BLOOD VESSELS
AND
LYMPHATICS
IN
ORGAN SYSTEMS

Contributors

D. I. Abramson

S. S. Adler

K. H. Albertine

R. J. Altiere

B. M. Altura

J. U. Balis

R. Beeuwkes III

D. R. Bell

D. Bergel

J. E. Bernstein

J. A. Bevan

L. R. Caplan

P. K. Chaudhuri

S. Chien

K. Christensen

K. E. Clark

C. E. Corliss

U. Desai

P. B. Dobrin

M. C. Fishbein

A. H. Friedman

C. N. Gillis

J. P. Gilmore

S. Glagov

I. Goldstein

D. N. Granger

H. J. Granger

C. V. Greenway

M. Hamaji

T. S. Harrison

D. D. Heistad

D. S. Hungerford

H. K. Jacobs

E. D. Jacobson

P. C. Johnson

H. A. Kontos

R. J. Krane

P. R. Kvietys

K. R. Larsen

M. Lawrence

L. V. Leak

D. W. Lennox

J. V. Lieponis

B. Linde

A. J. Lonigro

A. L. Lorincz

T. V. McCaffrey

R. S. McCuskey

J. M. McGreevy

J. A. McNulty

D. J. Martin

A. J. Miller

F. G. Moody

E. Nelson

G. Northrop

C. C. C. O'Morchoe

M. P. Owen

R. J. Paul

A. R. Pavlakis

M. A. Perry

E. M. Ramsey

W. C. Randall

A. M. Rappaport

F. D. Reilly

J. A. G. Rhodin

E. R. Ross

G. Ross

C. F. Rothe

M. M. Schwartz

S. S. Sobin

H. V. Sparks, Jr.

N. C. Staub

M. B. Stemerman

A. E. Taylor

B. L. Tepperman

C. Thomas

L. P. Thompson

S. F. Vatner

N. E. Warner

R. M. Weiss

J. S. Wheeler, Jr.

L. A. Whiteside

D. L. Wilbur

C. L. Witte

M. H. Witte

W. C. Worthington, Jr.

R. D. Wurster

BLOOD VESSELS
AND
LYMPHATICS
IN
ORGAN SYSTEMS

Edited by

DAVID I. ABRAMSON

Departments of Physical Medicine and Rehabilitation and Medicine
University of Illinois College of Medicine
Chicago, Illinois

PHILIP B. DOBRIN

Department of Surgery
Hines Veterans Administration Hospital
Hines, Illinois
and Departments of Surgery and Physiology
Loyola University Stritch School of Medicine
Maywood, Illinois

1984

ACADEMIC PRESS, INC.

(Harcourt Brace Jovanovich, Publishers)

Orlando   San Diego   San Francisco   New York   London
Toronto   Montreal   Sydney   Tokyo   São Paulo

ACADEMIC PRESS, INC.
Orlando, Florida 32887

United Kingdom Edition published by
ACADEMIC PRESS, INC. (LONDON) LTD.
24/28 Oval Road, London NW1 7DX

Library of Congress Cataloging in Publication Data
Main entry under title:

Blood vessels and lymphatics in organ systems.

    Includes bibliographies and index.
    1. Blood-vessels. 2. Lymphatics. 3. Organs (Ana-
tomy)- Blood-vessels. I. Abramson, David Irvin,
Date. II. Dobrin, Philip B. [DNLM: 1. Blood
circulation. 2. Blood vessels. 3. Lymphatic system. WG 103 B655]
QP102.B55   1984   612'.13   83-11860
ISBN 0-12--42520-3

PRINTED IN THE UNITED STATES OF AMERICA

84 85 86 87   9 8 7 6 5 4 3 2 1

Contents

QP102
.B55
1984

PART ONE

GENERAL PROPERTIES OF BLOOD VESSELS AND LYMPHATICS

CHAPTER 1

Morphologic and Biochemical Aspects of Arteries and Veins

CHAPTER 2

Biomechanics of Arteries and Veins

CHAPTER 6

Central Nervous System: Spinal Cord

CHAPTER 7

Central Nervous System: Special Senses

CHAPTER 8

Endocrine System: Pituitary, Thyroid, and Adrenal

CHAPTER 9

Endocrine System: Pancreas, Parathyroid, and Pineal

CHAPTER 10

Cardiopulmonary System: Heart

CHAPTER 11

Cardiopulmonary System: Lungs

CHAPTER 12

Digestive System: Esophagus and Stomach

CHAPTER 13

Digestive System: Small and Large Intestines

CHAPTER 14

Digestive System: Liver

CHAPTER 15

Genitourinary System: Kidney

CHAPTER 16

Genitourinary System: Male Reproductive Organs and Lower Urinary Tract

CHAPTER 17

Genitourinary System: Female Reproductive Organs

CHAPTER 18

Integumentary System: Skin and Adipose Tissue

CHAPTER 19

Locomotor System: Skeletal Muscle

CHAPTER 20

Locomotor System: Bones

CHAPTER 21

Hemopoietic System: Spleen and Bone Marrow

Contributors

Numbers in parentheses indicate the pages on which the authors' contributions begin.

D. I. ABRAMSON (602, 640, 656), *Departments of Physical Medicine and Rehabilitation and Medicine, University of Illinois College of Medicine, Chicago, Illinois 60680*

S. S. ADLER (705), *Department of Medicine, Section of Hematology, Rush University Medical School, Chicago, Illinois 60612*

K. H. ALBERTINE (363, 396), *Cardiovascular Research Institute, School of Medicine, University of California, San Francisco, California 94143*

R. J. ALTIERE (385), *Pharmacodynamics and Toxicology Division, College of Medicine, University of Kentucky, Lexington, Kentucky 40506*

B. M. ALTURA (113), *Department of Physiology, State University of New York Downstate Medical Center, Brooklyn, New York 11203*

J. U. BALIS (390), *Department of Pathology, College of Medicine, University of South Florida, Tampa, Florida 33612*

R. BEEUWKES III (507), *Department of Physiology and Biophysics, Harvard Medical School, Boston, Massachusetts 02115*

D. R. BELL (663), *Department of Physiology, Albany Medical College of Union University, Albany, New York 12208*

D. BERGEL (70), *University Laboratory of Physiology, Oxford OX1 3PT, England*

J. E. BERNSTEIN (596, 608, 614), *Department of Dermatology, Northwestern University Medical School, Chicago, Illinois 60611*

J. A. BEVAN (32), *Department of Pharmacology, University of California School of Medicine, Los Angeles, California 90024*

L. R. CAPLAN (196), *Department of Neurology, Michael Reese Hospital and Medical Center, and University of Chicago Pritzker School of Medicine, Chicago, Illinois 60637*

P. K. CHAUDHURI (461), *Department of Surgery, Loyola University Stritch School of Medicine, and Hines Veterans Administration Hospital, Maywood, Illinois 60153*

S. CHIEN (77), *Department of Physiology, College of Physicians and Surgeons, Columbia University, New York, New York 10032*

K. CHRISTENSEN (595, 639), *Department of Anatomy, St. Louis University School of Medicine, St. Louis, Missouri 63104*

K. E. CLARK (572, 573), *Departments of Obstetrics and Gynecology, Physiology, and Pediatrics, College of Medicine, University of Cincinnati, Cincinnati, Ohio 45267*

C. E. CORLISS (123, 175, 211, 259, 318, 362, 409, 437, 459, 506), *Department*

xiii

[1]Present address: First Department of Surgery, Osaka University Medical School, Fukushima-ku, Osaka 553, Japan.

H. K. JACOBS (212, 219), *Departments of Surgery and Physiology, Loyola University Stritch School of Medicine, Maywood, Illinois 60153*

E. D. JACOBSON (415, 419), *College of Medicine, University of Cincinnati, Cincinnati, Ohio 45267*

P. C. JOHNSON (106), *Department of Physiology, College of Medicine, University of Arizona, Tucson, Arizona 85724*

H. A. KONTOS (186, 191), *Department of Medicine, Medical College of Virginia, Richmond, Virginia 23298*

R. J. KRANE (544, 548, 552), *Department of Urology, Boston University School of Medicine, Boston, Massachusetts 02215*

P. R. KVIETYS (450), *Department of Physiology, College of Medicine, University of South Alabama, Mobile, Alabama 36688*

K. R. LARSEN (410), *Department of Surgery, University of Utah School of Medicine, Salt Lake City, Utah 84132*

M. LAWRENCE (233), *Department of Otorhinolaryngology, Kresge Hearing Research Institute, University of Michigan Medical School, Ann Arbor, Michigan 48109*

L. V. LEAK (134, 164), *Ernest E. Just Laboratory of Cellular Biology, College of Medicine, Howard University, Washington, D.C. 20059*

D. W. LENNOX (682, 686, 688), *Department of Orthopaedic Surgery, The Johns Hopkins University School of Medicine, Baltimore, Maryland 21205*

J. V. LIEPONIS (212, 222), *Department of Orthopedic Surgery, Loyola University Stritch School of Medicine, Maywood, Illinois 60153*

B. LINDE (622), *Department of Clinical Physiology, Karolinska Institute, Huddinge Hospital, S-141 86 Stockholm, Sweden*

A. J. LONIGRO (521), *Division of Clinical Pharmacology, Departments of Medicine and Pharmacology, St. Louis University School of Medicine, Veterans Administration Medical Center, St. Louis, Missouri 63125*

A. L. LORINCZ (596, 608, 614), *Department of Medicine, Section of Dermatology, University of Chicago Pritzker School of Medicine, Chicago, Illinois 60637*

T. V. McCAFFREY (241), *Department of Otorhinolaryngology, Mayo Medical School, Rochester, Minnesota 55901*

R. S. McCUSKEY (698), *Department of Anatomy, West Virginia University School of Medicine, Morgantown, West Virginia 26506*

J. M. McGREEVY (423), *Department of Surgery, University of Utah School of Medicine, Salt Lake City, Utah 84132*

J. A. McNULTY (308), *Department of Anatomy, Loyola University Stritch School of Medicine, Maywood, Illinois 60153*

D. J. MARTIN (618), *Laboratoire de Physiologie, Faculté de Médecine de Grenoble, 38700 La Tronche, France*

A. J. MILLER (348), *Department of Medicine, Northwestern University Medical School, Chicago, Illinois 60611*

F. G. Moody[2] *(410, 423), Department of Surgery, University of Utah School of Medicine, Salt Lake City, Utah 84132*

E. Nelson *(182), Department of Neurosciences, Rush Medical School, Chicago, Illinois 60612*

G. Northrop *(273, 304), Departments of Internal Medicine and Obstetrics and Gynecology, Rush Presbyterian–St. Lukes Medical Center, Chicago, Illinois 60612*

C. C. C. O'Morchoe *(126, 532), Department of Anatomy, Loyola University Stritch School of Medicine, Maywood, Illinois 60153*

M. P. Owen *(32), Department of Pharmacology, University of California School of Medicine, Los Angeles, California 90024*

R. J. Paul *(17), Department of Physiology, College of Medicine, University of Cincinnati, Cincinnati, Ohio 45267*

A. R. Pavlakis *(544), Department of Urology, Boston University School of Medicine, Boston, Massachusetts 02215*

M. A. Perry *(427), Department of Physiology, College of Medicine, University of South Alabama, Mobile, Alabama 36688*

E. M. Ramsey *(577), Department of Embryology, Carnegie Institution of Washington, Washington, D.C. 20025*

W. C. Randall *(319), Department of Physiology, Loyola University Stritch School of Medicine, Maywood, Illinois 60153*

A. M. Rappaport *(461), Sunnybrook Medical Center, University of Toronto, Toronto, Ontario M4N 3M5, Canada*

F. D. Reilly *(698), Department of Anatomy, West Virginia University School of Medicine, Morgantown, West Virginia 26506*

J. A. G. Rhodin *(97), Department of Anatomy, College of Medicine, University of South Florida, Tampa, Florida 33612*

E. R. Ross *(224), Department of Pathology, Loyola University Stritch School of Medicine, Maywood, Illinois 60153*

G. Ross *(340), Departments of Physiology and Medicine, University of California School of Medicine, Los Angeles, California 90024*

C. F. Rothe *(85), Department of Physiology, Indiana University School of Medicine, Indianapolis, Indiana 46223*

M. M. Schwartz *(528), Department of Pathology, Rush Medical School, Chicago, Illinois 60612*

S. S. Sobin *(368), Department of Physiology and Biophysics, University of California School of Medicine, Los Angeles, California 90033*

H. V. Sparks, Jr. *(647, 652), Department of Physiology, College of Human Medicine, Michigan State University, East Lansing, Michigan 48824*

[2]Present address: Department of Surgery, The University of Texas Health Science Center, 6431 Fannin, Houston, Texas 77025.

N. C. STAUB (363, 396), *Cardiovascular Research Institute, School of Medicine, University of California, San Francisco, California 94143*

M. B. STEMERMAN (25), *Department of Medicine, Beth Israel Hospital, Boston, Massachusetts 02115*

A. E. TAYLOR (618), *Department of Physiology, College of Medicine, University of South Alabama, Mobile, Alabama 36688*

B. L. TEPPERMAN (415, 419), *Department of Physiology, Health Sciences Centre, University of Western Ontario, London, Ontario N6A 5C1, Canada*

C. THOMAS (196), *Department of Pathology, Michael Reese Hospital and Medical Center, Chicago, Illinois 60616*

L. P. THOMPSON (647, 652), *Department of Physiology, College of Human Medicine, Michigan State University, East Lansing, Michigan 48824*

S. F. VATNER (326, 333), *Department of Medicine, Harvard Medical School, Boston, Massachusetts 02115*

N. E. WARNER (296), *Department of Pathology, University of Southern California School of Medicine, Los Angeles, California 90033*

R. M. WEISS (560, 562), *Department of Surgery, Section of Urology, Yale University School of Medicine, New Haven, Connecticut 06510*

J. S. WHEELER, JR. (548), *Section for Urology, Department of Surgery, Loyola University Stritch School of Medicine, Maywood, Illinois 60153*

L. A. WHITESIDE (674), *Department of Surgery, Section of Orthopedic Surgery, De Paul Hospital, St. Louis, Missouri 63044*

D. L. WILBUR (259), *Department of Anatomy, Medical University of South Carolina, Charleston, South Carolina 29403*

C. L. WITTE (496), *Department of Surgery, University of Arizona College of Medicine, Tucson, Arizona 85724*

M. H. WITTE (496), *Department of Surgery, University of Arizona College of Medicine, Tucson, Arizona 85724*

W. C. WORTHINGTON, JR. (259), *Associate Dean for Academic Affairs, Medical University of South Carolina, Charleston, South Carolina 29425*

R. D. WURSTER (219, 222), *Department of Physiology, Loyola University Stritch School of Medicine, Maywood, Illinois 60153, and Rehabilitation Research and Development Center, Veterans Administration Hospital, Hines, Illinois 60141*

Preface

The format followed in this volume is similar to that found in "Blood Vessels and Lymphatics," which was edited by one of us (D. I. A.) over two decades ago. Since then, knowledge in the cardiovascular sciences has advanced so rapidly as to warrant the development of another monograph on the same subject. Such an effort was given further impetus by the fact that the available information regarding the various topics is scattered throughout the literature in a variety of disciplines and, therefore, can be gathered only after time-consuming examination of a large number of original papers and reviews.

This volume deals with both the general and the specific characteristics of blood vessels and lymphatics in organ systems, including such rarely described vascular networks as those in the pineal, parathyroids, pancreas, adrenals, adipose tissue, and special senses. Also, the newly recognized functions of vascular endothelium are given special emphasis. Although the volume is intended as an advanced treatise, it does not present the exhaustive detailed information required by the specialist in any given field. Instead, it provides a structured, multidisciplinary approach to the broad field of vascular science, emphasizing both established and recent concepts.

In the preparation of the volume, the expertise of a large group of eminent investigators was enlisted. Each contributed one or several sections, in every instance limited to areas of research in which he or she is presently engaged. The submitted material was then judiciously edited for the purpose of organizing, integrating, and interrelating the topics so as to form a coordinated account with a consistent orderly style.

The volume is divided into two major areas. The first, consisting of four chapters, is devoted to general aspects of the arteries, veins, microcirculation, and lymphatic channels. The second and main segment, which is made up of seventeen chapters, deals with the embryologic, morphologic, physiologic, pharmacologic, pathophysiologic, and pathologic characteristics of blood and lymph circulations in each of the important organ systems.

In addition to orthodox concepts, promising, innovative, and provocative, but still unproved, viewpoints are presented, in some instances emphasizing topics that are of special interest to the authors. Wherever indicated, attention is directed to future areas for investigation, thus helping to focus interest on gaps in our knowledge in a particular field. Cross-references are found throughout the text for the purpose of unifying and consolidating the contents of the volume.* At the end of each chapter, an extensive current bibliography is

*To facilitate location of cross-references, each item is identified by a capital letter (referring to a main heading in the chapter), an Arabic number (a subheading preceded by the number), and a lowercase letter (a tertiary heading found on the first line of the section referred to).

found, including review articles, which should be helpful in providing sources of useful additional material to the reader.

It is hoped that this volume will be of value to the graduate student in the areas of blood and lymph circulation, acting as a foundation and catalyst for future development, and to the advanced research worker or clinician seeking sources of information on recent advances in the diverse aspects of cardiovascular science.

We wish to express our appreciation to Virginia Thiel and Shirley Zwiesler for their expert secretarial assistance, to Chris Capelle and William C. Gley for their technical help in the preparation of the manuscript, and to Joan J. Mathews and Louise F. Abramson for their meticulous proofreading of the contents. The editorial cooperation of the staff of Academic Press is gratefully acknowledged.

DAVID I. ABRAMSON
PHILIP B. DOBRIN

General Properties of Blood Vessels and Lymphatics

Chapter 1
Morphologic and Biochemical Aspects of Arteries and Veins

A. ANATOMY

By S. GLAGOV

1. GENERAL CONSIDERATIONS

The walls of arteries and veins are well-organized connective tissue structures composed of cells and matrix fibers arranged in three transmural zones or tunicae, the *intima*, the *media*, and the *adventitia*. The intima, the innermost subluminal layer, consists of endothelium and a variable quantity of immediately underlying cells and matrix elements. The media, which lies beneath the intima, is composed mainly of smooth muscle cells and closely associated elastin and collagen fibers. The adventitia, the outermost layer, consists mainly of collagen and elastin fibers, fibroblasts, small blood vessels, and other connective tissue elements, depending on location and nature of the immediately surrounding tissues. Although all arteries and veins share these general organizational features, the former may be distinguished from the latter by the composition of their walls and by the thickness of the media in relation to vessel diameter. Furthermore, many of the major specific homologous mammalian arteries and veins which comprise the systemic and pulmonary circuits have characteristic microarchitectural features which are independent of vessel diameter or species and appear to correspond in large measure to the mechanical stresses normally imposed upon the principal structural layers. The present section deals with the structure of the major arteries and veins and emphasizes both the similarities and distinctive differences between them.

2. ARTERIES

a. Intima: A continuous monolayer of polygonal flat cells, the *endothelium*, lines the luminal surface of all arteries. Since this structure is the immediate interface between the bloodstream and the underlying arterial wall, it is continuously subjected to shearing forces related to blood flow and to normal forces related to blood pressure. It is also directly exposed to circulating cells and plasma components. In relation to shearing stress, endothelial cells tend to be elongated in the direction of blood flow, particularly where the latter is rela-

3

Blood Vessels and Lymphatics in Organ Systems
Copyright © 1984 by Academic Press, Inc.

tively rapid, laminar, linear, and unidirectional (Flaherty *et al.*, 1972; Clark and Glagov, 1976a). Transverse cross-sectional and *en face* views of the endothelium are shown in Figs. 1.1 and 1.2. Where flow is slow, complex, turbulent, or relatively stagnant (especially at selected sites about branch ostia and

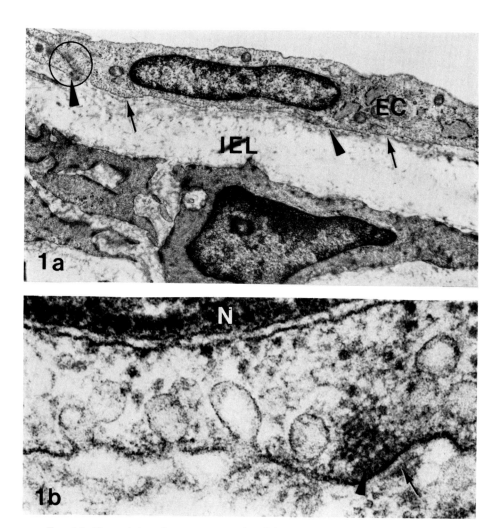

Fig. 1.1. Transmission electron micrographs of the intima of a normally distended rabbit abdominal aorta in longitudinal section. (a) Endothelial cells (EC) are separated from the internal elastic lamina (IEL) by the basal lamina (arrows). The edges of endothelial cells overlap (circle) and tight junctions are numerous. Peripheral dense bodies (arrowheads) on the abluminal aspect of the endothelial cells are sites of relatively firm attachment. ×14,000. (b) Higher magnification of a peripheral dense body (arrowhead) reveals increased electron density of the underlying basal lamina (arrow). N, nucleus. ×90,000.

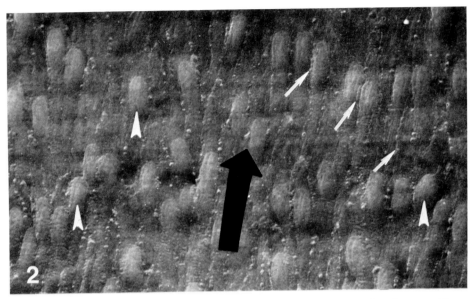

Fig. 1.2. Scanning electron micrograph of the luminal endothelial surface of monkey aorta. The endothelium is formed by a continuous sheet of closely associated cells, with overlapping edges (arrows). Both cells and nuclei (arrowheads) are oriented in the direction of flow (large arrow). ×3000.

bifurcations), the endothelial cells are less distinctly elongated or oriented (Bjorkerud and Bondjers, 1972; Gutstein *et al.*, 1973). In general, the edges of the cells overlap rather than abut, and each one tends to override its immediately distal or downstream neighbor. Tight (occluding) junctions are frequent between adjacent cells, whereas gap (communicating) junctions are less frequent and are most evident in large arteries (Simionescu *et al.*, 1976). A glycoprotein coat, demonstrable by special fixation and processing methods (Luft, 1966; Weber *et al.*, 1973), covers the luminal surface. Although villiform projections and other differences in endothelial luminal surface configuration have been reported in various locations (Gabbiani and Majno, 1969; Smith *et al.*, 1971; Robertson and Rosen, 1977), luminal surfaces prove to be largely smooth and regular when specimens are examined after controlled-pressure perfusion–fixation and under conditions designed to approach normal levels of temperature, pH, and ionic composition (Clark and Glagov, 1976a).

A basal lamina, which borders the endothelial cells on their abluminal or basal surfaces, forms a more or less continuous boundary between the endothelium and the immediately underlying intimal or medial structures. Although multilayered and complex in some locations, each basal laminal layer consists of two zones when viewed in usual electron micrographs. The inner zone is clear, whereas the outer finely fibrillar dense zone is rendered relatively electron opaque by the usual intensifying agents. In addition, discrete points of increased density and attachment have been noted between basal peripheral cytoplasmic dense bodies of endothelial cells and the subjacent internal elastic lamina (Ts'ao and Glagov, 1970). The basal lamina seems to provide a pliable, extensive bond which permits endothelial cells to comply with changes in configuration related to pulsation or vessel torsions and flexions, whereas the focal, rigid, dense body attachments probably prevent slippage, buckling, or excessive overriding of the lining cells when shearing stresses increase or change rapidly.

Shearing stresses exceeding 400 dynes/cm^2 have resulted in endothelial deformation, swelling, and sloughing in experimental models (Fry, 1968 and 1969). Nevertheless, in many areas where comparably high levels of shear stress normally prevail, there is little evidence for endothelial injury or detachment in carefully prepared specimens (Zarins et al., 1980). These findings suggest that, for some locations, departures exceeding usual shearing stresses may be injurious, whereas in other locations normally high absolute levels of shear may be associated with tighter, endothelial attachment. (For a discussion of cytologic and functional characteristics of endothelial cells, see Section C, below.)

Although the *intima* is defined as the tissue which extends from the lumen to the media, the endothelium and basal lamina are, in many locations, directly applied to the internal elastic lamina of the media and therefore comprise the entire intima. Where cells and/or matrix fibers intervene between endothelium and internal elastic lamina, the intima is thicker than the endothelial layer and in some sites is as thick as or thicker than the media. Intimal cells are predominantly smooth muscle (Geer, 1965), and both they and the associated matrix fibers, particularly those in immediate proximity to the internal elastic lamina, are frequently oriented in the longitudinal or axial direction (Buck, 1979).

Intimal fibrocellular thickening of an encircling and diffuse type appears to increase with age (Movat et al., 1958), but not to the same degree in all vessels. In man, the aorta and the coronary arteries, for example, tend to develop relatively thick intimas, but even this selective effect is not uniform in all populations (McGill, 1968; Persmen et al., 1975). Focal thickenings are prominent in relation to bends, bifurcations, and branches (Stehbens, 1981). These findings suggest that the development of such intimal pads may be related, at least in part, to local mechanical and/or metabolic factors and could conceivably play a functional mechanical role. (For a discussion of the general subject of intimal thickening and its relationship to atherogenesis, see Section E-2a, below.)

b. Media: The usual inner limit of the arterial media is the internal elastic lamina. In most locations it appears on transverse histologic sections as a more or less continuous hyaline layer of elastin. It is wavy or undulated in immersion-fixed histologic preparations but straightens to form a circular band when intraluminal pressure is maintained at normal levels during fixation. It is focally interrupted by gaps which correspond to clefts and fenestrations. In some locations, such as the aortic arch, a clearly defined internal elastic lamina may be absent, and elastic tissue appears instead as interrupted bars or islands. In the presence of marked degrees of intimal thickening, extensive defects, attenuations, or duplications of the internal elastic lamina are common. The outer limit of the media may be sharply demarcated when a well-defined external elastic lamina is present, as in the intraparenchymal pulmonary artery branches. However, in such vessels as the cerebral arteries, the external elastic lamina is commonly incomplete, interrupted, or absent. When the adventitia contains bands of smooth muscle cells, as in the case of some renal arteries, the precise outer boundary of the media is somewhat obscured. Otherwise, the distinction between media and adventitia is generally clear, for the latter is less compact than the media and contains fibroblasts, blood vessels, adipose tissue, and connective tissue fibers (see Section A-2c, below).

Smooth muscle cells, elastic fibers, and collagen fibers are the principal structural elements of the media, and arteries are normally classified as elastic or muscular types according to the relative proportions of these cellular and fibrous components found in the media. In *elastic arteries*, matrix fibers, in the form of well-defined elastic bands or lamellae and collagen bundles, are abundant and prominent in the media. The conducting vessels of relatively large diameter in close proximity to the heart, such as the aorta, the proximal brachiocephalic trunks, the iliac arteries, and the main pulmonary arteries, are examples of this type. These show relatively little variation in diameter in the course of normal function, and when removed from the body, their walls recoil and retract to some extent, but a prominent lumen and a relatively thin wall persist.

The media of a *muscular artery* contains relatively fewer connective tissue fibers than that of an elastic artery, smooth muscle being the morphologically predominant component. Except for the internal elastic lamina, which is prominent in muscular arteries, elastin appears in the form of branching strands. Some muscular arteries, such as the coronary, renal, and mesenteric vessels, branch directly from the aorta, whereas others arise by gradual transition from the proximal elastic arterial segments. Most, however, arise as second or third order branches of the elastic arteries. The predominantly muscular composition of these vessels corresponds to a greater capacity to change diameter actively in the course of normal function, under the influence of neurohumoral stimulation. When intraluminal pressures are markedly reduced or the vessels are excised, retraction and recoil may be considerable. *Small arteries*, i.e., those which comprise the distal subdivisions of the large muscular arteries, normally contain the lowest relative proportions of medial fibrous connective tissue.

Such vessels change diameter markedly in response to stimuli and are, there-fore, along with the arterioles, the principal regulators of peripheral resistance. [For a detailed discussion of the relation between cytologic structure and func-tion in medial smooth muscle cells, see Somlyo (1980).]

Since the media is the principal stress-bearing structure of the normal arterial wall, its thickness, as well as the relative abundance and arrangement of its structural components, would be expected to correspond to the imposed tensile stresses (Burton, 1954). In general, for adult mammalian arteries, the *thickness and composition* of the media at any particular location correspond closely to the magnitude of the tensile force normally exerted on the wall, whereas the *orientation* of medial cells and fibers appears to be related to the direction in which the tensile forces are applied. For example, when examined in transverse cross section in the undistended state, the media of the aorta of nearly all mammals consists of more or less distinct layers of wavy coaxial sheets or lamellae of elastic tissue, connected by intervening layers of radially or obliquely oriented cells, appearing to bridge from lamella to lamella. With redistention, diameter increases and wall thickness decreases as intraluminal pressure is raised. As diastolic pressure is approached, distensibility diminishes markedly, and further alterations in diameter or wall thickness are minimal when pressures are raised beyond systolic levels (Wolinsky and Glagov, 1964). During redistention, the large changes in aortic radius and wall thickness corre-spond to the straightening of the elastic lamellae, straightening and alignment of collagen fibers, and circumferential elongation of cells. At diastolic pressure, elastic fibers are no longer wavy, and the inextensible collagen fibers are drawn taut. Thus, the aorta, functioning normally at pressures ranging between di-astolic and systolic levels, is always nearly maximally distended. The circum-ferential elastic modulus of the media approaches that of collagen, thereby limiting further distention (Roach and Burton, 1957).

The relative proportions of elastin and collagen in the aortic media change with distance from the heart and also vary from vessel to vessel. The thoracic aorta normally contains a greater quantity of elastin than collagen, whereas the abdominal aorta has more collagen than elastin (Cleary, 1963). Yet, the total scleroprotein concentration, i.e., the sum of the collagen and elastin along the aorta, does not seem to change with distance from the heart. It has also been demonstrated that the anatomic position along the aorta at which relative pro-portions of elastin and collagen reverse differs from species to species. It would therefore seem that the accumulation of fibrous protein in the aortic media is normally governed by the prevailing mean tensile stress (Burton, 1954), where-as the relative proportions of elastin and collagen at particular locations along the aorta and in homologous arteries may be related to such factors as dif-ferences in the distribution and dampening of tensile stresses associated with the cardiac cycle and the reduction in diameter of the main trunk as major branches arise (Rodbard, 1970).

The integrated and coordinated function of the media is attributable to at least three modes of interconnection of its structural elements (Clark and Glagov, 1979), the first of which consists of flat sheets of interlacing collagen fibrils which run over and within the basal lamina investing the smooth muscle cells. Any pull tending to separate the overlapping elongated cells changes the orientation of the surrounding interlacing basketwork of collagen fibrils, so that the latter tighten about the cells and tend to hold groups of them together, thus preventing further stretching, separation, or slippage. The second binding arrangement consists of strong, focal attachment sites between peripheral dense bodies of the cells and adjacent elastic fibers. The tenacity of these cell–elastin attachment complexes may be greater than the cohesiveness of the cell body, for sudden experimental hyperdistention has been shown to result in focal disruption of cell projections near attachment points, as well as fracture and recoil of basal lamina, while preservation of the corresponding bond of cell to elastin persists. Aortic elastic fibers form interconnected bands or bars on either side of the cell layers which are oriented in the same direction as the cells. Large collagen fiber bundles, separate from the pericellular fine fibrils, are wedged between adjacent elastic fiber systems. The presence of interconnected elastic fibers in close association with the collagen fibers assures resiliency, as well as tensile strength (Wolinsky and Glagov, 1964). Such an arrangement tends to distribute tensile stresses uniformly through the aortic media, so that structural flaws and irregularities which develop with age tend to be bypassed rather than propagated. All of the medial structures are embedded in an ubiquitous, continuous, fine microfibrillar proteoglycan matrix, which may provide a third means of stress transmission and coordination. In general, a basal lamina investing smooth muscle cells of the media is more prominent and continuous in muscular than in elastic vessels. It is likely that the denser basal lamina–collagen matrix of muscular arteries corresponds to the high degree of intercellular coordination required for active contraction and changes in configuration during normal function, whereas the prominent elastic fibers and strong cell to elastic attachment sites of elastic arteries correspond to their primarily stress-bearing function.

The layers which comprise the aortic media of adult mammals are generally of similar thickness regardless of species, and the number of transmedial layers correspond closely to the total tension normally imposed upon the media. Since mean aortic pressure is similar for most adult mammals, and since there is a nearly linear relationship between aortic radius at mean blood pressure and the number of its fibrocellular layers, the average tension per layer proves to be nearly constant. Estimates made from a study of several species (Wolinsky and Glagov, 1967a) revealed that each increment of 0.017 mm in adult aortic luminal radius with increasing species size corresponds to an additional fibrocellular layer and that the average tension per aortic layer is about 2000 dynes/cm. On the basis of these findings, the aortic medial fibrocellular layer has been called a *lamellar unit* with the implication that it is the functional, as well as the structural, unit of aortic medial architecture. Similarly, an estimate of "layering" can be obtained for medias of muscular arteries and large pulmonary arteries by counting the number of cells traversed by a radius passing through the wall. Although the tension per transmedial layer is nearly constant for each of the several homologous mammalian arteries which have been studied in this manner (Glagov, 1972), both the estimated total tension and the average tension per layer at normal distending pressure are smaller for these vessels than for the aorta, corresponding in general to the lower content of matrix fibers.

More recent studies of normally distended elastic and muscular arteries, examined in semithin tissue sections and in fracture preparations by scanning electron microscopy, have revealed a further structural subdivision of arterial medial microarchitecture common to elastic and muscular arteries (Clark and Glagov, 1979). According to these findings, medias of both elastic and muscular arteries are composed of groups or fascicles of commonly oriented, elongated, smooth muscle cells lying mainly within tangential planes. Each fascicle is invested by a sheath of basal lamina and collagen fibrils, closely associated with a corresponding system of elastic fibers oriented in the same direction as the fascicle cells. Morphologic differences among arteries reside principally in the number, size, and composition of these musculoelastic structural groups. In straight portions of adult aortas and other elastic arteries, the fascicles are 1–3 cells thick in the radial direction, but they vary considerably in longitudinal width and circumferential length and in the prominence of their elastic fiber systems, depending on location across the media and on vessel diameter. The apparent layering of the aortic media would therefore appear to correspond to closely packed smooth muscle fascicles, with each cell layer bracketed on either side by elastin. The impression of a succession of single elastic lamellae on usual, relatively thick transverse cross sections of undistended preparations results from the juxtaposition and superimposition of the elastic fiber systems surrounding adjacent fascicles. Although medias of muscular arteries are also composed of such musculoelastic fascicles, the associated matrix fiber systems are less prominent than in the aorta. On the basis of anatomic appearances, the size and alignment of individual fascicles appear to correspond, respectively, to local differences in amplitude and resultant direction of tensile stress. The organization of the media consistent with the findings outlined above is shown semidiagrammatically in Fig. 1.3.

FIG. 1.3. Structure of the aortic media . (a) Light micrograph of monkey aorta, slightly oblique section near the arch. Elastic tissue lamina (arrows) bracket the smooth muscle groups (arrowheads). (b) Semidiagrammatic representation of the light microscopic appearance of the aortic media in a straight portion of the normally distended vessel. Smooth muscle cells (M) are elongated, overlapping, and oriented circumferentially. Intervening connective tissue consists of elastic fibers (arrows) closely associated with each cell layer and collagen bundles (arrowheads). (c) Ultrastructural infrastructure reveals that smooth muscle cells (M) are surrounded by sheaths (S) composed of basal lamina and fine collagen fibrils. Elastic fibers (E) consist mainly of branching arrays oriented in the same direction as immediately associated cells. Collagen bundles (C), separate from the pericellular fibrils, course mainly between adjacent elastic fiber layers.

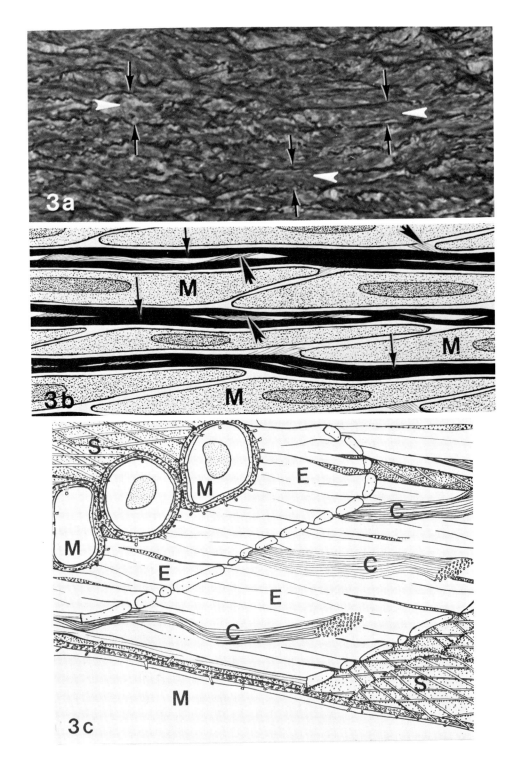

A close relationship among the changes in tangential tension, wall thickness, and medial composition and organization is apparent during postnatal growth as arteries establish adult patterns of medial organization. At birth, for example, the ascending aorta and pulmonary trunk are similar in diameter, wall thickness, composition, and morphologic appearance. During postnatal growth, wall thickness and total scleroprotein accumulation per cell rise much more rapidly for the aortic media than for the pulmonary trunk, as the transmural elastin lamellar pattern becomes increasingly prominent in the aorta and more obscure in the pulmonary artery. These differences closely parallel the developing differences in medial tangential tension or stress (Leung *et al.*, 1977). If elevated pulmonary artery pressure persists after birth, due to the presence of anomalous shunts, the neonatal pulmonary artery architecture tends to persist (Heath and Edwards, 1958). (For further discussion of mechanical aspects of the media, see Section A, Chapter 2.)

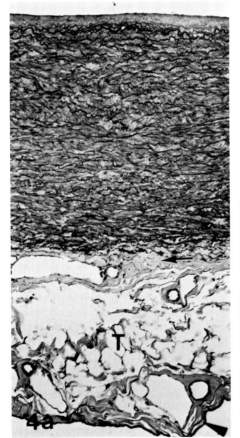
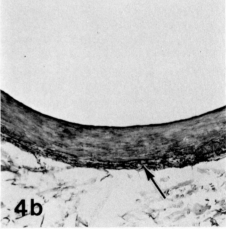

FIG. 1.4

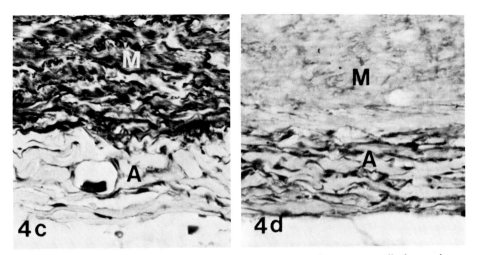

FIG. 1.4. Media and adventitia. Boundary between media and adventitia is usually distinct, but the structure and composition of the adventitia are quite variable. (a) Adventitia of the proximal thoracic aorta consists of a relatively thin band of collagen (arrow) and is bounded by adipose tissue (T) and the parietal pleura (arrowhead). (b) Adventitia of the coronary artery of a young monkey forms a prominent relatively thick layer (arrow). ×100. (c) Higher magnification of the media (M)-adventitia (A) boundary of the aorta reveals relatively dense collagen in the adventitia. (d) Higher magnification of the media (M)-adventitia (A) boundary of a coronary artery reveals alternating elastic and collagen fibers in the adventitia. ×500.

c. Adventitia: Although the distinction between media and adventitia rarely presents difficulties, the outer limit of the adventitia is often difficult to define, for it is usually contiguous and coextensive with the perivascular connective tissues. The aorta and pulmonary trunk have relatively slight adventitial condensations of fibrous connective tissue and may be surrounded by adipose tissue. In contrast, the adventitia of the large muscular visceral arteries, such as the renal and mesenteric branches, consists of prominent elastic and collagen fibers in well-organized layers, and in these locations it may be thicker than the media (Fig. 1.4). Compared with the media, however, adventitial cells are sparse and mainly fibroblasts. The characteristic laminated structure of the adventitia may be present initially, but in some vessels, such as the epicardial coronary artery branches, this may decrease in prominence with age. Given the range of adventitial structure, the contribution of the normal adventitia to the mechanical properties of the arterial wall is presumably variable. For the large elastic arteries, removal of this structure has little effect on static pressure–volume relationships (Wolinsky and Glagov, 1964). On the other hand, the structured, thick adventitia of most large muscular arteries may serve as a resilient mechanical buffer to prevent compression or restriction of pulsatile expansion by surrounding tissues and organs. A gradient in the direction of fiber orientation, with collagen fibers most circumferential in the deeper adventitial layers, would tend to support such a function (Smith *et al.*, 1981). Moreover, it may allow for changes in diameter and damping of pulsatile stresses, which might otherwise be transmitted to surrounding tissues. In any case, the adventitia of arteries is often strong enough to withstand tensions associated with normal levels of blood pressure in the event of medial rupture (see Section E-4a, below). It also contains vasa vasorum and nerves, the former providing nutrition to the adventitia and media and branches of the latter contributing to the regulation of medial smooth muscle function.

d. Artery wall nutrition: The aortic media is nourished by diffusion and convection from the endothelial surface and by vasa vasorum which supply it from the adventitial side. Ingress from the aortic lumen is apparently sufficient to nourish the inner 0.5 mm of the media (Geiringer, 1951). This zone contains no vasa vasorum and encompasses about 30 medial fibrocellular layers (Wolinsky and Glagov, 1967b). Aortic medias composed of more than 30 layers contain vasa vasorum between the thirtieth layer and the adventitia. Thus, aortas of small mammals, such as rats, rabbits, and small breeds of dogs, which have fewer than 30 medial layers are apparently nourished mainly by perfusion from the intimal side, whereas aortas of large mammals, such as pigs, sheep, large breeds of dogs, and humans, having more than 30 medial layers are also perfused by vasa from the adventitial side. The distinction between avascular and vascular aortic medias is apparently already drawn at birth, for animals which are destined to have more than 30 aortic medial layers as adults have more than this number at birth, and medial vasa are already present (Wolinsky, 1970a). Those aortas which have fewer than 30 layers at birth will not exceed 30 layers at full maturity and do not contain medial vasa at birth.

Two additional features are associated with the presence of medial vasa vasorum. First, the average tension per medial layer for aortas which contain medial vasa tends to be somewhat higher than for those without vasa, suggesting that the presence of medial vasa permits medial layers to function at a higher level of medial stress (Wolinsky and Glagov, 1969). Second, in very large mammals, such as horses and cattle, the outer vascularized zone may deviate somewhat from the usual uniformly lamellar architecture (Wolinsky and Glagov, 1967c). Elastic fibers are focally diminished in thickness in association with the occurrence of prominent smooth muscle bundles, suggesting that this mode of organization, possibly necessary to stabilize very large and long aortas, may depend on the presence of medial vasa.

The afferent vasa vasorum arise mainly from the large arterial branches almost immediately after the origin of the latter, arborize and anastomose in the adventitia, and enter the media, usually at right angles. Within the media, the vasa tend to be oriented axially at several branching levels. The details of microarchitecture which permit intramedial aortic vasa to remain open despite the compressive and shearing stresses within the arterial wall have not been explored. Lymphatic channels are demonstrable in the adventitia of most arteries (Johnson, 1969) and have been shown to drain to local lymph nodes, but they have not been identified within the media, and their role in clearance of substances which enter the vessel wall by way of the intima or by way of the afferent vasal channels has not been elaborated.

3. VEINS

Although major architectural features are similar for all veins, special details of microanatomy and composition can be identified for many specific vessels. Compared to those of arteries, vein walls are much thinner in relation to vessel diameter, more easily collapsed, and less extensible. These features correspond to a greater content of collagen and a lower content of elastin in vein walls.

a. Intima: The luminal surface of veins is covered by a continuous endothelium, but, in keeping with the normally lower rate of blood flow, individual endothelial cells are not as distinctly oriented as in arteries, and occluding or communicating junctions are not as frequent (Simionescu *et al.*, 1976). Endothelial cells have been reported to be thinner and to have larger surface areas than those of arteries, but such comparisons are difficult to interpret unless specimens have been distended at normal intraluminal pressures during fixation. Intimal thickening occurs in large veins such as the vena cava and locally in relation to ectasias and tortuosities, mainly on the inner aspect of the bends.

Valves are a prominent feature of most veins. Absent from the vena cavae, portal vein, coronary veins, and most veins not subjected to elevated hydrostatic pressure, venous valves are especially prominent in the vessels of the extremities, particularly the lower ones. They most often consist of cusps formed by crescentic extensions from the intima, usually just distal to the confluence of tributaries. They are covered by endothelium and contain loose connective tissue and a central plate of collagen fibers. There is no evidence that they develop an active component of tension, although some have scattered smooth muscle cells and are associated with an increase in medial smooth muscle at the angle of insertion.

b. Media: Regardless of species or vessel size, an internal elastic lamina is present in most major veins, although it is not as prominent as in arteries. The media usually contains a greater proportion of collagen than is seen in normal arteries, and this tissue tends to occur in the form of prominent fiber bundles between fascicles of smooth muscles. Thorough studies of vein walls in both the distended and undistended state indicate that smooth muscle cell bundles are oriented either circumferentially, axially, or obliquely in different locations, depending on the degree of distention (Kügelgen, 1955, 1956). Such findings indicate that venous smooth muscle probably provides a combination of longitudinal tethering and circumferential tensile support.

Medias of some large veins may have special architectural features. The media of the portal vein, for example, is organized in two distinct smooth muscle layers (Ts'ao *et al.*, 1970; Hammersen, 1972; Komuro and Burnstock, 1980). The internal layer is circular, whereas the external one is longitudinal. In addition to the usual subendothelial elastic lamina, the two layers may be separated by elastic fibers, with the result that the inner layer may be misinterpreted as a thickened intima. The segregation of these separate layers during early growth indicates, however, that this is actually a form of medial differentiation (Ts'ao *et al.*, 1971). Both nerve fibers and vasa vasorum are prominent in the media of the portal vein. The presence of nerves provides an anatomic basis for the periodic contractile properties which have been demonstrated for this vessel. The media of other large veins also contains vasa vasorum, suggesting that diffusion from the relatively hypoxic blood in the lumen is inadequate for maintainence of the venous media. Some veins, such as the main adrenal vein within the adrenal medulla, have a media which is markedly narrowed or absent at intervals about its circumference. The significance of this arrangement is not clear. Since the adrenal is supplied by several arteries and drained by a single vein, this morphologic feature may be related to a mechanism as yet undetermined for control of blood flow.

c. Adventitia: Although distinction of media from adventitia is usually possible for small veins, the adventitia of large veins is often thick and may contain bundles of collagen and smooth muscle. Collagen fiber bundles are frequently more prominent and the smooth muscle groups less prominent in the outer layers of the media than in the inner layers, particularly in large veins, and adventitial fiber bundles can often be followed into the media at some point on the vessel circumference if sequential sections are examined. As in muscular arteries, elastic fibers are more prominent in the adventitia than in the media and may help to demarcate the two tunicae. The matrix composition of the adventitia varies with location and vein diameter. Large veins, which sustain relatively high levels of wall tension, and veins in close proximity to arteries tend to have a relatively thick fibroelastic adventitia. In many species, the vena cava segments immediately adjacent to the right atrium are surrounded by a sheath of myocardial muscle fibers.

B. MECHANOCHEMISTRY OF VASCULAR SMOOTH MUSCLE

By R. J. Paul

1. General Considerations

The primary role of vascular smooth muscle (VSM) is the generation of the mechanical forces required for the control of vessel diameter. Like all muscle types, VSM requires chemical energy to perform mechanical work. This section deals with the processes involved in such energy transduction.

Current knowledge of the mechanochemistry of VSM is based largely on studies of skeletal muscle (Needham, 1971), for there are many similarities between the two types; nevertheless, VSM has evolved in a manner uniquely suited to its function. For example, a vasculature constructed with the most efficient skeletal muscle would be associated with an energy cost to an organism of some three to four times its basic metabolic rate, a task which VSM accomplishes at about one-hundredth the cost (Paul, 1980). Although maintenance of contraction by VSM requires little metabolic energy, the mechanism of energy production in this tissue renders it more immediately dependent on intermediary metabolism than skeletal muscle. Coordination of energy metabolism with contractility is critical to vascular function. [For detailed information regarding VSM, particularly with respect to the diversity of vascular tissues, see Bohr *et al.* (1980); for discussion of anatomic aspects of wall nutrition, see Section A-2d, above.]

2. Vascular Smooth Muscle Mechanics in Relation to Structure

a. VSM versus striated muscle: Although most concepts concerning VSM have been derived from skeletal muscle and in general appear valid, it must be kept in mind that the differences between the two types of muscle may be of more significance than the similarities. The contractile apparatus of VSM exhibits considerably less order than does skeletal muscle, but it seems to be constructed of morphologically similar protein elements. In both types, actin and myosin, organized into thin and thick filaments, are the central elements of their mechanochemical systems. Although some differences in electrophoretic

17

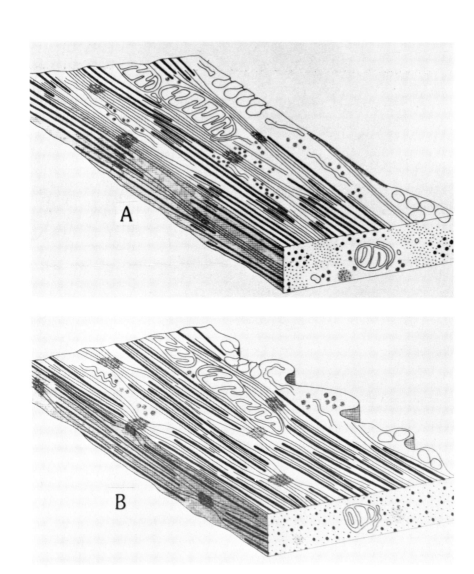

Fig. 1.5. Schematic sections of a smooth muscle cell in relaxed (A) and contracted (B) state. In the relaxed muscle, the thick, myosin-containing filaments are grouped together, separate from the thin, actin-containing filaments. The thin filaments are connected to the dense bodies. During contraction (B), the thin filaments slide and interdigitate between the thick filaments, forming a more regular pattern. The contraction of the filament units is transferred to the cell membrane, where dense bodies are present. At long cell lengths, the average interaction between thick and thin filaments would be reduced, presumably accounting for the decline in active isometric force and respiration rate at long tissue lengths, as shown in Fig. 1.6. (Adapted from Heumann, 1973, reproduced with permission from *Phil. Trans R. Soc.*)

properties have been noted, skeletal and smooth muscle actin can be used interchangeably in experiments concerned with actin-activated myosin ATPase. Vascular smooth muscle myosin is similar to that of skeletal muscle in molecular weight and appearance in electron micrographs, but it differs with regard to ATPase activity and to actin-binding characteristics. The organization of VSM actin and myosin into thin and thick filaments appears grossly comparable to skeletal muscle. However, the mechanical and energetic consequences of long filaments and sarcomeres, the characteristic of the contractile protein organization of some invertebrate smooth muscles (Rüegg, 1971), do not appear to be of major functional significance in mammalian smooth muscle. The control of actin–myosin interaction in VSM is the subject of much recent controversy (Adelstein and Eisenberg, 1980) and appears to differ qualitatively from the molecular mechanisms for regulation of skeletal muscle contractility (see Section B-4, below).

A schematized model for the arrangement of filament in smooth muscle is shown in Fig. 1.5, and the mechanical characteristics under isometric conditions as a function of muscle length are depicted in Fig. 1.6. Under relaxed conditions (in some cases requiring pharmacologic agents), isometric force progressively increases with tissue length. This exponential spring-like behavior is characteristic of all muscles and is operationally defined as a passive parallel element. Total force observed under activated conditions is shown in the upper record in Fig. 1.6. Subtracting the passive characteristics from the total force yields the length-active force relation. This is usually attributed operationally to the interaction of actin and myosin. The observation of a length-active force characteristic of VSM, similar to that found in skeletal muscle studies, forms the basis for postulating similar "sliding filament" mechanisms for both. The energy requirement for maintaining active force in VSM also displays a similar dependence on tissue length (Paul, 1980). Although a sliding filament mechanism for VSM is the simplest alternative, the evidence is as yet by no means as convincing as in the case of skeletal muscle.

At the optimum tissue length for force generation, VSM can generate isometric stresses of 200–300 mN/mm^2. When this is adjusted for cell content, the active stresses are at least as large as those generated by skeletal muscle. This is surprising in light of the fact that the myosin content of VSM is only 0.1 to 0.2 of that in skeletal muscle, thus yielding a higher force per mole of myosin in VSM than in skeletal muscle. However, since the actin content of VSM is similar to that of skeletal muscle, it must be concluded the ratio of thin to thick filament in smooth muscle (~15 : 1) is considerably greater than that in skeletal muscle (2 : 1). Although the filaments appear similar in size, it is still possible that some form of longer "effective" length (Paul and Rüegg, 1978), resulting from the higher thick to thin filament ratio in VSM, may provide a "mechanical advantage" to the force-generating apparatus. Models with longer "effective" sar-

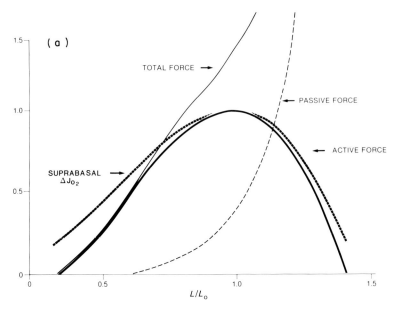

FIG. 1.6. Mechanical and metabolic characteristics of vascular smooth muscle. (a) Relations among isometric force, suprabasal oxygen consumption rate (ΔJ_{O_2}), and tissue length; bovine mesenteric vein characteristics. Ordinate: isometric force and suprabasal J_{O_2} normalized to the maximum values observed. Abscissa: tissue length normalized to the length at which maximum active isometric force is measured. Passive, total, and active forces are defined in text. (Adapted from Paul and Peterson, 1975, reproduced with permission from the *American Journal of Physiology*.) (b) Relation between force and velocity. Ordinate: Force. Maximum active isometric force for vascular muscle ranges from 25 to 250 mN/mm². Abscissa: Velocity of contraction. Unloaded shortening speeds range between 0.075 and 0.75 muscle lengths per second at 37°C. Experimental values can be fitted by the hyperbolic relation of Hill in which P_o is the value for maximal isometric force and a and b are constants.

comeres result in more force-generating elements (cross bridges) in parallel and hence higher possible forces per myosin content. The relation between contraction speed and force in smooth muscle can be described by the Hill hyperbolic equation, initially used to describe skeletal muscle, an example of which is shown in Fig. 1.6. Although this relation is of similar form for all muscle, contractile velocities of vascular tissue are 1 to 3 orders of magnitude slower than skeletal muscle. In recent years, experimental studies on VSM have analyzed the mechanical transients following the imposition of rapid (< 1 msec) changes in load or muscle length. At least part of this behavior may be ascribed to mechanisms at the level of the cross bridge itself, and hence this type of experimentation has attracted considerable interest (Hellstrand and Paul, 1982).

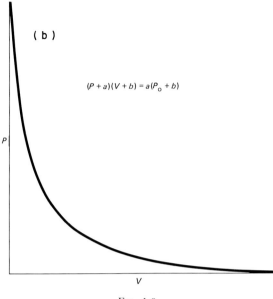

(b)

$$(P + a)(V + b) = a(P_0 + b)$$

FIG. 1.6

3. Vascular Metabolism and Energetics

The phosphagen pool (ATP + PCr) in VSM ranges from 1 to 4 μmole/gm, values which are approximately one-tenth of that in skeletal muscle. If endogenous phosphagen were the immediate source for mechanical activity in VSM, it would not provide sufficient substrate for more than a few minutes of contraction. Consequently, VSM is heavily dependent on intermediate metabolism for the maintenance of active tone. Oxidative metabolism provides the major portion of ATP synthesis, and respiration is closely coupled to mechanical activity (Paul, 1980). The basal rate of oxygen consumption (J_{O_2}) in VSM ranges between 0.05 and 0.5 μmole/min/gm, the rate being inversely related to the size of the animal and independent of the particular vessel. Vessels which show spontaneous mechanical activity, such as rat portal vein, are exceptions, in that they exhibit high basal respiratory rates. Stimulation of VSM causes J_{O_2} to increase to two to three times that of the unstimulated rate, while maintaining maximum isometric force. This phenomenon is essentially independent of the stimulus used to elicit contraction if compared at equivalent levels of isometric force. The increase in respiration probably is related to the increase in actomyosin ATPase because only a small increase in J_{O_2} is seen when a muscle is stimulated at lengths at which no isometric force (and presumably little actin–myosin interaction) occurs. About 20% of the increase in J_{O_2} can be attributed to processes involved with activation unrelated to actin–myosin interaction. It is of interest to note that under similar conditions, skeletal muscle respiration can increase by some 40-fold and yet be insufficient to maintain contractility.

Although metabolism in VSM is primarily oxidative, under fully oxygenated conditions, most of the glucose entering vascular tissue is catabolized only to lactate. On a molar basis, the rate of lactate production (J_{lac}) in VSM may exceed J_{O2} by a factor of 2. This aerobic glycolysis accounts for less than 30% of the cellular rate of synthesis of ATP. Because aerobic glycolysis of such magnitude is observed in relatively few types of cells, it has been suggested that aerobic glycolysis may be an index of vascular myopathy. However, recent evidence has shown that J_{lac} is correlated with Na-K transport in vascular tissue. By modifying contractile activity and Na-K transport conditions, respiration and lactate production can be varied independently in vascular tissue. This observation suggests that aerobic glycolysis reflects some form of enzymic compartmentation in vascular tissue rather than a metabolic defect. Evidence suggesting that fatty acids are the primary substrate for vascular oxidative metabolism, with little contribution from oxidation of carbohydrate (Chase and Odessey, 1981), reinforces such a possibility. Despite data documenting the presence of glycogen phosphorylase in VSM, glycogen does not appear to be a substrate for either oxidative metabolism under resting conditions or for the anomalous aerobic production of lactate.

Vascular muscle energetics, based largely on metabolic studies, indicate that the cost of maintaining tension, i.e., the rate of ATP synthesis per unit force maintained, is up to 300-fold less than that in skeletal muscle. Estimates of the actomyosin ATPase activities, based on *in vivo* tension-dependent J_{O2}, agree closely with the *in vitro* values for the isolated contractile proteins. This observation provides a basis for a model to explain both the low tension cost and contractile speed of VSM. Currently accepted models of muscle contraction involve the cyclic interaction of actin and myosin, where each cycle requires the hydrolysis of ATP allowing for the production of mechanical activity. While structural factors also may play a role, the lower vascular actomyosin ATPase activity and consequent actin–myosin cycle rate appear to account for both the decreased speed of contraction and the low metabolic cost in VSM. The molecular mechanism underlying the low actomyosin ATPase activity in this tissue is the focus of much muscle research.

4. CONTROL AND COORDINATION OF VASCULAR METABOLISM AND CONTRACTILITY

An increase in intracellular Ca^{2+} to micromolar levels generally is believed to be the signal initiating contraction in VSM. However, the mechanism whereby this increase in Ca^{2+} is translated into the contractile response is controversial. Current data indicate that Ca^{2+} binds to the ubiquitous calcium receptor protein, calmodulin, and that this complex activates a myosin light chain kinase (MLCK). In turn, activated MLCK phosphorylates a myosin light chain (MLC). Phosphorylation of the 20,000 dalton protein subunit of VSM myosin appears to be a prerequisite for actin–myosin interaction *in vitro* (Adelstein and Eisenberg, 1980). At this point, agreement ends, and there are currently at least three schools of thought concerning the regulation of VSM contractility. The first theory, which is based on *in vitro* studies of contractile proteins and of membrane-free smooth muscle, suggests that phosphorylation of myosin is central and acts as the switching mechanism for actin–myosin cycling (Hartshorne and Siemankowski, 1981). The second hypothesis, which is based on the correlation between changes in contraction velocity with MLC phosphorylation state in living VSM, suggests that while phosphorylation may be an initial switching mechanism, its primary role is to regulate the speed of the cross bridge cycling (Murphy *et al.*, 1983). In this scheme, an additional, yet unknown, mechanism of Ca^{2+} control is involved in regulation of unphosphorylated myosin, as observed in living VSM for contractions of long durations. The third theory holds that VSM is regulated by a protein called leiotonin, a protein believed to be analogous to troponin, the regulatory protein of skeletal muscle. Troponin acts by binding calcium and altering actin–myosin interaction on the thin filament (Ebashi, 1980). It is of interest in this regard that less than a decade ago, VSM was thought, by analogy to skeletal muscle, to be regulated by a troponin, thin filament-linked control system. Current evidence indicates that most smooth muscle probably lacks troponin, but that both thin and thick filament-linked regulatory processes may exist.

Although there is evidence to support each of the three theories, at present firm conclusions cannot be drawn concerning the mechanism of regulation of VSM contractility. However, three cautionary statements can be made: (1) The processes underlying relaxation in VSM may not simply be the reverse of the events leading to contraction. (2) Although Ca^{2+} plays a central role, isometric force cannot be used as an index of intracellular Ca^{2+}. Mechanisms controlling the activity of MLCK and phosphatase are known to alter the sensitivity of the actin–myosin interaction to Ca^{2+} in such a manner that an inhibition of isometric force in the presence of micromolar Ca^{2+} is a distinct possibility. (3) It is likely that no single mechanism, or even multiple mechanisms, can be generalized to include all varieties of VSM.

It has been argued that the synthesis of ATP by intermediary metabolism, rather than the vascular phosphagen pool, is crucial for the maintenance of contractile activity in VSM. Oxidative metabolism, the major source of ATP, has been observed to follow isometric force closely both in magnitude and time, but the control point for this coordination is unknown. An increase in free ADP, concomitant with the rise in actomyosin ATPase associated with contractile activity, is a mechanism postulated for stimulating mitochondrial oxidative phosphorylation in skeletal muscle. Changes in free ADP in phasically contracting rat portal vein are consistent with this hypothesis; however, in tonically contracting hog carotid artery, no change in ATP or PCr, and hence in free ADP, is measured during contraction, as compared with resting levels. Increased Ca^{2+} is known to activate glycogen phosphorylase, and by increasing substrate availability, this could potentially serve as a mechanism for coordination. However, as previously noted, carbohydrate does not appear to serve as a substrate for vascular oxidative metabolism. The mechanism for control of vascular metabolism and its coordination with contractility are not known with certainty, and, hence, analogies from skeletal muscle mechanisms are less than secure.

5. SUMMARY

Many of the structural features of the VSM contractile apparatus are grossly similar to those of skeletal muscle. The mechanical properties, in part reflecting structure, are characterized by force–length and force–velocity relations, similar in form to skeletal muscle relations. Vascular muscle differs from skeletal muscle in that (1) force per myosin is higher; (2) contraction velocities are slower; and (3) the cost of active tension is substantially lower. These characteristics can be related to the lower intrinsic actomyosin ATPase of VSM. Although the energy requirements for contraction are low, the lack of a large phosphagen pool renders VSM more dependent on intermediary metabolism, on a moment to moment basis, than skeletal muscle. Oxidative metabolism, with fatty acids the most likely substrate, is the primary source of ATP and is correlated with isometric force. However, VSM is also characterized by substantial aerobic glycolysis, which is correlated with Na-K transport. Glucose appears to be the primary source of aerobic lactate production. Intracellular Ca^{2+} plays a central role in control of vascular contractility. The molecular mechanisms underlying such a regulation are the subject of controversy and both thick and thin filament-linked mechanisms have been proposed. Factors involved in coordination of energy metabolism with contractility are also poorly understood and appear to be somewhat different from those in skeletal muscle.

C. VASCULAR ENDOTHELIUM

By M. B. STEMERMAN

The endothelial cell is the interfacing structure between the blood vessel lumen and the blood (Majno, 1965). This is an unique position which demands that the endothelium participate in the processes of hemostasis and vascular permeability. Three technical advances have provided important understanding of the biology of the vascular endothelium: electron microscopy, cell culture, and methods of inducing and assessing vascular injury. A comprehensive discussion of this rapidly growing and interesting field is not attempted here; instead recent reviews are listed for further information (Gimbrone, 1976, 1981; Schwartz *et al.*, 1981). (For discussion of morphologic and pharmacologic aspects of endothelium, see Section A-2a, above; Section D-5, below; Section C-2c, Chapter 3; Sections D-4 and E-1a, Chapter 11.)

1. GENERAL FEATURES OF VASCULAR ENDOTHELIUM

a. Structure of endothelial cells: The morphology of the endothelium provides the first insight into its uniqueness. When fixed under pressure and viewed from the lumen, endothelial cells of a large vessel appear to spread out and hug the vessel's surface. The outlines of the endothelium can be distinguished on staining the cells with silver nitrate. When the luminal surface is examined by light microscopy, the silver stain deposits on the borders of the cell give it the appearance of tiles laid upon a floor. By scanning electron microscopy, the cells are found to be more linear in appearance, with a raised central area, likely representing the nucleus (Clark and Glagov, 1976b).

2. CONSTITUENTS OF THE ENDOTHELIAL CELL

a. Pinocytotic vesicles: The endothelial cell shows many typical organelles, including mitochondria, microtubules, and microfilaments, but in keeping with a major functional purpose, namely, regulation of permeability, is the presence of abundant pinocytotic vesicles. The latter structures are responsible for the movement of material from the lumen of the vessel into its wall (Bruns and

Blood Vessels and Lymphatics in Organ Systems
Copyright © 1984 by Academic Press, Inc.

Palade, 1968). In such regions as the central nervous system, where the junctional complexes are fused, these vesicles provide the only means of transporting material from lumen to vessel wall (Hirano *et al.*, 1970). On the other hand, in organs in which the junctions are not fused, the junction apparatus may supply access of some of the luminal contents to the underlying vessel.

b. *Weibel–Palade bodies:* The endothelium contains a unique organelle, the rod-shaped tubular body or Weibel–Palade body (Weibel and Palade, 1964). This structure is approximately 0.1 μm in diameter and is up to 3 μm in length, giving it an oblong shape. The interior of the tube is occupied by cylindrical tubules, 20 nm in diameter, embedded in a dense matrix surrounded by a single unit membrane. Rod-shaped tubular bodies identify the endothelium and are seen throughout the vasculature (but rarely in capillaries), being most prominent in large veins. The origin of these highly organized structures is not clear, but they may be derived from the Golgi apparatus (Sengel and Stoebner, 1970). Their functions are also unknown, although it has been suggested that they are involved in hemostasis (Burri and Weibel, 1968). Rod-shaped tubular bodies appear to be plentiful in central nervous system endothelia in the case of vessels supplying cerebral tumors (Hirano, 1976).

c. *Junctional complexes:* Each endothelial cell is closely adjoined to its neighbor by different types of junctional complexes, depending upon the milieu of the vessel. These structures range from fused junctions, as seen in the cerebral vascular system, to loose junctional arrangements, noted in blood vessels of the kidney, bone marrow, liver, and spleen (Majno, 1965). In this regard, endothelia seem capable of undergoing morphologic change influenced by the organ in which they are located. For example, Hirano and Zimmerman (1972) described blood vessels growing from the central nervous system into a metastatic kidney tumor, in which the endothelium was transformed, taking on the appearance of the renal endothelium.

d. *Endothelial cell surfaces:* There are three surfaces to the endothelial cell: nonthrombogenic, adhesive, and cohesive. The one that faces the lumen is nonthrombogenic and is devoid of an electron-dense connective tissue, but it does possess a glycocalyx that can be seen after staining the endothelium with special stains, such as ruthenium red (Gerrity *et al.*, 1977). The adhesive inner or abluminal surface is attached to the connective tissue of the subendothelial zone. It is on the vessel side of the endothelial surface that dense condensations of filaments, called half-desmosomes, are found. It has been suggested that these organelles, together with filaments that extend from the abluminal surface of the vessel, are sites for attachment of endothelial cells to the subendothelium (Leak and Burke, 1968). The third or cohesive surface adjoins endothelial cells to each other by means of cell junctions.

3. Functions of the Endothelial Cell

Until recently, the physiology of the endothelium was poorly understood, and the cell was studied in a limited capacity as a permeability barrier and nonthrombogenic surface. With the current increased attention given to intravascular thrombus formation and to cellular dynamics involved in atherosclerotic plaque development, the function of the endothelial cell (EC) has attracted great attention. Of particular importance to the understanding of its physiology have been advances in cell culture (Jaffee *et al.*, 1973a,b). Prior to this approach physiologic studies of ECs were extremely difficult because of their position in the vasculature and their relative scarcity. With the advent of cell culture, the biologic nature of these cells has come under intensive study now that substantial quantities of ECs are available.

a. Role of endothelium: It is clear from numerous studies that a primary function of endothelium is that of a permeability barrier, screening out significant quantities of blood-borne materials. It is estimated that the endothelial cell allows only between 1 and 10% of the intraluminal concentration of proteins into the vessel wall (Bratzler *et al.*, 1977). In addition to this function, the endothelial cell appears to be a site of synthesis of a number of important materials. Included are the prostaglandins and especially prostacyclin (Weksler *et al.*, 1977), factor VIII antigen (Jaffee *et al.*, 1973b), fibronectin (Jaffee and Mosher, 1978), and histamine (Hollis and Rosen, 1972). It also participates in the metabolism of chylomicrons through the action of lipoprotein lipase (Schoefl and French, 1968), and it can process a number of vasoactive materials, such as bradykinin (Ryan and Ryan, 1977), serotonin (Small *et al.*, 1977), and norepinephrine (Hughes *et al.*, 1969). In addition, the ECs probably convert angiotensin I to angiotensin II (Johnson and Erdos, 1977) by means of the angiotensin converting enzyme. The EC also has unique immunologic characteristics, possessing ABO antigens and factor VIII antigen (Jaffee *et al.*, 1973a), α_2-macroglobulin antigen (Becker and Harpel, 1976), and tissue factor (thromboplastin) antigen (Stemerman *et al.*, 1976). The wide diversity of antigens found on the endothelial surface may explain, in part, the susceptibility of the cell to immunologic injury (Hardin *et al.*, 1973). Therefore, it appears that the EC not only serves a multiplicity of primary vascular functions, but it also is a highly diversified synthetic cell, the function of which may markedly influence the response of vessels to a number of agents or stimuli.

4. Factors Relating to Growth of Endothelium

Because the endothelium has such a diversification of function and is strategically located at the interface between the blood and the blood vessel, knowledge of its growth chracteristics is of considerable importance to the understanding of normal and pathologic conditions of the vessel wall. Of importance in this regard is the fact that *in vivo*, ECs can be induced experimentally to slough from the vessel wall under conditions as variable as blood chemical alterations to mechanical manipulation of the animal.

a. Rate of regrowth of endothelium: Following the use of a balloon catheter to remove large areas of aortic endothelium, the rate of recovery of ECs regrowing from branch sites of the otherwise denuded artery varies, depending upon the mammalian species studied. In rabbits, there is slow covering of the surface by new endothelium, the process taking up to 9 months for complete recovery (Stemerman *et al.*, 1977). In contrast, regrowth in the rat (Bettmann *et al.*, 1981) and pig (Scott *et al.*, 1980) is much faster, with aortic recovery being attained within 1 month. In the latter two animals, with injuries that denude only a few cells at a time, recovery of the deendothelialized region takes only 48 hr. Recovery under these circumstances is associated with replication of the endothelial cells; however, if the area of damage is kept to only 2–4 ECs, recovery can be attained by cellular spreading without replication (Reidy and Schwartz, 1981).

b. Importance of regrowth of endothelium: Endothelial regrowth has been considered necessary in order to reestablish normal vascular permeability, a normal functioning nonthrombogenic surface, and the synthetic function of these cells. Recovery by the endothelium quickly reestablishes a permeability barrier which is lost with its removal (Stemerman, 1981). Less crucial to this function, is the nonthrombogenic surface attributed to ECs. If smooth muscle cells (SMC) establish themselves at the luminal surface (Stemerman *et al.*, 1977), they show little ability to act as a permeability screen (Stemerman, 1981). However, they do appear to be a relatively effective nonthrombogenic surface, except in areas where electron-dense material may occupy the luminal surface of the SMC. In these limited regions, platelets may be seen to attach to the surface, although little is known about their state and whether they undergo release of their intracytoplasmic constituents; furthermore, no fibrin is found on the SMC surface.

c. Implications of regrowth of endothelium: Much more complex and potentially important is the effect that the regenerating ECs may have on the underlying neointima. With regrowth of the ECs over a thickened intima, there appears to be a change in the underlying connective tissue (Minick *et al.*, 1977). In this reendothelialized region, the extracellular connective tissue of the neointima stains with Alcian blue and, more importantly, is associated with the accumulation of lipid, in contrast to the non-reendothelialized regions. It appears, therefore, that the regenerating endothelium may exert an organizational effect on the underlying neointima. The supposition is that there is a change in the glycosaminoglycans of the neointima under the regenerated region, leading to retention of lipid. Such a feature of endothelial regrowth may have important implications for atherogenesis.

  d. *Role of tissue cultures:* The development of tissue culture as a means of study of endothelial cells may provide significant data regarding their growth and organization. Initial cultures consisted of human umbilical vein endothelium grown in high serum concentration as a starting material. However, only primary cultures of human cells could be used, and, as a result, many studies on EC growth were restricted to such species as bovine or porcine endothelium. Recently, Maciag *et al.* (1981) identified an endothelial cell growth factor (ECGF), obtained from bovine hypothalamus that allows repeated passage of human umbilical vein endothelial cells (HUVEC), using human fibronectin as a matrix for the cells. Furthermore, ECGF delays premature senescence of HUVEC and markedly reduces serum requirement for growth in culture. Thus far, ECGF has been demonstrated to be a potent EC mitogen, causing proliferation of HUVEC under quiescent conditions. Two forms of this growth factor have been identified: one with a high molecular weight (70,000) and one with a low molecular weight (17,000–25,000) (Maciag *et al.*, 1982). A material obtained from the culture medium of bovine ECs, which has been termed endothelial cell-derived growth factor (ECDGF) (Gajdusek *et al.*, 1980), has been found to act on mesenchymal cells other than EC to promote proliferation. Whether this substance is involved in reorganization of the neointima, as mentioned above, is unknown. It should also be noted that the growth of endothelial cells characteristically differs from the development of other mesenchymal cells, both *in vivo* and *in vitro*, since the endothelial cells do not require growth factor(s) from the platelet (platelet-derived growth factor) or pituitary factors, as is the case for other mesenchymal cells of the vessel wall (Maciag *et al.*, 1981).

  e. *Stability of endothelial cells:* Endothelial cells, if left undisturbed, are remarkably stable, with a low turnover rate *in vivo*. The labeling of endothelium by tritiated thymidine varies from 0.01% in the retina to 0.13% in the myocardium (Florentin *et al.*, 1969). However, there does appear to be a predilection for endothelial turnover in the area of the arteries most subjected to stress. Moreover, increased thymidine labeling has been demonstrated in growing animals (Spaet and Lejnieks, 1967). Also, increased labeling has been found at arterial branch sites, which has been attributed to local hemodynamic events, an observation which may help explain the predilection of such locations for intimal thickening. Hemodynamic effects exerted on the endothelium are thought to be caused by shear stress exerted by sudden alterations in flow as the blood exits to branches.

Increased incorporation of tritiated thymidine (Caplan and Schwartz, 1973) is also associated with a variety of vascular injuries produced by chemical or physical means. Gaynor *et al.* (1970) and Gaynor (1971) showed that a single intravenous injection of endotoxin in heparinized rabbits produces increased labeling as well as detachment of the endothelium. Blood smears from such animals contain circulating ECs, whereas smears from control animals do not. Such an observation indicates that accelerated labeling in this instance may be a product of endothelial sloughing and regrowth.

5. Role of Endothelium in Hemostasis

Initiation of thrombogenesis is usually associated with activation via a contact surface. However, the endothelium provides a surface for the vessel containing the blood without influencing activation of either the coagulation proteins or platelets. The reason the endothelium is able to function as a nonthrombogenic surface is not entirely clear, since it possesses features which are both pro- and antihemostatic.

The endothelium is known to contain factor VIII antigen (Jaffee *et al.*, 1973a,b) and tissue factor antigen (Stemerman *et al.*, 1977) and has been shown to synthesize both factor VIII antigen (Jaffee *et al.*, 1973a,b) and type IV collagen (Jaffee *et al.*, 1976). All these substances may contribute to thrombus formation, either by acting within the coagulation mechanism or by providing a surface for hemostatic activation.

On the other hand, the endothelium is a major source of prostacyclin (PGI_2), the most potent known inhibitor of platelet aggregation (Weksler *et al.*, 1977). In addition, it is known that ECs contain an inhibitor of coagulation, α_2-macroglobulin, and release plasminogen activator which promotes fibrin lysis (Loskutoff and Edgington, 1977). Also, the surface of the endothelium may be necessary for the activation of protein C, which when activated is a potent inhibitor of coagulation (Owen and Esmon, 1981). The dynamics of these presumably offsetting functions of the endothelium are difficult to explain. Nevertheless, the endothelium does provide a nonthrombogenic surface, and, apparently, even when injured *in vivo*, platelets will not adhere to it.

6. Injury of Endothelium

A central issue of endothelial biology and vascular pathobiology is the state of health of ECs. An example of this theme is the response to injury hypothesis of atherogenesis (Stemerman and Ross, 1972; Ross and Glomset, 1976). Implied in such a view is the notion that with endothelial injury comes EC loss and that, through the latter, a cascade of events follows, promoting SMC proliferation, intimal growth, and luminal narrowing. Several studies indicate that, in addition to mechanical means of denuding endothelium, changes in components of the blood, such as elevated lipoprotein levels or homocysteine levels, can induce EC sloughing (Harker *et al.*, 1976). Recent restudy of elevated levels of hypercholesterolemia in rabbits, however, do not demonstrate sloughing (Stemerman, 1981); previous studies showing large areas of endothelial loss were likely due to preparation artifact. When the endothelial surface is exposed to elevated levels of cholesterol, a functional change can be induced in focal areas of the endothelium, as demonstrated by increased uptake of horseradish peroxidase (Stemerman, 1981) or low density lipoprotein (LDL). Although the EC morphology in these foci show little substantial alteration, there is a change in function, termed endothelial dysfunction (Stemerman, 1981). It, therefore, appears that endothelial cells may remain attached to the luminal surface of the vessel wall and still undergo profound functional changes. These alterations may augment the passage of LDL or other materials into the vessel wall, thus promoting atherogenesis. Platelet-derived growth factor(s) and other stimulatory substances can also adversely affect vessel intima in involved focal regions. (For discussion of role of endothelium in thrombosis, see Section E-2b, Chapter 10.)

7. CONCLUSIONS AND FUTURE AVENUES FOR INVESTIGATION

The endothelial cell has increasingly gained attention. Close study of this structure has demonstrated that it is highly complex and often puzzling in its contrasting functions. Recent studies, however, have provided a series of important insights into explaining more fully the function of a cell that occupies such a critical position in the vasculature. In addition to its acknowledged permeability and nonthrombogenic functions, the EC has been shown to be an important synthetic site for many substances influencing the biology of the vessel wall. Further studies may provide a clearer understanding of these functions and also permit better insight into the control of pathologic entities of the vessel wall.

D. PHARMACOLOGY*

By J. A. Bevan and M. P. Owen

There are wide variations in the pharmacologic responses of blood vessels. Those of different size from the same vascular bed, those of the same size in different beds, arteries and veins lying within the same fascial sheath, and the same anatomically defined blood vessel in different species may exhibit remarkably different functional responses. Although some diversification appears to be of teleological benefit, the reason for most of the observed variations remains obscure. This section attempts to provide some generalizations concerning the pharmacologic attributes of the vasculature, although many exceptions may be found. Specific regional vascular pharmacology is found in appropriate sections throughout this volume.

1. Species Differences in Pharmacologic Properties

Differences in blood vessels are found among species (see, for example, Toda, 1977) in their anatomic organization; receptor populations and characteristics; innervation density, pattern, and ultrastructure; putative transmitters; relative dominance of constrictor and dilator innervation; extent and pattern of spontaneous tone; and excitation–contraction and excitation–relaxation coupling characteristics. On the other hand, there are many pharmacologic similarities between blood vessels. Present information, however, seems insufficient to allow meaningful generalizations to be made.

In a survey of the literature on the *in vitro* study of human blood vessels, Bevan (in press) concluded that adrenergic-related properties probably showed less variability than do other features, and virtually all the major attributes of the adrenergic-neuroeffector mechanism established in animal studies have been confirmed in one or more human vascular segments (Fig. 1.7). However, at present there is insufficient evidence to suggest which of the commonly used laboratory animal species is closest to man. The extent of species differences raises the question of advisability of drug testing on animal tissue if appropriate human material can be utilized.

*The new research described in this section was supported by the United States Public Health Service Grants HL 15805, 20581, and 26414.

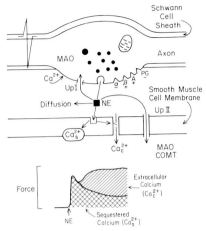

FIG. 1.7. Diagrammatic sequence of steps involved in the vascular adrenergic neuroeffector mechanism: (1) Action potential propagation along an adrenergic axon results in (2) Ca^{2+} entry into presynaptic axoplasm to effect release by (3) exocytosis of vesicles containing the transmitter [norepinephrine (NE)]. [Release has been shown to be modulated by both positive and negative feedback systems, including presynaptic α-adrenoceptor (negative), β-adrenoceptor (positive), prostaglandin (negative), and angiotensin II (positive)]. NE appears to be stored in both small and large dense core vesicles. (4) NE diffuses across the synaptic space to smooth muscle cells. (5) The effective concentration of NE at the receptor sites on smooth muscle is reduced as a result of (a) neuronal reuptake (UpI), (b) diffusion into surrounding cellular spaces, and (c) extraneuronal uptake (UpII) and subsequent muscle metabolism. (6) Transmitter (NE) binds to postjunctional smooth muscle adrenoceptors (both α and β). (7) The α-adrenoceptor is coupled to sequestered Ca^{2+} pools (Ca_s^{2+}) and Ca^{2+} channels (Ca_c^{2+}). (8) Two separate additive mechanisms [release of sequestered Ca^{2+} (Ca_s^{2+}) and influx of extracellular Ca^{2+}] are responsible for the biphasic contraction of the vascular smooth muscle to NE.

2. ADRENERGIC REGULATION OF BLOOD VESSELS

Some of the pharmacologic behavior of blood vessels may be related to their diameter (Bevan *et al.*, 1980). In general, the smaller the vessel, the narrower is the synaptic cleft (Rowan and Bevan, in press). When the cleft is wide, the norepinephrine concentrations that result from nerve stimulation are similar within, and immediately outside, the synapse (Bevan, 1977). A wide synaptic cleft favors a slow generalized accumulation of transmitter in the wall of the blood vessel and is consistent with a modest level and slow development of neurogenic tone. Adrenergic receptors are believed to be distributed throughout the width of the vascular muscle layer in tissues with wide synaptic clefts, although there appears to be a variation in sensitivity at different depths within the wall (see Section D-3, below). Wide clefts favor termination of the transmitter effect predominantly by diffusion and by uptake into smooth muscle cells and subsequent metabolism by the muscle (Gero and Gerova, 1971). These processes are slower than reuptake of transmitter into nerve varicosities.

With a narrow cleft, concentrations of norepinephrine within the synapse are higher than outside the synapse (Bevan, 1977). Such a gradient, as well as the location of an enveloping Schwann cell over the extra- but not the intra-synaptic surface of most varicosities, suggests that transmitter is released exclusively from the presynaptic membrane. The narrow synaptic cleft also favors rapid development of tone, which is dependent on myogenic spread. The importance of myogenic coordination in small blood vessels has been shown by electrophysiologic studies of the mesenteric arterioles in which excitatory junction potentials summate with the initiation of muscle action potentials (Hirst, 1977). The possibility of a restrictive location of adrenergic receptors in vessels with narrow clefts is derived from studies of the rat portal vein (Ljung, 1975), the vasculature of adipose tissue (Rosell and Belfrage, 1975), and the rabbit facial vein (Winquist and Bevan, 1981b). Thus, when the cleft is narrow, transmitter tends to remain within the cleft, to have an action within the cleft, and to have its effects terminated by neuronal reuptake. The latter is an efficient and rapid process.

Resistance vessels are believed to maintain active myogenic tone (Folkow, 1964). Two vessels with narrow synaptic clefts, the rat portal vein and the rabbit facial vein, have been used as models for phasic and tonic vascular activity, respectively. Tone has also been found in resistance vessels of the rabbit ear *in vitro* (Owen *et al.*, 1982, 1983). One consequence of the increased tone observed in small vessels is that vasodilator effects, by whatever mechanism, become functionally important in these vessels.

Two basic types of adrenergic receptors are recognized in blood vessel smooth muscle—α-adrenergic and β-adrenergic. Binding of agents to α-adrenergic receptors usually initiates events that lead to smooth muscle contraction and vasoconstriction, whereas binding to β-adrenergic receptors causes smooth muscle relaxation and vasodilatation. A number of investigators have proposed that α-adrenoceptors may be considered as "innervated" receptors, in contrast to the β-adrenoceptors, which may be more accessible to circulating catecholamines and are thus considered "humoral" receptors (Glick *et al.*, 1967; Hamilton and Feigl, 1976; Guimaraes and Paiva, 1977; Russell and Moran, 1980).

Studies of the rabbit facial vein, a vessel which develops myogenic tone *in vitro* and displays β-adrenoceptor-mediated neurogenic relaxation, imply that β-adrenoceptors in this vessel are in close proximity to the site of adrenergic transmitter release and neuronal uptake, whereas the α-receptors are more distant (Winquist and Bevan, 1981b). An investigation in the feline pulmonary vascular bed indicates that neuronally released norepinephrine can act on both α- and β-receptors (Hyman *et al.*, 1981). Moreover, vasodilator responses appear dependent on the existing level of tone in the pulmonary bed. β-Adrenergic-mediated vasodilation after α-adrenergic blockade has been observed in small resistance vessels (Lundvall and Järhult, 1974, 1976; Gründe, 1979), suggesting that vascular β-receptors may be of functional importance in the nervous control of vessels which display intrinsic myogenic tone. In any case, rapid disposition and termination of biologic effect are prerequisites for the rapid readjustments of vascular tone with changes in sympathetic outflow that are necessary in the small resistance blood vessels.

The actions of some sympathomimetic agents, which mimic at least a number of the actions of the sympathetic nervous system, and sympatholytic agents, which block sympathetic activity, are influenced by the innervation density of a blood vessel. The reason for this is apparent. All other neuroeffector parameters being equivalent, the greater the innervation density, the more marked is the influence of the sympathetic nervous system. This sympathetic control is effected by drugs which modify the adrenergic neuroeffector process locally. Clearly the consequence of this interference is a function of the level of tonic control. Some sympathomimetic agents—the "indirectly acting amines"—rapidly displace the transmitter from its storage site, the transmitter then reacting with the receptors. Others interfere with the disposition (reuptake) of transmitter by the axonal varicosity (uptake$_1$) and hinder the disposition of transmitter by smooth muscle uptake (uptake$_2$).

Sympatholytic agents can be divided into several groups. Some block biosynthesis of the transmitter; others interfere with its storage, resulting in its slow liberation and prevention of its reaccumulation. Yet other drugs hinder transmitter release, and a final group blocks the interaction of transmitter with postjunctional receptors. When two vessels have a similar density of innervation and differ only in their synaptic cleft width, the one with the narrower cleft can be expected to be affected to a greater extent by a drug which blocks nerve varicosity reuptake than by one which blocks smooth muscle uptake and metabolism. The reverse applies to a vessel with a wide synaptic cleft. One consequence of uptake$_1$ blockade in a vessel with a narrow cleft is that it is more difficult to block α-receptors.

3. ASYMMETRY OF CONTRACTILE RESPONSE

The characteristics of the contractile response of some arteries and veins appear to vary according to whether the vasoactive agent enters the media via the adventitia or the intima. This has been shown to be true at least for norepinephrine, histamine, and serotonin. For example, norepinephrine applied to the intimal surface of the rabbit ear artery causes a greater constriction than when placed on the adventitial surface (de la Lande et al., 1967; de la Lande and Jellet, 1972; Kalsner, 1972). This difference probably results from neuronal uptake mechanisms for norepinephrine (de la Lande *et al.*, 1967; de la Lande and Jellet, 1972). In the rabbit thoracic aorta, after blockade of neuronal uptake, differences in steady state contractile force observed with intimal and adventitial application of the drug seem to depend mainly on extraneuronal uptake and disposition mechanisms (Pascual and Bevan, 1980). The initial time course of responses to a drug differs according to the surface by which it enters the media. For example, contraction elicited by norepinephrine applied to the intima of a rabbit aortic strip takes place after a shorter latency and proceeds at a higher initial velocity than when the drug enters through the adventitia. Such effects are independent of amine uptake mechanisms (Pascual and Bevan, 1980). In the case of the rabbit ear artery, a vessel which responds to both norepinephrine and serotonin with a biphasic contraction, the initial phase (one that may be associated with a propagated electrical signal and the recruitment of intracellular Ca^{2+}) is greater when the amines are applied to the intimal than to the adventitial side of the vessel (McCalden and Bevan, 1980). This effect is also independent of the amine uptake mechanisms and may be the cause of the asymmetry seen after disposition blockade.

The differences between the inner and outer media may reflect a functional subdivision of the vessel wall. It is possible that the outer smooth muscle cells, being relatively insensitive to norepinephrine, may only react to this substance when released from nerves, under which conditions a high concentration is released to the smooth muscle cells. The inner cells, on the other hand, are relatively more sensitive and will respond to endogenous circulating factors. The mechanism of myogenic coupling, operating when the wall is exposed to a vasoactive agent, may be the method whereby excitation spreads rapidly into the deeper layers of the muscle wall.

4. MYOGENIC VASCULAR TONE

Analysis of pharmacologic vasodilatation frequently neglects to consider the level of initial vascular tone. Commonly, vasodilator agents are screened for their efficacy against extrinsic tone, despite the fact that a major determinant of vascular resistance in many beds is tone of intrinsic myogenic origin. In the absence of alternative evidence, it is tacitly assumed that vascular tone is essentially the same, irrespective of its origin, and hence that it has a similar pharmacology. However, there are several lines of evidence to suggest that this is not the case. For example, in the isolated rabbit facial vein, the concentration–inhibitory relationship of isoproterenol, salbutamol, and hydralazine (but not papaverine and sodium nitrite) has been shown to be altered when tone is initiated with histamine, as compared with the relationship when tone is initiated myogenically (Winquist and Bevan, 1980). In a study of second generation slow-calcium antagonists, such as diltiazem and nimodipine, it was found that drug- and potassium-induced tone was highly sensitive to the action of such agents, whereas this was not the case for intrinsically maintained tone, whether in large or small arteries or veins (Bevan, 1983). Such a finding was surprising, since both types of tone are exquisitely sensitive to alteration in external calcium concentration (Winquist and Bevan, 1981a). Nevertheless, the sensitivity of spontaneous myogenic tone, which develops via voltage-sensitive channels as occurs in the rabbit portal vein, was high. It must be concluded that the receptor for calcium antagonists associated with calcium channels in the smooth muscle cells is not identical for all types of channels. The above examples cast strong doubt on the general assumption that the pharmacology of vascular tone is the same irrespective of origin.

5. ENDOTHELIUM-DEPENDENT DILATATION

Furchgott and Zawadzki (1980) observed that the dilator action of acetylcholine, mediated through muscarinic receptors, was dependent on the integrity of the tunica intima. Experimental evidence obtained by these workers suggested that a substance released from the endothelial cells acts on the surrounding medial smooth muscle to cause relaxation. In the case of the rabbit aorta, it is probably a lipoxygenase derivative of arachidonic acid or some other unsaturated fatty acid.

A variety of substances in blood vessels owe part or all of their dilator capacity to an endothelial-dependent effect. These include the cyclic nucleotides; polypeptides, such as bradykinin and vasoactive intestinal polypeptide (VIP); some calcium ionophores; and vasodilator drugs, such as hydralazine. There is evidence that the tunica intima is also obligatory for the effect of some vasodilator nerves innervating cerebral and a number of extracerebral cranial arteries. Such nerves are believed to release acetylcholine and other substances, probably VIP into the walls of the vessels. The agents exert their vasodilator action through the release of a principle from the innermost layer of the arterial wall (Bevan et al., in press). (For additional discussion of endothelial cell function, see Section C, above; Section C-2c, Chapter 3; and Section D-4, Chapter 11.)

6. VASCULAR DIVERSITY AND EMBRYOGENESIS
OF THE CIRCULATION

Two groups of studies—one of the central arterial system and the other of thoracic and abdominal veins—have drawn attention to the remarkable similarity between the distribution of some vascular characteristics and the pattern of development of the circulation. Variability in the anatomy of the large central arteries and veins has been attributed to sequential events that take place when the simple segmental, bilaterally symmetric, arterial system of the embryo is transformed into the complex, asymmetric pattern of the mature animal. Anatomic variants are understood as developmental dissimilarities of this process (Wilson, 1961). Recent detailed functional studies of segments of the circulation reveal abrupt changes in the α-adrenergic properties of the vascular smooth muscle in the larger parts of the arterial tree (Ransom and Bevan, in preparation). In the rabbit, these transitional sites occur close to the roots of the celiac, superior and inferior mesenteric, renal, and lumbar arteries and at specific sites in the brachial, vertebral, and internal carotid arteries. Such abrupt transition sites demarcate a central arterial compartment in which the smooth muscle is more sensitive to norepinephrine—usually by several orders of magnitude—than in more distal vessels. There is no consistent change in adrenergic innervation density or pattern, and the blood vessels at these locations do not seem to possess common dimensions or hemodynamic significance.

It has been accepted for some time that smooth muscle cells of the circulation have a common origin from local mesodermal cells. However, Le Lievre and Le Douarin (1975) and Le Douarin (1980) have demonstrated that those in arteries that develop from the branchial arch system are derived at least in part from neural crest cells that have migrated laterally and ventrally from the neural folds. Thus, the cells of the central arterial compartment have a different origin from those of the peripheral vessels, because they arise from the same cells from which much of the peripheral nervous system, endocrine glands, and tissues of the head and face develop. There is remarkable similarity between the exact location of two functional transition sites and the limits of the branchial arch system, as defined by classic embryologic techniques (Bevan, 1979).

Shigei *et al.* (in press) have systematically surveyed many of the physiologic and pharmacologic characteristics of the central veins of the dog and have related these to venous embryology. Their findings show that the veins of the body wall exhibit a pattern of drug sensitivity that is different from those that drain the tissues of the digestive tube and its derivatives. However, there are two exceptions: (1) the middle segement of the inferior vena cava has characteristics similar to those of the veins of the digestive tube, and (2) the splenic vein behaves like the veins of the posterior abdominal wall rather than like the associated portal and mesenteric veins. In this regard, Shigei *et al.* (in press) argue that these two major groups of veins have different origins, those of the abdominal wall arising from primary ventral veins derived from body wall mesenchyme. In contrast, the muscle of the veins of the digestive tube, like that of the intestine itself, is derived from a dorsal group of primitive veins. These workers emphasize that the common origin of smooth muscle of the portal vein and of the gut may account for some of the functional similarities. The inferior vena cava is a developmental mosaic (Shigei *et al.*, 1978), in that the central part has many physiologic and pharmacologic similarities to the portal and mesenteric veins, and there is evidence that this particular segment may develop from the afferent veins of the liver (Takaoka, 1956). During embryogenesis, the splenic vein functions as a shunt between the veins of the abdominal wall and those of the mesentery (Miki, in press), a finding which may explain the seemingly anomalous functional properties of this vessel.

The location of abrupt junctions in veins can be visualized, and it has been suggested that, in the embryo, the development of the raised pressure in the artery segments causes changes in the blood vessel wall that tend to obscure embryologically determined structural differences (Shigei *et al.*, in press).

E. PATHOLOGY OF LARGE ARTERIES

By S. Glagov

1. General Considerations

Several conditions are specific affections of the large arteries, and their underlying pathogenetic mechanisms appear to be complex, involving the interaction of several factors over extended periods of time. The latter include exposure of the arterial wall to abnormal concentrations of normal blood components, variations in patterns of blood flow, elevation of blood pressure, and predispositions related to abnormalities of arterial cellular metabolism. In addition, exogenous substances which gain access to the bloodstream may affect endothelial permeability or integrity or otherwise modify the metabolism of arterial wall cells. Conditions which primarily affect the intima tend to narrow the lumen, thus interfering with volume flow or causing local instabilities of the flow field. Those states which involve the media are associated either with increased or decreased compliance and may therefore lead either to ectasia and/or disruption of the artery or to altered pulse pressure patterns. Pathologic processes which affect the adventitia may result in reduced compliance of the wall and/or interference with medial nutrition by involvement of the vasa vasorum. Regardless of the underlying initial process, however, most conditions eventually affect the structure and function of all three arterial wall coats. The common pathologic states which involve large arteries, i.e., the aorta and the principal proximal conducting and distributing branches, are outlined below, with particular emphasis on the relationship between the characteristic

Blood Vessels and Lymphatics in Organ Systems
Copyright © 1984 by Academic Press, Inc.

morphologic changes and the functional anatomy of the vessel wall. (For structural and mechanical aspects of maintenance of wall integrity, see Section A, above, and Section A, Chapter 2.)

2. ARTERIOSCLEROSIS

The category of arteriosclerosis includes a group of changes characterized mainly by stiffening, thickening, and deformation of the arterial wall, due to excessive or disproportionate accumulation of normal and/or abnormal connective tissue elements and/or degenerative debris. The common conditions in this group causing primarily intimal changes are *diffuse intimal thickening* and *atherosclerosis*. Nonspecific diffuse and local *sclerotic changes in the media* of large arteries, not directly related to intimal disease, are also frequent and are usually associated with relatively benign alterations in mechanical properties.

a. Diffuse intimal thickening: Relative to the width of the media, aortic intimal thickening appears to increase with the distance from the heart and may attain or exceed the thickness of the underlying media (Movat *et al.*, 1958), but the absolute degree of involvement may not be different in different aortic sites. Branches, bifurcations, inner aspects of curves, and areas of abnormally altered configuration and flow tend to be affected selectively. There is evidence that sites of intimal thickening may be predisposed to atherosclerosis (Stehbens, 1975; Thomas *et al.*, 1979) and that populations prone to diffuse intimal thickening manifest the same tendency (Tejada *et al.*, 1968). However, intimal thickening as such has not been shown to be either an inevitable or necessary precursor of atherosclerosis. The cells in diffuse intimal thickenings are mainly smooth muscle, with both collagen and elastic fibers being also present. The organization of these elements resembles, to a variable extent, that of the underlying media except that often there is relatively more collagen. Accumulations of lipid may be entirely absent. In contrast to the underlying media, cells and fibers, particularly those which are in close proximity to the intimal elastic lamina, tend to be oriented in the axial direction (Buck, 1979). The cells are presumed to arise from the underlying media as a result of mechanical and/or chemical stimulation to proliferation and migration. Similar changes can be induced experimentally by removal of the endothelium (see Section C-4, above). Such a procedure is thought to permit access to the media of a growth factor derived from thrombocytes or from other cells. Intimal proliferation can also be induced by altering local flow patterns (Stehbens, 1974). Whatever the causes or adaptive implications of diffuse intimal thickening, there is no evidence that as a result, the lumen is significantly narrowed, although the condition usually produces a thicker than normal arterial wall.

b. Atherosclerosis: This state is an obstructive and destructive, predominantly intimal, disease, occurring principally in the elastic and proximal muscular arteries of the systemic circulation. On the basis of human autopsy studies (Blumenthal, 1967) and several experimental models (Wissler *et al.*, 1974a), atherosclerotic lesions are thought to begin as focal intimal thickenings containing intracellular and extracellular lipid and accompanied by the accumulation of cells and the deposition of matrix macromolecules, including glycosaminoglycans, collagen, and elastin.

In preparations of fresh or fixed *undistended arteries*, atherosclerotic lesions viewed *en face* are present on the luminal surface as flat plaques, raised or elevated plaques, calcific encrustations, or ulcerated excoriations. Those which barely project from the luminal surface usually appear as yellow specks or streaks, often distributed in irregular axially oriented arrays. These so-called *fatty streaks* contain lipid-laden cells and variable quantities of matrix materials. Early experimental lesions, similar to human fatty streaks, show surface irregularities and crevices corresponding to the contours of the immediately underlying spherical intimal lipid-laden cells (Taylor *et al.*, 1978), but the overlying endothelium is usually intact. Other relatively narrow and flat intimal changes are composed mainly of amorphous intimal material and have been proposed as possible precursor sites of plaque formation (Haust, 1971; Smith *et al.*, 1979). Plaques which extend from the luminal surface as firm or rigid, raised plateaus or mounds are frequently white or translucent, well demarcated, and oval, with their long axes often oriented in the direction of flow. These are the so-called fibrous or pearly, *raised plaques.*

When arteries are fixed while distended by normal intraluminal pressures, transverse cross sections most often reveal lumens which are circular or slightly oval, regardless of the size of the intimal lesion. Under these conditions advanced experimental or human lesions appear as eccentric but symmetric widenings of the intima, with wedge-shaped edges and concave luminal surfaces (Glagov and Zarins, 1983). With more extensive disease, lesions tend to become confluent and nodular and develop irregular margins. A typical advanced raised plaque consists of a *center* or *core* and an overlying compact layer of tissue, the *fibrous cap.* The center may contain cell debris, formed matrix fibers, extracellular proteins derived from cells or plasma, intracellular and extracellular lipids in amorphous and/or crystalline form, and calcific deposits. When cell debris and amorphous material predominate, the core is considered to have originated by necrosis and decomposition of the component elements. The tissue surrounding such a lesion may contain intact cells and fibers, as well as vasa vasorum which arise from the underlying media and adventitia or, less often, from the vessel lumen. Extensive portions of the interior of lesions may also be myxomatous, gelatinous, or fibrous. In normally distended atherosclerotic arteries, the fibrous cap consists of oriented matrix fibers and cells forming a concave covering, the luminal curvature of which approximates that of the adjacent uninvolved wall. The fibrous cap is therefore usually a smooth-surfaced fibrous channel, continuous in contour with the adjacent artery wall and tending to sequester the dystrophic necrotic center from the lumen. Immediately beneath a lesion, the internal elastic lamina may be irregularly undulated, atrophic, and discontinuous, suggesting that the lesion, and particularly the fibrous cap, sustains a proportion of the mural stress associated with the distending pressure. Further support for such a splinting effect by the lesion is suggested by the presence in many instances of a straightened, neoformed elastic lamina beneath the endothelium covering the fibrous cap and by the atrophy, thinning, and outward bulging of the underlying media. Lesions are called fibrous, fibrocalcific, lipid-rich, cellular, necrotic, myxomatous, etc., depending upon their predominant or most impressive morphologic features.

Merging of adjacent plaques, secondary atheromatous deposits on fibrous caps of previous lesions, organization of thrombi upon lesions, and hemorrhages into plaques may all result in the development of complex architectural features. Calcification within lesions and in the immediately underlying media may be extensive and render the affected vessel rigid and nodular, particularly in older individuals (Rifkin *et al.*, 1979). The predominance of necrotic material would make a lesion *soft* in contrast to the *hard*, rigid, or brittle consistency of a mainly fibrocalcific lesion with an intact and prominent fibrous cap. Absence, minimal formation, or disintegration of a fibrous cap would tend to favor surface ulceration and exposure of the material in the necrotic center to the lumen, thus causing the lesion to appear as a friable, grumous excoriation, often surmounted by a thrombus.

Thrombi associated with plaques may be fresh or organizing and partially or totally occlusive. Organizing thrombi may appear on transverse sections as crescent-shaped deposits, similar in surface contour to the associated underlying lesion and the adjacent wall. Fresh thrombi are often associated with evidences of recent or remote lesion disruption, fragmentation, and/or hemorrhage, but well-organized thrombi may be difficult to distinguish from the lesions into which they have been incorporated. Thrombotic material, as well as debris from a disrupted lesion, may embolize and cause occlusions of distal arterial branches (see Section E-2, Chapter 19). Much of the symptomatology associated with carotid atherosclerosis, for example, is attributable to embolization rather than obstruction at the site of plaque formation. In the major coronary arteries, occlusive thrombosis in relation to atherosclerotic plaques has been considered to be the principal immediate cause of acute myocardial ischemia, resulting in sudden death or myocardial infarction. In many instances, occlusive thrombi have not been found on postmortem studies at various intervals after clinical myocardial infarction. Although such results have suggested that occlusive thrombi in epicardial vessels may follow rather than precede myocardial infarction (Spain and Bradess, 1970), most recent investigations indicate that an occlusive arterial thrombus on a plaque is probably the significant early event in many, if not most, instances of fatal myocardial infarction (Chandler et al., 1974; Bloor and Ashraf, 1977), (see Section E-2, Chapter 10). Prostacyclin (PGI_2), a potent inhibitor of platelet aggregation, is produced in arterial walls and from precursors present in platelets (Moncada and Amezcua, 1979). This substance is probably important in preventing thrombosis under conditions of normal arterial function and may be insufficiently elaborated at sites where the arterial wall is markedly altered by atherosclerosis. The mechanisms by which some atherosclerotic lesions become complicated by disruption, thrombosis, or hemorrhage whereas others are fibrotic, calcific, and smooth-surfaced have not been elucidated. Changes in plaque consistency due to differential resorption or accumulation of lesion components and marked differences in consistency between plaques and adjoining vessel segments would tend to favor plaque disruption, particularly under conditions of sudden alterations in mechanical stress. Blood may be presumed to enter a lesion from the arterial lumen by way of clefts created by mechanical stresses or by necrosis of the center or from the vasa vasorum which often extend into plaques and may at times cause them suddenly to bulge into the lumen (Paterson, 1952).

Narrowed and rigid vessels in which fibrocalcific lesions occupy the entire circumference may maintain in vivo luminal dimensions even in the absence of distending pressure. This effect may help to explain why correlations between angiograms and postmortem findings in undistended specimens are best for the most advanced lesions (Zarins et al., 1983), whereas reports of angiographic underestimation of narrowing by less advanced lesions may in many instances actually reflect postmortem overestimation due to measurements on collapsed vessels. Lesions may also vary widely in volume for any given decrease in lumen diameter, for vessels tend to dilate with age and possibly also in association with the formation of atherosclerotic lesions (Bond et al., 1981; Zarins et al., 1983). In any case, quantitative estimates of size and effect of lesions on vessel lumen size and configuration are likely to be most accurate in specimens fixed while distended.

A distinction has been drawn between atherosclerotic deposits which are "eccentric", i.e., associated with partially preserved arterial wall sectors and nonaxisymmetric lumens, and ones which are "concentric", i.e., completely encircling with or without axisymmetric lumens (Roberts, 1977). Inferences have been drawn that these configurations may have pathogenic significance and may represent different types or degrees of individual reaction to atherogenic stimuli. It has been suggested, for example, that concentric lesions with marked early medial involvement may result in part from an associated abnormal immune reaction (Wissler and Vesselinovitch, 1983). The extent to which coronary artery plaques and plaques in other locations are indeed "eccentric," as opposed to "concentric" or encircling in the course of lesion evolution, remains to be investigated.

Within any given arterial tree, vessels are not involved uniformly and some are commonly spared regardless of the severity of disease elsewhere (Roberts *et al.*, 1959; Glagov and Ozoa, 1968). The human abdominal aorta is nearly always more severely affected than the thoracic segment. Arteries to the lower extremities may be severely diseased, whereas branches to the upper extremities are nearly always spared. The carotid bifurcation is commonly the site of severe disease, particularly about the sinus of the internal carotid branch, whereas the adjacent common carotid and external carotid branch are relatively spared. The coronary arteries are more severely diseased than other arteries of similar size. Except for involvement of their ostia, the renal, celiac, and mesenteric arteries are nearly always spared. No completely adequate explanation has been furnished for these discrepancies. Special mechanical features which prevail at sites of preferential involvement (Glagov, 1972; Nerem, 1981) include alterations in shear stress due to changes in flow rate, turbulence of flow, and increased mural tensile stress due to geometric configuration and/or increased pressure. For the coronary arteries, marked excursions in flow rate and medial tension during the cardiac cycle, abundance of branchings, mechanical torsions and flexions of vessels associated with cardiac motion, and the special reactivity of coronary artery smooth muscle to vasoactive substances and nervous impulses have all been suggested as predisposing factors. Although the prevalence of atherosclerosis and the extent to which the coronary arterial tree is affected tend to increase with age (White *et al.*, 1950), characteristic complex lesions have been shown to occur by the second and third decades (Yater *et al.*, 1948), and intimal thickening, thought to be a precursor or potentiator of atherosclerosis, has been found to be present in coronary arteries of children within the first decade (Persmen *et al.*, 1975). Thus, individual differences in susceptibility may already be operative early in life, and the latent period between initiation of lesions and onset of clinical symptoms may be longer and more variable than previously realized. In some populations, however, comparatively few lesions are evident at any age (McGill, 1968). Furthermore, even in groups prone to develop severe atherosclerosis, many individuals have few lesions. Women develop less atherosclerosis than men, but this difference may be narrowed after menopause or oophorectomy, suggesting that hormonal status may be an important conditioning or protective factor (Rivin and Dimitroff, 1954; Wolinsky, 1973; Gordon *et al.*, 1978). Clinical data dealing directly with hormone levels and atherosclerosis are scant, however. Systemic hypertension also accelerates the disease in the major arteries and tends to reduce the differential involvement of the two aortic segments. The low pressure pulmonary arteries are largely spared, as compared with the aorta, except in the presence of pulmonary hypertension.

Because atherosclerotic lesions regularly contain lipid, occur earlier, and are more severe in persons with documented abnormalities of lipid metabolism, and because they can be induced experimentally by rendering animals hyperlipidemic by diets rich in fats, the ingress and/or persistence in the intima of abnormal quantities of lipid has been considered to be a major factor in atherogenesis (Duff and McMillan, 1951; Adams, 1981). Evidence has accrued which implicates the *low density, intermediate density,* and *very low density* fractions of the serum lipoproteins as the probable principal carriers of the cholesterol which forms such a prominent component of the lesions. There is also evidence that the *high density* lipoprotein fraction may act as a carrier which diverts or mobilizes cholesterol and inhibits lesion formation (Miller, 1981). Thus, the relative proportions of these fractions and the ability of cells to deal with various lipoprotein fractions may be determinants of lesion induction and progression.

Excessive intimal ingress and accumulation of lipoproteins and other blood components have been attributed to injury and/or to altered permeability of the endothelium. Such modifications of endothelial integrity and function have been considered to be due to a large number of endogenous and exogenous circulating agents, including low density lipoproteins and/or agents originating from tobacco smoke. Because the accumulation of cells is also a prominent feature of lesion formation and evolution, substances which induce cell proliferation have been assigned a putative role in atherogenesis. In particular, a growth factor isolated from blood platelets has been proposed as a major stimulant to arterial smooth muscle cell proliferation and lesion initiation. Such a mechanism requires disruption of endothelial continuity, with platelet deposition on the denuded surface, and resulting induction of focal intimal hypercellularity, with subsequent necrosis and accumulation of lipids if the injury is not resolved. This "response to injury" hypothesis for lesion formation, suggested many years ago by others, has recently been accorded considerable attention (Ross and Glomset, 1976). Other cells involved in lesion formation, such as macrophages, have also been shown to contain growth factors. Low density lipoproteins may likewise stimulate cell proliferation. Although many of these factors have been shown to induce a proliferative response in cultured smooth muscle cells, their role as inducers of an initiating proliferate phase in atherogenesis has not been demonstrated. Clusters of cells in advanced lesions appear to be monotypic with regard to certain isoenzymes, suggesting that at least some of the cell proliferation noted in lesions results from single cell transformations induced by circulating mitogens which gain access to the arterial wall (Benditt and Benditt, 1973; Benditt and Gown, 1980). Since fatty streaks appear to be reversible and their cells not monotypic, such lesions may be targets for the transforming events which lay the basis for the more persistent and complex forms of the disease (Pearson *et al.*, 1978).

The frequent occurrence of fibrin deposition and/or thrombus formation in association with atherosclerosis had led earlier investigators to propose a thrombogenic or "accretion" theory of atherogenesis (Duguid, 1960). Lesion formation and/or enlargement by this means would result in large measure from the focal deposition and incorporation of thrombus components into the intima. The normal inhibition of thrombus formation on the luminal surface has been attributed, in large part, to the presence of prostacyclin, a powerful antiaggregation agent for platelets (Moncada and Amezcua, 1979). In keeping with the suggested role of endothelial injury as an initiating event, it has been suggested that endothelial cells in zones of high shear stress are especially vulnerable to injury and atherogenesis due to increased flow velocity (Nerem, 1981). Evidence has been adduced, however, in both human and experimental material, that zones of relatively low shear stress, flow separation, and departure from laminar flow may be selectively involved (Caro *et al.*, 1971; Zarins *et al.*, 1982) and that plaques tend to form where endothelium is intact or restored and not absent (Minick *et al.*, 1979; Chidi *et al.*, 1979; Bomberger *et al.*, 1980).

The pathologic and experimental findings correspond with human epidemiologic findings to the degree that clinical risk factors which influence the induction and/or acceleration of the atherosclerotic process have been proposed. These include hypercholesterolemia related to inborn metabolic abnormalities or to excessive dietary intake, hypertension, smoking, and diabetes. It has been suggested, principally on the basis of diet manipulation in animal experimentation and on the reduction in ischemic heart disease during periods of human privation and undernutrition (Lopez *et al.*, 1966), that atherosclerotic lesions may regress or be partially resorbed under some conditions (Wissler *et al.*, 1974b). Control of risk factors would be expected to prevent, inhibit, or reverse lesions. Recent epidemiologic studies and clinical therapeutic trials tend to support such an approach (Wissler and Vesselinovitch, 1983). The possibility cannot be excluded that at least some of the focal intimal, fibrocellular thickenings in human arteries represent aborted, arrested, or regressed lesions, whereas others reflect adaptive changes related to hemodynamic stresses. However, convincing morphologic criteria for lesion involution in man remain to be established.

In summary, a good deal has been learned about the reaction of the vessel wall to endothelial injury, mechanical stress, chemical and mechanical stimuli, and exposure to hyperlipidemia, but the hierarchy and temporal relationships among these factors in the initiation, development, and regression of atherosclerotic lesions are still not clear.

c. Nonspecific sclerosis of media: Two principal forms of nonspecific connective tissue change in the media of large arteries have been described: fibrosis and calcification. Absolute quantities and relative proportions of collagen, elastin, and smooth muscle cells differ for different vessels (Fischer and Llaurado, 1966) but tend to be similar for homologous vessels, regardless of species. In general, the relative proportion of collagen increases and elastin content decreases with age. In the aorta, for example, an increase in the relative proportion of collagen occurs with age, but total scleroprotein may show little change, either with distance from the heart or across the wall (Feldman and Glagov, 1971); cellularity also decreases with age, with the overall result that the vessel wall becomes stiffer or less compliant. These changes are enhanced by hypertension. Experimental elevation of blood pressure in relatively young animals results in greater cross-sectional area of the aortic media, due initially to cell proliferation and subsequently to accumulation of collagen and elastin. The change is proportional to the rise in medial tension, although the relative proportions of the two fibrous proteins may not be altered, and the number of medial layers are not increased (Wolinsky, 1970b). These data suggest that some ideal level of stress is maintained by the elaboration of suitable quantities of matrix fibers in a fixed proportion. Although this adaptive potential may persist to some degree when maturation is complete, the changes related to growth and to hypertension in the adult human may be due to different mechanisms. Whether the greater tension per layer places a hypertensive aorta at some functional mechanical or biosynthetic disadvantage remains to be investigated.

Besides the deposition associated with atherosclerosis, *calcification* occurs extensively in arteries, strikingly within elastic fibers, and there are apparently zones of selective involvement which may correlate with flow patterns (Meyer, 1977) and possibly with preferential sites of atherogenesis. Calcification also occurs, mainly in peripheral muscular arteries, as a prominent medial band (Mönckeberg's sclerosis). This change is usually closely associated with intimal thickening which may be bland and myxomatous, similar to atherosclerosis, or it may form a well-organized laminated structure. The localization of the change and its close association with intimal thickening suggest that it is a form of dystrophic calcification, probably related to the relative ischemia of the midzone of the media.

3. ECTASIA AND TORTUOSITY

The considerations outlined in Section A, above, suggest that under normal conditions, the structure of the media of large arteries corresponds closely to vessel diameter and medial stress and that changes in vessel composition and structure during growth correlate with changes in tensile stress. Nevertheless, the aorta, as well as other major systemic arteries, tends to increase in diameter, elongate, and become less supple with increasing age after growth has ceased. This diffuse irreversible enlargement when marked is called *ectasia*. In some instances the expansions become clinically significant deformations. The common form of diffuse and extensive ectasia of the aorta and of the large arteries parallels the increase in matrix fiber accumulation (Mendez and Tejada, 1969) and the decrease in compliance of the wall (Busby and Burton, 1965). Since the enlargement usually includes increases in both diameter and length, the vessel tends to become *tortuous* or bowed. With greater diameter and the establishment of altered curvatures and branch configurations, the distribution of tensions is changed and flow patterns are modified. As a consequence, medial stresses may be focally augmented. Although equilibrium can be established when fibrosis of the media is sufficient to permit the artery wall to bear the increased stress without dilatation, small dystrophic flaws in the media may be propagated and lead to further local deformations. A change in configuration from a fusiform to a saccular or spherical deformation could also serve to reduce the effective medial stress and contribute to stabilization (see Section A-2, Chapter 2). Ectasias, whether diffuse or focal, are often not self-limiting, for the fibrous reaction may lag behind the increased stress associated with increased vessel radius and result in atrophy and further enlargement. Concavities of curvatures and obtuse branch angles are sites at which medial tensile stresses are relatively elevated and walls are normally thickest. Stresses are usually least at the acute angle or apex of branch points, and the media in this location is correspondingly comparatively thin. Depending on conditions of stress distribution and medial metabolic response, either location may be vulnerable to abnormal expansion, the former because of elevated tension and the latter in relation to an already thin wall. The diffuse and extensive form of artery enlargement is usually not associated, as such, with serious consequences. When complications occur, they are generally attributable to associated atherosclerosis or the formation of aneurysms.

4. ANEURYSMS

Aneurysms are local ectasias, outpouchings, or swellings of arterial walls which are associated with marked alterations in blood flow and a tendency to the formation of thrombi and/or rupture of the wall. Many of the anatomic and clinical features indicate that metabolic or structural defects of the media or alterations of this structure associated with atherosclerosis probably underly most acquired arterial aneurysms. These lesions can be classified according to three principal modes of pathogenesis. In the most common type, the *true aneurysm*, the arterial wall thins or stretches to form a segmental fusiform or globular enlargement or a focal saccular outpouching. The aneurysmal wall, although attenuated and fibrotic, is continuous with the wall of the adjacent uninvolved portion of the vessel. In a second type, the *dissecting aneurysm*, arterial wall structure is largely maintained but the cohesiveness of the media is inadequate. Hence, entry of blood into an extensive plane of separation within the media results in the formation of an intramural hematoma, with corresponding swelling of the vessel. Finally, *pseudoaneurysms* develop from the establishment of a communication between the bloodstream and a local periarterial sac, usually by way of a traumatic transmural tear.

a. True aneurysms: The abdominal aortic segment, below the level of the renal arteries, is particularly prone to the formation of true atherosclerotic aneurysms. The media of a large fusiform dilatation (often extending to the bifurcation) is usually reduced to a collagenous band, often less than 25% of normal mural thickness. The associated atherosclerotic plaques are frequently atrophic and calcific, so that the wall becomes a brittle shell, susceptible to fracture and rupture by compression or manipulation. The mural thrombi which deposit on the luminal surface are generally thick and laminated, with layers closest to the lumen showing the least organization. Frequently the clot consists of a dense, gelatinous, greenish mass of decomposing blood elements, with little evidence of organization. It may occupy most of the lumen, leaving a passage for blood flow of a normal or reduced diameter.

The close association of true aneurysms with advanced atherosclerotic disease suggests that local factors, peculiar to the abdominal aortic segment, may be of importance in the development of the lesion. Although the close quantitative relationship between the number of medial layers and the depth of penetration of vasa vasorum into the media applies to both the thoracic and abdominal aortic segments in most mammals, this is not the case for the human abdominal aorta (Wolinsky and Glagov, 1969). For example, midway between the renal arteries and the aortic bifurcation, the adult human aortic media is normally about 0.7 mm thick. For mammals in general, such thickness would be expected to correspond to about 40 layers, with the aortic media containing an outer vascularized zone of about 10 layers. Instead, the adult human abdominal aorta contains about 30 layers and is devoid of medial vasa vasorum. The dissociation between tension and complement of transmural layers at this level is reminiscent of the situation which occurs in animals exposed to experimentally induced hypertension (Wolinsky, 1970b), inasmuch as the thickness and cross-sectional area of the media correspond to the total tension across the media, but the number of transmural layers do not. Diffusion from the lumen alone may be inadequate to maintain the media in the face of the rapid development of intimal thickenings and atherosclerosis common to this segment. The tendency to medial atrophy may be further potentiated by atherosclerotic involvement of abdominal aortic branch ostia, with associated obstruction of vasal outflow. There is also evidence from human material that local increases in collagenase activity may be associated with abdominal aortic aneurysm formation (Busuttil *et al.*, 1980) and that elastase content increases with age in the aorta (Hornebeck *et al.*, 1978). The degree to which individual differences in medial architecture and composition, vasal supply, enzyme activity, and rate of atherogenesis in the abdominal aorta correspond to aneurysm formation remains to be determined. Although enlargement of an aneurysm and thinning of its wall raise mural stress and are likely to produce a vicious circle of increased stress and further enlargement, atrophy, and dilatation, the formation of increasingly dense collagenous tissue may strengthen the wall; moreover, the tendency toward spherical shape may reduce the magnitude of the tensile stress. (For a discussion of the mechanical aspects of aneurysms, see Section A-2, Chapter 2.)

Aneurysms of the aorta associated with *Treponema pallidum* infection are also true aneurysms. These occur mainly in the proximal thoracic aorta, and the changes in the media are different from those associated with atherosclerotic aneurysms of the distal aorta. Rather than replacement by a poorly vascularized fibrous band, the syphilitic aortic media often shows, at least initially, evidence of patchy disappearance of elastic fibers and scarification, largely of the outer half, i.e., in the region normally supplied mainly by medial vasa vasorum (Wolinsky and Glagov, 1967b). The changes in the media may be due, in part, to direct penetration by the spirochete via the vasa vasorum, but more likely they result from interference with blood supply to the outer media caused by syphilitic endarteritis of the vasa vasorum. The usual massive expansion of the most proximal ascending portion of the aorta suggests that the ectasia at this location is probably due to yielding of the damaged media at the site of greatest rate of change of tensile stress. The irregular scarification of the media is evidenced by wrinkling of the intima and may be associated with exceptionally severe atherosclerosis in this portion of the aorta.

Besides the thoracic and abdominal aorta, true aneurysms develop at other sites because of special local conditions. The splenic artery, for example, is the most frequent location for splanchnic aneurysm formation, and it is also commonly ectatic, tortuous, and atherosclerotic. Splenic artery aneurysms occur more often in women, particularly those who have had multiple pregnancies (Busuttil and Brin, 1980). This phenomenon may be related to changes in splenic flow associated with pregnancy, but it should be noted that both splenic and renal artery dysplasias tend to occur under similar conditions (Stanley and Fry, 1974). These and other findings suggest that hormonal status may modify medial composition, compliance, and cohesion by means of a direct effect on smooth muscle and/or matrix composition (Sirek *et al.*, 1977; Fischer and Swain, 1980).

In other locations, underlying medial defects, presumably congenital, have been incriminated. Typical examples are the aneurysms of the renal arteries in the kidney hilus (Hubert *et al.*, 1980) and those which form in the cerebral arteries (Suzuki *et al.*, 1979) (see Section E-3a, Chapter 5). In both of these locations, saccular aneurysms are generally situated at the angle of branch sites, where the arterial wall is normally thinnest. Such formations may be potentiated by hypertension, but the lesions likewise occur in the absence of hypertension. Whether these defects are due to congenital malformations or to limitations in the capacity of the vessel wall to remodel as vessels grow and angles change remains to be demonstrated.

Fusiform enlargements may also occur just distal to stenoses or focal compressions and have been called poststenotic dilatations. Under experimental conditions, marked increases in artery wall compliance have been demonstrated beyond constrictions, provided that bruits exist (Roach, 1963). This finding suggests that medial organization, cohesion, or composition may have been altered by the associated vibrations. The precise anatomic infrastructure of these poststenotic modifications has not been demonstrated.

b. *Dissecting aneurysms:* The predominant feature in the development of these lesions is the precipitous splitting of the media, which accounts for it being referred to as acute aortic dissection. One or several associated tears may extend between the split and the lumen, establishing continuity between the bloodstream and the disrupted media. Passage may also be established to the adventitial side, followed by catastrophic hemorrhage or the formation of a massive periarterial hematoma. The extensive circumferential and longitudinal propagation of the characteristic intramural cleft suggests that the underlying defect of the media is a loss or reduction of medial cohesion. The presence of a corresponding defect in medial matrix composition and/or organization is indicated by the following findings. The disease is a feature of the Marfan syndrome and of several other hereditary disorders of matrix fiber metabolism (McKusick, 1966). The incidence of dissection is increased in pregnant women (Pedowitz and Perell, 1957), and disruptions similar to those which occur in human dissection can be induced in experimental animals by administering β-aminopropionitrile, a compound which interferes with matrix fiber cross-linking (Barron *et al.*, 1974). The tendency of aortas prone to dissection to enlarge and/or to form true aneurysms in the hereditary disorders suggests that focal or segmental disturbances of medial cohesiveness may in some cases underlie both conditions. A precipitating role for increased medial stress in acute dissection is supported by the high frequency with which individuals with acute aortic dissection are hypertensive and the common occurrence of an acute hypertensive crisis immediately preceding the catastrophic event (Schlatmann and Becker, 1977b). The transverse intimal–inner medial disruption which provides access of the blood to the media commonly occurs in the ascending aorta, close to the inner aspect of the arch, i.e., in the region of greatest medial tension. The microscopic medial finding commonly associated with acute aortic dissection is cystic medial necrosis, a degenerative change involving mainly the medial smooth muscle cells, associated with focal accumulations of ground substance (Saruk and Eisenstein, 1977). However, it is necessary to point out that cystic medial necrosis is also noted in many individuals without manifest aortic disruption (Gore and Hirst, 1973; Schlatmann and Becker, 1977a).

The predominance of dissecting hematomas in the human thoracic aorta and the propagation of dissection in the outer layers of the media, i.e., in the region of the potentially ischemic watershed boundary zone between the vascular and avascular portions of the media (Wolinsky and Glagov, 1967b; Heistad *et al.*, 1978), suggest that inadequacy of perfusion may play a role in weakening the media. Also of interest is the finding that ligation of aortic vasa in experimental animals of appropriate size causes zonal ischemic necrosis of the media (Wilens *et al.*, 1965). Hypertension would be expected to expose the proximal aortic segment to relatively high levels of intramural shearing stress, which, under certain conditions, such as a sudden further rise in pressure, might compress, distort, or disrupt the vasa and interfere with vasal perfusion of the media (Sacks, 1975). Extensive aortic hemorrhagic dissections sometimes extend into the major arterial branches and interfere with flow by compression of the lumen. Dissections also occasionally originate in major branch arteries without preceding aortic dissection.

c. *Pseudoaneurysms:* When the wall of a large elastic artery is transected, the resulting hemorrhage may be contained by periarterial tissues, with the formation of a blood-filled cavity which communicates with the vessel lumen by way of the original mural defect. Since the abnormal outpouching is not encompassed by arterial wall and is caused by a transmural defect in an otherwise intact artery, it is called a *pseudoaneurysm.* This type of lesion occurs most commonly as a consequence of direct, penetrating injuries, such as those produced by bullet or knife wounds, or of transient forces which exceed the tensile strength of the media, such as those which are found in the thoracic aorta in association with severe crushing chest injuries. Whereas muscular arteries tend to retract and contract after local disruptive injury, thus sealing off a traumatic mural defect with a hemostatic occlusive plug that arrests flow entirely, large elastic arteries do not generally contract or recoil sufficiently to seal an extensive disruptive injury. Hence, if massive hemorrhage does not ensue, pseudoaneurysms, rather than thrombotic occlusions, are likely to occur. Despite interruption of the media at sites of transecting trauma, a tendency to splitting or separation between medial layers, with formation of extended intramural dissecting hematomas, is not a frequent complication. Although periarterial tissues may provide a surprisingly good barrier and prevent immediate catastrophic hemorrhage, the absence in the adventitia of an organized fibrocellular architecture designed to resist tensile stress may result in progressive enlargement of a pseudoaneurysm.

Pseudoaneurysms may also be associated with diagnostic or operative maneuvers, such as the insertion of arterial catheters, and can also develop in association with vascular grafting procedures, particularly at anastomotic sites, when sutures fatigue and break. This effect may be potentiated if mismatched compliance of vessel and prosthesis causes excessive motion at an anastomotic site. Unlike the strong, healed tissue bond which forms when the edges of a transected artery have been approximated and sutured together, the seal between a prosthetic graft and an artery remains dependent on the integrity of the sutures. Infection at graft anastomoses have the same grave implications as mycotic aneurysms, i.e., arterial disruptions formed by purulent destruction of an arterial wall where infected material has lodged. It should also be noted that sites of vessel wall devitalization by injury, the presence of sutures and foreign prosthetic material, and the occurrence of abnormal pressure and flow patterns are particularly prone to infection. Endarterectomy would also be expected to predispose to aneurysm formation, for the procedure involves removal of the entire media, in addition to the intimal plaques, leaving only an adventitial tube to contain arterial blood pressure. Yet, aneurysm formation under such conditions is unusual, for the adventitia surrounding advanced plaques is often thickened and fibrous. In this regard, it was found that in experimental tests performed on diseased human iliac artery segments, tensile strength of the arterial channel was not diminished by endarterectomy (Gryska, 1961; Sumner *et al.*, 1969). Although disruption of an atherosclerotic aneurysm may give rise to a superimposed pseudoaneurysm, fracture of plaques by transluminal angioplasty and catheter penetrations into intimal plaques rarely result in aneurysm formation.

REFERENCES*

Adams, C. W. M. (1981). Lipoprotein filtration in atherogenesis. Evidence from experimental and human pathology. *In* "Lipoproteins, Atherosclerosis and Coronary Heart Disease" (N. E. Miller and B. Lewis, eds.), pp. 91–106. Elsevier, Amsterdam. (E)

Adelstein, R. S., and Eisenberg, E. (1980). Regulation and kinetics of the actin myosin interaction. *Ann. Rev. Biochem.* **49**, 921–956. (B)

Barron, M. V., Simpson, C. F., and Miller, E. S. (1974). Lathyrism: A review. *Q. Rev. Biol.* **49**, 101–128. (E)

Becker, C. G., and Harpel, P. C. (1976). α_2-Macroglobulin on human vascular endothelium. *J. Exp. Med.* **144**, 1–9. (C)

Benditt, C. M., and Gown, A. M. (1980). Atheroma: The artery wall and the environment. *Int. Rev. Exp. Pathol.* **21**, 55–118. (E)

Benditt, E. P., and Benditt, J. M. (1973). Evidence for a monoclonal origin of human atherosclerotic plaques. *Proc. Natl. Acad. Sci. U.S.A.* **70**, 1753–1756. (E)

Bettmann, M. A., Stemerman, M. B., and Ransil, B. J. (1981). The effect of hypophysectomy on experimental endothelial cell regrowth and intimal thickening in the rat. *Circ. Res.* **48**, 907–912. (C)

Bevan, J. A. (1977). Some functional consequences of variation in adrenergic synaptic cleft width and in nerve density and distribution. *Fed. Proc., Fed. Am. Soc. Exp. Biol.* **36**, 2439–2443. (D)

Bevan, J. A. (1979). Sites of transition between functional systemic and cerebral arteries of rabbits occur at embryological junctional sites. *Science* **204**, 635–637. (D)

Bevan, J. A. (1983). Diltiazem selectively inhibits cerebrovascular extrinsic but not intrinsic myogenic tone. A review. *Circ. Res.* **52**, 104–109. (D)

Bevan, J. A. (in press). The human adrenergic neuromuscular mechanism. *In* "General Pharmacology," Proc. Symp. Pharmacol. Human Blood Vessels, Kuwait, 1982. (D)

Bevan, J. A., Bevan, R. D., and Duckles, S. P. (1980). Adrenergic regulation of vascular smooth muscle. *In* "Handbook of Physiology" (D. F. Bohr, A. P. Somlyo, and H. V. Sparks, Jr., eds.), Sect. 2, Vol. 2, pp. 325–351. Am. Physiol. Soc., Bethesda, Maryland. (D)

*In the reference list, the capital letter in parentheses at the end of each reference indicates the section in which it is cited.

Bevan, J. A., Buga, G. M., Snowden, A., and Said, S. I. (in press). Is the neural vasodilator mechanism to cerebral and extracerebral arteries the same? In "Proceedings of the Symposium on Cerebral Blood Flow: Effect of Nerves and Neurotransmitters" (D. D. Heistad and M. L. Marcus, eds.), Iowa City, Iowa, June 15–18, 1981. Elsevier, New York. (D)

Bjorkerud, S., and Bondjers, G. (1972). Endothelial integrity and viability in the aorta of the normal rabbit and rat as evaluated with dye exclusion tests and interference contrast microscopy. Atherosclerosis 15, 285–300. (A)

Bloor, C. M., and Ashraf, M. (1977). Pathogenesis of acute myocardial infarction. Adv. Cardiol. 23, 19–24. (E)

Blumenthal, H. T., ed. (1967). The pathology of human ateriosclerosis. In "Cowdry's Arteriosclerosis, A Survey of the Problem," Part IV, pp. 227–414. Thomas, Springfield, Illinois. (E)

Bohr, D. F., Somlyo, A. P., and Sparks, H. V., Jr., eds. (1980). "Handbook of Physiology," Sect. 2, Vol. 2. Am. Physiol. Soc., Washington, D.C. (B)

Bomberger, R. A., Zarins, C. K., and Glagov, S. (1980). Medial injury and hyperlipidemia in development of aneurysms or atherosclerotic plaques. Surg. Forum 31, 338–340. (E)

Bond, G. M., Adams, M. R., and Bullock, B. C. (1981). Complicating factors in evaluating coronary artery atherosclerosis. Artery 9, 21–29. (E)

Bratzler, R. L., Chisolm, M., Colton, C. K., Smith, K. A., Zilversmit, D. B., and Lees, R. S. (1977). A distribution of labeled albumin across the rabbit thoracic aorta in vivo. Circ. Res. 40, 182–190. (C)

Bruns, R. R., and Palade, G. (1968). Studies on blood capillaries. I. General organization of blood capillaries in muscle. J. Cell. Biol. 37, 244–277. (C)

Buck, R. C. (1979). The longitudinal orientation of structures in the subendothelial space of rat aorta. Am. J. Anat. 156, 1–14. (A,E)

Burri, P. H., and Weibel, E. R. (1968). Beeinflusung einer specifischen Cytoplasmatischen Organelle von Endothezellen durch Adrenalin. Zeit. Zell. 88, 426–440. (C)

Burton, A. C. (1954). Relation of structure to function of the tissues of the wall of blood vessels. Physiol. Rev. 34, 619–642. (A)

Busby, D. E., and Burton, A. C. (1965). The effect of age on the elasticity of the major brain arteries. Can. J. Physiol. Pharmacol. 43, 185–202. (E)

Busuttil, R. W., and Brin, B. J. (1980). The diagnosis and management of visceral artery aneurysms. Surgery 88, 619–624. (E)

Busuttil, R. W., Abou-Zamzam, A. M., and Machleder, H. I. (1980). Collagenase activity of the human aorta. Arch. Surg. 115, 1373–1378. (E)

Caplan, B. A., and Schwartz, C. J. (1973). Increased endothelial cell turnover in areas of in vivo Evans blue uptake in the pig aorta. Atherosclerosis 17, 401–417. (C)

Caro, C. G., Fitzgerald, J. M., and Schroter, R. C. (1971). Atheroma and arterial wall shear: Observation correlation and proposal of a shear dependent mass transfer mechanisms for atherogenesis. Proc. R. Soc. London (Biol.) 117, 109–159. (E)

Chandler, A. B., Chapmann, I., Erhardt, L. R., Roberts, W. C., Schwartz, C. J., Sinapius, D., Spain, D. M., Sherry, S., Ness, P. M., and Simon, T. L. (1974). Coronary thrombosis in myocardial infarction. Am. J. Cardiol. 34, 823–833. (E)

Chase, K. V., and Odessey, R. (1981). The utilization by rabbit aorta of carbohydrate, fatty acids, ketone bodies and amino acids as substrates for energy production. Circ. Res. 48, 850–858. (B)

Chidi, C. C., Klein, L., and DePalma, R. (1979). Effect of regenerated endothelium on collagen content in the injured artery. Surg. Gynecol. Obstet. 148, 839–843. (E)

Clark, J. M., and Glagov, S. (1976a). Luminal surface of distended arteries, eliminating configurational and technical artefacts. Br. J. Exp. Pathol. 57, 129–135. (A)

Clark, J. M., and Glagov, S. (1976b). Evaluation and publication of scanning electron micrographs. Science 192, 1360–1361. (C)

Clark, J. M., and Glagov, S. (1979). Structural integration of the arterial wall: I. Relationships and attachments of medial smooth muscle cells in normally distended and hyperdistended aortas. *Lab. Invest.* **40**, 587–602. (A)

Cleary, E. G. (1963). Correlative and comparative study of the non-uniform arterial wall. Ph.D. Thesis, University of Sydney, Sydney, Australia. (A)

de la Lande, I. S., and Jellet, L. B. (1972). Relationship between the roles of monoamine oxidase and sympathetic nerves in the vasoconstrictor response of the rabbit ear artery to noradrenaline. *J. Pharmacol. Exp. Ther.* **180**, 47–55. (D)

de la Lande, I. S., Frewin, D., and Waterson, J. G. (1967). The influence of sympathetic innervation on vascular sensitivity to noradrenaline. *Br. J. Pharmacol.* **31**, 82–93. (D)

Duff, G. L., and McMillan, G. C. (1951). Pathology of atherosclerosis. *Am. J. Med.* **18**, 92–108. (E)

Duguid, J. B. (1960). The thrombogenic hypothesis and its implications. *Postgrad. Med. J.* **36**, 226–229. (E)

Ebashi, S. (1980). Regulation of muscle contraction. *Proc. R. Soc. London Ser. B.* **207**, 259–286. (B)

Feldman, S. A., and Glagov, S. (1971). Transmedial collagen and elastin gradients in human aortas: Reversal with age. *Atherosclerosis* **13**, 385–394. (E)

Fischer, G. M., and Llaurado, J. G. (1966). Collagen and elastin content in canine arteries from functionally different vascular beds. *Circ. Res.* **19**, 394–399. (E)

Fischer, G. M., and Swain, M. L. (1980). Influence of contraceptive and other sex steroids on aortic collagen and elastin. *Exp. Mol. Pathol.* **33**, 15–24. (E)

Flaherty, J. T., Pierce, J. E., Ferrans, V. J., Patel, D. J., Tucker, W. K., and Fry, D. L. (1972). Endothelial nuclear patterns in the canine arterial tree with particular reference to hemodynamic events. *Circ. Res.* **30**, 23–33. (A)

Florentin, R. A., Nam, S. C., Lee, K. T., and Thomas, W. A. (1969). Increased ^{3}H-thymidine incorporation into endothelial cells of swine fed cholesterol for three days. *Exp. Mol. Pathol.* **10**, 250. (C)

Folkow, B. (1964). Description of the myogenic hypothesis. *Circ. Res.* **15**, (Suppl. 1–2), 279–285. (D)

Fry, D. L. (1968). Acute vascular endothelial changes associated with increased blood velocity gradients. *Circ. Res.* **22**, 165–197. (A)

Fry, D. L. (1969). Certain chemorheologic considerations regarding the blood vascular interface with particular reference to coronary artery disease. *Circulation* **40** (Suppl. 4), 38–59. (A)

Furchgott, R. F., and Zawadzki, J. V. (1980). The obligatory role of endothelial cells in the relaxation of arterial smooth muscle by acetylcholine. *Nature (London)* **288**, 373–376. (D)

Gabbiani, G., and Majno, G. (1969). Endothelial microvilli in the vessels of the rat gasserian ganglion and testis. *Z. Zellforsch.* **97**, 111–117. (A)

Gajdusek, C., DiCorleto, P., Ross, R., and Schwartz, S. (1980). An endothelial cell-derived growth factor. *J. Cell Biol.* **85**, 467–472. (C)

Gaynor, E. (1971). Increased mitotic activity in rabbit endothelium after endotoxin. An autoradiographic study. *Lab. Invest.* **24**, 318–327. (C)

Gaynor, E., Bouvier, C. A., and Spaet, T. H. (1970). Vascular lesions: Possible pathogenetic basis of the generalized Schwartzman reaction. *Science* **170**, 986–988.

Geer, J. C. (1965). Fine structure of aortic intimal thickening and fatty streaks. *Lab. Invest.* **14**, 1764–1783. (A)

Geiringer, E. (1951). Intimal vascularization and atherosclerosis. *J. Pathol. Bacteriol.* **6**, 201–211. (A)

Gero, J., and Gerova, M. (1971). *In vivo* studies of sympathetic control of vessels of different function. *In* "Physiology and Pharmacology of Vascular Neuroeffector Systems" (J. A. Bevan, R. F. Furchgott, R. A. Maxwell, and A. P. Somlyo, eds.), pp. 86–94. Karger, Basel. (D)

Gerrity, R. G., Richardson, M., Somer, J. B., Bell, F. P., and Schwartz, C. J. (1977). Endothelial cell morphology in areas of *in vivo* Evans blue uptake in the aorta of young pigs. II. Ultrastructure of the intima in areas of differing permeability to proteins. *Am. J. Pathol.* **89**, 313–326. (C)

Gimbrone, M. A. (1976). Culture of vascular endothelium. *In* "Progress in Hemostasis and Thrombosis" (T. H. Spaet, ed.), Vol. 3, pp. 1–28. Grune and Stratton, New York. (C)

Gimbrone, M. A. (1981). Vascular endothelium and atherosclerosis. *In* "Vascular Injury and Atherosclerosis" (S. Moore, ed), Vol. 9, pp. 25–52. Dekker, New York. (C)

Glagov, S. (1972). Hemodynamic risk factors: Mechanical stress, mural architecture, medial nutrition and the vulnerability of arteries to atherosclerosis. *In* "The Pathogenesis of Atherosclerosis" (R. W. Wissler and J. C. Geer, eds.), pp. 164–199. Williams & Wilkins, Baltimore, Maryland. (A,E)

Glagov, S., and Ozoa, A. K. (1968). Significance of the relatively low incidence of atherosclerosis in the pulmonary, renal and mesenteric arteries. *Ann. N. Y. Acad. Sci.* **149**, 940–955. (E)

Glagov, S., and Zarins, C. K. (1983). Quantitating atherosclerosis. Problems of Definition. *In* "Noninvasive Diagnosis of Atherosclerotic Lesions: Quantitative Evaluation of Morphology, Biochemistry and Pathophysiology" (M. G. Bond, W. Insull, Jr., S. Glagov, A. B. Chandler, and J. F. Cornhill, eds.), pp. 11–35. Springer-Verlag, Berlin and New York. (E)

Glick, G., Epstein, S. E., Wechsler, A. S., and Braunwald, T. (1967). Physiological differences between the effects of neuronally released and blood-borne norepinephrine on β-adrenergic receptors in the arterial bed of the dog. *Circ. Res.* **21**, 217–227. (D)

Gordon, T., Kannel, W. B., Hjortland, M. C., and McNamara, P. M. (1978). Menopause and coronary heart disease. The Framington Study. *Ann. Intern. Med.* **89**, 157–161. (E)

Gore, I., and Hirst, A. E., Jr. (1973). Dissecting aneurysm of the aorta: Clinical–pathologic correlations. *Cardiovasc. Clin.* **2**, 239–260. (E)

Grände, P.-O. (1979). Dynamic and static components in the myogenic control of vascular tone in cat skeletal muscle. *Acta Physiol. Scand.* (Suppl.) **476**, 1–44. (D)

Gryska, P. F. (1961). The physical properties of arteries after endarterectomy. *Surgery* **113**, 227–229. (E)

Guimaraes, S., and Paiva, M. Q. (1977). Differential influence of block of catechol-O-methyltransferase (COMT) activity and of neuronal uptake on α- and β-adrenergic effects. *J. Pharmacol.* **29**, 502–503. (D)

Gutstein, W. H., Farrell, G. A., and Armellini, C. (1973). Endothelial cell injury in pre-atherosclerotic swine. *Lab. Invest.* **29**, 134–149. (A)

Hamilton, F. N., and Feigl, E. O. (1976). Coronary vascular sympathetic β-receptor innervation. *Am. J. Physiol.* **230**, 1564–1576. (D)

Hammersen, F. (1972). On the fine structure of the portal vein in different rodents. *In* "Vascular Smooth Muscle" (E. Betz, ed.), pp. 113–115. Springer-Verlag, Berlin and New York. (A)

Hardin, N. J., Minick, C. R., and Murphy, G. E. (1973). Experimental induction of atherosclerosis by the synergy of allergic injury to arteries and lipid-rich diet. III. Role of earlier acquired fibromuscular intimal thickening in the pathogenesis of later developing atherosclerosis. *Am. J. Pathol.* **73**, 301–326. (C)

Harker, L., Ross, R., and Slichter, S. (1976). Homocystine-induced arteriosclerosis: The role of endothelial cell injury and platelet response in its genesis. *J. Clin. Invest.* **58**, 731–741. (C)

Hartshorne, D. J., and Siemankowski, R. F. (1981). Regulation of smooth muscle actomyosin. *Ann. Rev. Physiol.* **43**, 519–530. (B)

Haust, M. D. (1971). The morphogenesis and fate of potential and early atherosclerotic lesions in man. *Human Pathol.* **2**, 1–29. (E)

Heath, D., and Edwards, J. E. (1958). The pathology of pulmonary hypertensive disease. A description of six grades of structural changes in the pulmonary arteries with special reference to congenital cardiac defects. *Circulation* **18**, 533–547. (A)

Heistad, D. D., Marcus, M. L., Law, E. G., Armstrong, M. L., Ehrhardt, J. C., and Abboud, F. M. (1978). Regulation of blood flow to the aortic media in dogs. *J. Clin. Invest.* **62**, 133–140. (E)

Hellstrand, P., and Paul, R. J. (1982). Vascular smooth muscle: Relations between energy metabolism and mechanics. In "Vascular Smooth Muscle: Metabolic, Ionic, and Contractile Mechanisms" (M. F. Crass, III, and C. D. Barnes, eds.), pp. 1–35. Academic Press, New York. (B)

Heumann, H-G. (1973). Smooth muscle: Contraction hypothesis based on the arrangement of actin and myosin filaments in different states of contraction. Phil. Trans. R. Soc. London (Biol.) 265, 213–218. (B)

Hirano, A. (1976). Further observations of the fine structure of pathological reaction in cerebral blood vessels. In "The Cerebral Vessel Wall" (J. Cervos-Navarro, E. Betz, F. Matakas, and R. Wullenweber, eds.), pp. 41–50. Raven Press, New York. (C)

Hirano, A., and Zimmerman, H. M. (1972). Fenestrated blood vessels in a metastatic renal carcinoma in the brain. Lab. Invest. 26, 465–470. (C)

Hirano, A., Dembitzer, H. M., Becker, N. H., Levine, S., and Zimmerman, H. M. (1970). Fine structural alterations of the blood–brain barrier in experimental allergic encephalomyelitis. J. Neuropathol. Exp. Neurol. 29, 432–440. (C)

Hirst, C. D. S. (1977). Neuromuscular transmission in arterioles of guinea pig submucosa. J. Physiol. (London) 273, 263–275. (D)

Hollis, T. M., and Rosen, L. A. (1972). Histidine decarboxylase activity of bovine aortic endothelium and intima-media. Proc. Soc. Exp. Biol. Med. 141, 978–981. (C)

Hornebeck, W., Adnet, J. J., and Robert, L. (1978). Age dependent variation of elastin and elastase in aorta and human breast cancers. Exp. Gerontol. 13, 293–298. (E)

Hubert, J. P., Pairolero, P. C., and Kazmier, F. J. (1980). Solitary renal artery aneurysm. Surgery 88, 557–564. (E)

Hughes, J., Gilles, C. N., and Bloom, F. E. (1969). The uptake and disposition of dl-norepinephrine in perfused rat lung. J. Pharmacol. Exp. Ther. 169, 237–248. (C)

Hyman, A. L., Nandiwada, P., Knight, D. S., and Kadowitz, P. J. (1981). Pulmonary vasodilator responses to catecholamines and sympathetic nerve stimulation in the cat. Evidence that vascular B-2 adrenoreceptors are innervated. Circ. Res. 48, 407–415. (D)

Jaffee, E. A., and Mosher, D. F. (1978). Synthesis of fibronectin by cultured human endothelial cells. J. Exp. Med. 147, 1779–1791. (C)

Jaffee, E. A., Nachman, R. L., Becker, C. G., and Minick, C. R. (1973a). Culture of human endothelial cells derived from umbilical veins: Identification by morphologic and immunologic criteria. J. Clin. Invest. 52, 2745–2756. (C)

Jaffee, E. A., Hoyer, L. W., and Nachman, R. L. (1973b). Synthesis of antihemophilic factor antigen by cultured human endothelial cell. J. Clin. Invest. 52, 2757–2764. (C)

Jaffee, E. A., Minick, C. R., Adelman, B., Becker, C. G., and Nachman, R. (1976). Synthesis of basement membrane collagen by cultured human endothelial cells. J. Exp. Med. 144, 209–226. (C)

Johnson, A. R., and Erdos, E. G. (1977). Metabolism of vasoactive peptides by human endothelial cells in culture: Angiotensin I converting enzyme (kinase II) and angiotensinase. J. Clin. Invest. 59, 684–695. (C)

Johnson, R. A. (1969). Lymphatics of blood vessels. Lymphology 2, 44–56. (A)

Kalsner, S. (1972). Differential activation of the inner and outer muscle layers of the rabbit ear artery. Eur. J. Pharmacol. 20, 122–124. (D)

Komuro, T., and Burnstock, G. (1980). The fine structure of smooth muscle cells and their relationship to connective tissue in the rabbit portal vein. Cell Tissue Res. 210, 257–267. (A)

Kügelen, Von A. (1955). Über das Verhältnis von Ringmuskulatur und Innendruck in menschlichen grossen Venen. Z. Zellforsch. Mikroskop. Anat. 43, 168–183. (A)

Kügelen, Von A. (1956). Weitere Mitteilungen üden Wandbau der grossen Venen des Menschen unter besonderer Berücksichtigung ihrer Kollagenstrukturen. Z. Zellforsch. Mikroskop. Anat. 44, 121–174. (A)

Leak, L. U., and Burke, J. F. (1968). Ultrastructural studies on the lymphatics anchoring filaments. J. Cell Biol. 36, 129–249. (C)

Le Douarin, N. (1980). Migration and differentiation of neural crest cells. *Curr. Top. Dev. Biol.* **16**, 31–85. (D)

Le Lievre, C. S., and Le Douarin, N. M. (1975). Mesenchymal derivatives of the neural crest: Analysis of chimaeric quail and chick embryos. *J. Embryol. Exp. Morphol.* **34**, 125–154. (D)

Leung, D. Y. M., Glagov, S., and Mathews, M. B. (1977). Elastin and collagen accumulation in rabbit ascending aorta and pulmonary trunk during postnatal growth: Correlation of cellular synthetic response with medial tension. *Circulation Res.* **41**, 316–323. (A)

Ljung, B. (1975). Physiological patterns of neuroeffector control mechanisms. *In* "Vascular Neuroeffector Mechanisms" (J. A. Bevan, G. Burnstock, B. Johansson, R. A. Maxwell, and O. A. Nedergaard, eds.), pp. 143–155. Karger, Basel. (D)

Lopez, S. A., Krehl, W. A., Hodges, R. E., and Good, E. L. (1966). Relationship between food consumption and mortality from atherosclerotic heart disease in Europe. *Am. J. Clin. Nutr.* **19**, 361–369. (E)

Loskutoff, D. J., and Edgington, T. S. (1977). Synthesis of a fibrinolytic activator and inhibitor by endothelial cells. *Proc. Natl. Acad. Sci. U.S.A.* **74**, 3903–3907. (C)

Luft, J. H. (1966). Fine structure of capillary and endocapillary layer as revealed by ruthenium red. *Fed. Proc., Fed. Am. Soc. Exp. Biol.* **25**, 1773–1783. (A)

Lundvall, J., and Järhult, J. (1974). β-Adrenergic microvascular dilatation evoked by sympathetic stimulation. *Acta Physiol. Scand.* **92**, 572–574. (D)

Lundvall, J., and Järhult, J. (1976). β-Adrenergic dilator component of the sympathetic vascular response in skeletal muscle. *Acta Physiol. Scand.* **96**, 180–192. (D)

Maciag, T., Hoover, G. A., Stemerman, M. B., and Weinstein, R. (1981). Serial propagation of human endothelial cells *in vitro. J. Cell Biol.* **91**, 420–426. (C)

Maciag, T., Hoover, G. A., and Weinstein, R. (1982). High and low molecular weight forms of endothelial cell growth factor. *J. Biol. Chem.* **257**, 5333–5336. (C)

Majno, G. (1965). Ultrastructure of the vascular membrane. *In* "Handbook of Physiology" (W. Hamilton and P. Dow, eds.), Sect. 2, Vol. 3, pp. 2293–2362. Am. Physiol. Soc., Bethesda, Maryland. (C)

McCalden, T. A., and Bevan, J. A. (1980). Asymmetry of the contractile response of the rabbit ear artery to exogenous amines. *Am. J. Physiol.* **238**, H618–H624. (D)

McGill, H. C., Jr. (1968). "Geographic Pathology of Atherosclerosis." Williams & Wilkins, Baltimore, Maryland. (A,E)

McKusick, V. (1966). "Heritable Disorders of Connective Tissue." Mosby, St. Louis, Missouri. (E)

Mendez, J., and Tejada, C. (1969). Chemical composition of aortas from Guatemalans and North Americans. *Am. J. Clin. Pathol.* **51**, 598–602. (E)

Meyer, W. W. (1977). The mode of calcification in atherosclerotic lesions in atherosclerosis: Metabolic, morphologic and clinical aspects. *Adv. Exp. Med. Biol.* **82**, 786–792. (E)

Miki, S. (in press). The genesis of the splenic vein. *In* "Vascular Neuroeffector Mechanisms—IV," *Int. Symp. Vascular Neuroeffector Mechanisms, 4th, July 28–30, 1981, Kyoto, Japan.* Raven, New York. (D)

Miller, G. J. (1981). The epidemiology of plasma lipoprotein and atherosclerotic disease. *In* "Lipoproteins, Atherosclerosis and Coronary Heart Disease" (N. E. Miller and B. Lewis, eds.), pp. 59–71. Elsevier, Amsterdam. (E)

Minick, C. R., Litrenta, M. M., Alonso, D. R., Silane, M. F., and Stemerman, M. B. (1977). Further studies on the effect of regenerating endothelium on intimal lipid accumulation. *Prog. Biochem. Pharmacol.* **14**, 115–122. (C,E)

Minick, C. R., Stemerman, M. B., and Insull, W., Jr. (1979). Role of endothelium and hypercholesterolemia in intimal thickening and lipid accumulation. *Am. J. Pathol.* **95**, 131–158. (E)

Moncada, S., and Amezcua, J. L. (1979). Prostacyclin, thromboxane A$_2$ interactions in haemostasis and thrombosis. *Haemostasis*, **8**, 252–265. (D)

Movat, H. Z., More, R. H., and Haust, M. D. (1958). The diffuse intimal thickening of the human aorta with aging. *Am. J. Pathol.* **34**, 1023–1031. (A,E)

Murphy, R. A., Aksoy, M. O., Dillon, P. F., Gerthoffer, W. T., and Kamm, K. E. (1983). The role of myosin light chain phosphorylation in regulation of the cross bridge cycle. *Fed. Proc., Fed. Am. Soc. Exp. Biol.* **42**, 51–56. (B)

Needham, D. M. (1971). "Machina Carnis. The Biochemistry of Muscular Contraction in its Historical Development." Univ. Press, London, Cambridge. (B)

Nerem, R. M. (1981). Arterial fluid dynamics and interactions with the vessel wall. *In* "Structure and Function of the Circulation" (C. J. Schwartz, N. T. Werthessen, and S. Wolf, eds.), Vol. 2, pp. 719–835. Plenum, New York. (E)

Owen, M. P., Mason, M. F., Walmsley, J. G., Bevan, R. D., and Bevan, J. A. (1982). Vascular adrenergic control variation with artery diameter in the rabbit ear. *Fed. Proc., Fed. Am. Soc. Exp. Biol.* **41**, 1234. (D)

Owen, M. P., Walmsley, J. G., Mason, M. F., Bevan, R. D., and Bevan, J. A. (1983). Adrenergic control in three artery segments of diminishing diameter in rabbit ear. *Am. J. Physiol.* **245**, H320–H326. (D)

Owen, W. G., and Esmon, C. T. (1981). Functional properties of an endothelial cell co-factor for thrombin catalyzed activation of protein C. *J. Biol. Chem.* **256**, 5532–5535. (C)

Pascual, R., and Bevan, J. A. (1980). Asymmetry of consequences of drug disposition mechanisms in the wall of the rabbit aorta. *Circ. Res.* **46**, 22–28. (D)

Paterson, J. C. (1952). Factors in the production of coronary artery disease. *Circulation*, **6**, 732–739. (E)

Paul, R. J. (1980). The chemical energetics of vascular smooth muscle: Intermediary metabolism and its relation to contractility. *In* "Handbook of Physiology" (D. F. Bohr, A. P. Somlyo, and H. V. Sparks, Jr., eds.), Sect. 2, Vol. 2, pp. 201–236. Am. Physiol. Soc., Washington, D.C. (B)

Paul, R. J., and Peterson, J. W. (1975). Relations between length, isometric force, and O_2 consumption rate in vascular smooth muscle. *Am. J. Physiol.* **228**, 915–922. (B)

Paul, R. J., and Rüegg, J. C. (1978). Biochemistry of vascular smooth muscle: Energy metabolism and proteins of the contractile apparatus. *In* "Microcirculation" (B. M. Altura and G. Kaley, eds.), Vol. II, pp. 41–82. Univ. Park Press, Baltimore, Maryland. (B)

Pearson, T. A., Dillman, J. M., Solez, K., and Heptinstall, R. H. (1978). Clonal markers in the study of the origin and growth of human atherosclerotic lesions. *Circ. Res.* **43**, 10–18. (E)

Pedowitz, P., and Perell, A. (1957). Aneurysms complicated by pregnancy. I. Aneurysms of the aorta and its major branches. *Am. J. Obstet. Gynecol.* **73**, 720–735. (E)

Persmen, E., Norio, R., and Sarna, S. (1975). Thickenings in the coronary arteries in infancy as an indication of genetic factors in coronary heart disease. *Circulation* **51**, 218–225. (A,E)

Ransom, J. Y., and Bevan, R. D. The aorta and its ventral branches respond differently to norepinephrine (in preparation). (D)

Reidy, M. A., and Schwartz, S. M. (1981). Endothelial regeneration. III. Time course of intimal changes after small defined injury to rat aortic endothelium. *Lab. Invest.* **44**, 301–312. (C)

Rifkin, R. D., Parisi, H. F., and Follard, E. (1979). Coronary calcification in the diagnosis of coronary artery disease. *Am. J. Cardiol.* **44**, 141–147. (E)

Rivin, A. U., and Dimitroff, S. P. (1954). The incidence and severity of atherosclerosis in estrogen treated males and in females with a hypoestrogenic state. *Circulation* **9**, 533–538. (E)

Roach, M. R. (1963). An experimental study of the production and time course of poststenotic dilatation in the femoral and carotid arteries of adult dogs. *Circ. Res.* **13**, 537–551. (E)

Roach, M., and Burton, A. C. (1957). Reason for the shape of the distensibility curves of arteries. *Can. J. Biochem. Physiol.* **35**, 681–690. (A)

Roberts, J. C., Jr., Moses, C., and Wilkins, R. H. (1959). Autopsy studies in atherosclerosis. I. Distribution and severity of atherosclerosis in patients dying without morphologic evidence of atherosclerotic catastrophe, and II. Distribution and severity of atherosclerosis in patients dying with morphologic evidence of atherosclerotic catastrophe. *Circulation* **20**, 511–519, 520–526. (E)

Roberts, W. C. (1977). Coronary heart disease. A review of abnormalities observed in the coronary arteries. *Cardiovasc. Med.* **2**, 29–49. (E)

Robertson, A. L., and Rosen, L. A. (1977). The arterial endothelium: Characteristics and function of the endothelial lining of large arteries. In "Microcirculation" (G. Kaley and B. M. Altura, eds.), Vol. I, pp. 145–163. Univ. Park Press, Baltimore, Maryland. (A)

Rodbard, S. (1970). Negative feedback in the architecture and function of the connective and cardiovascular tissues. *Perspect. Biol. Med.* **13**, 507–527. (A)

Rosell, S., and Belfrage, E. (1975). Adrenergic receptors in adipose tissues and their relation to adrenergic innervation. *Nature (London)* **253**, 738–739. (D)

Ross, R., and Glomset, J. A. (1976). The pathogenesis of atherosclerosis. *N. Engl. J. Med.* **295**, 369–377 (C,E,)

Ross, R., and Harker, L. (1976). Hyperlipidemia and atherosclerosis. *Science* **193**, 1094–1100. (C)

Rowan, R. A., and Bevan, J. A. (in press). Distribution of adrenergic synaptic cleft width in vascular and nonvascular smooth muscle. In "Vascular Neuroeffector Mechanisms—IV," *Int. Symp. Vascular Neuroeffector Mechanisms, 4th, July 28–30, 1981, Kyoto, Japan.* Raven, New York. (D)

Rüegg, J. C. (1971). Smooth muscle tone. *Physiol. Rev.* **51**, 201–248. (B)

Russell, M. P., and Moran, N. C. (1980). Evidence for lack of innervation of B-2 adrenoceptors in the blood vessels of the gracilis muscle of the dog. *Circ. Res.* **46**, 344–352. (D)

Ryan, J. W., and Ryan, U. S. (1977). Pulmonary endothelial cells. *Fed. Proc., Fed. Am. Soc. Exp. Biol.* **36**, 2683–2691. (C)

Sacks, A. H. (1975). The vasa vasorum as a link between hypertension and arteriosclerosis. *Angiology* **26**(5), 385–390. (E)

Saruk, M., and Eisenstein, R. (1977). Aortic lesion in Marfan syndrome. *Arch. Pathol. Lab. Med.* **101**, 74–77. (E)

Schlatmann, T. J. M., and Becker, A. E. (1977a). Histologic changes in the normal aging aorta: Implications for dissecting aortic aneurysm. *Am. J. Cardiol.* **39**, 13–20. (E)

Schlatmann, T. J. M., and Becker, A. E. (1977b). Pathogenesis of dissecting aneurysm of aorta. Comparative histopathologic study of significance of medial changes. *Am. J. Cardiol.* **39**, 21–26. (E)

Schoefl, G. I., and French, J. E. (1968). Vascular permeability to particulate fat: Morphological observations on vessels of lactating mammary gland and of lung. *Proc. Royal Soc. London, Ser. B.* **169**, 153–165. (C)

Schwartz, S. M., Gajdusek, C. M., and Selden, S. C. (1981). Vascular wall growth control: The role of the endothelium. *Arteriosclerosis* **1**, 107–126. (C)

Scott, R. F., Imai, H., Makita, T., Thomas, W. A., and Reiner, S. M. (1980). Lining cell and intimal smooth muscle cell response and Evans blue staining in abdominal aorta of young swine after denudation by balloon catheter. *Exp. Mol. Pathol.* **33**, 185–202. (C)

Sengel, A., and Stoebner, P. (1970). Golgi origin of tubular inclusions in endothelial cells. *J. Cell Biol.* **44**, 223–231. (C)

Shigei, T., Ishikawa, N., Ichikawa, T., and Tsuru, H. (1978). Differences in the response of the three embryologically distinct segments of the isolated canine posterior vena cava to vasoactive substances. *Blood Vessels* **15**, 157–169. (D)

Shigei, T., Ichikawa, T., Ishikawa, N., Uematsu, T., and Tsuru, H. (in press). Functional venous characteristics in relation to embryology. In "Vascular Neuroeffector Mechanisms—IV" *Int. Symp. Vascular Neuroeffector Mechanisms, 4th, July 28–30, 1981, Kyoto, Japan.* Raven, New York. (D)

Simionescu, M., Simionescu, V., and Palade, G. E. (1976). Segmental differentiations of cell junctions in the vascular endothelium. Arteries and veins. *J. Cell Biol.* **68**, 705–723. (A)

Sirek, O. V., Sirek, A., and Fikar, K. (1977). The effect of sex hormones on glycosaminoglycan content of canine aorta and coronary arteries. *Atherosclerosis* **27**, 227–233. (E)

Small, R., Macaraki, E., and Fisher, A. B. (1977). Production of 5-hydroxyindoleacetic acid from serotonin by cultured endothelial cells. *J. Cell Physiol.* **90**, 225–231. (C)

Smith, E. B., Dietz, D. S., and Craig, I. B. (1979). Characterization of free and tightly bound lipoprotein in intima by thin layer isoelectric focussing. *Atherosclerosis* **33**, 329–342. (E)

Smith, J. F. A., Canham, P. B., and Starkey, J. (1981). Orientation of collagen in the tunica adventitia of the human cerebral artery measured with polarized light and the universal stage. *J. Ultrastruct. Res.* **77**, 133–145. (A)

Smith, U., Ryan, J. W., Michie, D. D., and Smith, D. S. (1971). Endothelial projections as revealed by scanning electron microscopy. *Science* **173**, 925–927. (A)

Somlyo, A. V. (1980). Ultrastructure of vascular smooth muscle. In "Handbook of Physiology" (D. F. Bohr, A. P. Somlyo, and H. V. Sparks, Jr., eds.), Sect. 2, Vol. 2, pp. 33–68. Am. Physiol. Soc., Bethesda, Maryland. (A)

Spaet, T. H., and Lejnieks, I. (1967). Mitotic activity of rabbit blood vessels. *Proc. Soc. Exp. Biol. Med.* **1125**, 1197–1198. (C)

Spain, D. M., and Bradess, V. A. (1970). Sudden death from coronary heart disease. Survival time, frequency of thrombi and cigarette smoking. *Chest* **58**, 107–110. (E)

Stanley, J. C., and Fry, W. J. (1974). Pathogenesis and clinical significance of splenic artery aneurysms. *Surgery* **76**, 898–909. (E)

Stehbens, W. E. (1974). Hemodynamic production of lipid deposition, intimal tears, mural dissection and thrombosis in the blood vessel wall. *Proc. R. Soc. London, Ser. B* **185**, 357–373. (E)

Stehbens, W. E. (1975). The role of hemodynamics in the pathogenesis of atherosclerosis. *Prog. Cardiovasc. Dis.* **18**, 89–103. (E)

Stehbens, W. E. (1981). Arterial structure at branches and bifurcations with reference to physiological and pathological processes, including aneurysm formation. In "Structure and Function of the Circulation" (C. J. Schwartz, N. T. Werthessen, and S. Wolf, eds.), pp. 667–694. Plenum, New York. (A)

Stemerman, M. B. (1981). Effects of moderate hypercholesterolemia on rabbit endothelium. *Arteriosclerosis* **1**, 25–32. (C)

Stemerman, M. B., and Ross, R. (1972). Experimental arteriosclerosis. I. Fibrous plaque formation in primates: An electron microscope study. *J. Exp. Med.* **136**, 769–789. (C)

Stemerman, M. B., Pitlick, F. A., and Dembitzer, H. B. (1976). Electron microscopic immunohistochemical identification of endothelial cells in the rabbit. *Circ. Res.* **38**, 146–156. (C)

Stemerman, M. B., Spaet, T. H., Pitlick, F., Cintron, J., Lejnieks, I., and Tiell, M. L. (1977). Intimal Healing: The pattern of re-endothelialization and intimal thickening. *Am. J. Pathol.* **87**, 125–137. (C)

Sumner, D. S., Hokanson, E., and Strandness, D. E. (1969). Arterial walls before and after endarterectomy. *Arch. Surg.* **99**, 606–611. (E)

Suzuki, K., Hori, S., and Ooneda, G. (1979). Electron microscopic study on the medial defect at the apex of human cerebral arterial bifurcations. *Virchows Arch. Path. Anat. and Histol.* **382**, 151–161. (E)

Takaoka, K. (1956). Development of blood vessels of pro- and opisthonephros of *Hynobius naevius* (a salamander). *Acta Anat. Nippon* **31**, 349–376. (D)

Taylor, K., Glagov, S., Lamberti, J., Vesselinovitch, D., and Schaffner, T. (1978). Surface configuration of early atheromatous lesions in controlled-pressure perfusion-fixed monkey aortas. *Scanning Electron Microsc.* **2**, 449–457. (E)

Tejada, C., Strong, J. P., Montenegro, M. R., Restrepo, C., and Solberg, L. A. (1968). Distribution of coronary and aortic atherosclerosis by geographic location, race and sex. *Lab. Invest.* **18**, 509–526. (E)

Thomas, W. A., Reiner, J. M., Florentin, R. A., and Scott, R. F. (1979). Population dynamics of arterial cells during atherogenesis. VIII. Separation of the roles of injury and growth stimulation in early aortic atherogenesis in swine originating in preexisting intimal smooth muscle cell masses. *Exp. Mol. Pathol.* **31**, 124–144. (E)

Toda, N. (1977). Actions of bradykinin on isolated cerebral and peripheral arteries. *Am. J. Physiol.* **232**, H267. (D)

Ts'ao, C. H., and Glagov, S. (1970). Basal endothelial attachment: Tenacity at cytoplasmic dense zone in the rabbit aorta. *Lab. Invest.* **23**, 510–516. (A)

Ts'ao, C. H., Glagov, S., and Kelsey, B. F. (1970). Special structural features of the rat portal vein. *Anat. Rec.* **166**, 529–540. (A)

Ts'ao, C. H., Glagov, S., and Kelsey, B. F. (1971). Structure of the mammalian portal vein; postnatal establishment of two mutually perpendicular medial muscle zones in the rat. *Anat. Rec.* **171**, 457–470. (A)

Weber, G., Fabbrini, P., and Resi, L. (1973). On the presence of a concanavalin A reactive coat over the endothelial aortic surface and its modifications during early experimental cholesterol atherogenesis in rabbits. *Virchows Arch. A: Pathol Anat.* **359**, 299–307. (A)

Weibel, E. R., and Palade, G. E. (1964). New cytoplasmic components in arterial endothelia. *J. Cell Biol.* **23**, 101–112. (C)

Weksler, B. B., Marcus, A. J., and Jaffee, E. A. (1977). Synthesis of prostaglandin I_2 (prostacyclin) by cultured human and bovine endothelial cells. *Proc. Natl. Acad. Sci. U.S.A.* **74**, 3922–3926. (C)

White, N. K., Edwards, J. E., and Dry, T. J. (1950). The relationship of the degree of atherosclerosis with age in men. *Circulation* **1**, 645–654. (E)

Wilens, S. L., Malcolm, J. A., and Vasquez, J. M. (1965). Experimental infarction (medial necrosis) of the dog's aorta. *Am. J. Pathol.* **47**, 695–711. (E)

Wilson, J. G. (1961). Embryology of the heart and major vessels. *In* "Development and Structure of the Cardiovascular System" (A. A. Luisada, ed.), pp. 1–12. McGraw-Hill, New York. (D)

Winquist, R. J., and Bevan, J. A. (1980). Relaxant properties of vasodilators on the myogenic tone of the rabbit facial vein. *Blood Vessels* **17**, 169 (Abstr.). (D)

Winquist, R. J., and Bevan, J. A. (1981a). *In vitro* model of maintained myogenic vascular tone. Brief Communications. *Blood Vessels* **18**, 134–138. (D)

Winquist, R. J., and Bevan, J. A. (1981b). Relative location of α- and β-adrenoceptors to sites of release of sympathetic transmitter in the rabbit facial vein. *Circ. Res.* **49**, 486–492. (D)

Wissler, R. W., and Vesselinovitch, D. (1974a). Differences between human and animal atherosclerosis. *In* "Atherosclerosis III" (G. Schettler and A. Weizel, eds.), pp. 319–325. Springer-Verlag, Berlin and New York. (E)

Wissler, R. W., and Vesselinovitch, D. (1974b). Evidence for prevention and regression of atherosclerosis in man and experimental animals at the arterial lesion level. *In* "Atherosclerosis III" (G. Schettler and A. Weizel, eds.), pp. 747–750. Springer-Verlag, Berlin and New York. (E)

Wissler, R. W., and Vesselinovitch, D. (1983). The complementary interaction of epidemiological and experimental animal studies. A key foundation of the preventive effort. *Prev. Med.* **12**, 84–99. (E)

Wissler, R. W., and Vesselinovitch, D. (1983). Atherosclerosis—relationship to coronary blood flow. *Am. J. Cardiol.* **52**, 2A–7A. (E)

Wolinsky, H. (1970a). Comparison of medial growth of human thoracic and abdominal aortas. *Circ. Res.* **27**, 531–538. (A)

Wolinsky, H. (1970b). Response of the rat aortic media to hypertension: Morphological and chemical studies. *Circ. Res.* **26**, 507–522. (E)

Wolinsky, H. (1973). Comparative effects of castration and antiandrogen treatment on the aortas of hypertensive and normotensive male rats. *Circ. Res.* **33**, 183–189. (E)

Wolinsky, H., and Glagov, S. (1964). Structural basis for the static mechanical properties of the aortic media. *Circ. Res.* **14**, 400–413. (A)

Wolinsky, H., and Glagov, S. (1967a). A lamellar unit of aortic medial structure and function in mammals. *Circ. Res.* **20**, 99–111. (A)

Wolinsky, H., and Glagov, S. (1967b). Nature of species differences in the medial distribution of aortic vasa vasorum in mammals. *Circ. Res.* **20**, 409–421. (A,E)

Wolinsky, H., and Glagov, S. (1967c). Zonal differences in modelling of the mammalian aortic media during growth. *Fed. Proc., Fed. Am. Soc. Exp. Biol.* **26**, 357. (A)

Wolinsky, H., and Glagov, S. (1969). Comparison of abdominal and thoracic aortic medial structure in mammals. Deviation of man from the usual pattern. *Circ. Res.* **25**, 677–686. (A,E)

Yater, W. M., Traum, A. H., Brown, W. G., Fitzgerald, R. P., Geisler, M. A., and Wilcox, B. B. (1948). Coronary artery disease in men 18 to 39 years of age. Report of 866 cases, 450 with necropsy examinations. *Am. Heart J.* **36**, 334–681. (E)

Zarins, C. K., Taylor, K. E., Bomberger, R. A., and Glagov, S. (1980). Endothelial integrity at aortic ostial flow dividers. *Scanning Electron Microsc.* **III**, 249–254. (A)

Zarins, C. K., Giddens, D. P., and Glagov, S. (1982). Atherosclerotic plaque distribution and flow velocity profiles in the carotid bifurcation. *In* "Cerebrovascular Disease" (J. J. Bergan, and J. S. T. Yao, eds.), pp. 19–30. Grune and Stratton, New York. (E)

Zarins, C. K., Zatina, M. A., and Glagov, S. (1983). Correlation of postmortem angiography with pathologic anatomy. *In* "Noninvasive Diagnosis of Atherosclerotic Lesions: Quantitative Evaluation of Morphology, Biochemistry, and Pathophysiology" (M. G. Bond, W. Insull, Jr., S. Glagov, A. B. Chandler, and J. F. Cornhill, eds.), Chap. 13, pp. 283–306. Springer-Verlag, Berlin and New York. (E)

Chapter 2
Biomechanics of Arteries and Veins

A. MECHANICAL PROPERTIES*
By P. B. DOBRIN

1. INTRODUCTION

The chief mechanical function of blood vessels is to transport blood. In addition, the arteries transiently store and transmit the pulsatile energy generated by contraction of the heart. The veins, especially the small venous channels, act as a reservoir for most of the blood volume, and the microcirculatory vessels provide both resistance and exchange functions. The present section briefly presents the mechanical behavior of relaxed and contracted blood vessels; more complete discussions of this subject are available in recent reviews (Dobrin, 1978, 1983; Fung *et al.*, 1979; McDonald, 1974; Murphy, 1980; Patel and Vaishnav, 1980).

2. CIRCUMFERENTIAL MECHANICS

a. Relaxed blood vessels: All blood vessels are compliant at low pressures, but become stiff when distended by high pressures (Bergel, 1961), for wall stiffening occurs as the connective tissue fibers are straightened with vessel distention. Stiffening takes place at about 80 mm Hg (Fig. 2.1) in normal arteries, at 35–50 mm Hg in normal veins, and at 3–7 mm Hg in large lymphatics (Ohhashi *et al.*, 1980) (see Section D-2 below). With age, stiffness in the human thoracic aorta appears at progressively lower pressures (Bader, 1967) so that under physiologic conditions these vessels are functionally less compliant. Enzymatic degradation experiments and histologic studies indicate that, in normal arteries, elastin bears circumferential loads at virtually all transmural pressures, whereas collagen does so only at 80 mm Hg pressures and above (Roach and Burton, 1957; Wolinsky and Glagov, 1964). Although stretched collagen fibers are several thousand times as stiff as elastin or the intact wall, quantitative estimates in a variety of dog vessels suggest that only 8 to 25% of

*Supported by a Veterans Administration Clinical Investigatorship and United States Public Health Service Grant HE-08682 from the National Institutes of Health.

64

ISBN 0-12-042520-3

the collagen present in arteries actually is load-bearing (Cox, 1978). Recent enzymatic degradation studies of dog carotid artery show that elastin bears loads in the circumferential, longitudinal, and radial directions, whereas collagen contributes to vessel stiffness almost solely in the circumferential direction (P. B. Dobrin, personal observations).

With advancing age, arteries become larger in diameter and develop thicker walls with widened media. They also exhibit increased collagen content. Such changes are accompanied by elevated vessel stiffness (Bader, 1967). Some of the greater stiffness is due to increased diameter, since vessels become stiffer as the connective tissues are stretched, but another portion can be attributed to the progressive unfolding and stiffening of elastin and collagen fibers at any given vessel diameter with aging (Roach and Burton, 1959; Samila and Carter, 1981). Finally, calcification of arteries with advancing age may similarly restrict the extensibility of the connective tissues.

Arteries also exhibit elevated stiffness with hypertension and advanced atherosclerosis, conditions associated with increased deposition of collagen. Although the intracellular signal triggering deposition of this material is uncertain, it seems likely that the physiologic stimulus is the mean or the oscillatory deformations and/or associated stresses, for *in vitro* oscillation causes smooth muscle cells to increase their synthesis of connective tissues (Leung *et al.*, 1976).

The development of arterial aneurysm is an example of pathophysiologic vascular mechanics. Although the genesis of abdominal aortic aneurysms usually is attributed to atherosclerosis, such lesions also may reflect an inherent connective tissue susceptibility to mechanical failure. Recently it has been proposed that endogenous circulating or mural collagenases and elastases may precipitate rupture of aneurysms (Swanson *et al.*, 1980; Busuttil *et al.*, 1980, 1982). Experimental treatment of normal vessels with high doses of these enzymes reveals that degradation of elastin leads to marked vessel dilatation and that sequential degradation of both elastin and collagen results in massive vessel dilatation resembling an aneurysm. Although degradation of elastin does not disrupt the wall, degradation of collagen, even with elastin intact and with minor vessel dilatation, produces frank arterial rupture (P. B. Dobrin, personal observations). Since the distending force in a cylindrical vessel increases directly with vessel diameter, one may question how an aneurysmal artery which cannot maintain equilibrium at normal dimensions is able to do so, even temporarily, at larger dimensions. Two mechanisms may be important: First, aneurysms recruit and stretch additional collagen fibers, some of which may have been unstretched and in the adventitia. Second, aneurysms often assume a fusiform shape, a configuration which is better described as a sphere than as a cylinder. The circumferential stress for a sphere is about one-half that for a cylinder of similar radius and wall thickness. Therefore, the wall stress in fusiform-shaped aneurysms increases much less rapidly than it would in cylindrical vessels of the same dimensions. (For morphologic aspects of aneurysms, see Section I-4a,b Chapter 1.)

   b. Contracted blood vessels: The large arteries possess considerable connective tissue and relatively little vascular muscle, and therefore seldom constrict actively; in fact, they usually are distended by elevations in arterial pressure. However, even the large conduit vessels may actively decrease in caliber under conditions of acute blood loss (Gerova and Gero, 1967). This indicates that the muscle cells in conduit arteries participate in pressor responses but normally are unable to overcome the distending force resulting from increased pressure.

   Small arteries and arterioles are capable of constricting under almost all conditions, and this response produces a rise in arterial pressure. Constriction of such vessels is favored by a geometric advantage; the circumferential force distending them is given by the product of transmural pressure, internal diameter, and vessel length. Therefore, because of the low transmural pressure and small diameter, small vessels are subject to lower distending forces than are large vessels. In addition, the arterioles and small arteries possess relatively thick walls with an abundance of vascular muscle, factors which also favor constriction.

   For both large and small vessels, maximum constriction occurs at their respective physiologic pressures. Active contraction of most vascular muscle occurs slowly and in some cases may require several minutes to achieve completion. This is in sharp contrast to skeletal and cardiac muscle in which contraction takes place within fractions of a second. The slow contraction velocity of vascular muscle results from (a) the long extracellular diffusion distance required of neurotransmitters in such tissue; (b) the long intracellular distance for calcium ion diffusion in sarcoplasmic reticulum-poor smooth muscle cells; and (c) the low ATPase activity of smooth muscle cells, since splitting of ATP is a requirement for cycling of actomyosin cross bridges. Mangel *et al.* (1982) reported that in the rabbit, aortic vascular muscle contracts synchronously with each pressure pulse and that the response is elicited by a pacemaker in the right, but not the left, atrium. This is a unique observation that requires confirmation.

   Vascular muscle is capable of markedly constricting arteries. In the case of small vessels, profound constriction is thought to cause "critical closure," i.e., the cessation of blood flow despite a positive pressure gradient (Burton, 1961). During constriction, the active retractive force provided by the vascular muscle adds to the passive retractive force exerted by the connective tissues. Burton (1961) proposed that when the sum of these two forces exceeds the distending force, then a disequilibrium results which leads to complete vessel closure. Burton's analysis was based on an application of the law of Laplace: $T = Pr$, where T is wall tension, P is transmural pressure, and r is radius. Azuma and Oka (1971) argued that whereas this expression applies to infinitely thin membranes, it neglects the finite wall thickness of blood vessels. These investigators offered an alternative analysis which accounted for wall thickness and concluded that the walls of most vessels actually are under compression, not tension. They further suggested that such compressive forces may be the cause of critical closure. However, neither the analysis of Burton nor that of Azuma and Oka accounts for the material properties of vessel walls at small diameters, for under such conditions the walls may resist compression. In addition, the aggregation of red blood cells at low shear rates may contribute to cessation of blood flow with low, but positive, pressure gradients (Schmid-Schoenbein, 1976). At present, the precise mechanism of critical closure remains to be determined.

If the muscle in a vessel is excited at the same time that the transmural pressure is elevated, contraction may proceed in an isometric fashion, i.e., pressure may rise with little or no change in vessel diameter. Maximally excited muscle of large arteries can tolerate about 100 mm Hg increase in pressure before the vessels are forced to distend. This corresponds to $2.2–3.5 \times 10^5$ mN/m^2 active isometric stress at optimum muscle length (L_0), with lesser values developed at shorter and longer lengths (Murphy, 1980). Such a range of values is comparable to that developed by skeletal muscle and is several times greater than that developed by nontetanizable cardiac papillary muscle. The shape of the length-active stress curve exhibited by smooth muscle resembles that of skeletal and cardiac muscles, but with a more precipitous decline in stress at long muscle lengths. This is due, at least in part, to the highly compliant series elastic component (SEC) of smooth muscle, i.e., the sum of compliant elements coupled in series with the force-generating apparatus. In vascular muscle, the SEC is extended 7–10% at peak active stress (Dobrin and Canfield, 1973; Murphy, 1980). Therefore, at L_0, the contractile element in smooth muscle is 7–10% shorter than that predicted by overall tissue length, and this may distort the shape of the manifest length-active stress curve. By contrast, the SEC in skeletal muscle is only 1–3%.

c. Vessel resistance to distention: In small arteries and arterioles, vasoconstriction results in a marked increase in circumferential stiffness (Hinke and Wilson, 1962) due, in part, to the low distending force in vessels of small caliber. In large arteries, vasoconstriction results in increased circumferential stiffness at small dimensions, but decreased stiffness at large dimensions (Dobrin and Rovick, 1969). This is seen most clearly when stiffness is examined as a function of vessel diameter or of strains computed with respect to a single reference value (Dobrin, 1983). However, increased stiffness also is evident simply upon inspection of pressure–radius curves (Fig. 2.1). With distention, the vessel exhibits increased stiffness at radii up to the peak in the length-active tension curve (L_0). At larger radii, the vessel exhibits decreased stiffness as the muscle is stretched to lengths greater than L_0; at these lengths, the muscle yields, thereby permitting vessel distention.

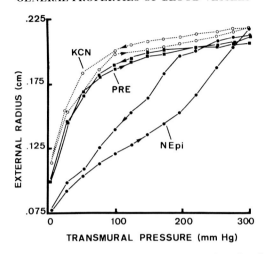

FIG. 2.1. Pressure–radius curves for a dog carotid artery in the relaxed, pretreatment state (PRE), after excitation of muscle with norepinephrine (NEpi) and after metabolic poisoning of muscle with potassium cyanide (KCN). The data illustrate several important features: (1) slopes of curves reflect vessel distensibility, with the relaxed vessel exhibiting marked distensibility at low pressures and increasing stiffness appearing at about 80 mm Hg; (2) the horizontal distance between NEpi and KCN curves reflects the unimodal, peaked length-active tension curve of vascular muscle; (3) activation of the muscle (NEpi) increases vessel stiffness at small and moderate dimensions of up to about 100 mm Hg, but it decreases vessel stiffness relative to the relaxed vessel at higher pressures as the constricted vessel gradually dilates; decreased stiffness occurs at dimensions greater than the peak in the length-active tension curve; (4) pressure–diameter curves obtained for the relaxed vessels (KCN) exhibit little hysteresis, whereas curves obtained for constricted vessels (NEpi) exhibit marked hysteresis. Direction of hysteresis indicates that vascular muscle, once excited, can bear more force than it can generate. (From Dobrin and Rovick, 1969, reproduced with permission of American Physiological Society.)

Finally, the smooth muscle cells in some vessels are myogenic, i.e., they actively contract when stretched (Winquist and Bevan, 1981). This behavior is seen most frequently in small arteries and some veins and may contribute to autoregulation. The characteristics of the mechanical stimulus and the identity of the intracellular sensor of myogenicity are important physiologic parameters that remain to be determined.

   d. Quantitative considerations: At equilibrium the diameter of a vessel is stable, i.e., it neither dilates nor constricts. Under these conditions, distending and retracting forces are equal, and the circumferential force exerted by the wall is given by

$$T = P_{T}r \tag{A.1}$$

where T is the tensile force exerted by the wall, P_{T} is transmural pressure, and r is mean radius. Dividing this tensile force by the thickness of the wall gives the circumferential stress (σ_{θ})

$$\sigma_\theta = P_T(r_i/h) \tag{A.2}$$

where r_i is the internal radius and h is the wall thickness. These expressions are determined from static equilibria and are independent of wall stiffness. Computation of stiffness, such as "elastic modulus," is much more complex for intact vessels because pressurization causes simultaneous deformation in all directions. (For detailed discussions of quantitative methods see Fung *et al.*, 1979; Patel and Vaishnav, 1980; and Dobrin, 1983.)

3. Longitudinal Mechanics

a. Relaxed blood vessels: Arteries and veins change length with motion of the body structure and viscera. For example, with extension of the knee joint, the popliteal artery is lengthened 9% in the dog and 20% in man (Browse *et al.*, 1979); with extension of the cervical spine, the dog carotid artery is lengthened from 10 to 60% (Dobrin, 1983). As the heart recoils during the ejection phase of the cardiac cycle, the ascending aorta and pulmonary arteries extend 5–11%. However, most vessels exhibit negligible length changes with the pressure oscillations that accompany each cardiac cycle. The descending thoracic aorta lengthens about 1% during each cardiac cycle, whereas the abdominal aorta actually *shortens* about 1% (Patel *et al.*, 1964). Perivascular tissues adherent to arteries, as well as branches originating from each artery, tether vessels at *in situ* length (Patel and Vaishnav, 1980). In addition, arteries are held in an extended position by traction, as exemplified by the marked shortening of vessels following excision. Most arteries exhibit little active muscle stress in the longitudinal direction, and the strong tethering forces mentioned above minimize any changes in length that might result from activation of the muscle.

b. Quantitative considerations: The tensile stress in the longitudinal direction due to pressure (σ_{Zp}) is

$$\sigma_{Zp} = \frac{P_T r_i^2}{\pi r_o^2 - \pi r_i^2} \cong P_T \frac{r_i}{2h} \tag{A.3}$$

where r_i is internal radius, r_o is external radius. Thus, σ_{Zp} is approximately one-half σ_θ. In addition, there is a strong traction component which extends vessels. The traction stress (σ_{ZT}) is given by the traction force (F_T) divided by the cross-sectional area of the wall $(\pi r_o^2 - \pi r_i^2)$

$$\sigma_{ZT} = \frac{F_T}{\pi r_o^2 - \pi r_i^2} \tag{A.4}$$

In addition, there is a small longitudinal component at the intima due to the shearing action of the blood (see Patel and Vaishnav, 1980). While this may disturb the integrity of the endothelium, it is several orders of magnitude smaller than the static stresses computed above and generally is disregarded in the evaluation of static wall mechanics. The longitudinal forces resulting from traction and from pressurization add algebraically to maintain nearly constant net longitudinal force and hence maintain nearly constant vessel length (Dobrin, 1978). However, with age, the traction force decreases, and this permits vessels to buckle at high pressure, a common finding in aged hypertensive patients.

4. RADIAL MECHANICS

Distention of large arteries during each cardiac cycle increases vessel diameter 8–10% and increases vessel length about 1%. These deformations are associated with narrowing of the wall thickness. In addition, the pressure exerted directly against the vessel wall produces a compressive stress in the radial direction. This radial stress is equal to blood pressure at the lumen but declines curvilinearly to vanish at the adventitial margin (Doyle and Dobrin, 1973; Vaishnav *et al.*, 1973) to give an average value that is approximately half the transmural pressure. Radial stress and the narrowing of the wall that accompanies circumferential deformation all compress the structures contained within the wall, including the vasa vasorum. Elevated blood pressure increases this compression but also raises perfusion pressure in the vasa. Nevertheless, experimental evidence indicates that hypertension decreases flow through the vasa (Heistad *et al.*, 1978). Although the interactions of mechanical factors in determining wall nutrition are complex (Dobrin, 1983), this area may be of pathophysiologic significance and deserves further investigation. (For the morphologic aspects of vasa vasorum and wall nutrition, see Section A-2d, Chapter 1.)

B. HEMODYNAMICS

By D. BERGEL

Hemodynamics has changed greatly in recent years due to improved instrumentation and the rigorous application of fluid dynamics to the circulation. What has been learned is well described in various texts, including those of Bergel (1972), McDonald (1974), Caro *et al.* (1978), Patel and Vaishnav (1980), and Milnor (1982). The following summarizes the basic principles of hemo-dynamics as they have been used to study events in large arteries.

1. ESTABLISHED FLOW IN A CYLINDRICAL CHANNEL

It is necessary to start with a statement of Poiseuille's expression, derived, it should be mentioned, from work directed toward understanding the circulation of blood

$$Q = (8/\pi)(\Delta P/\Delta L)(R^4/\eta) \tag{B.1}$$

Here the flow rate (Q) is related to three variables: the pressure gradient $(\Delta P/\Delta L)$; the tube radius (R); and a fluid property, the viscosity (η). This statement is true only within limits, but it does predict the minimum energy cost of flow. In the 1950s, Womersley, working with McDonald (see McDonald, 1974), produced linearized solutions for the flow due to an oscillatory pressure gradient P' given by $P' = P_0 \sin(\omega t)$. This is of the form

$$Q = (R^4/\eta) P_0 M \sin(\omega t + \epsilon) \tag{B.2}$$

which introduces an amplitude term (M) together with a phase lag (ϵ), both of which are frequency dependent so that Eq. (B.2) reduces to Eq. (B.1) when the frequency (ω) is zero, i.e., steady flow. The magnitudes of these modifying terms are determined by a nondimensional parameter, Womersley's α

$$\alpha = R(\omega\rho/\eta)^{1/2} \tag{B.3}$$

Thus α depends on tube radius, frequency, and the ratio of density (ρ) to fluid viscosity (η). The inverse ratio, η/ρ, is known as the kinematic viscosity (ν). When $\alpha > 10$, M and ϵ approach asymptotic values, the resulting flow is determined by fluid inertia, and it lags nearly 90° behind the pressure gradient, whereas for $\alpha > 3$, flow approximates that for the instantaneous Poiseuille solution. Values of α found in the human circulation are about 15 in the aorta and 3 in the femoral artery (McDonald, 1974; Milnor, 1982).

2. IN VIVO APPLICATIONS

a. Viscosity term: Blood is a non-Newtonian fluid (see Section C, below). Its rheological properties are of extreme importance in the microcirculation but have not been shown to be of significance in the large vessels (McDonald,

1974). Indeed, Taylor (1959) showed that the oscillatory pressure–flow relation would be in error by 2% if the high shear asymptotic value for viscosity were used. Of course, the actual number taken depends upon hematocrit. However, the viscous shear force exerted at the wall by the moving blood is importantly affected by the properties of blood.

  b. *Linearity and use of Fourier analysis:* Womersley's expressions deal with sinusoidal pressure and flow waves [Eq. (B.2)], but the waves actually seen are not truly sinusoidal. However, if pressure and flow are related linearly, i.e., twice the pressure gradient producing twice the flow, then one may make use of the principle of superposition. This can be stated as follows: Suppose that pressure gradient A (steady or oscillatory) drives a flow A' and that pressure gradient B drives flow B', etc., then the simultaneous application of complex gradient $A + B + C \cdots$ results in the flow $A' + B' + C' \cdots$. Since there are well established mathematical techniques to describe any pressure gradient as the sum of a series of sinusoidal waves, the flow due to each such component can be computed and combined to show the resulting complex flow wave. The validity of this approach is defended by McDonald (1974), and the method has been shown to be adequate as a first approach. However, when gross nonlinearities exist, the method cannot be used. For example, ventricular pressure and aortic flow are not related during diastole, and hence the use of Fourier analysis to relate aortic flow to the ventricular–aortic pressure gradient is not valid.

  c. *Oscillatory pressure gradient:* The simplest example of such a state is a pressure wave which propagates along an artery (distance x) at a constant velocity c. Since $c = dx/dt$, the pressure gradient, dp/dx, can be expressed as

$$dp/dx = (dp/dt)(1/c) \tag{B.4}$$

permitting the calculation of the flow from a single pressure measurement. While Eq. (B.4) does not apply exactly, it does show that the flow wave is expected to resemble the time differential of the pressure wave and not the pressure wave itself (McDonald, 1974).

  The pressure wave velocity is given by the Moens–Korteweg expression, $c^2 = Eh/2R\rho$, where E is the Young's elastic modulus of the vessel wall and h is the wall thickness. This expression must be modified to deal with the Poisson's ratio of the wall, i.e., the length changes and thinning that accompany circumferential distention and the tethering forces which restrain longitudinal wall motion (McDonald, 1974; Patel and Vaishnav, 1980; Dobrin, 1978). In addition, corrections are required to account for arterial wall viscoelasticity, and the value taken for E must be appropriate for the mean pressure level. However, even when all of this is done, it still is not established whether the computed wave velocity accurately describes measured values, for it is difficult to measure wave propagation (velocity and attenuation) because of the presence of reflected pressure waves in all arteries; moreover, opinions differ as to the best method of measurement (Milnor and Bertram, 1978; Busse *et al.*, 1979; Li *et al.*, 1981). At present, measured and computed values for pulse wave propagation must be considered to be somewhat different.

Therefore, the pressure gradient must be measured, which entails the use of two manometers a finite distance apart, i.e., $\Delta p/\Delta x$, not dp/dx. However, Δx may be made to be a very small fraction of a wavelength, since the velocities involved are several meters/second; then $\Delta p/\Delta x$ approximates dp/dx. The pressure waveform changes as it travels, and thus the various harmonic components show different velocities and attenuations. The chief reason for this is the presence of wave reflections, but changes in blood velocity due to alterations in vessel cross-sectional area also are important (Li *et al.*, 1981). In conclusion, the flow wave computed from the pressure differential agrees quite reasonably with measured flow wave (McDonald, 1974).

Following computation of hydraulic resistance as the ratio of mean pressure to mean flow, impedance may be computed from the oscillatory pressures and flows, i.e., the ratio of amplitudes of the pressure components to the amplitudes of the flow components. The result is not a single impedance but rather a spectrum of values showing variation with frequency. The impedance of each frequency is computed from the amplitude of the pressure and flow harmonic at that frequency and their phase relationship. The resulting data are expressed as (1) the impedance modulus and (2) the phase of the input impedance. Since the input flow wave is related to the pressure gradient, which, in turn, results from the interaction of the pressure wave and the propagation characteristics of the bed, the input impedance is strongly influenced by all of these variables. Indeed, the major determinants of the characteristic impedance (i.e., the impedance undistorted by the effects of wave reflection) are the Womersley frequency-dependent fluid parameters and the true wave velocity (McDonald, 1974). However, in the presence of reflections, the measured wave velocity exhibits frequency-related peaks and troughs, as does the measured impedance.

Impedance measurements have been reported for a variety of vascular beds in man and animals, and the general features of the spectra are well understood. They are of interest because they describe the flow characteristics of a vascular bed in exact and concise terms. In addition, the input impedances of the aorta and the pulmonary artery represent the loads presented to the left and right ventricles, respectively (Milnor, 1978).

Impedance plots reflect the frequency dependence of pressure and flow and are said to be in the frequency domain. Exactly the same information can be presented as a plot of the response of the system to a brief impulsive flow at the input. The response of a vascular bed to a flow impulse is a series of separated pressure waves demonstrating the reflected and re-reflected waves (Sipkema *et al.*, 1980). Both the impedance method and the flow impulse method make the same assumptions concerning linearity, but the time domain presentation is likely to prove superior when concerned with cardiac ejection; this can be represented as the sum of an appropriate series of flow impulses. Such a presentation is attractive when one tries to model the interaction between cardiac and vascular mechanics.

3. Turbulence and Other Nonlinear Effects

The description of pulsatile hemodynamics given above accounts satisfactorily for the major features of arterial flow properties, but with improved instrumentation and larger computers, it becomes feasible to look at more complex events.

a. Turbulence: Equation (B.1) deals only with streamline laminar flow. However, when the nondimensional Reynold's number (*Re*) is > 2000, turbulent flow can be expected. Reynold's number may be expressed as

$$Re = D\bar{v}\rho/\eta = 2R\bar{v}/\nu \qquad (B.5)$$

where D is tube or vessel diameter and $\bar{v}$ is average flow velocity ($Q/\pi R^2$). *Re* indicates the relative influence of viscous retardation and fluid inertia. When flow is turbulent, viscous forces are overcome and random three-dimensional velocity variations are superimposed on the mean flow. The excess energy dissipation is indicated by a discontinuity in the pressure–flow relation, but in the circulation, turbulence is best detected by the measurement of localized flow disturbances. Using a thin-film probe, Nerem and Seed (1972) showed transient turbulent disturbances during flow deceleration in the aortic root of dogs. In this region, flow ceases during diastole and the disturbance fades; therefore, when flow is pulsatile, the time available for the growth of turbulence is limited and flow tends to be stabilized. Moreover, Nerem and Seed found that the critical *Re* (based on peak velocity) was about 150α. Later Nerem (1977) extended the analysis with the use of a modified *Re* and confirmed that highly disturbed flow is common in the ascending aorta. The extra energy cost of turbulence is trivial, but the localized high shear forces can damage red cells (Sutera, 1977), and the higher frequency oscillations can affect the arterial wall (Roach, 1977).

b. Convective and local acceleration: In the idealized model of established oscillatory flow in a cylindrical vessel, all blood particles at the same radial location experience the same temporal pattern of velocity change. However, in the case of flow down a pipe of varying cross section, fluid particles undergo a velocity change as they are convected through the taper; the pressure difference across the region reflects this change in kinetic energy. The conversion between potential and kinetic energy is described by the Bernoulli equation, which can be written

$$1/2\rho v^2 + P = \text{constant} \tag{B.6}$$

The kinetic pressure represented by the first term is about 4 mm Hg for a velocity of 100 cm/sec; thus the kinetic terms are a small part of the work of cardiac pumping, but the effects on local flow mechanics may be highly significant (Li *et al.*, 1981). The pressure losses incurred in accelerating flow through a gently tapering vessel are commensurate with the transmission losses. The appearance of terms in v^2 in the equations of motion complicates matters enormously (Atabek, 1980).

c. Entrance length: Fluid passing from a reservoir at rest into a tube undergoes convective acceleration. Initially the fluid moves as a plug and the velocity profile across the tube is flat. Fluid near the wall suffers viscous retardation, and the viscous boundary layer grows in thickness with distance from the entrance. For laminar flow, the process is complete when the boundary occupies the whole lumen, and the entrance length over which this occurs is equal to $(0.03D)(Re)$. This leads to the conclusion that the steady flow entrance region extends over the whole human aorta. However, the established velocity profile is rather flat for turbulent and oscillatory flow, and the entrance lengths are considerably shorter, $(0.0693D)(Re)^{1/4}$ and $3.4\ U/\omega$ (where U is instantaneous core velocity), respectively (Caro *et al.*, 1978). Velocity profile measurements in dogs have, in fact, shown considerable profile development within the aorta (Schultz, 1972). Entrance effects arise at each branching point, although here the fluid is not accelerating from rest, and these inertial influences can be detected even in the microcirculation (Benis *et al.*, 1970).

d. Flow through a stenotic valve or artery: Blood entering a constricting channel is accelerated and loses pressure according to Eq. (B.5), but provided the constriction is not extremely sharp, flow is stable and laminar. However, on leaving a stenotic segment, deceleration occurs and an additional retarding pressure gradient is generated. This adverse force is most effective where the flow is slowest, i.e., near the wall. Unless the channel widens extremely gradually, a region will develop in the area distal and behind the obstruction in which flow near the wall is reversed. The fluid will circulate here, producing flow separation.

With increasing velocity, the separation zone increases in length and considerable extra energy losses occur. Moreover, flow in the emerging jet readily becomes turbulent, and the disturbance is projected a considerable distance downstream (Clark, 1980). Within the jet, the oscillations are not entirely random, but, instead, there is a characteristic frequency or frequencies which become greater as flow increases, giving rise to a murmur, the pitch of which varies with the flow. In addition to the considerable energy losses incurred in the presence of stenosis in large arteries (Clark, 1979), damage to red cells (Sutera, 1977) and alterations in wall properties also may occur.

4. Flow in Veins

Flow in the vena cava shows marked oscillations which result from the atrial pressure waves (Wexler et al., 1968). Venous hemodynamics appears to conform to what has been said above, but a major complicating factor arises from the high distensibility of veins. This is particularly true since the cross section is not circular at very low pressures, so that the vein collapses into a complex shape. Near the collapse point, marked instabilities arise and the system may oscillate (Lyon et al., 1981). The existing knowledge of hemodynamic events in veins is summarized by Caro et al. (1978).

5. Current Problems

a. Interactions between blood and vessel wall: Evidence continues to accumulate that flowing blood can affect the arterial wall. Roach (1977) has shown that turbulent vibrations below an arterial stenosis can reversibly modify the mechanics of the wall and lead to poststenotic dilatation. It has also been suggested that arterial endothelial cells orient themselves along flow lines, for example, around a branch orifice (Langille and Adamson, 1981), although the contribution of blood flow, as against that from some structural feature of the subintimal layer, remains to be determined (Jackman, 1982). It is also clear that high wall shear rates can damage endothelial cells (Fry, 1973). Furthermore, it appears that arterial wall disease has a predilection for certain sites, often occurring near bifurcations (Roach, 1977), and that vascular permeability to macromolecules is influenced by the arteries' flow history (Fry and Vaishnav, 1980). The questions that have to be asked are whether these are all instances of some common mechanism, and what exactly are the critical flow patterns which affect the endothelium. This has led to great interest in flow patterns in complex geometrics and, in particular, to studies of velocity profiles which would allow the prediction of shear rates and forces at the wall. Animal studies (Schultz, 1972; Nerem, 1977; Benson et al., 1980) give information on velocity profiles at a few arterial sites, but technical difficulties favor the use of models (Roach, 1977). From the latter, it appears that the details of geometry and of flow oscillations are of great significance (Lutz *et al.*, 1977), although they cannot be adequately specified.

b. Arterial fluid dynamics and cardiac ejection: The input impedance of the systemic bed represents the hydraulic load on the left ventricle (Milnor, 1978). Unfortunately, there is no accepted method to apply a general understanding of impedance toward a deduction of the state of a heart that is responsible for a particular flow pulse in a particular aorta. Various available techniques (Westerhof *et al.*, 1977; Sdougos *et al.*, 1982) illustrate the practical and conceptual difficulties.

c. Generation of cardiovascular sound: There is very little to add to the brief account of this subject by McDonald (1974). It was reiterated by him that turbulence alone does not generate sounds and that some solid structures must be set in vibration. From measurements on models of stenotic valves, Clark (1976) concluded that the kinetic energy in a jet was unlikely to be responsible for murmur generation, since the turbulent pressures at the wall were several orders of magnitude greater. The idea that sounds result from turbulent disturbances, rather than from a less random eddy shedding process, is supported by *in vivo* sound spectrum measurements. The question is not yet settled (Gross *et al.*, 1980), nor is it possible to predict stenosis geometry from analysis of the sound. Even further removed from a definitive answer is a quantitative evaluation of cardiac and Korotkoff sounds.

d. Blood vessel growth and development: Significant information regarding the influences that bring about the development of a particular vascular pattern is not available. It is known that the growth of collateral vessels involves changes in upstream arteries (Leon and Bloor, 1968) and that there is hypertrophy of arteries supplying a tumor or a pregnant uterus. Such anatomic changes could be due to the conduction of some vasodilatory process from the arteriolar region (Hilton, 1959), but it has been speculated (Rodbard, 1974) that local hemodynamic forces may also be involved.

C. BLOOD RHEOLOGY

By S. CHIEN

The rate of blood flow (Q) through an organ depends on the ratio of pressure drop (P) to flow resistance (R). R is equal to the product of blood viscosity (η_B)

and vascular hindrance (Z). For the flow of a simple fluid through a straight tube with length L and a uniform radius r, Poiseuille's law states that

$$P = Q(8\eta L/\pi r^4) \qquad\qquad (C.1)$$

i.e., $Z = 8L/\pi r^4$. Because blood vessels have complicated geometry with radii varying in space and time, Eq. (C.1) does not hold quantitatively for circulation *in vivo*. Nevertheless it is still true that Z is strongly dependent on r. Viscosity (η) is the ratio of shear stress (σ) to shear rate ($\dot{\gamma}$); in a tube with fixed geometry, η can be determined from P/Q ratio.

1. FACTORS DETERMINING BLOOD VISCOSITY

The viscosity of blood (η_B) varies inversely with temperature. At a given temperature, η_B is primarily a function of cell concentration, plasma viscosity (η_P), cell aggregation, and cell deformation (Chien, 1975; Schmid-Schoenbein, 1976). Because red blood cells (RBCs) occupy the largest volume fraction of blood cells, hematocrit (Hct) is a major determinant of η_B. White blood cells (WBCs) and platelets are much lower in their volume concentrations (sum < 1%) and normally have no significant influence on the flow behavior of blood. In very narrow vessels, WBCs may cause microvascular plugging due to their rigidity (Lichtman, 1973; Schmid-Schoenbein *et al.*, 1981), thus becoming a significant rheological factor in some pathologic conditions.

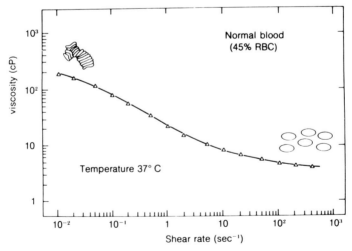

FIG. 2.2. Logarithmic plot showing the relationship between the apparent blood viscosity of normal human blood and shear rate. The inset sketches show the aggregation of red cells at lower shear rates and their disaggregation and deformation at higher shear rates. (Adapted from Chien, 1977.)

Plasma viscosity, η_P is a function of the concentration of plasma proteins, especially those with molecular asymmetry, i.e., fibrinogen and some globulins. The ratio η_B/η_P is the relative viscosity (η_r). The parameter η_r of a blood sample with Hct adjusted to a constant level serves to normalize variations in Hct and η_P among samples and reflects variations in RBC aggregation and RBC deformability. The shear-dependent variations in RBC aggregation and deformation are responsible for the non-Newtonian behavior of blood.

Under low flow conditions, RBCs aggregate to form rouleaux due to bridging of cell surfaces by fibrinogen and globulins. An increase in shear stress (σ) causes disaggregation and a decrease in η_B (Fig. 2.2).

Normal human RBCs are remarkably deformable in response to σ acting on the cell surface. Increases in $\dot{\gamma}$ and/or external fluid viscosity lead to a larger deforming σ, which causes the elongation of RBCs with their long axis aligned with flow. The rotation of RBC membrane can transmit σ to the internal fluid for its participation in flow (Fischer *et al.*, 1978). As a result, the viscous resistance of RBCs is reduced, and η_B becomes lower at high shear rates (Fig. 2.2).

2. ENERGY BALANCE IN RED CELL

AGGREGATION

Red blood cell aggregation involves a balance of energy at cell surfaces (Chien, 1980). The aggregating energy of blood (E_B) results from the bridging of the surfaces of two adjacent RBCs by the adsorbed or bound macromolecules. The aggregating energy, E_B, per unit RBC surface area is equal to the product of the aggregating energy per molecular bond, the number of molecular bonds per macromolecule, and the number of macromolecules adsorbed or bound per unit surface area. Variations in these factors can explain the changes in aggregating energy as a function of the type, size, and concentration of the macromolecule.

The primary disaggregating energies are those due to electrostatic repulsion (E_E) and mechanical shearing (E_S). Red blood cell surfaces have negative charges mainly due to the presence of sialic acid, and E_E is generated between two RBC surfaces brought into close range. Removal of sialic acid by neuraminidase or increased screening of surface charge by divalent cations leads to a decrease in E_E (Jan and Chien, 1973). This causes an enhancement of RBC aggregation except when E_B depends on electrostatic attraction between sialic acid and polycationic bridging macromolecules. A reduction in ionic strength decreases the screening of surface charge, thus enhancing E_E and reducing RBC aggregation (Jan and Chien, 1973; Brooks, 1976). High concentrations of neutral polymers (e.g., dextran) can reduce the effectiveness of counterion

screening of surface charge, thus enhancing E_E and causing RBC disaggregation (Brooks, 1976; Chien, 1980). Mechanical shearing causes RBC disaggregation due to the work done by the dispersing shear stress. A moderate degree of
shearing, however, may promote RBC aggregation by increasing the proba-
bility of cell to cell encounter or by affecting the cell membrane (Copley and
King, 1976; Greig and Brooks, 1981).

The net aggregation energy

$$E_A = E_B - E_E - E_S \qquad (C.2)$$

induces a change of strain energy in the RBC membrane and alters the cell
shape. The curvature of the end cell in rouleaux may change from the normal
concavity to a profile that is less concave and may even become convex as E_A
increases (Chien, 1980; Skalak *et al.*, 1981). E_A can be deduced from changes of
cell shape following aggregation (Skalak *et al.*, 1981; Evans and Buxbaum,
1981).

3. Microrheology of Individual Erythrocytes

Red blood cell deformability serves to lower η_B under physiologically high
flow conditions and makes possible RBC passage through narrow capillaries.
This deformability results from (a) the fluidity of internal hemoglobin-rich fluid,
(b) the favorable geometric relationship between membrane surface area and
cell volume, and (c) the viscoelastic properties of the cell membrane (Chien,
1975).

The internal fluid viscosity (η_i) increases with the mean corpuscular hemo-
globin concentration (MCHC), especially when MCHC approaches 40 gm/dl.
Deoxygenation of hemoglobin S leads to a marked elevation of η_i at any given
MCHC, causing the abnormal RBC deformability in sickle cell disease.

The normal discoid RBC has a surface area ($\simeq 140\ \mu m^2$) in excess of the ~ 90
μm^2 required to enclose a sphere with the same volume. This allows RBCs to
deform into a variety of shapes without changing its surface area (Fig. 2.3).

The rheology of RBC membrane has been the subject of many reviews
(e.g., Evans and Skalak, 1980). The deformation behavior of RBC membrane
depends on whether the surface area is stretched. Deformation of RBCs under
most physiologic conditions, e.g., during transit through a narrow capillary,
occurs at a constant surface area. The shear elastic modulus (μ_S) for constant
area deformation, as determined by micropipette aspiration, is approximately 4
$\times\ 10^{-3}$ dyn/cm (Evans and Hochmuth, 1976; Chien *et al.*, 1978). When deformation involves area expansion, e.g., RBC swelling in a hypotonic medium, the

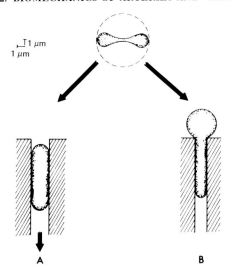

FIG. 2.3. Passage of normal human erythrocyte (top and side views of biconcave discoid shown at the top of the diagram) through cylindrical channels with diameter $(d_t) = 3$ μm in (A) and 2 μm in (B). Note that the deformation of the normal erythrocyte allows its passage through (A) but not (B).

areal elastic modulus is approximately 10^2 dyn/cm (Evans and Skalak, 1980), which is more than four orders of magnitude higher than μ_S. Such behavior is in keeping with mechanical models in which a network structure in the plane of the membrane is connected to posts running through the membrane. On the endoface of the RBC membrane, there is a network of proteins (formed by spectrin, actin, and band 4.1) which is connected to glycoproteins (e.g., band 3) spanning the membrane (Fig. 2.4). The μ_S of the RBC membrane is much more similar to that of protein monolayers than to that of lipids.

The membrane viscosity (η_m) during RBC deformation and recovery in micropipette tests is on the order of 10^{-4} dyn × sec/cm (Evans and Hochmuth, 1976; Chien *et al.*, 1978). This is comparable to the η_m deduced from the lateral diffusion coefficient of band 3 molecules in the RBC membrane (Koppel *et al.*, 1981), suggesting that η_m is related to the lateral motion of the glycoproteins and the spectrin–actin network to which they are connected.

The RBC membrane has a curved contour at rest, and cell deformation usually involves a change in curvature. The bending modulus has been estimated to be $\simeq 10^{-11}$ dyn × cm (Evans and Skalak, 1980), which is similar to that of lipid bilayers. The bending modulus of the RBC membrane may be influenced significantly by membrane lipids, especially their differential packing in the two halves of the bilayer. Various chemical agents can influence the lipid organization in the membrane and alter RBC shape and rheology (Bessis *et al.*, 1973; Meiselman, 1978); these changes are probably accompanied by alterations in bending modulus.

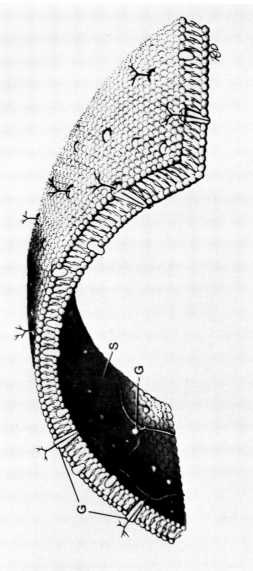

Fig. 2.4. Schematic drawing showing the hypothetical structure of red cell membrane. The lipid bilayer is penetrated by proteins located on the exoface and the endoface of the membrane. Some of the glycoproteins (G) span the entire thickness of the membrane. The spectrin–actin network (S) lining the endoface is shown to be connected with the glycoprotein molecules protruding from the membrane.

By using the elastic and bending moduli of the RBC membrane, the deformation of RBC from its resting biconcave discoid shape can be simulated under a variety of conditions, including osmotic swelling, rouleau formation, and flow through narrow capillaries (Skalak and Chien, 1981).

4. RELEVANCE OF BLOOD RHEOLOGY TO CIRCULATION *IN VIVO*

a. Blood viscosity and shear rate in vivo: Due to the complexity of vascular geometry and the non-Newtonian behavior of blood, it is difficult to determine $\dot{\gamma}$ in the circulation *in vivo*. A rough estimate of $\dot{\gamma}$ can be obtained as $4Q/\pi r^3$, which is the wall shear rate for a Newtonian fluid in Poiseuille flow.

In arteries $\dot{\gamma}$ is sufficiently high to cause RBC deformation and disaggregation, resulting in a low η_B. The highest $\dot{\gamma}$ is normally found in the capillaries, thus ensuring RBC deformation where it is required. The intravascular Hct in vessels smaller than 300 μm decreases with the vessel diameter (Fahraeus, 1929; Lipowsky *et al.*, 1980; Kanzow *et al.*, 1982); this further contributes to a reduction in η_B in small vessels (Fahraeus and Lindqvist, 1931). Therefore, the high Z in a narrow capillary due to the small r is compensated by a low η_B, thus preventing a markedly elevated resistance. It is only in capillaries with diameters smaller than 4–5 μm that the η_B rises because of the need for marked RBC deformation. Such an increase of η_B in small capillaries becomes more pronounced when RBC deformability is reduced, e.g., in sickle cell crisis, or when there is an increase in concentration of the more rigid WBCs, e.g., in leukemia.

The postcapillary venules and small veins have the lowest $\dot{\gamma}$ and are the most likely sites of RBC aggregation. In low flow states, the further lowering of $\dot{\gamma}$ in postcapillary segments would lead to a preferential elevation of postcapillary η_B, thus causing an increase in post- to precapillary resistance ratio and an elevation of capillary pressure. These changes would be more pronounced if the η_B vs. $\dot{\gamma}$ relationship is steepened due to hemoconcentration. Such a vicious cycle involving rheologically induced transcapillary fluid loss may play a significant role in some forms of shock.

b. Correlation between blood viscosity and flow resistance: When an isolated part of the body is perfused by blood with varying Hct, the flow resistance changes in the same direction as η_B. Alterations in η_B have been correlated with *in vivo* changes in total peripheral resistance in experimental endotoxic shock, hemorrhagic shock, and isovolumic hemoconcentration and hemodilution. A similar correlation between η_B *in vitro* and flow resistance *in vivo* has been observed in several clinical conditions, including acute myocardial infarction. By correlating the η_B measurements with *in vivo* resistance determinations, one can estimate alterations in vascular hindrance in disease states, e.g., trauma, sepsis, and polycythemia (Chien, 1981).

   c. *Optimum hematocrit for oxygen transport:* The rate of O_2 delivery (J_{O_2}) to tissues is equal to the product of blood flow and the arterial oxygen content (A_{O_2}). A_{O_2}, which is mainly carried with the hemoglobin, is proportional to the product of Hct, MCHC, and hemoglobin O_2 saturation. If arterial pressure, MCHC, and O_2 saturation are normal (Chien, 1975),

$$J_{O_2} = k \; \text{Hct}/\eta_B Z \tag{C.3}$$

where k is a constant. If Z is also unchanged, then J_{O_2} is proportional to the ratio Hct/η_B. In dogs subjected to Hct variations by isovolemic exchange, the

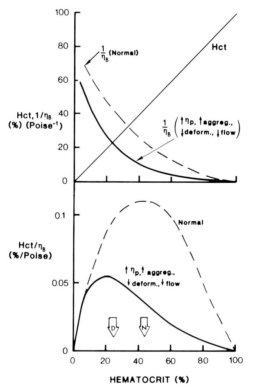

FIG. 2.5. Effects of variations in Hct on O_2 transport as reflected by the calculated Hct/η_B (bottom panel). Top panel shows the variations in Hct and $1/\eta_B$ with Hct. Dashed lines: results obtained when Hct is the only variable, with other hemorheological parameters remaining normal. Solid lines: results obtained when there are increases in η_p, RBC aggregation, or RBC rigidity or a decrease in flow rate. Arrow marked N shows normal Hct level, and arrow marked D shows Hct level found in patients with sickle cell disease or paraproteinemia.

rates of O_2 delivery and consumption in the systemic and coronary circulations remain essentially unchanged over a Hct range of 20–60% (Jan *et al.*, 1980). The optimum Hct range for O_2 delivery is narrower in some other organs (Fan *et al.*, 1980).

With Hct alterations, η_B and Hct vary in the same direction; hence the ratio Hct/η_B remains constant over a moderate range of Hct. When the increase in η_B is due to an increase in η_P and/or RBC aggregation (e.g., in paraproteinemias) or to a decrease in RBC deformability (e.g., in sickle cell disease), there is no concomitant increase in Hct, and J_{O_2} decreases. The decrease in Hct under these conditions serves to minimize the increase in η_B so that J_{O_2} would not decrease as much as that expected from changes in η_P, RBC aggregation, or RBC deformability per se (Fig. 2.5). Therefore, the decrease in Hct has a compensatory effect in minimizing the curtailment of J_{O_2} in these conditions. In low flow states, η_B is elevated at a given Hct (Fig. 2.2), and J_{O_2} is again better maintained at lower Hct levels (Fig. 2.5).

The above discussion is based on an unchanged vascular hindrance and J_{O_2} is considered only in terms of the ratio Hct/η_B. When there is a change in vascular hindrance, then Eq. (C.2) should be used. In hematologic diseases, compensatory vasodilatation may occur to counter the increase in η_B and minimize the decrease in J_{O_2}. When the primary disturbance is in the vascular system, however, the vasculature may lose its ability to undergo compensatory vasodilatation. Under these circumstances, any increase in η_B amplifies the detrimental effect of the increased vascular hindrance in reducing blood flow and O_2 delivery.

D. PROPERTIES OF VEINS

By C. F. ROTHE

1. FUNCTION OF VEINS

Veins provide a conduit for the return of blood from the exchange vessels to the heart. They also influence the distribution of blood between the thorax and periphery because they are distensible and have smooth muscle in their walls. This function is important, since reducing the mean central venous pressure by

Blood Vessels and Lymphatics in Organ Systems
Copyright © 1984 by Academic Press, Inc.

only 1 mm Hg causes cardiac output to decrease about 50 ml/min/kg body weight if the sympathetic tone is high and compensatory mechanisms are blocked (Guyton et al., 1973; Herndon and Sagawa, 1969). Thus, a 1% change in blood volume can potentially alter cardiac output by about 10%. In the standing position, nearly 0.5 liter of blood tends to accumulate in the lower limb veins, thus reducing cardiac output about 25%. However, reflex mechanisms (including active venoconstriction) act to maintain arterial blood pressure and minimize the decrease in cardiac output. Furthermore, a change in cardiac output induces passive compensatory alterations in blood volume distribution and hence alters central venous pressure so as to restore cardiac output toward normal. The concept of an equilibrium between cardiac output and venous return around a right atrial pressure operating point is of great help in understanding this passive influence of veins on cardiac output (e.g., Guyton et al., 1973). [For major reviews of the physiology of the venous system, see Franklin (1937), Brecher (1956), Shepherd and Vanhoutte (1975), Hainsworth and Linden (1979), Gow (1980), and Rothe (1983a,b)].

2. Vascular Capacitance

a. Definition: The *compliance* (C) of a blood vessel is defined as the ratio of volume change (ΔV) induced by a unit change in distending pressure (ΔP)

$$C = \Delta V / \Delta P \qquad (D.1)$$

Because the pressure–volume relationship may not be linear, the distending pressure or volume should be specified. Although units of ml/mm Hg are often used, normalization to a unit of tissue or body weight permits comparisons between organs, species, or studies. *Vascular distensibility* (D) is the fractional (or percentage) change in volume for each unit change in pressure

$$D = (\Delta V / V_o) \, \Delta P \qquad (D.2)$$

where V_o is the original total blood volume. Unfortunately, accurate measures of organ blood volume *in vivo* are usually difficult to obtain.

The total body vascular compliance is about 3 ml of blood per mm Hg change in transmural pressure for each kilogram of body weight (cf. Gow, 1980; Rothe, 1983a,b). The compliance of the heart and lungs is approximately 1 ml/mm Hg/kg of body weight and that of the systemic circulation is about 2 ml/mm Hg/kg. Skeletal muscle has both a lower compliance and lower blood volume per kilogram of tissue than the rest of the body, whereas the compliance of the liver is about 10 times that of the overall systemic circulation. The compliance of the splanchnic bed is approximately 0.7 ml/mm Hg/kg body weight (Karim and Hainsworth, 1976), a rather large fraction of the total.

Vascular capacitance concerns the magnitude of the contained volume of blood as related to various transmural pressures, the pressure outside the vessels usually being considered to be zero. The internal pressure of the capacitance vessels is assumed by many to be the mean circulatory filling pressure— "the pressure that would be measured at all points in the entire circulatory system if the heart were stopped suddenly and the blood were redistributed instantaneously in such a manner that all pressures were equal" (Guyton *et al.*, 1973). Blood vessels contain a significant volume at zero transmural pressure. Over part of the physiologic pressure range, the compliance of veins is constant (i.e., the pressure–volume relationship is linear), but at abnormally high pressures, these vessels become stiff, i.e., the compliance decreases, with the result that little further volume may be added as the pressure is increased. The *total vascular capacity* is the total blood volume at a specified capacitance vessel pressure (usually the mean circulatory filling pressure). The *unstressed volume* is that volume contained at zero transmural pressure. It may be estimated as the total vascular capacity minus the *stressed volume* (the product of compliance and mean circulatory filling pressure). The unstressed volume is thus a calculated volume. With a normal blood volume of about 75 ml/kg body weight, the total vascular stressed volume is approximately 25 ml/kg and the unstressed volume, 50 ml/kg.

b. Active capacitance vessel responses: These imply reflex activity and involve changes in the contractile activity of smooth muscle surrounding the vessel. Such responses affect wall tension and, in turn, induce alterations in transmural pressure at the same volume, changes in the contained volume at the same transmural pressure, or some combination of both. An active reaction involves a change in the pressure–volume relationship, not a move along a given capacitance curve. As emphasized by Shoukas and Sagawa (1973), reflex activity may produce a change in compliance, in unstressed volume, or in both. The arterial baroreceptor reflex, for example, causes changes in the unstressed volume but has little effect on compliance (Shoukas *et al.*, 1981).

c. Passive capacitance vessel volume changes: These follow alterations in transmural pressure and are related to vessel collapse, elastic recoil, or changes in flow. With vessels near the surface of the body, such as the jugular vein, changes in volume may occur with little alteration in transmural pressure if the vessels change from round to elliptical in cross section. The vessels collapse when the transmural pressure is negative. In the abdominal cavity, the highly pliable contents are suspended in fluid medium and gravitational influences are minimized. In this setting, vessel collapse does not generally occur. During straining (as in the Valsalva maneuver), both thoracic and abdominal pressures may rise dramatically. Retrograde flow into the limbs is impeded by the venous valves, and venous return from the limbs may be stopped, but flow continues between the heart and viscera. With inspiration, the veins just outside the thorax tend to collapse if the transmural pressure approaches zero. Near total occlusion persists until an increase in internal pressure from venous return opens the vessels, permitting flow to occur (e.g., Moreno *et al.*, 1969). Within organs, such as the liver, skeletal muscle, or kidney, the small veins are surrounded by tissue, and a large negative transmural pressure probably is required to reduce the contained volume below the unstressed volume. The quantitative importance of vessel collapse in influencing general cardiovascular homeostasis is not known.

    d. Effect of changes in transmural pressure: If the transmural pressure of capacitance vessels decreases, blood is expelled as a result of *passive elastic recoil* of the compliant vessels. Furthermore, a decrease in blood flow through tissue causes a reduction in pressure drop *along* the veins. Such a response leads to a fall in pressure in the small peripheral veins, a reduced transmural pressure, and a decreased contained blood volume—a *passive flow effect* related to flow and venous resistance. The response acts to transfer blood toward the heart. This raises central venous pressure somewhat, increases ventricular filling, and thus restores cardiac output—a physical negative feedback mechanism. About 65% of the blood mobilized during splanchnic nerve stimulation is a result of reduced blood flow (Brooksby and Donald, 1972). A 1 ml/min/kg change in blood flow through the body as a whole causes approximately 0.15 ml/kg change in blood volume of the tissue and 0.25 ml/kg for the total circulation, including the heart and lungs, provided that central venous pressure is held constant with a reservoir (e.g., Numao and Iriuchijima, 1977). Following hemorrhage, blood volume, cardiac output, and capacitance vessel pressures (closely related to the mean circulatory filling pressure) decrease. The reduction in peripheral venous pressures is about three times as great as the fall in central venous pressure (Rothe and Drees, 1976). Thus, in the presence of blood flow changes, alterations in central venous pressure are not a quantitative measure of active or passive changes of the transmural pressure of the compliant capacitance vessels. Furthermore, an increase in capacitance vessel transmural pressure will not be quantitatively transmitted to the heart if flow also changes.

    During exercise, contracting skeletal muscle compresses the veins, forcing blood toward the heart—the *muscle pump.* On relaxation, the reflux of blood is prevented by the venous valves, thus aiding venous return (flow) and minimizing blood pooling (e.g., Shepherd and Vanhoutte, 1975). If the valves become incompetent, pooling of blood occurs in the lower limbs, leading to enlarged (varicose) veins. Respiratory activity also influences venous return (e.g., Brecher, 1956).

   e. Transcapillary fluid shifts: Net water movement from the interstitial spaces provides a powerful compensatory mechanism for homeostatic maintenance of cardiac output in the presence of blood or water loss or blood pooling. Although the transcapillary fluid flow is a very small fraction of the blood flow through the system, over a period of even a few minutes the magnitude may become significant. These shifts in fluid volume confound most measurements of venous activity.

   Following a rapid, large hemorrhage, about one-third of the lost volume comes from the passive elastic recoil of the capacitance vessels within 5 to 15 sec. In the first minute or two, another one-third comes from reflex induced capacitance vessel constriction (see below), and transcapillary fluid shifts from the tissue supply the remainder by approximately 5 min (Rothe and Drees, 1976). In response to hemorrhage in the dog and cat, about 50% of the blood volume loss has been found to originate in the splanchnic bed (Carneiro and Donald, 1977b; Greenway and Lister, 1974). In the dog, the spleen is large and highly reactive to neural and hormonal stimuli, and hence it can provide a 2 to 5 ml/kg body weight transfer of blood. This accounts for about 60% of the active splanchnic capacitance vessel response (Carneiro and Donald, 1977a) and about one-third of the total response (Shoukas *et al.*, 1981).

3. NERVOUS CONTROL OF CAPACITANCE SYSTEM

   Smooth muscle is present in the walls of many venules and of all small veins. When stimulated, it acts to increase wall tension and reduce the contained volume. Vascular smooth muscle cells of the veins are innervated by the peripheral sympathetic nervous system and provide part of the reflex control of cardiovascular function. However, innervation is sparse as compared with that of arterioles. Nevertheless, a reflex-induced reduction in venous capacitance causes a 5–15 ml/kg transfer of blood toward the heart within 1 min of stimulation (e.g., Drees and Rothe, 1974; Numao and Iriuchijima, 1977; Rothe, 1983b; Shoukas and Sagawa, 1973; Shoukas *et al.*, 1981).

   There are numerous receptor inputs for the nervous control of the capacitance system, of which the *carotid sinus* and *aortic arch baroreceptors* are the most effective. A 25 mm Hg decrease in carotid sinus pressure causes about 3.5 ml of blood per kilogram of body weight to be moved from the peripheral veins toward the heart. In the dog, approximately 0.15 ml blood per kilogram of body weight is transferred with a 1 mm Hg change in carotid sinus pressure (Shoukas and Sagawa, 1973; Muller-Ruchholtz *et al.*, 1979). Over the total carotid sinus baroreceptor range, the capacitance vessel response is about 11 ml/kg body weight (Shoukas *et al.*, 1981), thus exerting a potentially large influence on cardiac output. In the dog, the carotid arterial pressure causing a 50% of maximum response of the capacitance vessels is about 15 mm Hg higher than that producing a 50% response in the resistance vessels (Hainsworth and Karim, 1976). In the cat, baroreceptors do not appear to influence the hepatic capacitance vessels (Lautt and Greenway, 1980). The aortic baroreceptors also are involved in changes in the capacitance vessel system (e.g., Karim *et al.*, 1978).

Stimulation of the *chemoreceptors* in the carotid or aortic bodies causes capacitance vessel constriction (Hainsworth *et al.*, 1980; Kahler *et al.*, 1962). However, the response is small, i.e., < 5 ml/kg body weight. The *cardiopulmonary receptors* also influence the capacitance vessels. Simultaneously blocking these and the carotid receptors causes the liver blood volume to decrease by about 130 ml/kg of liver (approximately 3 ml/kg body weight). The cardiopulmonary receptors contribute approximately 40% of this response (Carneiro and Donald, 1977b).

a. Variations in response of capacitance vessels: The capacitance vessels of the skin are little influenced by the baroreceptors or chemoreceptors. This may be due to the fact that cutaneous veins are under the strong influence of *thermoregulatory mechanisms* of the central nervous system (e.g., Rowell, 1977). Skeletal muscle capacitance vessels also seem to be little influenced by the cardiovascular reflexes. For example, studies of the capacitance vessels in the limbs suggest that there is little venous response to cardiovascular disturbances. The capacitance vessels in the abdomen provide most of the reflex compensation. Such capacitance vessel reflexes in people are difficult to assess because of the problem of measuring pressures and vascular volumes in the liver and intestine without causing injury to them.

4. MECHANISMS AVAILABLE FOR MAINTENANCE OF CARDIAC OUTPUT

Although reflex venoconstriction cannot fully compensate for conditions such as catastrophic hemorrhage, extreme pooling of blood in dependent limbs, massive cutaneous vasodilatation in a hot environment, or a debilitating loss of water, it does provide a rapidly acting means of aiding filling of the heart so as to provide adequate cardiac output. Only compensatory mechanisms in the kidneys or changes in fluid input can provide complete compensation for large alterations in blood volume. On standing, great stress is placed on compensatory vascular mechanisms, resulting in a reduced capacitance vessel reserve (Hainsworth and Linden, 1979). This accounts for the tendency toward fainting observed in debilitated patients or those who have lost blood. Because the right heart is exquisitely sensitive to filling pressure, reflex changes in capacitance vessel smooth muscle activity provide a powerful homeostatic mechanism for the cardiovascular system, the importance of which is neither fully understood nor appreciated.

References*

Atabek, H. B. (1980). Blood flow and pulse propagation in arteries. In "Basic Hemodynamics and Its Role in Disease Processes" (D. J. Patel and R. N. Vaishnav, eds.), pp. 255–362. Univ. Park Press, Baltimore, Maryland. (B)

Azuma, T., and Oka, S. (1971). Mechanical equilibrium of blood vessel walls. Am. J. Physiol. 221, 1310–1318. (A)

Bader, H. (1967). Dependence of wall stress in the human thoracic aorta on age and pressure. Circ. Res. 20, 354–361. (A)

Benis, A. M., Usami, S., and Chien, S. (1970). Effect of hematocrit and inertial losses on pressure-flow relation in the isolated hind paw of the dog. Circ. Res. 27, 1047–1068. (B)

Benson, T. J., Nerem, R. M., and Pedley, T. J. (1980). Assessment of wall shear stress in arteries, applied to the coronary circulation. Cardiovasc. Res. 14, 568–576. (B)

Bergel, D. H. (1961). The static elastic properties of the arterial wall. J. Physiol. (London) 156, 445–457. (A)

Bergel, D. H., ed. (1972). "Cardiovascular Fluid Dynamics." Academic Press, New York. (B)

Bessis, M., Weed, R. I., and Leblond, P. L., eds. (1973). "Red Cell Shape, Physiology, Pathology, Ultrastructure." Springer-Verlag, Berlin and New York. (C)

Brecher, G. A. (1956). "Venous Return." Grune & Stratton, New York. (D)

Brooks, D. E. (1976). Red cell interactions in low flow states. In "Microcirculation" (J. Grayson and W. Zingg, eds.), Vol. 1, pp. 33–52. Plenum, New York. (C)

Brooksby, G. A., and Donald, D. E. (1972). Release of blood from the splanchnic circulation in dogs. Circ. Res. 31, 105–118. (D)

Browse, N. L., Young, A. E., and Thomas, M. L. (1979). The effect of bending on canine and human arterial walls and on blood flow. Circ. Res. 45, 41–48. (A)

Burton, A. C. (1961). On the physical equilibrium of small blood vessels. Am. J. Physiol. 164, 319–329. (A)

Busse, R., Bauer, R., Schakert, A., Summa, Y., and Wetterer, E. (1979). An improved method for the determination of the pulse transmission characteristics of arteries in vivo. Circ. Res. 44, 630–636. (B)

Busuttil, R. W., Abou-Zamzam, A. M., and Machleder, H. I. (1980). Collagenase activity of the human aorta: A comparison of patients with and without abdominal aortic aneurysms. Arch. Surg. 115, 1373–1378. (A)

Busuttil, R. W., Heinrich, R., and Flesher, A. (1982). Elastase activity: The role of elastase in aortic aneurysm formation. J. Surg. Res. 32, 214–217. (A)

Carneiro, J. J., and Donald, D. E. (1977a). Blood reservoir function of dog spleen, liver, and intestine. Am. J. Physiol. 232, H67–H72. (D)

*In the reference list, the capital letter in parentheses at the end of each reference indicates the section in which it is cited.

Carneiro, J. J., and Donald, D. E. (1977b). Change in liver blood flow and blood content in dogs during direct and reflex alteration of hepatic sympathetic nerve activity. *Circ. Res.* **40**, 150–158. (D)

Caro, C. G., Pedley, T. J., Schroter, R. C., and Seed, W. A. (1978). "The Mechanics of the Circulation." Oxford Univ. Press, London and New York. (B)

Chien, S. (1975). Biophysical behavior of red cells in suspensions. *In* "The Red Blood Cell" (D. MacN. Surgenor, ed.), 2nd ed., Vol. 2, pp. 1031–1133. Academic Press, New York. (C)

Chien, S. (1977). Blood rheology in hypertension and cardiovascular diseases. *Cardiovasc. Med.* **2**, 356–360. (C).

Chien, S. (1980). Aggregation of red blood cells: An electrochemical and colloid chemical problem. *In* "Bioelectrochemistry: Ions, Surfaces, Membranes," Adv. Chem. Ser., pp. 1–32. Am. Chem. Soc., Washington, D.C. (C)

Chien, S. (1981). The Fahraeus Lecture: Hemorheology in disease, pathophysiological significance and therapeutic implications. *Clin. Hemorheol.* **1**, 419–442. (C)

Chien, S., Sung, K. L. P., Skalak, R., Usami, S., and Tozeren, A. (1978). Theoretical and experimental studies on viscoelastic properties of red cell membrane. *Biophys. J.* **24**, 463–487. (C)

Clark, C. (1976). Turbulent velocity measurements in a model of aortic stenosis. *J. Biomechanics* **9**, 677–687. (B)

Clark, C. (1979). Energy losses in flow through stenosed valves. *J. Biomech.* **12**, 737–746. (B)

Clark, C. (1980). The propagation of turbulence produced by a stenosis. *J. Biomech.* **13**, 591–604. (B)

Copley, A. L., and King, R. G. (1976). Erythrocyte sedimentation of human blood at varying shear rates. *Biorheology* **13**, 281–286. (C)

Cox, R. H. (1978). Passive mechanics and connective tissue composition of canine arteries. *Am. J. Physiol.* **234**, H533–H541. (A)

Dobrin, P. B. (1978). Mechanical properties of arteries. *Physiol. Rev.* **58**, 397–460. (A)

Dobrin, P. B. (1983). Vascular Mechanics. *In* "Handbook of Physiology" (J. T. Shepherd and F. M. Abboud, eds.), Sect. 2, Vol. 3, pp. 65–102. Am. Physiol. Soc., Bethesda, Maryland. (A)

Dobrin, P. B., and Canfield, T. R. (1973). Series elastic and contractile elements in vascular smooth muscle. *Circ. Res.* **33**, 454–464. (A)

Dobrin, P. B., and Rovick, A. A. (1969). Influence of vascular smooth muscle on contractile mechanics and elasticity of arteries. *Am. J. Physiol.* **217**, 1644–1652. (A)

Doyle, J. M., and Dobrin, P. B. (1973). Stress gradients in the walls of large arteries. *J. Biomech.* **6**, 631–639. (A)

Drees, J. A., and Rothe, C. F. (1974). Reflex venoconstriction and capacity vessel pressure–volume relationships in dogs. *Circ. Res.* **34**, 360–373. (D)

Evans, E. A., and Buxbaum, K. (1981). Affinity of red blood cell membrane for particle surfaces measured by the extent of particle encapsulation. *Biophys. J.* **34**, 1–12. (C)

Evans, E. A., and Hochmuth, R. M. (1976). Membrane viscoelasticity. *Biophys. J.* **16**, 1–11. (C)

Evans, E. A., and Skalak, R. (1980). "Mechanics and Thermodynamics of Biomembranes." Chem. Rubber Publ. Co., Cleveland, Ohio. (C)

Fahraeus, R. (1929). The suspension stability of blood. *Physiol. Rev.* **9**, 241–274. (C)

Fahraeus, R., and Lindqvist, T. (1931). The viscosity of the blood in narrow capillary tubes. *Am. J. Physiol.* **96**, 562–568. (C)

Fan, F.-C., Schuessler, G. B., Chen, R. Y. Z., and Chien, S. (1980). Effect of hematocrit alteration on the regional hemodynamics and oxygen transport. *Am. J. Physiol.* **238**, H545–H552. (C)

Fischer, T. M., Stoehr-Lissen, M., and Schmid-Schoenbein, H. (1978). The red cell as a fluid droplet: Tank tread-like motion of the human erythrocyte membrane in shear flow. *Science* **202**, 894–896. (C)

Franklin, K. J. (1937). "A Monograph on Veins." Thomas, Springfield, Illinois. (D)

Fry, D. L. (1973). Responses of the arterial wall to certain physical factors. *Ciba Found. Symp. Atherosclerosis*, pp. 93–125. (B)

Fry, D. L., and Vaishnav, R. N. (1980). Mass transport in the arterial wall. *In* "Basic Hemo-dynamics and Its Role in Disease Processes" (D. J. Patel and R. N. Vaishnav, eds.), pp. 425–482. Univ. Park Press, Baltimore, Maryland. (B)

Fung, Y. C., Fronek, K., and Patitucci, P. (1979). Pseudoelasticity of arteries and the choice of its mathematical expression. *Am. J. Physiol.* **237**, H620–H631. (A)

Gerova, M., and Gero, J. (1967). Reflex regulation of smooth muscle tone of conduit vessel. *Angiologica* **4**, 348–358. (A)

Gow, B. S. (1980). Circulatory correlates: Vascular impedance, resistance, and capacity. *In* "Hand-book of Physiology" (D. F. Bohr, A. P. Somlyo, and H. V. Sparks, Jr., eds.), Sect. 2, Vol. 2, pp. 353–408. Am. Physiol. Soc., Bethesda, Maryland. (D)

Greenway, C. V., and Lister, G. E. (1974). Capacitance effects and blood reservoir function in the splanchnic vascular bed during nonhypotensive haemorrhage and blood volume expansion in anaesthetized cats. *J. Physiol. (London)* **237**, 279–294. (D)

Greig, R. G., and Brooks, D. E. (1981). Enhanced concanavalin A agglutination of trypsinised erythrocytes is due to a specific class of aggregation. *Biochim. Biophys. Acta* **641**, 410–415. (C)

Gross, D. R., Lu, P. C., Dodd, K. T., and Hwang, N. H. C. (1980). Physical characteristics of pulmonary artery stenosis murmurs in calves. *Am. J. Physiol.* **238**, H876–H885. (B)

Guyton, A. C., Jones, C. E., and Coleman, T. G. (1973). "Circulatory Physiology: Cardiac Output and Its Regulation," 2nd ed. Saunders, Philadelphia, Pennsylvania. (D)

Hainsworth, R., and Karim, F. (1976). Responses of abdominal vascular capacitance in the anaes-thetized dog to changes in carotid sinus pressure. *J. Physiol. (London)* **262**, 659–677. (D)

Hainsworth, R., and Linden, R. J. (1979). Reflex control of vascular capacitance. *In* "Cardiovascu-lar Physiology III" (A. C. Guyton and D. B. Young, eds.), pp. 67–124. Univ. Park Press, Baltimore, Maryland. (D)

Hainsworth, R., Karim, F., McGregor, K. H., and Wood, L. M. (1980). Carotid body chemorecep-tors and abdominal vascular capacitance in the dog. *J. Physiol. (London)* **307**, 75P–76P. (D)

Heistad, D. D., Marcus, M. L., Law, E. G., Armstrong, M. L., Ehrhardt, J. C., and Abboud, F. M. (1978). Regulation of blood flow to the aortic media in dogs. *J. Clin. Invest.* **62**, 133–140. (A)

Herndon, C. W., and Sagawa, K. (1969). Combined effects of aortic and right atrial pressures on aortic flow. *Am. J. Physiol.* **217**, 65–72. (D)

Hilton, S. M. (1959). A peripheral arterial conducting mechanism underlying dilatation of the femoral artery and concerned in functional vasodilatation in skeletal muscle. *J. Physiol. (Lon-don)* **149**, 93–111. (B)

Hinke, J. A. M., and Wilson, M. L. (1962). A study of elastic properties of a 550-μ artery *in vitro*. *Am. J. Physiol.* **203**, 1153–1160. (A)

Jackman, R. W. (1982). Persistence of axial orientation cues in regenerating intima of cultured aortic explants. *Nature (London)* **296**, 80–83. (B)

Jan, K.-M., and Chien, S. (1973). Role of surface electric charge in red blood cell interactions. *J. Gen. Physiol.* **61**, 638–654. (C)

Jan, K.-M., Heldman, J., and Chien, S. (1980). Coronary hemodynamics and oxygen utilization after hematocrit variations in hemorrhage. *Am. J. Physiol.* **239**, H326–H332. (C)

Kahler, R. L., Goldblatt, A., and Braunwald, E. (1962). The effects of acute hypoxia on the systemic venous and arterial systems and on myocardial contractile force. *J. Clin. Invest.* **41**, 1553–1563. (D)

Kanzow, B., Pries, A. R., and Gaehtgens, P. (1982). Analysis of the hematocrit distribution in the mesenteric microcirculation. *Int. J. Microcirc.* **1**, 67–79. (C)

Karim, F., and Hainsworth, R. (1976). Responses of abdominal vascular capacitance to stimulation of splanchnic nerves. *Am. J. Physiol.* **231**, 434–440. (D)

Karim, F., Hainsworth, R., and Pandey, R. P. (1978). Reflex responses of abdominal vascular capacitance from aortic baroreceptors in dogs. *Am. J. Physiol.* **235**, H488–H493. (D)

Koppel, D. E., Sheetz, M. P., and Schindler, M. (1981). Matrix control of protein diffusion in biological membranes. *Proc. Natl. Acad. Sci. U.S.A.* **78**, 3576–3580. (C)

Langille, B. L., and Adamson, S. L. (1981). Relationship between blood flow direction and endothelial cell orientation at arterial branch sites in rabbits and mice. *Circ. Res.* **48**, 481–488. (B)

Lautt, W. W., and Greenway, C. V. (1980). Hepatic nerves: A review of their function and effects. *Can. J. Physiol. Pharmacol.* **58**, 105–123. (D)

Leon, A. S., and Bloor, C. M. (1968). Effect of exercise and its cessation on the heart and its blood supply. *J. Appl. Physiol.* **24**, 485–490. (B)

Leung, D. Y. M., Glagov, S., and Mathews, M. B. (1976). Cyclic stretching stimulates synthesis of matrix components by arterial smooth muscle cells *in vitro*. *Science* **191**, 475–477. (A)

Li, J.K-J., Melbin, J., Riffle, R. A., and Noordergraaf, A. (1981). Pulse wave propagation. *Circ. Res.* **49**, 442–452. (B)

Lichtman, M. A. (1973). Rheology of leukocytes, leukocyte suspensions and blood in leukemia. *J. Clin. Invest.* **52**, 350–358. (C)

Lipowsky, H. H., Usami, S., and Chien, S. (1980). *In vivo* measurements of "apparent viscosity" and microvessel hematocrit in the mesentery of the cat. *Microvasc. Res.* **19**, 297–319. (C)

Lutz, R. J., Cannon, J. N., Bischoff, K. B., Dedrick, R. L., Stiles, R. K., and Fry, D. L. (1977). Wall shear stress distribution in a model canine artery during steady flow. *Circ. Res.* **41**, 391–399. (B)

Lyon, C. K., Scott, J. B., Anderson, D. K., and Wang, C. Y. (1981). Flow through collapsible tubes at high Reynolds numbers. *Circ. Res.* **49**, 988–1002. (B)

Mangel, A., Fahim, M., and van Breeman, C. (1982). Control of vascular contractility by the cardiac pacemaker. *Science* **215**, 1627–1629. (A)

McDonald, D. A. (1974). "Blood Flow in Arteries." Williams & Wilkins, Baltimore, Maryland. (A,B)

Meiselman, H. J. (1978). Rheology of shape-transformed human red cells. *Biorheology* **15**, 225–237. (C)

Milnor, W. R. (1978). Influence of arterial impedance on ventricular function. *In* "The Arterial System" (R. D. Bauer and R. Busse, eds.), pp. 227–235. Springer-Verlag, Berlin and New York. (B)

Milnor, W. R. (1982). "Hemodynamics". Williams & Wilkins, Baltimore, Maryland. (B)

Milnor, W. R., and Bertram, C. D. (1978). The relation between arterial viscoelasticity and wave propagation in the canine femoral artery *in vivo*. *Circ. Res.* **43**, 870–879. (B)

Moreno, A. H., Katz, A. I., and Gold, L. D. (1969). An integrated approach to the study of the venous system with steps toward a detailed model of the dynamics of venous return to the right heart. *IEEE Trans. Biomed. Eng.* **BME-16**, 308–324. (D)

Muller-Ruchholtz, E. R., Grund, E., Hauer, F., and Lapp, E. R. (1979). Effect of carotid pressoreceptor stimulation on integrated systemic venous bed. *Basic Res. Cardiol.* **74**, 467–476. (D)

Murphy, R. A. (1980). Mechanics of vascular smooth muscle. *In* "Handbook of Physiology" (D. F. Bohr, A. P. Somlyo, and H. V. Sparks, Jr., eds.), Sect. 2, Vol. 2, pp. 325–351. Am. Physiol. Soc., Bethesda, Maryland. (A)

Nerem, R. M. (1977). Hot-film measurements of arterial blood flow and observations of flow disturbances. *In* "Cardiovascular Flow Dynamics and Measurements" (N. H. C. Hwang and N. A. Norman, eds.), pp. 191–215. Univ. Park Press, Baltimore, Maryland. (B)

Nerem, R. M., and Seed, W. A. (1972). An *in vivo* study of aortic flow disturbances. *Cardiovasc. Res.* **6**, 1–14. (B)

Numao, Y., and Iriuchijima, J. (1977). Effect of cardiac output on circulatory blood volume. *Jpn. J. Physiol.* **27**, 145–156. (D)

Ohhashi, T., Azuma, T., and Sakaguchi, M. (1980). Active and passive mechanical characteristics of bovine mesenteric lymphatics. *Am. J. Physiol.*, **239**, H88–H95. (A)

Patel, D. J., and Vaishnav, R. N. (1980). "Basic Hemodynamics and its Role in Disease Processes." Univ. Park Press, Baltimore, Maryland. (A,B)

Patel, D. J., Greenfield, J. C., and Fry, D. L. (1964). *In vivo* pressure–length–radius relationship of certain blood vessels in man and dog. *In* "Pulsatile Blood Flow" (E. O. Attinger, ed.), pp. 293–306. McGraw-Hill, New York. (A)

Roach, M. R. (1977). The effects of bifurcations and stenosis on arterial disease. *In* "Cardiovascular Flow Dynamics and Measurements" (H. N. C. Hwang and N. A. Norman, eds.), pp. 489–560. Univ. Park Press, Baltimore, Maryland. (B)

Roach, M. R., and Burton, A. C. (1957). The reason for the shape of the distensibility curves of arteries. *Can. J. Biochem. Physiol.* **35**, 681–690. (A)

Roach, M. R., and Burton, A. C. (1959). The effect of age on the elasticity of human arteries. *Can. J. Biochem. Physiol.* **37**, 557–569. (A)

Rodbard, S. (1974). Biophysical factors in vascular structure and calibre. *In* "Atherosclerosis" (E. Schottler and A. Wiezel, eds.), 3rd ed., pp. 44–63. Springer-Verlag, Berlin and New York. (B)

Rothe, C. F. (1983a). The venous system. Physiology of the capacitance vessels. *In* "Handbook of Physiology" (J. T. Shepherd and F. M. Abboud, eds.), Sect. 2, Vol. 3, pp. 397–452. Am. Physiol. Soc., Bethesda, Maryland. (D)

Rothe, C. F. (1983b). Reflex control of the veins and vascular capacitance. *Physiol. Rev.* (in press). (D)

Rothe, C. F., and Drees, J. A. (1976). Vascular capacitance and fluid shifts in dogs during prolonged hemorrhagic hypotension. *Circ. Res.* **38**, 347–356. (D)

Rowell, L. B. (1977). Reflex control of the cutaneous vasculature. *J. Invest. Dermatol.* **69**, 154–166. (D)

Samila, Z. J., and Carter, S. A. (1981). The effect of age on the unfolding of elastin lamellae and collagen fibers with stretch in human carotid arteries. *Can. J. Physiol. Pharm.* **51**, 1050–1057. (A)

Schmid-Schoenbein, H. (1976). Microrheology of erythrocytes, blood viscosity, and the distribution of blood in the microcirculation. *In* "Cardiovascular Physiology II" (A. C. Guyton and A. W. Cowley, eds.), Vol. 9, pp. 1–62. Univ. Park Press, Baltimore, Maryland. (C)

Schmid-Schoenbein, G. W., Sung, K. L. P., Tozeren, H., Skalak, R., and Chien, S. (1981). Passive mechanical properties of human leukocytes. *Biophys. J.* **36**, 243–256. (C)

Schultz, D. L. (1972). Pressure and flow in large arteries. *In* "Cardiovascular Fluid Dynamics" (D. H. Bergel, ed.), Vol. 1, pp. 287–314. Academic Press, New York. (B)

Sdougos, H. P., Schultz, D. L., Tan, L. P., Bergel, D. H., Rajagopalan, B., and Lee, G. de J. (1982). The effects of peripheral impedance and inotropic state on the power output of the left ventricle. *Circ. Res.* **50**, 74–85. (B)

Shepherd, J. T., and Vanhoutte, P. M. (1975). "Veins and Their Control." Saunders, Philadelphia, Pennsylvania. (D)

Shoukas, A. A., and Sagawa, K. (1973). Control of total systemic vascular capacity by the carotid sinus baroreceptor reflex. *Circ. Res.* **33**, 22–33. (D)

Shoukas, A. A., Macanespie, C. L., Brunner, M. J., and Watermeier, L. (1981). The importance of the spleen in blood volume shifts of the systemic vascular bed caused by the carotid sinus baroreceptor reflex. *Circ. Res.* **49**, 759–766. (D)

Sipkema, P., Westerhof, N., and Randall, O. S. (1980). The arterial system characterized in the time domain. *Cardiovasc. Res.* **14**, 270–279. (B)

Skalak, R., and Chien, S. (1981). Capillary flow: History, experiments and theory. *Biorheology* **18**, 307–330. (C)

Skalak, R., Zarda, P. R., Jan, K. M., and Chien, S. (1981). Mechanics of rouleau formation. *Biophys. J.* **35**, 771–781. (C)

Sutera, S. P. (1977). Flow-induced trauma to blood cells. *Circ. Res.* **41**, 2–8. (B)

Swanson, R. J., Littooy, F. N., Hunt, T. K., and Stoney, R. J. (1980). Laparotomy as a precipitating factor in rupture of intra-abdominal aneurysms. *Arch. Surg.* **115**, 299–304. (A)

Taylor, M. G. (1959). The influence of the anomalous viscosity of blood upon its oscillatory flow. *Phys. Med. Biol.* **3**, 273–290. (B)

Vaishnav, R. N., Young, J. T., and Patel, D. J. (1973). Distribution of stresses and of strain-energy density through the wall thickness in a canine aortic segment. *Circ. Res.* **32**, 577–583. (A)

Westerhof, N., Elzinga, G., Sipkema, P., and van den Bos, G. G. (1977). Quantitative analysis of the arterial system and heart by means of pressure–flow relations. *In* "Cardiovascular Flow Dynamics and Measurements" (N. H. C. Hwang, and K. A. Norman, eds.), pp. 403–438. Univ. Park Press, Baltimore, Maryland. (B)

Wexler, L., Bergel, D. H., Gabe, I. T., Makin, G. S., and Mills, C. J. (1968). Velocity of blood flow in normal human venae cavae. *Circ. Res.* **23**, 349–359. (B)

Winquist, R. J., and Bevan, J. A. (1981). *In vitro* model of maintained myogenic vascular tone. *Blood Vessels* **18**, 134–138. (A)

Wolinsky, H., and Glagov, S. (1964). Structural basis for the static mechanical properties of the aortic media. *Circ. Res.* **14**, 400–413. (A)

Chapter 3
Microcirculation

A. ANATOMY OF MICROCIRCULATION
By J. A. G. RHODIN

1. GENERAL CONSIDERATIONS

The term *microcirculation* refers to the smallest units of the cardiovascular system. The components are *arterioles, precapillary sphincter* areas, *capillaries* (arterial as well as venous), *postcapillary venules*, and *muscular venules*. Each segment of the microvascular bed has its own characteristic structure, and the length and the cross-sectional diameter, as well as the thickness of the wall of each microvascular segment, vary considerably.

There is a typical microvascular pattern for each organ, as well as for each tissue, and therefore no generalization can be made when referring to the term "microcirculation." Nevertheless, it is important to determine the kind of microvascular bed which lends itself best to the experimental study that one intends to pursue. From such an approach, comparison can then be made of the findings and a general, overall conclusion drawn relating to the correlation of structure and function of each particular segment of the microvascular bed.

In this section, the mesentery of the rat is used as an example, chiefly because it has a microvascular pattern spread out in two dimensions. It can, therefore, be studied easily *in vivo*.

2. TECHNICAL APPROACH TO THE STUDY OF ULTRASTRUCTURE OF THE MICROCIRCULATION

The architecture and the topography of microvascular beds in many organs and tissues have been studied extensively in sectioned tissues (Rhodin, 1974). However, structural analyses are of little value if they are not combined with *in vivo* studies in an arrangement which permits one to note the reactivity of the microvascular segment with cinemicrophotography or videotape recording (Wiedeman, 1980). The microvessels must also be accessible to micromanipula-

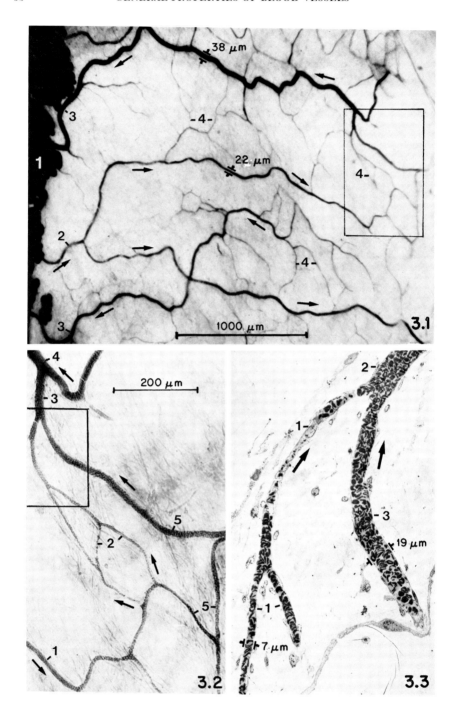

tion and to micropuncture, as well as to direct measurement of blood flow and velocity.

The findings reported in this section are based on the following *in vivo* techniques (Rhodin, 1967, 1968, 1973; Rhodin and Lim Sue, 1979): Careful videotape recording is obtained at a magnification of about ×40 of large areas of the microvascular bed in the mesentery of the rat (Fig. 3.1), followed by a detailed recording at magnifications ranging from ×100 to ×2400. Such an *in vivo* method establishes flow patterns and flow directions and makes possible positive identification of each segment of the microvascular bed. The procedure also permits the study of changes in vascular diameter (Figs. 3.8 and 3.9) and of the reactivity of each specific microvascular segment to drugs and chemicals applied either topically or by intravenous or intraarterial injection.

The second step of the procedure deals with fixation, embedding, sectioning, and microscopic analysis of the microvascular bed. Fixatives can be applied either topically by a superfusion technique (Figs. 3.1–3.3) or intraarterially by a perfusion technique (Figs. 3.4–3.7). The influence of the fixatives on the vascular diameter and/or components of the vascular wall can be recorded by videotape at the moment the fixatives reach the segment under observation. No major variations in vascular diameter have been found to occur as a result of either method of fixation. However, the endothelial cell surface is smoother in the perfused vessels, which indicates that there may be some slight rearrangement of the cytoplasm during the superfusion (Figs. 3.7 and 3.12). Following fixation and dehydration, the entire area under study is embedded in a flat position, which permits scanning and photographing of the microvessels in the light microscope (Figs. 3.1 and 3.2) before the sectioning procedure. Finally, selected areas are sectioned for electron microscopic analysis (Fig. 3.3). Because of the two-dimensional pattern of the mesenteric vessels, the vascular bed lends itself readily to sectioning in a plane which is parallel to the long axis of the microvessels.

FIG. 3.1. Whole mount of typical two-dimensional vascular bed in a mesenteric window of the rat fixed by superfusion of glutaraldehyde, postosmicated and embedded in epoxy resin. Arteries and veins, obscured by fat cells (1), give off arterioles (2) and receive venules (3). Capillary networks (4) are seen in between. Arrows indicate direction of blood flow, confirmed by intravital microscopy of this field. The density of the vessels is caused by their content of red blood cells, arrested seconds after the onset of the superfusion of fixative. Light micrograph. ×35.

FIG. 3.2. Components of the microcirculatory bed, enlarged from rectangle in Fig. 3.1. Precapillary (terminal) arteriole (1), capillary network (2), postcapillary venule (3), muscular venule (4), thoroughfare (5). Light micrograph. ×90.

FIG. 3.3. Segments of the microcirculatory bed seen in rectangle in Fig. 3.2. Venous ends of the capillaries (1), postcapillary venule (2), thoroughfare (3). This sequence of increasing magnifications demonstrates the accuracy of this approach to the study of each segment of a microvascular bed. Electron micrograph. ×310.

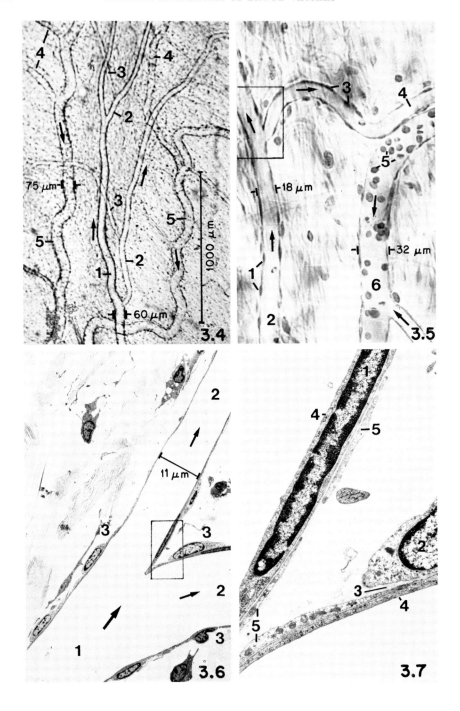

The advantage of the above-described approach to the study of the micro-vascular bed in the mesentery is that an identical segment can be studied *in vivo* and subsequently in gradually increasing magnifications by electron microscopy (Figs. 3.8–3.12). Such a method takes the guesswork out of the study of the different segments of the microvascular bed. For example, a venous end of a capillary can positively be identified, as compared with an arterial end, or a precapillary arteriole can be differentiated from a postcapillary venule. Technically, the superfusion is an easier approach, but it does not seem to fix the tissues as rapidly as perfusion; in the latter case the fixative reaches the vascular segments almost instantaneously. If the intent is to study the relationship of the formed elements of the blood to the endothelial surface, superfusion is necessary, because perfusion removes all the blood cells (Fig. 3.4) except those leukocytes which are adherent to the endothelial cell surface in the postcapillary venules (Fig. 3.5).

3. Vascular Patterns

The general architecture of the microvascular bed varies according to the organ and the tissue. However, in areas which are most commonly studied by *in vivo* methods, such as the rat cremaster muscle (Hutchins *et al.*, 1973), the

FIG. 3.4. Whole mount of mesenteric vascular bed, fixed by intraarterial perfusion of glutaraldehyde, postosmicated and embedded in epoxy resin. This method of fixation eliminates the red blood cells and enhances the delineation of the vascular walls. Arteriole (1), precapillary (terminal) arterioles (2), and capillaries (3) have a straighter course than postcapillary venules (4) and muscular venules (5). Arrows indicate flow direction. Light micrograph. ×40.

FIG. 3.5. Same method of preservation as in Fig. 3.4. At higher magnification nuclei of smooth muscle cells (1), of a precapillary arteriole (2), and pericytes (3) of an arterial capillary (4) can be seen, as well as leukocytes (5) sticking to endothelium of a postcapillary venule (6). Arrows indicate flow direction, confirmed by intravital microscopy, of the vascular segments. This preparation lends itself to a close analysis of different vascular segments before sectioning for electron microscopy. Light micrograph. ×250.

FIG. 3.6. Area similar to rectangle in Fig. 3.5 examined first *in vivo* for precise localization within the microvascular bed before electron microscope analysis. Terminal arteriole (1) gives rise to two arterial capillaries (2) sectioned longitudinally. Vascular lumena are devoid of blood cells and remain wide open following perfusion of glutaraldehyde. Luminal surface of endothelial cells is smooth, whereas pericytes (3) protrude from connective tissue aspect of capillary walls. Electron micrograph. ×1100.

FIG. 3.7. Enlargement of rectangle in Fig. 3.6. Endothelial nucleus (1) is long and does not protrude into vascular lumen after perfusion of fixative. Pericyte nucleus (2) is short and wrapped around endothelial tube. A basal lamina (3) may separate endothelial cells from pericytes. Luminal surfaces (4) of endothelial cells are smooth and straight. Pericyte processes are seen (5). Electron micrograph. ×7700.

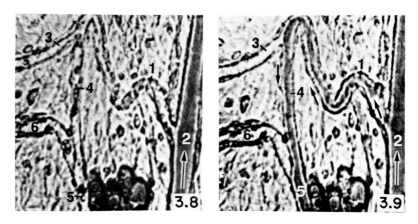

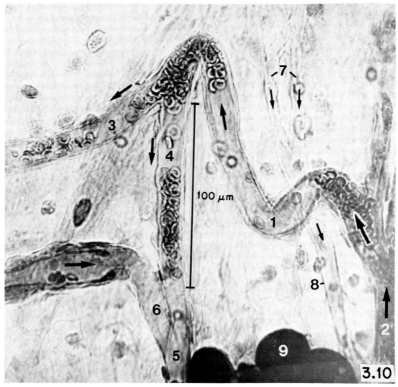

FIGS. 3.8–3.10. Precapillary sphincter area during contraction (Fig. 3.8), relaxation (Fig. 3.9), and after fixation by superfusion and epoxy embedding (Fig. 3.10). Area was observed for 1 hr, displaying typical vasomotion. It was fixed by superfusion of glutaraldehyde during a phase of relaxation. Precapillary sphincter area (1) temporarily did not direct flow into a growing capillary loop (3). Flow in main 30 μm arteriole (2) was continuous. Contraction (1) limited to a small part of precapillary sphincter area. It does not involve a short capillary segment (4) or postcapillary venule (5). Venous end of growing capillary loop (6) connects with postcapillary venule. Other structures are capillaries (7), postcapillary venule (8), and fat cells (9). Arrows indicate flow direction. Light micrograph. Whole mount. Figs. 3.8 and 3.9: ×256. Fig. 3.10: ×484.

hamster cheek pouch (Duling, 1973), and, to a certain extent, the mesentery (Frasher and Wayland, 1972), patterns have been recognized and mapped, although they are subject to variations. In the mesentery of the rat (Figs. 3.1 and 3.2), each major small artery gives rise to a number of arterioles as its diameter decreases toward the periphery of the mesentery.

a. Arterioles: It is generally accepted that these vessels possess a diameter of less than about 0.5 mm (500 μm). The muscular media contains 2 to 4 layers of circumferentially arranged smooth muscle cells. With decreasing arteriolar diameter, the number of smooth muscle layers is gradually reduced (Fig. 3.12). All arterioles have an inner lining of elongated endothelial cells. The periphery is invested by an incomplete layer of fibroblasts and a network of nonmyelinated nerves. *Precapillary (terminal) arterioles* have an inner, luminal diameter of 15 to 20 μm and usually have only one layer of smooth muscle cells. *Precapillary sphincter areas* vary in length and may extend for 10 to 15 μm along the vessel (Figs. 3.8–3.12), or instead they may be represented by one or two smooth muscle cells, forming a true sphincter around the entrance of a capillary.

b. Capillaries: These vessels may branch off from 50 μm arterioles, as well as from 15 μm precapillary arterioles (Figs. 3.10 and 3.11). In the rat mesentery, in the majority of cases the capillaries come off precapillary (terminal) arterioles (Fig. 3.6), forming interconnected networks which empty into postcapillary venules. However, there also may be *thoroughfare* vessels which function as arteriovenous anastomoses (Fig. 3.2). Structurally, the arterial end of these structures is a precapillary arteriole, and the venous end is a postcapillary venule or even a muscular venule. These thoroughfare vessels, discovered and described by Chambers and Zweifach (1944), are typical for the mesentery but are rarely found in other organs and tissues.

Capillaries consist of an endothelial tube, averaging 5 μm in diameter, a thin basal lamina, and occasional pericytes (Figs. 3.6 and 3.7). The thickness of the endothelium and its ultrastructure varies considerably with the tissue and organ. The capillaries are generally divided into the following categories according to the fine structure of their endothelium: (a) fenestrated, (b) continuous, and (c) discontinuous. Ordinarily, capillaries remain open even if blood flow has ceased. This fact and the lack of response to vasoactive drugs are considered to be evidence that endothelial cells are unable to contract.

c. Postcapillary venules: These vessels drain the capillary bed (Figs. 3.1 and 3.3). Their walls consist of an endothelial tube and several pericytes, but no smooth muscle cells. The postcapillary venules, the segment of the microvascular bed most reactive during inflammation, contain intercellular endothelial junctions which open up and allow large molecules to escape from the blood-

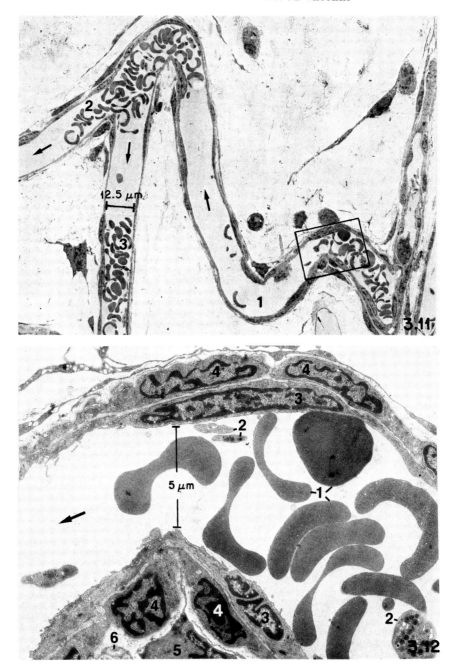

stream. There is also some margination of leukocytes in the venules, caused by a peculiar stickiness of these blood cells, possibly prior to diapedesis via the interendothelial junctions. The biochemical specifics of this endothelium– leukocyte interaction are not known. The pericyte is an amoeba-shaped cell, embracing the postcapillary venule with long slender cell processes (Figs. 3.6 and 3.7). Its function is also unknown. In skeletal muscle, the pericyte may be induced periodically to relax (Tilton *et al.*, 1979), from which it has been assumed that this cell possesses the ability to contract. Pericytes present in myocardial capillaries do not show such activity.

The diameter of postcapillary venules becomes larger as blood flow increases. Smooth muscle cells appear in the media of these vessels, at which point they are referred to as *muscular venules* (Fig. 3.1). It should be noted, however, that the venous end of a mesenteric capillary can join with a muscular venule, as well as with a postcapillary venule. Postcapillary venules may form a long or short meshwork of their own or may empty into a muscular venule. The smooth muscle cells of the muscular venules usually form one, or at the most, two layers around the endothelial tube. Peripherally, connective tissue cells (veil cells) are present, and nerves may be encountered along the vessel as well, but these are fewer in number than in the case of the arterioles.

4. FLOW PATTERNS

Because of the irregular vascular pattern found in the rat mesentery, it is necessary to comment briefly on the flow pattern in this microvascular bed. (For flow regulation in the microcirculation, see Section B-3, below.)

Under normal conditions and normal blood pressure, the flow direction is constant in the arterioles and the muscular venules. However, in the arterial and venous capillaries, as well as in the postcapillary venules, the flow pattern may change from time to time. Often the flow stops in a small arterial capillary

---

FIG. 3.11. Electron micrograph of microvessels in Figs. 3.8–3.10. A detailed analysis can now be made of the components, and corresponding segments and cells can be identified and traced in the *in vivo* recordings in Figs. 3.8 and 3.9 and in the fixed and embedded preparation in Fig. 3.10. Precapillary sphincter area (1) contains smooth muscle cells up to the point where the growing capillary loop begins (2). Short capillary (3) is devoid of smooth muscle cells. Red blood cells present in the growing capillary loop entered this vascular segment during last seconds of fixation. Electron micrograph. ×715.

FIG. 3.12. Enlargement of area indicated by a rectangle in Fig. 3.11, demonstrating details of the precapillary sphincter region: erythrocytes (1); platelets (2); endothelial nuclei (3); smooth muscle cell nuclei (4); fibroblast (5); nonmyelinated nerve axons (6). The sphincter region seems to be in a state of slight contraction judging from one narrow part of the lumen (5 μm), wrinkled nuclei of smooth muscle cells, and irregular luminal surface of endothelial cells. Electron micrograph. ×5400.

and then starts again after a short interval (intermittent flow), although there is no identifiable contraction at the entrance to the capillary near the arteriole nor at the inlet to the postcapillary venule. Frequently, flow is reversed in the venous capillaries and postcapillary venules. However, it must be remembered that in the mesentery, there are numerous anastomoses and cross connections between capillaries and postcapillary venules, facilitating a reversal of flow. There may also be a momentary stoppage of flow in one branch of the capillary network, whereas another permits unimpeded movement to continue. Thoroughfares, which establish a shortcut between an arteriole and a muscular venule, may be responsible for the reversal of flow in some segments of the adjoining capillary loops due to slight changes in intravascular pressure. Other, presently unknown, factors may likewise influence flow pattern changes in the mesentery. Intermittent flow has also been observed in skeletal muscle capillaries (Burton and Johnson, 1972). It was recently suggested by Tilton *et al.* (1979) that pericyte contraction may be responsible for a slight constriction of a small capillary segment, which momentarily prevents the flow of formed blood elements.

Most of the blood vessels and the microvascular beds present in the mesenteric membrane function as precursors of capillary networks involved in the development of adipose tissue (see Section G, Chapter 18). The majority of microvessels are therefore involved in growth, reflected in an intense capillary endothelial sprouting. The intermittent flow, as well as the reversal of flow direction, observed in the microvascular bed may be directly related to the elaboration of capillary networks for the delivery of lipids to the fat cell precursors.

B. PHYSIOLOGY OF MICROCIRCULATION

By P. C. JOHNSON

1. GENERAL CONSIDERATIONS

The physiology of the microcirculation is a matter of intensive study at the present time, and information on this topic is expanding rapidly. Several recent reviews and monographs (Duling and Klitzman, 1980; Gross and Popel, 1980; Johnson, 1978, 1980; Kaley and Altura, 1977, 1978, 1980; Wiedeman *et al.*,

Blood Vessels and Lymphatics in Organ Systems
Copyright © 1984 by Academic Press, Inc.

1981) are recommended for the reader interested in a more detailed exposition of recent experimental and theoretical investigations of this topic.

As described in Section A, above, the microcirculation consists of arterioles, capillaries, and venules which form a branching, tapered network of approximately circular tubes.

a. Arterioles: These vessels undergo active changes in diameter, approximately two- to threefold in the smallest branches and 20–40% in the large ones, depending on the initial state of vascular tone. They regulate flow to the capillary bed and also may act to regulate pressure at this level. The terminal arterioles give rise to a network of generally parallel capillaries which undergo a modest amount of additional branching.

b. Capillaries: The terminal branches of the arterioles appear to govern the flow to various regions of the capillary network. In some vascular beds, there is a specialized smooth muscle cell (precapillary sphincter) which regulates flow through individual capillaries, but in most beds (e.g., skeletal muscle), such a structure is not seen.

c. Venules: In general, these vessels are regarded as passive, distensible tubes which serve principally as conduits for returning the blood from the tissues to the central circulation and as part of the capacitance system which contains most of the blood volume. The organization of the venular network is similar to that of the arteriolar network except that the venules are two to three times wider and somewhat more numerous than the arterioles. Also, smooth muscle cells are less abundant in the venular wall; venules display less spontaneous vasomotion; and constriction causes smaller changes in venular diameter during excitation.

2. EXCHANGE PROCESS

a. Exchange of lipid-soluble substances: This process takes place between blood and tissue probably at all levels of the microcirculatory network, including arterioles and venules, although the capillary network provides the most favorable circumstances for it. Using O_2 as an example, there is a substantial drop in blood P_{O_2} from artery to terminal arteriole (68 to 25 mm Hg) in the hamster cheek pouch (Duling and Berne, 1970), and the hemoglobin oxygen (HbO_2) saturation in this same region falls from 92 to 41%. In exercising muscle, the transit time in arterioles may be substantially less, and hence P_{O_2} levels at the entrance to the capillary may be higher. Since the larger arterioles (above 30 μm) are paired with and adjacent to venules, a countercurrent flow situation may exist in which some of the O_2 lost from the arterioles may be taken up by the venules, thus leading to a situation in which venous O_2 levels may be higher than those in the capillary network.

   b. *Exchange of lipid-insoluble substances:* This process takes place between blood and tissue principally in the capillary network at specialized junctions between endothelial cells. Large molecules also may move through small numbers of rather large (0.5–1.0 μm) gaps in the small (10–15 μm) venules. These gaps become more numerous with histamine application. In addition, it is thought that pinocytotic vesicles in the endothelium may serve as a shuttle transport system (reviewed by Renkin, 1977). However, the role of vesicular transport is not clear at present, since a recent report suggests that the vesicles may be part of a stationary network of channels within the cell (Bundgaard, 1980).

   c. *Exchange of smaller molecular weight lipid-insoluble substances:* This process, especially in the case of H_2O, takes place principally in the capillary network at the specialized junctions between the endothelial cells. Figure 3.13 shows a plot of the filtration constant (Lp) in intestinal muscle capillaries (Gore, 1982). Lp is a measure of the conductance for water across the capillary mem-

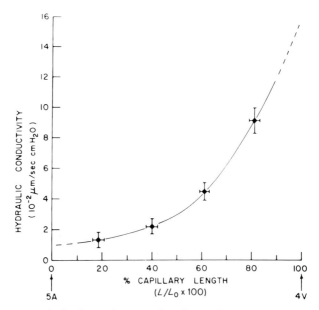

FIG. 3.13. Average hydraulic conductivity of capillary wall as a function of position between terminal (5A) arteriole and initial (4V) branch of the venular network. Note that conductivity increases by an order of magnitude along the capillary. (From Gore, 1982, reproduced with permission of the American Physiological Society.)

brane. Since conductance is much greater in the venous section of the capillary network, fluid movement may be more rapid in this region, depending on the magnitude of the forces which influence fluid exchange. In the renal glomerulus, the high Lp throughout the capillary bed allows the plasma water to filter rapidly and plasma colloid osmotic pressure to rise in the arterial section of the capillary. The latter change brings filtration to a halt (Beeuwkes and Brenner, 1978). In the intestinal capillary beds, this equilibration process may also occur but to a more limited degree (Johnson and Hanson, 1966). However, the equilibration may not be as significant in skeletal muscle capillaries because of the low Lp in these vessels.

   d. *Factors in fluid exchange process:* The colloid osmotic pressure in plasma, the colloid osmotic pressure in interstitial fluid, the hydrostatic pressure in blood, and the hydrostatic pressure in interstitial fluid all influence the fluid exchange process. Of these four forces, the capillary hydrostatic pressure has the greatest potential for change, and the alteration can occur rapidly. Despite the possibility of large changes, fluid balance in most tissues seems to be maintained within rather narrow limits, probably because physiologic control mechanisms appear to regulate capillary hydrostatic pressure (in some instances perhaps incidentally to regulation of flow) and because the other forces involved in filtration are altered by fluid movement. For example, when elevation of capillary hydrostatic pressure causes filtration, interstitial colloid osmotic pressure falls and tissue hydrostatic pressure rises; both of the latter changes tend to reduce filtration.

   Capillary hydrostatic pressure is sensitive to changes in venous pressure, with 60 to 80% of the increment in large vein pressure being transmitted to the capillaries. There is evidence that a rise in capillary pressure is attenuated, in part, by associated constriction of the arterioles and a fall in venous resistance (Johnson, 1980).

   In contrast, there is substantial evidence from whole organ studies that capillary pressure changes little over a wide range of arterial pressures. Jarhult and Mellander (1974), for example, found a change in mean capillary pressure (as measured by the isovolumetric technique) of only 4 mm Hg over an arterial pressure range of 30 to 170 mm Hg. Moreover, in the intestine and skeletal muscle, reduction of arterial pressure from 100 to 40 mm Hg was accompanied by a decrease in precapillary resistance (autoregulation) and, below the autoregulatory pressure range, by an increase in postcapillary resistance (Johnson, 1978). Both of these changes tend to maintain a constant capillary pressure. Direct pressure studies at the microcirculatory level, although incomplete, do not fully confirm the above observations. For example, dilatation of arterioles has been reported in cat mesentery with arterial pressure reductions (Johnson, 1980). The scant information available regarding the behavior of the venules in response to decreased arterial pressure indicates that, under such conditions, there is little change in venular dimensions. It is necessary to point out, however, that since blood shear rate in the venules is low (Lipowski *et al.*, 1978), a reduction in flow could lead to a rise in blood viscosity in these vessels. Both the dilatation of the arterioles and the increase in blood viscosity in the venules would tend to maintain a constant capillary pressure. In the kidney, direct measurement reveals that glomerular capillary pressure is independent of arterial pressure over the range 60–150 mm Hg, under which conditions, blood flow is also perfectly autoregulated. On the other hand, Bohlen and Gore (1977) found that during hemorrhage, pressure in pre- and postcapillary vessels in the denervated rat intestinal muscle fell in proportion to the drop in arterial pressure. Some of the precapillary vessels dilated, whereas postcapillary vessels generally constricted.

3. FLOW REGULATION IN THE MICROCIRCULATION

a. Autoregulation: Arterioles constrict when arterial pressure is elevated and dilate when pressure is reduced, thus tending to maintain constant flow (autoregulation). It has been suggested that autoregulation is due to an exquisite sensitivity of the arterioles to changes in the tissue level of vasodilator metabolites which, in turn, is responsive to blood flow, perhaps through alterations in tissue O_2 levels. Reactive vascular beds are, in fact, quite sensitive to changing O_2 levels in a suffusing solution passing over the tissue (Duling and Klitzman, 1980). However, in strongly autoregulating organs, vascular resistance can double when arterial pressure is doubled, with the result that large changes in resistance may be associated with no detectable alteration in blood flow and O_2 delivery. This circumstance would seem to provide no error signal for metabolic regulation of flow. However, though volume flow is held constant at reduced arterial pressure in the perfectly autoregulating organ, blood velocity is slower through the dilated arterioles, so that transit time is increased. Such a situation would allow greater opportunity for O_2 loss to the adjacent venule by a countercurrent exchange process. Thus, tissue P_{O_2} in the vicinity of the capillaries may be reduced at lower arterial pressure even though arterial blood flow and O_2 delivery remain constant.

b. Effect of metabolic needs or O_2 supply on flow regulation: This subject has been examined only to a limited extent by direct observation of the microcirculation. A variety of studies have shown that the arterioles dilate when tissue O_2 levels are lowered (Duling and Klitzman, 1980). When O_2 demand in exercising muscle rises, both volume flow and the number of capillaries carrying blood increase, raising O_2 delivery and reducing the average diffusion distance from blood to tissue. However, when tissue O_2 levels are high, excitation of skeletal muscle leads principally to an increase in the *number* of capillaries carrying blood, whereas when tissue O_2 levels are low, stimulation of striated muscle leads principally to arteriolar dilatation and increased volume flow (Duling and Klitzman, 1980). It is of interest that blood flow distribution in the capillary network becomes more homogeneous under conditions of increased metabolic demand.

The mechanism by which lowered O_2 leads to arteriolar dilatation is still a matter of active investigation. P_{O_2} changes in the immediate vicinity of the arterioles do not alter vascular tone; hence, it would appear that a direct effect of O_2 on the vessel itself is not responsible (Duling and Klitzman, 1980). It is more likely that when P_{O_2} falls, tissue metabolism changes, causing production of a vasodilator substance. However, mean tissue P_{O_2} is generally 20 mm Hg or more, whereas the critical P_{O_2} required by mitochondria for aerobic metabolism is only 1–2 mm Hg. Therefore, it would seem that drastic reduction of blood flow is required before critical P_{O_2} levels are reached. However, it is possible that localized tissue areas at sites near the venous end of the capillaries, where P_{O_2} is lowest, are vulnerable to O_2 deprivation. A modest fall in flow might lead to anaerobic metabolism in such areas.

c. Myogenic regulation of arterioles: The other local control mechanism which appears to be importantly involved in blood flow regulation is the myogenic response of the arterioles. According to this hypothesis, an increase in intravascular pressure stimulates the smooth muscle cells of the arterioles to contract. Evidence for this response has been obtained from whole organ and direct microcirculatory observations, the latter including *in situ* and *in vivo* studies (Johnson, 1980). Although the underlying mechanism is still a matter of speculation, there is evidence that circumferential wall tension or wall stress may be the regulated variable. According to the Laplace relationship ($T = Pr$), when intravascular pressure (P) is doubled, arteriolar radius (r) must decrease to half its initial value in order to restore tension (T) to its original level. Data from recent studies (Bouskela and Wiederheilm, 1979; Burrows and Johnson, 1981) indicate that tension may be regulated when intravascular pressure is altered.

By its nature, the myogenic mechanism would tend to maintain a constant intravascular pressure at the microcirculatory level. For example, when arterial pressure is elevated and capillary and arteriolar pressures rise, arterioles constrict, leading to a reduction in inflow and therefore an attenuation of the rise in capillary pressure. This may be seen from the relationship

$$P_c = P_v + F \times R_v \tag{B.1}$$

where P_c is capillary pressure, P_v is venous pressure, F is flow, and R_v is venous

resistance. At the same time, the constriction of arterioles would tend to maintain a normal pressure in the arteriolar and capillary network. By application of
the above equation, it can be seen that if flow is perfectly autoregulated,
changes in arterial pressure will have no effect on capillary pressure.

A modification of the myogenic response may provide essentially perfect
flow autoregulation without flow being the regulated variable. Consider the
hypothesis as shown in Fig. 3.14 (Johnson, 1980), in which several consecutive
branches of the arteriolar network are assumed to possess a myogenic response
but are still capable of acting independently of each other. When arterial
pressure is reduced by a single step, the pressure in the arterioles also falls.
However, as the large arterioles dilate due to the reduction of the pressure
stimulus, the pressure gradient in them is reduced and pressure downstream
tends to be restored. Since the process of dilatation occurs in the neighboring
downstream branches as well, a point will be reached in the network at which
pressure is normal. When such a condition occurs, flow in the network also will
return precisely to the initial level. With a larger reduction in pressure, the
vasodilatation will become more pronounced in the large arterioles and will
spread to the more distal vessels as well. A limiting point is reached when all
the arterioles are maximally dilated. This hypothesis predicts that pressure in
the larger arterioles is autoregulated only when arterial pressure is high,
whereas pressure in the distal arterioles and capillaries is autoregulated down

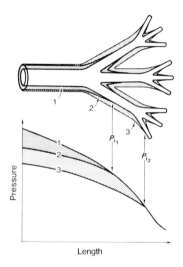

Fig. 3.14. Schema showing hypothetical behavior of an arteriolar network composed of series-
coupled myogenically active vessels. Upper panel shows consecutive vascular segments. Lower
panel shows pressure gradient in the control state (curve 1); following 20% reduction in arterial
pressure (curve 2); and following 40% reduction of arterial pressure (curve 3). (From Johnson, 1980,
reproduced with permission of the American Physiological Society.)

to a much lower arterial pressure. Such a mechanism could operate in concert with the sensitivity of arterioles to tissue metabolites to provide a versatile system in which flow would be autoregulated at a value determined by tissue metabolite levels.

C. PHARMACOLOGY OF MICROCIRCULATION*

By B. M. ALTURA

1. GENERAL CONSIDERATIONS

Much of the available information concerning the pharmacology of the microvasculature is derived from indirect whole-organ studies, perfusion experiments, and investigations of isolated arteries and veins. In this context, it often is assumed that isolated blood vessels, 150–300 μm i.d., can be used to gain information about the intact resistance and capacitance vessels. It is obvious, however, that such an approach may not reflect accurately the pharmacology of the *in situ* microvasculature. Furthermore, since microvessels from different regions in the circulation, as well as from within a single microvasculature, often display heterogeneity of responses and receptors, it is necessary to utilize direct *in situ* techniques in the study of these microscopic vessels. In addition, arterioles and venules are of various sizes, structures, and types within any vascular bed. (For a more complete discussion of these issues, as well as the experimental data cited in this paper, see Altura, 1971, 1978a,b, 1980, 1981a,b,c, 1982, 1983; B. M. Altura and Altura, 1976, 1977a,b,c, 1978, 1982a,b, 1983; Altura and Altura, 1978a,b; Altura and Chand, 1981; B. M. Altura *et al.*, 1980, 1982a,b, 1983; Chand and Altura, 1981a,b; Furchgott, 1981; Kaley and Altura, 1977, 1978, 1980; Weiss, 1981.)

Recently endothelial cells lining the intima of blood vessels have been demonstrated to play an important role in the regulation of vascular muscle reactivity and vasomotor tone. Since the intimal surfaces of endothelial and reticuloendothelial cells are vital to preserve (and protect) the integrity of organs and tissues, it must be entertained that the comparative pharmacology of these specialized microvascular structures also must be studied as precisely as possible.

*The original studies referred to herein were supported by Research Grant HL-18002, HL-18015, HL-29600, and DA-02339 from the United States Public Health Service.

2. Factors Influencing Arteriolar and Venular Smooth Muscle Tone and Reactivity

Comparison of studies from different laboratories reveals that the same agonists induce either similar, different, or no noticeable responses on identical types of microvessels. Several reasons may explain such findings, as noted in Table 3.I, in which are listed a number of physiologic, pharmacologic, and experimental factors known to modulate the responsiveness of microvessels to neurohumoral, hormonal, local metabolic, and physical influences. These factors are also elaborated upon below.

a. Effect of anesthetic and analgesic agents on tone and reactivity of microvessels: In order to study the intact microvasculature, most investigators first employ substances to anesthetize and/or narcotize the experimental animals. In this regard, it is necessary to point out that all intravenous and volatile general anesthetic and analgesic agents currently approved for clinical use, including ketamine, althesin, propanidid, and fentanyl, are known to compromise cardiovascular responses in intact animals and to affect the responsiveness of isolated vascular smooth muscles. In addition, many of these agents can influence plasma and tissue levels of endogenous vasoactive neurohumoral agents (Kaley and Altura, 1980).

High-resolution television image-splitting microscopy has revealed that most general anesthetic drugs which produce surgical anesthesia—barbiturates, α-chloralose, urethan, chloral hydrate, and diethyl ether—dilate arterioles and venules in a dose-dependent manner (Altura, 1981a; B. M. Altura *et al.*, 1980). Quantitative microcirculatory studies performed *in vivo* on the splanchnic, skeletal muscle, and cerebral (pial) microvasculatures of the rat indicate that both general and local anesthetic agents—procaine, lidocaine, mepivicaine, and tetracaine—affect not only the resting diameter of arterioles and venules but also their responsiveness to neurohumoral agents, the latter change often resulting in an attenuation of the action of constrictor agents. Studies of isolated vascular preparations and of microvessels suggest that such reactions are related to interference with the movements and/or translocation of calcium ions (Ca^{2+}) across and within the microvascular smooth muscle cells. Acute administration of morphine, meperidine, levorphanol, and codeine can exert excitatory, depressant, or no effect on vascular smooth muscle cells, depending upon the type of blood vessel and upon the vascular region studied. Irrespective of the route of administration, morphine (doses as little as 0.1 mg/kg), codeine, meperidine, and levorphanol cause dose-dependent dilations of up to 100% over control lumen size of terminal arterioles and precapil-

TABLE 3.I

Physiologic, Pharmacologic, and Experimental Factors Which Can Modify the
Tone and Actions of Vasoactive Agents on Microvessels

Physiologic	Pharmacologic and experimental
Heterogeneity of smooth muscle elements	Anesthetics
Endothelial cell integrity	Analgesics
Aging	Buffers
Sex hormones	Solvents, preservatives
Innervation	Presence of certain drugs
Ionic milieu	Route of drug administration
Physical	Purity of agonist
Length–tension (e.g., blood pressure, wall thickness, geometry, etc.)	Surgical manipulation
Temperature	
Tissue metabolism and osmolarity	
Local chemicals and hormones	
Species, strain	
Physiologic state of host	

lary sphincters. However, the lumen size of the mesenteric venules (30–70 μm i.d.) is not altered by any of the opiates tested. In the rat mesentery, intravenous injection of morphine, in doses used for anesthesia in man (2–4 mg/kg), produces an 80–90% increase in arteriolar lumen size, a response greater than that produced by any other known vasodilator agent. These microcirculatory actions of the opiates are abolished by intravenous administration of extremely low doses of the narcotic antagonist, naloxone (B. T. Altura et al., 1980), a drug which does not attenuate any other type of arteriolar dilator or anesthetic-induced dilator response in the rat mesenteric circulation. Thus, in the microvasculature of certain tissues, there are specific opiate receptors in arterioles and precapillary sphincters (but probably not in venular smooth muscle) which subserve relaxation.

In addition to the powerful effects that various drugs of abuse (alcohol, barbiturates, opiates) have on the CNS, they also exert potent influences on blood pressure and cardiac output. However, it is not known what peripheral vascular effects are evoked in persons who chronically self-medicate themselves with these drugs. Studies of blood vessels excised from rats given chronic doses of ethanol, barbiturates, and morphine indicate that arterial and arteriolar, but not venous or venular vessels, become supersensitive to constrictor vasoactive agents (Altura, 1981b; Altura and Altura, 1982a); barbiturate, alcohol, and opiate withdrawal is generally associated with elevated blood pressure. It would, therefore, be of considerable interest to determine the types of changes that occur in the arteriolar and venular levels of the microcirculation in different organ regions.

b. Influences of aging on reactivity: With aging, there appears to be a generalized increase in peripheral vascular resistance and a decrease in perfusion of a number of organs (Altura and Altura, 1977b). Data support the idea that, in old age, arterial, arteriolar, and precapillary sphincter smooth muscle cells become hyporeactive to contractile agents (e.g., catecholamines, serotonin, tyramine, angiotensin II) (Altura, 1981a; Altura and Altura, 1977b). However, certain venules (e.g., rat mesenteric), while displaying a loss of contractile sensitivity to serotonin and vasopressin on aging, do not demonstrate this type of change in the case of such constrictors as catecholamines, tyramine, or angiotensin II. Inhibitory responses of mesenteric terminal arterioles and precapillary sphincters (but not of muscular venules) to the β-adrenergic agonist, isoproterenol, also decrease with age (Altura, 1978a, 1981a; Altura and Altura, 1977b). Collectively, such data suggest that different blood vessels within a single vascular bed display heterogeneity which may be age dependent.

c. Role of endothelial cells in regulation of microvascular tone and reactivity: Endothelial cell (EC) damage has been implicated in several vascular diseases of major clinical importance, such as diabetes mellitus, primary pulmonary hypertension, pulmonary hypertension, arterial and venous thrombosis, acute renal failure, atherosclerosis, arterial spasm, anaphylaxis, and circulatory shock. Vascular ECs not only serve as a protective barrier in blood vessel walls, but they constitute an important active biochemical structure for the synthesis, metabolism, uptake, storage, and degradation of numerous vasoactive agents, among which are serotonin, catecholamines, angiotensins, and prostanoids. Endothelial cells are also known to possess binding (receptor?) sites for catecholamines, angiotensins, acetylcholine, histamine, serotonin, etc. It has been demonstrated that vascular ECs of intrapulmonary, intrarenal, hepatic, coronary, and splanchnic arteries appear to play an essential role in the relaxation of these blood vessels in response to several types of naturally occurring vasodilators, e.g., acetylcholine, bradykinin, arachidonic acid, substance P, ADP, and ATP. It is also clear that these vasorelaxants act on distinct EC receptors. On the other hand, relaxation of blood vessels induced by isoproterenol, impromidine, and dimaprit (H_2-histamine receptor agonists) and by prostanoids, prostacyclin, and papaverine, as well as other nonspecific vasodilator relaxants (e.g., glyceryl trinitrite, sodium nitrate, sodium nitrite), 5-AMP, and adenosine, does not depend upon the integrity of the EC, at least in the tissues so far investigated. Many agents act on ECs to cause relaxation of the underlying vascular smooth muscle cells. However, it is known that loss of responsiveness of ECs to the relaxants, bradykinin, acetylcholine, substance P, etc., or a loss of EC integrity, results in a transformation of these potent dilators into potent constrictor agents. If *in situ* regional arteries and arterioles suddenly or gradually were to lose their ability to dilate in response to such dilators, the end result could be a loss of microvascular patency in numerous organ regions. It is possible that such a sequence of events could play a role in several peripheral vascular disorders. The importance of ECs in controlling relaxation of blood vessels could also explain why investigators in the past have been perplexed by noting that acetylcholine, kinin, substance P, and other vasodilators often elicit contraction of excised vascular smooth muscle when studied *in vitro*. Extreme caution must thus be exercised in preparing blood vessels for *in vitro* study so as not to injure or destroy the intimal EC.

3. Responsiveness of Microvessels to Vasoactive Agents and Receptor Specificity

Peripheral arterioles, metarterioles, precapillary sphincters, and muscular venules are known to contain specific receptors for many neurohumoral substances and drugs which elicit contraction. These agents include catecholamines, neurohypophyseal peptides, angiotensins, prostanoids, dopamine, serotonin, cardiac glycosides, sympathomimetic amines, etc. The vasoconstrictor responses of different types of peripheral vessels to these substances usually display different orders of sensitivity and magnitude, depending upon organ region, segment within a region, and vasoactive substance. By contrast, it is not known whether all neurohumoral peripheral arteriolar dilators (e.g., histamine, acetylcholine, kinins, prostaglandins, purines, hydroperoxides, etc.) can actively dilate *all* types of muscular venules in different regions of the mammalian circulation. Whether this is due to a lack of receptor sites for certain neurohumoral dilators or to various physiologic and/or pharmacologic factors (Table 3.I) in the environment of the microscopic capacitance vessels is not known.

a. α-Adrenergic receptor specificity with regard to type and size of microvessel: Arterioles and precapillary sphincters are more sensitive to norepinephrine and epinephrine than are the venules. For example, in the rat mesentery, the precapillary sphincters (3–7 μm) are, on the average, 500–1000 times and 1000–10,000 times more sensitive to the vasoconstrictor action of epinephrine and norepinephrine, respectively, than are the arterioles and venules.

The smallest venules (30–40 μm) of the mesentery and cremaster (skeletal) muscle of the rat, which are invested with both smooth muscle cells and pericytes, are more responsive (lower threshold and greater maximal responses) to the constrictor effect of norepinephrine and epinephrine than are larger venules (40–75 μm). It has been suggested as one possibility that since there are fewer smooth muscle cells in the smallest venules, the pericytes may contribute to the overall responses. Vessel size and type must also perforce play an important role in a microvasculature's response to constrictor agents, such as catecholamines.

Overall, available evidence indicates that α-adrenergic receptors of the terminal arterioles, metarterioles, precapillary sphincters, and venules are more sensitive to epinephrine and norepinephrine than to dopamine and phenylephrine. These data support the idea that specific α-adrenergic receptors exist in all types of peripheral microvessels. The relative number and distribution of these receptors, however, differ with region and type of microvessel.

b. β-*Adrenergic receptors:* These structures, which subserve relaxation in vascular smooth muscle, exist in mammalian microvascular smooth muscle of numerous vascular beds (e.g., intestine, liver, pancreas, skeletal muscle, brain, adipose tissue, kidneys, lungs, heart, etc.). Although the β-adrenergic receptor sensitivity in arterioles, metarterioles, and precapillary sphincters depends on age, this is not the case for venules. In young rats, there is a preponderance of β-adrenergic, as compared with α-adrenergic, receptors in the muscular venules of the mesenteric vasculature (Altura, 1971). This implies that circulating epinephrine in physiologic concentrations may cause dilatation of the microscopic capacitance vessels in certain vascular beds.

4. INFLUENCE OF IONIC CALCIUM AND CALCIUM CHANNEL BLOCKERS ON MICROVASCULAR TONE

Since Ca^{2+} serves as the link between excitation and contraction of the vascular smooth muscle cells, its movements across the vascular cell membranes, even in the absence of neurohumoral substances, may serve to control directly the tone and caliber of muscular microvessels. If this is true, then removal of Ca^{2+} from the extracellular space or its binding to surface membranes should dilate arteries, veins, arterioles, and venules. In this regard, it has been found that in isolated arteries and veins which exhibit myogenic activity, withdrawal of extracellular Ca^{2+} or use of agents which block its entry into the cell causes rapid inhibition of spontaneous contractions. Blocking agents include Mg^{2+}, anesthetics, alcohol, artificial buffers, certain steroids, drugs, etc. However, information at the *in situ* microcirculatory level with respect to the actions of withdrawal of Ca^{2+} and of Ca^{2+} entry blockers on tone is sparse. The available data for the rat indicate that perfusion with the Ca^{2+} chelator, Na_2EDTA, produces only slight dilatation of splanchnic arterioles and venules, marked dilatation (30–40% increase in diameter) of cremasteric muscle, and no effect on pial arterioles and venules. Perfusion of splanchnic micro-vessels, as well as cerebral (pial) arterioles and venules, with the Ca^{2+} entry blocker, verapamil (1–100 μg), fails to influence the tone of the microvessels and does not affect capillary blood flow (Altura *et al.*, 1983). The perivascular application of verapamil, however, causes dose-dependent increases in lumen size and blood flow in terminal arterioles and muscular venules of rat crem-asteric muscle.

A new class of Ca^{2+} channel blockers, such as the 1,4-dihydropyridine derivatives, nimodipine and nitrendipine, exert potent (i.e., threshold equal to 10^{-9} to 10^{-8} M) dilator actions on splanchnic, skeletal muscle, and cerebral (pial) microscopic resistance and capacitance vessels in the intact rat, whereas other derivatives (e.g., nisoldipine) fail, like verapamil, to dilate cerebral ar-terioles and venules (Altura *et al.*, 1983).

Although the *in vitro* and the indirect studies with Ca^{2+} entry blockers suggest that these agents should nonspecifically inhibit vascular tone and in-crease blood flow, the above *in situ* microcirculatory studies do not unequivo-cally support this concept. Still several conclusions can be drawn from these data: (a) the influx of Ca^{2+} may be responsible for the maintenance of arteriolar and venular tone in skeletal muscle but not splanchnic and cerebral micro-vasculatures; (b) the apparent dilator actions reported for Ca^{2+} entry blockers in certain vascular beds may be the result of effects on small arteries rather than on either arterioles or venules; (c) the controlling Ca^{2+} in some microvessels may arise from an extracellular origin, provided it can reach the intracellular compartment by a Ca^{2+} channel blocker-insensitive mechanism, such as a Na^+, Ca^{2+} exchange system; (d) it is possible to design Ca^{2+} entry blockers which not only show different affinities for arterioles and venules in regional circulatory beds but which can yield varying potencies and degrees of maximal vasodilatation. Further studies along these lines should provide powerful tools in the treatment of arterial hypertension, vasospasm, and ischemia (especially of the brain).

REFERENCES*

Altura, B. M. (1971). Chemical and humoral regulation of blood flow through the precapillary sphincter. *Microvasc. Res.* **3**, 361–384. (C)

Altura, B. M. (1978a). Pharmacology of venular smooth muscle: New insights. *Microvasc. Res.* **16**, 91–117. (C)

Altura, B. M. (1978b). Humoral, hormonal, and myogenic mechanisms in microcirculatory regula-tion. *In* "Microcirculation" (G. Kaley and B. M. Altura, eds.), Vol. II, pp. 431–502. University Park Press, Baltimore, Maryland. (C)

Altura, B. M. (1980). "Vascular Endothelium and Basement Membranes." Karger, Basel. (C)

Altura, B. M. (1981a). Pharmacology of venules: Some current concepts and clinical potential. *J. Cardiovasc. Pharmacol.* **3**, 1413–1428. (C)

*In the reference list, the capital letter in parentheses at the end of each reference indicates the section in which it is cited.

Altura, B. M. (1981b). Pharmacology of venules. *Bibl. Anat.* **20**, 343–346. (C)

Altura, B. M. (1981c). Pharmacology of the microcirculation. *In* "Microcirculation: Current Physiologic, Medical, and Surgical Concepts" (R. M. Effros, H. Schmid-Schönbein, and J. Ditzel, eds.), pp. 51–105. Academic Press, New York. (C)

Altura, B. M. (1982). "Ionic Regulation of the Microcirculation." Karger, Basel. (C)

Altura, B. M. (1983). Endothelium, reticuloendothelial cells and microvascular integrity: Roles in host defense. *In* "Handbook of Shock and Trauma" (B. M. Altura, A. M. Lefer, and W. Schumer, eds.), Vol. 1, pp. 51–96. Raven, New York.

Altura, B. M., and Altura, B. T. (1976). Vascular smooth muscle and prostaglandins. *Fed. Proc., Fed. Am. Soc. Exp. Biol.* **35**, 2360–2366. (C)

Altura, B. M., and Altura, B. T. (1977a). Vascular smooth muscle and neurohypophyseal hormones. *Fed. Proc., Fed. Am. Soc. Exp. Biol.* **36**, 1853–1860. (C)

Altura, B. M., and Altura, B. T. (1977b). Ageing in vascular smooth muscle and its influence on reactivity. *In* "Factors Influencing Vascular Reactivity" (O. Carrier, Jr., and S. Shibata, eds.), pp. 169–188. Igaku-Shoin Ltd., Tokyo. (C)

Altura, B. M., and Altura, B. T. (1977c). Influence of sex hormones, oral contraceptives and pregnancy on vascular muscle and its reactivity. *In* "Factors Influencing Vascular Reactivity" (O. Carrier, Jr., and S. Shibata, eds.), pp. 221–254. Igaku-Shoin Ltd., Tokyo. (C)

Altura, B. M., and Altura, B. T. (1978). Interactions of locally produced humoral substances in regulation of the microcirculation. *In* "Mechanisms of Vasodilatation" (P. M. Vanhoutte and I. Leusen, eds.), pp. 98–106. Karger, Basel. (C)

Altura, B. M., and Altura, B. T. (1982a). Microvascular and vascular smooth muscle actions of ethanol, acetaldehyde, and acetate. *Fed. Proc., Fed. Am. Soc. Exp. Biol.* **41**, 2447–2451. (C)

Altura, B. M., and Altura, B. T. (1982b). Factors affecting responsiveness of blood vessels to prostaglandins and other chemical mediators of injury and shock. *In* "Role of Chemical Mediators in the Pathophysiology of Acute Illness and Injury" (R. McConn, ed.), pp. 45–63. Raven, New York. (C)

Altura, B. M., and Altura, B. T. (1983). Actions of oxytocin and synthetic analogs on cardiovascular smooth muscle. *Fed. Proc., Fed. Am. Soc. Exp. Biol.* (in press). (C)

Altura, B. M., and Chand, N. (1981). Bradykinin relaxation of renal and pulmonary arteries is dependent upon intact endothelial cells. *Br. J. Pharmacol.* **74**, 10–11. (C)

Altura, B. M., Altura, B. T., Carella, A., Turlapaty, P. D. M. V., and Weinberg, J. (1980). Vascular smooth muscle and general anesthetics. *Fed. Proc., Fed. Am. Soc. Exp. Biol.* **39**, 1584–1591. (C)

Altura, B. M., Altura, B. T., Carella, A., and Turlapaty, P. D. M. V. (1982a). Ca^{2+} coupling in vascular smooth muscle: Mg^{2+} and buffer effects on contractility and membrane Ca^{2+} movements. *Can. J. Physiol. Pharmacol.* **60**, 459–482. (C)

Altura, B. M., Altura, B. T., and Gebrewold, A. (1983). Comparative microvascular actions of Ca^{2+} antagonists. *In* "Ca^{2+} Entry Blockers, Adenosine and Neurohumors" (G. F. Merrill, ed.), pp. 155–175. Urban & Schwarzenberg, Baltimore, Maryland. (C)

Altura, B. T., and Altura, B. M. (1978a). Factors affecting vascular responsiveness. *In* "Microcirculation" (G. Kaley and B. M. Altura, eds.), Vol. 2, pp. 547–615. University Park Press, Baltimore, Maryland. (C)

Altura, B. T., and Altura, B. M. (1978b). Intravenous anesthetic agents and vascular smooth muscle function. *In* "Mechanisms of Vasodilatation" (P. M. Vanhoutte and I. Leusen, eds.), pp. 165–172. Karger, Basel. (C)

Altura, B. T., Gebrewold, A., and Altura, B. M. (1980). Are there opiate receptors in the microcirculation? *In* "Vascular Neuroeffector Mechanisms" (J. A. Bevan, T. Godfraind, R. A. Maxwell, and P. M. Vanhoutte, eds.), pp. 316–319. Raven, New York. (C)

Beeuwkes, R., and Brenner, B. M. (1978). Kidney. *In* "Peripheral Circulation" (P. C. Johnson, ed.), Chapter 6. Wiley, New York. (B)

Bohlen, H. G., and Gore, R. W. (1977). Comparison of microvascular pressures and diameters in the innervated and denervated rat intestine. *Microvasc. Res.* 14, 251–264. (B)

Bouskela, E., and Wiederhielm, C. A. (1979). Microvascular myogenic reaction in the wing of the unanesthetized bat. *Am. J. Physiol.* 237 (*Heart Circ. Physiol.* 6), H59–H65. (B)

Bundgaard, M. (1980). Transport pathways in capillaries—in search of pores. *Annu. Rev. Physiol.* 42, 325–336. (B)

Burrows, M. E., and Johnson, P. C. (1981). Diameter, wall tension and flow in mesenteric arterioles during autoregulation. *Am. J. Physiol.* 241 (*Heart Circ. Physiol.* 10), H829–H837. (B)

Burton, K. S., and Johnson, P. C. (1972). Reactive hyperemia in individual capillaries of skeletal muscle. *Am. J. Physiol.* 223(3), 517–524. (A)

Chambers, R., and Zweifach, B. W. (1944). Topography and function of the mesenteric capillary circulation. *Am. J. Anat.* 75, 173–205. (A)

Chand, N., and Altura, B. M. (1981a). Acetylcholine and bradykinin relax intrapulmonary arteries by acting on endothelial cells: Role in lung vascular diseases. *Science* 213, 1376–1379. (C)

Chand, N., and Altura, B. M. (1981b). Endothelial cells and relaxation of vascular smooth muscle cells: Possible relevance to lipoxygenases and their significance in vascular diseases. *Microcirculation* 1, 297–317. (C)

Duling, B. R. (1973). The preparation and use of the hamster cheek pouch for studies of the microcirculation. *Microvasc. Res.* 5, 423–429. (A)

Duling, B. R., and Berne, R. M. (1970). Longitudinal gradients in periarteriolar oxygen tension. *Circ. Res.* 27, 669–678. (B)

Duling, B. R., and Klitzman, B. (1980). Local control of microvascular function: Role in tissue oxygen supply. *Annu. Rev. Physiol.* 42, 373–382. (B)

Frasher, W. G., and Wayland, H. (1972). A repeating modular organization of the microcirculation of cat mesentery. *Microvasc. Res.* 4, 62–76. (A)

Furchgott, R. F. (1981). The requirement for endothelial cells in the relaxation of arteries by acetylcholine and some other vasodilators. *Trends Pharmacol. Sci.* 2, 173–176. (C)

Gore, R. W. (1982). Fluid exchange across single capillaries in rat intestinal muscle. *Am. J. Physiol.* 242 (*Heart Circ. Physiol.* 11), H268–H287. (B)

Gross, J. F., and Popel, A. (1980). "Mathematics of Microcirculation Phenomena." Raven, New York. (B)

Hutchins, P. M., Goldstone, J., and Wells, R. (1973). Effects of hemorrhagic shock on the microvasculature of skeletal muscle. *Microvasc. Res.* 5, 131–141. (A)

Jarhult, J., and Mellander, S. (1974). Autoregulation of capillary hydrostatic pressure in skeletal muscle during regional arterial hypo- and hypertension. *Acta Physiol. Scand.* 91, 32–41. (B)

Johnson, P. C., ed. (1978). "Peripheral Circulation." Wiley, New York. (B)

Johnson, P. C. (1980). The myogenic response. *In* "Handbook of Physiology" (D. F. Bohr, A. P. Somlyo, and H. V. Sparks, Jr., eds.), Sect. 2, Vol. 2, pp. 409–442, Chapter 15. Am. Physiol. Soc., Bethesda, Maryland. (B)

Johnson, P. C., and Hanson, K. M. (1966). Capillary filtration in the small intestine of the dog. *Circ. Res.* 19, 766–773. (B)

Kaley, G., and Altura, B. M., eds. (1977). "Microcirculation," Vol. I. University Park Press, Baltimore, Maryland. (B,C)

Kaley, G., and Altura, B. M., eds. (1978). "Microcirculation," Vol. II. University Park Press, Baltimore, Maryland. (B,C)

Kaley, G., and Altura, B. M., eds. (1980). "Microcirculation," Vol. III. University Park Press, Baltimore, Maryland. (B,C)

Lipowsky, H. H., Kovalcheck, S., and Zweifach, B. W. (1978). Distribution of blood rheological parameters in the microvasculature of cat mesentery. *Circ. Res.* 43, 738–749. (B)

Renkin, E. M. (1977). Multiple pathways of capillary permeability. *Circ. Res.* 41, 735–743. (B)

Rhodin, J. A. G. (1967). The ultrastructure of mammalian arterioles and precapillary sphincters. *J. Ultrastruct. Res.* **18**, 181–223. (A)

Rhodin, J. A. G. (1968). Ultrastructure of mammalian venous capillaries, venules and small collecting veins. *J. Ultrastruct. Res.* **25**, 452–500. (A)

Rhodin, J. A. G. (1973). Fixation of microvasculature for electron microscopy. *Microvasc. Res.* **5**, 285–291. (A)

Rhodin, J. A. G. (1974). "Histology." Oxford Univ. Press, London and New York. (A)

Rhodin, J. A. G., and Lim Sue, S. (1979). Combined intravital microscopy and electron microscopy of the blind beginnings of the mesenteric lymphatic capillaries of the rat mesentery. *Acta Physiol. Scand., Suppl.* **463**, 51–58. (A)

Tilton, R. G., Kilo, C., Williamson, J. R., Jr., and Murch, D. W. (1979). Differences in pericyte contractile function in rat cardiac and skeletal muscle microvasculatures. *Microvasc. Res.* **18**, 336–352. (A)

Weideman, M. P. (1980). *In vivo* methods for study of microcirculation. Direct and indirect methods. *In* "Microcirculation" (G. Kaley and B. M. Altura, eds.), Vol. III, pp. 277–302. University Park Press, Baltimore, Maryland. (A)

Weideman, M. P., Tuma, R. F., and Mayrovitz, H. N. (1981). "An Introduction to Microcirculation." Academic Press, New York. (B)

Weiss, G. B. (1981). "New Perspectives on Calcium Antagonists." Am. Physiol. Soc., Washington, D.C. (C)

Chapter 4
Lymphatics

A. EMBRYOLOGY OF LYMPHATIC SYSTEM

By C. E. CORLISS

1. GROWTH AND DEVELOPMENT

a. Origin: Primordia for the lymphatic system are five lymphatic sacs that develop in the human embryo during the first 4 months. The paired jugular lymph sacs first appear near the junctions of the anterior cardinal and posterior cardinal veins at about 6 weeks. Visible by 8 weeks are the midline cisterna chyli dorsal to the aorta and the retroperitoneal sac just ventral to it. The paired posterior sacs appear next in the angle between the femoral and iliac veins (Fig. 4.1A and B). The subclavian sac, (Fig. 4.1A) is not a discrete structure in the human embryo (Warwick and Williams, 1980).

b. Development of lymph sacs and lymph vessels: Lymphatics are lined with endothelium (von Recklinghausen, 1862a), but the embryonic origin of the inner layer has long been controversial, with two schools of thought prevailing.

In a series of India ink injection studies, Sabin (1916) produced strong evidence that primary lymph sacs bud off existing veins and that lymph vessels then sprout from these sacs and grow centrifugally to establish the lymphatic system (the "centrifugal" theory).

An opposing view is based on studies of serial sections and reconstructions (Kampmeier, 1960, 1969). According to this theory, primary lymph sacs are first formed among the numerous venous tributaries opening into the anterior cardinal and posterior cardinal veins near their junctions (d. ven. lymph., Fig. 4.2A). Formed at 1 month, the tributaries soon become the venolymphatic plexus (Kampmeier, 1969), coalescing into temporary precursors of the jugular lymph sacs [Fig. 4.2B (d. ven. plexus) and 4.2C]. With continued widening and merging, the plexus gradually loses continuity with the parent vein (Fig. 4.2C). Meanwhile, regional fluid-filled mesenchymal spaces begin to fuse and envelope the now isolated, degenerating venous plexus (Fig. 4.2C, shaded areas). Incorporating more and more of the peripheral mesenchymal spaces, by 8 weeks, lymph sacs materialize as discrete structures (Kampmeier, 1960). Under increasing fluid pressure, the mesenchymal cells enclosing the spaces become the typical endothelial cells of all lymphatic channels. The sacs, with their

Blood Vessels and Lymphatics in Organ Systems
Copyright © 1984 by Academic Press, Inc.

All rights of reproduction in any form reserved.
ISBN 0-12-042520-3

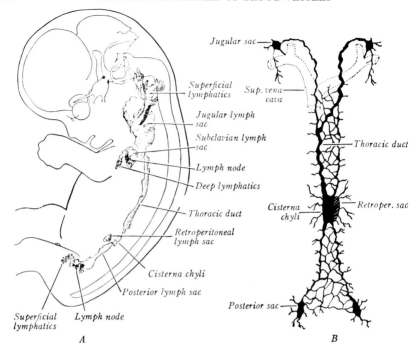

Fig. 4.1. Development of the primary lymphatics in man. (A) Profile reconstruction of the primitive lymphatic system in a human embryo of 9 weeks (after Sabin, 1916). ×3. (B) Diagram, in ventral view, of the definitive thoracic duct emerging from a lymphatic plexus. (From Arey, 1974; reproduced with permission from W. B. Saunders Company.)

endothelial walls, completely replace the venolymphatics and now reestablish the well-known connection with the jugular veins (Kampmeier, 1969). Kampmeier (1969) gives convincing arguments for this concept (Fig. 4.2). Because centripetal extensions grow back from early sacs and vessels to return lymph to the heart via veins (instead of extending out from the veins), this concept is called the "centripetal" theory (Gray and Skandalakis, 1972).

Both views have been challenged. Sabin's injections have been criticized because of the use of excessive pressures that could have forced India ink into nonlymphatic areas (Kampmeier, 1969). Also, it is possible that stained sections may not have revealed all of the lymphatics present or allowed accurate differentiation of small blood vessels from small lymphatics (Clark, 1911). The careful restudy done by Kampmeier (1969) of Sabin's 7-week human embryo (carefully injected with its heart still beating) supported the "centripetal" theory. Because all blood vessels in the specimen were filled with ink, with none found in lymphatics, any confluence of the two systems seemed impossible. Some authors (Gray and Skandalakis, 1972) feel that resolution of this controversy awaits new approaches and new methods; others (e.g., Warwick and Williams, 1980) state that the "balance of evidence" favors the view that the origin of all but the earliest lymphatics is independent of the venous system.

c. Development of the thoracic duct: The upper part of the thoracic duct begins in the cervical region at the sixth week as a series of mesenchymal spaces. These grow, envelop an existing venous plexus, and eventually merge into vessels draining into the newly formed jugular sacs. Thoracic and abdominal portions of the duct are similarly formed, originally being bilateral (Fig.

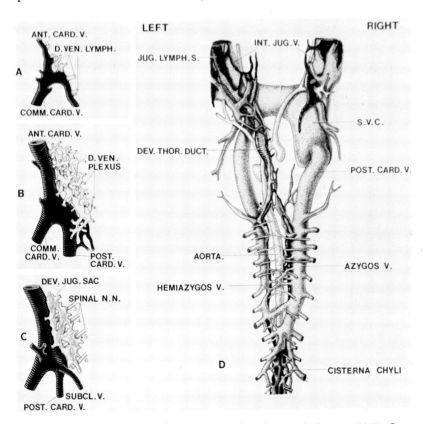

FIG. 4.2. Developmental stages of the left jugular lymph sac in the human. (A) Confluence of the cardinal veins (black), with dorsal venous tributaries (white, stippled), in a 4-week embryo [adapted after a reconstruction by Ingalls (1908)]. (B) Semischematic reconstruction of the dorsal venous plexus, the precursor of the jugular lymph sac. (C) Same of the developing sac replacing the degenerating plexus (shaded areas within the sac); spinal nerves (brachial plexus) drawn in to show topographical levels of the forming jugular lymph sac. (D) Reconstruction, dorsal view, of vessels in territory of the developing thoracic duct of an 8-week human embryo. Note the degenerating venous channel (shaded) within the lumen of the thoracic duct and its surviving connection, at *, with the left azygos (hemiazygos) vein. (Adapted from Kampmeier, 1969, with permission from Charles C. Thomas, Publisher, Springfield, Illinois.)

4.2D). The right one becomes the main channel of the thoracic duct, passing obliquely to join the left member at the level of T4–T6 (Fig. 4.2D). By the end of the ninth week, this main endothelial channel unites the sacs previously described, and, with the jugular sacs already opening into the jugular veins at the seventh week, the definitive pattern for human lymphatic drainage is now established. Valves develop early in lymphatic vessels, appearing by the end of the second month.

d. Development of the lymph nodes: Once formed, the lymph sacs and vessels provide primordia for their ubiquitous nodes. Lymph cells accumulate along the lymph vessels and sacs, and, as connective tissue invades the sacs, the cells are caught in the developing mesh. Their rapid proliferation forms the typical postnatal cortical nodules and medullary cords. All the sacs become nodes except for the upper part of the cisterna chyli, which usually remains a permanent dilatation.

B. ANATOMY OF LYMPHATIC SYSTEM*

By C. C. C. O'Morchoe

1. General Introduction

a. Organization: Lymphatics are thin-walled vessels that converge upon the neck where lymph gains entrance to the blood. The channels are classified according to their size and their position, but differences in nomenclature unfortunately exist. The smallest, which form the peripheral component, are commonly referred to as capillaries, although the terms "initial" and "terminal" lymphatics have also been advocated (Casley-Smith, 1977). The second component of the system is formed by collecting vessels that carry lymph from the capillary networks to the regional nodes. These are properly referred to as prenodal collecting lymphatics. Postnodal collecting lymphatics comprise the next component; they carry postnodal lymph directly to larger vessels of the system or indirectly between successive sets of nodes. Larger vessels, called trunks, which form the next component, are but a few named vessels that leave the final major groups of nodes and drain into the ducts which form the fifth

*For discussion of lymphatics in specific organs, see sections devoted to this subject in the chapters comprising Part Two of this volume.

Blood Vessels and Lymphatics in Organ Systems   All rights of reproduction in any form reserved.
Copyright © 1984 by Academic Press, Inc.   ISBN 0-12-042520-3

component. The ducts are two in number. One, the thoracic duct, ascends from the upper abdomen to the lower neck and ends where the left subclavian vein joins the internal jugular vein. It carries the major portion of the body lymph. The other, the right lymphatic duct, is short and ends like the thoracic duct, but on the right side of the neck. It carries lymph from the upper right segment of the body. One feature of the lymphatic system is the inconstancy of its arrangement, for, great variations exist from individual to individual. Although not considered as definitive, the foregoing description portrays the general format of the system.

b. Distribution: Lymphatics exist in most regions of the body, their presence being dependent upon local lymph formation. Since this process originates in interstitial fluid that is, in turn, derived from plasma transudate, lymphatics are only found in tissues that possess blood vessels. Thus, cartilage, as well as the optic cornea and lens, do not contain lymph vessels—although when blood capillaries grow into a damaged cornea, lymphatics will follow suit. The central nervous system is also without lymphatics, albeit for different reasons. In it, blood vessels are believed to restrict the escape of plasma proteins from their lumen and thereby eliminate the need for lymphatic drainage. Although this is a well-accepted view, it is possible that lymph from the central nervous system may yet exist. One controversial hypothesis is that prelymphatic pathways lie within the sheaths of arteries, running from the brain and spinal cord toward the lymph vessels that drain the head and neck. A second theory is that cerebrospinal fluid is drained, in part, by way of lymph. Support for the latter concept comes from tracer studies, for substances injected into the subarachnoid space have been recovered in deep cervical lymph (Bradbury and Cole, 1980). Whatever the truth of this controversy, there is no doubt that the brain and spinal cord do not contain discrete lymphatic vessels. The bone marrow is another important tissue which does not possess lymphatics, possibly because the high permeability of the marrow sinusoids eliminates the need for lymph formation.

2. LYMPHATIC CAPILLARIES

a. Organization: Lymphatic capillaries form networks within the interstitial spaces that they drain. The distal ones are closed at their peripheral ends, analogous to the fingers of a glove. More proximally they communicate and join to form early collecting vessels. In certain parenchymatous organs such as the liver, lymph vessels are confined to connective tissue septa; thus fluid predestined to be lymph may flow long distances to reach the earliest lymphatics. Accordingly, the concept of interstitial prelymphatic pathways has arisen (Casley-Smith, 1977). Such preferential channels, though evident in edema, are not so obvious in the normal interstitium and so are in dispute. Whether the space of Disse serves as a prelymphatic pathway for hepatic lymph is also controversial.

The lymphatic networks in most organs, including the liver, skin, and kidney, comprise a superficial and a deep capillary system. Such an arrangement appears to be of morphologic advantage rather than of functional significance.

  b. Structure: Lymphatic capillaries are difficult to recognize by light microscopy. Their walls are thin and so they commonly collapse and either disappear or simulate artifactual clefts in tissue sections. Alternatively, when the lumen does remain patent, the walls exhibit a single endothelial layer which mimics that of blood capillaries or veins. Identification may thus depend upon the use of serial sections, tracer substances, or ultrastructural evidence, of which the last is most reliable.

The ultrastructure of the endothelium has been studied in several investigations, largely because of interest in translymphatic pathways (see Section C-2c, below), (Leak, 1970; Casley-Smith, 1977; Albertine and O'Morchoe, 1980; Yang *et al.*, 1981). These transport pathways, though highly permeable even to macromolecules, are in considerable dispute. At issue is the relative importance of intercellular channels (Casley-Smith, 1977), in contrast to endocytotic vesicles or other possible routes across the cells (Albertine and O'Morchoe, 1980; Yang *et al.*, 1981). The internal structure of the cells themselves is not in question. The cells form a continuous nonfenestrated sheet of squamous cells and thus present three surfaces for study—luminal, abluminal, and intercellular. The luminal side is relatively simple. Occasional cytoplasmic processes extend for varying distances into the lumen, their function being quite unknown. Also present are caveolae, invaginations of the plasma membrane that form a part of the vesicular system.

Similarly the abluminal surface has caveolae, and, as in the case of the luminal surface, they vary in depth from mild depressions to almost complete vesicles that merely touch the plasma membrane. It is believed that they represent the formation or dissolution of endocytotic vesicles on the surface. Another feature that distinguishes the abluminal surface is that, unlike blood vascular endothelium, it lacks a continuous basal lamina, which may be completely missing or show as discontinuous patches. This is the one most useful finding by which lymphatic and blood capillaries may be differentiated from each other; another is the presence of anchoring filaments (Leak, 1970) which are attached to the cell surface and extend into the interstitium where they become anchored to the surrounding tissues. These filaments are vital to lymphatic function (see Section C-2i, below). Without them, the interstitial fluid, especially in the presence of edema, would compress the lymphatic vessels and occlude their lumen. With them, the anchored ends are separated and so exert a tension on the endothelial cells, thereby ensuring vessel patency. These filaments are often crowded at the margins of the cells. Therefore, when stretched, they separate the cells and thus expand the intercellular space. This may be important in normal lymph formation, as well as in edema.

The third face of the endothelial cell, the intercellular surface, borders the intercellular space. Adjacent cells may either touch by simple abutment of their cytoplasm, overlap each other to a varying degree, or interdigitate in complex ways (Albertine and O'Morchoe, 1980). Whatever the manner of their contact, the intervening space, with some exceptions, is about 20 nm wide and may have junctional complexes of the occludens or adherens type; desmosomes are rarely if ever seen. Sometimes the intercellular channel becomes wider in the center but narrows to normal width at either end. Only rarely do adjacent cells lie 30 nm or more apart, the spaces thus formed being called open junctions, for they serve as discontinuities in the vessel wall. Their incidence and significance are in dispute, partly because they vary with the site and status of the tissue. Subdiaphragmatic capillaries, for instance, have many open junctions that widen when the diaphragm is stretched; they thereby serve to drain the peritoneal space. In contrast, renal lymphatics rarely have open junctions (Albertine and O'Morchoe, 1980; Yang et al., 1981). However, whatever their incidence in normal tissues, it is now well known that their frequency increases in edema.

Internally, the fine structure of the lymphatic endothelial cells is similar to that of blood vascular endothelium. The usual organelles are present, including mitochondria, scattered rough endoplasmic reticulum, a small Golgi complex, a pair of centrioles, lysosomes, and endocytotic vesicles. The vesicular system has attracted the greatest interest because of its possible role in translymphatic transport (see Section C–1g, below) (Albertine and O'Morchoe, 1980; Yang et al., 1981). Occasional vesicles are of the coated type, but most are simple and smaller, with a diameter of about 75 nm. According to recent work, the majority of the vesicles that seem to lie within the cell retain a connection with the surface, but the importance of this finding for the transport process has yet to be elucidated.

3. PRENODAL COLLECTING LYMPHATICS

a. Organization: Collecting lymphatics begin where the capillary networks end, but the point of their transition is not clear. In some it may be where the wall begins to thicken, in others, where the valves appear. The functional change appears to be more gradual since lymph may be formed across both types of vessel. Collecting lymphatics divide and anastomose along their course so that lymph, formed at a localized site, becomes dispersed throughout the regional nodes. The immune reaction to a local challenge may thereby be intensified.

b. Structure: The structure of collecting lymphatics depends upon their site and on the region that they drain. Species differences also exist. Lymph vessels with thin walls are found in the dog and horse, whereas thicker-walled vessels occur in the goat and man. Regardless of these variations, collecting vessels show gradual increases in the thickness of their walls. First is the appearance of connective tissue beneath the endothelium; next, occasional smooth muscle cells are noted. Coincidentally the basal lamina becomes more evident, though rarely does it become complete. Closer to the lymph node, the wall grows thicker and the lumen becomes larger. The thickest walls reveal a trilaminar structure resembling that of blood vessels. The intima consists of endothelium that overlies a narrow layer of loose connective tissue. Ultrastructurally, the endothelium is similar to that which forms the walls of the lymphatic capillaries. The media contains smooth muscle cells disposed in one or two concentric layers, enabling the lymphatic vessel to contract. The adventitia is formed of loose connective tissue, consisting of collagen and elastin, as well as fibroblasts. Within the adventitia are small blood vessels and nerves that serve the needs of the lymphatic wall.

4. Lymph Nodes

a. Organization: Lymph nodes are considered here only as they form a part of the lymphatic circulation. All lymph normally traverses at least one set of nodes and some pass through several sets before entering the blood. The principal groups of nodes lie close to blood vessels, from which they commonly receive their names. [For details of the disposition of nodes and of their lymphatic drainage in man, see Rouviere (1938).]

The upper limb is drained by important groups of nodes that lie in the axilla around the parts or branches of the axillary artery. Of special clinical significance is that a major portion of the breast drains to this group of nodes. In a comparable manner, lymph from the lower limb passes to nodes beside the blood vessels of the groin. The major group, the external iliac nodes, receives lymph either directly from the lower limb or through an inguinal group adjacent to the great saphenous vein. Of clinical importance is that this group drains lymph from superficial elements of the perineum.

The lymphatic drainage of the upper part of the body, excluding the upper limbs, is more complex. Outlying groups of nodes encircle the junction of the head and neck, related to branches of the external carotid artery. They drain superficial regions and generally are palpable when enlarged. The deeper outlying nodes are primarily connected with the viscera—the pharynx, larynx, trachea, and esophagus. Lymph from both superficial and deep outlying groups drain to a central set, the deep cervical nodes, adjacent to the carotid sheath beneath the sternomastoid muscle.

The trunk contains outlying groups of nodes that drain its walls and viscera, as well as central groups related to major blood vessels. Lymph from the pelvic viscera flows mainly to nodes that lie along the internal iliac artery, although some vessels go directly to the external iliac group. As already mentioned, the superficial aspects of the perineal viscera drain to inguinal nodes. Lymphatic vessels from the gonads accompany gonadal blood vessels and run to nodes beside the abdominal aorta. Within the abdomen, outlying nodes lie close to viscera; hepatic, gastric, and mesenteric, as well as other groups, can be included in this category. Their efferent lymphatics pass on to central nodes that lie close to the aorta. Of these, the preaortic group drains viscera supplied by unpaired branches of the aorta. On either side, the paraaortic nodes receive the outflow from a common iliac group, to which, in turn, drain both internal and external iliac nodes. The paraaortic nodes also drain lymph derived from the abdominal wall and viscera supplied by paired branches of the abdominal aorta (kidneys, gonads, and suprarenal glands).

The thorax has outlying groups of nodes that drain the thoracic cage and diaphragm. Of special importance are the internal thoracic (parasternal) group, behind the margins of the sternum, for they receive lymphatics from the breast. The largest and most important groups of nodes that drain the thoracic viscera lie close to the tracheal bifurcation. Collectively they are called the tracheobronchial nodes, and they form the receiving group for lymph from the heart and lungs.

b. Structure: The pathway through the node begins at the point where incoming, afferent, lymphatics end. Each terminal branch divides upon the surface before penetrating the capsule, to enter the subcapsular sinus that surrounds the node. From there the lymph flows centrally through cortical and then medullary sinuses to reach the efferent vessels at the hilum of the node. These intranodal sinuses are lined by simple endothelium and lie between aggregates of lymphocytes that form the follicles and medullary cords. Thus, under normal states of flow, the lymph does not traverse the follicles. Within the sinuses are reticular fibers, as well as macrophages. Together they screen the lymph, showing impressive powers of filtration and phagocytic action.

5. POSTNODAL COLLECTING LYMPHATICS

a. Organization: Efferent lymphatics appear at the hilum of the node and there combine to form a few postnodal vessels that branch and reunite along their course. Their destination is either the next set in the chain of nodes or the larger lymphatic vessels of the system. The pathways that they take, with rare exceptions, follow the neighboring blood vessels.

b. Structure: The structure of postnodal vessels, like that of their prenodal counterparts, is highly variable. The lumina progressively enlarge and the walls increase in thickness. All possess an endothelial lining and variable amounts of connective tissue and smooth muscle. In the vessels with thicker walls, a muscular media can usually be distinguished from the fibrous adventitia, but in thinner vessels, such a distinction may not be realistic. Like all collecting lymphatics, they are well supplied with valves which have a tricuspid form; each cusp is pocket-shaped and is flanked by a dilatation of the vessel wall, the sinus. Evidence that the valves in pulmonary lymphatics are shaped like cones has yet to be confirmed.

6. LYMPHATIC TRUNKS

a. Organization: The term, lymphatic trunk, is often loosely used, being properly confined to those named vessels that drain the major regions of the body and form the larger tributaries of the ducts. The abdomen has three, one of which is unpaired, the intestinal trunk, transporting lymph of visceral origin from preaortic nodes to the dilated lower end (cisterna chyli) of the thoracic duct. The other two, a pair of lumbar trunks with similar destination, provide the outflow from the paraaortic nodes, thereby conveying lymph from the lower limbs, the pelvis, and the posterior abdominal viscera and walls. The remaining trunks are paired, and they converge upon the confluence of veins in the lower region of the neck. A subclavian trunk is formed by efferents of the axillary nodes. A jugular trunk transports lymph from deep cervical nodes. Vessels ascending from the tracheobronchial nodes unite to form three paired bronchomediastinal trunks, which may enter one of the two lymphatic ducts or open into the bloodstream independently.

b. Structure: The walls of lymphatic trunks are also thin, though thicker than their tributaries. The trilaminar structure is usually evident by light microscopy, and valves are plentiful. Nerves and blood vessels are found within their walls, primarily in the adventitial coat.

7. LYMPHATIC DUCTS

a. Organization: The right lymphatic duct stems from the confluence of jugular, subclavian, and bronchomediastinal trunks and rarely extends beyond a centimeter at most. It lies in front of the scalenus anterior muscle, close to the medial boundary of this structure, where it gains entrance to the venous system. However, often the duct is missing and one or all of the trunks end independently by entering the junction of the right subclavian and internal jugular veins.

The thoracic duct is the only other large lymphatic duct. It carries all of the circulating lymph except that from the right upper regions of the body. It starts in the upper abdomen at a small sac, the cisterna chyli. The latter receives the lumbar and intestinal trunks and is highly variable in form. Sometimes it cannot be discerned and is replaced by channels that unite the trunks with the beginning of the thoracic duct. When present, the cisterna chyli lies anterior to the upper two lumbar vertebrae, wedged in between the aorta and the right crus of the diaphragm. The thoracic duct ascends from the cysterna chyli to enter the lower region of the chest by passing through the aortic opening of the diaphragm. Thereafter it lies between the aorta to its left and the azygos vein to its right up to the level of the fifth thoracic vertebra. Then it crosses to the left side of the body, where it continues to ascend along the left side of the esophagus. On entering the neck, it arches to the left and finally descends to reach the internal jugular vein and left subclavian vein. Just before entering this venous junction, it may receive the three lymphatic trunks from the thorax, upper limb, and head and neck. Alternatively, these trunks may open into the venous system independently. In its thoracic course, the thoracic duct receives tributaries from nodes in the posterior mediastinum. The anatomy of the duct, like much of the lymphatic system, is subject to wide variation. It frequently divides and reunites along its course and sometimes fails to cross the midline in the thorax.

b. Structure: The morphology of the thoracic and right lymphatic ducts conforms with other large lymph vessels, although the walls are often thicker. The typical trilaminar structure is obvious. Both ducts are guarded at their central ends by valves that stop blood from entering them. The thoracic duct also has valves throughout its course, thus permitting upward flow of lymph in the erect position.

8. LYMPHATICOVENOUS ANASTOMOSES

According to the description given above, lymph enters the bloodstream only where the subclavian vein meets the jugular vein, although numerous studies suggest that other lymphaticovenous junctions may exist. The extent of such junctions is controversial in the case of the normal subject, and wide differences exist among individuals and species. Beyond dispute, however, is evidence that connections between blood vessels and lymphatics develop when major lymphatic vessels are occluded. Ligation of the thoracic duct, for example, induces rapid alternative routes by which lymph is shunted to the bloodstream.

C. PHYSIOLOGY OF LYMPHATIC SYSTEM*

By L. V. Leak

1. Historical Perspective

In some invertebrates, the cardiovascular system is open, thereby permitting its fluid content (hemolymph) to pass freely from the arterial side into the extracellular spaces, where individual cells are exposed to a medium in which substances can be exchanged. From the extracellular spaces, the fluids slowly diffuse into veins for return to the heart (Barnes, 1964). However, in vertebrates, a closed circulatory system is present, along with a complex pumping organ (the heart) which facilitates the circulation of larger quantities of fluids and cells throughout various organ systems of the body. In higher animals, the blood circulatory system becomes more specialized, developing an elevated pressure system which pumps blood through an extensive network of branching blood capillaries. With the high pressure, blood capillaries are permeable to fluids and other substances which enter the interstitium (Starling, 1896). In order to remove fluids and macromolecules which have leaked into the interstitium, a drainage system is required, and the lymphatics have evolved to remove extracellular fluids, large molecules, and cells from the interstitium. However, unlike the blood vascular system, the lymphatic system is a one-way drainage system, the major vessels of which (thoracic duct and right lymphatic duct) empty into the great veins of the neck (see Section B-7, below).

Although Hippocrates spoke of "White Blood," it was not until 1627 that Asellius first recognized lymphatic vessels. By the next half century, the large lymphatic channels were clearly described by Bartholin (1651) and several hundred years later by Ebert and Belajeff (1866). However, the function of this system was not appreciated until it was recognized that the lymphatics formed a closed system of vessels which directed interstitial fluids derived from the blood capillaries into the veins in the base of the neck (His, 1863). It was Starling (1896) who first clearly documented filtration from the blood capillaries as a major factor in the formation of interstitial fluid and lymph.

Early concepts of the function of the lymphatic system were based on the notion that the lymphatic vessels were connected to arteries and blood capillaries. This belief persisted until von Recklinghausen (1862b) put forth the idea of the intercellular space, which set the stage for considering the phenomenon of fluid exchange between blood capillaries and the connective tissue spaces containing cells and fibrous elements (interstitium). Lymphatic vessels were con-

*Work cited in this section was supported, in part, by the United States Public Health Grants NIAID-10639 and NHLBI-13901 of the National Institutes of Health and Department of Health and Human Services.

134

sidered to act as a system for drainage and transport of fluids from the interstitial spaces back into the systemic blood circulation. Incorporated in the
filtration theory was the concept that blood plasma and lymph had a similar
composition. However, Ludwig and Tomsa (1862) opposed this view, suggesting that the capillaries were the source of lymph. It was not until the early part
of the twentieth century that Starling (1909) put forth the concept of blood
capillary and lymphatic function. Prior to this it was believed that all of the fluid
which escaped from blood capillaries into the connective tissue space was
returned mainly by way of the lymphatic system. However, Starling proposed
that the major pathway for return of water and crystalloid was by way of the
venous limb of the blood capillary, whereas the pathway for colloid was mainly
via the lymphatics. Although the general idea of fluid flow from blood capillary
to interstitium and subsequent drainage by the lymphatics has persisted, the
precise mechanism for the movement of fluids and plasma proteins from the
interstitial spaces still remains controversial. Through the use of ultrastructural
studies, information has been obtained on possible mechanisms by which fluids
and large molecules may gain access to the lymphatic lumen (Casley-Smith and
Florey, 1961; Casley-Smith, 1964; Leak and Burke, 1965, 1966, 1968; Leak,
1970, 1972b). These morphologic studies provide the anatomic framework of
the lymphatic capillaries, which can be correlated with physiologic and biochemical functions (see Section D, below). However, despite recent advances
in ultrastructural, physiologic, and cytochemical techniques, a consistent interpretation of the three-dimensional relationship of blood capillary– interstitial–lymph interface in the formation of lymph has not yet evolved.

2. ULTRASTRUCTURAL BASIS FOR LYMPHATIC CAPILLARY FUNCTION

Since the early 1960s, ultrastructural studies have revealed much information on the structural arrangement of the lymphatic vascular wall; from such
data, concepts have been advanced to explain how fluids and large molecules
enter the lymphatic vessels. Therefore, in outlining the functional framework of
the lymphatic vascular system, it will be necessary to consider in some detail
the ultrastructural aspect of the lymphatic capillary wall.

a. General characteristics: The lymphatic capillary wall consists of a continuous endothelial lining that lacks a complete basal lamina. The intercellular
junctions permit free and rapid passage of fluids, large molecules, and cells
from the adjacent connective tissue. This segment of the lymphatic vascular
system constitutes the initial or primary element which drains the interstitium
of fluids in the maintenance of homeostasis. The vessels vary in width, ranging
from 20 to 60 μm in diameter. They appear as blind end tubes (Leak, 1972b) or
bulbous saccules (Cliff and Nicoll, 1970) (Fig. 4.3).

b. Lymphatic sinusoids and spaces: In the interstitial tissue of the mammalian testis, lymphatic vessels have been observed that are labyrinthine and pleomorphic in shape. Due to their irregular morphology, these vessels have been termed lymphatic sinusoids (Fawcett *et al.*, 1970). They form a continuous system of lymphatic channels which serve as the initial drainage site for

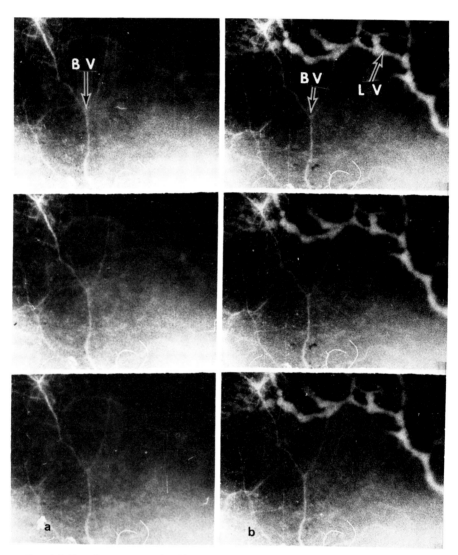

Fɪɢ. 4.3. Cinephotomicrographs of blood (BV) and lymphatic (LV) vessels. (a) An area of tissue prior to an interstitial injection of colloidal carbon and trypan blue. (b) The same tissue area after injection of the mixture which labels lymphatic capillaries. ×60. (From Leak, 1971; reproduced with permission from *Journal of Cellular Biology.*)

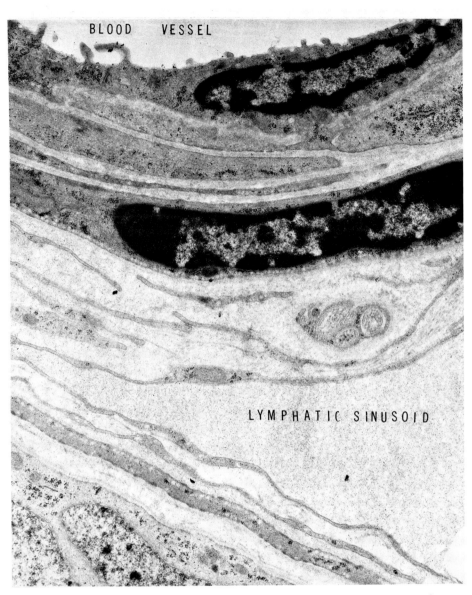

FIG. 4.4. Electron micrograph illustrating a lymphatic sinusoid within the interstitium of guinea pig testis. Tracer particles (colloidal ferritin) injected interstitially fill the connective tissue area and the lumen of lymphatic sinusoid. ×8400. (From Leak, 1981; reproduced with permission from *Anatomic Record.*)

interstitial fluid, as well as playing a role in the removal of hormone secretions by the Leydig cells of the testis (Lindner, 1963) (Fig. 4.4).

In some interstitial compartments of the mammalian testis, large intercommunicating cavities appear to be filled with protein-rich lymph; however, the walls give the impression of being only partially lined by endothelium. This arrangement provides a lumen which is continuous with the extracellular fluid phase of the surrounding tissue space. The boundary of the space is discontinuous in some areas, whereas in others, the endothelium gradually becomes continuous with a conventional type of lymphatic vessel. The term, discontinuous sinusoid, is also used to describe this type of vessel (Fawcett *et al.*, 1973). (For further discussion of the lymphatic drainage of the testis, see Section A-3, Chapter 16.)

c. Lymphatic capillary ultrastructure: In classifying blood vessels, the walls of which allow the passage of fluid, metabolites, and plasma proteins into the connective tissue spaces, morphologists and physiologists described these vessels as capillaries (Benninghoff, 1930; Chambers and Zweifach, 1947; Krogh, 1959). The term, capillary, has also been used to classify the smallest and thinner-walled lymphatic vessels (Hudack and McMaster, 1932). Such terms as initial, primary, and terminal lymphatics are also used to denote the more permeable portion of lymphatic vessels (Zweifach, 1973; Casley-Smith, 1977). However, from a functional standpoint, the segment of the lymphatic drainage system which removes interstitial fluids, proteins, and cells is located at the site of cellular metabolism and in close proximity to the blood capillaries and venules. The strategic location, coupled with the salient features which provide for a rapid and continuous drainage of interstitial fluids, proteins, and cells, makes the term, lymphatic capillary, more appropriate for these permeable and thinner-walled lymphatic vessels (Leak, 1972b, 1980). The application of improved techniques for electron microscopic studies of the lymphatic vessels has provided structural features which may now serve as criteria for the identification of lymphatic capillaries with some degree of certainty at the ultrastructural level. Lymphatic capillaries can thus be recognized by the following features.

1. The endothelial cells lack a continuous basal lamina.

2. Cytoplasmic projections extend from both luminal and abluminal surfaces to give an irregular contour when compared with the accompanying blood capillaries and venules.

3. The lymphatic capillary wall is held in close apposition to the surrounding interstitium by lymphatic anchoring filaments.

4. The terminal margins of adjacent endothelial cells extensively overlap and are loosely adherent over wide areas, with the result that separation of adjacent endothelial cells can easily occur.

5. The lymphatic capillary is irregular in its diameter, with an extremely attenuated endothelial cell cytoplasm except in the nuclear region.

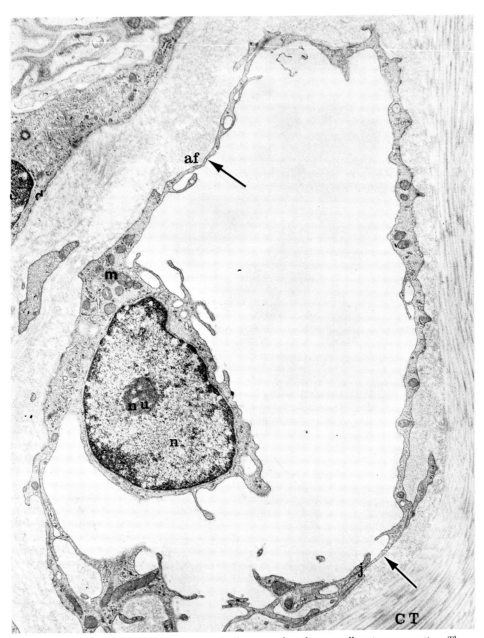

FIG. 4.5. A survey electron micrograph illustrating lymphatic capillary in cross section. The close association of the adjoining connective tissue components (CT) with the lymphatic wall is maintained by numerous anchoring filaments (af) which appear as a meshwork of fine filaments. The endothelium is extremely attenuated at various points (arrows), and the nucleus (n) with its nucleolus (nu) protrudes into the lumen. Several intercellular junctions (j) are observed. Mitochondria (m) appear in the juxtanuclear region, as well as in the thin cytoplasmic rims. ×11,000. (From Leak, 1970; reproduced with permission from Academic Press.)

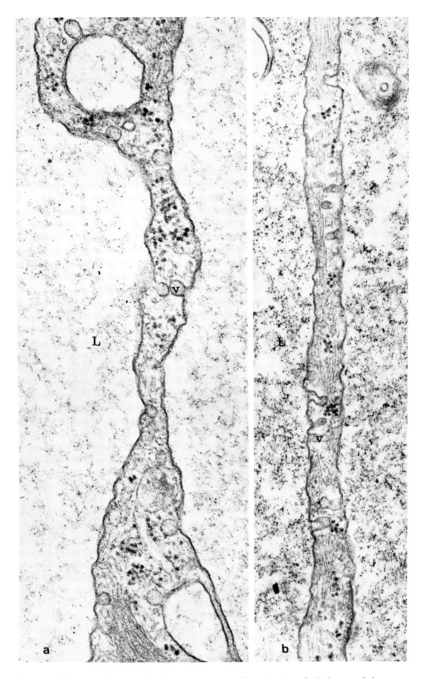

FIG. 4.6. Electron micrographs showing portions of lymphatic endothelium and the occurrence of plasmalemmal vesicles (v) in the thick and thin regions of the endothelial wall. (a) ×76,500. (b) ×68,000.

d. Lymphatic capillary wall: This structure consists of a single layer of endothelial cells which are extremely attenuated in thickness (50 to 100 nm) over large areas of the vessel wall (Fig. 4.5). The nucleus occupies a position in the center of the cell, which is also the thickest portion (up to 3 μm in width) and characteristically bulges into the lumen. The plasmalemma contains nu-

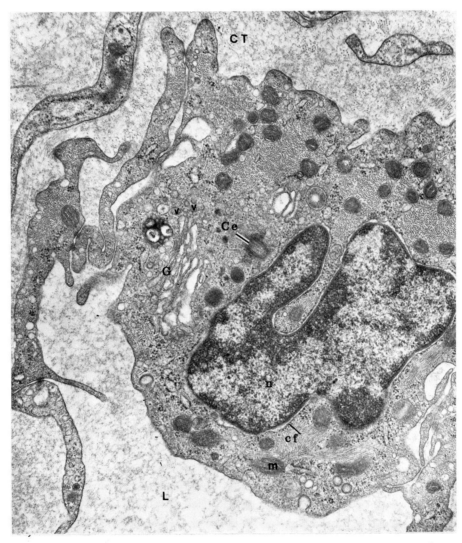

FIG. 4.7. Electron micrograph of perinuclear region of lymphatic endothelium. Figure shows lumen (L). The cytoplasm surrounding the nucleus (n) contains large numbers of cytoplasmic filaments (cf), mitochondria (m), a centriole (Ce), and Golgi complex (G), with numerous associated vesicles (v). ×28,000. (From Leak, 1981; reproduced with permission from *Anatomic Record.*)

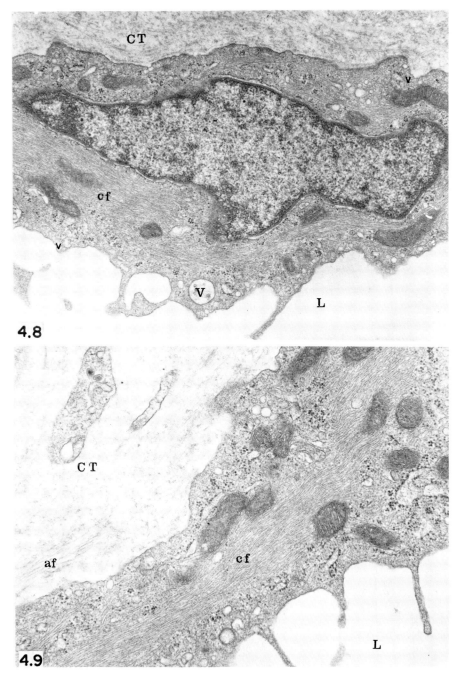

FIG. 4.8. Electron micrograph of a longitudinal section through the nuclear region of the lymphatic endothelium. Numerous cytoplasmic filaments (cf) appear in the perinuclear region, and

merous invaginations (caveoli or pinocytotic vesicles) along both luminal and abluminal surfaces. Many vesicles also appear free in the cytoplasm. Occasionally several vesicles appear connected in series to form transcapillary endothelial channels (Fig. 4.6). The perinuclear region contains a prominent Golgi apparatus. Closely associated with the forming and mature faces of the Golgi apparatus, there are numerous vesicles of varying sizes, the content of which shows varying degrees of electron densities (Fig. 4.7). Many of these represent lysosomes as indicated by the presence of acid phosphatase activity.

The endoplasmic reticulum is of the rough variety and is randomly distributed throughout the thicker regions of the cytoplasm and occasionally in the thin cytoplasmic rims of the endothelium (Figs. 4.6 and 4.7). Ribosomes also occur throughout the cytoplasm as clusters of polyribosomes. The distribution of the endoplasmic reticulum and free ribosomes is reminiscent of that found in the blood capillary endothelium.

The mitochondria are observed throughout the juxtanuclear areas, and small numbers are seen in other regions of the cytoplasm except in the very attenuated segments. They are oval to elongate in shape and display an internal structural arrangement like the mitochondria in blood capillary endothelial cells. The outer membrane exhibits a smooth contour whereas the inner membrane is invaginated into the matrix as cristae (Fig. 4.8).

Microtubules are found in close proximity to centrioles. They are also present throughout the cytoplasm and are generally aligned parallel to the long axis of the cell. They measure 25 nm in diameter and are of an indeterminate length due to their course in and out of the plane of section (Fig. 4.7).

  e. Role of filaments in endothelial cells: A salient feature of the lymphatic capillary endothelial cell is the presence of numerous cytoplasmic filaments measuring about 6 nm in diameter. They are often arranged parallel to the long axis of the cell and many appear in bundles throughout the cytoplasm. Those filaments appearing subjacent to the plasmalemma give the impression of dense plaques suggestive of possible attachment sites similar to those seen in smooth muscle cells (Figs. 4.7 and 4.9). While cytoplasmic filaments are observed in

---

some are oriented parallel to the long axis of the cell, while others are randomly dispersed. A large vesicle (V) appears near the luminal surface; the smaller plasmalemmal vesicles (v) are found along both luminal (L) and connective tissue (CT) fronts. Short segments of basal lamina (bl) are also illustrated. ×18,700. (From Leak, 1970; reproduced with permission from Academic Press.)

  FIG. 4.9. Electron micrograph depicting longitudinal section through the endothelial cell at some distance from the perinuclear region. Cytoplasmic filaments (cf) are arranged into bundles which exclude other cytoplasmic organelles. Vesicles (v) occupy both luminal (L) and connective tissue (CT) fronts of the endothelium. Anchoring filaments (af) are as indicated. ×25,500. (From Leak, 1970; reproduced with permission from *Microvascular Research*.)

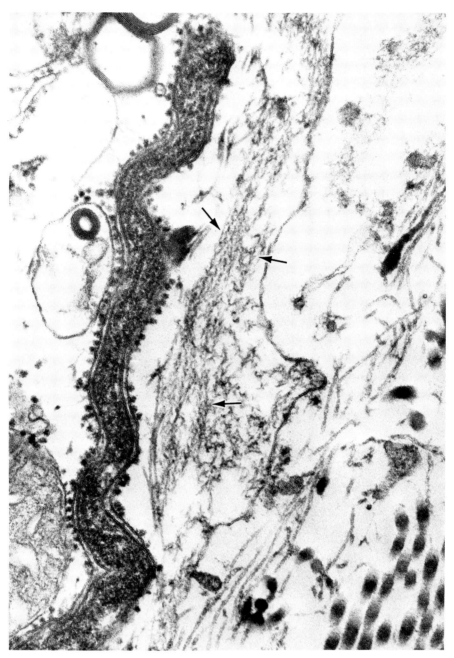

FIG. 4.10. Electron micrograph of part of a diaphragmatic lymphatic endothelial cell showing arrowhead formations (arrows). Cell was treated with glycerin prior to incubation in subfragment 1 of myosin. ×10,200.

the blood capillary endothelium (Majno *et al.*, 1969), they are not as abundant as those found in endothelial cells of the lymphatic capillary.

Both morphologic and biochemical data have been obtained from a wide range of nonmuscle cells regarding actin and myosin (Pollard and Weihing, 1974). The presence of actin filaments in lymphatic endothelial cells of the lung (Lauweryns *et al.*, 1976) and of the diaphragm (Fig. 4.10) is also demonstrated by the formation of arrowhead complexes after the endothelium is exposed to subfragment 1 of myosin. Such a contractile system within the lymphatic capillary endothelium suggests the presence of a mechanism for the active contraction and relaxation of the lymphatic capillary wall when appropriately stimulated. This would also provide a means for regulating the separation of the intracellular clefts between adjacent cells.

f. Mechanisms responsible for rhythmic contraction of lymphatic capillaries: The rhythmic contraction seen in large lymphatic collecting vessels is attributed to smooth muscle cells in the tunica media which are closely associated with nerve axons at various regions along the wall. In studies of the dermal lymphatic capillaries using cinephotographic methods, these vessels have been observed to undergo regular rhythmic contractions (Leak, 1971). The lack of smooth muscle cells in the lymphatic capillary wall and the presence of numerous cytoplasmic filaments that react with heavy meromyosin to form arrowhead complexes suggest that many of the latter represent the intrinsic contractile components responsible for the observed changes.

g. Vesicular transport: The lymphatic capillary endothelium also contains numerous plasmalemmal invaginations that extend into the cytoplasm from both the luminal and connective tissue fronts (Figs. 4.6 and 4.7). Many appear in the deeper regions of the cytoplasm as if in the process of moving across the lymphatic endothelium. They are similar to the micropinocytotic vesicles (75 nm in diameter) that are observed in endothelial cells of blood vessels (Bruns and Palade, 1968).

In a number of laboratories, studies have been made using various electron-dense tracer substances to follow the movement of both water-soluble and inert substances across the lymphatic endothelial cell (French *et al.*, 1960; Casley-Smith, 1964; Leak and Burke, 1966). It is now apparent that large molecules and particles that are removed from both the luminal and connective tissue fronts within endocytic-like vesicles may aggregate within large vacuoles in the lymphatic endothelium. This is indicated by the accumulation of inert particles, such as carbon and thorium dioxide, into large aggregates in vacuoles that can reach several micrometers in diameter. Such vacuoles (Fig. 4.11) may be observed 12 months after interstitial injection of the particles (Leak, 1971). These observations indicate that vesicular transport is not unidirectional in the lymphatic capillary endothelium; instead there is movement of large molecules

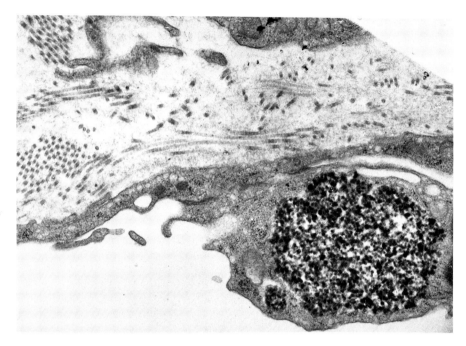

FIG. 4.11. Part of lymphatic capillary endothelium showing the accumulation of colloidal carbon in large vacuole at 6 months after interstitial injection. ×21,250.

and particles within vesicles which proceeds from both the luminal and connective tissue surfaces toward the perinuclear region of the cytoplasm (Leak, 1971, 1972a). It is in the latter region that the endocytosed vesicles fuse with lysosomal vesicles containing hydrolytic enzymes involved in the digestion of the engulfed substances into smaller components within 18–24 hr, presumably for utilization by the cell (Leak, 1968, 1972a). However, in the case of such inert substances as colloidal carbon, thorium, and latex spheres injected interstitially, the lytic activity has no effect. Therefore, these materials accumulate into very large vacuoles which remain in the cell for an indefinite time period (Leak, 1971). The rapid breakdown of such substances as ferritin and peroxidase suggests that the lymphatic capillary endothelium plays an important role in the intracellular digestion of endocytosed substances that may be removed from the luminal, as well as the connective tissue, fronts of the lymphatic capillary.

  h. Intercellular junctions: The lymphatic capillary wall is distinguished by a large number of intercellular junctions in which the adjacent cell margins are extensively overlapped (Fig. 4.12a and b). The distance between apposing cells may reach several micrometers and extend the total length of the intercellular

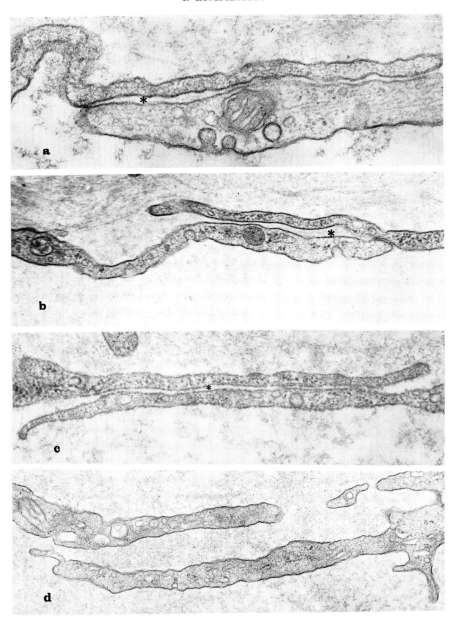

FIG. 4.12. Electron micrographs depicting extensive overlapping of adjacent endothelial cell margins. The width of the intercellular cleft (*) is quite variable, ranging from areas of close apposition to regions where adjacent cells are widely separated to form patent junctions, as in (d). (a) ×76,500. (b) ×21,250. (c) ×33,150. (d) ×23,800. (From Leak, 1971; reproduced with permission from *Journal of Cell Biology*.)

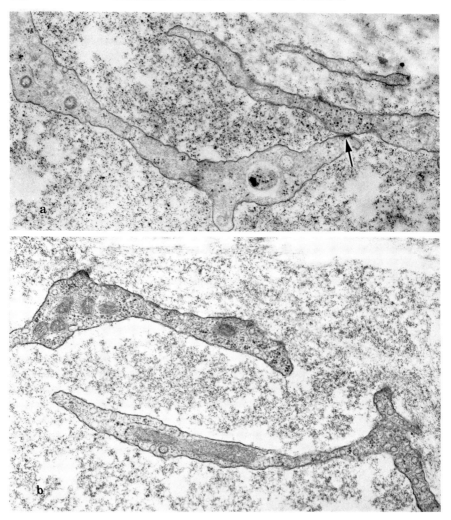

Fig. 4.13. (a) Close apposition between adjacent endothelial cells, maintained at one point (arrow) by a macula adherens. ×23,800. (b) Intercellular cleft is extremely wide throughout its length, forming an uninterrupted passageway from the connective tissue into the lymphatic lumen (patent junction). ×25,500.

cleft (Fig. 4.12c and d). These structures represent patent intercellular junctions which are unique to lymphatic capillaries. It is through such open clefts that large molecules, particulate substances, and cells enter the lymphatic lumen (Fig. 4.13a and b). However, in some of the overlapping junctions, there are foci in which the adjacent plasma membranes are closely approximated (Fig. 4.12a and b). When such sites are viewed at higher magnifications, it becomes immediately apparent that the apposed membranes are held together by *maculae adherentes* or desmosomes, with the distance between apposing cells ranging from 10 to 25 nm in width.

Occasionally there are areas of close apposition where short segments of the intercellular cleft are obliterated, thus preventing the passage of such molecules as lanthanum and peroxidase (Fig. 4.14). These intermittent seals within the intercellular cleft are formed by a fusion between the outer leaflets of adjacent cell membranes and are recognized as quintuple structures which represent *maculae ocludentes* (Fig. 4.15). Such specialized sites serve to maintain cell-to-cell adhesion without causing a complete obliteration along the total length of the intercellular cleft; instead this arrangement provides a spot-weld effect for maintaining a close continuity between cells along the total length of the lymphatic capillary wall. At the same time, many of the adjacent cell margins remain loosely apposed to each other and can be easily separated to form patent channels. Such areas would be readily available to accommodate the rapid passage of excessive amounts of interstitial fluids, particulate components, and cells into the lumina of lymphatic capillaries.

In the lymphatic capillaries of the diaphragm, the cell margins of the submesothelium lymphatic endothelial cells extend to the diaphragmatic surface, to make junctional contact with the surface mesothelial cells (Fig. 4.16). This arrangement between mesothelial and lymphatic endothelial cells forms circular pores that provide open channels between the peritoneal cavity and lymphatic vessels of the diaphragm. The openings represent the stomata of von Recklinghausen (Leak and Rahil, 1978). Similar structures have been described in the visceral pleura of the rat lung by Pinchon *et al.* (1980) and in the intercostal space of the lower thorax of the mouse by Wang (1975). These structures provide uninterrupted channels between the serous cavities and lymphatic lumina, thus allowing fluids and large substances to pass readily into the lymphatics.

i. *Lymphatic anchoring filaments:* The lymphatic capillary lacks a definitive basal lamina around its wall, a structural feature which serves as a major criterion for the identification of this vessel at the electron microscopic level (Leak and Burke, 1968). When noted, the short segments of a basal lamina are presumed to represent the transitional region between the lymphatic capillary and the larger lymphatic collecting vessel which contains a continuous basal lamina. Segments of newly formed regenerated blood capillaries also lack a continuous basal lamina, and they are likewise very permeable to large molecules (Schoefl, 1964).

The early histologists described the lymphatic wall as being closely connected to the surrounding connective tissue, but the significance of such an intimate association with the extracellular components was first proposed by Pullinger and Florey (1935). These workers observed that, in the presence of

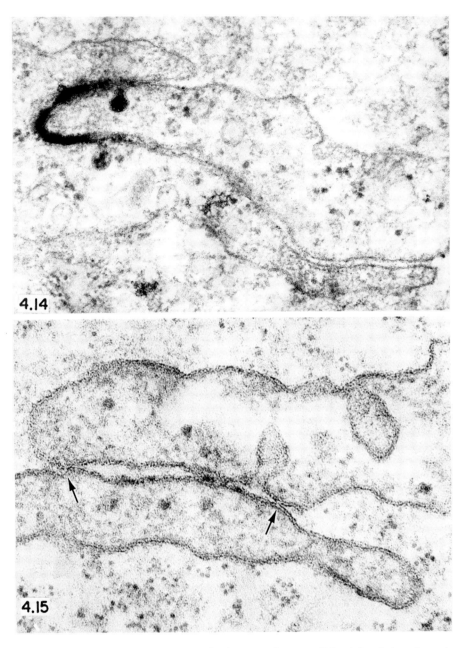

4.14

4.15

Fɪɢ. 4.14. Electron micrograph showing lanthanum in the intercellular cleft and plasmalemmal vesicles. ×70,720. (From Leak, 1971; reproduced with permission from *Journal of Cellular Biology.*)

Fɪɢ. 4.15. Portion of lymphatic capillary showing areas of close apposition between endothelial cells (arrows). ×146,200. (From Leak, 1971; reproduced with permission of *Journal of Cellular Biology.*)

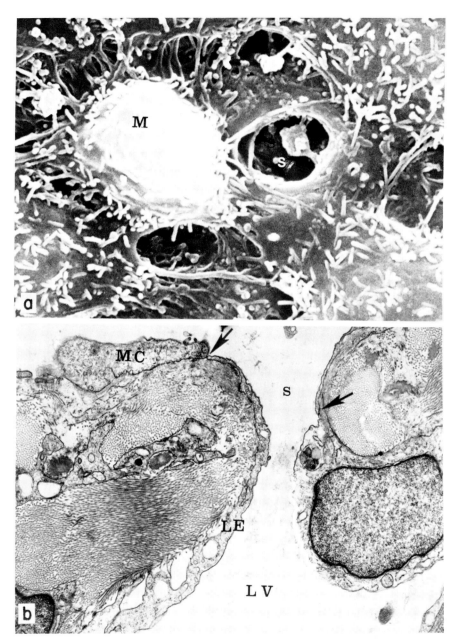

FIG. 4.16. (a) Scanning electron micrograph illustrating the appearance of stoma (S) overlying lymphatic vessel. Mesothelial (M) cell contains numerous microvilli on apical surface. ×11,560. (b) Electron micrograph showing a thin section of diaphragm, with surface mesothelium, stoma (S), and lymphatic vessel (LV). The lymphatic endothelial cells (LE) extend onto the peritoneal surface to form intercellular junctions (arrows) with the surface mesothelial cells (MC). There is close contact between surface mesothelium and underlying lymphatic vessels. ×15,300. (From Leak and Rahil, 1978; reproduced with permission from *American Journal of Anatomy*.)

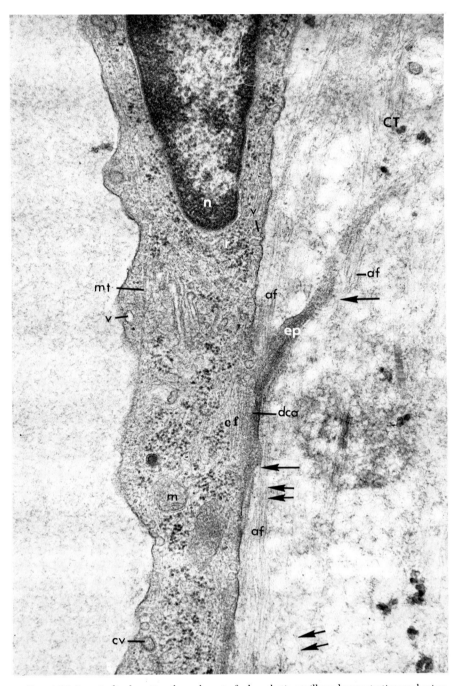

FIG. 4.17. Longitudinal section through part of a lymphatic capillary demonstrating anchoring filaments (af) coursing along lymphatic wall. Note the beading pattern throughout the long axis of anchoring filaments (double arrows). Some of the filaments insert on the outer leaflet of the trilaminar unit membrane of the lymphatic endothelium (single arrows). Endothelial projections

local edema, collagen and reticulum fibers were in close proximity to the lymphatic endothelium and suggested that such fibers provided a means for connecting the lymphatic capillaries to the surrounding tissue area. With the advent of electron microscopy, Casley-Smith and Florey (1961) reported that collagen bundles and small fibers did invest the lymphatic capillary wall, but the intimate association between the lymphatic wall and surrounding interstitial components described earlier (Pullinger and Florey, 1935) was not borne out by the new approach. With development of improved fixation and staining techniques, however, Leak and Burke (1965, 1966, 1968) and Burke and Leak (1968) demonstrated that fine filaments, ranging from 6 to 10 nm in diameter, insert within a densely staining substance on the connective tissue surface of the endothelial cell membrane and extend into the adjoining interstitium between collagen bundles and connective tissue cells (Fig. 4.17). These structures provide a means for holding the lymphatic capillary wall in intimate contact with the surrounding tissue areas. Because of this, the term, lymphatic anchoring filaments, has been used to describe them (Leak and Burke, 1968). The close topographical relationship provided by the anchoring filaments prevents collapse of the lymphatic capillary wall by stabilizing and anchoring it to the adjoining interstitium in which these filamentous structures are firmly embedded.

3. Physiology of Lymph Formation and Flow

Although both blood and lymphatic vessels are lined by a continuous layer of endothelial cells, striking differences have been noted in their anatomic features, as well as their topographical relationship with the connective tissue. For example, the blood capillary is able to maintain its regular size and shape by the structural rigidity of its wall, plus the maintenance of a high positive pressure (Landis and Pappenheimer, 1963; Intaglietta et al., 1970; Weiderhielm and Weston, 1973). On the other hand, the lymphatic capillary wall is less rigid, due to its intimate association with the adjoining connective tissue by way of lymphatic anchoring filaments (see Section C-2i, above). Thus, its lumen reflects the irregular form of the surrounding interstitial space and is influenced by changes in the connective tissue.

(ep) with closely associated filaments (af) extend into the adjoining connective tissue area (CT). Note dense cytoplasmic accumulation (dca) at periphery of endothelium. Pinocytotic vesicles (v) occur along both abluminal and luminal surfaces of endothelium. Numerous free ribosomes (r), in addition to microtubules (mt), cytoplasmic filaments (cf), and mitochondria (m), are found in the cytoplasm. Coated vesicles (cv) are also shown. Part of nucleus (n) is located in lower portion of cell. ×38,250. (From Leak, 1971; reproduced with permission from *Journal of Cellular Biology*.)

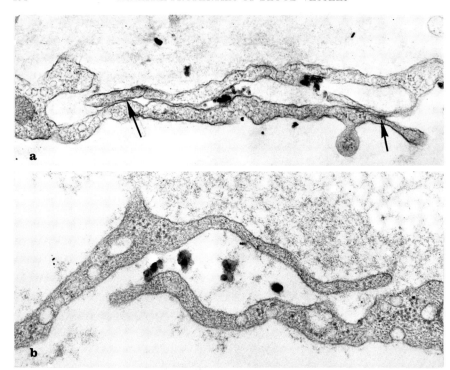

FIG. 4.18. Electron micrographs illustrating passage of colloidal carbon via the intercellular cleft. In (a), several areas of close apposition are found along the intercellular cleft (arrows), whereas in (b), the intercellular cleft forms an uninterrupted passageway. Specimens were injected with colloidal carbon 5 min before fixation. Lumina are at bottom of photographs. (a) ×24,480. (b) ×46,750. (From Leak, 1971; reproduced with permission from *Journal of Cellular Biology.*)

a. Basis for lymphatic capillary permeability: Information has now accumulated which provides a structural basis for lymphatic capillary permeability (French *et al.*, 1960; Casley-Smith and Florey, 1961; Casley-Smith, 1964, 1967; Leak and Burke, 1965, 1966, 1968; Leak, 1970, 1971, 1976). In this regard, the transport of intravenously injected protein tracers at the light microscopic level (Nicolaysen *et al.*, 1975) and of electron-dense tracer substances within the size range of plasma proteins has been monitored across the blood–tissue–lymph interface (Leak, 1972a, 1977; Dobbins and Rollins, 1970). The data suggest that the lymphatic capillaries serve as "active passageways" for the removal of excess fluid, plasma protein, and particulate substances from the interstitial spaces (Fig. 4.18).

b. Role of lymphatics as a transport system: In considering the overall physiology of the lymphatic system, it must be thought of as primarily a transport system, designed to maintain homeostasis of the interstitial environment by removing tissue fluids (lymph) and returning it to the bloodstream. The importance of this drainage system lies in the fact that normal blood capillaries are permeable to protein molecules, a mechanism which may disturb the normal Starling pattern of exchange between the blood capillary wall and the surrounding connective tissue space; hence, if the process were left unchecked, it would lead to edema formation (Mayerson, 1963). However, it is necessary to point out that although freely permeable to water and crystalloids, the blood capillary endothelium is semipermeable to colloid of protein molecular size, and thus only small amounts relative to the total plasma protein cross the blood capillary wall (Landis and Pappenheimer, 1963). Therefore, colloid is retained to a large extent in the blood capillary lumen, creating a colloid osmotic (oncotic) pressure gradient which acts to reabsorb water and crystalloid into venular blood (Starling, 1896). Except for the blood–brain barrier (Feder et al., 1969), proteins of molecular weight of approximately 40,000 or less (Karnovsky, 1967; Venkatachalom and Fahimi, 1969; Simionescu et al., 1975) readily cross the blood capillary wall into the interstitium. Under normal physiologic conditions, excess tissue fluids and plasma proteins are continuously being removed from the intercellular space by the lymphatic capillaries.

4. Mechanisms of Lymph Formation

With the exception of its protein content, normal interstitial fluid is similar in composition to plasma (Yoffey and Courtice, 1970; Vakiti et al., 1970). In addition, a number of studies have shown that both blood plasma and lymph contain proteins which belong to a single system of extracellular fluid proteins (Mayerson et al., 1960; Morris and Sass, 1966).

The term, lymph, is used to describe the fluid contained within lymphatic vessels. Although much controversy has existed over its source, there is no question that its formation begins within the lymphatic capillary as fluid, plasma protein, and cells removed from the connective tissue space (Figs. 4.13 and 4.19), the fluid being similar to the free fluid of the tissues (Gibson and Garr, 1970; Yoffey and Courtice, 1970; Taylor and Gibson, 1975).

a. Early work on lymphatic capillary permeability: If the lymphatic system is to serve as a unidirectional drainage system for interstitial fluid homeostasis, there must be preferential pathways within the wall of the lymphatic capillary to provide for a free and rapid transport from the interstitium into the vessel. In the earlier studies of lymphatic capillary permeability, vital dyes and colloidal carbon were shown to be removed from the interstitium with surprising rapidity (Hudack and McMaster, 1932). Likewise, blood proteins, such as

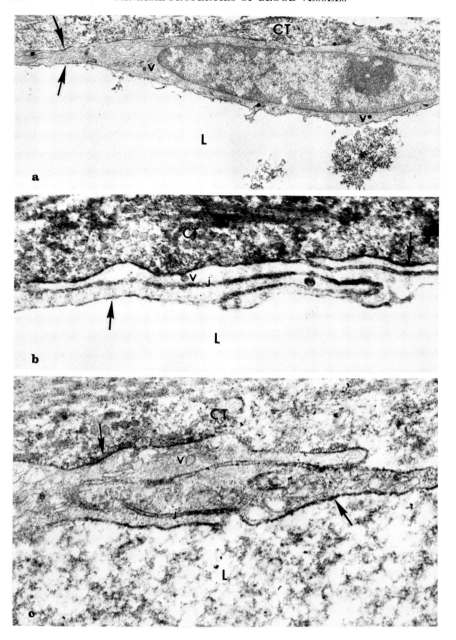

FIG. 4.19. Electron micrographs demonstrating movement of peroxidase across the blood–tissue–lymph interface. Figures show connective tissue (CT), vesicles (V), lumen (L), intercellular junction (j). Arrows point to peroxidase staining of endothelium. (a) 1 min, (b) 2 min, and (c) 5 min after intravenous injection of peroxidase. (a) ×12,155. (b) ×39,440. (c) ×35,360.

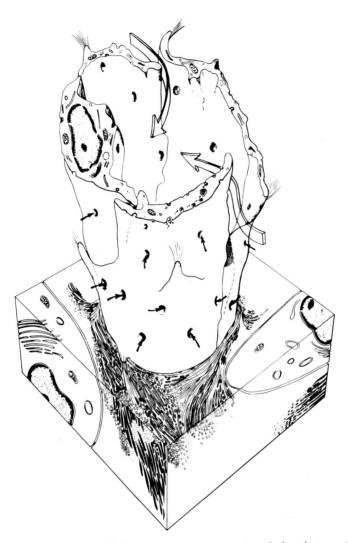

FIG. 4.20. Three-dimensional diagram representing a portion of a lymphatic capillary, reconstructed from collated electron micrographs. The major passageway for transport of fluids and large molecules from the interstitium into the lymphatic lumen is by way of the intercellular cleft (long white arrow). The uptake of large molecules from both the connective tissue and luminal fronts may occur within vesicles (small arrows) which move toward the central cytoplasm; they merge with autophagic vacuoles in which intercellular digestion occurs for subsequent utilization or discharge. In the case of inert particles, such as carbon and colloidal thorium, these components are not digested by the cell and remain aggregated into large vacuoles as indicated in Fig. 4.11. (From Leak, 1971; reproduced with permission from *Journal of Cellular Biology*.)

albumin and globulin, when coupled with radioactive iodine and injected intravenously, were demonstrated to cross the blood capillary wall into the connective tissue and leave by way of the lymphatic system (Mayerson *et al.*, 1962;
Nicolaysen *et al.*, 1975). While these investigations established that the general
flow pattern of fluids and proteins was from blood capillary lumen into interstitium, with subsequent removal by the lymphatic capillaries, the anatomic
basis for lymphatic capillary permeability was still unclear because the resolution provided by light microscopy did not allow sufficient delineation of the
lymphatic capillary endothelial wall.

b. Role of intercellular junction: With the advent of electron techniques
and improved tissue preservation methods, electron-dense tracer substances
have been used to show that the major pathway for uptake of interstitial fluids
and particulate substances is the intercellular junction (Figs. 4.13 and 4.18). In
addition, it is also clear that endocytosis occurs via vesicular uptake of interstitially injected particles by the lymphatic endothelial cells (Casley-Smith,
1964; Dobbins and Rollins, 1970; Leak, 1971) (Fig. 4.19). Whereas the movement of large molecules across the blood capillary endothelium is mainly from
the luminal to the connective tissue front (Simionescu *et al.*, 1975; Clementi
and Palade, 1969; Karnovsky, 1967), in the case of the lymphatic capillary
endothelium, large molecules and particulate materials proceed from both connective tissue and luminal fronts of the endothelium within vesicles which may
aggregate into large vacuoles (Leak, 1970, 1971) (Fig. 4.20).

In addition to the results obtained with interstitial injection of tracer substances for lymphatic capillary uptake, evidence for the rapid transport of fluids
and large molecules across the blood–interstitial–lymphatic interface has also
come from studies with tracers injected intravenously. The progress of these
substances across the blood capillary wall into the connective tissue space and
lymphatic capillaries was monitored at successive time periods using ultrastructural techniques (Dobbins and Rollins, 1970; Leak, 1971). Since with such
procedures the possibility of producing unphysiologic pressures within the
interstitium is eliminated, the results showing marker substance within the
clefts of intercellular junctions is presumed to represent the normal transport
route of fluids and plasma protein across the blood–tissue–lymphatic interface.
With intravenous injection of peroxidase, the greatest intensity of staining by
this protein trace is observed between 15 and 30 min (Fig. 4.19), a time period
which coincides with that of maximal concentration of peroxidase in the thoracic duct lymph after intravenous administration (Clementi and Palade, 1969).
Although there is also transport of the tracer by endocytosis, it is evident that
intense peroxidase staining occurs within the cleft of the intercellular junction,
indicating that this is a major passageway for the transport of interstitial fluid
and large molecules.

5. Mechanisms of Lymph Flow

a. Lymphatic capillary pump: A unidirectional flow process from the blood capillary to the tissue–lymph interface is assured by the presence in the lymphatic capillaries of specialized structures—endothelial cell junctional valves (Fig. 4.21), which facilitate the rapid removal of interstitial fluid, large molecules, and cells. They consist of extensively overlapping adjacent endothelial cell margins, coupled with lymphatic anchoring filaments which insert along the abluminal endothelial cell surface, except for the inner segment of the junction overlapped by its neighbor. The lack of adhesion devices in the latter portion of the junction permits it to move freely as a flap valve in response to pressure changes across the lymphatic capillary wall (Fig. 4.22). The wide overlapping of adjacent cell margins, combined with the stabilizing effect of the anchoring filaments, allows the unstabilized segment of the endothelial cell of each intercellular junction to act as a one-way flap valve (trap door). As pressure in the interstitial space exceeds that in the lymphatic lumen, the unstabilized

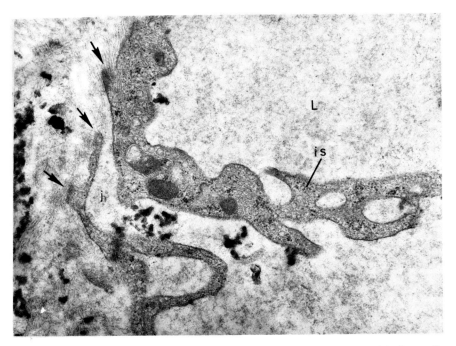

Fig. 4.21. Electron micrograph depicting the flap valve effect of inner segment (is) of intercellular junction (j) cell margin, which is free to swing into the lumen (L). Anchoring filaments are indicated by arrows. ×25,925. (From Leak and Burke, 1966; reproduced with permission from *American Journal of Anatomy.*)

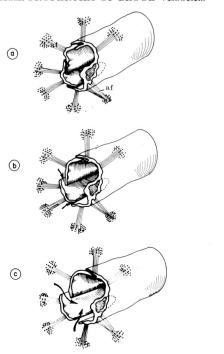

Fig. 4.22. Diagrammatic representation of lymphatic capillary tissue interface. The anchoring filaments (af), the sparsity of adhesive devices between the apposing endothelial cells, and their extensive overlap provide a mechanism which allows for expansion of the capillary lumen, as well as for regulating direction of bulk flow from the interstitial space into the lymphatic capillary. When the pressure in the interstitial space exceeds that in the lymphatic lumen, the unsupported portion of the endothelial cell is free to swing into the lumen (b) and (c), opening a wide direct channel from interstitial space into vessel. When the pressure within the lymphatic lumen is raised above that of the surrounding tissue, this flap is closed against the overlapping endothelial cell. Such a mechanism provides for a unidirectional flow, transporting large molecules, as well as cells. (From Leak, 1972b; reproduced with permission from Springer-Verlag.)

segment of the endothelial cell is free to swing into the lymphatic lumen (Fig. 4.22), thus providing a direct channel from the interstitial space into the lymphatic vessel. On the other hand, when the lymphatic capillary luminal pressure is raised above that of the surrounding connective tissue space, the flap valve is closed against the overlapping endothelial cell. Facilitating the opening and closing of the flap valve are the attached lymphatic anchoring filaments. These structures are in close contact with the lymphatic endothelial surface and extend into the surrounding connective tissue where they become firmly embedded. As a result, when interstitial exudate is increased in the tissue space, tension is placed on the collagen and elastic fibers in which the lymphatic anchoring filaments are firmly embedded, causing them to be separated by the increased interstitial fluid, a response which occurs during normal, as well as inflammatory, states. Regions of the lymphatic capillary wall to which anchoring filaments are attached would also be pulled along with the separating collagen and connective tissue fibers. Such a movement throughout the interstitial compartment would lead to a separation of loosely overlapped adjacent endothelial cells, causing a widening of the lymphatic capillary lumen, as well as producing patent intercellular junctions through which fluids and particulate substances could readily enter the lymphatic lumen (Fig. 4.22).

When hyaluronidase is used to digest components of the ground substance of edematous connective tissue, the firm connection within the interstitium is lost, and under such circumstances, instead of becoming dilated, the lymphatic capillaries now collapse and the wall becomes separated from the interstitium (Casley-Smith, 1967). It can, therefore, be concluded that the lymphatic anchoring filaments provide a structural basis for maintaining a firm connection between the lymphatic wall and the surrounding interstitium, as well as facilitating the opening and closing of the intercellular junctions (Leak and Burke, 1974). The response of the overlapped intercellular junctions to an increase in the interstitial fluid pressure during inflammation and edematous conditions is very pronounced, with the width of the intercellular cleft being greatly increased and the action of the flap valve greatly exaggerated (Fig. 4.23).

   b. Role of interstitial fluid pressure: The studies of Guyton (1963) and of Guyton *et al.* (1971, 1975, 1976) have demonstrated the significance of interstitial fluid pressure in determining the movement of fluids within the interstitial space and from the interstitium into the lymphatics. Using the perforated capsule technique, these workers demonstrated that the interstitial fluid pressure usually measures about -5 mm Hg in subcutaneous tissue, comparable negative pressure values being reported by other workers (Stromberg and Weiderhielm, 1970; Lucas and Floyer, 1973). Using the wick method, Schölander *et al.* (1968) and Prather *et al.* (1971) obtained similar negative pressure readings for the interstitial fluid pressure. With the needle method of measuring interstitial fluid pressure, both positive and negative (McMaster, 1946) readings have been reported. These data therefore suggest that the normal interstitial fluid pressure is usually subatmospheric. However, in inflammation it is more positive, and when this state is associated with edema, the pressure values measure above atmospheric (Guyton, 1963).

   c. Role of lymphatic capillaries in lymph transport: Although the earlier studies of McMaster (1946) on interstitial and lymphatic pressures supported the view that interstitial fluid moves down a pressure gradient into lymphatic capillaries to become lymph, more recent studies have found lymphatic capillary pressures to be near atmospheric (Weiderhielm and Weston, 1973; Zweifach and Prather, 1975). Under such conditions, in order for fluid to be

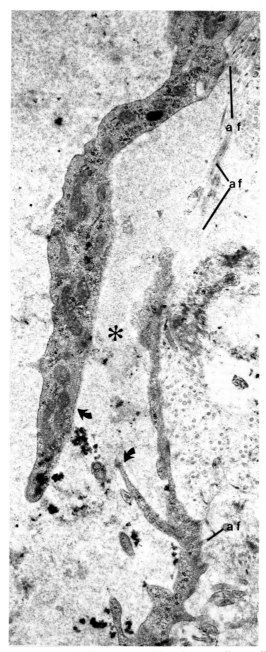

Fig. 4.23. Electron micrograph showing part of lymphatic capillary wall 1 hr after bacterial injection. The electron-dense flocculent material within the intercellular cleft (*) is continuous with the interstitium and lumen of the vessel. Anchoring filaments (af) are present along the connective tissue front of the vessel. Arrows point to overlapping cell margins. ×30,600. (From Leak and Burke, 1974; reproduced with permission from Academic Press.)

transported from a subatmospheric pressure into the lymphatic capillary lumen (which is at or above atmospheric pressure), the lymphatic capillary would have to play an active role in the process. In this regard, it has been demonstrated by Leak and Burke (1966, 1968) and Leak (1970, 1972b) that the lymphatic anchoring filaments and the extreme overlapping intercellular junctions provide the structural basis for postulating an active mechanism of lymph formation. Based on these morphologic findings, it has been suggested that lymph formation occurs as a result of pressure differentials in the lymphatic lumen and the interstitium (Leak, 1972b). Other possibilities that have been offered are the lymphatic suction theory (Reedy et al., 1975; Guyton et al., 1975) and the osmotic pull theory (Casley-Smith, 1977).

In regard to the lymphatic suction theory, it has been proposed that as interstitial fluid collects in the tissue spaces, it expands, along with the anchoring filaments (see above), causing the lymphatic capillary also to expand, thus creating a suction within its lumen. This would, in turn, pull fluid into the lymphatic capillary lumen as the one-way flap valve is also pulled open. Once filled, the content of the lymphatic capillary would be emptied of its contents (propelled upstream) by contraction of the endothelial cells or by external compression (i.e., by surrounding skeletal muscles, etc.). During relaxation the tissue recoil would place tension on the lymphatic wall which would start the filling or suction cycle over again. This explanation is similar to that suggested by Leak (1976).

In explaining the formation of lymph by the osmotic pull hypothesis, Casley-Smith (1977) has proposed that compression of the lymphatic capillaries (initial lymphatics) would express water from them into the interstitial space, thereby concentrating the protein within the lymphatic lumen. Upon relaxation, the concentrated lymph would exert an osmotic pull on the surrounding water, moving it into the lymphatic lumen.

Studies of Hogan (1979, 1980) on the interstitial fluid pressure and the intraluminal pressure of the contracting lymphatic bulb in the bat wing demonstrated that lymphatic capillary contraction could transfer fluid against a net pressure gradient which lowers the surrounding interstitial fluid pressure. Such a finding provides evidence in support of the lymphatic capillary as a pump (suction) system in the formation of lymph. In addition, the observation of actively contracting lymphatic capillaries (Cliff and Nicoll, 1970; Leak, 1971; Nicoll, 1975; Nicoll and Taylor, 1977; Hogan, 1979), coupled with the cytochemical demonstration of actin within their endothelium (Lauweryns et al., 1976; Leak, 1980), also suggests that these vessels possess a contractile apparatus to provide the pumping action necessary for the propulsion of lymph upstream into collecting vessels.

d. Role of collecting channels in lymph transport: The large collecting lymphatic vessels have valves which prevent the back flow of lymph, and in addition, the muscular coat in their tunica media provides the motile power within their wall for the continued flow of lymph upstream toward the lymphatic trunks. In some vertebrates (reptiles and birds), lymph is propelled along the collecting lymphatic vessels by rhythmically contracting lymph hearts. While there is no such mechanism found along the course of the adult mammalian lymphatic vasculature, rhythmic contractions have been observed by a number of investigators (Florey, 1927; Carlton and Florey, 1927; Kinmonth and Taylor, 1956; Hall et al., 1965).

Lymph propulsion has also been attributed to the effects of extrinsic factors on the collecting vessels, such as muscle contraction (Hudack and McMaster, 1933), respiratory movements (Morris, 1953; Rusznyak et al., 1960; Mayerson, 1963; Hall et al., 1965; Yoffey and Courtice, 1970), and movements of the intestines (Simmonds, 1957). However, more recent observations indicate that intrinsic rhythmic contractions of the valved lymphatic vessels are primarily responsible for the perfusion of lymph toward the thoracic duct, and the strength and frequency of the intrinsic contractions of these channels have been related to the rate of lymph flow (Hall et al., 1965).

Adrenergic nerve fibers have been demonstrated throughout the wall of large lymphatic vessels (Todd and Bernard, 1971). The close topographical relationship between nerve fibers and smooth muscle cells within the lymphatic vascular wall also provides morphologic data implicating neural regulation as a source for controlling the rhythmic contraction of the lymphatic collecting vessels and larger lymphatic ducts (Leak, 1972b).

D. PATHOPHYSIOLOGY AND PATHOLOGY OF THE LYMPHATIC SYSTEM

By L. V. LEAK

Studies by a number of investigators have demonstrated that the vasodilatation which occurs in response to injury exceeds the physiologic limit by causing an increased permeability of the blood capillary wall (Majno and Palade, 1961; Spector and Willoughby, 1965), thus permitting large amounts of plasma protein to escape into the interstitium. In addition, the number of blood capillaries in the inflamed area, carrying blood at an optimal flow rate, are also increased (Burke and Miles, 1958). The lymphatics respond by dilating their lumens in order to accommodate the greater fluid demand (Miles and Miles, 1958). The

Blood Vessels and Lymphatics in Organ Systems
Copyright © 1984 by Academic Press, Inc.

dilatation of the lymphatic capillaries is made possible by the attached anchoring filaments and the extensively overlapping intercellular junctions that are loosely apposed to each other. As described above, the increased interstitial fluid volume not only expands the interstitial space but also the lymphatic capillaries. Such structural changes explain why histologic examination of an intense inflammatory reaction demonstrates venules which are compressed, whereas lymphatics are greatly distended (Pullinger and Florey, 1935). Thus, the lymphatic capillaries are capable of responding to increased demands for fluid transport by a widening of their lumina. The vessels continue to function satisfactorily even though widely dilated until the junctions are pulled so widely apart that the junctional valve no longer seats on the adjacent cell and fluid continues to move in and out of the lymphatic lumen (Leak and Kato, 1970).

Information on lymph flow in both normal and pathologic condition has come from a large number of lymph cannulation studies. In the early experiments, glass catheters were used to collect lymph from inflamed tissues, to determine changes in protein concentration and the presence of inflammatory cells (Cameron and Courtice, 1946; Yoffey and Courtice, 1970). With development of improved plastic tubing, it became possible to cannulate the thoracic duct and large lymphatics in various regions and collect lymph over longer time periods (Bollman et al., 1948; Staub et al., 1975; Sprent, 1977). In addition to providing information on lymph flow and the characteristics of lymph in a wide variety of tissues and from a large number of animals, information on lymphocyte migration and the cell content of lymph has also been provided (Lascelles and Morris, 1961; Morris and Courtice, 1977).

Using combined lymph cannulation and the perforated capsule methods for interstitial fluid pressure measurements, Taylor et al. (1973) demonstrated that lymph flow increases as interstitial pressure is elevated from values of -7 to $+2$ mm Hg interstitial pressure. However, there is no further augmentation when interstitial fluid pressure is raised to more positive values. These observations suggest that the inability of the lymphatic system to accommodate excess fluids leading to edematous conditions manifests itself as the interstitial fluid pressure is increased from its negative value up to and approaching $+2$ mm Hg. The studies of Guyton (1965) and Taylor et al. (1973) also demonstrated that edema does not occur in subcutaneous tissue so long as the interstitial fluid pressure remains negative.

Chronic cannulations of lymph vessels in the sheep lung have provided much information on lung fluid and protein exchange in both normal and pathologic conditions (Staub, 1974; Staub et al., 1975). Using this technique, Vreim et al. (1976) demonstrated that during moderate (mild) interstitial edema in sheep lung, lymph and free-fluid proteins could be considered to be identical, thus dispelling the notion that lung lymph is concentrated as it is propelled through the collecting lymphatics.

Lymph fistulas have also provided information on changes in the flow rate, as well as cellular composition of lymph following the inflammatory reaction.

The degree to which the lymphatic system responds to various inflammatory stimuli can also be monitored in the flow rates and cellular composition of afferent lymph. Lymph in sheep following the injection of a suspension of larvae from the intestinal worm, *Haemonchus contortus*, contains polymorphonuclear leukocytes within 24 hr; this is followed by the appearance of eosinophils (Hay, 1979). The flow of lymph from the efferent vessels of a lymph node draining the injection site reaches maximal levels several days after the injection. At this time there is also a large increase in the number of lymphocytes (Hay, 1979). With the exception of eosinophils, a similar sequence of events is observed in the peritoneal cavity following the injection of bacterial toxin (Leak, 1981). (For a discussion of the pathologic changes in the lymphatics of the heart, see Section F-4c, Chapter 10.)

Cannulation of the thoracic duct and lymphatic vessels from various regions has provided a convenient means of sampling for changes in the traffic of fluids and cells removed by the lymphatics during an inflammatory response.

REFERENCES*

Albertine, K. H., and O'Morchoe, C. C. C. (1980). Renal lymphatic ultrastructure and translymphatic transport. *Microvasc. Res.* **19**, 338–351. (B)

Arey, L. B. (1974). "Developmental Anatomy: Textbook and Laboratory Manual of Embryology," p. 371. Revised 7th ed. Saunders, Philadelphia, Pennsylvania. (A)

Barnes, R. (1964). "Invertebrate Zoology," p. 345. Saunders, Philadelphia, Pennsylvania. (C)

Bartholin, T. (1651). "Ex caspari bartholini parentis intituitionibus, Omnius recentiorum, et propries observationibus tertium ad sanguinis circulationem reformata." Hack, Leyden. (C)

Benninghoff, A. (1930). Blutegefasse und Herz. *In* "Handbuck der mikroskopischen Anatomie des Menschen" (W. von Möllendorff, ed.), Vol. 6, Part 1. Springer-Verlag, Berlin and New York. (C)

Bollman, J. L., Cain, J. C., and Grindlay, J. H. (1948). Technique for collection of lymph from liver, small intestine or thoracic duct of the rat. *J. Lab. Clin. Med.* **33**, 1349–1352. (D)

Bradbury, M. B. W., and Cole, D. F. (1980). The role of the lymphatic system in drainage of cerebrospinal fluid and aqueous humor. *J. Physiol. (London)* **299**, 353–365. (B)

Bruns, R. R., and Palade, G. E. (1968). Studies on blood capillaries. I. General organization of blood capillaries in muscle. *J. Cell Biol.* **37**, 244–276. (C)

Burke, J. F., and Leak, L. V. (1968). Lymphatic function in normal and inflamed states. *In* "Progress in Lymphology" (M. Viamonte, P. R. Kochler, M. Witte, and C. Witte, eds.), Vol. II, pp. 81–85. Thieme, Stuttgart. (C)

Burke, J. F., and Miles, A. A. (1958). The sequence of vascular events in early infective inflammation. *J. Pathol. Bacteriol.* **76**, 1–19. (D)

Cameron, G. R., and Courtice, F. C. (1946). The production and removal of edema fluids in the lung after exposure to carbonyl chloride (phosgene). *J. Physiol. (London)* **105**, 175–185. (D)

Carlton, H. M., and Florey, H. W. (1927). The mammalian lacteal: Its histologic structure in relation to its physiological properties. *Proc. R. Soc. London, Ser. B* **102**, 110–118. (C)

Casley-Smith, J. R. (1964). An electron microscopic study of injured and abnormally permeable lymphatics. *Ann. N.Y. Acad. Sci.* **116**, 803–830. (C)

*In the reference list, the capital letter in parentheses at the end of each reference indicates the section in which it is cited.

Casley-Smith, J. R. (1967). Electron microscopical observation on the dilated lymphatics in edematous regions and their collapse following hyaluronidase administration. *Br. J. Exp. Pathol.* **48**, 680–686. (C)

Casley-Smith, J. R. (1977). Lymph and lymphatics. *In* "Microcirculation" (G. Kaley and B. M. Altura, eds.), Vol. I, pp. 423–508. University Park Press, Baltimore, Maryland. (B,C)

Casley-Smith, J. R., and Florey, H. W. (1961). The structure of normal small lymphatics. *Q. J. Exp. Physiol.* **46**, 101–106. (C)

Chambers, R., and Zweifach, B. W. (1947). Intercellular cement and capillary permeability. *Physiol. Rev.* **27**, 436–463. (C)

Clarke, E. R. (1911). An examination of methods used in the study of the development of the lymphatic system. *Anat. Rec.* **5**, 395. (A)

Clementi, F., and Palade, G. E. (1969). Intestinal capillaries. I. Permeability to peroxidase and ferritin. *J. Cell Biol.* **41**, 33–58. (C)

Cliff, W. J., and Nicoll, P. A. (1970). Structure and function of the bat's wing. *Q. J. Exp. Physiol. Cogn. Med. Sci.* **55**, 112–121. (C)

Dobbins, W. O., and Rollins, E. L. (1970). Intestinal mucosal lymphatic permeability: An electron microscopic study of endothelial vesicles and cell junctions. *J. Ultrastruct. Res.* **33**, 29–59. (C)

Ebert, C. J., and Belajeff, A. (1866). Über die lymphgefasse des herzens. *Virchows Arch. Pathol. Anat. Physiol.* **37**, 124–131. (C)

Fawcett, D. W., Leak, L. V., and Heidger, P. M. (1970). Electron microscopic observations on the structural components of the blood–testis barrier. *J. Reprod. Fertil.* **10**, Suppl., 105–122. (C)

Fawcett, D. W., Neaves, W. B., and Flores, M. N. (1973). Comparative observation on intertubular lymphatics and the organization of the interstitial tissue of the mammalian testis. *Biol. Reprod.* **9**, 500–532. (C)

Feder, N., Reese, T. S., and Brightman, M. W. (1969). Microperoxidase, a new tracer of low molecular weight. A study of the interstitial compartment of the mouse brain. *J. Cell Biol.* **43**, 35A. (C)

Florey, H. W. (1927). Observations on the contractility of lacteals. Part II. *J. Physiol. (London)* **63**, 1–18. (C)

French, J. E., Florey, H. W., and Morris, B. (1960). The absorption of particles by the lymphatics of the diaphragm. *Q. J. Exp. Physiol. Cogn. Med. Sci.* **45**, 88–103. (C)

Gibson, H., and Garr, K. A., Jr. (1970). Dynamics of the implanted capsule. *Fed. Proc., Fed. Am. Soc. Exp. Biol.* **29**, 319A. (C)

Gray, S. W., and Skandalakis, J. E. (1972). The lymphatic system. *In* "Embryology for Surgeons," pp. 695–714. Saunders, Philadelphia, Pennsylvania. (A)

Guyton, A. C. (1963). A concept of negative interstitial pressure based on pressures in implanted perforated capsules. *Circ. Res.* **12**, 399–414. (C)

Guyton, A. C. (1965). Interstitial fluid pressure. II. Pressure–volume curves of interstitial space. *Circ. Res.* **16**, 452–460. (D)

Guyton, A. C., Granger, H. J., and Taylor, A. E. (1971). Interstitial fluid pressure. *Physiol. Rev.* **51**, 527–563. (C)

Guyton, A. C., Taylor, A. E., and Granger, H. J. (1975). "Circulatory Physiology: Dynamics and Control of the Body Fluids." Saunders, Philadelphia, Pennsylvania. (C)

Guyton, A. C., Taylor, A. E., and Brace, R. A. (1976). A synthesis of interstitial fluid regulation and lymph formation. *Fed. Proc., Fed. Am. Soc. Exp. Biol.* **35**, 1881–1885. (C)

Hall, J. G., Morris, B., and Woolley, G. (1965). Intrinsic rhythmic propulsion of lymph in the unanaesthetized sheep. *J. Physiol. (London)* **180**, 336–349. (C)

Hay, J. B. (1979). Kinetics of the inflammatory response in regional lymph. *Curr. Top. Pathol.* **68**, 89–108. (D)

His, W. (1863). Über das epithel der lymphgefass-wurzeln und uber die V. Recklinghausenschen saftkanalchen. *Z. Wiss. Zool.* **13**, 455–473. (C)

Hogan, R. D. (1979). The initial lymphatics and interstitial fluid pressure. *In* "Tissue Fluid Pressure and Composition" (A. R. Hargens, ed.), pp. 155–163. Williams & Wilkins, Baltimore, Maryland. (C)

Hogan, R. D. (1980). Intralymphatic vs. tissue pressure in the edematous bat wing. *Adv. Physiol. Sci.* **7**, 193–200. (C)

Hudack, S. S., and McMaster, P. D. (1932). I. The permeability of the lymphatic capillary. *J. Exp. Med.* **56**, 223–238. (C)

Hudack, S. S., and McMaster, P. D. (1933). The lymphatic participation in human cutaneous phenomena. A study of minute lymphatics of the living skin. *J. Exp. Med.* **57**, 751–774. (C)

Ingalls, N. W. (1908). A contribution to the embryology of the liver and vascular system in man. *Anat. Rec.* **2**, 338–344. (A)

Intaglietta, M., Pawula, R. F., and Tompkins, W. R. (1970). Pressure measurements in the mammalian microvasculature. *Microvasc. Res.* **2**, 212–220. (C)

Kampmeier, O. F. (1960). The development of the jugular lymph sacs in the light of vestigial, provisional and definitive phases of morphogenesis. *Am. J. Anat.* **107**, 153. (A)

Kampmeier, O. F. (1969). "Evolution and Comparative Morphology of the Lymphatic System." Thomas, Springfield, Illinois. (A)

Karnovsky, M. J. (1967). The ultrastructural basis of permeability studies with peroxidase as a tracer. *J. Cell Biol.* **35**, 213–236. (C)

Kinmonth, J. B., and Taylor, G. W. (1956). Spontaneous rhythmic contractility in human lymphatics. *J. Physiol. (London)* **133**, 3P. (C)

Krogh, A. (1959). "The Anatomy and Physiology of Capillaries." Hafner, New York. (C)

Landis, E. M., and Pappenheimer, J. R. (1963). Exchange of substances through the capillary walls. *In* "Handbook of Physiology" (W. F. Hamilton, ed.), Sect. 2, Vol. II, Chapter 29, pp. 961–1035. Am. Physiol. Soc., Bethesda, Maryland. (C)

Lascelles, A. K., and Morris, B. (1961). Surgical techniques for the collection of lymph from unanaesthetized sheep. *Q. J. Exp. Physiol. Cogn. Med. Sci.* **46**, 199–205. (D)

Lauweryns, J. M., Baert, J. H., and DeLoecker, W. (1976). Fine filaments in lymphatic endothelial cells. *J. Cell Biol.* **68**, 163–167. (C)

Leak, L. V. (1968). Lymphatic capillaries in tail fin of amphibian larva. An electron microscopic study. *J. Morphol.* **125**, 419–466. (C)

Leak, L. V. (1970). Electron microscopic observations on lymphatic capillaries and the structural components of the connective tissue-lymph interface. *Microvasc. Res.* **2**, 361–391. (B,C)

Leak, L. V. (1971). Studies on the permeability of lymphatic capillaries. *J. Cell Biol.* **50**, 300–323. (C)

Leak, L. V. (1972a). The transport of exogenous peroxidase across the blood–tissue–lymph interface. *J. Ultrastruct. Res.* **39**, 24–42. (C)

Leak, L. V. (1972b). The fine structure and function of the lymphatic vascular system. *In* "Handbuch der allgemeinen Pathologie" (H. Meessen, ed.), pp. 149–196. Springer-Verlag, Berlin and New York. (C)

Leak, L. V. (1976). The structure of lymphatic capillaries in lymph formation. *Fed. Proc., Fed. Am. Soc. Exp. Biol.* **35**, 1863–1871. (C)

Leak, L. V. (1977). Pulmonary lymphatic and interstitial fluid. *Lung Biol. Health Dis.* **5**, 631–685. (C)

Leak, L. V. (1980). Lymphatic vessels. *Electron Microsc. Hum. Med.* **5**, Part 3, 157. (C)

Leak, L. V. (1981). Continuous pathways for cellular transport from the peritoneal cavity. *Anat. Rec.* **199**, 151A. (C,D)

Leak, L. V., and Burke, J. F. (1965). Ultrastructure of lymphatic capillaries. *J. Appl. Physiol.* **36**, 2620A. (C)

Leak, L. V., and Burke, J. F. (1966). Fine structure of the lymphatic capillary and the adjoining connective tissue area. *Am. J. Anat.* **118**, 785–810. (C)

Leak, L. V., and Burke, J. F. (1968). Ultrastructural studies on the lymphatic anchoring filaments. *J. Cell Biol.* **36,** 129–149. (C)

Leak, L. V., and Burke, J. F. (1974). Early events of tissue injury and the role of the lymphatic system in early inflammation. *In* "The Inflammatory Process" (B. W. Zweifach, L. Grant, and R. I. McCluskey, eds.), Vol. 3, Chapter 4, pp. 163–235. Academic Press, New York. (C)

Leak, L. V., and Kato, F. (1970). Electron microscopic studies of lymphatic capillaries during early inflammation. *Lab. Invest.* **61,** 572–588. (D)

Leak, L. V., and Rahil, K. (1978). Permeability of the diaphragmatic mesothelium: The ultrastructural basis for "stoma." *Am. J. Anat.* **151,** 557–594. (C)

Lindner, H. R. (1963). Partition of androgen between the lymph and venous blood of the testis in the ram. *J. Endocrinol.* **25,** 483–494. (C)

Lucas, J., and Floyer, M. A. (1973). Renal control of changes in the compliance of the interstitial space: A factor in the aetiology of renoprival hypertension. *Clin. Sci.* **44,** 397–416. (C)

Ludwig, C., and Tomsa, W. (1862). Die Lymphwege des hodens und ihr verhältnis zu den Blut- und Samengefassen. *Sitzungsber. Akad. Wiss. Wien, Math.-Naturwiss. Kl. Abt. 2* **46,** 221–237. (C)

Majno, G., and Palade, G. E. (1961). Studies on inflammation. I. The effect of histamine and serotonin on vascular permeability. An electron microscopic study. *J. Cell Biol.* **11,** 571–605. (D)

Majno, G., Shea, S. M., and Leventhal, M. (1969). Endothelial contraction induced by histamine type mediators. An electron microscopic study. *J. Cell Biol.* **42,** 647–672. (C)

Mayerson, H. S. (1963). The physiological importance of lymph. *In* "Handbook of Physiology" (J. Field, ed.), Sect. 2, Vol. III, pp. 1035–1073. Am. Physiol. Soc., Bethesda, Maryland. (C)

Mayerson, H. S., Wolfram, G. C., Shirley, H. A., and Wasserman, K. (1960). Regional differences in capillary permeability. *Am. J. Physiol.* **198,** 155–160. (C)

Mayerson, H. S., Patterson, R. M., McKee, A., LeBru, S. L., and Mayerson, P. (1962). Permeability of lymphatic vessels. *Am. J. Physiol.* **203,** 98–106. (C)

McMaster, P. D. (1946). The pressure and interstitial resistance prevailing in the normal and edematous skin of animals and man. *J. Exp. Med.* **84,** 473–494. (C)

Miles, A. A., and Miles, E. M. (1958). The state of lymphatic capillaries in acute inflammatory lesions. *J. Pathol. Bacteriol.* **76,** 21–35. (D)

Morris, B. (1953). The effect of diaphragmatic movement on the absorption of protein and of red cells from the peritoneal cavity. *Anat. J. Exp. Biol. Med. Sci.* **31,** 239–246. (C)

Morris, B., and Courtice, F. C. (1977). Cells and immunoglobulins in lymph. *Lymphology* **10,** 62–70. (D)

Morris, B., and Sass, M. B. (1966). The formation of lymph in the ovary. *Proc. R. Soc. London, Ser. B* **164,** 577–591. (C)

Nicolaysen, G., Nicolaysen, A., and Staub, N. C. (1975). A quantitative radioautographic comparison of albumin concentration in different sized lymph vessels in normal mouse lungs. *Microvasc. Res.* **10,** 138–152. (C)

Nicoll, P. A. (1975). Excitation contraction of single vascular smooth muscle cells and lymphatics *in vivo. Immunochemistry* **12,** 511–515. (C)

Nicoll, P. A., and Taylor, A. E. (1977). Lymph formation and flow. *Annu. Res. Physiol.* **39,** 73–95. (C)

Pinchon, M. C., Bernaudin, J. F., and Bignon, J. (1980). Pleural permeability in rat. I. Ultrastructural basis. *Biol. Cell.* **37,** 269–272. (C)

Pollard, T. D., and Weihing, R. R. (1974). Cytoplasmic actin and myosin and cell motility *CRC Crit. Rev. Biochem.* **2,** 1–65. (C)

Prather, J. W., Bowes, D. N., Warrell, D. A., and Zweifach, B. W. (1971). Comparison of capsule and wick techniques for the measurement of interstitial pressure. *J. Appl. Physiol.* **31,** 942–945. (C)

Pullinger, B. D., and Florey, H. W. (1935). Some observations on structure and function of lymphatics. Their behavior in local oedema. *Br. J. Exp. Pathol.* **16**, 49–61. (C,D)

Reedy, N. P., Krouskap, T. A., and Newell, P. H. (1975). Biomechanics of a lymphatic vessel. *Blood Vessels* **12**, 261–278. (C)

Rouviere, H. (1938). "Anatomy of the Human Lymphatic System" (trans. by M. J. Tobias). Edwards, Ann Arbor, Michigan. (B)

Rusznyak, I., Foldi, M., and Szabo, G. (1960). "Lympatics and Lymph Circulation," 2nd ed. Pergamon, Oxford. (C)

Sabin, F. R. (1916). The origin and development of the lymphatic system. *Johns Hopkins Hosp. Rep.* **17**, 347. (A)

Schoefl, G. F. (1964). Electron microscopic observation on the regeneration of blood vessels after injury. *Ann. N.Y. Acad. Sci.* **116**, 789–802. (C)

Schölander, P. F., Hargens, A. R., and Miller, S. L. (1968). Negative pressure in the interstitial fluid of animals. *Science* **161**, 321–328. (C)

Simionescu, M., Simionescu, N., and Palade, G. E. (1975). Segmental differentiation of cell junction in the vascular endothelium: The microvasculature. *J. Cell Biol.* **67**, 863–885. (C)

Simmonds, W. J. (1957). The relationship between intestinal motility and the flow and rate of fat output in thoracic duct lymph in anaesthetized rats. *Q. J. Exp. Physiol. Cogn. Med. Sci.* **42**, 205–221. (C)

Spector, W. G., and Willoughby, D. A. (1965). *In* "The Inflammatory Process" (B. W. Zweifach, L. Grant, and R. T. McCluskey, eds.), 1st ed., pp. 427–448. Academic Press, New York. (D)

Sprent, J. (1977). Recirculating lymphocytes. *In* "The Lymphocyte Structure and Function" (J. H. Marchalonis, ed.), pp. 43–112. Dekker, New York. (D)

Starling, E. H. (1896). On the absorption of fluids from the connective tissue spaces. *J. Physiol. (London)* **19**, 312–326. (C)

Starling, E. H. (1909). "The Fluids of the Body. The Herter Lectures" W. T. Keuer & Co., Chicago, Illinois. (C)

Staub, N. C. (1974). Pulmonary edema. *Physiol. Rev.* **54**, 678–811. (D)

Staub, N. C., Bland, R. D., Brigham, K. L, Demling, R. H., Erdmann, A. J., III, and Woolverton, W. C. (1975). Preparation of chronic lung lymph fistula in sheep. *J. Surg. Res.* **19**, 315–320. (D)

Stromberg, D. D., and Weiderhielm, C. A. (1970). Effects of oncotic gradients and enzymes on negative pressures in implanted capsules. *Am. J. Physiol.* **219**, 928–932. (C)

Taylor, A. E., and Gibson, W. H. (1975). Concentrating ability of the lymphatic vessels. *Lymphology* **8**, 43–49. (C)

Taylor, A. E., Gibson, W. H., Granger, H. J., and Guyton, A. C. (1973). The interaction between intracapillary forces in the overall regulation of interstitial fluid volume. *Lymphology* **6**, 192–208. (D)

Todd, G. L., and Bernard, G. R. (1971). Functional anatomy of the cervical lymph duct of the dog. *Anat. Rec.* **169**, 443A. (C)

Vakiti, C., Ruiz-Ortiz, F., and Burke, J. R. (1970). Chemical and osmolar changes of inflammatory states. *Surg. Forum* **21**, 227–228. (C)

Venkatachalam, M. A., and Fahimi, H. D. (1969). The use of beef liver catalase as a protein tracer for electron microscopy. *J. Cell Biol.* **42**, 480–489. (C)

von Recklinghausen, F. (1862a). "Die Lymphgefasse und ihre Beziehung zum Bindegewebe." Hirschwald, Berlin (cited by Yoffey and Courtice, 1970). (A)

von Recklinghausen, F. T. (1862b). Zur fettesorption. *Arch. Pathol. Anat.* **26**, 172–208. (C)

Vreim, C. E., Snashall, P. D., Demling, R. H., and Staub, N. C. (1976). Lung lymph and free interstitial fluid protein composition in sheep with edema. *Am. J. Physiol.* **230**, 1650–1653. (D)

Wang, N. S. (1975). The preformed stomas connecting the pleural cavity and the lymphatics in the parietal pleura. *Am. Rev. Respir. Dis.* **111**, 12–20. (C)

Warwick, R., and Williams, P. L., eds. (1980). "Gray's Anatomy," 36th ed. Saunders, Philadelphia, Pennsylvania. (A)

Weiderhielm, C. A., and Weston, B. V. (1973). Microvascular, lymphatic, and tissue pressures in the unanesthetized mammal. *Am. J. Physiol.* **225**, 992–996. (C)

Yang, V. V., O'Morchoe, P. J., and O'Morchoe, C. C. C. (1981). Transport of protein across lymphatic endothelium in the rat kidney. *Microvasc. Res.* **21**, 75–91. (B)

Yoffey, J. M., and Courtice, F. C., eds. (1970). "Lympatics, Lymph and Lymphomyeloid Complex." Academic Press, New York. (C,D)

Zweifach, B. W. (1973). Micropressure measurements in the terminal lymphatics. *Bibl. Anat.* **12**, 361–365. (C)

Zweifach, B. W., and Prather, J. W. (1975). Micromanipulation of pressure in terminal lymphatics in the mesentery. *Am. J. Physiol.* **228**, 1326–1335. (C)

PART TWO

Blood Vessels and Lymphatics in Specific Organ Systems

Chapter 5
Central Nervous System: Brain

A. EMBRYOLOGY OF BLOOD CIRCULATION

By C. E. Corliss

1. Development of Arteries

Two arterial systems are responsible for the blood supply to the human brain: the internal carotid and the basilovertebral systems.

a. Internal carotid system: The primitive internal carotid artery begins in the fourth week as a composite vessel—a persistent third aortic arch, plus a portion of the dorsal aortic root—and extends rostrally to the developing brain (Fig. 5.1A). After 4 weeks, it splits into a rostral primitive olfactory branch and a caudal primitive posterior communicating artery (Fig. 5.3). When the first two aortic arches degenerate, blood supply to the branchial arches is assumed temporarily by the ventral pharyngeal artery from the aortic sac. It is, in turn, replaced by the stapedial artery extending from the stem of the second aortic arch (Fig. 5.3). The stapedial artery then supplies the branchial arches of the area, along with the orbit, via the supraorbital artery (Fig. 5.4).

By the middle of the sixth week, the internal carotid artery possesses the following branches: the primitive dorsal ophthalmic, the primitive anterior choroidal, the middle cerebral, and the anterior cerebral, the last three originating from its rostral branch (Fig. 5.3). At the same time, the external carotid, arising from the base of the third aortic arch, takes over the region once served by the transient ventral pharyngeal and stapedial arteries, including supply to the middle meningeal and supraorbital arteries (Fig. 5.4).

b. Basilovertebral system: The vertebral artery, formed by anastomosis of 7 rostral intersegmental arteries (Figs. 5.1A and 5.3 right) supplies blood to the longitudinal neural arteries forming along the ventral brain wall. These early bilateral plexiform vessels (Fig. 5.3) consolidate into the primitive midline basilar artery by the seventh week (Figs. 5.1–5.3) and give off several branches to the cerebellum. Located in their final pattern, the latter vessels are the posterior and anterior inferior cerebellars, the pontine, and the superior cerebellar (Figs. 5.1B, 5.2, and 5.4). Immediately rostral to this complex is the cerebral arterial circle of Willis, described below.

175

Blood Vessels and Lymphatics in Organ Systems
Copyright © 1984 by Academic Press, Inc.
ISBN 0-12-042520-3

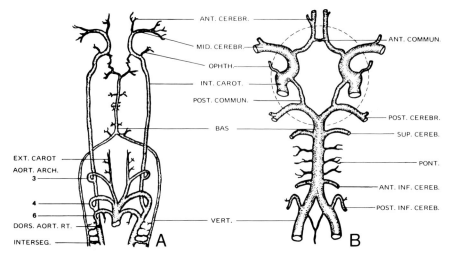

ANT. CEREBR.
MID. CEREBR.
OPHTH.
INT. CAROT.
POST. COMMUN.
BAS
EXT. CAROT
AORT. ARCH.
3
4
6
DORS. AORT. RT.
INTERSEG.
A

ANT. COMMUN.
POST. CEREBR.
SUP. CEREB.
PONT.
ANT. INF. CEREB.
POST. INF. CEREB.
VERT.
B

FIG. 5.1.

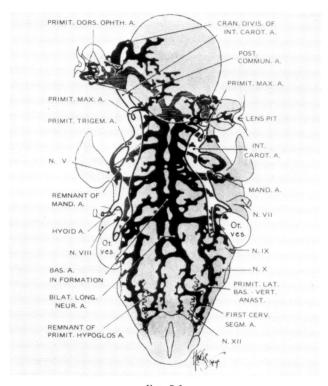

PRIMIT. DORS. OPHTH. A.
CRAN. DIVIS. OF INT. CAROT. A.
POST. COMMUN. A.
PRIMIT. MAX. A.
PRIMIT. MAX. A.
PRIMIT. TRIGEM. A.
LENS PIT
INT. CAROT. A.
N. V
MAND. A.
REMNANT OF MAND. A.
N. VII
HYOID A.
Ot. ves.
N. VIII
Ot. ves.
N. IX
N. X
BAS. A. IN FORMATION
PRIMIT. LAT. BAS. · VERT. ANAST.
BILAT. LONG. NEUR. A.
FIRST CERV. SEGM. A.
REMNANT OF PRIMIT. HYPOGLOS A.
N. XII

FIG. 5.2.

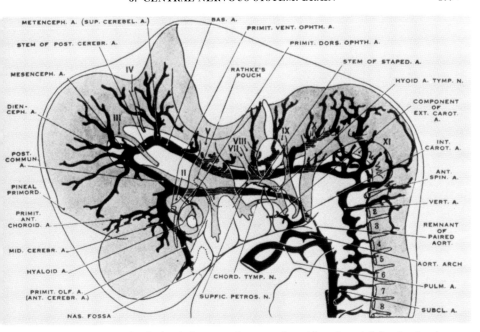

FIG. 5.3. Drawing, left side, head of 12.5 mm human embryo (about 6 weeks) showing developing cranial arteries. Bifurcation of the internal carotid (INT. CAROT. A.) into the rostral primitive olfactory artery (PRIMIT. OLF. A.) and caudal posterior communicating artery (POST. COMMUN. A.) is clearly indicated at the left. (From Padget, 1948; reproduced with permission from Carnegie Institution of Washington, Baltimore, Maryland.)

FIG. 5.1. (*opposite*) (A) Schematic diagram (dorsal aspect) of human aortic arch system at about 5 weeks. The internal carotid (INT. CAROT.) is forming by the combination of the persistent third aortic arch (3) and rostral part of the dorsal aortic root (DORS. AORT. RT.) Arches 3, 4, and 6 are so numbered. (B) Schematic drawing (dorsal aspect) of the adult cerebral arterial circle of Willis, encircled with dashed lines in figure, to show union of internal carotid (INT. CAROT.) and basilovertebral (BAS.) systems. (Adapted from Patten, 1968; reproduced with permission from McGraw-Hill Book Company, New York.)

FIG. 5.2. (*opposite*) Drawing, ventral view, to show formation of the basilar artery (BAS. A. IN FORMATION) in a 5.5 mm human embryo (about 4 weeks) from the plexiform bilateral longitudinal neural arteries (BILAT. LONG. NEUR. A.). Note the posterior communicating (POST. COMMUN. A.) upper left, shaded, and the internal carotid (INT. CAROT. A.) overlain in white. (From Padget, 1948; reproduced with permission of the Carnegie Institution of Washington, Baltimore, Maryland.)

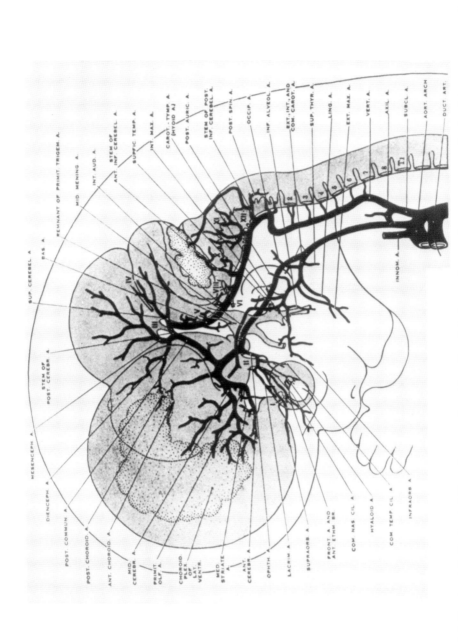

FIG. 5.4. Drawing, left side, head of 43 mm human embryo (about 9 weeks) showing the cranial arteries in essentially the adult pattern with the basilar (BAS. A.) and the left internal carotid arteries (INT. CAROT. A.) joined rostrally by the relatively large posterior communicating artery (POST. COMMUN. A.), in the center of the drawing. (From Padget, 1948; reproduced with permission from Carnegie Institution of Washington, Baltimore, Maryland.)

c. Cerebral arterial circle of Willis: This structure is the conjunction of the internal carotid and basilovertebral systems at the base of the brain. Caudally, it begins where the paired posterior cerebrals join the larger internal carotids via the posterior communicating arteries (Fig. 5.1B). The carotids, in turn, send forward the anterior cerebrals, which are linked by a single anterior communicating artery to close the circle rostrally (Fig. 5.1B). [For more details, see classic works of Congdon (1922) and Padget (1957).]

2. DEVELOPMENT OF VEINS

The venous circuit of the brain is completed in the fifth embryonic week when the primary head sinus empties into the anterior cardinal vein. This sinus is fed by three dural plexuses: the anterior dural from the fore- and midbrain, the middle dural from the metencephalon, and the posterior dural from the myelencephalon (Fig. 5.5). The branchial arches are served by the primitive

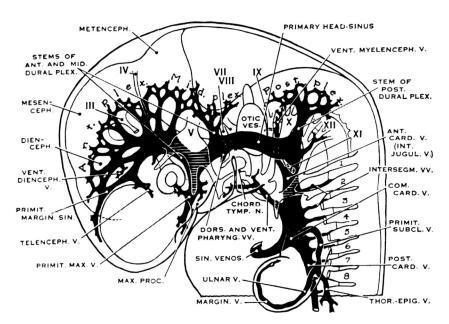

Fig. 5.5. Drawing, left side, head of 10 mm human embryo (about 5 weeks) showing formation of the large (horizontal) primary head sinus (PRIMARY HEAD-SINUS) from confluence of stems of the anterior, middle, and posterior dural plexuses (STEMS OF ANT., MID., POST. DURAL PLEX.). In the lower middle part of the drawing can be seen the primitive maxillary vein (PRIMIT. MAX. V.) and the ventral pharyngeal vein (VENT. PHARYNG. V.), draining branchial arch region and emptying into the large (horizontal) primary head sinus (PRIMARY HEAD-SINUS) and the (vertical) anterior cardinal vein (ANT. CARD. V.), respectively. (From Padget, 1957; reproduced with permission from Carnegie Institution of Washington, Baltimore, Maryland.)

maxillary vein rostrally and by the ventral pharyngeal veins caudally; the latter vessels empty into the anterior dural plexus (primary head sinus) and anterior cardinal vein, respectively (Fig. 5.5).

By 7 weeks the anterior and middle dural plexuses have anastomosed just dorsal to the otocyst, to form the sigmoid sinus, with a contribution from the stem of the posterior dural sinus. The primitive transverse sinus, too, is being formed by the rostral part of the middle dural sinus and the primitive marginal sinus. The superior sagittal sinus begins from the anterior dural plexus and the nearby tentorial plexus (Figs. 5.5 and 5.6). At the same time, the stem of the middle dural plexus becomes the prootic sinus (Figs. 5.6 and 5.7), which drains the ophthalmic region and contributes to the cavernous sinus (Fig. 5.7, shaded zone).

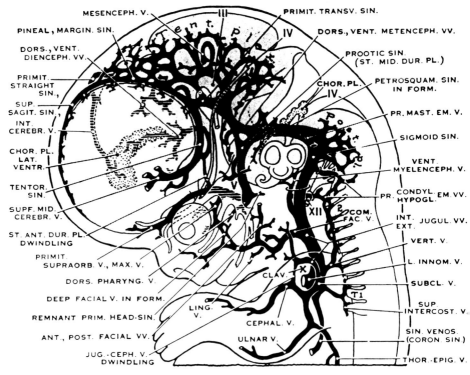

FIG. 5.6. Drawing, left side, head of 24 mm human embryo (about 7 weeks) showing formation of major cranial sinuses. The large sigmoid sinus (SIGMOID SIN.) formed by anastomosis of anterior, middle, and stem of posterior dural plexuses (see Fig. 5.6) is shown on right. The primitive transverse sinus (PRIMIT. TRANSV. SIN.) and the developing superior sagittal (SUP. SAGIT. SIN.) appear in the upper part of the drawing. The prootic sinus (PROOTIC SIN.) is seen as a vertical vessel in the center. (From Padget, 1957; reproduced with permission from Carnegie Institution of Washington, Baltimore, Maryland.)

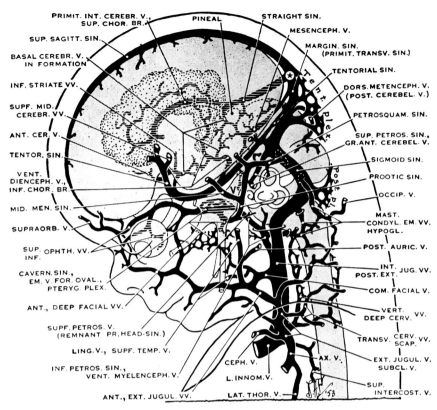

FIG. 5.7. Composite drawing of head of 60–80 mm human embryo (about 12 weeks) showing what is essentially the adult plan of the cerebrovenous system. The cavernous sinus (CAVERN. SIN., shaded here) drains the ophthalmic system and empties into the inferior petrosal sinus (INF. PETROS. SIN.). The transverse (PRIMIT. TRANSV. SIN.), superior sagittal (SUP. SAGITT. SIN.), and straight sinuses (STRAIGHT SIN.) meet in the confluence of sinuses. (From Padget, 1957; reproduced with permission from Carnegie Institution of Washington, Baltimore, Maryland.)

After 9 weeks the sigmoid sinus is well formed, and the cavernous sinus encloses part of the internal carotid artery in the hypophyseal region. The superior sagittal sinus is still developing at this time, and the straight sinus arises to drain the choroid plexus of the hemisphere ("PRIMIT. STRAIGHT SIN.", Fig. 5.6). By about 12 weeks, the basic plan of the definitive cerebrovenous system is established. The transverse, superior sagittal, and straight sinuses join in the confluence of the sinuses (Fig. 5.7, torcular herophili, at the black star in the white circle), and the ophthalmic system drains into the cavernous sinus and thence to the inferior petrosal (Fig. 5.7). At this time, the basal cerebral veins form (Fig. 5.7, shaded, near choroid plexus) but do not yet join the primitive internal cerebrals to form the great vein (of Galen) [for more details, see Padget (1957)].

B. ANATOMY OF BLOOD CIRCULATION

By E. Nelson

With good reason, the circulation of the brain has probably been subjected to more casual observations and scientific scrutiny than blood vessels of any other organ. This reflects the relative complexity of the vasculature itself, the heterogeneity of the brain supplied, the importance of blood flow to this organ, and the myriad of symptoms, some of which often directly relate to disease of specific vessels.

In this section, the gross anatomy of brain circulation will be briefly mentioned; light and electron microscopic observations on arteries, capillaries, and veins described; and the still controversial question of vascular innervation of the brain considered.

1. Gross Anatomy

The blood supply to the brain comes from the internal carotid arteries (anterior circulation) and the vertebral arteries (posterior circulation). The former are branches from the common carotid artery while the latter ordinarily arise from the subclavian artery, pass cephalad through or in close relation to the cervical spine and, unlike vessels elsewhere in the body, unite to form a larger vessel, the basilar artery. Branches from vertebral and basilar arteries supply the brainstem and cerebellum, with the basilar artery usually terminating as the two posterior cerebral arteries. From the latter, there commonly arise the posterior communicating arteries that merge with the two carotid arteries which, themselves, are joined by the anterior communicating artery, creating the polygon at the base of the brain known as the circle of Willis. A significant feature with clear clinical ramifications is the collateral circulation which potentially occurs from at least four sources: (1) the circle of Willis, (2) meningeal arterial anastomoses between internal and external carotid branches, (3) major branches of the anterior and posterior circulation, and (4) capillary anastomoses within the brain parenchyma. The combination of intracerebral vascular anastomoses and possible channels of collateral circulation are important determinants in cerebral vascular disease. [For details of the origin, course and branches of the major arterial and venous channels, their important anatomic variations, and the portions of the brain supplied or drained, see Stehbens (1972, 1979) and Toole and Patel (1974).]

Blood Vessels and Lymphatics in Organ Systems
Copyright © 1984 by Academic Press, Inc.

2. LIGHT AND ELECTRON MICROSCOPY OF BRAIN VESSELS

Cerebral arteries contain the usual successive layers: tunica intima, media, and adventitia; but they differ from muscular arteries in other organs in that the media and adventitia are thinner and the elastic tissue is concentrated in the internal elastic lamina, with no clear-cut external elastic lamina and relatively little elastic tissue within the media. In the normal intracranial arteries, true vasa vasorum have not been demonstrated within the media or adventitia (Stehbens, 1972; Mérei, 1974; E. Nelson, unpublished electron microscopic observations), probably because the relatively thin walls of the vessels and their location in the subarachnoid space render vasa vasorum unnecessary.

The existence of gaps (discontinuities or defects) in the media have been recognized for several decades (Stehbens, 1960). They occur at branches or forks in the artery and consist of discontinuities of the muscle layer which are filled by adventitial elements. Such gaps are present in systemic arteries as well, and, contrary to earlier speculations, probably represent an area of decreased rather than increased vulnerability of the wall, acting to reinforce the wall at a branch point (Stehbens, 1979).

It has long been recognized that foci of intimal proliferation may occur at the branches or forks of both large and small arteries in various species and at all ages. Ultrastructural studies (Takayanagi et al., 1972) of these pads or "cushions" in the larger cerebral vessels of the cat show the proliferative cells to be identical to the smooth muscle cells of the media and specifically fail to find innervation of these pads. Their existence in infancy and childhood suggests a physiologic genesis; however, their occurrence at the preferential location for atherosclerotic plaques in adult animals and man has led to the hypothesis that they are a stage in atherosclerosis (Stehbens, 1979). However, it seems most rational to consider them to be the morphologic reaction to "normal" rheological stress which, under appropriate external and internal circumstances, become the focus for an atherosclerotic plaque in some instances. These pads are focal alterations which may coalesce, be accompanied by changes in the internal elastic lamina, and eventually be associated with increased metachromatic ground substance within the intima. Such changes have been described in "normal" aging (Flora et al., 1967).

As in other organ systems, transmission and scanning electron microscopy have complemented gross and light microscopic observations in human and other species and in normal and abnormal situations (Maynard et al., 1957; Dahl et al., 1965; Nelson et al., 1973, 1974; Hazama et al., 1979; Povlishock et al., 1980). [For morphologic features of the cerebral vessel wall utilizing a variety of techniques, including electron microscopy and relevant pathophysiology, see Cervós-Navarro *et al.* (1974, 1976), Cervós-Navarro (1979), and Cervós-Navarro and Fritshke (1980).]

With scanning electron microscopy, the endothelial surface appears as a continuous lining, with individual cells varying in shape from ovoid to polygonal, probably related to rheological factors. The portion of the cell beneath which the nucleus lies protrudes slightly into the lumen, and cell boundaries are clearly marked by overlapping marginal folds. Endothelium of both arteries and veins appears to be continuous and nonfenestrated. Except for an unusually thin wall with few or no muscle fibers and variable amounts of loosely arranged collagen in the adventitia, cerebral veins appear to have no special characteristics. The abluminal portion of the internal elastic lamina of the cerebral arteries of man and animals stains with phosphotungstic acid or uranyl acetate and often shows interruptions or "fenestra." Occasional isolated smooth muscle cells are present between the internal elastic lamina and the endothelium. The tunica media consists of smooth muscle cells with the usual complement of organelles and fibrillar material recognizable as contractile protein. Contacts between smooth muscle cells themselves and between smooth muscle and endothelium are not uncommon. Within the adventitial layer, there are varying amounts of collagen, nerves which may be organized into bundles especially in the larger arteries, and an external border consisting of overlapping processes of fibroblasts. The smaller penetrating arteries and arterioles enter the brain parenchyma within the Virchow–Robin space (Jones, 1970). The boundaries of this perivascular space consist of the basement membrane of the superficial astrocyte processes of the brain parenchyma and the basement membrane (basal lamina) surrounding the smooth muscle cells of the penetrating vessel. Within this space, there are fibroblasts and nerves, and its extent defines the perivascular "cuff" of inflammatory cells in pathological conditions. As the penetrating arteriole loses its muscle layers, the basement membrane of endothelial cells or of pericytes coalesces with that of the parenchymal astrocyte, the vessel now being properly termed a true capillary (Nelson *et al.*, 1961).

Features of the cerebral capillary deserve special mention because of the importance of this vessel in metabolism and its relationship to blood–brain barrier. Its endothelium differs from that of capillaries in most other organs in that it is both continuous and has a tight junction between adjacent endothelial cells with fusion of the opposing plasma membranes. The only exceptions are capillaries of the choroid plexus, area postrema, portions of the hypothalamus, and a few other regions (Hirano, 1974). Another ultrastructural feature characteristic of most normal brain capillaries is the paucity of pinocytotic vesicles. Either or both of these features are thought to be the major morphologic constituents of the blood–brain barrier. The close apposition of the basal lamina to the astrocytic foot processes without intervening extracellular space is another peculiarity of cerebral capillaries (with the exceptions noted above), as is the increased number of mitochondria in the endothelial cells (Oldendorf *et al.*, 1980). Pericytes, which do not appear different from similar cells in other capillary beds, are often seen, surrounded, as usual, by basement membrane which is then contiguous with the basement membrane of astrocytic processes. Of interest is the observation that actin and myosin exist in brain capillaries (Owman *et al.*, 1977).*

3. Innervation of Cerebral Blood Vessels

Although the existence of intracranial perivascular nerves has been recognized at least since 1664 when Thomas Willis described them on the circle bearing his name (quoted in Mitchell, 1953), detailed studies using modern methodology have been relatively recent and related to a possible influence on cerebral blood flow and metabolism. A transmission electron microscopic study on innervation of human brain arteries by Dahl and Nelson (1964) has been followed by a number of ultrastructural studies designed to identify both the location and type of nerve (Nelson and Rennels, 1969, 1970; Nelson *et al.*, 1972; Iwayama *et al.*, 1970; Dahl, 1973; Cervós-Navarro, 1977).

In 1973, Hartman, using fluorescent techniques, described central noradrenergic innervation of small cerebral blood vessels, and in 1975, Rennels and Nelson published ultrastructural evidence suggesting existence of central innervation of cerebral capillaries. Heistad (1981) in a summary of a symposium, has updated the information on innervation of the cerebral vessels, including cellular mechanisms involved, neurotransmitters, the emerging evidence for a functional significance of cerebral vascular innervation, the continuing problem of true vasodilator nerves and their transmitter substance, the role of "vasoactive peptides," the significance of the demonstrated morphologic innervation of cerebral capillaries, and the origin of this innervation. [For further discussion of the neurogenic control of the brain circulation, see Owman and Edvinsson (1977, 1978).]

---

*Pertinent to the subject is the detailed study by Forbes *et al.* (1977) of pericytes in heart capillaries, associated with arrangements of microfilaments that closely resemble contractile elements of smooth muscle cells.

C. PHYSIOLOGY OF BLOOD CIRCULATION

By H. A. KONTOS AND D. D. HEISTAD

1. GENERAL CONSIDERATIONS OF CEREBRAL BLOOD FLOW

Because of its high rate of aerobic metabolism, the brain is heavily dependent on a continuous supply of blood at a high rate. This requirement is fulfilled by the development of specialized features of the cerebral circulation. The most important of these are relatively high sensitivity to local metabolic influences, relatively low responsiveness to reflex neurogenic influences, and elimination or limitation of the action of most circulating vasoactive substances by reducing their access to the vascular smooth muscle of cerebral vessels by virtue of the presence of the blood–brain barrier.

Cerebral blood flow is influenced by the following factors: (1) arterial blood pressure, (2) cerebral venous pressure, (3) intracranial pressure, (4) blood viscosity, and (5) length and caliber of the cerebral vessels. As in other vascular beds, regulation of the cerebral circulation under physiologic conditions is carried out mainly via alterations in the caliber of the cerebral vessels, secondary to changes in the activity of their vascular smooth muscle.

2. NEUROGENIC REGULATION OF CEREBRAL BLOOD FLOW

Cerebral vessels are supplied by extrinsic adrenergic, cholinergic, and peptidergic nerves (Owman and Edvinsson, 1977). It has also been suggested that adrenergic nerves originating in the brainstem innervate cortical cerebral vessels, but this is not fully established.

The influence of adrenergic vasoconstrictor fibers on cerebral circulation is much less pronounced than in other vascular beds. First, these nerves do not seem to exert resting vasoconstrictor tone, since their elimination causes little change in blood flow (Heistad *et al.*, 1978). Second, maximum stimulation of the fibers by high-frequency electrical current causes only modest vasoconstriction. For example, in the cat, pial vessels constrict only 7–12% (Wei *et al.*, 1975; Kuschinsky and Wahl, 1975). It is important to note that the effects of sympathetic nerve stimulation on cerebral blood flow depends on the species studied. In cats and dogs, the vasoconstrictor response is minimal (Heistad *et*

Blood Vessels and Lymphatics in Organ Systems
Copyright © 1984 by Academic Press, Inc.

al., 1978), whereas in rabbits (Sercombe *et al.*, 1978) and in primates (Heistad *et al.*, 1978), a more pronounced effect has been observed.

Although considerable controversy exists, the preponderant evidence suggests that, under normal conditions, the cerebral vessels are not very responsive to reflex regulatory mechanisms which, in other vascular beds, have pronounced vasoconstrictor effects. For example, cerebral blood flow remains invariant during alterations in activity of arterial baroreceptors (Rapela *et al.*, 1967) and of arterial chemoreceptors (Heistad *et al.*, 1976; Traystman *et al.*, 1978).

Under abnormal conditions, however, the responses may be different. For example, the effect of activation of sympathetic adrenergic nerves on cerebral blood flow is more pronounced during marked arterial hypertension (Heistad *et al.*, 1978). Also, these nerves are physiologically activated and exert small, but significant vasoconstriction of the larger pial vessels during CO_2 inhalation (Wei *et al.*, 1980b) and increase cerebral vascular resistance during severe hypotension (Fitch *et al.*, 1975).

Sympathetic nerves seem to be important in protecting the cerebral circulation from breakdown of the blood–brain barrier and from the occurrence of stroke in stroke-prone spontaneously hypertensive rats (Sadoshima *et al.*, 1981). It would appear from these studies that the main function of sympathetic adrenergic nerves is to induce constriction of the larger vessels under certain conditions. The mechanism of activation of cholinergic and peptidergic nerves and their potential physiologic functions are not known.

3. Chemical Regulation of Cerebral Blood Flow

Most vasoactive agents circulating in the blood do not have access to the vascular smooth muscle of cerebral vessels because of the presence of the blood–brain barrier. However, two agents, carbon dioxide and oxygen, do have a most important action on cerebral blood flow.

a. Carbon dioxide: This gas has a pronounced action, with arterial hypercapnia dilating cerebral vessels and increasing blood flow and arterial hypocapnia having the reverse effect. Such responses to CO_2 are readily reproducible and frequently are used as a measure of the reactivity of the cerebral circulation.

The primary mechanism of the vasoactive effect of arterial blood CO_2 on cerebral vessels is a local action (Kontos *et al.*, 1977a) mediated via change in extracellular fluid pH (Kontos *et al.*, 1977b). The blood–brain barrier is freely permeable to molecular CO_2 but relatively impermeable to hydrogen ion or bicarbonate ion. Therefore, the pH in the vicinity of cerebral vascular smooth muscle is determined by the local bicarbonate ion concentration and by the P_{CO_2}, which, in turn, is dependent on diffusion of CO_2 from the blood and from cerebral tissue. These features explain the marked influence on cerebral vessels of changes in arterial blood CO_2 and their relative insensitivity to alterations in blood pH (Lassen, 1968).

It has been proposed that the vasodilator effect of arterial hypercapnia is mediated by release of vasodilator prostaglandins. Such a view is based on the observation that indomethacin, a cyclooxygenase inhibitor, reduces the response to CO_2 inhalation in baboons (Pickard and MacKenzie, 1973) and rats (Sakabe and Siesjö, 1979). However, it is unlikely that this suggestion is correct, since subsequent experiments on cats failed to modify the vasodilator effect of CO_2 on pial arterioles (Wei *et al.*, 1980a) or on the cerebral circulation as a whole (Busija and Heistad, 1983), despite the fact that the effectiveness and specificity of indomethacin were demonstrated.

Bicarbonate ion concentration in cerebrospinal fluid (CSF) remains relatively constant during acute alterations in systemic pH. However, with prolonged exposure to abnormal atmospheres, such as protracted inhalation of CO_2, exposure to high altitude, or prolonged hypercapnia as a result of disease, there is an alteration in the concentration of bicarbonate ion in the CSF and in the cerebral extracellular fluid. This response seems to be related to changes in the rates of generation of bicarbonate, which, in turn, are dependent on the prevailing P_{CO_2} (Maren, 1979). During hypercapnia, extracellular fluid bicarbonate ion concentration increases, whereas during hypocapnia (as might occur as a result of hyperventilation induced by hypoxia), bicarbonate ion concentration decreases. This alteration then is reflected in a changed reactivity of the cerebral vessels to CO_2 inhalation, with the result that responsiveness is decreased following prolonged hypercapnia and is increased following hypocapnia (Levasseur *et al.*, 1979). Other mechanisms involving reflex or other remote effects of CO_2 seem to be of minor importance in determining the response of the cerebral circulation to alterations in Pa_{CO_2}.

b. Arterial hypoxia: This state dilates cerebral arterioles (Kontos *et al.*, 1978b) and increases cerebral blood flow (Shapiro *et al.*, 1970). The vasodilator effect of hypoxia on cerebral vessels is mediated mainly by local mechanisms, since it can be counteracted completely by supplying sufficient oxygen via local application of oxygenated fluorocarbons. The effect of hypoxia seems to depend on induction of reduced oxygen tension within the tissue and not on a direct action of low oxygen on vascular smooth muscle (Kontos *et al.*, 1978b). Tissue hypoxia, in turn, results in a release of vasoactive agents which diffuse to the vascular smooth muscle and cause relaxation. A prime candidate as mediator of this response is adenosine, the concentration of which in brain rises markedly during hypoxia (Winn *et al.*, 1981).

4. METABOLIC REGULATION OF CEREBRAL BLOOD FLOW

Changes in neuronal activity of portions of the brain are accompanied by alterations in metabolism and by corresponding alterations in regional cerebral blood flow. These variations are restricted to the active areas and may not be apparent from measurements of total cerebral blood flow (Ingvar, 1976). The coupling between cerebral blood flow, on the one hand, and cerebral metabolism and cerebral function, on the other, has been well demonstrated (Sokoloff, 1981a).

a. Mode of action: The mechanism by which the coupling between blood flow and cerebral metabolism is accomplished is unclear. It is believed to involve the release of vasoactive substances from neural cells. These diffuse through the extracellular fluid space and exert their action directly on vascular muscle. It is likely that more than one vasoactive agent are involved. The most promising agents for consideration are adenosine and potassium ions. Adenosine, which dilates cerebral vessels (Wahl and Kuschinsky, 1976), originates in the breakdown of its phosphorylated derivatives and is very quickly released into the extracellular fluid space in response to ischemia, hypoxia, hypotension, or increases in metabolism during seizures (Winn *et al.*, 1981). The common denominator in release of adenosine may be cellular P_{O_2}, but the exact link between the latter and adenosine is not known. Potassium also is a dilator of cerebral arterioles (Kuschinsky *et al.*, 1972), and it is likewise released during increased activity (Lothman *et al.*, 1975), but not during moderate changes in arterial blood pressure (Wahl and Kuschinsky, 1979). Other potential mediators are hydrogen ion, which may be involved in the later stages of metabolic vasodilation (Kuschinsky and Wahl, 1979), and increased osmolarity or decreased calcium ion concentration.

5. AUTOREGULATION OF CEREBRAL BLOOD FLOW

Alterations in perfusion pressure exert both passive and active changes in the cerebral circulation. In the absence of responsive vascular smooth muscle, decreases in arterial blood pressure cause passive reductions in vascular caliber, whereas increases have the opposite effect (Kontos *et al.*, 1981).

Passive behavior of cerebral vessels is modified by changes in the activity of vascular smooth muscle. Increases in arterial blood pressure in an active vascular bed induce arteriolar constriction and rises in cerebral vascular resistance, whereas decreases produce arteriolar dilatation and reductions in cerebral vascular resistance (Lassen, 1964; Kontos *et al.*, 1978a; MacKenzie *et al.*, 1976b, 1979). These compensatory responses, which tend to maintain cerebral blood flow relatively constant over a wide range of arterial blood pressure, constitute the phenomenon of autoregulation. The range of blood pressure over which autoregulation is dominant varies considerably depending on the conditions of testing. In dilated vascular beds, it is narrow, and the relation between flow and pressure is steep. The reverse features characterize relatively constricted vascular beds. Outside the range of control by autoregulation, cerebral blood flow changes more rapidly per unit variation in perfusion pressure. At high pressures, it increases and cerebral blood vessels dilate, the latter response possibly becoming irreversible at very high pressures. At low pressures, cerebral blood flow falls as the result of passive collapse of some vessels. Vasodilator responses are also seen during increases in intracranial pressure (Miller *et al.*, 1973). The changes produced by elevated venous pressure have been more variable, with both vasodilator (Moyer *et al.*, 1954) and vasoconstrictor reactions (Ekström-Jodal, 1970) having been observed.

Autoregulation is the result of the various local regulatory mechanisms. The action of vasomotor nerves is not necessary, although these structures influence pressure–flow relations as a result of alteration in the state of the vascular bed. Two mechanisms have been proposed to explain autoregulation: (1) the myogenic mechanism, which is based on the premise that vascular muscle responds to increased stretch or tension with contraction and to decreased stretch or tension with relaxation; and (2) the metabolic mechanism, which holds that adjustments of blood flow, in response to alterations in pressure, are due to passive changes in flow, which then affect the concentration of metabolites in the vicinity of vascular smooth muscle. At present, the evidence is not conclusive regarding which mechanism plays the dominant role. In pial arterioles, however, there is strong evidence that metabolic mechanisms are predominant, as, for example, in the case of the cerebral vascular responses to arterial hypotension (Kontos *et al.*, 1978b), a state in which there is tissue hypoxia, with secondary release of metabolites.

6. FUTURE AVENUES OF INVESTIGATION

In the past several years, rapid progress has been made in the understanding of the physiologic mechanisms involved in regulation of the cerebral circulation. Several aspects, however, remain controversial or are not fully understood. It would appear that future investigations directed toward the following goals would be fruitful: (1) a clearer definition of the physiologic function of the cerebral adrenergic nerves; (2) identification of the mechanism of activation of the cerebral vasodilator nerves and the understanding of their physiologic role; (3) gathering of more conclusive evidence, anatomic as well as functional, about the potential existence of a central adrenergic vasomotor pathway; (4) clarification of the role of various vasoactive agents in the coupling between cerebral metabolism and blood flow, including identification of the mechanisms which regulate release and disposal of these agents; (5) more complete understanding of the mechanisms responsible for cerebral vascular autoregulatory responses to changes in perfusion pressure.

D. PHARMACOLOGY OF BLOOD CIRCULATION*

By D. D. HEISTAD AND H. A. KONTOS

Cerebral vessels are very responsive to changes in brain metabolism, arterial pressure, blood gases, and pH of arterial blood. In contrast, many humoral stimuli and drugs that produce large changes in resistance in other vascular beds have little effect on cerebral vessels. Several specialized features of the cerebral circulation account for the relative insensitivity to humoral and pharmacologic stimuli. These unusual adaptive features protect the brain from fluctuations in blood constituents and facilitate normal function of neurons.

1. DETERMINANTS OF CEREBROVASCULAR RESPONSES

a. Vascular muscle: Cerebral vessels *in vitro* are able to generate strong contractions in response to some stimuli but not to others. For example, the effective concentration sufficient to elicit 50% response (EC_{50}) for serotonin is similar in the basilar and the saphenous arteries (Bevan *et al.*, 1975). On the other hand, cerebral arteries are almost 100 times less sensitive to norepinephrine than are extracranial segments of these vessels (Bevan, 1979).

*Supported by Research Grants HL16066, HL21851, and NS12587, Program Project Grant HL14388, and SCOR Grant HL14251 from the National Institutes of Health, by a Medical Investigatorship and Research Funds from the Veterans Administration, and by a Contract from the United Army Research and Development Command.

One factor that may contribute to variations in reactivity of cerebral as compared with other systemic arteries may relate to differences in dependence on intracellular and extracellular calcium. Sustained contractile responses to norepinephrine, but not to serotonin, are almost exclusively dependent on influx of extracellular calcium in the basilar artery (McCalden and Bevan, 1981). In the ear artery, contractile responses to both norepinephrine and serotonin are produced by release of intracellular calcium, as well as by influx of extracellular calcium. It therefore is possible that the relative insensitivity of cerebral vessels to norepinephrine is related to the unusual receptor-contraction coupling mechanism.

b. *Blood–brain barrier:* Entry of proteins and polar substances from the blood into extracellular fluid of the brain is effectively prevented by the blood–brain barrier (BBB). In addition, an enzymatic barrier limits extraction of catecholamines during one pass through the cerebral circulation to approximately 5% (Oldendorf, 1971). High levels of monoamine oxidase activity apparently degrade catecholamines that pass the endothelium (Hardebo and Owman, 1980). Thus, not only are cerebral vessels relatively insensitive to catecholamines, but the BBB limits access of these agents.

The BBB is absent in several sites in the brain that have a major role in cardiovascular regulation. These include the area postrema and some portions of the hypothalamus. As a result, although circulating catecholamines, angiotensin, and other vasoactive substances are excluded from most areas of the brain, they still may have important central effects through their access to the area postrema and hypothalamus.

Active transport mediates the entry into the brain of amine precursors (Edvinsson and MacKenzie, 1977), such as L-dopa. This mechanism is the basis for repletion of brain dopamine in the treatment of Parkinson's disease. A number of drugs, such as amphetamine, propranolol, and atropine, rapidly enter the brain by a carrier-mediated process or because they are lipid soluble, and have important central effects.

Permeability of the BBB can be increased by several stimuli, such as acute hypertension (Johansson *et al.*, 1970) and hyperosmotic solutions (Rapoport, 1970), both of which produce transient and reversible disruption of the BBB. Drugs that normally have no direct effect on cerebral vessels may alter cerebral blood flow after disruption of the barrier (MacKenzie *et al.*, 1976a).

c. *Indirect effects mediated by changes in cerebral metabolism:* The net effect of hormones and drugs on cerebral vessels is a summation of direct action and indirect effects mediated by changes in cerebral metabolism. The resultant of direct and indirect effects is similar to the physiologic responses in the coronary circulation.

2. METHODS FOR STUDY OF PHARMACOLOGIC EFFECTS ON BRAIN VESSELS

a. Vascular segments in vitro: By studying strips or segments of cerebral vessels *in vitro*, experimental conditions can be controlled carefully and changes in cerebral metabolism can be avoided. These studies have been criticized because only relatively large arteries, and not arterioles, have been investigated. However, because large cerebral vessels account for a major portion of total cerebral vascular resistance (Heistad and Kontos, 1983), it is appropriate to consider them as resistance vessels and to examine their responses.

b. Observation of pial vessels: Direct observation of vessels on the surface of the brain is a useful method for the study of the cerebral circulation. Substances can be applied locally by micropipettes (Kuschinsky and Wahl, 1975) or by superfusion (Levasseur *et al.*, 1975). The method allows examination of direct effects on cerebral vessels of hormones and drugs that normally cannot penetrate the BBB.

c. Measurement of cerebral blood flow: Among the numerous methods that have been used to measure cerebral blood flow, the appropriate one depends on the experimental conditions and objectives. First, to measure cerebral blood flow continuously, it is necessary to study cerebral venous outflow (Traystman *et al.*, 1978) or velocity of flow through a pial artery in which diameter is determined simultaneously (Busija *et al.*, 1981). Second, to measure distribution of cerebral blood flow with great spatial resolution, use of [14C]iodoantipyrine is the preferred method (Sakurada *et al.*, 1978). Third, to investigate regional cerebral blood flow repeatedly, the microsphere method is appropriate (Marcus *et al.*, 1976). Fourth, determination of xenon-133 clearance after injection into the internal carotid artery is the most accurate method for the investigation of cerebral blood flow in humans. The method is not used widely, however, because intracarotid injections are not commonly made. New, accurate methods are needed for measurement of cerebral blood flow in humans.

d. Measurement of cerebral metabolism: Cerebral oxygen consumption can be determined with the Fick principle by measuring cerebral blood flow and the arteriovenous oxygen difference. An important advance is the introduction of the [14C]deoxyglucose method for the study of rate of glucose uptake (Sokoloff, 1981b). This approach allows determination of metabolic rate in various parts of the brain, with excellent spatial resolution.

3. Responses to Humoral Stimuli and Drugs

a. Amines: Norepinephrine produces only modest contraction in cerebral vessels *in vitro* (Bevan *et al.*, 1975; Bevan, 1979; Harder *et al.*, 1981) and modest constriction of pial arteries *in vivo* (Kuschinsky and Wahl, 1975; Wei *et al.*, 1975). Intravenous infusion of norepinephrine constricts cerebral vessels, but such constriction is an autoregulatory response to the resulting hypertension rather than due to a direct effect on cerebral vessels. Intracarotid infusion of norepinephrine has little effect on cerebral blood flow (MacKenzie *et al.*, 1976a), presumably because only a minimal amount of the agent passes the BBB (Oldendorf, 1971). When the latter is disrupted, the drug increases flow by raising cerebral metabolism (MacKenzie *et al.*, 1976a).

Stimulation of β-adrenergic receptors produces relaxation of cerebral vessels *in vitro* when the vessels have "tone." *In vivo*, β-adrenergic agonists also produce cerebral vasodilatation (Edvinsson and MacKenzie, 1977), although the responses are relatively small. These studies indicate that cerebral vessels have β receptors.

Dopamine, like norepinephrine, penetrates the BBB only minimally (Oldendorf, 1971). Stimulation of dopaminergic receptors by apomorphine, a drug which readily crosses the BBB, significantly increases cerebral blood flow (McCulloch and Harper, 1977). The augmentation in flow is secondary to a rise in cerebral metabolism, rather than to a direct effect on cerebral vascular muscle.

Serotonin (5-hydroxytryptamine) produces marked contraction of cerebral vessels *in vitro* (Bevan *et al.*, 1975), and when applied to pial vessels, the same reaction is noted *in vivo*. There are numerous studies on the responses to intracarotid infusion of serotonin (see Edvinsson and MacKenzie, 1977) which indicate that the changes are complex. It appears that the drug constricts large cerebral arteries at the same time that small vessels dilate, so that the reductions in blood flow are usually small. When cerebral vessels are already dilated, serotonin produces larger decreases in blood flow.

Application of histamine to cerebral vessels *in vitro* produces contraction mediated by H_1 receptors and relaxation mediated by H_2 receptors (Heinemann *et al.*, 1977). Intracarotid infusion of the drug does not affect cerebral blood flow when the BBB is intact. When the latter is disrupted, infusion of the drug increases cerebral blood flow, apparently by stimulating vascular H_1 and H_2 receptors (Gross *et al.*, 1981).

b. Vasodilators: Intravenous infusion of vasodilator drugs results in dilatation of cerebral vessels. The response is primarily an autoregulatory reaction to hypotension, however, and not a direct effect of the drugs on cerebral vessels (Kontos *et al.*, 1978a).

Recent studies indicate that polypeptides affect cerebral vessels. Vasoactive intestinal peptide (VIP) relaxes them *in vitro* and dilates pial arterioles *in vivo* (Traystman *et al.*, 1982). It appears that dilator effects of VIP on cerebral arterioles are mediated by increased prostaglandin synthesis. Intracarotid infusion and intraventricular administration of VIP produce moderate elevations in cerebral blood flow.

Infusions of adenosine and acetylcholine into the carotid artery have been reported to have no effect on cerebral blood flow. It was suggested that the BBB prevents access of these agonists to cerebral vascular muscle. However, the brain uptake of adenosine is 9% (Cornford and Oldendorf, 1975) and that of acetylcholine is 4% (Oldendorf, 1971), so that infusion of high doses might be expected to have appreciable effects on cerebral vessels. In this regard, recent studies have demonstrated that intracarotid infusions of large amounts of either drug produce approximately twofold increases in cerebral blood flow in rabbits (Heistad *et al.*, 1980, 1981). Previous negative results in cats and dogs may be explained by the fact that vasodilator agents produce more vasodilatation in skeletal muscle than in brain and that common carotid injections of vasodilator drugs in these two species are diverted to extracranial tissue, so that only minimal doses of the drugs reach the brain.

  c. Prostaglandins: Naturally occurring prostaglandins, except for PGF_2, dilate small cerebral vessels (Heistad and Kontos, 1983). The predominant prostaglandins produced by brain parenchyma vary in different species, but the predominant one formed by brain vessels is PGI_2.

  d. Calcium entry blockers: Responses to some vasoconstrictor agents, such as norepinephrine, are more dependent on influx of extracellular calcium in cerebral vessels than is the case for other vessels (McCalden and Bevan, 1981). Thus, it might be anticipated that some calcium entry blockers would have pronounced effects on cerebral vessels. Studies *in vitro* indicate that nimodipine, a calcium blocker, is a more potent inhibitor of amine-induced contractions in basilar artery than in saphenous artery (Towart, 1981). Some specificity of this effect has been demonstrated by the observation that nimodipine is equipotent in inhibiting potassium-induced contraction in basilar and saphenous arteries. Intravenous or intracarotid infusion of nimodipine increases cerebral blood flow by its direct effect on cerebral vessels, without any rise in cerebral metabolism (Harper *et al.*, 1981).

4. FUTURE AVENUES FOR INVESTIGATION

The recent development of new methods to measure cerebral blood flow and metabolism in various regions of the brain will permit examination of the effects of drugs under normal conditions and in disease states characterized by focal abnormalities, such as brain tumors and strokes. In addition, application of methods to measure membrane potential (Harder *et al.*, 1981) and to inhibit calcium entry (Towart, 1981) should provide important new insights into cellular mechanisms that mediate responses of cerebral vessels to pharmacologic agents.

E. PATHOGENESIS AND PATHOLOGY OF BLOOD
CIRCULATION

By L. R. Caplan and C. Thomas

1. Introduction

Stroke has long been a well-known clinical phenomenon. Only within the last few decades, however, has clarification of its morbid anatomy, pathophysiology, and pathogenesis been possible. Advances in the understanding of cerebrovascular disease has followed the widespread application of newer diagnostic techniques, able to image the cerebral vasculature and its function during life. Among these are cerebral angiography with magnification and angiotomography, computerized axial tomography, positron emission tomography, and methods which quantitate cerebral blood flow. Advances also relate to the increased number of clinicians, investigators, and morphologists studying cerebrovascular disease.

2. Vascular Pathology Underlying Brain
Infarction

a. Atherosclerosis: The vessels supplying the brain are morphologically heterogeneous, as is the degenerative process, atherosclerosis, which affects them. In the larger extracranial vessels, such as the carotid and vertebral arteries, the major pathologic alteration involves the intima. In the superficial subintimal regions, collagenous and fibrous tissue proliferate, whereas in the deeper layers lipid-filled foam cells accumulate (Baker and Ianone, 1959a). The internal elastic lamina often is folded, frayed, and split. When the vessel is observed from the luminal side, these changes are reflected as flat, raised, ulcerated plaques, frequently demonstrating calcification. Often, platelet aggregates, thrombin, or fresh clot is deposited on the surface of the plaque, leading to further obliteration of the lumen. Fresh clot may extend rostrally to the next branch, e.g., a thrombus occluding the internal carotid artery may propagate into the intracranial ophthalmic artery branch, or, if it is loosely adherent, it may be liberated and embolize distally. With time, the clot becomes more adherent to the vessel wall and is less likely to become free in the bloodstream. Another mechanism for occlusion of large extracranial vessels is hemorrhage into a plaque, a common finding in the morphologic study of Imparato *et al.* (1979) (Fig. 5.8). When larger extracranial vessels occlude, the decrease in distal perfusion pressure and the local metabolic changes caused by the ischemia, such as acidosis, act to promote collateral circulation. Survival of ischemic brain depends on the adequacy of collateral circulation, the resistance of the particular brain tissues to hypoxia, and extension or embolization of the thrombus.

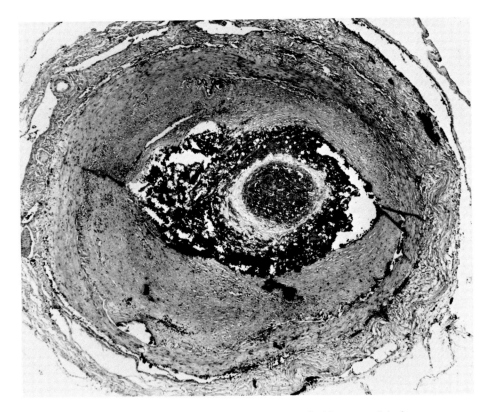

FIG. 5.8. Hemorrhage into an atheromatous plaque with obliteration of the lumen.

The atheromatous process maximally involves the carotid bifurcation, often extending 15–30 mm along the internal carotid artery (Fisher et al., 1965). The internal carotid artery above this level is relatively free of atherosclerosis up to the carotid siphon. Ulceration is infrequent in the internal carotid artery within the siphon, but this portion of vessel often is calcified. The inner surface may contain bony, hard, yellow-brown excresences, pits, and shiny smooth heavily calcified regions (Fisher et al., 1965). The vertebral artery is most severely diseased at its origin, but it may have plaques scattered throughout its nuchal course especially adjacent to spondylitic bars. The distal vertebral artery, beginning within a few millimeters of dural penetration, also is often the site of severe atherosclerotic narrowing. The basilar artery is frequently severely diseased, especially the proximal third.

The atherosclerotic process is morphologically different in the various large branch vessels. For example, in the anterior and middle cerebral arteries, the vascular media is primarily involved. There is some increase in connective tissue within the intima where thin strands of collagen fibers are deposited between the endothelium and elastica interna. Usually ulcerated plaques or subintimal lipid deposits are not noted. The internal elastic membrane may reveal segmental swelling or irregularity as demonstrated by elastic stains. In the media, collagenous fibers increase, replacing the muscle elements and eventually producing heavy bands of collagen fibers which become acellular, hyalinized, and calcified. Adventitial connective tissue may encroach on the media. Ultimately the muscle of the media is replaced by fibrous elements. Atherosclerosis is most severe at the origin and horizontal segments of the anterior, middle, and posterior cerebral arteries, before these vessels branch, whereas the more distal branches are spared. The smaller intracerebral arteries (150–500 μm) may demonstrate an increase in collagen usually beginning at the adventitial surface and compressing the vascular media. Though this process results in replacement of the media with connective tissue, it ordinarily does not encroach on the lumen or thicken the wall (Baker and Ianone, 1959b). Calcium deposits are often observed in the media of small arteries and capillaries in the basal ganglia and cerebellar dentate nuclei, generally without associated ischemia or an inflammatory or glial reaction. (For a detailed discussion of the pathogenesis and pathology of arteriosclerosis, see Section E-2, Chapter 1.)

b. Emboli: A large body of evidence supports the notion that occlusion of the smaller more superficial cerebral arteries is due to embolization from more proximal sites. In one autopsy study, 48% of cerebral occlusions arose from a cardiac source, most not recognized during life (Blackwood et al., 1969). Newer cardiac noninvasive imaging and arrhythmia monitoring techniques have clearly documented mitral valve prolapse, calcified mitral annulus, myocardiopathy, and chronic or intermittent atrial arrhythmias as sources of cerebral embolization. In addition, ulcerated plaques within the proximal extracranial arteries form the nidus for platelet clumps on their surfaces, with subsequent cerebral embolization. Only one-sixth of occlusions within the middle cerebral artery in the study of L'hermitte *et al.* (1970) could be attributed to local atherosclerosis; the others were caused by emboli from proximal vessels or the heart. Castaigne *et al.* (1973) studied the posterior circulation and found that nearly all occlusions within the posterior cerebral arteries or their branches represented either emboli or direct extension from basilar artery thrombus. In some individuals who intravenously inject drugs manufactured only for oral use, such as methylphenidate (Ritalin), pentazocine hydrochloride (Talwin), or pyribenzamine, the ophthalmic and smaller cerebral vessels, as well as others, may contain birefringent matter which can be seen with the polarizing microscope. This material is microcrystalline cellulose or magnesium silicate and originates from the filler material used in the manufacture of the drug; it frequently incites a granulomatous response within the vessel wall (Caplan *et al.*, 1982).

c. Vasculopathy in superficial vessels within arterial boundary zones: Some pial vessels over the convexal surface of the brain may contain segments of white hyaline material. The white external appearance, due to chalky white platelet clumps within the vessel, is found only in the pia overlying cerebral infarcts within the arterial border zones, that is, the regions of the brain lying between arterial boundaries. Romanul and Abramowicz (1964) defined the pathogenesis of this phenomenon as due to stasis related to low perfusion pressures with *in situ* aggregation of platelets. Later, the platelet clumps become translucent and hyalinized. Previously this change had been attributed by others to cerebral Buerger's disease, a conclusion that has no basis.

d. Fibromuscular dysplasia: This disorder, initially described in the renal arteries of hypertensive young women, is now known to affect the cerebral vasculature. The lesions are most commonly found in the pharyngeal portion of the internal carotid artery and the nuchal vertebral artery. The carotid siphon and the middle and posterior cerebral arteries are occasionally involved. Several morphologic subtypes are recognized: (1) pure intimal fibroplasia with fibroelastic thickening of the intima; (2) medial fibroelastosis, in which segmental fibrosis alternates with areas of deficient medial tissue, with the development of aneurysmal bulges; (3) subadventitial fibroplasia in which a cuff of collagen replaces the outer layers of the media; and (4) fibromuscular hyperplasia, in which the essential changes are a disordered arrangement of muscle fibers and medial fibrous tissue (Pollock and Jackson, 1971). The lesions are frequently mixed, and all may be related to a fibroblast-like transformation of smooth muscle cells. Cerebral angiography has added information about the pathophysiology of the condition. Long regions of smooth concentric arterial narrowing, aneurysmal bulges, and alternating bands of medial contraction give the vessel a beaded appearance. The mechanism of ischemia distal to the lesions is uncertain. There is an increased incidence of berry aneurysms in patients with fibromuscular dysplasia.

e. Arterial dissection: The most common locus of spontaneous dissection is the internal carotid artery where the lesion usually begins 2–3 cm distal to the origin or 2–4 cm proximal to the skull base. Dissections of the basilar and middle cerebral arteries have also been described. Blood, having dissected the lumen, usually reenters it in the form of a thrombus which may then embolize distally. With time, the dissection heals. Occult trauma, congenital loose connective tissue in the media, and edema of the media due to migraine have all been proposed as causes of dissection.

f. Arteritis: Acute necrotizing vasculitis may be caused by some gram-negative and gram-positive bacteria. *Aspergillus, Mucor,* and *Candida* produce vascular necrosis with superimposed thrombosis of small arteries. In tuberculous meningitis, the adventitia, media, and often intima of the meningeal arteries contain foci of necrosis, granulomatous inflammation, and frequent thrombosis (Fig. 5.9). Subintimal fibrous tissue proliferates. The lesions usually involve the base of the brain, especially the brainstem, penetrating vessels, and branches of the middle cerebral artery (Dastur and Udani, 1966). In syphilis, inflammatory cells, chiefly lymphocytes and plasma cells, surround the walls of vessels as thick cuffs in the Virchow–Robin spaces. Meninges are thickened. Intimal proliferation may be concentric or eccentric and is collagenous. The media is thin, and the adventitia is thick and infiltrated by lymphoplasmacytic cells. In the larger meningeal arteries, the lesions especially involve the vasa vasorum. Intimal thickening and fibrosis, possibly due to destruction of the media and adventitia, lead to luminal encroachment.

Giant cell (temporal) arteritis is a subacute inflammatory disease which especially affects small branches of the external carotid artery, such as the temporal and occipital arteries. Of special importance is involvement of the ophthalmic artery and its branches, producing blindness in one or both eyes in about 25% of cases. The internal carotid and vertebral arteries may be affected in sites located just before the vessels pierce the dura. The lesions are usually

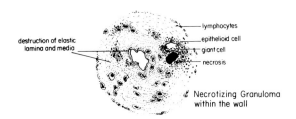

FIG. 5.9. Arteritis in tuberculosis.

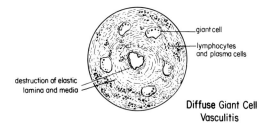

FIG. 5.10. Giant cell arteritis.

segmental, with thickening of the wall and eccentric narrowing of the lumen. Intimal thickening; destruction of the internal elastic lamina and media; diffuse inflammatory cell infiltration with lymphocytes, plasma cells, and a few neutrophils; and epithelioid and giant cells characterize the lesion (Fig. 5.10). The adventitia may also be involved and may contain giant cells.

Sarcoidosis of the brain is generally limited to the basal meninges and pituitary–hypothalamic regions. Characteristic noncaseating granulomas are situated along the adventitia of arteries (Fig. 5.11) and especially along veins where gray-white sheaths surround the vessels and lead to dilatation and focal destruction of the walls. Arterial walls are usually not necrosed or thrombosed. There are small ischemic lesions and hemorrhage from injured veins, as well as diffuse meningeal infiltration.

Central nervous system arteritis may complicate polyarteritis nodosa, Wegener's granulomatosis, lymphomatoid granulomatosis, systemic lupus erythematosus, and rheumatoid arthritis. The vascular lesions are usually focal necrotizing arteritis characterized by subendothelial edema, fibrinoid necrosis of the media, destruction of the internal elastic lamina, and infiltration of the wall by neutrophils, lymphocytes, and occasionally eosinophils. In the chronic stage there are subendothelial fibrosis and thickened media and adventitia. In systemic lupus erythematosus immune complex deposits may be demonstrable.

In some of the inherited disorders of connective tissue, as for example, pseudoxanthoma elasticum, Ehlers–Danlos syndrome, homocystinuria con-

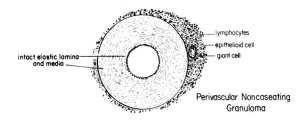

FIG. 5.11. Sarcoid arteritis.

genital complement deficiency, and Marfan's syndrome, collagen and elastic tissue of cerebral vessels degenerate, with frequent aneurysmal formation and vascular occlusions.

3. Vascular Pathology Associated with Cerebral and Subarachnoid Hemorrhage

Intracerebral hemorrhage is most commonly caused by leakage of blood from lipohyalinotic penetrating vessels in hypertensive patients. The initial accumulation of blood leads to compression of satellite vessels, which themselves undergo pressure damage and leak. Alteration of blood clotting factors (as for example, in patients taking warfarin or heparin or suffering from hemophilia, thrombocytopenia, or hypoprothrombinemia) and trauma commonly cause hemorrhage from normal appearing vessels.

a. Aneurysms: The most common cause of nontraumatic subarachnoid hemorrhage is leakage of blood under arterial pressure from an aneurysm. Such an arterial lesion (variously called congenital, saccular, or berry aneurysm) is produced by a combination of congenital and acquired factors. Aneurysms form at bifurcation points, especially at the posterior communicating–internal carotid artery junction, anterior cerebral–anterior communicating artery junction, and the trifurcation of the middle cerebral artery. Congenital medial defects probably underlie the condition, as supported by the work of Smith and Windsor (1961) who found such medial defects in 30.1% of 509 arterial bifurcations in infants at sites corresponding to the usual distribution of aneurysms. However, Stehbens (1975) reviewed the ultrastructure of early aneurysms and identified other changes in addition to the medial defect: lamination, thickening, redundancy, and separation of basement membranes; absence of elastica; and abundant cellular debris and extracellular lipid and lipophages. With increasing age and arterial hypertension, these medial defects enlarge, leading to aneurysm formation. Usually only the larger ones (greater than 8 mm) rupture. Thrombus is often found within the larger aneurysms and may form the nidus for embolization to distal vessels and subsequent ischemic stroke (Duncan *et al.*, 1979). Occasionally sclerotic changes within the wall of the vertebral, basilar, and carotid arteries can lead to fusiform enlargement, but such an abnormality rarely leads to hemorrhage. Mycotic aneurysms are due to embolization of infected vegetations which produce microinfarction of the vessel wall, with subsequent aneurysmal dilatation. A similar phenomenon is seen in patients with cardiac myxomas. Mycotic and myxoma aneurysms usually are more distally located in the arterial tree and are less common along the circle of Willis, the usual site of berry aneurysms. (For further discussion of the morphology and mechanics of aneurysms, see Section I-4a, Chapter 1; Section A-2, Chapter 2.)

b. Vasospasm: Recently, clinical investigators have identified a syndrome of "vasospasm" in patients with subarachnoid hemorrhage. Narrowing of cerebral arteries, especially in areas of maximal bleeding, can be visualized angiographically; functional spasm can be induced in the experimental animal by placing blood on the middle cerebral or basilar artery. The accompanying clinical features in the human include headache, stupor, and focal cerebral ischemia. Crompton (1964) first emphasized the frequent finding of cerebral infarction in patients coming to autopsy with aneurysmal subarchnoid hemorrhage. In humans, functional vasospasm may lead to subsequent pathologic changes in the involved vessels, a type of constrictive endarteropathy (Conway and McDonald, 1972). In monkeys subjected to experimental subarachnoid hemorrhage (Clower *et al.*, 1981), subintimal swelling and thickening are present within 3 days of the operative procedure, and subintimal proliferation, fibrous replacement of the medial smooth muscle layer, and interruption of the internal elastic membrane become progressively more severe during the next month.

c. Vascular malformations: Although arteriovenous and capillary malformations occur in many organs, they are rarely as problematic as in the brain, where they are an important cause of intracerebral, intraventricular, and subarachnoid hemorrhage and seizures. Telangiectases occur throughout the nervous system but are especially common in the pontine base, cerebellar cortex, and deep cerebral white matter. Grossly they are ill defined, nonencapsulated zones in which vessels may be closely compact or separated by fields of intervening neural tissue. Microscopically, telangiectasis consists of irregular saccular capillaries or venules, with relatively acellular walls devoid of muscular fibers and made up of collagen fibers or reticulin. Although draining veins may be enlarged, there are no direct arterial feeding vessels. Arteriovenous malformations, in contrast, occur outside or within brain parenchyma and consist of dilated hypertrophic arterial channels (which may show segmental dilatation, intimal sclerosis, and mural fibrosis) and veins (which make up the bulk of the lesions and contain more collagen fibers than usual). Pure venous angiomas also occur but are rare. (For discussion of vascular malformations of the spinal cord, see Section E-1, Chapter 6.)

4. VASCULOPATHIES ASSOCIATED WITH BOTH INFARCTION AND HEMORRHAGE

a. Lipohyalinosis and disease of small penetrating arteries: Small penetrating arteries (less than 200 μm), which originate from larger arteries to perforate the deeper structures in the brainstem and basal ganglia, are subject to a different type of vasculopathy, namely, lipohyalinosis. The lenticulostriate, thalamostriate, thalamoperforating, and basilar penetrating vessels are often the sites of medial hyaline and lipid deposition and fibrinoid change (Fisher, 1969). This process can lead to segmental arterial disorganization with obliteration of the lumen and ischemia distally. At times, lipohyalinosis is followed by obliteration of the lumen and at other times, by outpouchings from the lumen called Charcot–Bouchard aneurysms. Bleeding from the microaneurysms causes intracerebral hemorrhage (Cole and Yates, 1967). Lipohyalinosis is usually found in hypertensive patients and is the arteriopathy which underlies lacunar infarction and hypertensive intracerebral hemorrhage. Also atheroma within a larger vessel may at times block small vessels or extend into the orifice of a branch. For example, in the basilar artery, where pontine penetrating branches exit perpendicularly from the main vessel, atheroma in the latter may block such channels or extend into them as a junctional plaque (Fisher and Caplan, 1971).

b. *Congophilic (amyloid) angiopathy:* In some aged patients, small cerebral arteries and arterioles undergo degeneration, with the walls consisting of pale metachromatic material. The intima is spared, but elastic fibers are destroyed by the process and the lumen is narrowed. The arterial wall has an affinity for Congo red and is argyrophilic and PAS positive. At times amyloid material is prominent around the artery and almost forms a reduplication of the vessel. These vascular lesions are most prominent in the occipital lobes but are scattered through many cerebral regions, generally sparing penetrating vessel territory. They are often accompanied by an increased number of senile plaques and produce a clinical picture of cerebral infarctions and hemorrhage in elderly patients. The lesions are limited to the brain and have no connection with systemic amyloidosis.

c. *Vascular lesions associated with hematologic disease:* Sickle cell anemia affects both large and small arteries (Boros *et al.*, 1976). Large vessel disease takes the form of an occlusive arteriopathy, with endothelial proliferation and gradual luminal encroachment. Ectasia and aneurysm formation of smaller arteries can also lead to hemorrhages and infarcts in the brain. In acute myelogenous leukemia, there may be leukostatic thrombi with multiple petechiae and sometimes larger hemorrhages. Thrombosis of intracerebral arteries and veins is occasionally due to hemostasis in relation to abnormalities of the blood clotting system rather than as a result of intrinsic abnormalities of the vessels themselves. Under such circumstances, a relatively normal appearing vessel is filled with thrombus. Venous thrombosis may also be a sequel of polycythemia.

d. *Drug-induced vasculopathy:* Abuse of drugs may lead to a vasculopathy and cerebral hemorrhage or infarction. Citron *et al.* (1970) described fibrinoid necrosis of the arterial intima and media, with perivascular inflammatory infiltrates and nodose-like segmental swellings, in a group of multiple-drug abusers, all admitting to amphetamine abuse. The same changes can be produced in the experimental animal by injection of methamphetamine. Vasculopathy may result from a variety of drugs and is probably due to an immunogenic mechanism produced by repeated injection of antigenic material. Some drugs, especially amphetamine, cocaine, and phencyclidine, also cause hemorrhage by a sympatheticomimetic effect on the cerebral vessels without morphologic change.

References*

Baker, A., and Iannone, A. (1959a). Cerebrovascular disease. I. The large arteries of the circle of Willis. *Neurology* **9**, 321–332. (E)

Baker, A., and Iannone, A. (1959b). Cerebrovascular disease. II. The smaller intracerebral arterioles. *Neurology* **9**, 391–396. (E)

Bevan, J. A. (1979). Sites of transition between functional systemic and cerebral arteries of rabbits occur at embryological junctional sites. *Science* **204**, 635–637. (D)

Bevan, J. A., Duckles, S. P., and Lee, T. J. F. (1975). Histamine potentiation of nerve- and drug-induced responses of rabbit cerebral artery. *Circ. Res.* **36**, 647–653. (D)

Blackwood, W., Hallpike, J., Kocen, R., and Mair, W. (1969). Atheromatous disease of the carotid arterial system and embolization from the heart in cerebral infarction: A morbid anatomical study. *Brain* **92**, 897–910. (E)

Boros, L., Thomas, C., and Weiner, W. J. (1976). Large vessel disease in sickle cell anemia. *J. Neurol., Neurosurg. Psychiatry* **39**, 1236–1239. (E)

Busija, D. W., and Heistad, D. D. (1983). Effects of indomethacin on cerebral blood flow during hypercapnia in cats. *Am. J. Physiol.* **244**, H519–H524. (C)

Busija, D. W., Heistad, D. D., and Marcus, M. L. (1981). Continuous measurement of cerebral blood flow in anesthetized cats and dogs. *Am. J. Physiol.* **241**, H228–H234. (D)

Caplan, L., Thomas, C., and Banks, G. (1982). Central nervous system complications of Ts & Blues addiction. *Neurology* **32**, 623–628. (E)

Castaigne, P., L'hermitte, F., Gautier, J., Escourolle, R., Derouesne, C., Der Agopian, P., and Popa, C. (1973). Arterial occlusions in the vertebro-basilar system. *Brain* **96**, 133–154. (E)

Cervós-Navarro, J. (1977). The structural basis of an innervatory system of brain vessels. *In* "Neurogenic Control of Brain Circulation" (C. Owman and L. Edvinsson, eds.), pp. 75–89. Pergamon, Oxford. (B)

Cervós-Navarro, J. (1979). "Pathology of Cerebral Microcirculation," Adv. Neurol., Vol. 20. de Gruyter, Berlin. (B)

Cervós-Navarro, J., and Fritshke, E., eds. (1980). "Cerebral Microcirculation and Metabolism." Raven Press, New York. (B)

Cervós-Navarro, J., Matakas, F., Grčević, N., and Waltz, A. G., eds. (1974). "Pathology of Cerebral Microcirculation." de Gruyter, Berlin. (B)

Cervós-Navarro, J., Betz, E., Matakas, F., and Wüllenweber, R., eds. (1976). "The Cerebral Vessel Wall." Raven Press, New York. (B)

Citron, B., Halpern, M., McCarron, M., Lundberg, G., McCormick, R., Pincus, I., Tatter, D., and Havenback, B. (1970). Necrotizing angiitis associated with drug abuse. *N. Engl. J. Med.* **283**, 1003–1011. (E)

*In the reference list, the capital letter in parentheses at the end of each reference indicates the section in which it is cited.

Clower, B., Smith, R., Haining, J., and Lockard, J. (1981). Constrictive endarteropathy following experimental subarachnoid hemorrhage. *Stroke* **12**, 501–508. (E)

Cole, F., and Yates, P. (1967). Intracerebral microaneurysms and small cerebrovascular lesions. *Brain* **90**, 759–768. (E)

Congdon, E. D. (1922). Transformations of the aortic arch system during the development of the human embryo. *Carnegie Inst. Washington Publ.* **277**, 47–110. (A)

Conway, L., and McDonald, L. (1972). Structural changes of the intradural arteries following subarachnoid hemorrhage. *J. Neurosurg.* **37**, 715–723. (E)

Cornford, E. M., and Oldendorf, W. H. (1975). Independent blood-brain barrier transport systems for nucleic acid precursors. *Biochim. Biophys. Acta* **394**, 211–219. (D)

Crompton, M. R. (1964). The pathogenesis of cerebral infarction following the rupture of cerebral berry aneurysms. *Brain* **87**, 491–510. (E)

Dahl, E. (1973). The innervation of cerebral arteries. *J. Anat.* **115**, 53–63. (B)

Dahl, E., and Nelson, E. (1964). Electron microscopic observations in human intracranial arteries. II. Innervation. *Arch. Neurol. (Chicago)* **10**, 158–164. (B)

Dahl, E., Flora, G., and Nelson, E. (1965). Electron microscopic observations on normal human intracranial arteries. *Neurology* **15**, 132–140. (B)

Dastur, D. K., and Udani, P. M. (1966). The pathology and pathogenesis of tuberculous encephalopathy. *Acta Neuropathol.* **6**, 311–326. (E)

Duncan, A., Rumbaugh, C., and Caplan, L. (1979). Cerebral emboli disease: A complication of carotid aneurysms. *Radiology* **133**, 379–384. (E)

Edvinsson, L., and MacKenzie, E. T. (1977). Amine mechanisms in the cerebral circulation. *Pharmacol. Rev.* **28**, 275–348. (D)

Ekström-Jodal, B. (1970). Effect of increased venous pressure on cerebral blood flow in dogs. *Acta Physiol. Scand., Suppl.* **350**, 51–61. (C)

Fisher, C. (1969). The arterial lesions underlying lacunes. *Acta Neuropathol.* **12**, 1–15. (E)

Fisher, C., and Caplan, L. (1971). Basilar artery branch occlusion: A cause of pontine infarction. *Neurology* **21**, 900–905. (E)

Fisher, C., Gore, I., Okabe, N., and White, P. (1965). Atherosclerosis of the carotid and vertebral arteries—extracranial and intracranial. *J. Neuropathol. Exp. Neurol.* **24**, 455–476. (E)

Fitch, W., MacKenzie, E. T., and Harper, A. M. (1975). Effects of decreasing arterial blood pressure on cerebral blood flow in the baboon. *Circ. Res.* **37**, 550–557. (C)

Flora, G., Dahl, E., and Nelson, E. (1967). Electron microscopic observations on human intracranial arteries—changes seen with aging and atherosclerosis. *Arch. Neurol. (Chicago)* **17**, 162–173. (B)

Forbes, M. S., Rennels, M. L., and Nelson, E. (1977). Ultrastructure of pericytes in mouse heart. *Am. J. Anat.* **149**, 47–69. (B)

Gross, P. M., Harper, A. M., and Teasdale, G. M. (1981). Cerebral circulation and histamine. 1. Participation of vascular H_1 and H_2 receptors in vasodilatory responses to carotid arterial infusion. *J. Cereb. Blood Flow Metab.* **1**, 97–108. (D)

Hardebo, J. E., and Owman, C. (1980). Barrier mechanisms for neurotransmitter monoamines and their precursors at the blood–brain interface. *Ann. Neurol.* **8**, 1–11. (D)

Harder, D. R., Abel, P. W., and Hermsmeyer, K. (1981). Membrane electrical mechanism of basilar artery constriction and pial artery dilation by norepinephrine. *Circ. Res.* **49**, 1237–1242. (D)

Harper, A. M., Craigen, L., and Kazda, S. (1981). Effect of the calcium antagonist, nimodipine, on cerebral blood flow and metabolism in the primate. *J. Cereb. Blood Flow Metab.* **1**, 349–356. (D)

Hartman, B. K. (1973). The innervation of cerebral blood vessels by central noradrenergic neurons. *In* "Frontiers in Catecholamine Research" (E. Usdin and S. H. Snyder, eds.), pp. 91–96. Pergamon, Oxford. (B)

Hazama, F., Ozaki, T., and Amano, S. (1979). Scanning electron microscopic study of endothelial cells from spontaneously hypertensive rats. *Stroke* **10**, 245–252. (B)

Heinemann, U., Lux, H. D., and Gutnick, M. J. (1977). Extracellular free calcium and potassium during paroxysmal activity in the cerebral cortex of the cat. *Exp. Brain Res.* **27**, 237–243. (D)

Heistad, D. (1981). Summary of symposium on cerebral blood flow: Effect of nerves and neurotransmitters. *J. Cereb. Blood Flow Metab.* **1**, 447–450. (B)

Heistad, D. D., and Kontos, H. A. (in press). Regulation of cerebral circulation. In "Handbook of Physiology" (J. Shepherd and F. M. Abboud, eds.), Sect. 2, Vol. 4. Am. Physiol. Soc., Bethesda, Maryland. (D)

Heistad, D. D., Marcus, M. L., Ehrhardt, J. C., and Abboud, F. M. (1976). Effect of stimulation of carotid chemoreceptors on total and regional cerebral blood flow. *Circ. Res.* **38**, 20–25. (C)

Heistad, D. D., Marcus, M. L., and Gross, P. M. (1978). Effects of sympathetic nerves on cerebral vessels in dog, cat, and monkey. *Am. J. Physiol.* **235**, H544–H552. (C)

Heistad, D. D., Marcus, M. L., Said, S. I., and Gross, P. M. (1980). Effect of acetylcholine and vasoactive intestinal peptide on cerebral blood flow. *Am. J. Physiol.* **239**, H73–H80. (D)

Heistad, D. D., Marcus, M. L., Gourley, J. K., and Busija, D. W. (1981). Effect of adenosine and dipyridamole on cerebral blood flow. *Am. J. Physiol.* **240**, H775–H780. (D)

Hirano, A. (1974). Fine structural alterations of small vessels in the nervous system. In "Pathology of Cerebral Microcirculation" (J. Cervós-Navarro, F. Matakas, N. Grčević, and A. G. Waltz, eds.), pp. 203–217. de Gruyter, Berlin. (B)

Imparato, A., Riles, T., and Gorstein, F. (1979). The carotid bifurcation plaque: Pathological findings associated with cerebral ischemia. *Stroke* **10**, 238–245. (E)

Ingvar, D. H. (1976). Functional landscapes of the dominant hemisphere. *Brain Res.* **107**, 181–197. (C)

Iwayama, T., Furness, J. B., and Burnstock, G. (1970). Dual adrenergic and cholinergic innervation of the cerebral arteries of the rat. *Circ. Res.* **26**, 635–646. (B)

Johansson, B., Li, C.-L., Olsson, Y., and Klatzo, I. (1970). The effect of acute arterial hypertension on the blood–brain barrier to protein tracers. *Acta Neuropathol.* **16**, 117. (D)

Jones, E. G. (1970). On the mode of entry of blood vessels into the cerebral cortex. *J. Anat.* **106**, 502–520. (B)

Kontos, H. A., Wei, E. P., Raper, A. J., and Patterson, J. L., Jr. (1977a). Local mechanism of CO_2 action of cat pial arterioles. *Stroke* **8**, 226–229. (C)

Kontos, H. A., Raper, A. J., and Patterson, J. L., Jr. (1977b). Analysis of vasoactivity of local pH, P_{CO_2} and bicarbonate on pial vessels. *Stroke* **8**, 358–360. (C)

Kontos, H. A., Wei, E. P., Navari, R. M., Levasseur, J. E., Rosenblum, W. I., and Patterson, J. L., Jr. (1978a). Responses of cerebral arteries and arterioles to acute hypotension and hypertension. *Am. J. Physiol.* **234**, H371–H383. (C, D)

Kontos, H. A., Wei, E. P., Raper, A. J., Rosenblum, W. I., Navari, R. M., and Patterson, J. L., Jr. (1978b). Role of tissue hypoxia in local regulation of cerebral microcirculation. *Am. J. Physiol.* **234**, H582–H591. (C)

Kontos, H. A., Wei, E. P., Dietrich, W. D., Navari, R. M., Povlishock, J. T., Ghatak, N. R., Ellis, E. F., and Patterson, J. L., Jr. (1981). Mechanism of cerebral arteriolar abnormalities after acute hypertension. *Am. J. Physiol.* **240**, H511–H527. (C)

Kuschinsky, W., and Wahl, M. (1975). Alpha-receptor stimulation by endogenous and exogenous norepinephrine and blockade by phentolamine in pial arteries of cats. *Circ. Res.* **37**, 168–174. (C, D)

Kuschinsky, W., and Wahl, M. (1979). Perivascular pH and pial arterial diameter during bicuculline induced seizures in cats. *Pfluegers Arch.* **382**, 81–85. (C)

Kuschinsky, W., Wahl, M., Bosse, O., and Thurau, K. (1972). Perivascular potassium and pH as determinants of local pial arterial diameter in cats. *Circ. Res.* **31**, 240–247. (C)

Lassen, N. A. (1964). Autoregulation of cerebral blood flow. *Circ. Res.* **14–15,** Suppl. I, 201–204. (C)

Lassen, N. A. (1968). Brain extracellular pH: The main factor controlling cerebral blood flow. *Scand. J. Clin. Lab. Invest.* **22,** 247–251. (C)

Levasseur, J. E., Wei, E. P., Raper, A. J., Kontos, H. A., and Patterson, J. L., Jr. (1975). Detailed description of a cranial window technique for acute and chronic experiments. *Stroke* **6,** 308–317. (D)

Levasseur, J. E., Wei, E. P., Kontos, H. A., and Patterson, J. L., Jr. (1979). Responses of pial arterioles after prolonged hypercapnia and hypoxia in the awake rabbit. *J. Appl. Physiol.* **46,** 89–95. (C)

L'hermitte, F., Goutier, J., and Derouesne, C. (1970). Nature of occlusions of the middle cerebral artery. *Neurology* **20,** 82–88. (E)

Lothman, E., Lamanna, J., Cordingly, G., Rosenthal, M., and Somjen, G. (1975). Responses of electrical potential, potassium levels, and oxidative metabolic activity of the cerebral neocortex of cats. *Brain Res.* **88,** 15–36. (C)

McCalden, T. A., and Bevan, J. A. (1981). Sources of activator calcium in rabbit basilar artery. *Am. J. Physiol.* **241,** H129–H133. (D)

McCulloch, J., and Harper, A. M. (1977). Cerebral circulation: Effect of stimulation and blockade of dopamine receptors. *Am. J. Physiol.* **233,** H222–H227. (D)

MacKenzie, E. T., McCulloch, J., O'Keane, M., Pickard, J. D., and Harper, A. M. (1976a). Cerebral circulation and norepinephrine: Relevance of the blood–brain barrier. *Am. J. Physiol.* **231,** 483–488. (D)

MacKenzie, E. T., Strandgaard, S., Graham, D. I., Jones, J. V., Harper, A. M., and Farrar, J. K. (1976b). Effects of acutely induced hypertension in cats on pial arteriolar caliber, local cerebral blood flow, and the blood–brain barrier. *Circ. Res.* **39,** 33–41. (C)

MacKenzie, E. T., Farrar, J. K., Fitch, W., Graham, D. I., Gregory, P. C., and Harper, A. M. (1979). Effects of hemorrhagic hypotension on the cerebral circulation. *Stroke* **10,** 711–718. (C)

Marcus, M. L., Heistad, D. D., Ehrhardt, J. C., and Abboud, F. M. (1976). Total and regional cerebral blood flow measurement with 7–10-, 15-, 25-, and 50-μm microspheres. *J. Appl. Physiol.* **40,** 501–507. (D)

Maren, T. H. (1979). Effect of varying CO_2 equilibria on rates of HCO_3 formation in cerebrospinal fluid. *J. Appl. Physiol.* **47,** 471–477. (C)

Maynard, E. A., Schultz, R. L., and Pease, D. C. (1957). Electron microscopy of the vascular bed of rat cerebral cortex. *Am. J. Anat.* **100,** 409–422. (B)

Mérei, F. T. (1974). Fine structure of the cerebral arteries with special reference to their functional role. *In* "Pathology of Cerebral Microcirculation" (J. Cervós-Navarro, F. Matakas, N. Grčević, and A. G. Waltz, eds.), pp. 39–44. de Gruyter, Berlin. (B)

Miller, J. D., Stanek, A. E., and Langfitt, T. W. (1973). Cerebral blood flow regulation during experimental brain compression. *J. Neurosurg.* **39,** 186–196. (C)

Mitchell, G. A. G. (1953). "Anatomy of the Autonomic Nervous System." Livingstone, Edinburgh. (B)

Moyer, J. H., Miller, S. I., and Snyder, H. (1954). Effect of increased jugular pressure on cerebral hemodynamics. *J. Appl. Physiol.* **7,** 245–247. (C)

Nelson, E., and Rennels, M. (1969). Neuromuscular contacts in intracranial arteries of the cat. *Science* **167,** 301–302. (B)

Nelson, E., and Rennels, M. (1970). Innervation of intracranial arteries. *Brain* **93,** 475–490. (B)

Nelson, E., Blinzinger, K., and Hager, H. (1961). Electron microscopic observations on sub-arachnoid and perivascular spaces of the Syrian hamster brain. *Neurology* **11,** 285–295. (B)

Nelson, E., Takayanagi, T., Rennels, M. L., and Kawamura, J. (1972). The innervation of human intracranial arteries: A study by scanning and transmission electron microscopy. *J. Neuropathol. Exp. Neurol.* **31,** 526–534. (B)

Nelson, E., Sunaga, T., and Shimamoto, T. (1973). Microvasculature in focal cerebral ischemia and infarction. Scanning transmission electron microscopy of ischemic endothelium in the monkey carotid artery. In "Eighth Princeton Conference on Cerebral Vascular Diseases" (P. McDowell and S. Whisnant, eds.), pp. 30–44. Grune & Stratton, New York. (B)

Nelson, E., Kawamura, J., Sunaga, T., Rennels, M. L., and Gertz, S. D. (1974). Scanning and transmission electron microscopic study of endothelial lesions following ischemia with special attention to ischemic and "normal" branch points. In "Pathology of Cerebral Microcirculation" (J. Cervós-Navarro, F. Matakas, N. Grčević, and A. G. Waltz, eds.), pp. 267–273. de Gruyter, Berlin. (B)

Oldendorf, W. H. (1971). Brain uptake of radiolabelled amino acids, amines, and hexoses after arterial injection. Am. J. Physiol. **221**, 1629–1638. (D)

Oldendorf, W. H., Cornford, M. E., and Brown, W. J. (1980). Some unique ultrastructural characteristics of rat brain capillaries. In "Cerebral Microcirculation and Metabolism" (J. Cervós-Navarro and E. Fritschke, eds.). Raven, New York. (B)

Owman, C., and Edvinsson, L. (1977). Histochemical and pharmacological approach to the investigation of neurotransmitters, with particular regard to the cerebrovascular bed. In "Neurogenic Control of Brain Circulation" (C. Owman and L. Edvinsson, eds.), pp. 15–38. Pergamon, Oxford. (B,C)

Owman, C., and Edvinsson, L. (1978). Histochemical and pharmacological approach to the investigation of neurotransmitters, with particular regard to the cerebrovascular bed. Ciba Found. Symp. **56**, 275–311. (B)

Owman, C., Edvinsson, L., Hardebo, J. E., Groschel-Stewart, U., Unsicker, K., and Walles, B. (1977). Immunohistochemical demonstration of actin and myosin in brain capillaries. Acta Neurol. Scand. **56**, Suppl. 64, 384–385. (B)

Padget, D. H. (1948). The development of the cranial arteries in the human embryo. Contrib. Embryol. Carnegie Inst. **575**, 205–261. (A)

Padget, D. H. (1957). The development of the cranial venous system in man from the viewpoint of comparative anatomy. Contrib. Embryol. Carnegie Inst. **611**, 70–140. (A)

Patten, B. M. (1968). "Human Embryology," 3rd ed. McGraw-Hill, New York. (A)

Pickard, J. D., and MacKenzie, E. T. (1973). Inhibition of prostaglandin synthesis and the response of baboon cerebral circulation to carbon dioxide. Nature (London) New Biol. **245**, 187–188. (C)

Pollock, M., and Jackson, B. (1971). Fibromuscular dysplasia of the carotid artery. Neurology **21**, 1226–1230. (E)

Povlishock, J. T., Kontos, H. A., Rosenblum, W. L., Becker, D. P., Jenkins, L. W., and DeWitt, D. S. (1980). A scanning electron microscopic analysis of the intraparenchymal brain vasculature following experimental hypertension. Acta Neuropathol. **51**, 203–213. (B)

Rapela, C. E., Green, H. D., and Denison, A. B., Jr. (1967). Baroreceptor reflexes and autoregulation of cerebral blood flow in the dog. Circ. Res. **21**, 559–568. (C)

Rapoport, S. I. (1970). Effect of concentrated solutions on blood–brain barrier. Am. J. Physiol. **219**, 270–274. (D)

Rennels, M. L., and Nelson, E. (1975). Capillary innervation in the mammalian central nervous system: An electron microscopic demonstration. Am. J. Anat. **144**, 233–241. (B)

Romanul, F., and Abramowicz, A. (1964). Changes in brain and pial vessels in arterial border zones. Arch. Neurol. (Chicago) **11**, 40–65. (E)

Sadoshima, S., Busija, D., Brody, M., and Heistad, D. D. (1981). Sympathetic nerves protect against stroke in stroke-prone hypertensive rats: A preliminary report. Hypertension **3**, Suppl. I, I124–I127. (C)

Sakabe, T., and Siesjö, B. K. (1979). The effect of indomethacin on blood flow-metabolism couple in the brain under normal, hypercapnic and hypoxic conditions. Acta Physiol. Scand. **107**, 283–284. (C)

Sakurada, O., Kennedy, C., Jehle, J., Brown, J. D., Carbin, G. L., and Sokoloff, L. (1978).

Measurement of local cerebral blood flow with [^{14}C]iodoantipyrine. *Am. J. Physiol.* **3**, H59–H66. (D)

Sercombe, R., Lacombe, P., Aubineau, P., Mamo, H., Pinard, E., Reynier-Rebuffel, A. M., and Seylaz, J. R. (1978). Is there an active mechanism limiting the influence of the sympathetic system on the cerebral vascular bed? Evidence for vasomotor escape from sympathetic stimulation in the rabbit. *Brain Res.* **164**, 81–102. (C)

Shapiro, W., Wasserman, A. J., Baker, J. P., and Patterson, J. L., Jr. (1970). Cerebrovascular response to acute hypocapnic and eucapnic hypoxia in normal men. *J. Clin. Invest.* **49**, 2362–2368. (C)

Smith, D. E., and Windsor, R. (1961). Embryologic and pathogenic aspects of the development of cerebral saccular aneurysms. *In* "Pathogenesis and Treatment of Cerebrovascular Disease" (W. S. Fields, ed.), p. 367. Thomas, Springfield, Illinois. (E)

Sokoloff, L. (1981a). Relationships among local functional activity, energy metabolism, and blood flow in the central nervous system. *Fed. Proc., Fed. Am. Soc. Exp. Biol.* **40**, 2311–2316. (C)

Sokoloff, L. (1981b). Localization of functional activity in the central nervous system by measurement of glucose utilization with radioactive deoxyglucose. *J. Cereb. Blood Flow Metab.* **1**, 7–36. (D)

Stehbens, W. E. (1960). Focal intimal proliferation in the cerebral arteries. *Am. J. Pathol.* **36**, 289–301. (B)

Stehbens, W. E. (1972). "Pathology of the Cerebral Blood Vessels." Mosby, St. Louis, Missouri. (B)

Stehbens, W. E. (1975). Ultrastructure of aneurysms. *Arch. Neurol. (Chicago)* **32**, 798–807. (E)

Stehbens, W. E. (1979). "Hemodynamics and the Blood Vessel Wall." Thomas, Springfield, Illinois. (B)

Takayanagi, T., Rennels, M., and Nelson, E. (1972). An electron microscopic study of intimal cushions in intracranial arteries of the cat. *Am. J. Anat.* **133**, 415–429. (B)

Toole, J., and Patel, A. (1974). "Cerebrovascular Disorders." McGraw-Hill, New York. (B)

Towart, R. (1981). The selective inhibition of serotonin-induced contractions of rabbit cerebral vascular smooth muscle by calcium-antagonistic dihydropyridines. An investigation of the mechanism of action of nimodipine. *Circ. Res.* **48**, 650–657. (D)

Traystman, R. J., Fitzgerald, R. S., and Loscutoff, S. C. (1978). Cerebral circulatory responses to arterial hypoxia in normal and chemodenervated dogs. *Circ. Res.* **42**, 649–657. (C, D)

Traystman, R. J., Kontos, H. A., and Heistad, D. D. (1982). Vasodilator action of VIP on cerebral vessels. *In* "Vasoactive Intestinal Peptide: Advances in Polypeptide Hormone Research" (S. Said, ed.), pp. 161–168. Raven, New York. (D)

Wahl, M., and Kuschinsky, W. (1976). The dilatatory action of adenosine on pial arteries of cats and its inhibition by theophylline. *Pfluegers Arch.* **362**, 55–59. (C)

Wahl, M., and Kuschinsky, W. (1979). Unimportance of perivascular H$^+$ and K$^+$ activities for the adjustment of pial arterial diameter during changes of arterial blood pressure in cats. *Pfluegers Arch.* **382**, 203–208. (C)

Wei, E. P., Raper, A. J., Kontos, H. A., and Patterson, J. L., Jr. (1975). Determinants of response of pial arteries to norepinephrine and sympathetic nerve stimulation. *Stroke* **6**, 654–658. (C, D)

Wei, E. P., Ellis, E. F., and Kontos, H. A. (1980a). Role of prostaglandins in pial arteriolar response to CO_2 and hypoxia. *Am. J. Physiol.* **238**, H226–H230. (C)

Wei, E. P., Kontos, H. A., and Patterson, J. L., Jr. (1980b). Dependence of pial arteriolar response to hypercapnia on vessel size. *Am. J. Physiol.* **238**, H697–H703. (C)

Winn, H. R., Rubio, R., and Berne, R. M. (1981). The role of adenosine in the regulation of cerebral blood flow. *J. Cereb. Blood Flow Metab.* **1**, 239–244. (C)

Chapter 6
Central Nervous System: Spinal Cord

A. EMBRYOLOGY OF BLOOD CIRCULATION
By C. E. CORLISS

1. DEVELOPMENT OF THE ARTERIAL SUPPLY
TO THE SPINAL CORD

a. Spinal arteries: Several authors (Evans, 1912; Hoskins, 1914; Strong, 1961) have studied these vessels in vertebrate embryos. In his study of injected spinal vessels of pig embryos from 6.2 to 240 mm, Hoskins found the vasculature to be similar to that of human embryos, including the pattern of radicular arteries. However, there was also a fourth main arterial channel, the median dorsal, which was not found in the human fetus at term.

In the human, dorsal segmental arteries supplying the spinal branches begin as offshoots from the aorta at 3 weeks, 11 of them having been formed by 3½ weeks (Evans, 1912). The spinal branches, in turn, divide into anterior and posterior radicular arteries which contribute to the anterior and posterior spinal arteries, respectively. The primitive anterior spinal artery in the 5-week embryo is a double vessel derived from the paired vertebrals of the area. By 7 weeks it has merged into a typical midline artery. At 6 weeks, the central arteries branching off the anterior spinals to nourish the cord are seen as a double row of vessels in the anterior fissure.

Hoskins' pig embryos provide information on later phases in the vasculogenesis of the vertebrate spinal cord. At the 12 mm stage (reached by 7 weeks), a lateral plexus appears just anterior to the exit of the posterior root and covers the lateral aspect of the cord and the adjacent ganglia. The posterior part of the cord is devoid of vasculature for a short time, but then it, too, receives a blood supply.

In the 18 mm stage of the pig embryo, an irregular vessel coalesces from the vascular meshwork to become a definite posterolateral artery by the time the embryo measures 75 mm. Radicular branches, similar to those in the human, develop and reinforce the anterior spinal and, later, the posterior spinal arteries. The median dorsal (posterior) artery is the last to appear (at the 30 mm stage) and does not become the typical midline vessel until the 45 mm stage.

The account by Hoskins (1914) is in accord in general with studies on other vertebrates, except that he does not believe that the spinal artery arises from

211

the vertebral arteries. Instead, he is of the opinion that it forms from segmental spinal arteries and then anastomoses with, or is reinforced by, the vertebral arteries.

b. Spinal veins: As in the case of the arterial system, the spinal venous complex begins in the human embryo as a capillary plexus on the anterolateral aspect of the cord. At the 6.2 mm stage, it extends over the dorsum and then, by 7.5 mm, over the related ganglia. The longitudinal anterior spinal vein develops by the time the embryo is 30 mm long but fails to reach the size of the anterior spinal artery. The posterolateral veins, appearing in the 50–60 mm human embryo, closely resemble the definitive vessels seen at term (240 mm) in the pig.

B. ANATOMY OF BLOOD CIRCULATION

By J. V. LIEPONIS AND H. K. JACOBS

The anatomy of the spinal cord vasculature remains obscure and controversial because of multiple and variable vascular sources, as well as confusing nomenclature. Variability exists among different species, different individuals of the same species, and in the same individual during different periods of development and in different areas of the cord.

1. EXTRASPINAL ARTERIAL SUPPLY

In the adult, only about one-fourth of the 31 fetal segmental vessels actually reaches the spinal cord. The anterior and posterior radiculomedullary arteries, branches of the radicular arteries, run with the spinal nerves. After penetrating the dura and arachnoid, the 6–10 anterior radicular branches and between 10 and 23 posterior radicular arteries form the anterior and posterior longitudinal spinal arterial systems, respectively (Lazorthes, 1972). Not only do the number of radicular arteries vary among individuals, but the extraspinal arterial supply varies according to the region of the spinal cord supplied. The regional blood supply to the cord consists of three major arterial divisions, as presented below.

a. Cervicothoracic region: The cervicothoracic region, composed of the cervical portion of the spinal cord and the first two thoracic segments, receives a rich vascular supply from arteries all of which are branches of the subclavian artery: the vertebral artery (via a C3 radicular artery), the deep cervical artery (via a C6 radicular artery), and the superior intercostal artery (via a C8 radicular artery) (Fig. 6.1). An almost equal number of anterior radiculomedullary arteries arise from both sides of the aorta. The cervical spinal cord also derives some blood supply from the posterior radiculomedullary arteries. Although of smaller caliber, these vessels are more numerous than their anterior counterparts. The cervical enlargement of the spinal cord is noted to have a significantly richer blood supply than the rostral third.

b. Midthoracic region: This segment of the spinal cord (T3–T8) receives only one to three anterior radiculomedullary arteries and very few posterior radiculomedullary arteries. All these vessels are segmental branches of the aorta with no functional extraspinal anastomoses. Of all the regions of the spinal cord, the midthoracic area has the least developed collateral flow, both in size and number of vessels. It therefore is more susceptible to injury in the event of an obstruction to one of the branches supplying it.

c. Thoracolumbar region: This segment includes the thoracic segments caudal to T8, the lumbar region, and the conus medullaris. The blood supply to this area usually consists of a single radiculomedullary vessel, sometimes called the "great anterior radicular artery," the "lumbar enlargement artery," (Lazorthes, 1972), or the "artery of Adamkiewicz." In over 75% of cases, it enters the spinal canal from the left side, traveling with the T10, T11, or T12 nerve root. Some accessory supply also comes from small arteries accompanying the sacral roots and the filaments of the cauda equina. Although small and

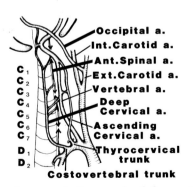

FIG. 6.1. The arteries and communicating vessels of the cervical cord. (Modified from Lazorthes, 1972; reproduced with permission from American Elsevier.)

few in number, these vessels connect with the anastomotic ring around the conus medullaris. They provide significant blood flow to the cord, particularly if the larger radiculomedullary arteries are located at relatively high levels or become narrowed (Lazorthes, 1972). The spinal cord in the thoracolumbar region also receives blood supply from the posterior radiculomedullary arteries originating predominantly from the left side. In addition, the latter vessels provide small arterial branches to the nerve roots and the pial plexus.

2. Longitudinal Spinal Arterial Systems

The anterior and posterior radiculomedullary arteries supply the two major arterial networks found on the surface of the spinal cord: the ventral and dorsal longitudinal systems.

a. Ventral longitudinal system: Rostrally, the vertebral arteries send two branches to form the anterior median longitudinal arterial trunk (AMLAT) (Dommisse, 1975), just caudal to the olive of the medulla. In the upper cervical cord, the AMLAT is often doubled. It runs caudally, ventral to the ventral median sulcus, and eventually anastomoses with the dorsal longitudinal system (see below) about 1.5 cm proximal to the conus terminalis. In some cases, the AMLAT may become greatly narrowed or actually discontinued in the mid-thoracic region, resulting in a relatively avascular, vulnerable region (Lazorthes, 1972). The anatomic, though not the functional, existence of this discontinuity has been questioned (Dommisse, 1975).

b. Dorsal longitudinal system: Rostrally, the paired posterolateral longitudinal arterial trunk (PLLAT) originates from branches of the vertebral or inferior cerebellar arteries. The vessels are found weaving a tortuous course along the posterolateral aspect of the spinal cord, just ventral to the entrance points of the dorsal roots of the spinal nerves. Multiple horizontal anastomoses across the midline are found interconnecting the dorsal longitudinal trunk with the ventral longitudinal trunk, particularly in the cervical and lumbar regions. These horizontal, transverse, and oblique vessels make up an arterial arcade known as the perimedullary coronary plexus. Although present in the adult, this plexus is greatly diminished in size and functional significance, the only significant connection between the PLLAT and AMLAT being the terminal anastomosis of the two posterior branches with the single anterior vessel, to form the cruciate arch of the conus terminalis.

c. Overview of longitudinal arterial systems: The extrinsic circulatory system to the spinal cord can be characterized by its extremely variable nature, even among different individuals of the same species. Dommisse (1975), in all of his dissections, did not find any two identical patterns, his work documenting the immense amount of variation in the medullary supply of the two longitudinal trunks. The only identifiable pattern was that of increased vascularity to more metabolically active regions of the cord. These, consisting of areas of predominantly gray matter in the cervical and lumbosacral enlargements, had a significantly larger number of medullary vessels than did regions of primarily white matter, such as the thoracic cord.

3. INTRINSIC MEDULLARY ARTERIAL VASCULAR SUPPLY

The intrinsic arterial vascular supply of the spinal cord arises from the AMLAT, the PLLAT, and their branches. The medullary circulation can be divided into two unequal arteriolar beds: the anterior medullary system supplying approximately two-thirds of the spinal cord; and the posterior medullary system providing vascularity to the remaining posterior one-third of the cord (Dommisse, 1975).

a. Anterior medullary arterial system: The AMLAT gives off central perforator arteries along its course. These vessels pass through the anterior medial sulcus up to the anterior white commissure, where they travel to one side of the midline and distribute blood supply unilaterally. Alternating central perforator arteries generally supply one side and then the other. Embryologically, these arteries originate from the two paired anterior primitive arterial trunks. When the tracts fuse to form the AMLAT, the perforating vessels remain separate and, thus, retain their regional integrity. Occasionally, central perforator arteries supply both sides of the cord.

Approximately 250–300 central perforator arteries are found branching from the AMLAT at regular intervals. However, regional differences do exist. Central perforator arteries are not only larger but also more numerous in the areas of cervical and lumbar enlargements. Also, they are essentially end arteries, for they do not form significant functional anastomoses with adjacent perforator arteries, perforating branches from the medullary arteries, or perforating branches from the pial plexus (Craigie, 1972). The central perforator arteries supply the gray and white matter in the deeper parts of the lateral columns (Fig. 6.2). There are five orders of branching of vessels of diminishing size, ultimately ending in a rich capillary plexus.

The pial plexus is formed by branches of the posterior and anterior radiculomedullary arteries and both major longitudinal arterial trunks. It gives off perforating arteries which run perpendicular to the white matter columns, contributing substantially to the vascularity of the outer portion of the cord, particularly the anterior two-thirds (Fig. 6.3).

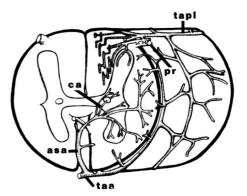

Fig. 6.2. Diagrammatic representation of the arteries and their distribution in the spinal cord, indicating early orders of branching of the perforator arteries. asa, anterior sulcal artery; ca, paracentral artery; tapl, the posterolateral arterial trunk; pr, posterior root; taa, truncus arteriosis anterior. (From Herren and Alexander, 1939; reproduced with permission from *Archives of Neurological Psychiatry.*)

b. Posterior medullary arterial system: The posterior one-third of the spinal cord is supplied by the posterior medullary system, the paired PLLAT, and the perforating pial vessels. Blood flow to this portion of the cord is supplied almost exclusively by penetrating rami from the PLLAT and that portion of the posterior pial plexus which interconnects these two large tracts. Some of the penetrating pial rami supply the posterior columns. Others extend into the posterior tip of the dorsal horn (Gillilan, 1958).

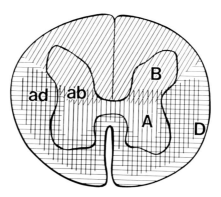

Fig. 6.3. Diagrammatic representation of the distribution of the intrinsic arteries of the spinal cord. A, area of distribution of the central artery; D, area served by the penetrating arteries from the lateral and ventral pial plexus; ad, area which may be supplied by either the central or lateral and ventral penetrating arteries; B, region served by the penetrating arteries from the posterior plexus; ab, the area which may be covered by either the central or by the posterior penetrating arteries. (From Gillilan, 1958; reproduced with permission from *Journal of Comparative Neurology.*)

4. TERMINAL VASCULATURE

The perforating arteries give off terminal arterioles which divide into multiple, segmental precapillary beds. From the latter, an extensive continuous interlocking capillary system develops and ultimately traverses the entire length of the spinal cord. It forms, albeit indirectly, from interconnections between adjacent segments, from the longitudinal arterial systems, and from the terminal twigs of the pial plexus. Vascular density, determined by metabolic requirements, is much greater in gray matter than in white matter. In the adult, the interconnecting capillary network is extensive and well developed, particularly in areas with high cell body density (Fazio and Agnoli, 1970).

5. VENOUS SYSTEM

The venous system of the spinal cord is characterized as multiple, irregular, and vastly interconnecting. The intrinsic veins are found scattered among the arteries in the medulla of the cord. These small vessels do not follow their arterial counterparts, but instead run in random directions without any segmental distributions. Thus, blood entering a particular capillary bed through any given central perforating artery in all likelihood leaves by a vein at a different level. These small venous channels ultimately form central veins. Unlike the central arteries, the central veins run in a radial pattern, emerging from the cord at various points. They form multiple longitudinal channels on

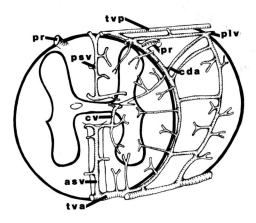

Fig. 6.4. Diagrammatic representation of the veins and their distribution in the spinal cord. Only higher level branching is included. asv, anterior sulcal vein; cda, centrodorsolateral anastomosis; cv, paracentral vein; plv, posterolateral venous trunk; pr, posterior root; psv, posterior septal vein; tva, truncus venosus anterior; tvp, posterior venous trunk. (From Herren and Alexander, 1939, reproduced with permission of *Archives of Neurological Psychiatry.*)

the surface of the spinal cord. Also unlike the arterial system, there are multiple interconnections among these longitudinal channels, between these channels and central veins, and between one central vein and another (Fig. 6.4).

Not all of the longitudinal veins described traverse the entire length of the cord. They may run for varying distances and then coalesce to form two major longitudinal tracts. The anterior spinal vein is found in the ventral medial sulcus, posterior to the AMLAT. The larger posterior spinal vein follows along the dorsal medial sulcus. Large channels interconnect these two trunks. Radiculomedullary veins provide outflow from the trunks at various points. Like the radiculomedullary feeder arteries, there is no characteristic segmental venous pattern. There are 14 principal radiculomedullary veins, 7 anterior and 7 pos-

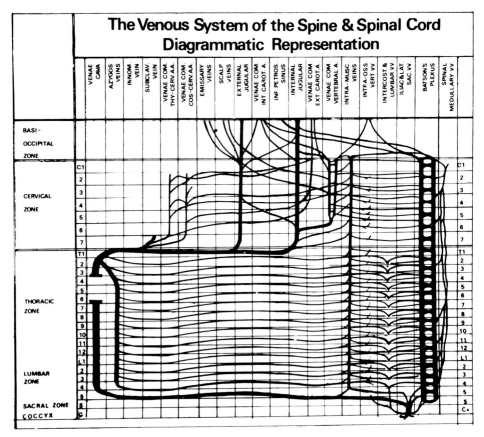

FIG. 6.5. Diagrammatic representation of the veins of the spine and spinal cord indicating relative capacities and anastomotic channels. (From Dommisse, 1975, reproduced with permission from Churchill-Livingstone.)

terior. Their distribution is unpredictable, and they are not necessarily found at the same levels as the arteries.

Ultimately, the radiculomedullary veins empty into Batson's plexus. This series of intermingling anastomotic venous channels in the bony spinal canal extends from the occiput to the coccyx. In addition to their role in draining the spinal cord, the vessels behave as capacitance channels for multiple vascular beds. Thus, spinal cord venous outflow intermixes with venous blood from various vascular beds in Batson's plexus and then follows a circuitous course back to the heart (Dommisse, 1975) (Fig. 6.5).

C. PHYSIOLOGY OF BLOOD CIRCULATION

By H. K. JACOBS AND R. D. WURSTER

Until the 1970s understanding of the physiology of spinal cord blood flow (SCBF) was based upon inference from the known data on cerebral blood flow. Moreover, due to scant human studies, most of the current understanding of the physiology of SCBF comes from investigations on a variety of experimental animals, and hence caution must be exercised in applying such conclusions directly to man.

1. BLOOD FLOW

Spinal cord blood flow has been measured using the distribution of [14C]antipyrene, radiolabeled microspheres, and hydrogen clearance technique in different animal models. Flow values reported for white matter have generally been consistent, but those for gray matter have varied (Hales *et al.*, 1981; Jacobs *et al.*, 1982). The flow for the entire spinal cord is in the range of 11 to 30 ml/min/100 gm, with the readings for the gray matter in the range of 30 to 60 ml/min/100 gm and those for the white matter varying between 9 and 15 ml/min/100 gm. Thus, white matter flow is about 20–30% of gray matter flow. The flow for whole spinal cord varies depending upon the level studied, with the highest readings being reported for the lumbosacral segments. The thoracic and cervical cord flows are about 70 and 45%, respectively, of that of the lumbosacral flow. Stimulation of peripheral nerves increases whole spinal

Blood Vessels and Lymphatics in Organ Systems
Copyright © 1984 by Academic Press, Inc.

cord flow by about 35 to 50%, a change thought to be secondary to increased metabolic activity (Marcus *et al.*, 1977; Kobrine *et al.*, 1978).

2. AUTOREGULATION

Available data indicate that autoregulation exists in the spinal cord of all mammalian species studied, but it is severely blunted or absent following spinal cord trauma (Kobrine *et al.*, 1976; Griffiths *et al.*, 1979). Simply stated, SCBF remains relatively constant at levels of about 50 to 135 mm Hg of systolic arterial pressure. Above and below this range, a passive relationship exists between flow and pressure. The implication of the broad zone of near constant flows with widely varying pressures is that the caliber of the resistance vessels must change in response to variations in pressures. As intraluminal pressures increase, constriction ensues and as they decrease, vasodilatation occurs. Several theories have been proposed to explain autoregulation: metabolic, myogenic, neural, and tissue pressure. The metabolic and myogenic hypotheses are often implicated in SCBF autoregulation but lack direct evidence. However, the metabolic theory is supported by the observation of increased blood flow associated with presumed elevated metabolic need due to increased neural activity (Marcus *et al.*, 1977).

According to the neural theory, the central nervous system exerts a direct neural control over vessel caliber in autoregulation. However, since autoregulation remains intact after high cervical cord lesions (Kobrine *et al.*, 1976), this argues against the neurogenic hypothesis, at least as far as supraspinal mediation of SCBF autoregulation is concerned. Nevertheless, one cannot rule out all neurally induced vasomotion on SCBF control. After α-adrenergic blockade, autoregulation is apparently absent (Kobrine *et al.*, 1977).

The tissue pressure theory contends that autoregulation occurs in the low-pressure vessels due to the compression that tissue fluids exert on them. According to such a view, an increase in arterial pressure raises capillary filtration, thus resulting in an elevation in tissue pressure in the relatively nondistensible anatomic locus of the spinal cord, i.e., its leptomeninges and the vertebral canal. The increased tissue pressure would reduce the microvascular transmural pressure and, hence, elevate blood flow resistance. With low perfusion pressure, the reverse would occur. This hypothesis is supported by the observed decreased SCBF with increased cerebrospinal fluid pressure (Griffiths *et al.*, 1979); nevertheless, it is not universally accepted (Kobrine *et al.*, 1976).

At low and high arterial pressure, SCBF autoregulation has limits. With perfusion of the spinal cord at low perfusion pressures, the degree of maximal vasodilatation is apparently the limiting factor. At the limits for high perfusion pressures, breakthrough occurs at levels over 160 mm Hg perfusion pressure, under which circumstances damage of the capillary endothelium may take place (Brightman, 1976).

3. CO_2 RESPONSE

Similar to cerebral blood flow, SCBF is altered by changes in arterial partial pressure of CO_2 (Pa_{CO_2}). In the range of 30 to 50 mm Hg of arterial Pa_{CO_2}, SCBF sensitivity is in the range of 0.5 to 1 ml/min/100 gm/mm Hg Pa_{CO_2} (Griffiths, 1975; Marcus et al., 1977; Jacobs et al., 1982). Also reported is a decreased CO_2 sensitivity under hypotensive conditions (Griffiths et al., 1978), as well as persistence of sensitivity (Jacobs et al., 1982). Such variations may be due to differences in anesthesia and arterial pH and in the extent of hypotension. Carbon dioxide sensitivity is probably a local response, since it is still present after cervical cord section (Kindt, 1971). In man, no increased or decreased CO_2 sensitivity has been reported in patients with intramedullary tumors (Jellinger, 1972).

4. COUNTERPRESSURE

Counterpressure is an increase of extravascular pressure, from whatever cause, which results in a decreased transmural pressure, increased resistance, and decreased SCBF. Elevations in cerebrospinal fluid pressure, particularly above 50 mm Hg, will result in reduced SCBF (Griffiths et al., 1978). Venous stasis may also produce a counterpressure and decrease SCBF (Assenmacher and Ducker, 1971), as do certain hypotensive drugs (Rogers and Traystman, 1979).

Spinal edema formation following spinal injury leads to counterpressure and decreased SCBF and even to ischemic cord transection (Albin et al., 1967). This process commonly occurs locally within minutes after the trauma is sustained (Yashon et al., 1973; Green and Wagner, 1973) and has a tendency to spread diffusively to adjacent cord areas. If the compressive injury is mild, edema occurs primarily with the removal of compression. Beggs and Waggener (1975) found that the endothelial barrier is injured within 6 hr of compression, remains so for days, and begins to recover its normal function at about 14 days. This observation provides a likely explanation for the presence of edema for roughly a 2-week period following injury (Yashon et al., 1973).

5. "STEAL" PHENOMENON

If a portion of the spinal cord is maximally dilated, e.g., due to cord trauma, interventional factors which cause vasodilatation may result in a greater blood flow to the normal adjacent zones of the cord at the expense of the injured area, a manifestation of the "steal" phenomenon. The reverse also occurs, i.e., if vascular resistance to the normal bed is increased, flow may be augmented in the already maximally dilated traumatized area ("reverse steal"). The "steal" phenomenon depends upon changes in the arterial pressure to the traumatized spinal cord region. If the steal is to occur strictly from one spinal cord region to another without a change in aortic pressure, the two portions must be supplied by a common spinal artery. In addition, there must be a sufficient upstream resistance in the common supply artery so that changes in flow to the normal region cause a pressure change through it. Such a situation is most likely present in the midthoracic region, although it could also occur in other regions. It is possible to consider the "steal" phenomenon as beneficial, since it reduces hemorrhagic necrosis or vascular leakage at the site of trauma. The phenomenon has likewise been used to explain a decreased flow to an injured area of spinal cord in response to an elevated Pa_{CO_2} (Palleske *et al.*, 1970; Griffiths, 1975).

6. SYMPATHETIC CONTROL OF SCBF

Although the spinal cord vasculature receives a rich catecholamine innervation, baroreceptor or chemoreceptor reflexes fail to change SCBF (Marcus *et al.*, 1977). Thus, the role of the presumed sympathetic innervation is unknown. Many questions remain unanswered. Does stimulation of the sympathetic chain alter blood flow; what type of autoregulation exists after sympathectomy; and do other physiologic conditions reflexly alter SCBF?

D. PHARMACOLOGY OF BLOOD CIRCULATION

By R. D. WURSTER AND J. V. LIEPONIS

Understanding of the pharmacology of spinal cord vessels is largely assumed indirectly from investigations of the effect of various drugs on cerebral blood flow, or it has its origin principally in studies on spinal trauma and the resultant circulatory problems. The following is a brief review of the therapeutic approach to spinal trauma and the presumed role of spinal cord blood flow (SCBF).

1. CATECHOLAMINES

Catecholamines, particularly norepinephrine, may play an important role in spinal cord dysfunction following trauma. Catecholamine release from the spinal cord (Osterholm, 1974) or elevated circulating catecholamines sequestered at the injury site (Vise et al., 1974) have been considered to produce spinal vasospasm, hypoxia, and eventual tissue necrosis. Of interest in this regard are the findings that pretreatment or treatment shortly following the trauma with catecholamine depletors or α-blocking drugs limited the development of post-trauma necrotic cord lesions (Osterholm, 1974; Vise et al., 1974; Hedeman et al., 1974). Hedeman and Sil (1974) have also examined the hypothesis that the release of dopamine is responsible for spinal vascular dysfunction rather than, or in addition to, norepinephrine. There is support for this contention (Naftchi et al., 1974). Still other workers have failed to observe changes in either of these drugs (de la Torre et al., 1974). Thus, the effectiveness of catecholamine depletors or blockers in spinal trauma is controversial, and the clinical applicability of such drugs remains unproved.

2. GLUCOCORTICOID STEROIDS

As with other areas of the central nervous system (CNS), steroid therapy has been utilized to improve the incidence of clinical recovery of the spinal cord following trauma. Several reports have indicated the effectiveness of intramuscularly or intrathecally administered steroids (Hansebout et al., 1975), but the unfortunate aspect of the treatment is that it must be given either prophylactically or within minutes of the trauma to be fully effective. Steroid-induced reversal of histopathologic changes caused by cord injury has also been reported when the steroid treatment is coupled with hypothermia (Richardson and Nakamura, 1971). The mechanism for the therapeutic action of steroids remains open to question. It could be through their antiinflammatory/antiedema effects on spinal vasculature or through their action in maintaining cellular integrity (Lewin et al., 1974).

3. NALOXONE

Intravenous administration of naloxone, an opiate antagonist, has been reported to improve reflex recovery after spinal trauma (Goldfarb and Hu, 1976), probably, in part, by reversing the hypotension associated with spinal shock (Faden et al., 1981) and, in part, by improving recovery from CNS ischemia (Hosobuchi et al., 1982). The drug also produces higher SCBF in the injured area (Young *et al.*, 1981). Thus, the mechanisms of the beneficial effects of naloxone may be via an improved oxygen supply/demand ratio involving vascular opiate receptors, which are in high concentration in the spinal cord.

4. Other Pharmacologic Agents

There are other therapeutic modalities affecting spinal cord vasculature which have been proposed for the prevention of spinal cord lesions. For example, histaminergic blockade has been shown to minimize the postlesion hyperemic response and resulting edema (Kobrine and Doyle, 1976), and hyperbaric oxygen has been found to improve the recovery from spinal cord contusions (Kelly *et al.*, 1972). Recently, some speculative interest has been shown in the use of free radical scavengers, e.g. selenium or glutathione, to minimize posttraumatic damage (Chen and Fishman, 1980).

5. Summary

Controversy remains in the area of SCBF following injury, particularly with regard to the benefits of a wide variety of treatments. Although local gray matter necrosis may occur due to spinal trauma, the major focus of attention should be placed upon preserving the integrity of white matter function, i.e., long tract conduction. Thus, the maintenance of white matter blood flow is of major concern. There exists considerable controversy over the use of single mechanisms as therapy in posttraumatic dysfunction. However, since this abnormal state may be due to a concert of mechanisms, multiple therapeutic approaches may be necessary.

E. PATHOGENESIS AND PATHOLOGY OF BLOOD CIRCULATION

By E. R. Ross

1. General Considerations of Vascular Malformations

Vascular malformations of the spinal cord, or so-called spinal angioma, include those developmental anomalies of the vasculature which may involve the spinal cord, leptomeninges, or dura. It is necessary to point out, however, that the distinction between these disorders and benign neoplasm of the vasculature has frequently been disregarded. In fact, review of even recent literature still reveals studies in which both vascular malformations and vascular neoplastic lesions (hemangioblastomas, hemangiopericytomas, etc.) are grouped together as "angioma" of the central nervous system, despite the fact that Cushing and Bailey as early as 1928 clearly differentiated one from the other. The term angioma has been defined as a tangle of vessels which forms an abnormal communication between the arterial and venous systems, the condition falling into the category of a developmental abnormality and not a neoplastic tumor (Ad Hoc Committee, 1958). It is used in this context in the present discussion.

a. Types: According to McCormick (1966), the complex nomenclature of vascular malformations of the spinal cord can generally be reduced to five types. Excluding the berry aneurysm, these are telangiectasis, varix, cavernous malformation, arteriovenous malformation (AVM), and the purely venous malformation. Odom (1962) reported a rare clinical case of a telangiectatic lesion at T5, as well as a mixed telangiectasis–cavernous angiomatous lesion of the spinal cord. Although venous angiomas have also been found in the spinal region, it is the AVM that is the most common and clinically most important type of malformation affecting the spinal cord. Hence, the following section will be limited to a discussion of this entity, which varies in size from a few millimeters to several centimeters. Because the vessels contributing to the AVM are often so altered that they no longer exhibit their usual morphologic characteristics, there is no advantage to the use of such terms as arterial, venous, or capillary angiomas.

b. Incidence: Although vascular anomalies of the brain have been well documented (see Section E-3c, Chapter 5), those involving the spinal cord have received little attention. The incidence of spinal cord AVMs has been reported to vary from 3.3% (Krayenbuehl and Yasargil, 1963) to 11.5% (Newman, 1959). In Wyburn-Mason's classic work (1943), vascular anomalies made up 3–4% of the total cases of spinal neoplasia, and Yasargil (1971) reported similar findings (4.35% of 961 spinal tumors). The figures reported by Newman (1959) appear to be unusually high.

Between 1935 and 1945, Newman (1959) reported only 2 cases of spinal angioma, whereas in the following decade, 19 additional ones were diagnosed. For example, Epstein *et al.* (1949) found 6 such cases in the 77 patients operated upon for suspected spinal cord neoplasia. Ten others were reported by Gross and Ralston (1959) in the decade following the introduction of selective arteriography (Hanson and Croft, 1956; Höök and Lidvall, 1958). Preoperative diagnosis has not only become more reliable with the use of this technique, but it has permitted the identification of smaller anomalies which might otherwise have been missed. By 1971, Yasargil had collected approximately 500 vascular anomalies of the spinal cord from the literature.

  c. Age and sex distribution for spinal cord angioma: The lesion is found in patients ranging from 1 to 80 years old, with most of the cases first being diagnosed at ages 30 to 60. Aminoff (1976) reported that females most frequently develop symptoms between the ages of 20 and 40 years, whereas in males the clinical manifestations are first noted between 40 and 60 years of age.

  d. Anatomic considerations: Vascular malformations tend to show a preferential topographic distribution with regard to three developmental vascular territories: the upper cervicothoracic portion supplying the entire cervical area, as well as T1, T2, and possibly T3; the middle thoracic segment supplying T3 through T8; and the great anterior radicular artery of Adamkiewicz supplying T9 through the lumbar levels. Cervical involvement varies from 10 to 18%; thoracic involvement, from 28 to 30%; and thoracolumbar involvement, from 53 to 60%. (For embryology and gross anatomy of extraspinal arterial blood supply, see Sections A-1a and B-1, respectively, above.)
  Through selective arteriography, it has been possible to identify and localize the AVM and determine the characteristics of the feeding and draining vessels. The lesions may be limited to the leptomeninges or they may extend into the spinal cord, but the majority are extramedullary. They may be supplied by one or more unilateral or bilateral arterial pedicles. Most occur on the dorsal surface of the spinal cord (Wyburn-Mason, 1943; Houdart *et al.*, 1966; Béraud, 1972), venous lesions being more frequent in the thoracolumbar region (Luessenhop and Cruz, 1969). The reported common finding of AVM in the anterior and lateral aspects of the cervical and thoracic segments (Luessenhop and Cruz, 1969) has not been substantiated by other studies (Wyburn-Mason, 1943; Brion *et al.*, 1952; Aminoff and Logue, 1974; Yasargil, 1971).
  Béraud (1972) has reported that a single feeding arterial vessel is present in 30% of cases, whereas multiple feeding vessels (as many as 10) are found in the remaining 70% of lesions. The abnormally distended, tortuous, and often coiled vessels may be limited to a few spinal segments or may extend over much of the length of the spinal cord. Aneurysmal out-pouchings are occasionally present. Although some vessels may remain thin, most are thickened, sclerotic, and no longer recognizable as arteries or veins. The main feeding arteries of the malformation generally enter the upper segments, with the numerous draining veins proceeding for a considerable distance both cranially or caudally from the mass of distended, tortuous, malformed vessels. The draining veins may be even larger than the malformation itself, thus leading to compression symptoms distant from the main lesion. Although distended vascular channels that feed or drain the angioma may be prominent within the cord, they should not be considered as components of the malformation itself. Bleeding from aneurys-

mal dilatations or regions of marked thinning and rupture may lead to sub-arachnoid hemorrhage or hemorrhage within the spinal cord.

Although some workers (Wyburn-Mason, 1943; Bergstrand et al., 1964; McCormick, 1966) believed venous angiomas and AVMs to be distinct and separate entities, others (Turner and Kernohan, 1941; Houdart et al., 1966) expressed the view that all vascular malformations are in reality AVMs, with one or another component predominating. Attempts to distinguish venous from arterial angiomas macroscopically, on the basis of the presence of bright red or dark blood, has proved to be inaccurate. For example, Aminoff (1976) has pointed out that when the size of an AVM permits a large volume of blood to flow from arterial to venous channels, the blood seen in the latter vessels may be bright red, and the venous channels themselves may even pulsate, indicating the presence of a large shunt. On the other hand, nonpulsating draining channels, containing only dark blood, are considered to be low volume shunts, the type classified as venous angioma by early observers.

2. PATHOGENESIS OF VASCULAR MALFORMATIONS

a. Congenital origin: Most widely accepted is the view that the AVM is the result of a congenital defect (Kaplan et al., 1961; Houdart et al., 1966; Djindjian et al., 1977), which occurs in early embryonic life when there is a persistence of embryonic arteriovenous shunts without normal development of an intervening capillary network. This hypothesis is supported by the fact that the vascular malformation is sometimes associated with other congenital disorders, such as Kartagener (Hayakawa et al., 1981) and Klippel–Trenaunay–Weber syndromes; intracranial aneurysms and AVMs; cutaneous, lymphatic, and osseous dysplasias; and angiomas of the paraspinal muscles, epidural space, vertebrae, retina, choroidae, and viscera. The primitive vasculature may remain latent and only later enlarge and become functional. Because of the comparatively few reports of familial occurrences, genetic factors have been largely discounted.

b. Origin from arteriovenous shunts: The possibility that the small arteriovenous shunts found on the surface of the human brain is the origin of cerebral arteriovenous angioma (Rowbotham and Little, 1965) has led to the belief that a similar cause may be the basis for the spinal cord AVM. However, thus far, the presence of arteriovenous shunts on the surface of the cord has not been reported. Such structures have been found in tissues of the human ear, indicating that they are not unique to the brain.

The mechanism whereby the embryonic AV shunts or the ones demonstrated by Rowbotham and Little (1965) could slowly develop into distinct vascular anomalies remains unclear. A mechanical theory has been offered, based on the observation that most anomalies occur at the caudal end of the spinal cord, thus suggesting that there may be interference with normal venous return. However, the presence of arterial blood within some of the anomalous vessels (large volume shunts) argues against such a theory.

c. Origin from infection and other factors: Kaydi (1889) and Foix-Ala-jouanine (1926) have proposed that infection may be a possible mechanism for the development of the vascular malformation. In this regard, it should be pointed out that the vascular changes described by Foix-Alajouanine (1926), as well as by others (Wyburn-Mason, 1943; Greenfield and Turner, 1939), appear to be those of thrombophlebitis, with sparing of arterial channels. Nor is there any clear-cut evidence to indicate that the pathologic process responsible for the development of the AVM can result from interference with normal flow through the capillary network and shunting through the arteriovenous communication. Trauma and hormonal influences also have been reported in sporadic cases as possible factors in the development of malformations.

d. Mechanisms responsible for clinical manifestations: Houdart *et al.* (1966) expressed the view that draining venous channels undergo dynamic alterations with progressive distention. They and others (Doppman *et al.*, 1969; Shephard, 1963) point out that the enlarging mass, which is found more often dorsally and in the pia, may become large enough to compress the subadjacent spinal cord, giving rise to clinical symptoms. This may occur at sites remote from the AVM itself. Aminoff and Logue (1974), on the other hand, noted that clinical symptomatology resulting from cord compression is relatively rare. The myelomalacia found accompanying spinal AVMs is believed to result from diversion of blood from adjacent normal arteries, where the pressure is normal, to the malformed vessels, where it is decreased. This "steal" phenomenon results in local chronic tissue ischemia, followed by gliotic scarring and frank necrosis of the surrounding tissue. Probably because symptoms tend to occur early, before spinal cord AVMs reach a large size, no clinical hemodynamic impairment of the type noted in such lesions located in the brain and in the limbs has been reported. However, bruits have been noted (Matthews, 1959). Aminoff and Logue (1974) believe that the arteriovenous shunt gives rise to an elevated pressure within draining coronal veins, as well as secondarily in the intramedullary veins. In their view, the eventual resulting diminished intramedullary blood flow leads to ischemia of the tissues, with the development of gliosis; increased vascularity (including hyalinization of small vessels); neuronal, myelin, and axonal degeneration; and even eventual cystic cavitation.

3. Pathology of Vascular Malformations

a. Gross pathology: Spinal AVMs may originate in the vertebral body, the epidural space (extradurally), intradurally, extramedullary, and intramedullary, or as a combination of these types. The following discussion will be limited to leptomeningeal (extramedullary) and intramedullary lesions.

In the presence of an AVM, the spinal cord may be atrophic or frankly necrotic, secondary to compression, or it may be swollen due to the angioma or to its complications, i.e., edema or hemorrhage. The changes often reach beyond the distribution of a particular arterial blood supply. Aminoff *et al.* (1974) have stated that the affected region has a territorial pattern similar to that of the draining intramedullary veins.

b. Microscopic pathology: Histologic examination of an AVM reveals numerous vascular channels of different sizes and structure. Large, middle, and small calibered vessels are seen, the walls of which vary considerably in thickness. In those arteries in which the internal elastic lamina is still identifiable, this structure may be fragmented, duplicated, and thickened. Within the media of both arteries and veins, connective tissue often replaces the muscle cells. In addition, the vessel wall may be completely or partially hyalinized, and focal calcification may also occur. The width of the vessel wall varies, with some segments appearing markedly thickened and others, extremely thinned and even demonstrating aneurysmal outpouchings. Within the perivascular connective tissue, increased fibrosis may be found, along with occasional lymphocytic, mononuclear, and plasma cell aggregates. Partial or complete occlusion by thrombus may also be noted within the lumen of some vessels. The accompanying leptomeninges generally are thickened by fibrotic connective tissue which is sometimes adherent to the underlying spinal cord. Histiocytes, some of which contain hemosiderin and round cell infiltrates, are also scattered within the thickened leptomeningeal covering. With involvement of the spinal cord, loss of myelin, axons, and even gray matter with its neurons may occur. Secondary reactive astrocytosis and microgliosis, some containing hemosiderin, are seen in the surrounding, as well as within, tissues between the distorted vessels.

References*

Ad Hoc Committee for Advisory Council for the National Institute of Neurological Diseases and Blindness, Public Health Service (1958). *Neurology* **8**, 401–433. (E)

Albin, M. S., White, R. J., Locke, G. S., Massopust, L. C., Jr., and Kretchmer, H. E. (1967). Localized spinal cord hypothermia. Anesthetic effects and application to spinal cord injury. *Anesth. Analg. (Cleveland)* **46**, 8–15. (C)

Aminoff, M. J. (1976). "Spinal Angiomas," 1st ed. Blackwell, Oxford. (E)

Aminoff, M. J., and Logue, V. (1974). Clinical feature of spinal vascular malformations. *Brain* **97**, 197–210. (E)

Aminoff, M. J., Barnard, R. O., and Logue, V. (1974). The pathophysiology of spinal vascular malformations. *J. Neurol. Sci.* **23**, 255–263. (E)

*In the reference list, the capital letter in parentheses at the end of each reference indicates the section in which it is cited.

Assenmacher, D. R., and Ducker, T. B. (1971). Experimental traumatic paraplegia. The vascular and pathological changes seen in reversible and irreversible spinal-cord lesions. *J. Bone Jt. Surg., Am.* Vol. **53-A**, 671–680. (C)

Beggs, J. L., and Waggener, J. D. (1975). Vasogenic edema in the injured spinal cord: A method of evaluating the extent of blood–brain barrier alteration to horseradish peroxidase. *Exp. Neurol.* **49**, 86–96. (C)

Béraud, R. (1972). Vascular malformations of the spinal cord. In "Handbook of Clinical Neurology" (P. J. Vinken, and G. W. Bruyn, eds.), Vol. 12, Chapter 21. Am. Elsevier, New York. (E)

Bergstrand, H., Höök, O., and Lidvall, J. (1964). Vascular malformations of spinal cord. *Arch. Neurol. Scand.* **40**, 169–183. (E)

Brightman, M. (1976). Some attempts to open the blood-brain barrier to protein. In "Head Injuries" (R. L. McLaurin, ed.), pp. 107–113. Grune and Stratton, New York. (C)

Brion, S., Netsky, M. G., and Zimmerman, H. M. (1952). Vascular malformations of the spinal cord. *Arch. Neurol. Psychiatry* **63**, 339–359. (E)

Chen, R. H., and Fishman, R. A. (1980). Transient formation of superoxide radicals in polyunsaturated fatty acid-induced brain swelling. *J. Neurochem.* **35**, 1004–1007. (D)

Craigie, E. H. (1972). Vascular supply of the spinal cord. In "The Spinal Cord; Basic Aspects and Surgical Considerations" (G. M. Austin, ed.), 2nd ed., pp. 57–87. Thomas, Springfield, Illinois. (B)

Cushing, H., and Bailey, P. (1928). "Tumors Arising from the Blood Vessels of the Brain." Thomas, Springfield, Illinois. (E)

de la Torre, J. C., Johnson, C. M., Harris, L. H., Kajihara, K., and Mullan, S. (1974). Monoamine changes in experimental head and spinal cord trauma: Failure to confirm previous observations. *Surg. Neurol.* **2**, 5–11. (D)

Djindjian, M., Djindjian, R., and Hurth, M. (1977). Spinal cord arteriovenous malformations and the Klippel-Trenaunay-Weber syndrome. *Surg. Neurol.* **8**, 229–237. (E)

Dommisse, G. F. (1975). "The Arteries and Veins of the Human Spinal Cord from Birth." Churchill-Livingstone, Edinburgh and London. (B)

Doppman, J. L., Di Chiro, G., and Ommaya, A. K. (1969). "Selective Arteriography of the Spinal Cord," 1st ed. Warren H. Green, St. Louis, Missouri. (E)

Epstein, J. A., Beller, A. J., and Cohen, I. (1949). Arterial anomalies of the spinal cord. *J. Neurosurg.* **6**, 45–56. (E)

Evans, H. M. (1912). Development of the vascular system. In "Manual of Human Embryology" (F. Keibel and F. P. Mall, eds.), Vol. 2, pp. 570–623. Lippincott, Philadelphia, Pennsylvania. (A)

Faden, A. I., Jacobs, T. P., Mougey, E., and Holaday, J. W. (1981). Endorphins in experimental spinal injury. *Ann. Neurol.* **10**, 326–332. (D)

Fazio, C., and Agnoli, A. (1970). The vascularization of the spinal cord: Anatomical and pathophysiological aspects. *Vasc. Surg.* **4**, 245–257. (B)

Foix-Alajouanine, T. (1926). La myelite necrotique subaigue. *Rev. Neurol.* **33**, 1–42. (E)

Gillilan, L. A. (1958). The arterial blood supply of the human spinal cord. *J. Comp. Neurol.* **110**, 75–103. (B)

Goldfarb, J., and Hu, J. W. (1976). Enhancement of reflexes by naloxone in spinal cats. *Neuropharmacology* **15**, 785–792. (D)

Green, B. A., and Wagner, F. C. (1973). Evolution of edema in the acutely injured spinal cord: A flourescence microscopic study. *Surg. Neurol.* **1**, 98–101. (C)

Greenfield, J. A., and Turner, J. W. (1939). Acute and subacute necrotic myelitis. *Brain* **62**, 227–252. (E)

Griffiths, I. R. (1975). Spinal cord blood flow after impact injury. In "Blood Flow and Metabolism in the Brain" (A. M. Harper, W. B. Jennett, J. D. Miller, and J. Rowan, eds.), pp. 427–429. Churchill-Livingstone, Edinburgh and London. (C)

Griffiths, I. R., Pitts, L. H., Crawford, R. A., and Trench, J. G. (1978). Spinal cord compression and blood flow. I. The effect of raised cerebrospinal fluid pressure on spinal cord blood flow. *Neurology* **28**, 1145–1151. (C)

Griffiths, I. R., Trench, J. G., and Crawford, R. A. (1979). Spinal cord blood flow and conduction during experimental cord compression in normotensive and hypotensive dogs. *J. Neurosurg.* **50**, 353–360. (C)

Gross, S. W., and Ralston, B. L. (1959). Vascular malformations of the spinal cord. *Surg., Gynecol. Obstet.* **108**, 673–678. (E)

Hales, J. R. S., Yeo, J. D., Stabback, S., Fawcett, A. A., and Kearns, R. (1981). Effects of anesthesia and laminectomy on regional spinal cord blood flow in conscious sheep. *J. Neurosurg.* **54**, 620–626. (C)

Hansebout, R. R., Kuchner, E. F., and Romero-Sierra, C. (1975). Effects of local hypothermia and of steroids upon recovery from experimental spinal cord compression injury. *Surg. Neurol.* **4**, 531–536. (D)

Hanson, R. A., and Croft, P. B. (1956). Spontaneous spinal subarachnoid hemorrhage. *Q. J. Med.* **25**, 53–66. (E)

Hayakawa, T., Kondoh, T., and Watanabe, M. (1981). Spinal arteriovenous malformation with Kartagener's syndrome. *Surg. Neurol.* **13**, 463–467. (E)

Hedeman, L. S., and Sil, R. (1974). Studies in experimental spinal cord trauma. Part 2. Comparison of treatment with steroids, low molecular weight dextran, and catecholamine blockade. *J. Neurosurg.* **40**, 44–51. (D)

Hedeman, L. S., Shellenberger, M. K., and Gordon, J. H. (1974). Studies in experimental spinal cord trauma. Part 1. Alterations in catecholamine levels. *Neurosurgery* **40**, 37–43. (D)

Herren, R. Y., and Alexander, L. (1939). Sulcal and intrinsic blood vessels of human spinal cord. *Arch. Neurol. Psychiatry* **41**, 678–687. (B)

Höök, O., and Lidvall, H. (1958). Arteriovenous aneurysms of the spinal cord. *J. Neurosurg.* **15**, 84–91. (E)

Hoskins, E. R. (1914). On the vascularization of the spinal cord of the pig. *Anat. Rec.* **8**, 371. (A)

Hosobuchi, Y., Baskin, D. S., and Woo, S. K. (1982). Reversal of induced ischemic neurologic deficit in gerbils by the opiate antagonist naloxone. *Science* **215**, 69–71. (D)

Houdart, R., Djindjian, R., and Hurth, M. (1966). Vascular malformations of the spinal cord. *J. Neurosurg.* **24**, 583–594. (E)

Jacobs, H. K., Lieponis, J. V., Bunch, W. H., Barber, M. J., and Salem, M. R. (1982). The influence of halothane and nitroprusside on canine spinal cord hemodynamics. *Spine* **7**, 35–40. (C)

Jellinger, K. (1972). Circulation disorders of the spinal cord. *Acta Neurochir.* **26**, 327–337. (C)

Kaplan, H. A., Aronson, S. N., and Browder, E. J. (1961). Vascular malformations of the brain: An anatomical study. *J. Neurosurg.* **18**, 630–635. (E)

Kaydi, H. (1889). "Ueber die Blutgefaesse des menschlichen rueckenmarkes." Gubrynowicz & Schmidt, Lemberg. (E)

Kelly, D. L., Jr., Lassiter, K. R. L., Vongsvivut, A., and Smith, J. M. (1972). Effects of hyperbaric oxygenation and tissue oxygen studies in experimental paraplegia. *J. Neurosurg.* **36**, 425–429. (D)

Kindt, G. W. (1971). Autoregulation of spinal cord blood flow. *Eur. Neurol.* **6**, 19–23. (C)

Kobrine, A. I., and Doyle, T. F. (1976). Role of histamine in posttraumatic spinal cord hyperemia and the luxury perfusion syndrome. *J. Neurosurg.* **44**, 16–20. (D)

Kobrine, A. I., Doyle, T. F., and Rizzoli, H. V. (1976). Spinal cord blood flow as affected by changes in systemic arterial blood pressure. *J. Neurosurg.* **44**, 12–15. (C)

Kobrine, A. I., Evans, D. E., and Rizzoli, H. V. (1977). The effects of beta adrenergic blockade on spinal cord autoregulation in the monkey. *J. Neurosurg.* **47**, 57–63. (C)

Kobrine, A. I., Evans, D. E., and Rizzoli, H. V. (1978). The effect of sciatic nerve stimulation on spinal cord blood flow. *J. Neurol. Sci.* **38**, 435–439. (C)

Krayenbuehl, H., and Yasargil, M. G. (1963). Die Varicosis Soinalis und ihre Behanllung. *Schweiz. Arch. Neurochir. Psychiatr.* **92**, 74–92. (E)

Lazorthes, G. (1972). Pathology, classification and clinical aspects of vascular diseases of the spinal cord. *In* "Handbook of Clinical Neurology" (P. J. Vinken and G. W. Bruyn, eds.), Vol. 12, Part II, pp. 492–506. Am. Elsevier, New York. (B)

Lewin, M. G., Hansebout, R. R., and Pappius, H. M. (1974). Chemical characteristics of traumatic spinal cord edema in cats. *J. Neurosurg.* **40**, 65–75. (D)

Luessenhop, A. J., and Cruz, T. D. (1969). The surgical excision of spinal intradural vascular malformations. *J. Neurosurg.* **30**, 552–559. (E)

McCormick, W. F. (1966). The pathology of vascular ("arteriovenous") malformations. *J. Neurosurg.* **24**, 807–816. (E)

Marcus, M. L., Heistad, D. D., Ehrhardt, J. C., and Abboud, F. M. (1977). Regulation of total and regional spinal cord blood flow. *Circ. Res.* **41**, 128–134. (C)

Matthews, W. B. (1959). The spinal bruit. *Lancet* **2**, 1117–1118. (E)

Naftchi, N. E., Demeny, M., DeCrescito, V., Tomasula, J. J., Flamm, E. S., and Campbell, J. B. (1974). Biogenic amine concentrations in traumatized spinal cords of cats. Effects of drug therapy. *J. Neurosurg.* **40**, 52–57. (D)

Newman, M. J. D. (1959). Racemose angioma of the spinal cord. *Q. J. Med.* **28**, 97–108. (E)

Odom, G. L. (1962). Vascular lesions of the spinal cord: Malformations, spinal and extradural hemorrhage. *Clin. Neurosurg.* **8**, 196–236. (E)

Osterholm, J. L. (1974). The pathophysiological response to spinal cord injury. *J. Neurosurg.* **40**, 3–33. (D)

Palleske, H., Kivelitz, R., and Loew, F. (1970). Experimental investigations on the control of spinal cord circulation. IV. The effect of spinal or cerebral compression on the blood flow of the spinal cord. *Acta Neurochir.* **22**, 29–41. (C)

Richardson, H. D., and Nakamura, S. (1971). An electron microscopic study of the spinal cord edema and the effect of treatment with steroids, mannitol and hypothermia. *Proc. Vet. Adm. Spinal Cord Injury Conf.* **18**, 10–16. (D)

Rogers, W. C., and Traystman, R. J. (1979). Nitroglycerin effects on cerebral hemodynamics. *Physiologist* **22**, 107 (abstr.). (C)

Rowbotham, G. F., and Little, E. (1965). The discovery of surface arteriovenous shunts. *Br. J. Surg.* **52** (7), 539–542. (E)

Shephard, R. H. (1963). Observations on intradural spinal angioma: Treatment by excision. *Neurochirurgia (Stuttgart)* **6**, 58–74. (E)

Strong, L. H. (1961). The first appearance of vessels within the spinal cord of the mammal: Their developing patterns as far as partial formation of the dorsal septum. *Acta Anat.* **44**, 80. (A)

Turner, O. A., and Kernohan, J. W. (1941). Vascular malformations and vascular tumors involving the spinal cord. *Arch. Neurol. Psychiatry* **46**, 444–463. (E)

Vise, W. M., Yashon, D., and Hunt, W. E. (1974). Mechanism of norepinephrine accumulation within sites of spinal cord injury. *J. Neurosurg.* **40**, 76–82. (D)

Wyburn-Mason, R. (1943). "The Vascular Abnormalities in Tumors of the Spinal Cord and Its Membranes." Kimpton, London. (E)

Yasargil, M. G. (1971). Intradural spinal arteriovenous malformations. *In* "Handbook of Clinical Neurology" (P. J. Vinken and G. W. Bryun, eds.), Vol. 20, Chapter 13, pp. 481–524. Am. Elsevier, New York. (E)

Yashon, D., Bingham, W. G., Faddoul, E. M., and Hunt, W. E. (1973). Edema of the spinal cord following experimental impact trauma. *J. Neurosurg.* **38**, 693–697. (C)

Young, W., Flamm, E. S., Demopoulos, H. B., Tomasula, J. J., and DeCrescito, V. (1981). Effect of naloxone on posttraumatic ischemia in experimental spinal contusion. *J. Neurosurg.* **55**, 209–219. (D)

Chapter 7
Central Nervous System: Special Senses

A. BLOOD CIRCULATION TO THE INNER EAR*

By M. LAWRENCE

1. GROSS ANATOMY OF BLOOD CIRCULATION

a. Vertebral artery: This vessel and its branches are the sole supply for the membranous labyrinth, and for this reason the inner ear remains unaffected by vascular changes occurring in the more peripheral ear or the bone of the otic capsule.

b. Internal auditory artery: This vessel, which passes to the inner ear through the internal acoustic meatus, originates from either the basilar artery or as a branch of the anterior inferior cerebellar artery (Fig. 7.1). In either case, there seems to be little consistency in the form that it takes. Most often it arises as a double vessel, each branch of which may again divide, with not all branches ending up in the inner ear. Some, of various sizes, supply areas of the petrous bone (Mazzoni, 1972), and others enter the arachnoid at the base of the cerebellum (Konashko, 1927; Walker, 1965). A single internal auditory artery is rare.

c. Branches of the internal auditory artery: Regardless of the variability of the internal auditory artery, it divides into the various labyrinthine branches: the anterior vestibular, the vestibular cochlear, and the cochlear ramus. Further branching then becomes quite complicated. The anterior vestibular branch divides into arterioles which supply the macula of the utricle, the cristas of the lateral and superior ampullas and the membranous canals, and the superior surfaces of the utricle and saccule. The vestibulo-cochlear branch divides into the posterior vestibular and the cochlear ramus, with the former supplying the macula of the saccule, the nerve and crista of posterior ampulla and membranous canals, and the inferior surfaces of utricle and saccule. Other structures making up the auditory system, such as the cecum vestibulare, modiolus, and lower one-third of the basal coil of the cochlea, are served by the cochlear ramus branch (Hawkins, 1967).

*This research was supported by funds provided by the United States Public Health Service Grants NS-05785 and NS-11731.

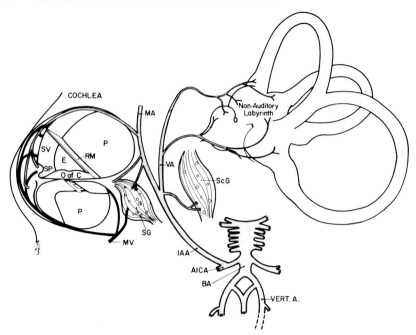

FIG. 7.1. Distribution of blood vessels to the inner ear. AICA, anterior inferior cerebellar artery; BA, basilar artery; E, endolymph; IAA, internal auditory artery; MA, modiolar artery; MV, modiolar vein; P, perilymph; RM, Reissner's membrane; ScG, Scarpa's ganglion; SG, spiral ganglion; SP, spiral prominence; SV, stria vascularis; O of C, organ of Corti; VA, vestibular artery; Vert. A, vertebral artery.

2. MICROSCOPIC ANATOMY OF BLOOD CIRCULATION

The capillary nets supplying the various membranes of the nonauditory labyrinth obviously facilitate tissue metabolism. At the same time, the vessels to the planum semilunatum at the base of an ampullar crista may also provide the plasma from which specialized cells control the contents of vestibular endolymph (Dohlman, 1967).

The capillaries of the auditory labyrinth consist of several complicated networks, both in architectural arrangement and in function. The modiolar artery at its entrance into the modiolus is about 83 μm in diameter (Kimura and Ota, 1975), and it soon twists and loops to form what has been called the cochlear plexus (Balogh and Koburg, 1965; Musebeck, 1965; Mootz and Musebeck, 1970), a structure for which no function has been determined. It was described as early as 1887 by Schwalbe, who likened it to glomeruli and suggested that it might serve to reduce the pressure of blood before entering the cochlea. Interestingly, unmyelinated nerve fibers have been observed among the fenestrated vessels making up the cochlear plexus.

As the modiolar artery coils around the nerve (Fig. 7.2), it gives off smaller arterioles (16 μm in diameter) at regular intervals. In these, the smooth muscle cells appear less prominent and the endothelial cells are flat, occasionally bulging into the vessel lumen. Some nerve fibers have been reported in the vicinity of the vessels (Terayama *et al.*, 1966).

a. Vessels of the spiral ligament: Branches without smooth muscle cells pass through the bony partition between the scala vestibuli of one turn and the scala tympani of a more apical turn to supply the structures of the spiral ligament. Here they develop into essentially four parallel networks. One supplies an area of connective tissue which is part of the spiral ligament that extends above the attachment of Reissner's membrane into the scala vestibuli (Fig. 7.2). Another set continues into the spiral ligament to supply it, while a third group enters the stria vascularis. A distinct structure, the spiral prominence, near the Claudius cells, also has its separate supply of capillaries (the fourth network). All of these groups of vessels eventually drain into the spiral modiolar vein and the cochlear aqueduct vein.

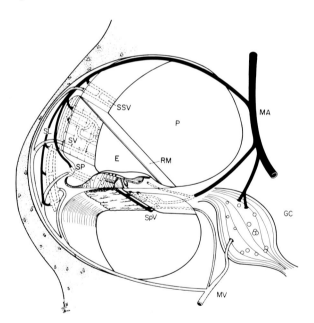

FIG. 7.2. The cochlear duct in cross section showing the various capillary areas. E, endolymph; GC, ganglion cells; MA, modiolar artery; MV, modiolar vein; P, perilymph; RM, Reissner's membrane; SL, spiral ligament; SP, spiral prominence; SpV, spiral vessels; SSV, suprastrial vessels; SV, stria vascularis.

As the arterioles from the modiolar artery reach the spiral ligament, they give rise to a network of capillaries that runs around the cochlear turn above Reissner's membrane. The vessels appear to be surrounded by well-defined pericapillary spaces, often connected by channels (Hawkins, 1967). Such an anatomic arrangement suggests that filtration and reabsorption of fluid could take place, with capillary pressure high enough to cause fluid to move out of the vessels, thus acting as a source of perilymph. In this regard, 78% of perilymph production has been reported to be derived from cochlear blood flow (Kellerhals, 1979).

The capillaries of the spiral ligament lie deep in the latter structure and serve no special purpose other than to supply the tissues with blood. The spiral prominence, a structure lying along the wall of the spiral ligament (Fig. 7.2), near the organ of Corti, receives blood from arterioles that pass through the spiral ligament, as well as from a branch extending from a vessel supplying the stria vascularis (Fig. 7.2). The arterioles entering this region are relatively large, branching into smaller capillaries that spiral around, in short courses, with the prominence. Beneath the bulge of the spiral prominence is the external sulcus, in which are found cells with long slender processes that extend up into and behind the spiral prominence, with vessels weaving among them. As yet no experimental evidence has revealed a function for the spiral prominence.

The cellular structure of the stria vascularis (Fig. 7.2) is unique and is no doubt related to its specialized function. As the arterioles enter it, they take many varied pathways. Some may anastomose directly with venous capillaries, whereas others may run longitudinally for short or long distances, branching in great complexity before draining into spiral veins.

The cells of the stria vascularis, among which the capillaries pass, have been studied by both light and electron microscopy. Kimura and Schuknecht (1970) described three types in man. Those which face endolymph and have nuclei which are close to the surface are called marginal or dark cells because of their staining properties. On the inner surface of these structures, the membrane is deeply corrugated, interlocking with intermediate and deeper basal cells. The luminal surface of the marginal cells shows varying numbers of microvilli and pinocytotic invaginations. An extensive network of tubules filled with a diffuse substance lies close to the luminal surface; marginal cells adjacent to each other are interlocked with numerous infoldings. The intermediate or light cells, containing many mitochondria, are fewer in number than the marginal cells with which they interlock. The basal cells lie next to the spiral ligament, are flat and long, and have cell processes that extend a short distance toward the marginal cells. Jahnke (1975), using freeze-fracture techniques in the guinea pig, has found that the marginal cells appear to provide a barrier between the endolymphatic space and the intracellular space. However, the intracellular spaces are very tightly sealed at the spiral ligament so that they form a compartment.

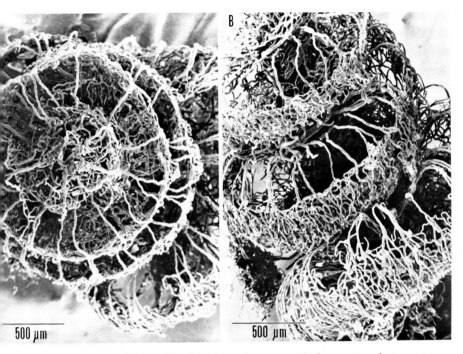

FIG. 7.3. Epoxy cast of the cochlear blood vessels as seen with the scanning electron microscope. (A) and (B) are two views of the same casting. (From Lawrence, 1980, reproduced with permission from *American Journal of Otolaryngology.*)

The local capillaries enter through the basal cell layer and pass among the intermediate and marginal cells. Vascular smooth muscle cells and neural elements are absent, the walls of the vessels being extremely thin but not fenestrated. The complexity of the capillary arrangement of the stria vascularis can be seen in Fig. 7.3 which is an epoxy cast of the cochlear blood vessels, as studied with the scanning electron microscope (Miodonski *et al.*, 1978), an approach which gives an impressive view of the dense capillary supply. The radiating arterioles are clearly visualized, and the capillary network of the stria demonstrates the intricate pattern of vessels. When flow in the vessels is observed in the living state, the hematocrit is found to be extremely high, with the rate being slower than that in the spiral ligament (Perlman and Kimura, 1955).

b. Vessels of the osseous spiral lamina: When Corti (1851) described the stria vascularis, he suggested that it might be the source of nutrients for the sensory cells of the organ which now bears his name. However, there is another group of capillaries beneath the osseous spiral lamina (SpV in Fig. 7.2) and occasionally beneath the tunnel of Corti that experiments have shown to be important in maintaining the fluid environment of the hair cells. What gave support to such a possibility is the belief that if endolymph with its high potassium content were to surround the unmyelinated nerve fibers present throughout the organ of Corti, the fibers could not function. Therefore, peri-lymph is considered the more likely fluid to occupy the spaces within the organ of Corti. For such a function, it is necessary to have a blood supply capable of bringing about a rapid exchange of oxygen, carbon dioxide, and nutrients, and this is provided by the spiral vessels (Lawrence, 1971).

After leaving the spiral modiolar artery, some of the arterioles extend along the dendrites coming from the organ of Corti, whereas others pass along the scala tympani surface of the osseous spiral lamina to form a series of loops at the bony lip. Still others may extend out onto the basilar membrane beneath the tunnel of Corti. A protection to the system lies in the fact that, should the flow in a section be impeded or stopped by a microthrombus, flow will continue in other sections or even reverse, since the capillaries are terminal vessels. In Fig.

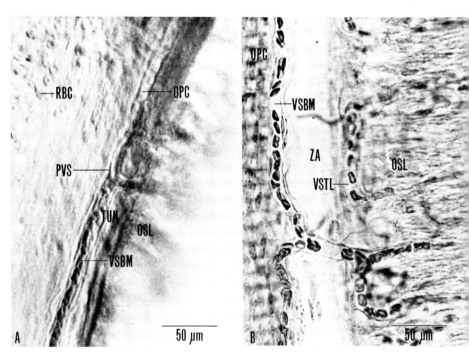

FIG. 7.4. Views of the spiral vessels as seen from beneath the organ of Corti. (A) In the living guinea pig. (B) In a fixed specimen from guinea pig. OPC, outer pillar cell; OSL, osseous spiral lamina; PVS, perivascular space; RBC, red blood cell floating in perilymph; TUN, tunnel of Corti; VSBM, spiral vessel of basilar membrane; VSTL, spiral vessel of tympanic lip; ZA, zona arcuata (beneath tunnel). (From Lawrence, 1980; reproduced with permission from *American Journal of Otolaryngology*.)

7.4A is pictured the living spiral vessel as seen from the underside of the basilar membrane. In Fig. 7.4B, there is a better view of the consistent looping of the vessels, noted in a fixed specimen. The capillaries of the lip of the osseous spiral lamina are also shown. A large white cell can be seen in the branch crossing the tunnel which appears as a clear strip.

The basilar membrane is divided into two clearly distinguished sections: the thin zona arcuata, which is beneath the tunnel from the lip of the osseous spiral lamina to the feet of the outer pillar cells, and the zona pectinata, which extends to the spiral ligament (Corti). Figure 7.4 clearly shows the two zones, particularly the zona arcuata. Apparently perilymph filling the scala tympani beneath the basilar membrane easily passes through this region. The fluid within the tunnel and around the sensory cells differs from perilymph by the presence of oxygen and nutrients that diffuse from the spiral capillaries. This reconstituted fluid has been given the name "cortilymph" (Engström and Wersall, 1953).

3. AUTONOMIC INNERVATION OF BLOOD VESSELS

In addition to the control of capillary flow by local products of metabolism, there has been speculation about the role of the autonomic nervous system. However, determining the presence of autonomic nerve fibers in the inner ear has been a difficult task. The cochlea is embedded in bone and bone surrounds the spiral ganglion and entering nerve fibers. Also, in an attempt to obtain the best preparation, various animal species have been used by different investigators so that interpretation of results is difficult. There is, however, general agreement that the origin of the adrenergic innervation is the ipsilateral cervical sympathetic ganglion.

Spoendlin and Lichtensteiger (1966), using a histochemical method in which the catecholamines and certain tryptamines are caused to fluoresce when treated with formaldehyde, found two different groups of adrenergic fibers. One consisted of a rich perivascular plexus around the internal auditory artery and the main arterial branches within the modiolus, but extending no further. The second was a much more impressive system, made up of fibers that extended between the regular nerve fibers, independent of blood vessels, and formed a plexus in the area of the habenula perforata (the cribiform area at the basilar membrane end of the osseous spiral lamina, where the dendrites leave the organ of Corti and pick up their myelin sheaths). Densert (1974), using the same technique in the rabbit as mentioned above, also described the adrenergic fibers terminating at the habenula perforata. In addition, he reported an extensive innervation of the spiral vessels of the osseous spiral lamina. No autonomic innervation of the stria vascularis capillaries has ever been reported.

Studies of the function of the adrenergic system are probably more meaningful than anatomic investigations. It is possible that under certain conditions, autoregulation, if such exists in the ear, competes with autonomic control. However, definite evidence that there is autonomic control of blood flow through the ear does not exist. For example, Perlman and Kimura (1955), in the course of taking motion pictures of the blood flow through the stria vascularis of guinea pigs and cats, saw no changes in the vessels or in the rate of blood flow after they cut the cervical trunk above the stellate ganglion or electrically stimulated the stellate ganglion, cervical trunk, superior cervical ganglion, vertebral artery, basilar artery, and the anterior inferior cerebellar artery. Using a radioactive microsphere technique in the dog, Todd *et al.* (1974) found that electrical stimulation of the stellate, caudal cervical, and superior cervical ganglia and vertebral and cervical sympathetic nerves had no effect on blood flow. Lawrence (1980) pointed out that the microcirculation of the ear is an extension of that in the brain, and hence both should respond similarly in demonstrating resistance to neurogenic control of blood flow, while reacting more readily to local metabolic demands. (For a discussion of the neurogenic regulation of cerebral blood flow, see Section C-2, Chapter 5; for a discussion of the metabolic regulation of cerebral blood flow, see Section C-4, Chapter 5.)

4. Pharmacology of Blood Circulation

Suga and Snow (1969a,b) measured blood flow in guinea pigs by electrical impedance plethysmography, using electrodes implanted in the scala vestibuli and scala tympani of the basal turn of the cochlea. They determined that intravenous administration of epinephrine and ephedrine increased blood flow, as indicated by a change in impedance accompanied by a rise in systemic blood pressure. Other α-receptor stimulants, such as norepinephrine, phenylephrine, and methoxamine, produced a transient decrease in blood flow followed by an increase, along with an elevation in systemic blood pressure. Isoproterenol and isoxsuprine (β-receptor stimulants) caused an increase in cochlear blood flow, quickly overcome by a marked fall of systemic blood pressure. α-Adrenergic blocking agents elicited an increase, while β-adrenergic blocking agents produced a decrease in blood flow.

Further research by Suga and Snow (1969a,b) revealed that such cholinomimetic agents as acetylcholine, bethanechol, and pilocarpine may cause increased blood flow, but this was apparently counteracted by a fall in blood pressure and bradycardia. Anticholinesterase agents, such as neostigmine and edraphonium, produced small increases, whereas the cholinolytic agents, atropine and scopolamine, elicited decreases. Because of the associated blood pressure changes resulting from the injection of the above drugs, it is difficult to draw conclusions concerning autonomic control.

5. FUTURE AVENUES FOR INVESTIGATION

Blood pressure, autoregulation, and the autonomic nervous system are obviously complexly interrelated, and primarily because of the difficulty of working with the capillaries of the inner ear, definitive experiments have not been done. However, surgical exposure of these vessels is becoming more precise so that better information should be forthcoming (Lawrence, 1971).

B. BLOOD CIRCULATION TO THE NOSE

By T. V. McCaffrey

1. ANATOMY OF BLOOD CIRCULATION

a. Arteries: The nasal mucosa has an abundant blood supply. In man and most animal species, two main groups of vessels supply the nose: the sphenopalatine artery and its branches and the ethmoidal group. The course of these vessels within the nose is similar for various species. The sphenopalatine artery enters the nose through the sphenopalatine foramen. It then gives off a nasopalatine branch to the septum and continues along the lateral wall of the nose. The ethmoidal vessels enter the superior part of the nasal cavity and supply the cribriform plate and superior portion of the lateral nasal wall. Within the nasal mucosa, the arteries form a latticework pattern close to the periosteum or perichondrium. Branches from the main arterial network supply the capillary plexuses in the subepithelial layer, around nasal glands, and in the periosteum or perichondrium (Dawes and Prichard, 1953).

b. Capillaries: One group, the superficial capillary network of the nasal mucosa, has a characteristic fine structure, with the vessels being supported by an incomplete layer of pericytes. The capillary endothelium is characterized by extremely attenuated areas devoid of cell organelles. These thin areas are found opposite the gaps between pericytes, and they contain circular fenestrae approximately 500 Å in diameter. The center-to-center spacing of the fenestrae varies between 0.1 and 0.5 μm. Another group, the submucosal capillaries, is smaller than the superficial capillaries and is without fenestrae. The third group, the deep capillaries enclosing the acini of the nasal glands, demonstrates fenestrations, like the superficial capillaries (Cauna and Hinderer, 1969).

c. Arteriovenous anastomoses: These structures have been identified in the nasal mucosa of man and in a large number of animal species (Dawes and Prichard, 1953; Bugge, 1968). In the nasal mucosa of the dog, cat, and rabbit, arteriovenous anastomoses are found in large numbers in the swell body, the nasoturbinal region, and the portion of the septum exposed to the inspiratory airstream. They arise from the latticework arteries. Small vessels originating from the anastomotic vessels supply the glandular tissue. When an arteriovenous anastomosis is fully open, the diameter of its lumen is the same at the arterial end as the artery from which it arises; it then gradually increases in size as the vessel approaches the vein. Arteriovenous anastomoses open either directly into thick-walled veins of the deep venous plexus or into a superficial venous plexus just before the latter communicates with the thick-walled veins. They have thick walls, the media containing both typical smooth muscle cells and cells of an epithelioid type with large pale nuclei. Arteriovenous anastomoses are from 12 to 60 μm in luminal diameter and from 100 to 150 μm in length (Dawes and Prichard, 1953).

d. Veins: These vessels drain by two routes: Anteriorly, they communicate with the palatine and anterior facial veins. Posteriorly, two main groups are involved. The ethmoidal veins pass through the cribriform plate and into the sagittal sinus, and the sphenopalatine veins are collected into a large trunk that passes through the sphenopalatine foramen to enter the infraorbital vein.

Animals can be divided into two major groups on the basis of the relationship of the arteries and veins in the nasal mucosa (Swindle, 1937). In animals designated "V/A" by Swindle, the major venous network lies superficial to the major arterial network. Into this category fall man, monkey, dog, cat, bear, rabbit, rat, porcupine, opossum, armadillo, mink, and many others. Animals designated "A/V" by Swindle have an inverted relationship between arteries and veins. The ox, antelope, sheep, goat, deer, and, probably, giraffe form this group. The functional significance of these artery–vein relationships is not known, and since only animals of the V/A type have been extensively studied, the rest of this discussion will concern only this group.

The nasal mucosa contains two major venous plexuses, both of which overlie the arteries. One, the superficial plexus, is composed of small veins arranged in a fine network. The other, the deeper plexus, consists of larger veins in a coarse network. Both groups of veins have a winding course and freely communicate with each other. The large veins of the deep plexus run at roughly right angles to the arteries but in a much closer-knit latticework. Some of the veins of the deep venous plexus possess unusually thick muscular walls.

The muscular veins of the nasal mucosa receive blood from the capillary bed and from arteriovenous anastomoses. They drain through the cushion veins, which can reduce or close their lumens in response to various stimuli.

The wall of cushion veins generally has the same histologic structure as that of regular nasal veins of similar size. The subendothelial cushion from which this vessel takes its name occupies only a fraction of the circumference of the wall and protrudes into the lumen in the form of a longitudinal ridge. The cushion consists mainly of loosely arranged smooth muscle cells longitudinally aligned; collagen and elastic fibers are also present. Contraction of the cushion musculature can lead to changes in the shape of the cushion (Cauna and Cauna, 1975), this action being capable of regulating drainage of the muscular veins.

 e. Summary: The nasal circulation is highly adapted for regulation of the volume of blood in the large muscular veins. Inflow to these vessels is controlled by the circulation through the capillary system and arteriovenous anastomoses, opening of the latter permitting a rapid flow of blood into the veins and congestion of the mucosa. Outflow from the muscular veins is regulated by the cushion veins, as well as by the thick muscular coat of the veins themselves, which allows rapid decongestion and emptying of the venous system. Since the nasal mucosa is contained within a rigid bony and cartilagenous cavity, mucosal congestion narrows the airway and leads to increased resistance to airflow.

2. VASOMOTOR INNERVATION OF BLOOD VESSELS

 a. Neuroanatomy: The autonomic innervation of about three-fourths of the nasal mucosa reaches the nose by way of the vidian nerve. The rest is supplied by autonomic fibers within the medial and lateral internal nasal branches of the ophthalmic division of the trigeminal nerve. Vasoconstrictor impulses reaching the nasal blood vessels are conveyed by preganglionic sympathetic nerves that emerge from the spinal cord at the T1 to T5 level and make synapse in the superior cervical ganglion. From there the postglionic fibers travel by way of the internal carotid plexus to the vidian nerve (Franke and Bramante, 1964). Preganglionic parasympathetic vasodilator and secretory nerves, which originate in the superior salivatory nucleus, travel by way of the greater superficial petrosal branch of the facial nerve to the sphenopalatine ganglion, where they make synapse with postganglionic nerves (Malcomson, 1959). The latter also run in the vidian nerve.

 b. Adrenergic nerves: Using the fluorescence method for localization of monoamines, Dahlström and Fuxe (1965) first reported on the distribution of adrenergic nerve terminals within the nasal mucosa. Norepinephrine was found within these structures, which ran in anastomosing strands surrounding and directly superimposed on the muscle layer of arteries, arterioles, and veins. The nerve terminals were rarely seen penetrating the muscular layer of the vessels. The investigations identified no specialized innervation of vascular sphincters.

Electron microscopic studies have confirmed the distribution of adrenergic nerve terminals within the nasal mucosa (Cauna, 1970; Änggård and Densert, 1974). Extensive adrenergic innervation of the blood vessels has been demonstrated by the presence of terminals containing dense-cored vesicles about 500 Å in diameter. Rows of adrenergic nerve terminals are located on the outside of the smooth muscle layer around the arteries, arterioles, and venous sinusoids. The terminals are partly enfolded in a Schwann cell, with their free surface facing the muscle fibers. A rich adrenergic innervation is also present in the cushion veins, but unlike arteries, arteriovenous anastomoses, and cavernous veins, the venous cushions have nonmyelinated nerve fibers inside the cushion tissue (Cauna and Cauna, 1975). The fenestrated capillaries of the respiratory epithelium and the nasal glands are devoid of any innervation.

c. Cholinergic nerves: These structures were identified in the nasal mucosa by histochemical detection of acetylcholinesterase activity (Ishii and Toriyama, 1972). They were identified near arteries and arterioles, but few fibers were seen near veins. The specificity of acetylcholinesterase has been questioned, however (Lehmann and Fibiger, 1979). It is generally agreed that a high acetylcholinesterase content is a prerequisite for location of cholinergic nerves but not definite proof that acetylcholine is present.

Ultrastructurally, the cholinergic nerve ending is considered to be characterized by many small clear vesicles and some large dense-cored vesicles. However, because of the absence of specific vesicles in the terminals of parasympathetic cholinergic neurons, electron microscopy cannot distinguish parasympathetic fibers from sensory fibers in the nasal mucosa. However, various investigators have described nerve endings with a "cholinergic" appearance around blood vessels and glandular acini in the nasal mucosa (Cauna, 1970; Änggård and Densert, 1974).

d. Peptidergic nerves: Recently, several peptides have been identified in both the peripheral and the central nervous systems. One of these—vasoactive intestinal polypeptide (VIP)—has been implicated in autonomic neurotransmission. Vasoactive intestinal polypeptide-immunoreactive nerves have been identified in the nasal mucosa (Lundberg *et al.*, 1981). They apparently originate in the sphenopalatine ganglion, since removal of this structure results in the disappearance of the VIP nerves; however, sympathectomy or sensory denervation produces no change in their activity. The distribution of the VIP neurons around blood vessels and glandular acini is similar to that of the acetylcholinesterase-positive nerves previously described. Because of the parallel reduction of VIP and choline acetyltransferase activity in nasal mucosa after extirpation of the sphenopalatine ganglion, the suggestion has been made that at least some of the VIP neurons of the nasal mucosa are cholinergic (Fig. 7.5).

Another biologically active polypeptide found in nervous tissue is avian pancreatic polypeptide (APP). Studies have demonstrated that APP coexists with norepinephrine in sympathetic nerves around arteries and arterioles in the nasal mucosa (Lundberg *et al.*, 1980a). APP-containing cell bodies have been found in the trigeminal and sphenopalatine ganglia. The APP nerves of the nasal mucosa disappear after sympathectomy.

Substance P is a peptide with potent vasodilatory and secretory actions. In the nasal mucosa, nerve fibers containing substance P surround blood vessels and glandular acini (Uddman *et al.*, 1981). In addition, nerves containing substance P are seen in the subepithelial layer and sometimes within the epithelium. In other tissues, there is evidence that fibers containing substance P are primary sensory neurons, and although most such fibers in the nasal mucosa surround blood vessels and glands, some, especially those in the epithelium, may have a sensory function. A few nerve bodies containing substance P have

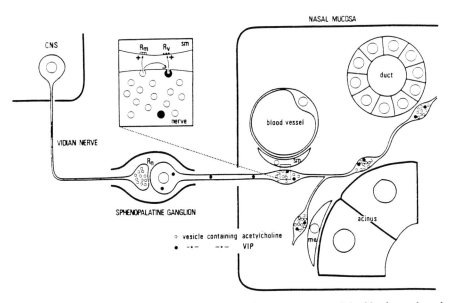

FIG. 7.5. Schematic illustration of the parasympathetic innervation of the blood vessels and exocrine tissue of the cat nasal mucosa. Preganglionic fibers originate in the central nervous system (CNS) and run in the vidian nerve. Upon activation they release acetylcholine, which acts via nicotinic receptors (R_n) on postganglionic neurons in the sphenopalatine ganglion. These neurons project via the posterior nasal nerves to the nasal mucosa. They may influence vascular smooth muscle (sm) and secretory elements [ducts, acini, and myoepithelial cells (me)] via release of both acetylcholine and VIP stored in large dense cored vesicles. Acetylcholine and VIP stimulate (+) postsynaptic muscarinic (R_m) and VIP (R_v) receptors, respectively. Experimental observations also suggest the existence of muscarinic autoreceptors, which may regulate (inhibit) VIP release. (From Lundberg *et al.*, 1981.)

been identified in the sphenopalatine ganglion by immunohistochemical techniques. Since these cell bodies also contain acetylcholinesterase, substance P, like VIP, may coexist with acetylcholine in cholinergic nerves.

3. Vasomotor and Pharmacologic Control of Blood Vessels

a. Sympathetic control: The blood vessels of the nasal mucosa are normally under sympathetic vasoconstrictor tone. Sectioning of the cervical sympathetic chain produces an immediate dilatation of nasal veins, reducing the patency of the nasal airway (Malcomson, 1959). Electrical stimulation of the cervical sympathetic nerve constricts both the resistance and the capacitance vessels of the nasal mucosa. However, capacitance vessels appear to have a lower threshold than resistance vessels (Malm, 1973). In addition, stimulation of the sympathetic trunk results in a greater reduction in arteriovenous shunt flow than in capillary flow, thus producing a redistribution of the flow from shunt to exchange vessels (Änggård, 1974). The effect of sympathetic nerve stimulation on the nasal blood vessels is reduced, but not completely abolished, by sectioning the vidian nerve, suggesting that at least a portion of the sympathetic innervation of the nasal mucosa does not travel in the vidian nerve (Jackson and Rooker, 1971). Although there have been reports of sympathetic vasodilator nerves (Blier, 1930), these have not been confirmed. However, stimulation of β-adrenergic receptors on nasal vessels does produce vasodilatation (Malm, 1974b; Hiley *et al.*, 1978).

b. Parasympathetic control: The effect of parasympathetic nerve stimulation of the nasal blood vessels has been studied by stimulation of the greater superficial petrosal nerve and the facial nerve proximal to the geniculate ganglion (Malcomson, 1959). In addition, the parasympathetic fibers of the vidian nerve have been stimulated after removal of the superior sympathetic ganglion or after blocking of the effects of sympathetic nerve stimulation by α-blocking drugs (Malcomson, 1959; Malm, 1973). These investigations showed dilatation of both capacitance and resistance vessels and an increase in venous volume and blood flow in the nasal mucosa. Although the parasympathetic vasodilator fibers have been assumed to be cholinergic, the vasodilating effect could not be completely blocked with atropine (Malm, 1973).

c. Other vasomotor mechanisms: Since VIP has been demonstrated in parasympathetic nerve endings in the nasal mucosa, this peptide with vasodilating effects may play a role in parasympathetic vasodilatation in the nasal mucosa. During parasympathetic nerve stimulation, the VIP output of the nasal mucosa has been shown to be increased. The response is not blocked by atropine and, in fact, appears to be enhanced (Lundberg *et al.*, 1981).

Avian pancreatic polypeptide has been shown reversibly to block atropine-resistant parasympathetic vasodilatation in the nasal mucosa (Lundberg *et al.*, 1980b). This polypeptide, found in sympathetic nerve endings, appears to inhibit selectively the parasympathetic response, with the sympathetic vaso-constrictor action not being reduced.

Nasal vessels have receptors for other biologically active compounds, but the role of these receptors in the normal regulation of nasal blood flow is not well understood. β-Adrenergic receptors have been identified in both the re-sistance and the capacitance vessels of the nose of the cat. Their stimulation increases nasal blood flow, an effect that is blocked by propranolol. The potency ratio of isoproterenol and salbutamol suggests that these receptors are primarily of the β_2 type (Malm, 1974b; Hiley *et al.*, 1978).

Vasodilatation of nasal blood vessels also is produced by both H_1 and H_2 receptor agonists, a response which is appropriately antagonized by specific receptor antagonists (Hiley *et al.*, 1978). Such a mechanism may be important in local vasomotor control during immunologic stimulation.

When injected into the nasal circulation, angiotensin produces intense vasoconstriction in both resistance and capacitance vessels. The effect of this substance is more marked than that of catecholamines on resistance vessels but less marked on capacitance vessels (Malm, 1974a).

4. Future Avenues for Investigation

The action of the classic sympathetic and parasympathetic pathways in the regulation of nasal vessels is now well understood; but further clarification of the role of peptide neurotransmitters in nasal vascular control is needed. Inves-tigation into this area of nasal physiology may show that vasomotor control in the nose is more complex and subtle than was suspected.

C. BLOOD CIRCULATION TO THE EYE

By A. H. Friedman

1. General Considerations

The invention of the ophthalmoscope by von Helmholtz in 1851 opened a new era, permitting visualization of blood vessels of intact living eyes. Since that time, increasingly sophisticated methods have been developed to study specific areas of the eye. Uveal and retinal blood vessel visualization has been aided by the technique of fluorescein angiography (Novotny and Alois, 1960), a method which allows resolution of 5 μm and delineates the fine details of circulation and the integrity of the blood–retinal barrier. Electron microscopic and electron-opaque markers have been used in the study of small blood vessel permeability (Shiose, 1970, 1971; Henkind *et al.*, 1979). Laser Doppler anemometry (Abbiss *et al.*, 1974) and laser speckle photography (Briers and Fercher, 1982) are promising noncontact, noninvasive methods for qualitative evaluation of retinal blood flow. A recent invasive approach involves the injection of labeled microspheres of various sizes into the retinal capillary circulation (Törnquist *et al.*, 1979). (For more extensive reviews of the circulation of the eye, see Duke-Elder and Wybar, 1961; Alm, 1972a,b; Bill, 1975, 1981; Henkind *et al.*, 1979.)

2. Embryology of Blood Circulation

The central retinal vascular system originates at about the seventh week of embryogenesis from mesoderm surrounding the fetal fissure. It extends into the optic cup and gradually vascularizes the globe as far as the ora serrata. The choroidal circulation, which extends from the optic nerve to the ciliary body, also is derived from mesoderm at about the first month of development, maturing throughout the fetal period and beyond birth. The hyaloid vascular system of the vitreous body (Reeser and Aaberg, 1979), which derives from the primitive dorsal ophthalmic artery, also is of mesodermal origin. A connection with the choroid retina is lost with closure of the fetal fissure. Vascularization of the primary vitreal cavity is maximal at about the ninth week. Subsequently the system atrophies and is seen in the adult as two remnants: a Mittendorf dot on the nasal side of the posterior surface of the lens and Bergmeister's papilla on the optic disc, which marks the exit of the hyaloid vessels.

3. Anatomy of Blood Circulation

The eye has two anatomically and physiologically distinct blood circulations: the choroidal blood supply which delivers oxygen and nutrients to the pigment epithelium and the uveal tract, and the central retinal artery, which supplies the inner portion of the retina, extending as far as the inner nuclear layer (Fig. 7.6). Retinal arteries are end vessels, in that they do not anastomose. While anastomoses are found at the capillary level, it is controversial as to whether they function in living tissue (Tso *et al.*, 1981). An avascular area separates the regions supplied by the two sets of arteries.

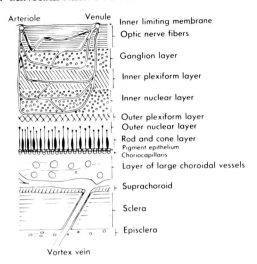

Fig. 7.6. Retinal capillaries, distributed in the inner layers of retina. Avascular outer layer, 130 μm thick, nourished mainly by choroidal vessels. (From Bill, 1981; reproduced with permission from C. V. Mosby Co.)

a. Gross anatomy: The blood circulation of the eye originates from branches of the ophthalmic artery, the latter vessel arising from the internal carotid artery as it emerges from the cavernous sinus. One of the branches, the central retinal artery, enters the subarachnoid space via the dural and arachnoid sheaths (Fig. 7.7). It penetrates the optic nerve, divides into two branches

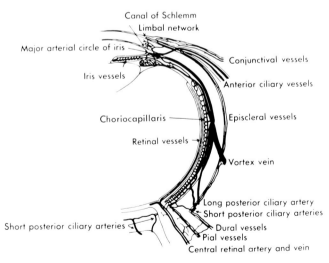

Fig. 7.7. Arteries and veins supplying the human eye. (From Bill, 1981, reproduced with permission from C. V. Mosby Co.)

on the optic disc, and then branches into arterioles and capillaries in the inner retinal layer. Drainage takes place into the cavernous sinus via retinal venules and veins and the orbital sinus.

Ciliary arteries supply the uvea. The major arterial circle of the iris, which itself gives rise to the ciliary processes and iris arteries, is formed in the ciliary body by a branch of the anterior ciliary artery and the long posterior ciliary arteries (Fig. 7.7). The short posterior ciliary arteries give rise to the choriocapillaris (lamina choroidocapillaris, Ruysch's membrane), the capillary net in the choroid, the anterior part of which is also supplied by arteries from the iris circle. The intrascleral circle of Zinn is formed by the short posterior ciliary arteries. It, together with intrascleral, retinal, and choroidal arterioles, supplies branches to the optic nerve head. Drainage from the choroid and parts of the anterior uvea is into the four vortex veins, which, in turn, empty into the superior and inferior ophthalmic veins. The latter vessels drain into the cavernous sinus. An intrascleral venous plexus receives drainage from other parts of the anterior uvea, as well as from the scleral vessels.

b. *Macular area:* There is a capillary-free zone about 120 μm in thickness surrounding arterioles (Michaelson and Campbell, 1940). The macula, originally vascularized when retinal vessels developed from mesenchyme in the fetus, becomes capillary-free by remodeling. This avascular layer comprises the total thickness of the retina at the center of the fovea. The avascular nature of the macular area permits discrimination of fine detail and color by the foveola, wherein 7×10^6 cone photoreceptors are concentrated among 1.2×10^8 rod photoreceptors, the latter acting as an intensity range finder.

c. *Microscopic anatomy:* Retinal arterioles and venules have a diameter of less than 20 μm, and retinal capillaries are generally less than 6 μm. The capillaries of the uvea are somewhat wider, those forming the choriocapillaris having a diameter of 20–30 μm. However, it has been suggested that the choriocapillaris is not truly composed of capillaries but is basically a sinusoid (Tso *et al.*, 1981). In a recently reopened area of controversy regarding this structure, the suggestion has been made that interpretation of its true nature may depend on the difference between the morbid and physiologic anatomy of the eye, on species differences, and possibly on the interpretation of angiograms (Tso *et al.*, 1981; Hayreh, 1981; Shimizu and Ujiie, 1981; Krey, 1981; Amalric, 1981).

4. PHYSIOLOGY OF BLOOD CIRCULATION

Ocular blood circulation acts to maintain the constancy of the *milieu intérieur* of the eye and to assure stable operation of the neural retina. The system requires precise regulation inasmuch as it must operate over a range of 7 or more log units of intensity, adapting rapidly to changes in contrast, contour, and color. Several barriers protect the chemical stability of this system: the blood–aqueous barrier, which maintains geometry of the visual pathway by regulating intraocular pressure, and the blood–retinal barrier (see Section C-5b, below), which regulates access to and from the neural retina, a system functioning beyond simple vision (Laties, 1973; Wurtman, 1978; Friedman and Marchese, 1978b). Constancy of neural retinal function is assured by properties inherent in the ocular microcirculation, as well as by a buffering system, the choriocapillaris, with its sinusoids which have an enormous rate of blood flow.

a. Autoregulation: This function assures relative constancy of blood flow over a wide range of perfusion pressures. Two mechanisms operate in the globe to achieve this: (1) altered muscle tone of the arterioles and (2) metabolic changes in the local environment of the blood vessels.

The vasculature in the retina, iris, and ciliary body autoregulates blood flow, whereas that in the uvea does not. Nevertheless, oxygen extraction in the uveal capillary bed is adequate, despite wide swings in rates of flow. In the autoregulated vessels of the retina, oxygen consumption is stable, about a third of the globe's oxygen requirements in the monkey being supplied by retinal vessels. The absence of precapillary sphincters in choroidal retinal capillaries has been thought to contribute to constancy of blood flow through this bed (Friedman and Oak, 1965; Friedman *et al.*, 1964).

b. Nervous innervation: The retina is not innervated by the autonomic nervous system. On the other hand, the choroid, with the exception of the choriocapillaris, contains both sympathetic and parasympathetic nerve terminals, but their role in regulating ocular blood flow is uncertain.

5. PHARMACOLOGY OF BLOOD CIRCULATION

Drugs alter normal physiologic processes, possibly acting on the blood–retinal and/or blood–aqueous barriers, or by directly affecting the various components of the vascular system. In assessing their action on ocular vessels, it is therefore necessary to differentiate the direct effects from those mediated through the systemic vasculature (Chandra and Friedman, 1972). For example, choroidal blood flow, choroidal vascular resistance, and mean arterial pressure changes produced by norepinephrine, epinephrine, isoproterenol, and acetylcholine vary depending upon the mode of administration. When given in a range of doses by close arterial injection via the lateral long posterior ciliary artery, blood flow to the eye is altered more profoundly than when the drugs are injected systemically via the femoral artery. Alm (1972a), observing the effects of vasoactive drugs (including angiotensin, nicotinic acid, di-hydroergotamine, papaverine, and histamine) on Pvr_{O_2} (oxygen concentration at the vitreoretinal junction as a measure of retinal oxygen consumption), found that only papaverine consistently increased Pvr_{O_2}. He concluded that the effect of this drug is due to its physicochemical characteristics which permit blood–retinal barrier penetration. Changing mean arterial pressure affects the hemodynamics of blood flow to the globe indirectly, because ocular perfusion pressure is approximately equal to the difference between mean arterial and intraocular pressures. Other compensatory mechanisms reflect mechanical properties of the vessels, as well as autonomic activity. (For excellent summaries of *in vivo* drug studies in a variety of species including man, see Bill, 1981; Henkind *et al.*, 1979.)

a. *Isolated extracorporeal eyes:* Such preparations (bovine, feline, amphibian), perfused through their own circulation with blood or enriched blood substitutes, circumvent problems arising in *in vivo* experiments (Seaman *et al.*, 1965; Gouras and Hoff, 1970; Marchese and Friedman, 1973; Marchese, 1979; Friedman and Marchese, 1981; Niemeyer, 1981). These studies show that extracorporeal eyes are electrophysiologically and otherwise similar to those *in situ* (Friedman and Marchese, 1978a, 1981; Niemeyer, 1981). The isolated perfused mammalian eye requires more stringent regulation than does the poikilothermic preparation, but both can be maintained 12 hr or more. Retinal detachment in the cat and seasonal changes in the frog become the limiting factors. In the frog preparation, two types of perfusion are used: constant pressure, mimicking frog arterial pressure, to assess vasoactive modification of flow, and constant volume, to overcome vascular resistance in order to assess neuroactive drug effects.

Flow rates in the frog preparation are reduced by norepinephrine in a dose-dependent fashion, as they are by γ-aminobutyric acid (GABA). Acetylcholine and other cholinergic substances, as well as dopamine, consistently increase flow rates in fresh preparations (Marchese, 1979; Friedman and Marchese, 1981). Results are confounded when several drugs are used in a single preparation. Vasodilatory dopamine receptors have not been described for ocular vessels (Goldberg, 1974), although results obtained with dopamine in the perfused frog eye suggest the presence, but not the location, of such receptors (A. H. Friedman and A. L. Marchese, unpublished observations).

DeVries and Friedman (1978) were unable to demonstrate any inhibitory neural activity after dopamine administration to an isolated retinal preparation. Dowling *et al.* (1976) had earlier suggested that the interplexiform dopaminergic neuron was an inhibitory regulator of neural retinal function. Dopamine, used clinically to produce vasodilatation in renal shutdown (Goldberg, 1974), also dilates brain vessels. When the pigment epithelium is in contact with the neural retina, i.e., as in an intact, perfused isolated globe (Marchese, 1979;

Friedman and Marchese, 1981), inhibition of some electroretinographic components is clearly and consistently shown. γ-Aminobutyric acid, on the other hand, exerts its inhibiting effects on either preparation. The importance of the intact system for full physiologic function and the effect of retinoschesis on ocular function and blood flow are areas for future investigation.

b. *Blood–retinal barrier:* Such a concept is important in the understanding of the ocular vascular responses to the administration of drugs. The juxtaposition of the choroid to the pigment epithelium and the impermeability of the tight junctions regulate flow of electrolytes, nutrients, oxygen, and other substances in and out of the eye. The requirement of lipophilicity determines whether the organ will be penetrated. This, like the blood–brain barrier (see Section D-1b, Chapter 5), affords a certain amount of protection and control of intraocular movement of undissociated lipophilic drugs.

There is also a complicating factor, in that the affinity of the melanin granules of pigment epithelium can alter transfer of vital substances. Blood–retinal barrier modification by pharmacologic agents can therefore affect function, as for example, the accumulation of drugs in uveal pigment, with subsequent ocular damage, as demonstrated *in vitro* by Potts (1962, 1964). Although chorioretinotoxic effects of phenothiazine derivatives were reported as early as 1956 by Verrey, the assessment of the melanin affinity of new drugs has not generally been made before drugs are used clinically. Chloroquine, which is used in the treatment of rheumatoid arthritis, can produce ocular lesions and chronic retinopathy many years after it has been discontinued (Burns, 1966). Ullberg *et al.* (1970) and Lindquist and Ullberg (1972) demonstrated that chorioretinotoxic drugs also accumulate in the fetal eye, indicating that the ocular system of the fetus is at risk long after medication is terminated. Many drugs likewise accumulate in ocular melanin located in the iris (Atlasik *et al.*, 1980), as well as in the blood–retinal barrier, thereby modifying visual function (Salazar and Patil, 1976). An example is tranexamic acid, an antifibrinolytic agent which disrupts the ocular vessels and neural elements in a time- and dose-dependent fashion (H.-E. Johansson, S. Ekvärn, M. Jönsson, N.-G. Lindquist, and K. Nerfström, personal communication).

6. PATHOLOGY OF BLOOD CIRCULATION

Because the literature on ocular vascular pathology is so vast, only some examples are selected for discussion.

a. *Pathologic changes in choriocapillaris:* The dense network and rapid circulation of the choriocapillaris, as well as the variability of density and diameter of its constituent vessels, make it a difficult area of investigation. Numerous disease entities are associated with inflammation, vascular degeneration, neovascularization, and neoformation of this structure. Hemorrhages originating in the choriocapillaris, subsequent to injury or surgery, may lead to blindness if the macular area is affected. Tumors of various origins may damage it by mechanical interference with local circulation. Choroidal folds may be formed as a result of neovascularization. (For a review of the problems affecting the choriocapillaris, see Amalric, 1981.)

b. Causes of visual loss: Neovascularization of whatever etiology can produce loss of vision as the result of vitreal hemorrhage and subsequent retinal detachment or by infiltration of the macular area. In the latter case, distortion of Amsler grids can be premonitory of serious pathology.

c. Glaucoma: Weinstein (1979) considered glaucoma patients to be vasoneurotic (labile), a state reflected by changes in systemic, as well as ocular, blood flow. Earlier, Mezaros and Toth (1933) demonstrated that peripheral capillaries of glaucomatous patients were abnormally formed, possessing a zigzag configuration that delays peripheral reactive hyperemia, a characteristic which lends itself to a noninvasive method for assessing the presence of glaucoma (A. H. Friedman, personal observations).

d. Summary: When vascular regulation breaks down and the physiologic compensatory processes are insufficient to cope with such a situation, then pathologic changes can occur. As a result, such disease entities as macular degeneration, fibroproliferative disease, and open angle glaucoma may develop.

REFERENCES*

Abbiss, J. B., Chubb, T. W., and Pike, E. R. (1974). Laser Doppler anemometry. *Opt. Laser Technol.* **6**, 249. (C)
Alm, A. (1972a). Effects of norephinephrine, angiotensin, dihydroergotamine, papaverine, isoproterenol, histamine, nicotinic acid, and xanthinol nicotinate on retinal oxygen tension in cats. *Ophthalmologica* **50**, 707–719. (C)
Alm, A. (1972b). Aspects of pharmacological regulation of blood flow through retina and uvea. A study in cats and monkeys. *Acta Univ. Ups.* **137**, 1–28. (C)
Amalric, P. (1981). Macular choriocapillaris pathology. *Ophthalmologica* **183**, 24–31. (C)
Änggård, A. (1974). Capillary and shunt blood flow in the nasal mucosa of the cat. *Acta Oto-Laryngol.* **78**, 418–422. (B)
Änggård, A., and Densert, O. (1974). Adrenergic innervation of the nasal mucosa in cat: An histological and physiological study. *Acta Oto-Laryngol.* **78**, 232–241. (B)
Atlasik, B., Stepien, K., and Wilczok, T. (1980). Interaction of drugs with ocular melanin *in vitro*. *Exp. Eye Res.* **30**, 325–331. (C)

*In the reference list, the capital letter in parentheses at the end of each reference indicates the section in which it is cited.

Balogh, K., and Koburg, E. (1965). Der plexus cochlearis. *Arch. Klin. Exp. Ohren-, Nasen-Kehlkopfheilkd.* **185**, 638–645. (A)

Bill, A. (1975). Blood circulation and fluid dynamics in the eye. *Physiol. Rev.* **55**, 383–417. (C)

Bill, A. (1981). Ocular circulation. *In* "Adler's Physiology of the Eye: Clinical Applications" (R. A. Moses, ed.), pp. 184–203. Mosby, St. Louis, Missouri. (C)

Blier, Z. (1930). Physiology of the sphenopalatine ganglion. *Am. J. Physiol.* **93**, 398–406. (B)

Briers, J. D., and Fercher, A. F. (1982). Retinal blood-flow visualization by means of laser speckle photography. *Invest. Ophthalmol. Visual Sci.* **22**, 255–259. (C)

Bugge, J. (1968). The arterial supply of the rabbit nose and oral cavity. *Acta Anat.* **70**, 169–183. (B)

Burns, R. P. (1966). Delayed onset of chloroquine retinopathy. *N. Engl. J. Med.* **275**, 693–696. (C)

Cauna, N. (1970). Electron microscopy of the nasal vascular bed and its nerve supply. *Ann. Otol., Rhinol., Laryngol.* **79**, 443–450. (B)

Cauna, N., and Cauna, D. (1975). The fine structure and innervation of the cushion veins of the human nasal respiratory mucosa. *Anat. Rec.* **181**, 1–16. (B)

Cauna, N., and Hinderer, K. H. (1969). Fine structure of blood vessels of the human nasal respiratory mucosa. *Ann. Otol., Rhinol., Laryngol.* **78**, 865–879. (B)

Chandra, S. R., and Friedman, E. (1972). Choroidal blood flow. II. The effect of autonomic agents. *Arch. Ophthalmol.* **87**, 67–69. (C)

Corti, A. (1851). Recherches sur l'organe de l'ouie des mammiferes. *Z. Wiss. Zool.* **3**, 109–169. (A)

Dahlström, A., and Fuxe, K. (1965). The adrenergic innervation of the nasal mucosa of certain mammals. *Acta Oto-Laryngol.* **59**, 65–72. (B)

Dawes, J. D. K., and Prichard, M. M. L. (1953). Studies of the vascular arrangements of the nose. *J. Anat.* **87**, 311–322. (B)

Densert, O. (1974). Adrenergic innervation in the rabbit cochlea. *Acta Otolaryngol.* **78**, 345–356. (A)

DeVries, G. W., and Friedman, A. H. (1978). GABA, picrotoxin and retinal sensitivity. *Brain Res.* **148**, 530–535. (C)

Dohlman, G. F. (1967). Secretion and absorption of endolymph. *NASA [Spec. Publ.] SP* **NASA SP-152.** (A)

Dowling, J. E., Ehinger, B., and Hedden, W. (1976). The interplexiform cell: A new type of retinal neuron. *Invest. Ophthalmol.* **15**, 916–926. (C)

Duke-Elder, S., and Wybar, K. C. (1961). The anatomy of the visual system. *In* "System of Ophthalmology" (S. Duke-Elder, ed.), Vol. 2, pp. 339–382. Mosby, St. Louis, Missouri. (C)

Engström, H., and Wersall, J. (1953). Is there a special nutritive cellular system around the hair cells of the organ of Corti? *Ann. Otol., Rhinol., Laryngol.* **62**, 507–512. (A)

Franke, F. E., and Bramante, P. O. (1964). Spinal origin of nasal vasoconstrictor innervation in the dog. *Proc. Soc. Exp. Biol. Med.* **117**, 769–771. (B)

Friedman, A. H., and Marchese, A. L. (1978a). Electrical activity of isolated intact perfused eye of *Rana catesbeiana* in response to various neurotransmitters, drugs and cations. *Fed. Proc., Fed. Am. Soc. Exp. Biol.* **37**, 861. (C)

Friedman, A. H., and Marchese, A. L. (1978b). Positive afterimage PAI: Early erasure by saccadic eye movement or Jendrassik maneuver. *Experientia* **34**, 71–73. (C)

Friedman, A. H., and Marchese, A. L. (1981). The isolated perfused frog eye: A useful preparation for the investigation of drug effects on retinal function. *J. Pharmacol. Methods* **5**, 215–234. (C)

Friedman, E., and Oak, S. M. (1965). Choroidal microcirculation *in vivo*. *Bibl. Anat.* **7**, 129–132. (C)

Friedman, E., Smith, T. R., and Kuwabara, T. (1964). Retinal microcirculation *in vivo*. *Invest. Ophthalmol.* **3**, 217–226. (C)

Goldberg, L. I. (1974). Dopamine–clinical uses of an endogenous catecholamine. *N. Engl. J. Med.* **291**, 707–710. (C)

Gouras, P., and Hoff, M. (1970). Retinal function in an isolated, perfused mammalian eye. *Invest. Ophthalmol.* **9**, 388–399. (C)

Hawkins, J. E. (1967). Vascular patterns of the membranous labyrinth. *NASA [Spec. Publ.] SP* **NASA SP-152.** (A)

Hayreh, S. S. (1981). Controversies on submacular choroidal circulation. *Ophthalmologica* **183,** 11–19. (C)

Henkind, P., Hansen, R. I., and Szalay, J. (1979). Ocular circulation. In "Physiology of the Human Eye and Visual System" (R. E. Records, ed.), pp. 98–155. Harper & Row, Hagerstown, Maryland. (C)

Hiley, C. R., Wilson, H., and Yates, M. S. (1978). Identification of β-adrenoceptors and histamine receptors in the cat nasal vasculature. *Acta Oto-Laryngol.* **85,** 444–448. (B)

Ishii, T., and Toriyama, M. (1972). Acetylcholinesterase activity in the vasomotor and secretory fibers of the nose. *Arch. Klin. Exp. Ohren- Nasen- Kehlkopfheilkd.* **201,** 1–10. (B)

Jackson, R. T., and Rooker, D. W. (1971). Stimulation and section of the vidian nerve in relation to autonomic control of the nasal vasculature. *Laryngoscope* **81,** 565–569. (B)

Jahnke, K. (1975). The fine structure of freeze-fractured intercellular junctions in the guinea pig inner ear. *Acta Otolaryngol., Suppl.* **336,** 1–40. (A)

Kellerhals, B. (1979). Perilymph production and cochlear blood flow. *Acta Otolaryngol.* **87,** 370–374. (A)

Kimura, R. S., and Ota, C. Y. (1975). Ultrastructure of the cochlear blood vessels. *Acta Otolaryngol.* **77,** 231–250. (A)

Kimura, R. S., and Schuknecht, H. F. (1970). The ultrastructure of the human stria vascularis. Part I. *Acta Otolaryngol.* **69,** 415–427. (A)

Konashko, P. I. (1927). Die Arteria auditiva interna des Menschen und ihre Labyrinthaste. *Z. Gesamte Anat.* **83,** 241–268. (A)

Krey, H. F. (1981). Distribution of arterioles, capillaries and venules in the equatorial choroid of the human eye. *Ophthalmologica* **183,** 20–23. (C)

Laties, A. M. (1973). The visual connection: Enter nucleus suprachiasmaticus. *Invest. Ophthalmol.* **12,** 237–239. (C)

Lawrence, M. (1971). The function of the spiral capillaries. *Laryngoscope* **81,** 1314–1322. (A)

Lawrence, M. (1980). Control of inner ear microcirculation. *Am. J. Otolaryngol.* **1,** 324–333. (A)

Lehmann, J., and Fibiger, H. C. (1979). Acetylcholinesterase and the cholinergic neuron. *Life Sci.* **25,** 1939–1947. (B)

Lindquist, N. G., and Ullberg, S. (1972). The melanin affinity of chloroquine and chlorpromazine studied by whole body autoradiography. *Acta Pharmacol. Toxicol.* **31** (II), 1–32. (C)

Lundberg, J. M., Hökfelt, T., Änggård, A., Kimmel, J., Goldstein, M., and Markey, K. (1980a). Coexistence of an avian pancreatic polypeptide (APP) immunoreactive substance and catecholamines in some peripheral and central neurons. *Acta Physiol. Scand.* **110,** 107–109. (B)

Lundberg, J. M., Änggård, A., Hökfelt, T., and Kimmel, J. (1980b). Avian pancreatic polypeptide (APP) inhibits atropine resistant vasodilation in cat submandibular salivary gland and nasal mucosa: Possible interaction with VIP. *Acta Physiol. Scand.* **110,** 199–201. (B)

Lundberg, J. M., Änggård, A., Emson, P., Fahrenkrug, J., and Hökfelt, T. (1981). Vasoactive intestinal polypeptide and cholinergic mechanisms in cat nasal mucosa: Studies on choline acetyltransferase and release of vasoactive intestinal polypeptide. *Proc. Natl. Acad. Sci. U.S.A.* **78,** 5255–5259. (B)

Malcomson, K. G. (1959). The vasomotor activities of the nasal mucous membrane. *J. Laryngol. Otol.* **73,** 73–98. (B)

Malm, L. (1973). Stimulation of sympathetic nerve fibres to the nose in cats. *Acta Oto-Laryngol.* **75,** 519–526. (B)

Malm, L. (1974a). Responses of resistance and capacitance vessels in feline nasal mucosa to vasoactive agents. *Acta Oto-Laryngol.* **78,** 90–97. (B)

Malm, L. (1974b). β-Adrenergic receptors in the vessels of the cat nasal mucosa. *Acta Oto-Laryngol.* **78,** 242–246. (B)

Marchese, A. L. (1979). A neuropharmacological analysis of retinal function in the isolated perfused frog eye. Dissertation for degree of Doctor of Philosophy, Loyola University, Chicago, Illinois. (C)

Marchese, A. L., and Friedman, A. H. (1973). Time-lapse recorded ERGs in the isolated perfused poikilotherm and homoiotherm eye. *Fed. Proc., Fed. Am. Soc. Exp. Biol.* **32,** 327. (C)

Mazzoni, A. (1972). Internal auditory artery supply to the petrous bone. *Ann. Otol., Rhinol., Laryngol.* **81,** 13–21. (A)

Mezaros, K. T., and Toth, Z. (1933). Uber das periphere Gefässystem von Glaukomkranken. *Klin. Monatsbl. Augenheilkd.* **90,** 67–72. (C)

Michaelson, I. C., and Campbell, A. C. P. (1940). The anatomy of the finer retinal vessels. *Trans. Ophthalmol. Soc. U. K.* **60,** 71–112. (C)

Miodonski, A., Hodde, K. C., and Kus, J. (1978). Scanning electron microscopy of the cochlear vasculature. *Arch. Otolaryngol.* **104,** 313–317. (A)

Mootz, W., and Musebeck, K. (1970). Die Ultrastructur des Plexus cochlearis. *Arch. Klin. Exp. Ohren- Nasen- Kehlkopfheilkd.* **196,** 301–306. (A)

Musebeck, K. (1965). Licht mikroskope Untersuchungen uber den Plexus cochlearis. *Arch. Klin. Exp. Ohren- Nasen- Kehlkopfheilkd.* **184,** 550–559. (A)

Niemeyer, G. (1981). Neurobiology of perfused mammalian eyes. *J. Neurosci. Methods* **3,** 317–337. (C)

Novotny, H. R., and Alois, D. L. (1960). A method of photographing fluorescence in circulating blood in the human retina. *Circulation* **24,** 82–86. (C)

Perlman, H. B., and Kimura, R. S. (1955). Observations of the living blood vessels of the cochlea. *Ann. Otol., Rhinol., Laryngol.* **64,** 1176–1192. (A)

Potts, A. M. (1962). Uveal pigment and phenothiazine compounds. *Trans. Am. Ophthalmol. Soc.* **60,** 517–552. (C)

Potts, A. M. (1964). Further studies concerning the accumulation of polycyclic compounds on uveal melanin. *Invest. Ophthalmol.* **3,** 349–404. (C)

Reeser, F. H., and Aaberg, T. M. (1979). Vitreous humor. *In* "Physiology of the Human Eye and Visual System" (R. E. Records, ed.), pp. 261–295. Harper & Row, Hagerstown, Maryland. (C)

Salazar, M., and Patil, P. N. (1976). An explanation for the long duration of mydriatic effect of atropine in eye. *Invest. Ophthalmol.* **15,** 671–673. (C)

Schwalbe, C. (1887). Ein Beitrag zur Kenntnis der Circulationsverhaltnisse in der Gehorschnecke. *In* "Beitrage zur Physiologie. C. Ludwig Festschrift," pp. 200–220. Vogel, Leipzig. (A)

Seaman, A., Rullman, D., Lutcher, C., and Moffat, C. (1965). The living extracorporeal eye. *Scand. J. Clin. Lab. Invest.* **17,** Suppl. 84, 101–108. (C)

Shimizu, K., and Ujiie, K. (1981). Morphology of the submacular choroid: Vascular structure. *Ophthalmologica* **183,** 5–10. (C)

Shiose, Y. (1970). Electron microscopic studies on blood–retinal and blood–aqueous barriers. *Jpn. J. Ophthalmol.* **24,** 73–90. (C)

Shiose, Y. (1971). Morphological study on permeability of the blood–aqueous barrier. *Jpn. J. Ophthalmol.* **15,** 17–26. (C)

Spoendlin, H., and Lichtensteiger, W. (1966). The adrenergic innervation of the labyrinth. *Acta Otolaryngol.* **61,** 423–434. (A)

Suga, F., and Snow, J. B. (1969a). Adrenergic control of cochlear blood flow. *Ann. Otol., Rhinol., Laryngol.* **78,** 358–374. (A)

Suga, F., and Snow, J. B. (1969b). Cholinergic control of cochlear blood flow. *Ann. Otol., Rhinol., Laryngol.* **78,** 1081–1090. (A)

Swindle, P. F. (1937). Nasal blood vessels which serve as arteries in some mammals and as veins in some others. *Ann. Otol., Rhinol., Laryngol.* **46,** 600–628. (B)

Terayama, Y., Holz, E., and Beck, C. (1966). Adrenergic innervation of the cochlea. *Ann. Otol., Rhinol., Laryngol.* **75,** 69–86. (A)

Todd, N. W., Dennart, J. E., Clairmont, A. A., and Jackson, R. T. (1974). Sympathetic stimulation and otic blood flow. *Ann. Otol., Rhinol., Laryngol.* **83**, 84–91. (A)

Törnquist, P., Alm, A., and Bill, A. (1979). Studies on ocular blood flow and retinal capillary permeability to sodium in pigs. *Acta Physiol. Scand.* **106**, 343–350. (C)

Tso, M., Shimizu, K., Hayreh, S. S., Bird, A., Gass, D., and Cunha-Vas, I. (1981). Discussion. *Ophthalmologica* **183**, 32–33. (C)

Uddman, R., Malm, L., and Sundler, F. (1981). Peptide containing nerves in the nasal mucosa. *Rhinology (Utrecht)* **19**, 75–79. (B)

Ullberg, S., Lindquist, N. G., and Sjöstrand, S. E. (1970). Accumulation of chorio-retinotoxic drugs in the fetal eye. *Nature (London)* **227**, 1257–1258. (C)

Verrey, F. (1956). Degenerescence pigmentaire de la retine d'origine medicamenteuse. *Ophthalmologica* **131**, 296–303. (C)

von Helmholtz, H. L. F. (1851). "Beschreibung eines Augen-Spiegels zur Untersuchung der Netzhaut im lebenden Auge." Förstner, Berlin. (C)

Walker, E. A. (1965). The vertebro-basilar arterial system and internal auditory angiography. *Laryngoscope* **75**, 369–407. (A)

Weinstein, P. (1979). Hemodynamics of glaucoma. *Glaucoma* **1**, 33–34. (C)

Wurtman, R. J. (1978). "The Pineal Gland." Springer-Verlag, Berlin and New York. (C)

Chapter 8
Endocrine System: Pituitary, Thyroid, and Adrenal

A. BLOOD CIRCULATION OF THE PITUITARY
GLAND

By C. E. Corliss,* D. L. Wilbur,† AND

W. C. Worthington, Jr.†

1. Embryology of Blood Circulation

The study by Wislocki (1937) of the vascular supply of the developing human hypophysis, based on three injected human embryos (21, 52, and 60 mm, sitting height), still remains the most useful reference on the subject. More recent papers consist of reports on the embryonic development of the chick hypophysis (Hammond, 1974), on the hypophyseal–portal system (Nemineva, 1950; Glydon, 1957), on the anterior lobe cells (Falin, 1961), on the hormonal aspects (Conklin, 1968; Siler-Khodr *et al.*, 1974), and on the origin of the pituitary capsule (Ciric, 1977; Chi and Myung, 1980). A well-illustrated paper on the vascularity of the developing pituitary–median eminence of the rat embryo has recently appeared (Szabo and Csanyi, 1982).

a. Interrelations of Rathke's pouch, brain, and pial plexus: In the earliest human embryos (21 mm), Wislocki found a plexus in the loose mesenchyme interposed between the evaginating Rathke's pouch (the future anterior pituitary) and the overlying diencephalon. Conklin (1968), in an even earlier stage (11.5 mm CR, i.e., crown-to-rump length), noted a comparable vascular concentration around Rathke's pouch, the primordia for the pial–hypophyseal plexus. Flanking the pouch and beneath it are capillaries and veins that are confluent with the anterior cardinal veins. The pial–hypophyseal plexus is supplied by branches of the internal carotid or posterior communicating artery and is drained by specific pial vessels. However, vessels of the future sella turcica, site of the definitive pituitary gland, drain separately into dural sinuses.

By the 25 mm stage, Rathke's pouch embraces the developing posterior lobe and continues upward as the pars tuberalis (Conklin, 1968). Mesenchyme,

---

*Author of Section A-1.

†Authors of Sections A-2 to A-8, inclusive.

Blood Vessels and Lymphatics in Organ Systems
Copyright © 1984 by Academic Press, Inc.

sandwiched between the developing epithelial pituitary gland and the brain, undergoes "intercresence" (Wislocki, 1937) with the parenchyma of the gland, i.e., the mesenchymal stroma invades the gland, bringing along a rich capillary blood supply from the pial–hypophyseal plexus. In the chick embryo, it is apparently on the vascularity of this plexus that continued differentiation of the walls of Rathke's pouch depends (Hammond, 1974). According to Wislocki and King (1936), a pial "surface condensation" on the brain at first seems to envelop the entire neurohypophysis, but later the infundibular part pushes into the mesenchyme of the sella turcica to fuse with the dura and lose its pial covering. However, Ciric (1977) disagrees, and without investigative support, proposes that a capsule is derived from pia mater. Chi and Myung (1980), with data from 56 human embryos and fetuses, state that the capsule is derived from neither source but arises from the original covering over Rathke's pouch. This concept does not necessarily contradict the accuracy of Wislocki's view that the superior hypophyseal arteries are deflected by the incorporation of pial mesenchyme by glandular parenchyma.

  b. Vessels of the posterior lobe: Most of these originate from the developing dural mesenchyme of the sella turcica (or of Rathke's pouch, see above) by way of the "free pole" portion of the pituitary lobe (Wislocki, 1937). This segment of the lobe receives its blood supply completely independently of the rest of the gland but only after it becomes attached to sellar connective tissue. Thus, it is always the last part to be vascularized. Blood for the "dural" capillary bed comes from the inferior hypophyseal artery, and the venous drainage is via the collecting veins that empty into the cavernous sinuses. The stalk is a relatively avascular structure despite the pial–hypophyseal plexus between it and the pouch-derived tuberalis (Szabo and Csanyi, 1982). As the pial–hypophyseal plexus continues to grow, its rich vascular net contributes to the primary capillary plexus. Plexiform tufts seen postnatally are not yet present in the early embryos studied by Wislocki.

  c. Portal trunks: Although these structures also develop too late to be present in the embryos described by Wislocki (1937), nevertheless he proposes their derivation from primary venules of the original pial—hypophyseal plexus after the latter has lost contact with its venous drainage. Furthermore, he believes that the trunks drain blood ". . . centrifugally from the brain (infundibular stalk) to the pituitary" (Wislocki, 1937, p. 114). Niemineva (1950) confirmed the presence of a late-forming human portal system but saw none at midterm, from which he deduced that the portal system is not indispensible for hypothalamic regulation before late pregnancy. However, a paper by Szabo and Csanyi (1982) on the rat embryo states that ". . . at a very early stage of embryonic development (E, 12), there is a direct vascular connection between brain and adenohypophysis." These authors feel that via this route, substances needed for differentiation of pituitary cells may reach the anterior lobe, even if in amounts not yet detectable by current immunohistochemical techniques.

A study of human fetal adenohypophysis, *in vitro* from 5–40 prenatal weeks by Siler-Khodr *et al.* (1974), detected the release of pituitary hormone in tissues from a 5-week-old human embryo. This provides the first evidence for cell differentiation before the pituitary gland is completely formed. Much more data on human embryos are needed to clarify this complex area of prenatal endocrinology.

d. Physiologic significance of embryologic findings: Glydon's (1957) paper on the development of the hypophyseal vessels in the rat emphasized that the portal vessels are not seen until 3 days prior to birth and that the primary capillary plexus itself is not visible until 5 days after birth. In the same paper, Glydon cited works of others indicating that the anterior lobe functions prenatally despite the tardy development of portal trunks and a primary capillary plexus. These data, then, suggest a potential for adenohypophyseal function without the unique vascular arrangement found in the adult median eminence. However, considering the most recent evidence of Szabo and Csanyi (1982), there may actually be a neurohypophyseal vascular connection present in the early rat embryo.

Several matters should be considered in evaluating the foregoing information on the developing vascular–hormonal relationships of the pituitary gland. (1) The intratuberalis vessels and those of the "pial–hypophyseal plexus" of Wislocki (1937) and of Hammond (1974) or the "perihypophyseal plexus" of Glydon (1957) might represent the actual beginnings of the primary capillary plexus. (2) The discovery of a "direct vascular connection between the brain and adenohypophysis" (Szabo and Csanyi, 1982) as early as 12 days in embryonic rats seems to obviate the necessity to consider that this gland can function without such a "vascular connection". (3) Although new information continues to appear concerning hypothalamic humoral mediators and their transport to the anterior lobe, the actual chemistry and physics of these substances are still unclear. (4) The tiny size of the embryonic pituitary gland, with the loose organization of surrounding tissue, should be kept in mind when pondering functional vascular requirements.

2. GROSS ANATOMY OF ARTERIAL CIRCULATION*

In the human hypophysis, the main blood supply is derived from paired right and left superior hypophyseal arteries and from paired right and left inferior hypophyseal arteries.

*By D. L. Wilbur and W. C. Worthington, Jr.

a. Superior hypophyseal arteries: These paired vessels derive from the internal carotid arteries after the latter have passed through the cavernous sinuses and dura mater. The vessels divide into anterior and posterior branches, both of which supply the pituitary stalk. The anterior branch also provides blood to the optic nerve and chiasma and the supraoptic and infundibular portions of the hypothalamus. The posterior branch likewise perfuses the optic tract, as well as the tuber cinereum. The anterior and posterior branches communicate and send small branches up and down the pituitary stalk. An important twig of the superior hypophyseal artery is the "loral artery" (McConnell, 1953; Stanfield, 1960) or the "artery of the trabecula" (Xuereb *et al.*, 1954a,b). This vessel passes caudad and anteriorly in the subarachnoid space, near the hypophyseal stalk, to enter the anterior hypophyseal lobe, usually in a position anterior or lateral to the stalk. (See Section A-4c, below, for a description of further ramifications of this vessel.)

b. Inferior hypophyseal arteries: These paired vessels, like their superior counterparts, arise from the intercavernous carotid artery, to supply the inferolateral aspects of the hypophysis. Each artery sends a branch to the dural covering of the anterior lobe and then gives off medial and lateral branches. The medial branch passes toward the midline to anastomose with the medial branch of the opposite side. The lateral branch courses upward between the anterior and posterior lobes of the pituitary to anastomose with the lateral branch of its contralateral equivalent posterior to the hypophyseal stalk. Thus, an arterial ring is produced around the posterior lobe which gives rise to branches that supply the posterior lobe and the dura covering the posterior and anterior lobes.

c. Arterioarterial anastomoses: The superior and inferior hypophyseal arteries are interconnected by anastomoses (Xuereb *et al.*, 1954a), the most constant of which are (1) the artery of the trabecula interconnecting with the ipsilateral lateral branch of the superior hypophyseal artery; (2) the artery of the trabecula interconnecting with the contralateral lateral branch of the inferior hypophyseal artery; and (3) the artery of the trabecula, with a vessel arising from the inferior hypophyseal arterial ring and supplying the intraglandular portion of the stalk. The latter vessel has been termed "the inferior artery of the lower infundibular stem" (Xuereb *et al.*, 1954a).

3. Gross Anatomy of Venous Circulation*

Blood draining from the anterior hypophyseal lobe exits via lateral, superior, and anteroinferior groups of veins. The lateral group drains into the superi-

*By D. L. Wilbur and W. C. Worthington, Jr.

or plexiform sinus, located on the anterior and posterior lobes superiorly, and into the anteroinferior plexiform sinuses found on the anteroinferior surface of the anterior lobe (H. T. Green, 1957). The superior group of veins, originating from the anterior pituitary lobe, enters the superior plexiform sinus. The anteroinferior group enters into the anteroinferior plexiform and inferior intercavernous sinus, the latter structure passing across the floor of the sella turcica just anterior to the groove between the anterior and posterior lobes. The small veins of the posterior lobe pass peripherally toward the anterior part of the lateral surface, emerging close to the branches of the inferior hypophyseal artery. They drain either into the posterior intercavernous sinus, which lies in front of the posterior clinoid plate, or into a posterior venous network, which is located posterior and lateral to the posterior lobe of the pituitary (H. T. Green, 1957). There appears to be no systemic venous drainage from the median eminence or from the upper and lower portions of the pituitary stalk other than the hypophyseal portal system (see below).

4. GROSS ANATOMY OF HYPOPHYSEAL PORTAL SYSTEM*

a. General considerations: The single most important fact concerning the vasculature of the pituitary gland and surrounding hypothalamic region is the existence of a third portal system of vessels in all animals thus far studied: the frog (J. D. Green, 1947; Lametschwandtner *et al.*, 1977), the rat (Green and Harris, 1949; Page and Bergland, 1977a); the cat (Morato, 1939; Akmayev, 1971a,b), the rabbit (Murakami, 1975b; Page *et al.*, 1976), the sheep (Daniel and Prichard, 1957; Bergland *et al.*, 1977), the mouse (Morin and Bottner, 1941; Worthington, 1955, 1963, 1964), the goat (Prichard and Daniel, 1958; Daniel and Prichard, 1975), and man (Fumagalli, 1942; McConnell, 1953; Xuereb *et al.*, 1954a; Stanfield, 1960; Holmes and Ball, 1974; Daniel and Prichard, 1975). The first published description of this system of vessels in vertebrate animals is that of Popa and Fielding (1930, 1933) in man. Although prior accounts of the pituitary vasculature existed, their work was the opening of a 30-year period of detailed morphologic and deductive anatomic study of this subject.

The hypophyseal portal system is composed of a series of portal trunks (the term "vein" being avoided purposely) which connects to a group of capillary-sized vessels, generally referred to as the primary capillary network. The system is located in the neurohypophysis, with sinusoids in the anterior lobe. In most animals, it is the sole significant source of blood to the anterior lobe. The specific features of the portal system—the configuration of the capillaries and

*By D. L. Wilbur and W. C. Worthington, Jr.

their relation to the surrounding tissues—are essentially the same in all animals, irrespective of the species or size.

 b. *Methods of study:* Investigators of the hypothalamico-hypophyseal axis have employed India ink or dye injections into the vascular system (Green, 1947; Torok, 1964; Akmayev, 1971a,b; Dierickx *et al.*, 1974; Holmes and Ball, 1974; Lametschwandtner and Simonsberger, 1975; Daniel and Prichard, 1975; Wilbur *et al.*, 1974, 1975, 1978; Wilbur, 1974; Wilbur and Spicer, 1980). Others have used latex-like resins and unsaturated polyester resins (Batson, 1955; Dollinger and Armstrong, 1974; Page *et al.*, 1976). Murakami (1971) employed a resin mixture of methyl methacrylate, injected into delicate vessels, to obtain a three-dimensional visualization of the microcirculation when viewed with the scanning electron microscope. Subsequently, the same investigator (Murakami, 1975a) and Lametschwandtner *et al.* (1976) used a casting substance that produced a pliable replica of the vasculature which demonstrated no leakage or discontinuities, defects often encountered with latex resins. By combining this new approach with microdissection and scanning electron microscopy, the authors were able to replicate the micromeshes of the finest capillaries in the hypothalamico–hypophyseal axis.

 c. *Vessels of portal system:* The ascending and descending branches of the superior hypophyseal artery surrounding the hypophyseal stalk, the ascending branches of the artery of the trabecula ("loral artery"), and the branches from the anastomoses between the superior and inferior hypophyseal arterial arteries all give off twigs to form the primary capillaries of the portal system (Xuereb *et al.*, 1954a,b; Daniel and Prichard, 1975). Many of the supplying vessels run with the pars tuberalis and in the area between the latter and the hypophysis, ultimately to enter the pituitary gland and subdivide into capillaries. Xuereb *et al.* (1954a) described the capillary system as "simple loops," "compact tufts of convoluted loops," and "intricate formations in which the convoluted loops are arranged in the form of a long spike." These authors also noted that the capillaries of the neurohypophysis are highly concentrated as compared with the surrounding neural tissue. Kobayashi and Matsui (1969) and Halasz (1972) found the capillaries of the median eminence to be fenestrated in character. Wislocki and King (1936) and Xuereb *et al.* (1954b) reported few connections between the capillary systems of the two types of tissue. However, Page *et al.* (1976) did note vascular connections in the rabbit in the median eminence, infundibular stem, and posterior pituitary. In this regard, it is noteworthy that the capillary beds of the intraglandular stalk are derived from the artery of the trabecula and arterial branches of the inferior hypophyseal artery.

 The capillary beds of the pituitary drain into the hypophyseal portal vessels which may also receive additional tributaries during their course through the pituitary stalk. Anastomoses between parallel hypophyseal portal trunks have not been reported in man but have been found in some lower animals. Electron microscopic studies of the portal system in the rat and rabbit have revealed that such vessels may be fenestrated (Clementi and Ceccarelli, 1971; Knigge and Scott, 1970; Duffy and Menefee, 1965).

According to Xuereb et al. (1954a,b), the hypophyseal portal vessels branch and reanastomose to form a network of sinusoids. However, the studies of Daniel and Prichard (1956) suggest that they are virtually end-arteries and that the sinusoidal network in man may not be freely anastomotic. The investigation in living pituitaries by Worthington (1966) supports the belief that the pressure gradient in the sinusoids of the anterior lobe is very low, so that even if there are anatomic connections, there may not be effective physiologic continuity.

Daniel and Prichard (1956, 1975) reported that two groups of portal vessels have their own areas of distribution in the anterior pituitary: the long portal vessels, which supply the bulk of the lobe, and their short counterparts, which supply only the region adjacent to the lower infundibular stem and the infundibular process. These observations were subsequently confirmed by the same investigators in tissue obtained from patients subjected to surgical removal of the pituitary stalk. Similar conclusions were reached by Porter et al. (1967) based on blood flow measurements in the rat. These investigators estimated that 30% of the blood reaching the anterior pituitary is supplied by the short portal vessels and 70% by the long portal vessels.

In man, rat, goat, and sheep, the anterior pituitary receives no direct arterial supply (Xuereb et al., 1954a,b; Daniel and Prichard, 1956, 1975; Page and Bergland, 1977a). This absence of direct arterial supply may render the gland especially susceptible to ischemia and infarction. Bergland and Page (1978) described an arterial complex in the monkey which appears to provide both direct and indirect anastomotic connections among the superior hypophyseal arteries, trabecular artery, inferior hypophyseal arteries, and paired carotid arteries. Stanfield (1960) reported that a number of small vessels originating from the trabecula artery and the capsule between the lobes may directly supply the anterior lobe.

A question arises concerning the extent of the direct arterial blood supply which supposedly is present in the rabbit pituitary. It is described by Harris (1947) and Daniel and Prichard (1975) as arising from the internal carotid artery and passing directly to the anterior lobe; the data presented by these authors, however, are not elaborate. Furthermore, it is difficult to develop any concept from them concerning the extent of direct arteriolar connections in the anterior pituitary, whether these connections supply their own capillary network or join the sinusoids and, if they do, in precisely what manner this takes place. It would appear that in considering a system so constant phylogenetically in its essential detail, great care should be exercised in interpreting morphologic data which imply a radically different vascular arrangement in a single species. In a recent study of rabbit vascular casts by scanning electron microscopy, Page et al. (1976) were unable to find a direct arterial supply to the anterior pituitary.

5. Microscopic and Submicroscopic Anatomy of Blood Vessels and Nervous Innervation*

a. Arteries: According to Green (1948), the superior hypophyseal arteries possess elastic lamina and a discernible smooth muscle coat which persist as the vessels subdivide into arterial branches in the pars tuberalis. Similar structures have been noted in the inferior hypophyseal arterial system (W. C. Worthington, unpublished observation).

b. Primary capillaries and portal trunks: The capillaries of the pituitary have been described as "tufted vessels," "glomerular skeins," and "complex vascular formations" (Green, 1948; Holmes, 1967; Page *et al.*, 1978). Green (1948) reported that the primary capillaries studied in human tissues varied from 5.8 to 39.7 μm in diameter, findings which have also been reported in living preparations (Worthington, 1955; Wilbur and Spicer, 1980). These vessels, as well as the venules that drain them, have been shown to possess a collagenous connective tissue sheath with a variable number of smooth muscle cells (Green, 1948). The connecting venules join together in the pars tuberalis to form portal channels of large caliber (see Section A-4c, above). In a number of human pituitaries stained with hematoxylin and eosin or periodic acid–Schiff reaction, the hypophyseal portal trunks have been seen as wide diameter, thin-walled structures, consisting of a very thin endothelium and comparatively small amounts of collagenous or reticular connective tissue, as well as some scattered cells resembling smooth muscle (Wilbur, 1974). Although it has been suggested that the smooth muscle cells are contractile in function (Murakami, 1975a), contraction of the walls of the portal trunks has never been observed in living animals (Worthington, 1960).

c. Neurovascular zones: There is a morphologically distinct, highly specialized neurovascular zone on the posterior aspect of the human pituitary stalk which may extend posteriorly almost to the mammillary bodies (Green, 1948). It is described as having a sharp line of demarcation between it and the tuber cinereum, and, besides collagen and islands of glandular cells, it apparently consists primarily of blood vessels and nerve fibers. The latter are arranged in extremely complex and highly individual perivascular plexuses which are unusually thick. Several authors have described a subependymal network of blood vessels, intercalated between the plexus of the median eminence and the hypothalamus, as a portal route from the former to the latter (Torok, 1964; Holmes, 1967; Knigge and Scott, 1970; Porter *et al.*, 1973; Oliver *et al.*, 1977;

---

*By D. L. Wilbur and W. C. Worthington, Jr.

Bergland and Page, 1979). Murakami (1975a) was unable to support this view and instead stated that "our micrographs, however, prove that the plexus of the median eminence not only converges into the portal venules to supply the anterior lobe, but also direct communications at its periphery with the systemic venous twigs surround the eminence." As attractive as the finding of very definite connections between the neurovascular zone and the tuberohypophyseal tract might be, in the light of current understanding of the neurovascular link between the hypothalamus and the anterior pituitary, such structures are an interesting possibility which, if they do exist in the human, require further study.

d. Perivascular nerve endings of hypophyseal stalk: In the hypophyseal stalk and in the pars tuberalis, Green (1948) found nerve endings to be common in the vessel sheaths, taking a variety of forms. Regarding the origin of these nerve fibers, he made the cautious statement that "many seem to come from the tractus hypophyseus proper." He believed that some of the neurovascular zone fibers gave rise to vasomotor-type endings which perhaps were supplying some of the vessels in the neural stalk.

An interpretation that at least some of the fibers ending around blood vessels in the primary capillaries and portal trunks in the human neurohypophysis are derived from the tractus hypophyseus is consistent with the findings in other animals and is the only view in accord with the uniform application of the neurohumoral hypothesis. The ever-present problem of whether silver-stained fibers are neural or reticular might cause considerably more concern if it were not for Green's extremely judicious handling of material of this type, both in previous and subsequent work. He demonstrated, by the painstaking use of silver stains and auxiliary techniques, that while there may be some vasomotor fibers accompanying the blood vessels which enter the anterior lobe, there are no nerve fibers ending directly as secretomotor fibers on cells of the anterior lobe. This work (Green, 1951a,b) is quite convincing and lends credence to the interpretations concerning the presence and sources of origin of nerve fibers in the vicinity of the blood vessels in the neurohypophysis and the special neurovascular zone.

The organization of neurohypophyseal perivascular structures has been studied by Liss (1958), Page and Bergland (1977a,b), Page *et al.* (1978), and Domokos *et al.* (1981). These investigators described nerve plexuses which lay directly on the vessel walls and were thought to derive from either the hypothalamic–hypophyseal tract or from autonomic plexuses. "Pericytes" also have been noted in the same plane as the nerve fibers. Pituicytes were found near the perivascular glial tissue, with processes extended toward, as well as away from, the blood vessels. Some of the processes terminated on blood vessels with a type of sucker foot. Perivascular connective tissues were found to form a dense sheet containing all of the other perivascular structures. It might be noted that Liss (1958) suggested that the pituicytes in this location were a type of modified astrocyte.

6. Ultrastructure of Microcirculation*

The study of the pituitary microcirculation presents many peculiar problems not encountered in general microcirculatory investigations. Because these are so fundamental and because the difficulties in exposing the pituitary for *in vivo* examination with a microscope are so formidable, there is some question as to whether experimental study in this field can be productive of useful information.

a. Neurohypophysis: Capillaries in this structure in the rat were studied electron microscopically by Hartmann (1958), and those of the posterior lobe were investigated in the cat and dog by Bargmann and Knoop (1957) and Palay (1957) and in the rat by Page and Bergland (1977a,b) and Page *et al.* (1978). It was found that the such vessels in the posterior lobe are surrounded by two layers of basement membrane, separated by a space of variable width. The inner layer is located immediately adjacent to the capillary endothelium, and the outer layer borders the pituicyte processes and nerve terminals. The formed space is usually optically empty, but it may contain occasional connective tissue fibers or fibroblasts. The endothelium of the capillaries possesses fenestrations (Hartmann, 1958; Palay, 1957), bridged by a single, very thin membrane, thus resembling fenestrations found in thyroid capillaries (Ekholm, 1957) (see Section B-1c, below). Hartmann (1958) found that, as in the thyroid, there were thickenings of the endothelial cytoplasm at the points of junction between adjacent endothelial cells. He noted widening of endothelial cells in the region of the nucleus, a site which also contained mitochondria, Golgi material, and some elements of endoplasmic reticulum. In areas distal to the nucleus, the endothelial cytoplasm was found to be very thin, measuring as little as 150 Å. Vesicles in the endothelial cytoplasm were likewise present, with their number increasing in experimental animals that had been treated with histamine, an agent known to raise the titer of posterior lobe hormones in the circulating blood. No neurosecretory vesicles or granules were found in the pericapillary spaces or the capillary lumen. Interestingly, the general arrangement of endothelial cells, endothelial discontinuities of basement membranes in the posterior pituitary, and the relationship of these structures to the parenchyma resembled those in the thyroid.

*By D. L. Wilbur and W. C. Worthington, Jr.

b. Adenohypophysis: Rinehart and Farquhar (1955) studied the ultrastructure of the sinusoids of the anterior pituitary in the rat and noted an inner layer of endothelial cells, a thin basement membrane closely applied to the outer margin of the endothelial cells, a connective tissue space, and a second basement membrane more closely applied to the cells of the parenchyma. The endothelial layer was found to be essentially continuous, but extremely thin in some areas. The cytoplasm of the endothelial cells was usually observed to contain mitochondria, vesicles, Golgi apparatus, and endoplasmic reticulum. The perisinusoidal space between the two basement membranes was of varying width, was of low electron density, and contained substances relatively high in water content. Connective tissue fibrils also were present. Rinehart and Farquhar (1955) noted that, "Granules and small segments of granule-containing cytoplasm derived from anterior pituitary parenchymal cells are frequently found lying free in the perisinusoidal and intersinusoidal spaces." They also found that rats treated with trypan blue exhibited very little dye in the cytoplasm of the endothelial cells, but considerable dye in the perisinusoidal spaces. Moreover, these investigators showed that in rats treated with trypan blue, the phagocytic cells of the anterior pituitary, which lie outside the endothelium, did not constitute a "reticuloendothelium," unlike that of the liver, spleen, and bone marrow.

Farquhar (1961) provided additional information, which makes it clear that the basic pattern of organization of the sinusoidal wall and its relationship to the surrounding parenchyma are similar to that found in the posterior pituitary, thyroid, parathyroid, and adrenal glands. For example, he described fenestrations in the endothelium that are of the same order of magnitude, are bridged by the same narrow membrane, and have the same relationship to the underlying basement membrane and to the points of junction between endothelial cells as are found in other endocrine glands. There are no discontinuities in the basement membrane in the location of the fenestrations, each of the latter being separate and distinct from the points of junction between neighboring endothelial cells. Farquhar also described granules, with their membranes in continuity with the cell membrane at the vascular poles of the parenchymal cells and with parenchymal cell pseudopodia projecting into the perisinusoidal spaces.

7. PHYSIOLOGY OF BLOOD CIRCULATION*

Despite a large amount of information indicating the importance of the vascular link between the hypothalamus and the anterior pituitary and among

*By D. L. Wilbur and W. C. Worthington, Jr.

various parts of the pituitary gland for endocrine functioning in general, very little is known about the physiology of the vessels or the mechanisms by which they are controlled (Worthington, 1955, 1960, 1963, 1964; Porter *et al.*, 1967, 1971, 1973, 1981).

First, despite all the experimental procedures which have been carried out on the pituitary (transplantations, stalk sections, and electrolytic lesions, etc.), it would still be desirable to know what constitutes a satisfactory blood supply for adequate functioning of all the different parts of this gland. Such information cannot be obtained from experiments employing injected fixed specimens, since an open capillary or an open vessel of any kind gives no information on how much blood is flowing through it or whether blood is flowing through it at all. Moreover, it is necessary to know what constitutes vascular insufficiency in any part of the pituitary. Infarction of the pituitary tissue (Sheehan, 1937; Sheehan and Stanfield, 1961; Daniel and Prichard, 1975) is easily demonstrable but is an extreme degree of vascular insufficiency. Consideration should be given to the possibility that rates of perfusion to parts of the pituitary sufficient to prevent infarction may, however, still cause functional deficiencies. Finally, it is necessary to obtain information on how much variation in blood flow to any given part of the gland may alter physiologic functions.

a. Direction of portal flow: The thesis that neurohumoral materials from the hypothalamus are transported by way of the portal system from the median eminence to the anterior pituitary depends for its validity on establishing the direction of blood flow in the portal vessels. Many of the purely anatomic studies have been directed primarily toward settling this question. While minute details of ramifications of the parts of the portal system have been reported, almost all of the available data concerning the vascular function of the pituitary gland consist of information deduced from the study of dead tissues or extrapolated from what is known regarding reactions of blood vessels and blood flow elsewhere in the brain case. Information so obtained cannot be expected to provide a sound basis for understanding the vascular physiology of the pituitary.

A large volume of data demonstrates that the anterior pituitary is supplied by portal vessels arising from the capillaries of the primary plexus in the median eminence of the hypothalamus and that the primary direction of blood flow in these vessels is hypothalamic–hypophyseal, i.e., from the median eminence descending into the anterior pituitary (Houssay *et al.*, 1935; Wislocki and King, 1936; Harris, 1947; Green and Harris, 1947; Hasegawa, 1953, 1960; Green, 1966; Daniel and Prichard, 1975; Akmayev, 1971a; Lametschwandter *et al.*, 1975, 1977; Worthington, 1955, 1960, 1963; Murakami, 1975a,b; Flerkó, 1980; Wilbur, 1974; Wilbur *et al.*, 1974, 1975; Wilbur and Spicer, 1980). Another theory dealing with possible routes of circulation holds that some blood flow may ascend in a retrograde manner from the pituitary gland toward the median eminence (Popa and Fielding, 1930; Ofuji, 1949; Szentágothai *et al.*, 1962; Torok, 1964; Negm, 1971; Akmayev, 1971a; Oliver *et al.*, 1977; Page and Bergland, 1977a,b; Page *et al.*, 1978; Porter *et al.*, 1981). Changes in direction of blood flow have been seen in almost any of the capillaries of the system (except those that flow directly into the collecting vessels) and in the arterial anastomoses, including the large arterial loop anterior to the median eminence (Worthington, 1962). Reversal of flow has, however, never been seen in the collecting vessels and portal trunks, for movement in these vessels and portal trunks has always been downward toward the anterior lobe, unless the pituitary gland was manually manipulated (Torok, 1964; Oliver *et al.*, 1977), in which case there was a pressure gradient reversal.

b. Vasomotion, vascularity, and capillary flow: Descriptive microcirculatory studies of the pituitary stalk of the mouse (Worthington, 1955, 1960) and of the frog (Worthington, 1963) have shown that the small arteries and arterioles supplying the primary capillary system of the median eminence and stalk are contractile. Intermittency of flow has been inferred from the observation that a number of primary capillaries were lost from view for varying periods of time. In addition, there usually were found to be fewer open capillaries in the stalk following superior cervical ganglionectomy (Blakely and Worthington, 1960). However, no evidence of active contractility of the hypophyseal portal trunks proper was observed.

c. Physical stimuli: Application of painful stimuli induces arterial and arteriolar vasoconstriction of the vessels around the median eminence and the stalk (Worthington, 1960). Conversely, systemic hypoxia and hypercapnea produce marked arterial vasodilatation, with an increase in portal blood flow. Responses observed following hemorrhage are variable, with small losses of blood causing slowing of portal blood flow, followed by acceleration, whereas large hemorrhage results in almost immediate and persistent reduction in portal blood flow and simultaneous arterial dilatation. In mice, bilateral common carotid artery ligation results in slowing of blood flow in the portal vessels and primary capillaries, but not complete cessation of flow in these vessels, probably because of intact vertebral vessels.

d. Superior cervical ganglionectomy: Experiments by Blakely and Worthington (1960) have suggested that the hypophyseal arterial vessels receive vasomotor fibers by way of the superior cervical ganglia. When the latter were removed bilaterally from mice, dilatation of the arterial vessels of the stalk and acceleration of portal vessel flow resulted, and the arteries demonstrated increased sensitivity to epinephrine.

e. Summary: It is necessary to point out that the many experimental procedures which have been applied to the pituitary and to the intact animal, with the view of elucidating the relationship between the hypothalamus and anterior pituitary, all involve the possibility of vascular interference. While the direct neurohumoral mechanisms of control of pituitary function have been clearly established, it still is possible that vasomotion has effects on the trophic function of the anterior pituitary, a response which has not been satisfactorily ruled out and which merits further investigation. [For a review of this subject, see Worthington (1960).]

While it is apparent that the blood vessels of the pituitary stalk have an active vasomotor system and react quite clearly and, on occasions, dramatically to peripheral stimuli, the available microcirculatory data are insufficient to prove that vasomotor responses have significant effects on any trophic function. It can only be hoped that refinements in technique will make it possible to advance our knowledge of the physiology of the pituitary blood vessels. At present, microcirculatory techniques have provided the only information regarding the physiology of these vessels.

8. PHARMACOLOGY OF BLOOD CIRCULATION*

a. Anesthetics: Sodium pentobarbital and morphine sulfate are two commonly employed anesthetic agents, both of which have been considered to block adrenocorticotropic responses to stress. According to Worthington (1960), these substances reduce the rate of blood flow to the hypophyseal portal vessels, although, at the same time, they may cause vasodilatation of the superior hypophyseal arteries and arterioles.

b. Sympathomimetic agents: Topical application of epinephrine to the pituitary stalks in dilutions as small as $1:5,000,000$ results in constriction of superior hypophyseal arteries and arterioles, with a reduction of the rate of blood flow through the portal system. Parenteral doses of epinephrine are followed by gradual acceleration of portal blood flow, with transient narrowing of arterial vessels and subsequent return to the original, or slightly larger, arterial diameters. Both local and systemic administration of *l*-norepinephrine bitartrate in small doses gives inconsistent results. Low concentrations sometimes produce arteriolar constriction, with reduction of portal blood flow. Large pharmacologic doses given systemically produce temporary cessation of blood flow in the portal trunks and primary capillaries, with extreme slowing of flow in the superior hypophyseal arteries. These changes are followed by intense arteriolar constriction and extremely rapid portal blood flow. Mecholyl or

*By D. L. Wilbur and W. C. Worthington, Jr.

acetylcholine, applied topically to the stalk in dilutions of 1 : 100 or 1 : 100,000, consistently produce arterial and arteriolar dilatation and acceleration of portal blood flow.

9. FUTURE AVENUES FOR INVESTIGATION

Although both physiologic and pharmacologic stimuli may alter blood flow through the pituitary gland, it has not been demonstrated that alterations in local circulation have significant effects on the endocrine function of this organ. Further studies of the microcirculation of the pituitary and of the influence of this unique vascular bed on the production and release of hormones from the glandular tissue remain to be undertaken.

B. BLOOD VESSELS AND LYMPHATICS OF THE
THYROID GLAND

By G. NORTHROP

1. ANATOMY OF BLOOD CIRCULATION

a. Extraglandular arterial anatomy: In human beings, arterial blood is supplied to the thyroid gland via two paired arteries. One pair, the right and left superior thyroid arteries, most commonly arises from the external carotid arteries (46%) or from the carotid bifurcation (36%) on the ipsilateral side. However, in 18% of patients, the common carotid artery is the site of origin (Faller and Scharer, 1947). As the superior thyroid artery enters the thyroid gland at its superior pole, it provides a posterior branch, which supplies the posterior aspect of the gland and anastomoses with the inferior thyroid artery, and a lateral branch, which descends along the lateral border of the gland. The superior thyroid artery continues as the anterior branch and follows the medial border of the upper lobe to the superior margin of the isthmus, where it anastomoses with its counterpart from the opposite side.

The second pair of vessels that supplies blood to the thyroid gland, the right and left inferior thyroid arteries, commonly arises from the thyrocervical trunk of the subclavian artery on the corresponding side of the neck. The course of the inferior thyroid artery is variable and in close association with important structures (recurrent laryngeal nerve and parathyroid gland). At varying distances from the inferior pole of the thyroid gland, the artery divides into a lower branch, which enters the inferior pole of the thyroid gland, and a superior branch, which supplies the posterior aspect of the gland. The superior branch anastomoses with the posterior branch of the superior thyroid artery on the same side. There may be a single thyroid artery, the thyroidea ima artery (reported in 12.2% of patients), which has been found to arise from the aortic arch or from the innominate, common carotid, or internal mammary arteries (Faller and Scharer, 1947). Vascular communications between the thyroid gland and the trachea are varied, both in frequency and in size. In the rabbit, Ichev (1967) noted their functional potential after bilateral ligation of the single thyroid artery that normally supplies each thyroid lobe in this species.

b. Intraglandular arterial anatomy: The physiologically functional unit of the thyroid gland is the follicle, a spheroid structure composed of a core of colloid-containing iodinated protein (thyroglobulin), surrounded by a single layer of epithelial cells. Twenty to 40 of the follicles encased in a fine connective tissue sheath are termed a thyroid lobule (Major, 1909; Johnson, 1953). The lobules, bound together with connective tissue, form functional lobes, many of which, in turn, are covered by the capsule to form the anatomic lobe (Johnson, 1955; Booth and Ghoshal, 1979).

As the terminal vessels of the superior and inferior thyroid arteries penetrate the thyroid gland, they are supported by connective tissue septa. They become progressively smaller as they pass from the capsule to between the structural lobes to form lobar arteries which subsequently split into lobular arteries and course in the interlobular septa. Lobular arteries divide into interfollicular arteries, each of which usually terminates as two follicular arteries.

Controversy still exists regarding the number of follicular arteries that supply one follicle, with some investigators (Johnson, 1953; Booth and Ghoshal, 1979) favoring a solitary follicular vessel breaking up into capillaries surrounding a single follicle, as found in the dog and man. On the other hand, Major (1909) and Ichev (1965) maintain that in the cat, rabbit, dog, and man, several follicular arteries may contribute to the capillary network surrounding a single follicle. Ichev (1970) resolved the problem by injecting fluorescent substances into anesthetized rabbits both before and after ligating the thyroid arteries. He discovered that in normal conditions (before ligation), the anatomic distribution of the artery agrees with the blood flow pattern, one follicular artery serving one follicle. However, when blood pressure is reduced to an area, as by ligation, anastomotic connections become functional so that additional follicular arteries can serve a single follicle. Thus, under different physiologic stimuli, local blood flow could be altered to a given follicle. Anastomosis between arteries on the surface of the gland are both obvious and frequent. Using injection techniques, Johnson (1953) demonstrated a similar arrangement within the gland.

An anatomic mechanism for altering blood flow to regions of the thyroid gland may involve occlusion of the arterial wall by cushions, as described by Modell (1933). These structures are composed of bundles of muscle cells that extend from the circular muscle layer of the artery into the lumen at a point where the vessel branches from a large trunk. The muscle bundles do not encircle the artery; instead, they form a cushion on either side of the vessel under the endothelial lining. The cushions appear large enough to occlude the lumen.

c. Microvasculature: Capillaries originate from twigs off the follicular artery and form an encompassing meshlike basket about the follicle (Fujita and Murakami, 1974). They progress in a parallel manner over the follicle, between the afferent and efferent connections (Rienhoff, 1931). The endothelial cell of the capillary has an intimate relationship with the epithelial cell of the follicle, the basal lamina closely following the plasma membrane.

The ultrastructure of the capillary endothelial cell is similar to that in most endocrine tissue (Ekholm, 1957), in that there are numerous fenestrations, 400–500 Å in diameter, covered by a diaphragm (Klinck *et al.*, 1970; Fujita, 1975). In the center of the diaphragm is a ring about 200 Å in diameter which, in turn, contains a light area with a diameter of about 70 Å, possibly representing the actual size of the opening (Fujita, 1975). The fenestrated endothelial wall is also present in the venular collecting system and in veins without and with a few smooth muscle cells in their walls (Luciano and Koch, 1975). Bestetti *et al.* (1977) have described numerous cytoplasmic processes, different from microvilli, that project into the capillary lumen in the thyroid gland of domestic fowl. These processes have fenestrations closed by diaphragms, some of which possess a thickening assumed to be a central knob. Ichev (1969) described capillary wall evaginations of several shapes in nonthyroid and thyroid tissues from cats and rabbits. He believes that the function and number of evaginations relate to the extent of exchange of substances between the capillary lumen and endothelial cell. Protrusions are infrequently noted on the basal surface of the endothelial cells.

Unique to thyroid capillaries is their response to thyroid-stimulating hormone (TSH), which causes the number of fenestrations in the endothelial cell to increase in parallel with the elevation in activity in the adjoining follicular cell (Bobkov and Zubarik, 1977). It seems likely that this response produces the frequent association of large vesicles along the follicular cell surface, within the perivascular space, and inside the endothelial cells of the capillaries (Klinck *et al.*, 1970).

d. Intraglandular venous anatomy: Venular egress occurs on the opposite side of the follicle from the entrance of the follicular arteriole. The veins follow in parallel fashion the pathway of the arteries, first forming twigs of the follicular vein, then follicular vein, lobular vein, lobar vein, and finally subcapsular vein (Johnson, 1953). Valves are reported in the larger veins in dogs but not in man (Modell, 1933). Veins anastomose frequently, both deep inside and upon the surface of the thyroid gland. Anastomoses between arteries and veins have been clearly noted in the thyroid gland of the dog, where they appear to be able to shunt blood past follicles; however, only indirect evidence for arteriovenous anastomoses in the thyroid gland of human beings has been reported (Modell, 1933; Johnson, 1953; Silva *et al.*, 1980).

e. Extraglandular venous anatomy: There is considerably more inconsistency in venous drainage than in the arterial blood supply to the thyroid gland, with three paired veins conducting blood away. A superior thyroid vein collects blood from the upper pole of each lobe and conveys it to the internal jugular vein on the ipsilateral side. The paired middle thyroid veins, which have no counterpart in the arterial system, leave the lateral border of the thyroid gland on each side and terminate in the internal jugular vein at a site inferior to that of the superior thyroid veins. The right and left inferior thyroid veins originate primarily on the anterior surface of the thyroid gland, travel a varied and complex course, and terminate in their respective innominate veins. Great variability in venous drainage from the lower poles of the thyroid gland has been illustrated by Krausen (1976), who found the classic right and left inferior thyroid veins in only three of ten cadavers. In seven dissections, there was fusion of the two veins into a thyroidea ima vein, which in six instances drained into the left innominate and in one into the right innominate vein. Other variations included large veins connecting the inferior with the superior thyroid veins in six dissections; in three, a pyramidal lobe vein coursed down the trachea to join the inferior thyroid venous plexus.

2. Physiology of Blood Circulation

a. Blood flow: Total thyroid blood flow has been measured in 75 euthyroid human beings, employing electromagnetic flowmetry at the time of neck surgery, with the average reading being reported as 31 (range 9 to 109) ml/min (Tegler *et al.*, 1981a). Thyroid gland weight was estimated so that flow rate per gram of thyroid tissue, 1.2 (range 0.4 to 3.8) ml/min/gm, could be calculated (Gillquist *et al.*, 1979; Tegler *et al.*, 1981a). Although excellent reviews (Sodenberg, 1959; Kapitola, 1973) of blood flow data in thyroid glands have long been available, specific physiologic interpretation remains unclear. Methodology, animal species, use of various anesthetic agents, and inadequate control data are just a few of the factors that, at best, make crude generalizations the maximal information that can be secured from these blood flow studies.

b. Vascular responses to autonomic stimulation: Microcirculation in the thyroid gland is influenced by vasoactive amines that can be released from adrenergic and muscarinic nerve fiber terminals found in close proximity to smooth muscle cells in the walls of arteries and arterioles (Melander *et al.*, 1974a; Melander and Sundler, 1979). In man and mouse, stimulation of sympathetic terminals usually results in increased blood flow in the thyroid gland (a response that can be abolished by sympathectomy), whereas cholinergic nerve stimulation generally causes decreased blood flow (Melander *et al.*, 1974b; Melander and Sundler, 1979).

3. Pharmacology of Blood Circulation

Substances responsible for an increase in blood flow through the thyroid gland may act by altering systemic circulation or, as discussed below, by dilating thyroid vasculature, thus permitting more blood to pass through the gland. Augmenting the number of new blood vessels, enlarging the lumen size of vessels already carrying blood, or opening up channels already present but not carrying blood all the time are ways to increase the blood flow rate. A chemical substance may directly increase blood flow or it may potentiate or inhibit the vascular effect of another agent.

Infusion of bovine or human TSH into the inferior thyroid artery in euthyroid patients during surgery results in a twofold increase in thyroid gland blood flow above the basal rate (Tegler *et al.*, 1981b). Thyrotropin-releasing hormone (TRH) infused systemically also causes an augmentation in thyroid gland blood flow (Tegler *et al.*, 1981b). The most likely reason for such a response is the release of TSH. However, a direct effect upon the central nervous system cannot be excluded. For example, in the rat, intraventricular injection of TRH results in a profound increase in thyroid gland blood flow, a response which is prevented by prior vagotomy (Tonoue and Nomoto, 1979). Additional substances reported to alter blood flow in the thyroid gland are antithyroid agents (goitrogens) (such as methylthiouracil and carbimazole) and perchlorate, pentobarbital, angiotensin II, theophylline, diazoxide, and iodine (Kapitola, 1973).

Mast cells found in the perivascular spaces in the thyroid gland contain granules of histamine and 5-hydroxytryptamine. These amines can be released by TSH, resulting in an increase both in blood flow and in follicular cell permeability. In contrast, mast cells located in juxtathyroidal muscle tissue do not respond to TSH with degranulation, and blood flow does not change (Melander *et al.*, 1975; Clayton and Szego, 1967).

4. Lymphatic System

The anatomy of the lymphatic pathways in the thyroid gland was accurately described in dog and man by Rienhoff in 1931. He found that injection of India ink into thyroid parenchyma on one side caused visualization of the lymphatic vessels in the opposite lobe in only 4 of the 40 dogs studied (Fig. 8.1). In these animals, examination revealed a connecting lymphatic channel in the pre-tracheal fascia whereby India ink particles, injected under a steady pressure into one lobe, could be forced through the channel and dispersed into the lymph vessels throughout the opposite lobe. Similar India ink injections were performed on thyroid lobes obtained from cadavers; however, lymph and blood vessel relationships with the follicular cells could be studied only in anatomically preserved regions of the lobe distant from the site of injection. Except for size, the anatomy of the lymphatic system in dog and man appears similar; thus a single description follows.

Lymph vessels originate in the perifollicular space, exterior to the perifollicular capillary network, as blind, irregularly shaped endothelial-lined sacs

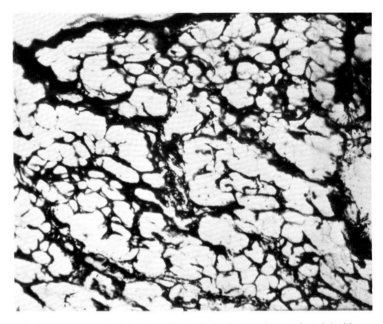

FIG. 8.1. Photomicrograph of the thyroid gland of a dog in whom India ink had been injected into the contralateral lobe. The interfollicular lymphatic plexus contains India ink and surrounds the thyroid follicles which appear as clear areas. Human thyroid tissue injected with India ink has a similar alveolar appearance. ×3. (From Rienhoff, 1931, reproduced with permission from the American Medical Association.)

or pouches. Tissue pressure and available space govern the size and shape of these pathways, which may vary from a thin ribbon to a wide, irregular sheet. The lymphatic plexus surrounds the perifollicular capillary network and forms the follicular lymph vessels, which, in turn, pass out of the lobule in the same connective tissue septa which support the lobular arteries and veins. Lymphatic vessels form an intraglandular plexus in which frequent anastomoses occur with other vessels as they pass along the septa toward the extraglandular plexus, located on the surface of the thyroid gland. Large lymph trunks, formed from the subcapsular plexus, pass out of the gland in a distribution that is species dependent. Tzinas *et al.* (1976) have described four areas of drainage in man, whereas only one or two areas are found in dogs (Booth and Ghoshal, 1979). The medial aspect of the upper two lobes drains either superiorly to the Delphian nodes or more deeply to the pretracheal nodes. Lymph leaving the medial aspect of the lower portion of the two lobes of the thyroid passes to the pretracheal and brachiocephalic nodes. Multiple lymphatic trunks arise from the lateral aspects of both lobes and drain into chains along the internal jugular veins. Posterior drainage is into lymph nodes located in the vicinity of the recurrent laryngeal nerves (Rienhoff, 1931; Tzinas *et al.*, 1976; Booth and Ghoshal, 1979). In most dogs, cats, and guinea pigs, lymph drainage passes superiorly to the cervical nodes and posteriorly to the pretracheal nodes which, in turn, drain into the mediastinal nodes (Rienhoff, 1931).

A study of radioactive iodinated substances in the thyroid gland of baboons pretreated with radioactive iodine has revealed higher concentrations of both thyroxine and iodoprotein in thyroid lymph than in venous blood. However, the higher rate of flow makes venous blood the major pathway for thyroxine (not iodoprotein) egress from the gland (Daniel *et al.*, 1967). Stimulation with TSH results in a prolonged increase of thyroxine in venous blood and a rise in iodoproteins in lymph.

5. AVENUES FOR FUTURE INVESTIGATION

Blood flow in the thyroid gland is increased following administration of TSH. It is well known from studies with thyroid slices and cell suspensions that TSH activity is mediated via cyclic AMP, the second messenger. There are no reports in which the mechanism of TSH-related increase in blood flow has been investigated. Other "trophic" hormones from the pituitary are also reported to modify blood flow in their target organ (as for example, ACTH increases blood flow to the adrenal cortex). However, it is unclear as to whether trophic hormones act on the central nervous system to influence release of amines, endorphins, etc., or directly affect blood vessel receptors, cushions, or smooth muscle to alter flow rates.

C. BLOOD VESSELS AND LYMPHATICS OF
THE ADRENAL GLAND

By M. Hamaji and T. S. Harrison

1. Anatomy of Blood Circulation

a. Extraglandular arterial anatomy: The architecture of the arteries supplying the adrenal gland differs from one species to another, as well as in any one species. The adrenal arterial supply is composed of numerous branches originating from large-named vessels. A unique feature of adrenal vasculature lies in the centripetal distribution of cortical sinusoids containing a corticomedullary portal venous system. This results in a dual blood supply to the adrenal medulla, which also receives both arterial and venous blood from the adrenal cortex.

Adrenal arteries usually originate from three main vessels: the superior adrenal artery arising from the inferior phrenic artery; the middle adrenal artery originating directly from the aorta; and the inferior adrenal artery arising either from the aorta or the renal artery. The major blood supply is thought to be from the superior and inferior adrenal arteries. In fact, in some instances, the middle adrenal artery is even absent and the inferior phrenic artery or inferior adrenal artery substitutes. The main arteries approach the upper, medial, and lower aspects of the adrenal gland and give off numerous threadlike twigs, which, in man, may number more than 50 tiny vessels.

b. Intraglandular arterial anatomy: From the subcapsular plexus, capillaries arise and perfuse the zona glomerulosa of the adrenal cortex. The arterioles run longitudinally along the column of cells in the zona fasciculata and terminate at a larger sinusoidal network, with rich anastomosis in the zona reticularis. The boundary zone in the adrenal medulla is referred to as the corticomedullary vascular dam.

Besides the arteries described above, a different adrenal arterial distribution (the medullary arteries) is seen in a variety of animals and in man. The vessels in this system originate from the subcapsular cortical plexus, pass radially through the cortical tissue without giving off any branches, and reach the corticomedullary vascular dam, where they arborize in the adrenal medulla. Flint (1900) found 50 medullary arteries among a total of 580 adrenal arteries in the dog. The number of these vessels vary, being proportional to the mass of medullary tissue; they have a relatively thick wall, particularly in large animals. The anastomosis between adjacent arteries in the subcapsular plexus varies

Blood Vessels and Lymphatics in Organ Systems
Copyright © 1984 by Academic Press, Inc.
All rights of reproduction in any form reserved.
ISBN 0-12-042520-3

between species, but it is rich in the cat, dog, monkey, and man (Harrison and Hoey, 1960; Merklin, 1962). Arteries reaching the surface of the adrenal gland are actually end-arteries, as demonstrated by the fact that their experimental ligation induces focal necrosis of the inner layer of the adrenal cortex (Harrison and Hoey, 1960). The adrenal medulla may escape such ischemic damage because it is independently supplied by medullary arteries.

In the human fetus, the pattern of blood vessel distribution within the adrenal gland is different from that in the adult (Ivemark et al., 1967). At the beginning of the second trimester, arterial blood is centrally supplied and spreads out centrifugally to the outer zone. This type of vascular arrangement gradually is replaced by a centripetal blood supply from capsular arterioles with radial branching. The change of the vascular pattern into the adult type seems to occur with differentiation of the cortical zone (Swinyard, 1943). In the mature organ, blood supply relative to gland weight is markedly diminished, as compared with the situation in the fetus, with fewer arterial branches and a meager subcapsular plexus (Merklin, 1962).

c. *Intraglandular venous anatomy:* Adrenal venous architecture topographically is divided into three regions (Coupland, 1975): an extraglandular vein extending from the hilum to the inferior vena cava or left renal vein; a longitudinal vein located within the adrenal gland; and medullary veins constituting treelike radicles which arborize throughout the adrenal medulla.

Adrenal venous drainage begins at the corticomedullary junction in dilated zona reticularis sinusoids. The medullary veins, arborizing in five orders of tributaries, terminate in the central vein. Interestingly, the latter and the greatest length of the medullary vein are not surrounded by medullary tissue but by an invaginated cuff of cortical tissue. These vessels, which course through the adrenal cortex, are especially prominent in the head and body of the adrenal gland (Dobbie and Symington, 1966).

Commonly a communicating vessel, varying in diameter from 1 to 2 mm, is noted between the superficial and deep veins. Large connecting venous channels are present in more than 80% of human adrenals. They arise from the central vein or its tributaries and traverse the entire adrenal in reverse direction, receiving no tributaries in their course (Merklin and Eger, 1961; Miekos, 1976). Although the exact distribution of venous blood from the adrenal gland is unclear, adrenal venous drainage is to some extent bidirectional, coursing through the central vein and superficial veins.

Currently, it is widely held that most of the venous blood from the adrenal cortex perfuses medullary tissue through venous sinusoids before entering the main adrenal vein. This vascular organization supports the concept of an adrenal portal venous system and is the anatomic basis for the strong functional dependence of the adrenal medulla on the cortex (Coupland, 1965; Merklin and Eger, 1961). However, Dempster (1974) has proposed an opposite opinion, suggesting that there are two distinctly independent adrenal circulations: one, cortical, consisting of arteries passing through the corticomedullary interface and draining directly into a large central vein, and the other, medullary, formed by medullary arteries which also ultimately drain into the central vein. Dempster stressed that there is no evidence that cortical venous blood bathes the chromaffin cells before joining the central vein, and he speculated that the two vascular systems originate from different embryologic sources. According to him, the fact that the central vein is surrounded by a cortical cuff within the adrenal medulla suggests that it was derived originally from the adrenal cortex. The concept of an intraadrenal portal system, however, is strongly supported by a number of outstanding morphologic and biochemical data (see Section C-2g, below). Which view is correct must await improved methods for collecting more definitive data on the intraglandular adrenal circulation.

d. Extraglandular venous anatomy: The main adrenal vein is large and it possesses a striking anatomic feature—a longitudinal muscle in its adventitia. Fewer anatomic variations are present than in the case of the adrenal arteries. The right adrenal vein empties directly into the inferior vena cava, although variations in number and in the entry site are seen. Rarely, it communicates with one of the accessory hepatic veins or the portal vein. The left adrenal vein invariably drains into the left renal vein, forming a common or separate channel with the left inferior phrenic vein. An alternative venous collateral, which develops in the presence of inferior vena caval obstruction, is the azygous system; another is the emissary veins, which form in response to central adrenal vein thrombosis (Dobbie and Symington, 1966). Superficial adrenal veins empty into superior and inferior phrenic veins and communicate with the renal capsular vein through a connecting trunk present on both sides, an arrangement found in the cat, dog, and man. The superficial veins allow bidirectional venous outflow, and their functional significance is discussed in Section C-3a, below.

As already mentioned, the main adrenal vein has the striking feature of an adventitial longitudinal muscle, a finding present only in certain species, including man, kangaroo, elephant, rhinoceros, chimpanzee, and hippopotamus. It is not observed in many domestic and laboratory animals. In the extraglandular portion of the adrenal vein, the muscular layer surrounds the entire wall and is continuous with the muscle at the junction of the inferior vena cava and left renal vein. It becomes eccentric along the peripheral vein and is arranged in muscular bundles. Peripheral veins join the main vessel only between the muscular bundles, the latter protruding into the lumen of third or fourth tributaries of medullary veins (Dobbie and Symington, 1966).

Most of the muscle cells develop in the first year of postnatal life. In the adult, the adrenal vein is surrounded by circular muscle in the media and by longitudinal muscle in the adventitia. Similar musculature is observed in the central vein during childhood, and from the age of 10–20 years, it develops further. As adrenal veins pass peripherally, the circular muscle disappears, but an irregular longitudinal muscle layer persists at the junction of venous tributaries (Symington, 1969).

2. PHYSIOLOGY OF BLOOD CIRCULATION

a. Arterial blood flow: The adrenal is densely vascular and the total blood flow per gland can be as much as 10 ml/min/gland in man (Horton *et al.*, 1966) and 5 ml/min/gland or 2.1 ml/min/gm tissue weight in the dog and cat (Hartman *et al.*, 1955; Houck and Lutherer, 1981). Adrenal circulation may be regulated by labile neurohumoral influences which facilitate the powerful secretory ability in the adrenal cortex and medulla independently. Porter and Klaiser (1964) doubted the functional importance of an accurate measurement of total adrenal flow and concluded that corticosteroid output in the rat is a function of two variables: ACTH reaching cortical cells and adrenal venous flow (Porter and Klaiser, 1965). Many workers have measured the rate of corticosteroid and catecholamine release and found an intermittent pattern.

Adrenal blood flow is usually determined by direct measurement of venous outflow through an adrenal vein cannula (Hume and Nelson, 1958) or by the radioactive microsphere method (Heymann *et al.*, 1977). The latter procedure provides repetitive measurements of regional distribution throughout the body with little operative manipulation and hence is suitable for small animals; however, it is not applicable to human study. Direct cannulation is believed to lack accuracy in view of the fact that expected pharmacologic effects sometimes are not seen (Hoechter *et al.*, 1955); moreover, it has the possibility of altering resistance to adrenal venous outflow *in situ*. Two other methods, [131I]albumin and a fractionation technique with 86Rb or [131I]antipyrine, also have been used to study adrenal blood flow (Kramer and Sapirstein, 1967; Maier and Staehelin, 1968), but, with their application, the possibility of a resulting change in cellular permeability after pharmacologic intervention cannot be ruled out.

Adrenal blood flow increases following ACTH infusion (Geber and Nies, 1979), endotoxin infusion (Wyler *et al.*, 1969), and during cardiopulmonary bypass (Baucher *et al.*, 1974); it decreases in experimental diabetes (Lucas and Foy, 1977). The experimental results in hemorrhagic shock are not consistent, with adrenal blood flow being reported as decreasing (Hume and Nelson, 1958; Walker *et al.*, 1959) or remaining constant (Houck and Lutherer, 1981; Zenner *et al.*, 1977). The disparity appears to derive not only from the different methods used in the various studies, but also from the complexity of the adrenal vascular response to hypotension. The dependence of adrenal blood flow upon systemic blood pressure is unclear. Urquhart (1965) reported a linear relationship between adrenal flow and systemic pressure in an adrenal perfusion preparation, whereas Houck and Lutherer (1981) pointed out that the adrenal vascular bed autoregulates through a wide range of blood pressure. Various kinds of humoral and neural mechanisms are also believed to modulate the adrenal circulation as well as secretion; however, the details remain unsettled.

b. *Regional distribution of blood within the adrenal gland:* The adrenal medulla receives a double blood supply: venous blood rich in corticosteroid through the cortical sinusoids and arterial blood directly from medullary arteries. Such a vascular arrangement regulates the regional distribution of blood flow within the gland and makes possible the functional dependence and independence of adrenal medulla and cortex.

Using Thorotrast microangiography in the rabbit, Harrison and Hoey (1960) noted vasoconstriction of medullary arteries and marked changes in adrenal blood distribution following intramuscular administration of epinephrine. Similar changes followed norepinephrine and histamine administration and stimulation of the splanchnic nerve in the rat, rabbit, and monkey. The response was termed "cortical vascularization," and since ACTH had no effect, it was suggested that medullary arteries regulate the regional distribution of adrenal circulation by diverting blood from the medulla into the cortex in response to vasoconstriction. However, there has been no verification of such a view.

A contrary hypothesis has been presented by Mack *et al.* (1969), who observed that during hemorrhagic shock, patchy areas of cortical nonperfusion appeared without any noticeable change in the medulla. To explain such findings, these workers offered the possibility that active vasoconstriction of precapillary arterioles occurred in the capillary plexus. Whether the studies of Harrison and Hoey and of Mack *et al.* are compatible and whether either accurately portrays intraadrenal hemodynamic changes is uncertain.

c. *Question of intramedullary regulation of blood flow:* Medullary arteries show zonal distribution in close apposition to medullary veins. Bennet and Kilham (1940) found that these vessels are distributed to definite groups of chromaffin cells. A cells (epinephrine-synthesizing cells) are located in closer association with the cortical tissue than the NA cells (norepinephrine-synthesizing cells) (Coupland, 1975), a cellular distribution favorable to epinephrine synthesis, which is promoted by a high concentration of glucocorticoid. If actually present, the vascular arrangement appears to be functionally significant; however, no selective distribution of medullary arteries to a given group of chromaffin cells has been demonstrated (Coupland and Selby, 1976). A recent radioautographic study suggests that chromaffin cells (A cells) close to the cortex are functionally more active than those in the central zone (Kent and Coupland, 1981).

d. *Role of muscular structure of adrenal vein:* As already mentioned, in man and some other mammals, adrenal venous tributaries have been found to have the striking feature of a longitudinal muscle in their adventitia. Many histologic studies have been performed evaluating the pathophysiologic significance of this structure, but, unfortunately, all the hypotheses proposed are still controversial.

A restrictive action of the muscle was considered, with contraction producing closure of the lumen of the tributaries and prevention of venous outflow. The resulting increase in intraglandular pressure was believed to lead to the opening of the vascular channels through emissary veins. Under stress, sympathetic stimuli would provoke contraction of the muscle, thus diverting part of the secreted epinephrine to the liver.

An opposing hypothesis is that contraction of the longitudinal muscle promotes venous outflow from the gland (Henderson, 1927). On the basis of the dynamic geometry of the muscular bundles existing between the junctions of the tributaries, contraction has the potential of widening the lumen at the anastomotic sites and decreasing the venous pressure within the gland. Valican (1948) speculated that contraction of both smooth muscle layers in the media could eject blood actively into the systemic circulation. Symington (1962) considered the possibility that muscular contraction could obstruct blood flow from cortical capillaries without interfering with the total adrenal circulation, thereby controlling adrenal cortical blood flow. Stagnation of cortical blood might permit longer contact of ACTH with cortical cells, permitting increased glucocorticoids to augment epinephrine biosynthesis. Barenthin (1975) suggested that the longitudinal muscular structure might exist for protection from venous spasm by the excessive contractile effect of catecholamines released into adrenal venous blood. In this regard, some mechanism involving adrenomedullary hyperfunction may have instigated the abnormality of venous musculature observed at autopsies of patients with hypertension and contracted kidneys.

e. Adrenal corticomedullary portal venous connections: It is possible that adrenal venous outflow can be bidirectional through numerous superficial veins, as well as through the central vein. These venous channels are distributed to the liver through the right portal vein. Because of the glycogenolytic action of epinephrine on the liver, direct drainage of epinephrine into the portal vein could be uniquely effective for glucose homeostasis. However, aside from potential physiologic significance, the anatomic validity of adrenoportal venous connections is doubtful because they occur inconstantly.

f. Role of the adrenorenal venous network: The function of adrenorenal venous connections was considered by Cow (1917), who speculated that epinephrine-rich blood could drain directly into the kidney, producing a significant decrease in urinary volume. Recently, Katholi *et al.* (1979) reported a postarrhythmic decrease of renal blood flow, caused by cortical vasoconstriction of the kidney, which coincided with increased epinephrine and norepinephrine content in renal subcapsular venous blood. It is possible that splanchnic and vagus nerves, together, modulate the close relationship between the adrenal gland and the kidney, also through this directly connecting venous network. The humoral and neural interaction of the sympathetic nervous system and the renin–angiotensin system, centrally or peripherally, has been increasingly clear.

g. *Functional dependence of adrenal medulla on cortex:* The close relationship between the medulla and cortex in mammals is based not only on their anatomic continuity, but also on their special vascular arrangement, especially the intraadrenal portal venous system.

Several excellent studies have led to the discovery of glucocorticoid induction of catecholamine biosynthesis. These deal with the phylogenic spectrum of catecholamine content, fetal and postnatal distribution, and the development of chromaffin tissue, as well as with postnatal regression of extraadrenal chromaffin cells. Phylogenetically, the distribution of chromaffin tissue varies widely, the anatomic relation of the medulla to cortex not being constant among species (Coupland, 1965). In adult mammals, most of the chromaffin cells are surrounded by the adrenal cortex, which has been suggested to be essential for epinephrine biosynthesis (Coupland, 1965). Adrenal chromaffin tissue develops during fetal life and after birth, whereas extraadrenal chromaffin tissue, as well as fetal cortical tissue, shows rapid regression after birth. The fetal adrenal medulla is rather immature compared with the fetal adrenal cortex.

Wurtman and Axelrod (1965) reported that hypophysectomy resulted in a profound decline of adrenal PNMT* activity and a concomitant decrease of epinephrine in the adrenal medulla. ACTH or large amounts of dexamethasone reverse the depression of PNMT activity and restore epinephrine biosynthesis. Recognition of this adrenal corticomedullary interaction has stimulated the study of phylogenic and ontogenic development of the adrenal medulla. Postnatal development of medullary tissue and epinephrine biosynthesis parallel the course of increasing PNMT activity. The regression of extraadrenal chromaffin tissue after birth is prevented by glucocorticoid administration, and, in some species, epinephrine synthesis is induced. Epinephrine content and PNMT activity rise progressively during fetal life, and, with fetal decapitation, adrenal epinephrine fails to increase.

Extensive biochemical studies have been performed on the enzymatic characteristics of PNMT, molecular basis of enzymatic induction by glucocorticoid, and PNMT isozyme. The humoral modulation of glucocorticoid effects on catecholamine biosynthesis has proved not to be specific to PNMT, since the activity of all of the enzymes participating in catecholamine biosynthesis falls after hypophysectomy. There is no glucocorticoid effect on enzyme levels in normal animals (Weiner, 1975).

*PNMT, phenyl ethanolamine *N*-methyltransferase, an enzyme promoting the last step of epinephrine biosynthesis.

Recent studies suggest that neural and hormonal stimuli regulate the adrenal medulla through other mechanisms. Neural stimuli increase PNMT and dopamine β-hydroxylase (DBH) levels by inducing *de novo* enzyme synthesis, whereas glucocorticoid inhibits *in vivo* enzyme proteolysis by controlling the levels of respective cofactor S-adenosylmethionine and ascorbic acid (Ciarranello, 1980).

Additionally, adrenal medullary function is modulated not only through splanchnic nerves, but also by ACTH which can act on central nuclei supplying innervation to splanchnic sympathetic neuronal cell bodies; therefore, ACTH may be a potentially important extracortical modulator of medullary function (Ciarranello, 1980). The concept of humoral and neural control of adrenomedullary function requires constant revision because of rapid progress in understanding catecholamine release.

3. Lymphatic System in Adrenal Glands

The adrenal lymphatics have not been studied as intensively as blood vessels, probably because of the limited methodology available. Only a few studies exist in several different species. The results are diverse and the physiologic role of adrenal lymphatics remains uncertain. In part, such a situation is due to the fact that for many years lymphatic anatomy has been investigated by injection techniques, which often produced misleading results because of incomplete filling of the channels and/or coincidental filling of blood vessels (Merklin, 1966).

Rich lymphatic plexuses are present on the surface and subcapsular region of the adrenal gland. Furthermore, a capillary network coursing throughout the parenchyma empties into the lymphatic vessels running parallel to the central vein. According to the meticulous work of Merklin (1966), no lymphatics are present in the adrenal cortical or medullary parenchyma, although lymphatic channels are prominent in the capsule and around the central vein. The finding that there are no adrenal lymphatic channels in the rat (Verhofstad and Lensen, 1973) suggests that other species may also have no adrenal parenchymal lymphatics.

Most adrenal lymphatic channels run to the thoracic duct, often directly without passing through regional lymph nodes, which are usually small in size and number. Some adrenal lymphatics course inferiorly toward the kidney and superiorly toward the diaphragm. In 12 of 68 cases studied, Merklin (1966) found that the capsular lymphatic channels of the right adrenal gland continue to the wall of the inferior vena cava. Histologically, no direct lymphaticovenous communications are seen, and the lymphatics end abruptly near the vasa vasorum in the caval wall (Merklin, 1966).

4. Future Avenues for Investigation

Vascular anatomy of the adrenal gland is functionally important, not only for the metabolism of the endocrine cells, but also because of its humoral regulation of catecholamine biosynthesis, particularly in mammals. However, there have been few studies on the adrenal microcirculation to verify the existing biochemical results or functional alterations during pathophysiologic conditions. Another important avenue for study is the role of the unique longitudinal muscle in the central adrenal vein.

References*

Akmayev, I. G. (1971a). Morphological aspects of the hypothalamic-hypophysial system. II. Functional morphology of pituitary microcirculation. Z. Zellforsch. Mikrosk. Anat. 116, 178–194. (A)

Akmayev, I. G. (1971b). Morphological aspects of the hypothalamic-hypophyseal system. III. Vascularity of the hypothalamus, with special reference to its quantitative aspects. Z. Zellforsch. Mikrosk. Anat. 116, 195–204. (A)

Barenthin, J. (1975). Über den Wandbau der Vena Centralis der Nebenniere des Menschen. Z. Mikrosk.-Anat. Forsch. 89, 647–664. (C)

Bargmann, W., and Knoop, A. (1957). Electron microkopische Beobachtungen an der Neurohypophyze. Z. Zellforsch. Mikrosk. Anat. 46, 242–251. (A)

Batson, O. V. (1955). Corrosion specimens prepared with a new material. Anat. Rec. 121, 425A. (A)

Baucher, J. K., Rudy, L. W., Jr., and Edmunds, L. H., Jr. (1974). Organ blood flow during pulsatile cardiopulmonary bypass. J. Appl. Physiol. 36, 86–90. (C)

Bennet, H. W., and Kilham, L. (1940). The blood vessels of the adrenal gland of the cat. Anat. Rec. 77, 447–471. (C)

Bergland, R. M., and Page, R. B. (1978). Can the pituitary secrete directly to the brain? (Affirmative anatomical evidence). Endocrinology 102, 1325–1338. (A)

Bergland, R. M., and Page, R. B. (1979). Pituitary–brain vascular relations: A new paradigm. Science 204, 18–24. (A)

Bergland, R. M., Davis, S. L., and Page, R. B. (1977). Pituitary secretes to brain. Lancet 2, 276–278. (A)

Bestetti, G., Canese, M. G., and Rossi, G. L. (1977). The ultrastructure of the blood capillary endothelium in the thyroid gland of the domestic fowl. J. Submicrosc. Cytol. 9, 23–30. (B)

Blakely, J. L., and Worthington, W. C., Jr. (1960). Vascular changes in the pituitary stalk following superior cervical ganglionectomy. Anat. Rec. 136, 165–166. (A)

Bobkov, V. M., and Zubarik, S. A. (1977). The effect of thyrotropic hormone on the ultrastructure of the endothelium of the thyroid perifollicular capillaries. Anat., Anthropol., Embryol. Histol. 31, Abstr. No. 2568. (B)

Booth, K. K., and Ghoshal, N. G. (1979). Angioarchitecture of the canine thyroid gland. Anat. Anz. 145, 32–51. (B)

Chi, Je G., and Myung, H. L. (1980). Anatomical observations of the development of pituitary capsule. J. Neurosurg. 52, 667–670. (A)

Ciarranello, R. D. (1980). Regulation of adrenal catecholamine biosynthetic enzymes: Integration of neural and hormonal stimuli in response to stress. In "Catecholamines and Stress" (E.

*In the reference list, the capital letter in parentheses at the end of each reference indicates the section in which it is cited.

Usdin, R. Kvetnanský, and I. J. Kopin, eds.), pp. 317–327. Elsevier/North-Holland, New York. (C)

Ciric, I. (1977). On the origin and nature of the pituitary gland capsule. *J. Neurosurg.* **46**, 596–600. (A)

Clayton, J. A., and Szego, C. M. (1967). Depletion of rat thyroid serotonin accompanied by increased blood flow as an acute response to thyroid-stimulating hormone. *Endocrinology* **80**, 689–697. (B)

Clementi, F., and Ceccarelli, B. (1971). Fine structure of rat hypothalamic nuclei. *In* "The Hypothalamus" (L. Martini, M. Motta, and F. Fraschini, eds.), pp. 17–44. Academic Press, New York. (A)

Conklin, J. L. (1968). The development of the human fetal adenohypophysis. *Anat. Rec.* **160**, 79–92. (A)

Coupland, R. E. (1965). "Natural History of the Chromaffin Cell." Longmans, Green, New York. p. 279. (C)

Coupland, R. E. (1975). Blood supply of adrenal gland. *In* "Handbook of Physiology" (H. Blaschko, G. Sayers, and A. D. Smith, eds.), Sect. 7, Vol. VI, pp. 283–294. Am. Physiol. Soc., Washington, D.C. (C)

Coupland, R. E., and Selby, J. E. (1976). The blood supply of mammalian adrenal medulla, a comparative study. *J. Anat.* **122**, 539–551. (C)

Cow, D. (1917). The suprarenal bodies and diuresis. *J. Physiol. (London)* **48**, 443–452. (C)

Daniel, P. M., and Prichard, M. M. L. (1956). Anterior pituitary necrosis. Infarction of the pars distalis produced experimentally in the rat. *Q. J. Exp. Physiol. Cogn. Med. Sci.* **41**, 215–229. (A)

Daniel, P. M., and Prichard, M. M. L. (1957). The vascular arrangements of the pituitary gland of the sheep. *Q. J. Exp. Physiol. Cogn. Med. Sci.* **42**, 237–248. (A)

Daniel, P. M., and Prichard, M. M. L. (1975). Studies of the hypothalamus and the pituitary gland. *Acta Endocrinol. (Copenhagen), Suppl.* **201**, 1–216. (A)

Daniel, P. M., Plaskett, L. G., and Pratt, O. E. (1967). The lymphatic and venous pathways for the outflow of thyroxine, iodoprotein and inorganic iodide from the thyroid gland. *J. Physiol. (London)* **188**, 25–44. (B)

Dempster, W. J. (1974). The nature of venous system in the adrenal gland. *Tohoku J. Exp. Med.* **112**, 63–77. (C)

Dierickx, K., Goossens, N., and De Woele, G. (1974). The vascularization of the neural isolated pars ventralis of the tuber cinereum-hypophysis of the frog, *Rana temporaria. Cell Tissue Res.* **149**, 431–436. (A)

Dobbie, J. W., and Symington, T. (1966). The human adrenal gland with special reference to the vasculature. *J. Endocrinol.* **34**, 479–489. (C)

Dollinger, R. K., and Armstrong, P. (1974). Scanning electron microscopy of injection replicas of the chick embryo circulatory system. *J. Microsc. (Oxford)* **102**, 179–181. (A)

Domokos, I., Laszló, F. A., Bilbao, J. M., and Kovács, K. (1981). Fine structural study of the posterior pituitary after destruction of the hypophysial stalk in the rat. *Acta Morphol. Acad. Sci. Hung.* **29** (2–3), 195–202. (A)

Duffy, P. E., and Menefee, M. (1965). Electron microscopic examination of neurosecretory granules, nerve and glial fibers, and blood vessels in median eminence of the rabbit. *Am. J. Anat.* **117**, 251–286. (A)

Ekholm, R. (1957). The ultra-structure of the blood capillaries in the mouse thyroid gland. *Z. Zellforsch. Mikrosk. Anat.* **46**, 139–146. (A, B)

Falin, L. I. (1961). The development of the human hypophysis and differentiation of cells of its anterior lobe during embryonic life. *Acta Anat.* **44**, 188–205. (A)

Faller, A., and Scharer, O. (1947). Über die variabilitat der arteriae thyroideae. *Acta Anat.* **4**, 119–122. (B)

Farquhar, M. G. (1961). Fine structure and function in capillaries of the anterior pituitary gland. *Angiology* **12**, 270–292. (A)

Flerkó, B. (1980). The hypophysial portal circulation today. *Neuroendocrinology* **30**, 56–63. (A)

Flint, J. M. (1900). The blood supply, angiogenesis, organogenesis, reticulum and histology of the adrenal. *Johns Hopkins Hosp. Rep.* **4**, 154–229. (C)

Fujita, H. (1975). Fine structure of the thyroid gland. *Int. Rev. Cytol.* **40**, 197–280. (B)

Fujita, H., and Murakami, T. (1974). Scanning electron microscopy on the distribution of the minute blood vessels in the thyroid gland of the dog, rat, and rhesus monkey. *Arch. Histol. Jpn.* **36**, 181–188. (B)

Fumagalli, Z. (1942). La vascolarizzazione dell'ipofisi umana. *Z. Anat. Entwick-lungsgesch.* **111**, 266–306. (A)

Geber, J. G., and Nies, A. S. (1979). The failure of indomethacin to alter the ACTH-induced adrenal hyperemia in the anesthetized dog. *Br. J. Pharmacol.* **67**, 217–220. (C)

Gillquist, J., Lundström, B., Larsson, L., Sjödahl, R., Brote, L., and Anderberg, B. (1979). Preoperative estimation of the size of the thyroid remnant. *Acta Chir. Scand.* **145**, 459–461. (B)

Glydon, R. St.J. (1957). The development of the blood supply of the pituitary in the albino rat with special reference to the portal vessels. *J. Anat.* **91**, 237–244. (A)

Green, H. T. (1957). The venous drainage of the human hypophysis cerebri. *Am. J. Anat.* **100**, 435–470. (A)

Green, J. D. (1947). Vessels and nerves of amphibian hypophyses. A study of the living circulation and of the histology of the hypophyseal vessels and nerves. *Anat. Rec.* **99**, 21–54. (A)

Green, J. D. (1948). The histology of the hypophyseal stalk and median eminence in man with special reference to blood vessels, nerve fibers and a special neurovascular zone in this region. *Anat. Rec.* **100**, 273–296. (A)

Green, J. D. (1951a). The comparative anatomy of the hypophysis, with special reference to its blood supply and innervation. *Am. J. Anat.* **88**, 225–312. (A)

Green, J. D. (1951b). Innervation of the pars distalis of the adenohypophysis studied by phase microscopy. *Anat. Rec.* **109**, 99–108. (A)

Green, J. D. (1966). The comparative anatomy of the portal vascular system and of the innervation of the hypophysis. *In* "The Pituitary Gland" (G. W. Harris and B. J. Donovan, eds.), Vol. 1, pp. 127–146. Butterworth, London. (A)

Green, J. D., and Harris, G. W. (1947). The neurovascular link between the neurohypophysis and adenohypophysis. *J. Endocrinol.* **5**, 136–146. (A)

Green, J. D., and Harris, G. W. (1949). Observation of the hypophysio-portal vessels of the living rat. *J. Appl. Physiol.* **108**, 358–361. (A)

Halasz, B. (1972). Hypothalamic mechanisms controlling pituitary function. *Top. Neuroendocrinol.* **38**, 97–122. (A)

Hammond, W. S. (1974). Early hypophysial development in the chick embryo. *Am. J. Anat.* **141**, 303–315. (A)

Harris, G. W. (1947). The blood vessels of the rabbit's pituitary gland and the significance of the pars and zona tuberalis. *J. Anat.* **81**, 343–351. (A)

Harrison, R. G., and Hoey, M. J. (1960). "Adrenal Circulation." Thomas, Springfield, Illinois. (C)

Hartman, F. A., Brownell, K. A., and Liu, T. Y. (1955). Blood flow through the dog adrenal. *Am. J. Physiol.* **180**, 375–377. (C)

Hartmann, J. F. (1958). Electron microscopy of the neurohypophysis in normal and histamine-treated rats. *Z. Zellforsch. Mikrosk. Anat.* **48**, 291–308. (A)

Hasegawa, K. (1953). On the vascular supply of the hypophysis and of the hypothalamus of the cat. III. On the hypophysio-portal vessels. *Shikoku Acta Med.* **4**, 212–221. (A)

Hasegawa, K. (1960). Comparative anatomy of the vascular supply in the hypothalamo-hypophysial system. *Kyushu J. Med. Sci.* **11**, 147–193. (A)

Henderson, E. F. (1927). The longitudinal smooth muscle of the central vein of the suprarenal gland. *Anat. Rec.* **36,** 69–78. (C)

Heymann, M. A., Payne, B. D., Hoffman, J. I. E., and Rudolph, A. M. (1977). Blood flow measurement with radionuclide-labelled particles. *Prog. Cardiovasc. Dis.* **20,** 55–79. (C)

Hoechter, O., Macchi, I. A., and Korman, H. (1955). Quantitative variations in the adrenocortical secretion of dogs. *Am. J. Physiol.* **182,** 29–34. (C)

Holmes, F. L., and Ball, J. N. (1974). "The Pituitary Gland: A Comparative Account." Cambridge Univ. Press, London and New York. (A)

Holmes, R. L. (1967). The vascular pattern of the median eminence of the hypophysis in the macaque. *Folia Primatol.* **7,** 216–230. (A)

Horton, R., Romanoff, E., and Walker, J. (1966). Androstenedione and testosterone in ovarian venous and peripheral plasma during ovariectomy for breast cancer. *J. Clin. Endocrinol. Metab.* **26,** 1267–1269. (C)

Houck, P. C., and Lutherer, L. O. (1981). Regulation of adrenal blood flow: Response to hemorrhagic hypotension. *Am. J. Physiol.* **241,** H872–H877. (C)

Houssay, B. A., Biasotti, A., and Sammartino, R. (1935). Modifications fonctionelles de l'hypophyse après les lesions infundibulo-tubériennes chez le crapaud. *C. R. Seances Soc. Biol. Ses Fil.* **120,** 725–727. (A)

Hume, D. M., and Nelson, D. H. (1958). Adrenal cortical function in surgical shock. *Surg. Forum* **5,** 568–575. (C)

Ichev, K. (1965). Intraparenchymal distribution of the blood vessels of the thyroid gland. *Nauchni Tr. Vissh. Med. Inst., Sofia* **44,** 51–57. (B)

Ichev, K. (1967). Blood circulation in the thyroid gland of rabbits, studied by means of biological fluorescent substances. *Dokl. Bolg. Akad. Nauk* **20,** 983–985. (B)

Ichev, K. (1969). Certain features of the ultrastructure of terminal vascular ramifications. *Dokl. Bolg. Akad. Nauk* **22,** 1091–1094. (B)

Ichev, K. (1970). Anastomoses and territories of arterial blood supply. *Nauchni Tr. Vissh. Med. Inst., Sofia* **49,** 1–4. (B)

Ivemark, B., Ekström, T., and Lagergren, C. (1967). The vasculature of the developing and mature human adrenal gland. *Acta Paediatr. Scand.* **56,** 601–606. (C)

Johnson, N. (1953). The blood supply of the thyroid gland. 1. The normal gland. *Aust. N. Z. J. Surg.* **23,** 95–103. (B)

Johnson, N. (1955). The blood supply of the human thyroid gland under normal and abnormal conditions. *Br. J. Surg.* **42,** 587–594. (B)

Kapitola, J. (1973). Blood flow through the thyroid gland in rats. *Acta Univ. Carol., Med., Monogr.* **55,** 6–99. (B)

Katholi, R. E., Oparil, S., Urthaler, F., and James, T. N. (1979). Mechanism of postarrhythmic renal vasoconstriction in the anesthetized dog. *J. Clin. Invest.* **64,** 17–31. (C)

Kent, C., and Coupland, R. E. (1981). On the uptake of exogenous catecholamines by adrenal cells and nerve endings. *Cell Tissue Res.* **221,** 371–383. (C)

Klinck, C. H., Oentel, J. E., and Winship, T. (1970). Ultrastructure of normal human thyroid. *Lab. Invest.* **22,** 2–22. (B)

Knigge, K. M., and Scott, D. E. (1970). Structure and function of the median eminence. *Am. J. Anat.* **129,** 223–244. (A)

Kobayashi, H., and Matsui, T. (1969). Fine structure of the median eminence and its functional significance. *In* "Frontiers in Neuroendocrinology" (W. F. Ganong and L. Martini, eds.), pp. 3–46. Oxford Univ. Press, London and New York. (A)

Kramer, R. J., and Sapirstein, L. A. (1967). Blood flow to the adrenal cortex and medulla. *Endocrinology* **81,** 403–405. (C)

Krausen, A. S. (1976). The inferior thyroid veins—the ultimate guardians of the trachea. *Laryngoscope* **86,** 1849–1855. (B)

Lametschwandtner, A., and Simonsberger, P. (1975). Light and scanning electron microscopic studies of the hypothalamo-adenohypophysial portal vessels of the toad *Bufo bufo* (L). *Cell Tissue Res.* **162**, 131–139. (A)

Lametschwandtner, A., Simonsberger, P., and Adam, H. (1976). Scanning electron microscopical studies of corrosion casts. The vascularization of the paraventricular organ (organon vasculosum hypothalami) of the toad *Bufa bufo* (L.). *Mikroskopie* **32**, 195–203. (A)

Lametschwandtner, A., Simonsberger, P., and Adam, H. (1977). Vascularization of the *Pars distalis* of the hypophysis in the toad, *Bufo bufo* (L.). *Cell Tissue Res.* **179**, 1–10. (A)

Liss, L. (1958). Die perivascularen structuren der menschlichen Neurohypophyse. *Z. Zellforsch. Mikrosk. Anat.* **48**, 283–290. (A)

Lucas, P. D., and Foy, J. M. (1977). Effect of experimental diabetes and genetic obesity on regional blood flow in the rat. *Diabetes* **26**, 786–792. (C)

Luciano, L., and Koch, A. (1975). The finer structure of venules and lymphatics of the dog thyroid gland. *Acta Anat.* **92**, 101–109. (B)

McConnell, E. M. (1953). The arterial supply of the human hypophysis cerebri. *Anat. Rec.* **115**, 175–204. (A)

Mack, E., Wyler, D. J., and Egdahl, R. H. (1969). Adrenal microcirculation in hemorrhagic shock. *Surg., Gynecol. Obstet.* **129**, 511–518 (C)

Maier, F., and Staehelin, M. (1968). Adrenal hyperemia caused by corticotropin. *Acta Endocrinol. (Copenhagen)* **58**, 613–618. (C)

Major, R. H. (1909). Studies on the vascular system of the thyroid gland. *Am. J. Anat.* **9**, 475–492. (B)

Melander, A., and Sundler, F. (1979). Presence and influence of cholinergic nerves in the mouse thyroid. *Endocrinology* **105**, 7–9. (B)

Melander, A., Ericson, L. E., Sundler, F., and Ingbar, S. H. (1974a). Sympathetic innervation of the mouse thyroid and its significance in thyroid hormone secretion. *Endocrinology* **94**, 959–966. (B)

Melander, A., Ericson, L. E., Ljunggren, J.-G., Norberg, K.-A., Persson, B., Sundler, F., Tibblin, S., and Westgen, V. (1974b). Sympathetic innervation of the normal human thyroid. *J. Clin. Endocrinol. Metab.* **39**, 713–718. (B)

Melander, A., Westgren, V., Sundler, F., and Ericson, L. E. (1975). Influence of histamine and 5-hydroxytryptamine containing thyroid mast cells on thyroid blood flow and permeability in the rat. *Endocrinology* **97**, 1130–1137. (B)

Merklin, R. J. (1962). Arterial supply of the adrenal gland. *Anat. Rec.* **144**, 359–371. (C)

Merklin, R. J. (1966). Suprarenal gland lymphatic drainage. *Am. J. Anat.* **119**, 359–374. (C)

Merklin, R. J., and Eger, S. A. (1961). The adrenal venous system in man. *J. Int. Coll. Surg.* **35**, 572–585. (C)

Miekos, E. (1976). Connection of superficial and internal adrenal veins. *Int. Urol. Nephrol.* **8**, 277–281. (C)

Modell, W. (1933). Observations on the structure of the blood vessels within the thyroid of the dog. *Anat. Rec.* **55**, 251–269. (B)

Morato, M. J. X. (1939). The blood supply of the hypophysis. *Anat. Rec.* **74**, 297–320. (A)

Morin, F., and Bottner, V. (1941). Contributi alla conoscenza della irroazione sanguina dell'ipofisi e dell'ipotalama di alcuni mammiferi. *Morphol. Jahrb.* **85**, 470–504. (A)

Murakami, T. (1971). Application of the scanning electron microscope to the study of the fine distribution of the blood vessels. *Arch. Histol. Jpn.* **32**, 445–454. (A)

Murakami, T. (1975a). Pliable methacrylate casts of blood vessels: Use in a scanning microscope study of the microcirculation in rat hypophysis. *Arch. Histol. Jpn.* **38**, 151–168. (A)

Murakami, T. (1975b). Injection replica scanning electron microscope method for studying the detailed vascular arrangement. *Cell* **7**, 11–18. (A)

Negm, I. M. (1971). The vascular blood supply of the pituitary and its development. *Acta Anat.* **80**, 604–619. (A)

Nemineva, K. (1950). Observations on the development of the hypophysial-portal system. *Acta Paediat.* **39**, 336–377. (A)

Ofuji, T. (1949). A study of vascular architecture in the hypothalamo-hypophyseal system of the mouse. III. Hypophyseal portal vessels. *Okayama Igakkai Z.* **61**, 207. (A)

Oliver, C., Mical, R. S., and Porter, J. C. (1977). Hypothalamic–pituitary vasculature: Evidence for retrograde blood flow in the pituitary stalk. *Endocrinology* **101**, 598–604. (A)

Page, R. B., and Bergland, R. M. (1977a). Pituitary vasculature. *In* "The Pituitary: A Current Review" (M. B. Allen, Jr. and V. B. Mahesh, eds.), pp. 9–17. Academic Press, New York. (A)

Page, R. B., and Bergland, R. M. (1977b). The neurohypophyseal capillary bed. I. Anatomy and arterial supply. *Am. J. Anat.* **148**, 345–358. (A)

Page, R. B., Munger, B. L., and Bergland, R. M. (1976). Scanning microscopy of pituitary vascular casts. *Am. J. Anat.* **146**, 273–302. (A)

Page, R. B., Leure-Dupree, A. E., and Bergland, R. M. (1978). The neurohypophyseal capillary bed. II. Specializations within median eminence. *Am. J. Anat.* **153**, 33–66. (A)

Palay, S. L. (1957). The fine structure of the neurohypophysis. *In* "Ultrastructure and Cellular Chemistry of Neural Tissue" (H. Waelsch, ed.), Chapter II. Harper (Hoeber). New York. (A)

Popa, G. T., and Fielding, U. (1930). A portal circulation from the pituitary to the hypothalamic region. *J. Anat.* **65**, 88–91. (A)

Popa, G. T., and Fielding, U. (1933). Hypophysio-portal vessels and their colloid accompaniment. *J. Anat.* **67**, 227–232. (A)

Porter, J. C., and Klaiser, M. S. (1964). Relationship of input of ACTH to secretion of corticosterone in rats. *Am. J. Physiol.* **207**, 789–792. (C)

Porter, J. C., and Klaiser, M. S. (1965). Corticosterone secretion in rats as a function of ACTH input and adrenal blood flow. *Am. J. Physiol.* **209**, 811–814. (C)

Porter, J. C., Hines, M. F. M., Smith, K. R., Repass, R. L., and Smith, A. J. K. (1967). Quantitative evaluation of local blood flow of the adenohypophysis in rats. *Endocrinology* **80**, 583–598. (A)

Porter, J. C., Kamberi, I., and Grazia, Y. R. (1971). Pituitary blood flow and portal vessels. *In* "Frontiers in Neuroendocrinology" (L. Martini and W. F. Ganong, eds.), pp. 145–175. Oxford Univ. Press, London and New York. (A)

Porter, J. C., Mical, R. S., Ben-Jonathan, N., and Ondo, J. G. (1973). Neurovascular regulation of the anterior hypophysis. *Recent Prog. Horm. Res.* **29**, 169–198. (A)

Porter, J. C., Gudelsky, G. A., Nansel, D. D., Pilotte, N. S., Burrows, G. H., and Tilders, F. J. H. (1981). Pituitary and hypothalamic hormones in pituitary stalk blood. *Front. Horm. Res.* **8**, 139–148. (A)

Prichard, M. M. L., and Daniel, P. M. (1958). The effects of pituitary stalk section in the goat. *Am. J. Pathol.* **34**, 433. (A)

Rienhoff, W. F. (1931). The lymphatic vessels of the thyroid. *Arch. Surg. (Chicago)* **23**, 783–804. (B)

Rinehart, J. A., and Farquhar, M. G. (1955). The fine vascular organization of the anterior pituitary gland. An electron microscopic study with histochemical correlations. *Anat. Rec.* **121**, 207–240. (A)

Sheehan, H. L. (1937). Post-partum necrosis of the anterior pituitary. *J. Pathol. Bacteriol.* **45**, 189–214. (A)

Sheehan, H. L., and Stanfield, J. P. (1961). The pathogenesis of post-partum necrosis of the anterior lobe of the pituitary gland. *Acta Endocrinol. (Copenhagen)* **37**, 479–510. (A)

Siler-Khodr, T. M., Morgenstern, L. L., and Greenwood, F. C. (1974). Hormonal synthesis and release from human fetal adenohypophysis *in vitro. J. Clin. Endocrinol. Metab.* **39**, 891–905. (A)

Silva, Z., Pinto, E., Silva, P., Campos, V. J. M., Silva, D. A. O., and Fernandes, W. A. (1980). Study of the venous drainage of the thyroid gland in dogs (*Canis familiaris*). *Anat. Anz.* **148**, 245–251. (B)

Söderberg, U. (1959). Temporal characteristics of thyroid activity. *Physiol. Rev.* **39**, 777–810. (B)

Stanfield, J. P. (1960). The blood supply of the human pituitary gland. *J. Anat.* **94**, 257–273. (A)

Swinyard, C. A. (1943). Growth of the human suprarenal glands. *Anat. Rec.* **87**, 141–150. (C)

Symington, T. (1962). Morphology and secretory cytology of the human adrenal cortex. *Br. Med. Bull.* **18**, 117–121. (C)

Symington, T. (1969). "Functional Anatomy of the Human Adrenal Gland." Williams & Wilkins, Baltimore, Maryland. (C)

Szabo, K., and Csanyi, K. (1982). The vascular architecture of the pituitary-median eminence complex in the rat. *Cell Tissue Res.* **224**, 563–577. (A)

Szentágothai, J., Flerkó, B., Mess, B., and Halasz, B. (1962). "Hypothalamic Control of the Anterior Pituitary." Akadémiai Kladó, Budapest. (A)

Tegler, L., Gillquist, J., Anderberg, B., Lundström, B., and Johansson, H. (1981a). Thyroid blood flow in man: Electromagnetic flowmetry during operation in euthyroid normal gland, nontoxic goiter, and hyperthyroidism. *J. Endocrinol. Invest.* **4**, 335–341. (B)

Tegler, L., Gillquist, J., Anderberg, B., Jacobson, G., Lundström, B., and Roos, P. (1981b). Human thyroid blood flow response to endogenous, exogenous human, and bovine thyrotrophin measured by electromagnetic flowmetry. *Acta Endocrinol. (Copenhagen)* **98**, 540–548. (B)

Tonoue, T., and Nomoto, T. (1979). Central nervous system mediated stimulation by thyrotropin-releasing hormone of microcirculation in thyroid gland of rat. *Endocrinol. Jpn.* **26**, 749–752. (B)

Torok, B. (1964). Structure of the vascular connections of the hypothalamo-hypophysial region. *Acta Anat.* **59**, 84–99. (A)

Tzinas, S., Droulias, C., Harlaftis, N., Akin, J. T., Gray, S. W., and Skandalakis, J. E. (1976). Vascular patterns of the thyroid gland. *Am. Surg.* **42**, 639–644. (B)

Urquhart, J. (1965). Adrenal blood flow and the adrenocortical response to corticotropin. *Am. J. Physiol.* **202**, 1162–1168. (C)

Valican, C. (1948). Le dispositif sphinctèropropulseur de la surrènale. *Arch. Anat. Mikrosk. Morphol. Exp.* **37**, 28–40. (C)

Verhofstad, A. A. J., and Lensen, W. F. J. (1973). On the occurrence of lymphatic vessels in the adrenal gland of the white rat. *Acta Anat.* **84**, 475–483. (C)

Walker, W. F., Zileli, M. S., Reutter, F. W., Shoemaker, W. C., and Moore, F. D. (1959). Adrenal medullary secretion in hemorrhagic shock. *Am. J. Physiol.* **197**, 773–780. (C)

Weiner, N. (1975). Control of the biosynthesis of adrenal catecholamines by the adrenal medulla. *In* "Handbook of Physiology" (H. Blaschko, G. Sayers, and A. D. Smith, eds.), Sect. 7, Vol. VI, pp. 357–366. Am. Physiol. Soc. Washington, D.C. (C)

Wilbur, D. L. (1974). An ultrastructural study of anterior pituitary somatotrophs following hypophysial portal vessel infusion of growth hormone secretagogues. Ph.D. Dissertation, Medical University of South Carolina Press, Columbia. (A)

Wilbur, D. L., and Spicer, S. S. (1980). Pituitary secretory activity and endocrinophagy. *In* "Biochemical Endocrinology: Synthesis and Release of Adenohypophyseal Hormones" (M. Jutisz and K. W. McKerns, eds.), pp. 167–186. Plenum, New York. (A)

Wilbur, D. L., Worthington, W. C., Jr., and Markwald, R. R. (1974). Ultrastructural observations of anterior pituitary somatotrophs following pituitary portal vessel infusion of dibutyryl cAMP. *Am. J. Anat.* **141**, 139–146. (A)

Wilbur, D. L., Worthington, W. C., Jr., and Markwald, R. R. (1975). An ultrastructural and radioimmunoassay study of anterior pituitary somatotrophs following pituitary portal vessel infusion of growth hormone releasing factor. *Neuroendocrinology* **19**, 12–27. (A)

Wilbur, D. L., Yee, J. A., and Raigue, S. E. (1978). Hypophysial portal vessel infusion of TRH in the rat: An ultrastructural and radioimmunoassay study. *Am. J. Anat.* **151**, 277–294. (A)

Wislocki, G. B. (1937). The meningeal relations of the hypophysis cerebri. II. An embryological study of meninges and blood vessels of the human hypophysis. *Am. J. Anat.* **61**, 95–129. (A)

Wislocki, G. B., and King, L. S. (1936). The permeability of the hypophysis and hypothalamus to vital dyes with a study of the hypophyseal vascular supply. *Am. J. Anat.* **58**, 421–472. (A)

Worthington, W. C., Jr. (1955). Some observations on the hypophyseal portal system in the living mouse. *Bull. Johns Hopkins Hosp.* **97**, 343–357. (A)

Worthington, W. C., Jr. (1960). Vascular responses in the pituitary stalk. *Endocrinology* **66**, 19–31. (A)

Worthington, W. C., Jr. (1962). The blood vessels of the pituitary and the thyroid. *In* "Blood Vessels and Lymphatics" (D. I. Abramson, ed.), pp. 428–464. Academic Press, New York. (A)

Worthington, W. C., Jr. (1963). Functional vascular fields in the pituitary stalk of the mouse. *Nature (London)* **199**, 461–465. (A)

Worthington, W. C., Jr. (1964). The pituitary microcirculation. *Eur. Conf. Microcirc. 2nd, 1962* pp. 449–454. (A)

Worthington, W. C., Jr. (1966). Blood samples from the pituitary stalk of the rat. Method of collecting and factors determining volume. *Nature (London)* **210**, 710–712. (A)

Wurtman, R. J., and Axelrod, J. (1965). Adrenaline synthesis: Control by the pituitary gland and adrenal glucocorticoid. *Science* **150**, 1464–1465. (C)

Wyler, F., Forsyth, R. P., Nies, A. S., Neutze, J. M., and Melmon, K. L. (1969). Endotoxin-induced regional circulatory changes in the unanesthetized monkey. *Circ. Res.* **24**, 777–786. (C)

Xuereb, G. P., Prichard, M. M. L., and Daniel, P. M. (1954a). The arterial supply and venous drainage of the human hypophysis cerebri. *Q. J. Exp. Physiol. Cogn. Med. Sci.* **39**, 199–218. (A)

Xuereb, G. P., Prichard, M. M. L., and Daniel, P. M. (1954b). The hypophyseal portal system of vessels in man. *Q. J. Exp. Physiol. Cogn. Med. Sci.* **39**, 219–230. (A)

Zenner, M. J., Gurll, N. J., and Reynolds, D. G. (1977). The effect of hemorrhagic shock and resuscitation on regional blood flow in cynomulgus monkey. *Circ. Shock* **4**, 291–296. (C)

Chapter 9
Endocrine System: Pancreas, Parathyroid, and Pineal

A. BLOOD VESSELS AND LYMPHATICS OF THE PANCREAS

By N. E. WARNER

1. ANATOMY OF BLOOD CIRCULATION

a. Extraglandular arterial anatomy: The human pancreas derives its extrinsic blood supply from the celiac and superior mesenteric arteries (Gray, 1973; Edwards, 1980). Both the common hepatic and splenic branches of the celiac artery furnish major arteries to the gland. The gastroduodenal branch of the hepatic artery gives rise to anterior and posterior superior gastroduodenal arteries that supply the pancreatic head and neck and contribute to the anterior and posterior superior pancreaticoduodenal arcades. The splenic artery supplies the pancreatic body and tail via the dorsal pancreatic artery, the pancreatica magna, and the caudae pancreatis, together with smaller unnamed branches. The first branch of the superior mesenteric artery ascends to form the anterior and posterior inferior pancreaticoduodenal branches that anastomose with corresponding pancreaticoduodenal arcades. The extrapancreatic arteries then enter the organ and divide to form interlobular branches within its septa.

b. Intraglandular arterial anatomy: The intrapancreatic vasculature is less well known than the extraglandular blood supply. The following description is based upon investigations in various species by Kühne and Lea (1882), Beck and Berg (1931), Wharton (1932), Tohyama (1935), Ferner (1952), Thiel (1954), Brunfeldt *et al.* (1958), Bunnag *et al.* (1963, 1964, 1966), Bunnag (1966), Heisig (1968), McCuskey and Chapman (1969), Hayasaka and Sasano (1970), Fujita (1973), Fujita and Murakami (1973), Henderson and Daniel (1979), Fraser and Henderson (1980), Lifson and Lassa (1981), and Henderson *et al.* (1981).

With regard to comparative anatomy, in many species, including humans, monkeys, dogs, and cats, the pancreas is a compact organ with lobules closely apposed, whereas in others (rodents, rabbits), the organ is diffusely dispersed within mesenteries and the lobules are thin and separated by loose septa. The latter type of architecture is particularly favorable for observation of the microvasculature *in vivo,* and most investigators have used animals possessing such an anatomic arrangement for this purpose (rabbits, rats, and mice). In the diffuse pancreas, islets are found in three locations: along the ducts, within the pancreatic lobules, and in the interstitium. Bensley (1911) classified these islets into four types: ductal (class 1), lobular (class 2 and 3), and interstitial (class 4). This distinction is very important because the lobular islets are an integral part of the insuloacinar portal system, a vascular arrangement that permits modulation of acinar function by insular hormones (Henderson, 1969; Fujita, 1973; Fujita and Murakami, 1973; Fujita and Watanabe, 1973; Henderson and Daniel, 1979; Fraser and Henderson, 1980; Henderson *et al.*, 1981).

c. Anatomy of microvasculature: Pancreatic microvascular connections have been studied by three techniques: direct microscopic observation *in vivo*, injection of dyes and examination of histologic sections, and analysis of microvascular casts by light or scanning electron microscopy (Fraser and Henderson, 1980). Thus it has been found that interlobular arteries give rise to intralobular arteries, one for each lobule. Arteriovenous anastomoses between interlobular arteries and veins have been described only in dogs (Dal Zotto, 1950; Papp *et al.*, 1969). Intralobular arteries, in turn, supply one or more afferent arterioles for each islet and may also give separate branches that directly supply the capillary bed of the acinar parenchyma. An islet located near the periphery of a lobule may receive its afferent arteriole directly from a nearby interlobular artery or even an extrapancreatic artery. Some small lobules have no islets, and the intralobular arteries supply only acinar parenchyma. Afferent arterioles connect with the insular capillary plexus at the periphery or the center of an islet, depending upon the species. Insular capillaries are wider than acinar capillaries, and they form a distinctive pattern resembling the network of a renal glomerulus, with convoluted sinuous channels separating small groups of cells.

d. Intraglandular venous anatomy: The venous drainage of ductal (class 1) islets differs from lobular (class 2 and 3) islets of rodents and rabbits. A lobular islet drains into capillary channels that connect directly (e.g., in series) with the capillary net of the acinar parenchyma, forming the insuloacinar portal system. The acinar plexus, in turn, empties into an intralobular vein which drains into an interlobular vein. Insuloacinar connections have been observed by many investigators, but their significance was recognized only recently by Henderson (1969) and Fujita (1973). A ductal islet, on the other hand, empties directly into one or more venules that merge and ultimately empty into the portal vein (Kühne and Lea, 1882; Berg, 1930; Beck and Berg, 1931; Thiel, 1954; Brunfeldt *et al.*, 1958; Bunnag *et al.*, 1963). Thus the capillary bed of the ductal islets is not connected with the capillaries of acinar parenchyma, but instead it lies in parallel to them.

The exocrine capillary plexus forms a meshwork surrounding the acini. It drains into intralobular veins, which empty, in turn, into interlobular veins and ultimately into the portal vein. In humans, the extrapancreatic veins (the pancreatic and pancreaticoduodenal) empty into the splenic and superior mesenteric veins. These unite to form the portal vein, and thus all venous blood from the pancreas ultimately is transported to the liver.

A second type of pancreatic portal system of unknown significance has been described in the rabbit by Lifson and Lassa (1981), who found an acinoductal portal bed lying in series with the insuloacinar system, described above.

2. Physiology of Blood Circulation

a. Blood flow in macrocirculation: The relationship of pancreatic blood flow to exocrine secretion has been studied for more than a century, ever since Claude Bernard, in 1856, observed plethora of the gland in dogs during digestion and pallor during fasting. Although it generally is agreed that circulation of blood is essential to sustain the cells, the question of whether the rate of flow serves as a primary stimulus to secretion continues to be debated. At present, the evidence points to a close relationship between blood flow and secretion, both exocrine and endocrine (for pertinent reviews, see Tankel and Hollander, 1957; Grim, 1963; Jacobson, 1967; Lanciault and Jacobson, 1976). Methods for study of whole organ blood flow include collection of venous outflow; the use of plethysmography, flowmeters, electrical conductance, and photocell measurement of transmitted light; clearance of hydrogen gas or radioactive elements; and analysis of injected radioactive microspheres. Blood flow measurements in unstimulated pancreas vary from 25 to 172 ml/min/100 gm (Delaney and Grim, 1966; Jacobson, 1967; Aune and Semb, 1969; Eliassen *et al.*, 1973; Bor *et al.*, 1974; Horowitz and Myers, 1982). Such variations almost surely reflect different experimental conditions. Effects of anesthesia are unclear. Although Aune and Semb (1969) found a reduction in flow due to pentobarbital, rates reported by Ercan *et al.* (1974) in dogs under this anesthetic were 25% higher than those in conscious dogs studied by Aune and Semb (1969). Hypoxia and hypercarbia decrease blood flow (Broadie *et al.*, 1979; Delaney and Grim, 1966).

Various types of hormones have different vascular effects on the macrocirculation. For example, gastrointestinal hormones, including secretin, pancreozymin, cholecystokinin-pancreozymin, caerulein, and pentagastrin, all induce secretion of pancreatic juice and, according to most reports, are associated with increased pancreatic blood flow (Grim, 1963; Jacobson, 1967; Goodhead *et al.*, 1970; Lanciault and Jacobson, 1976; Papp *et al.*, 1977; Beijer *et al.*, 1979; Takeshima *et al.*, 1981) and increased red cell velocity through individual capillaries (Heisig, 1968). Vasopressin reduces pancreatic circulation in normal dogs (Delaney *et al.*, 1966) but maintains, or slightly increases, blood flow in those with hemorrhagic pancreatitis (Pissiotis *et al.*, 1972). Vasopressin also increases survival in dogs with hemorrhagic pancreatitis. Epinephrine and nor-epinephrine cause vasoconstriction and decreased blood flow, and both reduce the rate of exocrine secretion. Insulin elicits a reduction in blood flow (Semb and Aune, 1971), and glucagon has the same effect (Delaney and Grim, 1966; Takeshima *et al.*, 1981). According to most investigators, acetylcholine, bradykinin, and histamine increase blood flow. In humans, angiotensin, although decreasing blood flow to most abdominal viscera, has no affect on the pancreas (Kaplan and Bookstein, 1972). The basis for such a response is unclear.

With regard to the pathophysiology of the macrocirculation, vascular changes and abnormal blood flow are characteristic of experimental pancreatitis, both local circulation and perfusion being seriously impaired. There is little question that such abnormal reactions play an important role in the pathogenesis of the disease (Morris, 1964; Papp *et al.*, 1966; Anderson *et al.*, 1967; Goodhead, 1969; Boijsen and Tylén, 1972; Pissiotis *et al.*, 1972; Donaldson *et al.*, 1978).

b. Responses of the microcirculation: The circulation in the minute vessels of the living pancreas has been described by a number of investigators using epi- or transillumination (Kühne and Lea, 1882; Mathews, 1899; Brunfeldt *et al.*, 1958; Palmer, 1959; Bunnag *et al.*, 1963, 1964, 1966, 1967; Bunnag, 1966; Heisig, 1968; McCuskey and Chapman, 1969; Fraser and Henderson, 1980). Blood flow through the islets is so rapid that individual red cells cannot be distinguished (Berg, 1930; Bunnag *et al.*, 1963). Heisig (1968) found the average velocity of red cells in exocrine capillaries to be 848 μm/sec. Although some observers have reported constant flow in islets, others describe fluctuation, interruption, or even temporary reversal. Intermittent capillary flow also has been observed in exocrine parenchyma. McCuskey and Chapman (1969) found independently contractile, bulging endothelial cells that influenced velocity and volume of flow in insular and exocrine capillaries of the mouse and could temporarily block the lumen. They also noted flow of blood from islets to acini via capillary anastomoses of the insuloacinar portal system. Fraser and Henderson (1980) conclusively demonstrated the passage of dye *in vivo* from insular plexuses to exocrine capillaries through these islet-to-insular connections. Some intralobular arterioles supply acinar parenchyma directly, and these have precapillary sphincters that locally control volume and rate of flow to acini (McCuskey and Chapman, 1969). Thus, not all afferent blood is delivered to the insuloacinar portal system. By injecting microspheres of various sizes, Lifson *et al.* (1980) calculated that less than one-quarter of total pancreatic blood flow goes directly to islets, but that nearly all efferent blood from the islets is delivered to acinar capillaries. These workers concluded that flow through islets is large enough to permit significant local action of insular hormones on the acini.

Demonstration of blood flow from lobular islets to acini via direct capillary connections strongly supports the hypothesis that insular hormones in high concentration can be transported directly to the exocrine pancreas for modulation of its metabolic activities (for further support for this view, see Henderson, 1969; Fujita and Watanabe, 1973; Henderson *et al.*, 1981; Editorial, 1981). Thus, islet hormones furnished to acinar cells could sustain synthesis of protein and amylase (insulin), influence the synthesis and release of enzymes and bicarbonate (glucagon), and inhibit production of pancreatic juice (somatostatin) as needed.

Concerning other aspects of the physiology of pancreatic microcirculation, anoxia causes cessation of insular flow (Palmer, 1959). With respect to the action of digestive hormones, only secretin and pancreozymin-cholecystokinin (Cecekin-Vitrum) have been studied, and Heisig (1968) found that both cause average corpuscular velocity to rise from 848 to 1195 μm/sec and zymogen granules to disappear from acinar cells. Capillary dimensions are unchanged, but arterioles and venules become dilated. Unquestionably, hormonal stimulation is associated with increased velocity of pancreatic acinar blood flow, but whether the increased flow stimulates or results from augmented secretion remains to be settled. Pituitrin causes constriction of arterioles and blanching and interruption of circulation in islets (Berg, 1930). Local application of vasopressin causes cessation of islet flow (Palmer, 1959), and intravenous administration of the drug produces slowing of flow in arterioles, islets, and venules (Bunnag *et al.*, 1963). Intravenous and local epinephrine constricts arterioles and stops circulation in islets (Berg, 1930; Palmer, 1959; Bunnag *et al.*, 1963), and intravenous norepinephrine causes segmental constriction of interlobular arteries and veins and temporarily slows but does not stop islet circulation. Insulin and glucagon have no effect on insular blood flow (Bunnag *et al.*, 1963). Serotonin produces temporary interruption of flow in islets, followed by increased circulation and transient dilatation of arterioles, whereas histamine elicits a gradual slowing of insular flow, followed by brief cessation and then a period of sustained rapid flow (S. C. Bunnag, N. E. Warner, and S. Bunnag, unpublished results).

3. PHARMACOLOGY OF BLOOD CIRCULATION

a. Responses of macrocirculation: The pharmacology of pancreatic macrocirculation has been studied mainly to shed light on the relationship of blood flow to secretion and to elucidate the pathogenesis of acute hemorrhagic pancreatitis. Isoproterenol, a strong β-adrenergic stimulating agent, increases blood flow and insulin output in dogs (Mandelbaum and Morgan, 1969). Phenoxybenzamine, an α-adrenergic blocking agent, prevents the decrease in blood flow produced by epinephrine (Barlow *et al.*, 1968) and stimulates exocrine secretion (Solomon *et al.*, 1974). Bethanechol chloride (Urecholine) mimics the effect of acetylcholine, causing increased blood flow and a small rise in exocrine secretion. Atropine augments local circulation but does not alter insulin secretion (Mandelbaum and Morgan, 1969). Papaverine raises blood flow in cats; it has no effect on basal exocrine secretion but augments secretin-stimulated flow of juice (Lenninger, 1973). Mannitol increases circulation but does not alter insulin output (Fischer *et al.*, 1976), whereas tolbutamide has no effect on blood flow but increases insulin secretion (Mandelbaum and Morgan, 1969). Intravenous ethanol reduces pancreatic blood flow by one-third, an effect that is reversed by treatment with 25% mannitol (Horwitz and Myers, 1982).

b. Responses of microcirculation: The pharmacology of pancreatic microcirculation has been relatively neglected. Ephedrine, a sympathomimetic amine, has effects like epinephrine, causing temporary interruption of insular blood flow (Bunnag *et al.*, 1963). Dihydroergotamine, an α-receptor blocker, and methacholine chloride (Mecholyl) both cause transient interruption of blood flow in islets (S. C. Bunnag, N. E. Warner, and S. Bunnag, unpublished results). Pilocarpine in lethal doses first increases insular flow, then causes agonal oscillations of circulation in the capillaries (O'Leary, 1930). Tolbutamide, a sulfonylurea hypoglycemic agent, produces transient vascular dilatation and rapid flow in islets, coinciding with maximal hypoglycemic effect of the drug (Bunnag *et al.*, 1967). Alloxan, a diabetogenic drug, has little effect on islet circulation until after about 24 hr when blood flow slows or stops (Ruangsiri, 1949; Bunnag *et al.*, 1967); by the third day, it invariably ceases, and by day 4, it disappears. Parathion, a cholinesterase inhibitor, has effects resembling methacholine (Hruban *et al.*, 1963), producing marked and prolonged dilatation of arterioles and venules of ductal islets, with concomitant hypoglycemia.

4. LYMPHATIC SYSTEM

a. Anatomy: It is generally accepted that the lymphatics of the pancreas (1) occur, together with blood vessels, in interlobular spaces; (2) unite to form successively larger channels and plexuses that accompany blood vessels and ducts in an axial arrangement; and (3) form plexuses on the surface of the organ that give rise to lymphatic channels passing to regional nodes (Hoggan and Hoggan, 1881; Grau and Taher, 1965; Cubilla *et al.*, 1978). In humans, the pancreatic lymphatics drain into five principal groups of nodes: superior, inferior, anterior, posterior pancreatic, and splenic (Cubilla *et al.*, 1978). These structures, in turn, drain into the cisterna chyli (Mackie and Moosa, 1980).

The distribution of lymphatic capillaries is disputed. One view is that islets are devoid of lymphatics and that the latter are present only in interstitial connective tissue (Bartels, 1909, Grau and Taher, 1965). However, some workers have reported finding networks of lymphatic capillaries surrounding the acini (Hoggan and Hoggan, 1881; Rusznyák *et al.*, 1967), and Hoggan and Hoggan (1881) clearly demonstrated in rodents that club-shaped endothelial-lined lymphatic capillaries form loops and meshes over the external aspect of each pancreatic lobule and are drained by efferent lymphatic channels. Studies cited in Rusznyák *et al.* (1967) confirm the presence of lymphatic capillaries closely applied to exocrine acini, separated from acinar cells only by acinar basement membranes, but without any lymphatic capillaries in the islets. However, Jdanov (1960) found lymphatic capillaries surrounding both islets and acinar tissue. He described the origin of pancreatic lymphatics in tridimensional networks of lymphatic capillaries surrounding acini and islets. These vessels, in turn, emptied into smaller interlobular lymphatics that formed a plexus around pancreatic lobules.

Lymphatic communications between gallbladder and pancreas have been postulated in humans and demonstrated in dogs (Weiner *et al.*, 1970), lending credence to an old theory of lymphogenous spread of inflammatory diseases involving both organs.

b. Physiology: Information on the functional aspects of lymphatic circulation is scanty, and most studies have dealt only with exocrine parenchyma.

Direct observations *in vivo* have been reported by Duprez *et al.* (1963), Godart (1965), Heisig (1968), and Reynolds (1970). Duprez *et al.* and Godart injected fluorescein or patent blue dye under gentle pressure into the pancreatic duct of rabbits, rats, and dogs and observed patterns of fluorescence or the passage of dye grossly and microscopically. Fluorescein passed from the duct system into the interstitium, outlining the acini to form a mosaic pattern, and then gradually disappeared over the next 20 min. Sometimes efferent lymphatics also were rendered fluorescent. Patent blue dye passed rapidly from ducts into parenchyma, and within 5 to 20 min, efferent pancreatic lymphatics were colored blue. In dogs, lymph from the thoracic duct was stained blue. Heisig (1968) studied interlobular lymphatics in living rabbits and found that lymph was clear in normal animals but contained erythrocytes after trauma to the area.

Reynolds (1970) injected patent V dye into the human pancreas and observed its transport. In the normal gland, selective uptake demonstrated an interlacing network of fine lymphatic channels, with flow being cephalad from head and body of the organ. Removal was rapid, with clearing completed in about 45 min. In patients with recent pancreatitis, uptake was prolonged and clearing was only partial, suggesting lymphatic blockade. In chronic pancreatitis, uptake was minimal and clearing failed to occur.

Although a lymphatic pathway for transport of insulin from pancreas to thoracic duct has been demonstrated, its significance is disputed. Biedl (1923) proposed that insulin passes from islets directly into the lymphatic circulation, and he demonstrated that systemic administration of canine thoracic duct lymph lowers blood sugar in rabbits. Daniel and Henderson (1966) measured insulin in thoracic duct lymph in rabbits and found a concentration comparable to that in blood from inferior vena cava. However, Pepin *et al.* (1970) and Henderson (1974) concluded that secretion of islets directly into pancreatic lymphatics has only a negligible role in transport of insulin from islets to systemic circulation.

Lymphatics also transport exocrine secretions of the pancreas. In this regard, pancreatic enzymes have been repeatedly demonstrated in thoracic duct lymph in normal animals and humans (Dumont and Mulholland, 1960; Dumont *et al.*, 1960; Bartos *et al.*, 1966; Sim *et al.*, 1966).

Various hormones have been investigated for their effect upon lymph flow from the pancreas. Although the action of secretin in animals is controversial, the consensus is that, in normal humans, administration of the purified product slightly increases the concentration of pancreatic enzymes in thoracic duct lymph, without having much influence on volume of lymph flow (Dumont *et al.*, 1960; Bartos *et al.*, 1966). Vega *et al.* (1967) found the same effect in dogs. Dreiling (1970) investigated cholycystokinin and pancreozymin in dogs and found that they increased lymph flow from the thoracic duct. Pepin *et al.* (1970) reported a similar response following injection of epinephrine and glucagon, but they noted no change in the concentration of amylase or lipase in the duct.

Increased pressure in the pancreatic duct, produced by ligation, cannulation, or injection, has a profound effect upon lymphatic drainage. Administration of morphine sulfate, which causes contraction of the sphincter of Oddi, has the same result (Dumont *et al.*, 1960), with fluid rich in enzymes immediately beginning to accumulate in the interstitium (Popper, 1940; Duprez *et al.*, 1963; Godart, 1965; Dumont and Martelli, 1968). By retrograde drainage, flow of lymph from the thoracic duct increases, and the content of pancreatic enzymes in the lymph rises (Dumont and Martelli, 1968; Dreiling, 1970; Dumont *et al.*, 1971). These effects are augmented by stimulation with secretin, pancreozymin, or cholecystokinin-pancreozymin (Popper and Necheles, 1942; Dumont *et al.*, 1960; Dumont and Mulholland, 1960; Bartos *et al.*, 1966; Figarella *et al.*, 1978). Thus the thoracic duct functions as a "safety valve" for the pancreas in case of obstruction of the pancreatic duct (Dumont *et al.*, 1960, 1971). In accord with such a view is the finding that if the thoracic and pancreatic ducts are ligated simultaneously, focal necrosis of pancreatic parenchyma and of peripancreatic fat ensues (Papp *et al.*, 1958), presumably as a consequence of stasis of enzyme-rich lymph in congested lymphatics (Papp, 1976). However, acute hemorrhagic pancreatitis does not occur since other factors are required to initiate this disease (Papp, 1976).

c. Pharmacology: Information on drugs affecting lymphatic circulation of the pancreas is sparse. Morphine sulfate has been shown to contract the sphincter of Oddi and raise pressure in the pancreatic duct and its radicles. Administration of this drug, together with secretin, causes increased flow of thoracic duct lymph and a higher concentration of lipase and amylase due to augmented drainage of lymph and enzymes from the interstitium of the pancreas, as described above (Dumont *et al.*, 1960). Alcohol given orally (but not intravenously) elicits increased flow and raises the concentration of amylase and lipase in thoracic duct lymph in humans (Bartos and Brzek, 1981). Such an effect probably is mediated through a local action of alcohol on the sphincter of Oddi and the duodenum, increasing retrograde drainage into pancreatic lymphatics (Baum and Iber, 1973).

Benzopyrones, a class of nonsteroidal antiinflammatory compounds, stimulate lymph flow (Casley-Smith, 1976), increase the contractility of lymphatics (Mislin, 1971), and enhance the phagocytic activity of macrophages (Piller, 1980). In normal humans, benzopyrones increase lipase and amylase in thoracic duct lymph but have no effect on exocrine secretion into the duodenum (Bartos and Brzek, 1979). These findings have been attributed to enhanced lymphatic transport of enzymes from the interstitium of the pancreas (Brzek and Bartos, 1969). The use of benzopyrones in humans and dogs with acute pancreatitis suggests that these substances may improve survival (Preissich, 1977; Bartos *et al.*, 1979). In dogs, lymph flow from the thoracic duct is increased, together with a rise in amylase and trypsin levels; edema of the pancreas is less severe; fat necrosis is minimal; and hemorrhagic necrosis is absent (Bartos *et al.*, 1979).

B. BLOOD VESSELS AND LYMPHATICS OF THE PARATHYROID GLANDS

By G. NORTHROP

1. ANATOMY OF BLOOD CIRCULATION

a. Extraglandular vascular anatomy: Each parathyroid gland is supplied by a single parathyroid artery, which furnishes blood only to this structure (Welsh, 1898; Halsted and Evans, 1907; Jander *et al.*, 1980). Variations in blood supply to the parathyroid glands occur primarily because of developmental factors that influence the migration of the superior and inferior parathyroid glands from their anlagen in branchial pouches IV and III, respectively, to their definitive adult locations (Gilmour, 1937; Norris, 1937). Data on the variability in location of the parathyroid glands are presented in Fig. 9.1. [For further

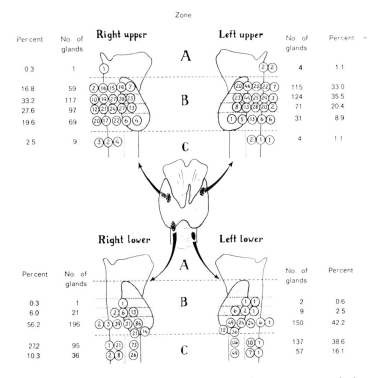

FIG. 9.1. The positions of 1405 parathyroid glands identified at 354 autopsies. The locations of the upper and lower glands on the right and left sides are indicated in separate schematic drawings. A lateral view of the larynx and trachea is shown, with the thyroid gland mobilized and dislocated ventrally and medially. Three zones (A, B, C) are indicated (dashed lines) in relation to the thyroid gland. In zone B, which represents the level of the thyroid, the dotted horizontal lines indicate the midline of the zone and the boundaries of its upper, middle, and lower thirds. The ventral and dorsal extents of the trachea below the thyroid (C) are separated (dotted vertical lines). The number of glands in various areas of the zones are noted in circles. Parathyroids situated more than 2 cm below the lower pole of the thyroid are included in the most caudal circles. (From Alveryd, 1968; with permission to reproduce photograph requested from *Acta Chirurgia Scandinavica*.)

discussion of this subject and of the number and vascular supply of the parathyroid glands, see Alveryd (1968).]

The inferior parathyroid glands travel a greater distance caudally with the thymus anlage than do the superior pair of parathyroid glands, which arise with the ultimobranchial body (Gilmour, 1937). Thus the artery to each superior parathyroid glands tends to be short, 4–5 mm in length, with less variability in location than the artery to the inferior parathyroid gland, which can be 2–3 cm long (Halsted and Evans, 1907). Alveryd (1968) reported that the inferior thyroid artery supplied both parathyroid glands on the right side in 86.1% of cases and both glands on the left side in 76.8% of cases. Arterial branches from the

TABLE 9.I

NUMBER OF PARATHYROID GLANDS REPORTED PER CASE IN THREE DIFFERENT
AUTOPSY INVESTIGATIONS

Reference	Number of cases	Number of parathyroid glands				
		2	3	4	5	6
Gilmour (1938)	428	1	26	374	25	2
		0.2%	6.1%	87%	6%	0.5%
Alveryd (1968)	354	2	18	319	13	—
		0.6%	5.1%	96.6%	3.7%	—
Wang (1976)[a]	160	—	—	156	3	1
				97.5%	1.9%	0.6%

[a]Cases with less than four glands were deleted from this series.

larynx, esophagus, trachea, and mediastinum, as well as the thyroidea ima or the superior thyroid artery, supplied the parathyroid glands in the remaining instances. Because of variation in parathyroid gland number (Table 9.I) and location (Fig. 9.1), it can be concluded that there will be both variability and inconsistency in arterial and venous circulation.

The venous effluent leaves the parathyroid gland from the hilus and drains into the venous plexus that surrounds the thyroid gland, as described in Section B-1e, Chapter 8. [For further discussion of the many variations of the arterial and venous circulations, see Dunlop et al. (1980) and Mallette et al. (1981).]

  b. Intraglandular vascular anatomy: Each parathyroid gland is enclosed in a connective tissue capsule from which septa penetrate the parenchyma, dividing it into irregular sheets, nests, or cords of cells which are, in fact, folds of unistratified epithelium (Krstic, 1980). The parathyroid artery enters the gland via the hilus and continues in connective tissue septa as a central artery, from which branches are given off at acute and right angles. The latter vessels continue in the septa and become successively smaller, until they appear as capillaries surrounded only by a delicate reticulum (Curtis, 1930). These vessels have round fenestrations of variable size distributed irregularly throughout the endothelium (Krstic, 1980).

Blood from the capillary network passes into venules and then into increasingly larger veins contained in the connective tissue septa until it finally flows from the hilus in a solitary vein in company with the entering artery (Gilmour, 1939).

2. PHYSIOLOGY OF BLOOD CIRCULATION

Although Welsh (1898) provided the first detailed description of the human parathyroid glands, their function remained in dispute until about a decade later (MacCallum and Voegtlin, 1909). With the development of the electron microscope, emphasis has been placed on the intracellular ultrastructure of the parathyroid hormone secreting (chief) cell, and, consequently, there is a plethora of information on intracellular synthesis and release of parathyroid hormone. In contrast, data concerning the gross structure of this gland remain lacking. Thus, it is not surprising that studies on blood flow have yet to be performed under normal *in vivo* conditions.

Morphologic evidence for both vagal and adrenergic nerve innervation of smooth muscle in the arterial wall and of the chief cells has been reported (Raybuck, 1952; Altenähr, 1971; Norberg *et al.*, 1975; Atwal, 1981). Mast cells containing 5-hydroxytryptamine are found in the perivascular spaces (Gilmour, 1939; Zawistowski, 1967). Thus the mechanisms for regulation of blood flow and parathyroid hormone secretion have been identified, although actual data on blood flow have not been reported.

3. LYMPHATIC SYSTEM

There is no detailed information available concerning intraparathyroid lymph vessels or lymph composition in human beings. Paloyan *et al.* (1973) have stated that lymph leaving the parathyroid glands drains into the internal jugular chain of nodes. Balashev and Ignashkina (1965) have described a double layer basketlike network of lymph vessels surrounding each parathyroid gland. From this plexus, vessels branch and penetrate the parenchyma via connective tissue septa, in company with blood vessels. The same investigators found frequent anastomoses between the subcapsular and glandular lymph vessels. Lymph drainage from the gland is either by large vessels, 0.15 to 0.75 mm in diameter, that accompany the parathyroid vein or by anastomosis with the thyroid lymph plexus. Senior (1974) described channels in the tropical lizard which he considered to be lymph vessels. They were smaller than capillaries and were without an apparent endothelial lining. They coursed inside the cords of cells in the parathyroid glands and contained no red blood cells and only occasional white blood cells.

4. FUTURE AVENUES FOR INVESTIGATION

Lymph vessels in man have been described as leaving the parathyroid gland at the hilus; however, a description of the origin of lymphatic ducts is entirely lacking. Localization of parathyroid glands in the hypercalcemic individual with elevated parathyroid hormone levels remains difficult or impossible almost 10% of the time. Investigation into methods for identifying this small gland should be vigorously undertaken because of the increasing frequency in which its localization is essential.

C. BLOOD CIRCULATION OF THE PINEAL
GLAND

By J. A. McNulty

1. Anatomy of Blood Circulation

a. Extrinsic morphology: The principal blood supply to the pineal gland of both rodents and primates is from branches of the posterior cerebral artery (von Bartheld and Moll, 1954; Quay, 1974; Hodde and Veltman, 1979). Arterioles enter the gland along all of its surface and often run in connective tissue septa. Venous drainage is into vessels that course beneath the capsule of the gland and eventually into the great cerebral vein (Galen) that in man empties into the sinus rectus. In rodents, the pineal gland is located in a superficial position directly beneath the superior sagittal sinus. From 12 to 16 veins (40–60 μm in diameter) exit the gland and most enter the distal part of the great cerebral vein, which drains directly into the superior sagittal sinus (von Bartheld and Moll, 1954; Hodde, 1979).

b. Intrinsic morphology: An extensive network of capillaries is found within the parenchyma of the pineal gland. Tikhomirov (1970) estimated by a photometric method that the vascular compartment has a volume of 5.5 mm³ and comprised about 30% of the total volume of the human pineal gland. Calculations of blood content (expressed as number of erythrocytes per cubic millimeter of tissue) in the pineal gland of the rat by Quay (1958) have indicated that vascularization in the cortex of the gland is 1.2 to 2.0 times greater than that in the medulla. This worker further noted that blood content of the pineal in the rat is one-third of that of the posterior lobe of the hypophysis and one-fifth of that of the anterior lobe.

c. Ultrastructure of capillaries: The capillary endothelial cells of most rodent species are generally attenuated and fenestrated. Matsushima and Reiter (1975) found in the rat, mouse, and ground squirrel that the fenestrae measure about 70 nm in width and are bridged by a thin diaphragm. The cell cytoplasm contains all of the usual organelles and numerous uncoated vesicles. Some of the latter, which are between 70 and 120 nm in diameter, abut the cell membrane. Adjacent endothelial cells are connected by junctions having a gap between 4 and 20 nm and by occasional tight junctions.

Exceptions to the above ultrastructural features are found in capillaries of the pineal gland of the chinchilla. Endothelial cells in this species are thicker,

308

Blood Vessels and Lymphatics in Organ Systems
Copyright © 1984 by Academic Press, Inc.

lack fenestrae, and are characterized by abundant lysosomelike bodies and filaments (Matsushima and Reiter, 1975). Numerous vesicles are present in the cytoplasm, but these are smaller (40–70 nm in diameter) than those in the rat, mouse, and ground squirrel. Finally, adjacent endothelial cells frequently overlap for long distances and are connected by tight junctions. The functional significance of the species differences in the fine structure of pineal capillaries is still not clear.

Capillaries in the pineal gland of primates resemble those in the chinchilla, with the endothelial cells being continuous and lacking fenestrae (Moller, 1974; Wartenberg, 1968; J. A. McNulty, personal observations). In the human fetus, tight junctions are found between endothelial cells (Moller, 1974). The absence of fenestrae suggested to Moller the existence of a blood–brain barrier. This view is also based on unpublished results derived from the use of different dyes which penetrate pineal capillaries of both rats and rabbits but not those of the cat, a species which has continuous, unfenestrated endothelial cells (Wartenberg, 1968).

2. PHYSIOLOGY OF BLOOD CIRCULATION

The pineal gland of mammals is an endocrine organ that synthesizes and secretes active compounds, including melatonin. Enzymatic activity of pinealocytes is stimulated by sympathetic fibers which reach the gland from the superior cervical ganglia. The release of norepinephrine from the sympathetic nerve terminals is regulated by environmental light: dark cycles via retinal pathways and the suprachiasmatic nucleus of the hypothalamus. In accordance with its endocrine functions, the pineal gland exhibits alterations in blood flow related to its physiologic state.

  a. Question of role of a blood–brain barrier: The dynamics of blood flow and the permeability characteristics of endothelial cells are of great importance in understanding the secretory functions of the pineal gland. Because pineal capillaries of some rodent species are permeable to dyes and silver nitrate (Moller, 1974; Wislocki and Leduc, 1952), Moller *et al.* (1978) investigated the mechanisms by which substances cross the microvasculature by injecting the tail vein of mice with the tracers, horseradish peroxidase and microperoxidase, which mimic proteins and peptides, respectively. They found that both tracers crossed the capillary wall via several pathways, including (1) fenestrae, (2) intercellular junctions, (3) cytoplasmic vesicles, and (4) "channels" formed by vesicles which open toward both the luminal and abluminal surfaces of the endothelium. Labeling of cytoplasmic vesicles with both tracers was most pronounced after 1 min elapsed time of circulation. The vesicles containing the reaction product were uncoated and most exhibited diameters between 50 and

110 nm. This latter size range agrees closely with the population of vesicles described by Matsushima and Reiter (1975). Welsh and Beitz (1981) utilized the intravenous injections of the same tracers to ascertain transendothelial uptake by the pineal gland of the Mongolian gerbil, a species in which the endothelial cells are not fenestrated but are permeable to some extent to trypan blue. After circulation times of 15 and 60 min, the label crossed the capillary wall with the reaction product being located in vesicles along the abluminal surface of the cell and in the intercellular clefts. These findings indicate that the absence of fenestrations in the endothelial cells are not a definitive morphologic criterion for a blood–brain barrier to proteins and peptides in the pineal gland. Although such a barrier has been implicated in the pineal gland of primate species (Moller, 1974), this does not necessarily exclude a secretory function, since the pineal hormone, melatonin, is a lipophilic molecule capable of penetrating the barrier (Wurtman and Axelrod, 1965).

   b. *Relative quantity of blood flow to pineal gland:* Although anatomic evidence in the rat has suggested that the vascular supply to the pineal gland is considerably less than that to the pituitary gland (Quay, 1958), physiologic data do not support such a view. For example, by measuring the uptake of radioactive isotopes of rubidium (^{86}Rb) and of potassium (^{42}K), Goldman and Wurtman (1964) estimated that blood flow to the pineal gland was approximately 4 ml/min/gm of tissue, a rate which equals or exceeds that of many other endocrine organs.

   c. *Role of sympathetic innervation in enzymatic activity:* The enzymatic activity of the pineal gland depends upon sympathetic innervation. In this regard, Goldman (1967) compared the uptake of rubidium isotope in intact rats with that in an experimental group in which the pineal gland had been denervated by either superior cervical ganglionectomy or section of the cervical sympathetic trunk. The resulting decrease in pineal blood flow in the second operated group was approximately 70% that of intact controls. This change corresponds to the lowered metabolic activities of the organ which follows denervation.

   d. *Effects of light and dark phases on metabolic activity:* The consistent circadian rhymicity in metabolic activity of the pineal gland prompted Quay (1972) to examine blood content in this organ in the rat over a 24-hr period. He reported that the number of erythrocytes per unit volume of tissue declined by about 50% in both the cortex and the medulla of the gland during the later part of the dark phase and early light. Presumably, this response was related to norepinephrine levels, which are high at the end of the dark period and decrease during the light phase. A functional relationship among pineal gland enzymatic activity, light, and blood flow has been substantiated by Rollag *et al.*

(1978). In their study, radioactive isotopes of strontium (^{85}Sr) and cerium (^{141}Ce) were infused over a 1-min interval through the brachiocephalic artery of ewes 2 min before lights off and 5 min after lights on. There was a statistically significant reduction in the uptake of radiolabeled microspheres by the pineal gland of those animals sacrificed after the lights were turned on. Uptake by the choroid plexus and other parts of the brain did not differ. These data correlate well with the rapid decline in enzymatic activity of the pineal gland following a light stimulus.

3. Pathology of Blood Circulation

Reports of pathologic conditions associated with the vasculature of the pineal gland usually involve arteriovenous malformations caused by local pineal tumors or masses (Lazar and Clark, 1974). A true venous malformation of the pineal region was reported by Ventureyra and Ivan (1979) in a 6-year-old girl and described as a varicose dilatation of the vein of Galen.

References*

Altenáhr, E. (1971). Electron microscopical evidence for innervation of chief cells in human parathyroid glands. *Experientia* **27**, 1077. (B)

Alveryd, A. (1968). Parathyroid glands in thyroid surgery. *Acta Chir. Scand., Suppl.* **389**, 1–120. (B)

Anderson, M. C., Schoenfeld, F. B., Iams, W. B., and Suwa, M. (1967). Circulatory changes in acute pancreatitis. *Surg. Clin. North Am.* **47**, 127–140. (A)

Atwal, O. S. (1981). Myelinated nerve fibers in the parathyroid gland of the dog: A light and electron microscopic study. *Acta Anat.* **109**, 3–12. (B)

Aune, S., and Semb, L. S. (1969). The effect of secretin and pancreozymin on pancreatic blood flow in the conscious and anesthetized dog. *Acta Physiol. Scand.* **76**, 406–411. (A)

Balashev, V. N., and Ignashkina, M. S. (1965). The lymphatic system of the human parathyroid gland. *Anat., Anthropol., Embryol. Histol.* **19**, Abstr. No. 3232. (B)

Barlow, T. E., Greenwell, J. R., Harper, A. A., and Scratcherd, T. (1968). Factors influencing pancreatic blood flow. *In* "Blood Flow Through Organs and Tissues" (W. H. Bain and A. M. Harper, eds.), pp. 469–484. Williams & Wilkins, Baltimore, Maryland. (A)

Bartels, P. (1909). "Das Lymphgefassystem." Fischer, Jena. (A)

Bartos, V., and Brzek, V. (1979). Exocrine secretion of the pancreas and enzymes in the lymph after administration of a combination of coumarin and rutin sulphate. *Arzneim.-Forsch.* **29**, 548–549. (A)

Bartos, V., and Brzek, V. (1981). Pancreatic enzymes in thoracic duct lymph after ethanol administration. *Lymphology.* 14, 29–31. (A)

Bartos, V., Brzek, V., and Groh, J. (1966). Alterations in human thoracic duct lymph in relation to the function of the pancreas. *Am. J. Med. Sci.* **252**, 31–38. (A)

Bartos, V., Kolc, J., Vanacek, R., and Malek, P. (1979). Lymph and blood enzymes and pathologic

*In the reference list, the capital letter in parentheses at the end of each reference indicates the section in which it is cited.

alterations in canine experimental pancreatitis after administration of benzopyrones. *Scand. J. Gastroentol.* **14**, 343–347. (A)

Baum, R., and Iber, F. L. (1973). Alcohol, the pancreas, pancreatic inflammation and pancreatic insufficiency. *Am. J. Clin. Nutr.* **26**, 347–351. (A)

Beck, J. S. P., and Berg, B. N. (1931). The circulatory pattern in the islands of Langerhans. *Am. J. Pathol.* **7**, 31–36. (A)

Beijer, H. J. M., Brouwer, F. A. S., and Charbon, G. A. (1979). Time course and sensitivity of secretin-stimulated pancreatic secretion and blood flow in the anesthetized dog. *Scand. J. Gastroenterol.* **14**, 295–300. (A)

Bensley, R. R. (1911). Studies on the pancreas of the guinea pig. *Am. J. Anat.* **12**, 297–388. (A)

Berg, B. N. (1930). A study of the islands of Langerhans *in vivo* with observations on the circulation. *Am. J. Physiol.* **95**, 186–189. (A)

Bernard, C. (1856). "Memoire sur le pancréas." Balliere, Paris. (A)

Biedl, A. (1923). Ueber die Abfuhrwege des Pankreasinkretes und die Bedeutung des Insulins für die Theorie des Pankreasdiabetes. *Dtsch. Med. Wochenschr.* **29**, 937–938. (A)

Boijsen, E., and Tylén, U. (1972). Vascular changes in chronic pancreatitis. *Acta Radiol. Diagn.* **12**, 34–38. (A)

Bor, N. M., Ercan, M. T., Alvur, M., Bekdik, C., and Oner, G. (1974). Calculation of net insulin secretion and pancreatic blood flow. *Pfluegers Arch.* **352**, 179–188. (A)

Broadie, T. A., Devedas, M., Rysavy, J., Leonard, A. S., and Delaney, J. P. (1979). The effect of hypoxia and hypercapnia on canine pancreatic blood flow. *J. Surg. Res.* **27**, 114–118. (A)

Brunfeldt, K., Hunhammar, K., and Skouby, A. P. (1958). Studies on the vascular system of the islets of Langerhans in mice. *Acta Endocrinol. Copenhagen* **29**, 473–480. (A)

Brzek, V., and Bartos, V. (1969). Therapeutic effect of the prolonged thoracic duct lymph fistula in patients with acute pancreatitis. *Digestion* **2**, 43–50. (A)

Bunnag, S. (1966). Postnatal neogenesis of islets of Langerhans in the mouse. *Diabetes* **15**, 480–491. (A)

Bunnag, S. C., Warner, N. E., and Bunnag, S. (1963). Microcirculation in the islets of Langerhans of the mouse. *Anat. Rec.* **146**, 117–123. (A)

Bunnag, S. C., Warner, N. E., and Bunnag, S. (1964). Microvasculature of the pancreatic ducts. *Bibl. Anat.* **4**, 142–149. (A)

Bunnag, S. C., Warner, N. E., and Bunnag, S. (1966). Effect of tolbutamide on postnatal neogenesis of the islet of Langerhans in mouse. *Diabetes* **15**, 597–603. (A)

Bunnag, S. C., Warner, N. E., and Bunnag, S. (1967). Effect of alloxan on the mouse pancreas during and after recovery from diabetes. *Diabetes* **16**, 83–89. (A)

Casley-Smith, J. R. (1976). The actions of the benzopyrones on the blood–tissue–lymph system. *Folia Angiol.* **24**, 7–22. (A)

Cubilla, A. L., Fortner, J., and Fitzgerald, P. (1978). Lymph node involvement in carcinoma of the head of the pancreas area. *Cancer* **41**, 880–887. (A)

Curtis, G. M. (1930). The blood supply of the human parathyroids. *Surg., Gynecol. Obstet.* **51**, 805–809. (B)

Dal Zotto, E. (1950). Anastomosi artero-venose e dispositivi arteriosi di blocco nel pancreas. *Boll. Soc. Ital. Biol. Sper.* **25**, 1322–1323; cited in *Anat., Anthropol., Embryol. Histol.* **5**, 245 (1951) (Abstr. No. 837). (A)

Daniel, P. M., and Henderson, J. R. (1966). Insulin in the lymph of the thoracic duct of the rabbit. *J. Physiol. (London)* **184**, 36P–37P. (A)

Delaney, J. P., and Grim, E. (1966). Influence of hormones and drugs on canine pancreatic blood flow. *Am. J. Physiol.* **211**, 1398–1402. (A)

Delaney, J. P., Goodale, R. L., Jr., Chang, J., and Wangensteen, O. H. (1966). The influence of vasopressin on upper gastrointestinal blood flow. *Surgery* **59**, 397–400. (A)

Donaldson, L. A., Williams, R. W., and Schenk, W. G., Jr. (1978). Experimental pancreatitis: Effect of plasma and dextran on pancreatic blood flow. *Surgery* **84**, 313–321. (A)

Dreiling, D. A. (1970). The lymphatics, pancreatic ascites and pancreatic inflammatory disease. *Am. J. Gastroenterol.* **53**, 119–131. (A)

Dumont, A. E., and Martelli, A. B. (1968). Pathogenesis of pancreatic edema following exocrine duct obstruction. *Ann. Surg.* **168**, 302–309. (A)

Dumont, A. E., and Mulholland, J. H. (1960). Measurement of pancreatic enzymes in human thoracic duct lymph. *Gastroenterology* **38**, 954–956. (A)

Dumont, A. E., Doubilet, H., and Mulholland, J. H. (1960). Lymphatic pathway of pancreatic secretion in man. *Ann. Surg.* **152**, 403–409. (A)

Dumont, A. E., Witte, C. L., and Witte, M. H. (1971). Studies of lymph in certain disorders of the liver, pancreas and small intestine. *Am. J. Gastroenterol.* **56**, 346–351. (A)

Dunlop, D. A. B., Papapoulos, S. E., Lodge, R. W., Fulton, A. J., Kendall, B. E., and O'Riordan, J. L. H. (1980). Parathyroid venous sampling: Anatomic considerations and results in 95 patients with primary hyperparathyroidism. *Br. J. Radiol.* **53**, 183–191. (B)

Duprez, A., Godart, S., Platteborse, R., Litvine, J., and Dupont, J. M. (1963). La voie de dérivation interstitielle et lymphatique de la secrétion du pancréas. *Bull. Acad. R. Med. Belg.* **7**, 691–706. (A)

Editorial. (1981). Pancreatic islet-acinar interactions. *Br. Med. J.* **283**, 570–571. (A)

Edwards, E. A. (1980). Abdominal visceral circulation in man. In "Structure and Function of the Circulation" (C. J. Schwartz, N. T. Werthessen, and S. Wolf, eds.), pp. 392–393. Plenum, New York. (A)

Eliassen, E., Folkow, B., and Hilton, S. (1973). Blood flow and capillary filtration capacities in salivary and pancreatic glands. *Acta Physiol. Scand.* **87**, 11A–12A. (A)

Ercan, M. T., Bor, N. M., Bekdik, C. F., and Oner, G. (1974). Measurement of pancreatic blood flow in dog by ^{133}Xe clearance technique. *Pfluegers Arch.* **348**, 51–57. (A)

Ferner, H. (1952). "Das Inselsystem des Pankreas." Thieme, Stuttgart. (A)

Figarella, C., de Caro, A., Oi, I., and Sahel, J. (1978). Lipase activity in blood following endoscopic pancreatography: Demonstration of its pancreatic origin and existence of ductal or acino-venous pathways in man. *Scand. J. Gastroenterol.* **13**, 393–399. (A)

Fischer, U., Hommel, H., and Salzsieder, E. (1976). Pancreatic blood flow in conscious dogs after oral administration of glucose. *Diabetologia* **12**, 133–136. (A)

Fraser, P. A., and Henderson, J. R. (1980). The arrangement of endocrine and exocrine pancreatic microcirculation observed in the living rabbit. *Q. J. Exp. Physiol. Cogn. Med. Sci.* **65**, 151–158. (A)

Fujita, T. (1973). Insulo-acinar portal system in the horse pancreas. *Arch. Histol. Jpn.* **35**, 161–171. (A)

Fujita, T., and Murakami, T. (1973). Microcirculation of monkey pancreas with special reference to the insulo-acinar portal system. *Arch. Histol. Jpn.* **35**, 255–263. (A)

Fujita, T., and Watanabe, Y. (1973). The effects of islet hormones upon the exocrine pancreas. In "Gastro-entero-pancreatic Endocrine System—A Cell Biological Approach" (T. Fujita, ed.), pp. 164–173. Igaku Shoin, Tokyo. (A)

Gilmour, J. R. (1937). The embryology of the parathyroid glands, the thymus, and certain associated rudiments. *J. Pathol. Bacteriol.* **45**, 507–522. (B)

Gilmour, J. R. (1938). The gross anatomy of the parathyroid glands. *J. Pathol. Bacteriol.* **46**, 133–149. (B)

Gilmour, J. R. (1939). The normal histology of the parathyroid glands. *J. Pathol. Bacteriol.* **48**, 187–222. (B)

Godart, S. (1965). Lymphatic circulation of the pancreas. *Bibl. Anat.* **7**, 410–413. (A)

Goldman, H. (1967). The nervous control of blood flow to the pineal body. *Life Sci.* **6**, 2071–2077. (C)

Goldman, H., and Wurtman, R. J. (1964). Flow of blood to the pineal body of the rat. *Nature (London)* **203**, 87–88. (C)

Goodhead, B. (1969). Acute pancreatitis and pancreatic blood flow. *Surg., Gynecol. Obstet.* **129**, 331–340. (A)

Goodhead, B., Himal, H. S., and Zanbilowicz, J. (1970). Relationship between pancreatic secretion and pancreatic blood flow. *Gut* **11**, 62–68. (A)

Grau, H., and Taher, E. (1965). Histologische Untersuchungen über das innere Lymphgefässystem von Pankreas und Milz. *Berl. Muench Tieraerztl. Wochenschr.* **78**, 147–151. (A)

Gray, H. (1973). "Anatomy of the Human Body" (C. M. Goss, ed.), 29th Am. ed., pp. 629–636. Lea & Febiger, Philadelphia, Pennsylvania. (A)

Grim, E. (1963). The flow of blood in mesenteric vessels. In "Handbook of Physiology" (W. F. Hamilton and P. Dow, eds.), Sect. 2, Vol. II, pp. 1439–1456. Am. Physiol. Soc., Washington, D.C. (A)

Halsted, W. S., and Evans, H. M. (1907). The parathyroid glandules. Their blood supply and their preservation in operation upon the thyroid gland. *Ann. Surg.* **46**, 489–506. (B)

Hayasaka, N., and Sasano, N. (1970). Vascular and ductal patterns of pancreas by microradiography and their relation to lesions of pancreatitis. *Tohoku J. Exp. Med.* **100**, 327–347. (A)

Heisig, N. (1968). Functional analysis of the microcirculation in the exocrine pancreas. *Adv. Microcirc.* **1**, 89–151. (A)

Henderson, J. R. (1969). Why are the islets of Langerhans? *Lancet* **2**, 469–470. (A)

Henderson, J. R. (1974). Insulin in body fluids other than blood. *Physiol. Rev.* **54**, 1–22. (A)

Henderson, J. R., and Daniel, P. M. (1979). A comparative study of the portal vessels connecting the endocrine and exocrine pancreas, with a discussion of some functional implications. *Q. J. Exp. Physiol. Cogn. Med. Sci.* **64**, 267–275. (A)

Henderson, J. R., Daniel, P. M., and Fraser, P. A. (1981). The pancreas as a single organ: The influence of the endocrine upon exocrine part of the gland. *Gut* **22**, 158–167. (A)

Hodde, K. C. (1979). The vasculature of the rat pineal organ. *Prog. Brain Res.* **52**, 39–44. (C)

Hodde, K. C., and Veltman, W. A. M. (1979). The vascularization of the pineal gland (epiphysis cerebri) of the rat. *Scanning Electron Microsc.* **3**, 369–374. (C)

Hoggan, G, and Hoggan, F. E. (1881). On the lymphatics of the pancreas. *J. Anat. Physiol. Norm. Pathol. Homme Anim.* **15**, 475–495. (A)

Horwitz, L. D., and Myers, J. H. (1982). Ethanol-induced alterations in pancreatic blood flow in conscious dogs. *Circ. Res.* **50**, 250–256. (A)

Hruban, Z., Schulman, S., Warner, N. E., Dubois, K. P., Bunnag, S., and Bunnag, S. C. (1963). Hypoglycemia resulting from insecticide poisoning. *JAMA, J. Am. Med. Assoc.* **184**, 590–593. (A)

Jacobson, E. D. (1967). Secretion and blood flow in the gastrointestinal tract. In "Handbook of Physiology" (C. F. Code and A. W. Heidel, eds.). Sect. 6, Vol. II, pp. 1056–1058. Williams & Wilkins, Baltimore, Maryland. (A)

Jander, H. P., Dietheim, A. G., and Russinovich, N. A. E. (1980). The parathyroid artery. *AJR, Am. J. Roentgenol.* **135**, 821–828. (B)

Jdanov, D. A. (1960). Nouvelles données sur la morphologie fonctionnelle du système lymphatique des glandes endocrines. *Acta Anat.* **41**, 240 259. (A)

Kaplan, J. H., and Bookstein, J. J. (1972). Abdominal visceral pharmacoangiography with angiotensin. *Radiology* **103**, 79–83. (A)

Krstic, R. (1980). Three-dimensional organization of the rat parathyroid glands. *Z. Mikrosk.-Anat. Forsch.* **94**, 445–450. (B)

Kühne, W., and Lea, A. S. (1882). Beobachtungen über die Absonderung des Pankreas. *Unters. Physiol. Inst. Univ. Heidelberg* **2**, 448–487. (A)

Lanciault, G., and Jacobson, E. D. (1976). The gastrointestinal circulation. *Gastroenterology* **71**, 851–873. (A)

Lazar, M. L., and Clark, K. (1974). Direct surgical management of masses in the region of the vein of Galen. *Surg. Neurol.* **2**, 17–21. (C)

Lenninger, S. (1973). Effects of acetylcholine and papaverine on the secretion and blood flow from the pancreas of the cat. *Acta Physiol. Scand.* **89**, 260–268. (A)

Lifson, N., and Lassa, C. V. (1981). Note on the blood supply of the ducts of the rabbit pancreas. *Microvasc. Res.* **22**, 171–176. (A)

Lifson, N., Kramlinger, K. G., Mayrand, R. R., and Lender, E. J. (1980). Blood flow to the rabbit pancreas with special reference to the islets of Langerhans. *Gastroenterology* **79**, 466–473. (A)

MacCallum, W. G., and Voegtlin, C. (1909). On the relation of tetany to the parathyroid glands and to calcium metabolism. *J. Exp. Med.* **11**, 118–151. (B)

McCuskey, R. S., and Chapman, T. M. (1969). Microscopy of the living pancreas in situ. *Am. J. Anat.* **126**, 395–403. (A)

Mackie, C. R., and Moosa, A. R. (1980). Surgical anatomy of the pancreas. *In* "Tumors of the Pancreas" (A. R. Moosa, ed.), p. 16. Williams & Wilkins, Baltimore, Maryland. (A)

Mallette, L. E., Gomez, L., and Fisher, R. G. (1981). Parathyroid angiography: A review of current knowledge and guidelines for clinical application. *Endocr. Rev.* **2**, 124–135. (B)

Mandelbaum, I., and Morgan, C. R. (1969). Pancreatic blood flow and its relationship to insulin secretion during extracorporeal circulation. *Ann. Surg.* **170**, 753–758. (A)

Mathews, A. (1899). The changes in structure of the pancreas cell. *J. Morphol. Suppl.* **15**, 171–216. (A)

Matsushima, S., and Reiter, R. J. (1975). Ultrastructural observations of pineal gland capillaries in four rodent species. *Am. J. Anat.* **143**, 265–282. (C)

Mislin, H. (1971). Die Wirkung von Cumarin aus *Melilotus officinalis* auf die Funktion des Lymphangions. *Arzneim.-Forsch.* **21**, 852–853. (A)

Moller, M. (1974). The ultrastructure of the human fetal pineal gland. I. Cell types and blood vessels. *Cell Tissue Res.* **152**, 13–30. (C)

Moller, M., van Deurs, B., and Westergaard, E. (1978). Vascular permeability to proteins and peptides in the mouse pineal gland. *Cell Tissue Res.* **195**, 1–15. (C)

Morris, R. E., Jr. (1964). Studies on the development of pancreatic necrosis in the living mouse. *Bull. Johns Hopkins Hosp.* **114**, 212–229. (A)

Norberg, K.-A., Pensson, B., and Granberg, P.-O. (1975). Adrenergic innvervation of the human parathyroid glands. *Acta Chir. Scand.* **141**, 319–322. (B)

Norris, E. H. (1937). The parathyroid glands and the lateral thyroid in man: Their morphogenesis, histogenesis, topographic anatomy and prenatal growth. *Contrib. Embryol. Carnegie Inst.* **159**, 247–294. (B)

O'Leary, J. L. (1930). An experimental study of the islet cells of the pancreas in vivo. *Anat. Rec.* **45**, 27–58. (A)

Palmer, A. A. (1959). A study of blood flow in minute vessels of the pancreatic region of the rat with reference to intermittent corpuscular flow in individual capillaries. *Q. J. Exp. Physiol. Cogn. Med. Sci.* **44**, 149–159. (A)

Paloyan, E., Lawrence, A. M., and Straus, F. H. (1973). "Hyperparathyroidism," pp. 56–76. Grune & Stratton, New York. (B)

Papp, M. (1976). Pathogenesis of acute pancreatitis: Pancreatic ductal–interstitial–vascular and lymphatic pathways. *Acta Med. Acad. Sci. Hung.* **33**, 191–206. (A)

Papp, M., Németh, E., Feuer, I., and Fodor, I. (1958). Effect of an impairment of lymph flow on experimental acute "pancreatitis." *Acta Med. Acad. Sci. Hung.* **11**, 203–208. (A)

Papp, M., Makara, G. B., Hajtman, B., and Csáki, L. (1966). A quantitative study of pancreatic blood flow in experimental pancreatitis. *Gastroenterology* **51**, 524–528. (A)

Papp, M., Ungvári, G., Németh, P. E., Munkácsi, I., and Zubek, L. (1969). The effect of bile-induced pancreatitis on the intrapancreatic vascular pattern in dogs. *Scand. J. Gastroenterol.* **4**, 681–689. (A)

Papp, M., Németh, E. P., and Horvath, E. J. (1971). Pancreaticoduodenal lymph flow and lipase activity in acute experimental pancreatitis. *Lymphology* **4**, 48–53. (A)

Papp, M., Feher, S., Varga, B., and Folly, G. (1977). Humoral influences on local blood flow and external secretion of the resting pancreas. *Acta Med. Acad. Sci. Hung.* **34**, 185–198. (A)

Pepin, J., Singh, H., Pairent, F. W., Appert, H. E., and Howard, J. M. (1970). A study of insulin secretion in thoracic duct lymph of the dog. *Ann. Surg.* **172**, 56–60. (A)

Piller, N. B. (1976). Conservative treatment of acute and chronic lymphedema with benzopyrones. *Lymphology* **9**, 132–137. (A)

Piller, N. B. (1980). Lymphedema, macrophages and benzopyrones. *Lymphology* **13**, 109–119. (A)

Pissiotis, C. A., Condon, R. E., and Nyhus, L. M. (1972). Effect of vasopressin on pancreatic blood flow in acute hemorrhagic pancreatitis. *Am. J. Surg.* **123**, 203–207. (A)

Popper, H. L. (1940). Enzyme studies in edema of the pancreas and acute pancreatitis. *Surgery* **7**, 566–570. (A)

Popper, H. L., and Necheles, H. (1942). Edema of the pancreas. *Surg., Gynecol. Obstet.* **74**, 123–124. (A)

Preissich, P. (1977). Zur medikamentösen Therapie der akuten Pankreatitis. *Leber, Magen, Darm* **5**, 26–29. (A)

Quay, W. B. (1958). Pineal blood content and its experimental modification. *Am. J. Physiol.* **195**, 391–395. (C)

Quay, W. B. (1972). Twenty-four-hour rhythmicity in carbonic anhydrase activities of choroid plexus and pineal gland. *Anat. Rec.* **174**, 279–288. (C)

Quay, W. B. (1974). "Pineal Chemistry." Thomas, Springfield, Illinois. (C)

Raybuck, H. E. (1952). The innervation of the parathyroid glands. *Anat. Rec.* **112**, 117–120. (B)

Reynolds, B. M. (1970). Observations of subcapsular lymphatics in normal and diseased human pancreas. *Ann. Surg.* **171**, 559–566. (A)

Rollag, M. D., O'Callaghan, P. L., and Niswender, C. D. (1978). Dynamics of photo-induced alterations in pineal blood flow. *J. Endocrinol.* **76**, 547–548. (C)

Ruangsiri, C. (1949). Changes in islets of Langerhans in living mice after alloxan administration. *Anat. Rec.* **105**, 399–419. (A)

Rusznyák, I., Földi, M., and Szabó, G. (1967). "Lymphatics and Lymph Circulation: Physiology and Pathology," 2nd ed., p. 147. Pergamon, Oxford. (A)

Semb, L. S., and Aune, S. (1971). The effect of glucose and insulin on pancreatic blood flow in the anesthetized dog. *Scand. J. Clin. Lab. Invest.* **27**, 105–111. (A)

Senior, W. (1974). Parathyroid gland structure of some tropical lizards. *J. Morphol.* **142**, 91–108. (B)

Sim, D. N., Duprez, A., and Anderson, M. C. (1966). Alterations of the lymphatic circulation during acute experimental pancreatitis. *Surgery* **60**, 1175–1182. (A)

Solomon, T. E., Solomon, N., Shanbour, L. L., and Jacobson, E. D. (1974). Direct and indirect effects of nicotine on rabbit pancreatic secretion. *Gastroenterology* **67**, 276–283. (A)

Takeshima, R., Miyamoto, J., Iwasaki, Y., and Dreiling, D. (1981). Relationships between pancreatic blood flow and secretion in the dog. *Mt. Sinai J. Med.* **48**, 1–6. (A)

Tankel, H. I., and Hollander, F. (1957). The relation between pancreatic secretion and local blood flow: A review. *Gastroenterology* **32**, 633–641. (A)

Thiel, A. (1954). Untersuchungen über das Gefäss-System des Pankreasläppchens bei verschiedenen Säugern mit besonderer Berucksichtigung der Kapillarknäuel der Langerhansschen Inseln. *Z. Zellforsch. Mikrosk. Anat.* **39**, 339–342. (A)

Tikhomirov, Y. L. (1970). Determination of the capillary network volume of the human pineal gland by a photometric method. *Arkh. Anat., Gistol. Embriol.* **58**, 98–102. (C)

Tohyama, M. (1935). The blood supply of the pancreatic islets of the dog. *Jpn. J. Med. Sci.* **5**, 61–67. (A)

Vega, R. E., Appert, H. E., and Howard, J. M. (1967). Effects of secretin in stimulating the output of amylase and lipase in the thoracic duct of the dog. *Ann. Surg.* **166**, 995–1001. (A)

Ventureyra, E. C. G., and Ivan, L. P. (1979). Venous malformation of the pineal region. *Surg. Neurol.* **11**, 225–228. (C) von Bartheld, F., and Moll, J. (1954). The vascular system of the mouse epiphysis with remarks on the comparative anatomy of the venous trunks in the epiphyseal area. *Acta Anat.* **22,** 227–235. (C)

Wang, C.-A. (1976). The anatomic basis of parathyroid surgery. *Ann. Surg.* **183,** 271–275. (B)

Wartenberg, H. (1968). The mammalian pineal organ: Electron microscopic studies on the fine structure of pinealocytes, glial cells and on the perivascular compartment. *Z. Zellforsch. Mikrosk. Anat.* **86,** 74–97. (C)

Weiner, S., Gramatica, L., Voegle, L. D., Hauman, R. L., and Anderson, M. C. (1970). Role of the lymphatic system in the pathogenesis of inflammatory disease in the biliary tract and pancreas. *Am. J. Surg.* **119,** 55–61. (A)

Welsh, D. A. (1898). Concerning the parathyroid glands. A critical anatomical and experimental study. *J. Anat. Physiol. Norm. Pathol. Homme Anim.* **2,** 380–402. (B)

Welsh, M. G., and Beitz, A. J. (1981). Modes of protein and peptide uptake in the pineal gland of the Mongolian gerbil: An ultrastructural study. *Am. J. Anat.* **162,** 343–355. (C)

Wharton, G. K. (1932). The blood supply of the pancreas, with special reference to that of the islands of Langerhans. *Anat. Rec.* **53,** 55–81. (A)

Wislocki, G. B., and Leduc, E. H. (1952). Vital staining of the hematoencephalic barrier by silver nitrate and trypan blue, and cytological comparisons of the neurohypophysis, pineal body, area postrema, intercolumnar tubercle and supraoptic crest. *J. Comp. Neurol.* **96,** 371–413. (C)

Wurtman, W. J., and Axelrod, J. (1965). The formation, metabolism and physiological effects of melatonin in mammals. *Prog. Brain Res.* **10,** 520–528. (C)

Zawistowski, S. (1967). Fluorescence studies on the occurrence of catecholamines in the parathyroid gland in the white rat. *Folia Morphol. (Prague)* **26,** 193–198. (B)

Chapter 10
Cardiopulmonary System: Heart

A. EMBRYOLOGY OF BLOOD CIRCULATION
By C. E. CORLISS

1. SUPERFICIAL CIRCUITS

a. Vessel formation: Human angiogenesis begins early in the fourth week when clumps of mesenchymal cells, termed blood islands, form on the developing yolk sac. These soon become a vascular plexus that spreads, not only over the entire yolk sac, but also to other extraembryonic and embryonic structures, including the epicardium of the developing heart.

b. Development of the coronary arteries: By the end of the sixth week, the coronary arteries have developed as endothelial sprouts emerging from the right and left sides of the bulbus cordis. Both these new vessels quickly establish connections with the original surface plexus, mentioned above, and, by exploiting this union, create the overall pattern of coronary circulation. The human coronary branches arise in the following sequence: first, the anterior interventricular; then the posterior interventricular; and finally the circumflex branch of the left coronary artery.

c. Development of the cardiac veins: The cardiac veins arise in the middle of the sixth week, shortly before the arteries, with most of them originating as endothelial sprouts from the wall of the sinus venosus (coronary sinus). The middle cardiac vein appears first, followed by the posterior vein of the left ventricle. These vessels, as well as the great and small cardiac veins which appear next, rapidly anastomose with the plexus that earlier had spread from the yolk sac. The anterior cardiac veins sprout off the wall of the right atrium itself. All these superficial vessels anastomose with the original epicardial plexus to set up complete drainage for the imminent arrival of the coronary artery system. Some branches also penetrate the myocardium to communicate with the smallest cardiac veins of the intramural group (see Section A-2b, below).

2. INTRAMURAL VASCULARIZATION

a. Early formation: The early tubular heart is nourished by blood traversing it, this intramural mechanism being continued in later stages by endothelial

Blood Vessels and Lymphatics in Organ Systems
Copyright © 1984 by Academic Press, Inc.
All rights of reproduction in any form reserved.
ISBN 0-12-042520-3

extensions from the cardiac cavity into the myocardium. The latter vessels penetrate between and behind the rapidly growing muscular trabeculae as intertrabecular spaces, thereby initiating intramural vascularization.

b. Development of the intramural vascular channels: The developing intertrabecular spaces are compressed by the growing trabeculae into very small myocardial sinusoids. These connect with the epicardial vessels via penetrating branches and with the cardiac lumen via the intertrabecular spaces and Thebesian veins. Most of the blood reaching the myocardial plexus returns through the coronary veins, whereas some is carried through sinusoids and intertrabecular spaces into the cardiac cavity (Patten, 1968). Late in gestation, certain epicardial vessels, arterial in structure, pass directly to the cardiac lumen without traversing the intervening capillary bed, thus supplying an efficient by-pass mechanism.

3. Circulation to the Conduction System

a. Sinoatrial node: This structure appears early in the fifth week (O'Rahilly, 1971) and is supplied by the right nodal artery, an atrial branch of the right coronary artery. Sometimes a left nodal artery from the circumflex branch of the left coronary artery reaches the node via the superior surface of the atrium (Walmsley and Watson, 1978).

b. Atrioventricular node and bundle of His: The atrioventricular node and bundle of His can be seen in a human embryo by 37 days (O'Rahilly, 1971). Late in the second month, the node and posterior division of the left bundle branch are served by a perforating twig of the posterior interventricular branch of the right coronary artery. During the third month, the anterior interventricular artery sends perforating branches into the interventricular septum, one of which supplies the region of the atrioventricular bundle (Licata, 1962).

B. ANATOMY OF BLOOD CIRCULATION

By W. C. Randall

1. Anatomy of Main Coronary Arteries and Their Principal Branches

Two major coronary arteries, a right and a left, arise from the base of the aorta within the sinus of Valsalva, behind the anterior and posterior cusps of the aortic valve. During cardiac ejection, the ostia are shielded from stenosis by the valve leaflets.

   a. Left coronary artery: This vessel, which is 0.5–2.0 cm long, bifurcates and occasionally trifurcates to furnish the primary blood supply to the left ventricle and to the right heart as well. It shows considerable variability in its axis of exit from the aorta. In a horizontal plane, the axis may be anterior, transverse, or posterior, whereas in a frontal plane, it may be superior, horizontal, or inferior. Knowledge of the axis is important, as it pertains to the radiographic plane of examination when in search of occlusive arterial disease (McAlpine, 1975). The orifice of the main left branch often shows a knife-like incisura at its inferior or lateral margin, a finding which has functional significance in that it may contribute to turbulence downstream, impair flow due to skimming, or act as a flap valve. Any of these variables could partially account for the reduction in coronary flow as pressure rises with distention of the aortic wall. McAlpine (1975) suggests that such anatomic variations may account for certain sudden, unexplained deaths during exercise in the absence of coronary atherosclerosis.

   The left coronary artery passes anteriorly through epicardial areolar tissue and to the left in the auriculoventricular groove, between the pulmonary artery and the left auricular appendage, and branches after traversing only 10–15 mm in man and 2.0–5.0 mm in dogs and smaller mammals. In some mammals (dog and rabbit), but not in others (man and monkey), a septal artery arises close to the bifurcation into the anterior descendens and circumflex arteries. Small branches from the left coronary artery pass to the pulmonary conus and left atrium and in some species supply the vasa vasorum of the pulmonary artery (Sobin *et al.*, 1962). In dog and man, a particularly interesting and potentially important branch of the proximal segment of the left coronary artery supplies a small group of chemoreceptors which is capable of being maximally excited with serotonin to elicit profound hypertension (James *et al.*, 1975, 1980).

   The *anterior descendens* branch of the left coronary artery follows the anterior interventricular sulcus toward, and terminates in, the region of the apex. There are a variable number of branches, several of them large, distributed epicardially and angling anteriorly and to the left toward the apex. Smaller branches destined for the right ventricle cross from the interventricular groove to supply a narrow band of muscle and frequently anastomose with branches from the right coronary artery. Anastomoses interconnect with branches of the left circumflex and, at the apex, with the marginal and posterior descending branch of the circumflex or right coronary artery. Septal branches penetrate deeply all along the anterior interventricular sulcus.

   The *left circumflex* branch of the left coronary artery follows the auricular groove leftward along the left atrium to end in the vicinity of the posterior longitudinal sulcus. It is primarily epicardial, terminating in the posterior descendens artery in some species (dog) but not in others (pig). Branching to atrium and ventricle occurs in man and dog. Posteriorly the left circumflex shows frequent anastomoses via branches from the posterior descendens or marginal arteries. In the dog, an important branch supplies the A-V nodal and His bundle region (Halpern, 1954).

b. Right coronary artery: This vessel arises from its ostium in the right anterior aortic cusp, sometimes, as in dogs and primates, with smaller ostia of accessory branches. The proximal segment of the main vessel is about 1.5 cm in length and extends in a perpendicular manner from the right aortic sinus, commonly by way of a smooth orifice; this is in contrast to the finding observed in the case of the left coronary artery (see Section B-1a, above). The primary right coronary artery passes anteriorly behind the pulmonary artery and follows the auriculoventricular groove, eventually coursing toward the apex. In some species (dog and rabbit), it commonly terminates as the marginal branch, but in others (pig and man), it generally extends posteriorly to become the posterior descendens artery. It gives off branches to the atrium, one of which is a major supply to the sinoatrial node in both human and canine hearts. In man and pig, posteriorly a branch to the A-V node is given off at the crux, corresponding to the vessel from the circumflex in the dog. A branch to the pulmonary conus frequently arises from one of the accessory ostia.

c. Species differences: Variability is common in descriptions of the distribution of coronary arteries, particularly when the hearts of different species are compared. In all mammals, the entire anterior and lateral aspects of the left ventricle are supplied by the left coronary branches, whereas the free ventricular wall obtains its circulation from the right coronary artery. The posterior ventricular surface is nourished by either right or left, or both, and thus the designation "dominant pattern" becomes meaningful. Canine hearts generally are left coronary dominant, the left circumflex artery distributing to the posterior surface of the left and right ventricles, posterior septum, and A-V node, this pattern being least common in man and pig. The pig heart is right coronary dominant in about 50% of specimens. In about 67% of human hearts, the right coronary artery is dominant, with the right coronary artery crossing the crux and supplying part of the left ventricular wall and the ventricular septum. In 15%, the left coronary artery is dominant (as in the case of the dog), with the circumflex artery crossing the crux and giving off a posterior interventricular branch which supplies all of the left ventricle, the septum, and part of the right ventricular wall. In the remaining 18%, both coronary arteries reach the crux— the so-called "balanced" coronary arterial pattern. Sometimes there is no real posterior interventricular branch, but instead the posterior septum is penetrated by a large number of branches from either right or left coronary arteries (Netter, 1969).

d. Intramuscular arteries: The left main conduit artery courses immediately subepicardially, with secondary branches passing at right angles to penetrate deeply into the underlying muscle. In contrast, secondary vessels from the right coronary artery split off obliquely to supply the relatively thinner musculature. Once the arteries penetrate the myocardium, they tend to lose their tortuosity and follow the orientation of the muscle layers (Gregg, 1950; James, 1960).

2. COLLATERAL CHANNELS

Traditional teaching, derived from studies of histology and pathology, maintained that the coronary arteries are end vessels without anastomoses. As recently as 1945, Wiggers reaffirmed this observation and stated that collateral channels ("if indeed, they exist at all") are of little use in protection against acute coronary occlusion. He asked, "How else can one explain the hemodynamic effects, incidence of fibrillation, and immediate diminution in contractile force in the affected area of heart muscle?" If extensive collateral circulation has not developed prior to total occlusion, muscle in the affected zone is not likely to survive. However, the presence of anastomoses, with diameters of 35–500 μm which interconnect arterial branches, now seems well established (Berne and Rubio, 1979), the terms "anastomoses" and "collaterals" being used interchangeably.

a. Characteristics of anastomoses: Anastomoses behave more or less as passive blood conductors, dictated by prevailing pressure gradients, and they respond with enlargement during tissue hypoxia. They arise directly from the arterial trunks, proximal to the resistance vessels, and connect with neighboring arterial vessels, again proximal to the distal resistance bed. Initially they appear as thin, endothelial tubes, but with time, they become enlarged during a state of ischemia and develop into muscular vessels. Mautz and Gregg (1937) reported that they are overstretched arterioles lacking a muscular coat, but Schaper *et al.* (1969) felt this description was inadequate and detailed the progressive transition from thin-walled to apparently normal muscular vessels.

Postmortem injection of the coronary arteries with suitable injectate has revealed the epicardial interconnecting network in dogs and the predominantly endocardial network in normal human and pig hearts. The epicardial connections of the dog heart are usually larger and more numerous than those in the pig heart (Schaper, 1971). Some uncertainty still exists with regard to the interconnecting network in the nonischemic human heart. James (1961) stated that collaterals of the human heart may be even larger than those in dogs, whereas Schaper *et al.* (1971) found that in hearts of patients dying from causes unrelated to ischemic heart disease, collaterals usually were not visible on

arteriograms. However, slowly developing experimental coronary arterial occlusion (using Ameroid constrictors) in animals results in arteriographically
demonstrable collateral vessels. These are usually quite large and have the
same localization as the interconnecting networks in normal hearts, i.e., in the
epicardial system. Schaper (1971) believes that these collaterals probably arise
from enlargement of preexisting interconnections, but newly formed vessels
may also contribute to the revascularization of an ischemic bed.

b. *Collateral vessels in ischemic human heart:* A predominantly subendocardial network of interconnecting collaterals expands in the human heart
under the influence of ischemic disease and becomes arteriographically demonstrable (Schaper, 1971). However, this is not the only source of collateral blood
supply to the ischemic myocardium. Schaper also includes the following: (1)
collaterals from preexistent subendocardial network (most frequent and extensive); (2) endomural connections between the left anterior descendens and
posterior descendens arteries, over the interventricular septum; (3) occasional
large epicardial collaterals, particularly in the apical region; (4) bridge-type
collaterals, bypassing the coronary artery occlusion and possibly stemming
from an enlarged vasa vasorum; and (5) extracoronary anastomoses, connecting
the bronchial arterial system with the coronary arteries. Thus, it is clear that
human hearts are capable of developing collaterals without principal re-
strictions in location. The vessels are found throughout the full thickness of the
ventricular walls, with lowest density in the epicardium (Fulton, 1965). Measurement of collateral blood flow in ischemic myocardium supports the view
that an extensive collateral network is found throughout the myocardial wall
(Downey *et al.*, 1974). Intramural arteries, about 200 μm in diameter, empty
into a subendocardial network of intercommunicating arterial anastomoses
(Berne and Rubio, 1979).

c. *Role of exercise in production of collaterals:* Considerable controversy is
found in the literature concerning the role of exercise in developing collateral
circulation. The pioneering experiments of Eckstein (1957) suggested better
collateralization in exercised dogs with coronary occlusion than in nonrunning
dogs. However, Schaper (1971) was unable to demonstrate improved collateral
circulation in dog hearts with exercise. Scheel *et al.* (1981) employed a carefully
controlled isolated beagle heart preparation to show that, while exercise did not
stimulate collateral growth in normal animals, after coronary occlusion, it doubled collateral conductance to the impaired vascular bed. Coronary reserve in
the trained animal returned to 52% of initial (preocclusion) capacity, versus
34% in the untrained animal. The development of collateral vessels was in the
region of ischemia, collateral flow being directed primarily toward the left heart
following circumflex occlusion and toward the right heart after right coronary
artery occlusion. Although dominant collateralization was via epicardial vessels,

intramyocardial septal collaterals also strongly participated in growth development (Scheel and Ingram, 1981).

3. ATRIAL CIRCULATION

Because the atrial walls are relatively thin, the circulatory system runs parallel to the surface. There is some deeper branching into such structures as the crista terminalis, and there are special arterial vessels to the sinoatrial and atrioventricular nodal regions. The sinus node artery provides branches to Bachman's bundle, the crista terminalis, and both right and left atrial free walls (James, 1961). A contralateral left anterior atrial artery anastomoses with the sinus node artery (James, 1977). The latter vessel is especially important, since it supplies both the normal and subsidiary pacemakers of the heart, thus exerting a critically important influence upon the initiation of each heart beat and associated electrical events (Hardie *et al.*, 1981; Randall *et al.*, 1981). In the dog there are two nutrient arteries to the A-V junctional region: one, the A-V nodal artery, branching from the circumflex, and the other arising from the septal artery. In man, in 90% of specimens, the A-V nodal artery branches from the right coronary artery just as it crosses the crux of the heart; in the remaining 10%, it is a branch of the circumflex at the same location (James, 1961).

4. NERVOUS INNERVATION OF CORONARY
ARTERIES AND ARTERIOLES

In most species, the small coronary arteries and arterioles possess a distinct media of densely packed smooth muscle cells of uneven thickness and contour (Berne and Rubio, 1979). Autonomic nonmyelinated nerve fibers and varicosed nerve endings are present only in the adventitial layer, just outside the media. At least in the case of the arterioles, most of the smooth muscle cells are innervated, and the fractional number of innervated fibers decrease as the number of smooth muscle cell layers increase with vessel diameter. A major question concerns the relative amount of innervation of small, as compared with large, vessels (Zuberbuhler and Bohr, 1965). It is not known how the inner muscle layers in the vessels are activated. Thus, important questions remain as to whether sympathetic nerve stimulation activates all of the muscle cells and how major vascular muscle spasms, observed in large conduit arteries, occur. Definitive information on how arteriolar smooth muscle cells contract and on how they are activated by neural or humoral signals also is lacking. The local release of transmitter agent, its excitation of specific receptors, and intercommunications with adjacent cells are almost totally beyond present understanding. Still, it is clear that adrenergic α- and β-receptor mechanisms, as well as cholinergic receptor mechanisms, are functionally involved in both normal and abnormal coronary artery behavior.

5. CAPILLARIES

As the arterioles become narrow, the muscularis becomes discontinuous and muscle cells decrease in number, to form metarterioles. The latter are continuous with simple endothelial tubes, the capillaries. It is generally thought that not all capillaries are open at all times (Provenza and Scherlis, 1959). The metarterioles and precapillary sphincters can totally close off the capillary lumen, thus suggesting a varying state of patency and function. Anastomotic connections among arterioles, metarterioles, precapillaries, and venules suggest arteriovenous shunting as an integral part of the myocardial capillary circulation (Gregg and Fisher, 1963). The ratio of capillaries to cardiac muscle fibers approaches 1 : 1, thus providing an enormous diffusing area for distribution and collection of gases, nutrients, and metabolites. A maximum diffusing distance of 8 μm has been computed. Electron microscopic studies have revealed a continuous capillary lining, without evidence of intercellular or intracellular pores (Palade, 1961). Pericytes and pericapillary cells with long branching processes that surround the capillary wall are reported around a majority of the cardiac capillaries. These cells contain microfilaments that may provide for modulation of the capillary lumen diameter (Forbes *et al.*, 1977).

6. VEINS

a. Anatomic features: There are twice as many veins as arterial channels in the heart, their density being much greater in the left than in the right ventricle; they may be subdivided into superficial and deep systems (Truex and Schwartz, 1951; Truex and Angulo, 1952). The superficial left ventricular veins parallel the arterial branches and pass toward the base of the heart, to empty into the great cardiac vein anteriorly and into its continuation in the left auricular groove, the coronary sinus, posteriorly. The latter vessel empties into the right atrium in the posterior–inferior interatrial septum, located between the medial end of the inferior vena cava and the A-V ring. It receives subsidiary trunks up to its orifice (Hellerstein and Orbison, 1951). The anterior cardiac veins, which consist of small trunks emptying into the right atrium just above the A-V valves, drain the right ventricle (Gregg, 1950).

The deeper venous circuit communicates with both atrial and ventricular cavities via Thebesian and sinusoidal channels (Gregg, 1950). The venous sinusoids are lined by a single layer of endothelium and range in width from 40 to

75 μm in dog hearts and 60 to 90 μm in human hearts. In the dog and pig, there is a massive formation of sinuses in the left ventricular wall which directly communicates with the left ventricular cavity (Christensen and Campeti, 1959). Venous collateral channels freely communicate over the surface of the heart (Gregg, 1950). These include connections between the anterior cardiac veins of the right ventricle and the left ventricular coronary sinus system.

b. *Role of coronary venous pressure:* Little has been published on the role of venous pressure as a determinant of coronary blood flow. Yet the venous portion of the coronary vascular system represents a series of collapsible tubes in which the back pressure opposing flow is the surrounding myocardial tissue pressure, not pressure at the outflow end of the system (Bellamy *et al.*, 1980). It has been suggested that the behavior of the veins may be comparable to that of a natural waterfall, in which hydraulic conditions downstream from the brink of the fall are not determinants of flow over the fall. Bellamy *et al.* (1980) studied the effect of transient occlusion of the coronary sinus on the relationship between aortic pressure and circumflex blood flow during prolonged diastole. The coronary pressure–flow relation was linear, and flow stopped at arterial pressures (zero-flow pressure) that always exceeded coronary venous pressure. Coronary flow was noted to be exquisitely sensitive to changes in venous pressure, and no circumstance was found in which it was not involved in determination of coronary flow. The dynamic participation of the coronary venous system in the ultimate determination of overall coronary vascular resistance has been neglected. It is conceivable that either neural or humoral mechanisms of control may influence this segment of the vascular system (Armour and Klassen, 1981).

C. PHYSIOLOGY OF BLOOD CIRCULATION

By S. F. VATNER

1. INTRODUCTION

In order to understand the extent to which the coronary circulation is regulated, either by autonomic factors or by pharmacologic agents, it is important to account for the mechanical and metabolic factors that are involved in the control of the circulation as well. For example, a major determinant of coronary blood flow is the coronary driving pressure. While it is generally assumed that this equates with aortic pressure, recently there has been considerable dispute regarding the exact driving pressure in the coronary vascular bed. The controversy stems from work by Bellamy (1978) and Ellis and Klocke (1980), showing that when blood flow falls to zero in the coronary circulation, considerable pressure remains, i.e., under such conditions, coronary arterial pressure does not equate with right atrial pressure, as had been assumed previously. Thus, the extent to which an indication of the pressure drop across the coronary bed can be gained from measurements of aortic and coronary arterial pressures remains controversial at this time. It is still generally agreed, however, that rises in aortic pressure will tend to increase coronary blood flow and falls in that variable will tend to decrease coronary blood flow. The coronary bed, like other regional beds, also autoregulates in response to changes in arterial pressure. It has been shown in cardiac preparations where metabolic factors are held constant that with an increase in arterial pressure, coronary blood flow tends to return toward control and vice versa when driving pressure is lowered. However, the coronary bed is unique in that with each heart beat, the arterioles, capillaries, and venules undergo severe compression. The compressive forces are important in determining the distribution of blood flow across the ventricular wall, as well as the vascular resistance in the entire bed. Finally, the heart is exquisitely dependent on oxygen supply, due to the fact that almost all of the oxygen delivered to the coronary bed is normally extracted as the blood passes to the venous circulation. Thus, changes in oxygen delivery due to changes in arterial oxygen content or hematocrit affect the coronary circulation. Moreover, changes in the myocardial metabolic demands, and consequently in the oxygen requirements of the myocardial tissue, exert a major effect in regulating coronary blood flow and vascular resistance. The most important determinants of myocardial metabolic demands are changes in the frequency of contraction, ventricular wall tension, and myocardial contractility. Once all of these factors enumerated above are either held constant or accounted for, only then can the effects of an autonomic intervention or pharmacologic agent under study be evaluated.

Thus, regulatory control of the coronary circulation is an integrated, multifactorial process. Recent evidence supports the concept that it is not only an intrinsic phenomenon, consisting of alterations in extravascular compressive forces and the release of vasoactive metabolites from myocardial cells, but it also involves the autonomic nervous system. An understanding of the extent to which neural control of the coronary vasculature can modulate the critical balance between nutrient supply and metabolic demand of the heart is not only of basic physiologic interest, but it may have important clinical implications. Currently, interest in autonomic regulation of the coronary circulation has intensified as a result of the present emphasis on coronary vasospasm and its role in the etiology of angina pectoris and myocardial ischemia.

The goal of this section is to summarize current understanding of the role of the autonomic nervous system in mediating changes in coronary vascular tone, with particular emphasis on studies conducted in intact, conscious animals and man.

2. Parasympathetic Activation

Because of the multiple influences of vagal nerve stimulation on various determinants of coronary blood flow (e.g., coronary perfusion pressure, extra-vascular compression, and myocardial metabolism), it has been only recently that the direct effects of parasympathetic activation on the coronary vasculature have been established by the work of Feigl (1969). In his study, an experimental design was utilized to account for the known determinants of coronary blood flow, with the result that changes in flow associated with vagal stimulation could be directly ascribed to activation of parasympathetic innervation of the coronary vasculature. With heart rate held constant in anesthetized, open-chest dogs, pretreated with propranolol, vagal stimulation elicited a peak increase in coronary blood flow of 30% and a calculated decrease in late diastolic coronary resistance of 42%. This coronary vasodilator action of vagal stimulation could be blocked by atropine, indicating that direct parasympathetic cho-linergic vasodilatation resulted from vagal stimulation. Feigl (1975) also demonstrated parasympathetic coronary vasodilatation in response to reflex stimulation of receptors in the ventricular myocardium (Bezold–Jarisch reflex). In this regard, Hackett et al. (1972) reported parasympathetically mediated coronary vasodilatation following carotid chemoreceptor stimulation and carotid baro-receptor hypertension.

3. Sympathetic Activation

The effects of electrical stimulation of sympathetic nerves on the coronary circulation have been assessed by a number of investigators (Berne et al., 1965; Feigl, 1967). Electrical stimulation of the stellate ganglion results in marked positive chronotropic and inotropic responses, with concomitant increases in coronary blood flow. The response is not a direct effect of sympathetic nerve stimulation on the coronary vasculature, since neither sympathetic cholinergic nor β-adrenergic coronary vasodilatation has been documented convincingly in response to nerve stimulation. Most likely it is a secondary effect, resulting from metabolic processes produced by the associated increases in heart rate and myocardial contractility. When such metabolically induced elevations in coronary blood flow are blocked by pretreatment with a β-adrenergic receptor antagonist, a direct coronary vasoconstrictor effect of sympathetic nerve stimulation is unmasked, which, in turn, can be abolished by α-adrenergic receptor blockade (Feigl, 1967).

Gerova *et al.* (1979) examined the diameter changes of the large coronary arteries in dogs with arrested hearts and with the coronary vasculature perfused via an extracorporeal circulation. These investigators found that sympathetic nerve stimulation in the absence of autonomic blockade increased coronary artery pressure and diameter. However, if the distending pressure was returned to baseline, coronary dimensions fell below control, indicating a direct vasoconstrictor effect. The sympathetically mediated vasoconstriction was eliminated by pretreatment with α-adrenergic receptor blockade.

4. α-ADRENERGICALLY MEDIATED CORONARY VASODILATATION

Activation of the carotid chemoreceptor and pulmonary inflation reflexes provides the most striking example of α-adrenergically mediated coronary vasodilatation (Vatner and McRitchie, 1975). For example, in conscious, spontaneously breathing dogs, carotid chemoreceptor reflex stimulation, accomplished by injections of minute quantities of either nicotine or cyanide into the common carotid artery just proximal to the carotid body, elicited a marked rise in ventilation, which was rapidly followed by a striking (2- to 3-fold) increase in coronary blood flow (Fig. 10.1). The latter response occurred in the face of little change in arterial pressure or alteration in heart rate; moreover, it took place prior to any significant changes in ventricular wall tension and myocardial contractility. Furthermore, the increase in coronary blood flow was accompanied by a rise in coronary sinus O_2 content and a fall in A-V O_2 difference, thus indicating that the dilatation was clearly not mediated by metabolic mechanisms.

To confirm that the afferent pathway was neural in the above studies, experiments were conducted after section of the ipsilateral carotid sinus nerve. Under these conditions absolutely no effect was observed from intracarotid injection of either nicotine or cyanide. To obtain further support for the view that the efferent pathway was neurally mediated, experiments were conducted in the presence and absence of selective and combined autonomic blockade. Since the peak coronary vasodilatation following carotid chemoreceptor reflex stimulation preceded inotropic changes, it was anticipated that β-adrenergic blockade would not modify the coronary vasodilatation, and this type of response did occur. Atropine increased cardiac rate and consequently partially dilated the coronary bed on a metabolic basis. However, carotid chemoreceptor reflex stimulation in the presence of cholinergic blockade still induced significant vasodilatation, attaining peak levels similar to those obtained in the absence of cholinergic blockade. The addition of phentolamine almost abolished the coronary vasodilatation resulting from carotid chemoreceptor reflex stimulation. Thus, it can be concluded that carotid chemoreceptor reflex stim-

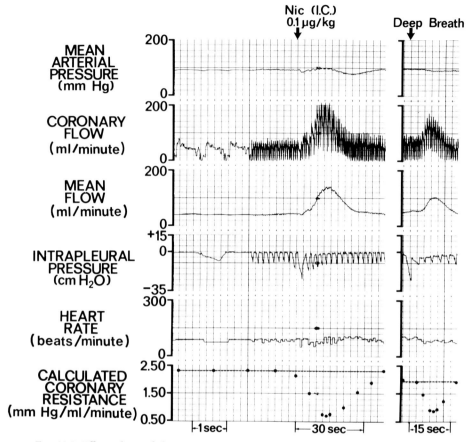

FIG. 10.1. Effects of carotid chemoreceptor stimulation, induced by intracarotid (ic) administration of nicotine (Nic) in a conscious dog with spontaneous rhythm (left section), on mean arterial blood pressure, phasic and mean coronary blood flows, intrapleural pressure, heart rate, and calculated coronary resistance. Carotid chemoreceptor stimulation elicits a striking increase in coronary flow and a reduction in coronary resistance that occurred immediately following an increase in respiratory depth (indicated by the change in intrapleural pressure). The effects of a spontaneous deep breath in the same conscious dog are shown on the right-hand side. Note that a spontaneous deep breath elicits a period of coronary dilatation similar to that which occurs with administration of nicotine into the carotid artery in the conscious state. (From Vatner and McRitchie, 1975; reproduced with permission from *Circulation Research*.)

ulation in the conscious dog elicits striking coronary vasodilatation. The efferent mechanism in the response appears to involve withdrawal of α-adrenergic tone, since the vascular change is blocked by phentolamine.

It is important to keep in mind that the above experiments were carried out in conscious dogs respiring spontaneously and that the carotid chemoreceptor reflex stimulation-induced coronary vasodilatation was always preceded by an increase in depth of respiration. However, it is conceivable and quite likely that the afferent mechanism responsible for the vascular response was the pulmonary inflation reflex and not the carotid chemoreceptor reflex. A similar type of vasodilator response can be evoked by increasing the depth of ventilation either spontaneously (deep breath) (Fig. 10.1) or mechanically (increasing the volume of a respirator) (Vatner and McRitchie, 1975). Such alterations are eliminated by α-adrenergic receptor blockade or attenuated markedly by vagotomy (Vatner and McRitchie, 1975). In contrast, carotid chemoreceptor reflex-induced coronary vasodilatation can be blocked by atropine (Hackett *et al.*, 1972; Vatner and McRitchie, 1975).

5. α-ADRENERGICALLY MEDIATED CORONARY VASOCONSTRICTION

Most evidence for α-adrenergically mediated vasoconstriction is derived from experiments in which calculated coronary driving pressure rises and coronary blood flow remains constant, thereby indicating an increase in coronary vascular resistance. Such increases in calculated coronary vascular resistance can be eliminated by α-adrenergic receptor blockade (Vatner *et al.*, 1974; Mudge *et al.*, 1976; Kelley and Feigl, 1978; Mohrman and Feigl, 1978; Murray and Vatner, 1981a).

Recently reported is an example of coronary vasoconstriction characterized by an increase in arterial pressure and absolute reduction in coronary blood flow (Murray and Vatner, 1981b). This type of response occurs following carotid chemoreceptor reflex stimulation, but only after subsidence of the initial period of coronary vasodilatation (Fig. 10.2). The late period of vasoconstriction in the intact, conscious dog is much more prominently displayed in the right coronary circulation than in the left. This is probably due to the fact that carotid chemoreceptor reflex stimulation is accompanied by a moderate increase in afterload for the left, but not for the right, ventricle, and in the dog, the right coronary artery supplies only the right ventricle. Thus, the latter vessel would not be expected to demonstrate an increase in blood flow due to enhanced myocardial metabolic demands faced by the left ventricle. The period of coronary vasoconstriction is associated with a decrease in coronary sinus O_2 content and a widening of A-V O_2 difference across the heart. The later period of coronary vasoconstriction is not affected by either β-adrenergic or cholinergic blockade, but can be virtually eliminated by α-adrenergic blockade.

The afferent mechanism of the late period of coronary vasoconstriction is most likely the carotid chemoreceptor reflex, since the response is even more intense in the presence of controlled ventilation (Murray and Vatner, 1981b), i.e., when the complicating influences of the pulmonary inflation reflexes are absent. Moreover, under these conditions, atropine eliminates the early vaso-

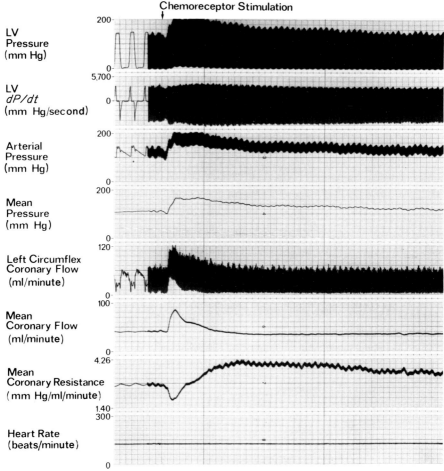

FIG. 10.2. Effects of intracarotid administration of nicotine in a conscious dog with controlled ventilation on left ventricular (LV) pressure, LV dP/dt, phasic and mean arterial pressure and left circumflex coronary blood flow, calculated mean coronary resistance, and heart rate. Chemo-receptor stimulation elicits an initial striking increase in coronary flow and a reduction in coronary vascular resistance; this is followed by a period of coronary vasoconstriction, characterized by a reduction in coronary blood flow in the face of sustained elevation in arterial pressure. The period of vasoconstriction is also typified by a fall in coronary sinus O_2 content, which could be eliminated by prior α-adrenergic receptor blockade.

dilatation but leaves intact the later period of vasoconstriction. Therefore, the latter change is not necessarily linked to the early vasodilator response. These experiments demonstrate that, in the same animal, stimulation of reflex pathways is sufficiently powerful to reduce coronary blood flow and coronary sinus O_2 content, despite an elevation of arterial pressure.

6. α-Adrenergically Mediated Coronary Vasoconstriction During Exercise

Severe exercise is associated with intense sympathetic drive to the heart. It is conceivable that the latter also affects the coronary vessels during severe exercise. To test this hypothesis, responses in dogs to near maximal exercise were studied in the presence and absence of α-adrenergic receptor blockade (Murray and Vatner, 1979). In the presence of α-adrenergic receptor blockade, coronary vascular resistance fell to a significantly lower level than in its absence. To obviate the criticism that, in these experiments, heart rate and myocardial contractility were uncontrolled, exercise was studied with heart rate held constant and after pretreatment with β-adrenergic receptor blockade, but in the presence and absence of α-adrenergic blockade. Again, exercise of similar intensity elicited significantly greater coronary vasodilatation in the presence of α-adrenergic receptor blockade than in its absence (Fig. 10.3) (Murray and Vatner, 1979). It is important to note that subsequent studies by Gwirtz and Stone (1981) and Heyndrickx *et al.* (1982) also demonstrated a component of α-adrenergic coronary vasoconstriction during exercise, which was associated with a widening of A-V O_2 difference across the heart. All these

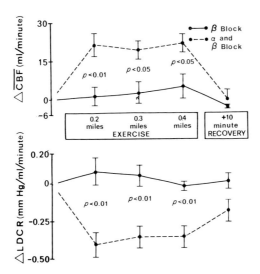

Fig. 10.3. Average ($\pm$ SEM) changes in mean left circumflex coronary blood flow (ΔCBF) (top panel) and late diastolic coronary resistance (ΔLDCR) (bottom panel) during spontaneous exercise at 0.2, 0.3, and 0.4 miles and 10 minutes after cessation of exercise, following pretreatment with propranolol, but with α-receptor activity intact (β block; solid lines) and following blockade of α-receptor activity (α and β block; dashed lines). Heart rate held constant at approximately 190 beats/minute by electrical pacing. (From Murray and Vatner, 1979; reproduced with permission from *Circulation Research*.)

investigations suggest that α-adrenergically mediated vasoconstriction attenuates the metabolic vasodilatation associated with exercise.

D. PHARMACOLOGY OF BLOOD CIRCULATION

By S. F. VATNER

1. INTRODUCTION

A large amount of information is available on the pharmacology of the coronary circulation. The majority of *in vivo* work has been conducted in experimental animals and man by measuring coronary blood flow and arterial pressure and calculating coronary vascular resistance. In addition, there are numerous *in vitro* studies dealing with the effects of drugs on isolated coronary vascular strips. Relatively few experiments have been performed in which the effects of pharmacologic agents have been examined on large coronary arteries *in vivo*, and still fewer yet in which intact or conscious preparations have been utilized. There is evidence to suggest that large coronary arteries react differently from the small resistance vessels in response to pharmacologic stimuli. Recently, techniques have become available to measure large coronary arterial dimensions instantaneously and continuously in the intact animal, as well as in man.

The goal of this section is to summarize the newer information on regulation of large coronary arteries by pharmacologic agents. Specifically, two aspects are dealt with: autonomic regulation of large coronary arteries and pharmacologic regulation, with particular emphasis on therapy of coronary vasospasm. The latter point is of particular interest, since it is being recognized with increasing frequency as a clinical entity that primarily involves large coronary arteries and that is amenable to pharmacologic therapy.

2. AUTONOMIC REGULATION

a. α-Adrenergic: Intravenous (Vatner *et al.*, 1974) or intracoronary (Mohrman and Feigl, 1978) infusions of norepinephrine have a demonstrable α-receptor-mediated vasoconstrictor effect on the coronary circulation, even with β-adrenergic activity intact and in the presence of secondary increases in myocardial oxygen consumption, which would ordinarily tend to oppose a direct vasoconstrictor action. For example, in healthy, conscious dogs, with heart rate held constant and with adrenergic activity intact, an intravenous bolus injection of norepinephrine elicits a transient decrease, followed by a sustained increase,

Blood Vessels and Lymphatics in Organ Systems
Copyright © 1984 by Academic Press, Inc.

in coronary vascular resistance during the steady state. The latter response cannot be explained entirely on the basis of coronary autoregulation, since when smaller doses of the drug are administered, coronary blood flow decreases, with arterial blood pressure remaining at or above control levels. Moreover, in the presence of incomplete α-adrenergic receptor blocking doses of phentolamine, norepinephrine still increases arterial pressure but only elicits vasodilatation (Vatner *et al.*, 1974). In this regard, Mohrman and Feigl (1978) demonstrated convincingly that α-adrenergic constrictor mechanisms can compete with metabolic vasodilatation during sympathetic activation associated with intracoronary infusion of norepinephrine. The net effect of the α-adrenergic constriction during norepinephrine infusion is to restrict the metabolically related increases in coronary blood flow by approximately 30% (i.e., the constriction reduces the extent of oxygen delivery associated with the norepinephrine infusion by approximately 30% for a given increment in oxygen consumption). This has the effect of raising oxygen extraction and lowering coronary venous oxygen content. Such norepinephrine-induced limitations in myocardial oxygen delivery are abolished by intracoronary administration of α-adrenergic receptor antagonists.

In the above studies, changes in coronary vascular cross-sectional area were assessed indirectly by a calculation of coronary resistance from measurements of coronary blood flow and arterial pressure. Recently, however, the effects of methoxamine, a specific α-adrenergic agonist, have been examined on large vessel coronary dimensions in conscious dogs, using an ultrasonic dimension gauge for the direct and continuous measurement of large coronary artery diameter (Vatner *et al.*, 1980). When given intravenously, this drug induces a striking, sustained decrease in coronary diameter, despite the opposing influence of increased arterial distending pressure. Such findings are consistent with the results of Kelley and Feigl (1978) in anesthetized dogs, showing norepinephrine induced vasoconstriction of conduit, as well as of, resistance coronary vessels. Thus, α-adrenergic stimulation of large coronary arteries appears to be sufficiently powerful to produce active changes in coronary vascular smooth muscle tone, which reduces cross-sectional area despite the opposing elevation in distending pressure (Fig. 10.4).

b. β-Adrenergic: Isoproterenol, a potent agonist of both β_1 and β_2-adrenergic receptors, increases myocardial O_2 consumption and consequently dilates resistance coronary vessels on a metabolic basis. In addition, it dilates coronary vessels by its action on vascular β-adrenergic receptors (Zuberbuhler and Bohr, 1965; Klocke *et al.*, 1965; Berne and Rubio, 1979; Parratt, 1980; Vatner *et al.*, 1982), independent of changes in myocardial metabolic demands. The results from most studies performed on either anesthetized or conscious animals indicate that isoproterenol stimulation of vascular receptors in coronary resistance vessels is primarily β_2-mediated, since it still occurs after selective β_1-blockade (Adam *et al.*, 1970; Ross and Jorgensen, 1970; McRaven *et al.*,

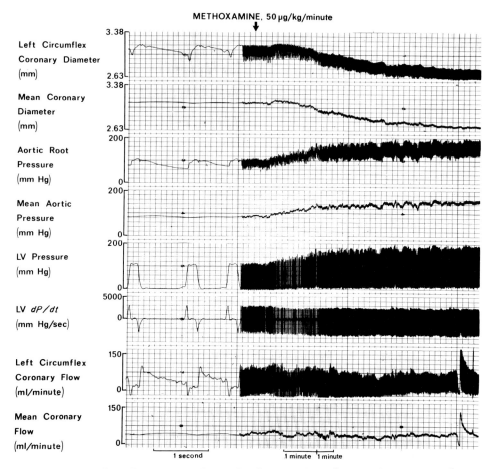

FIG. 10.4. Effects of a 10-minute infusion of methoxamine (50 μg/kg/minute) in a conscious dog are depicted on simultaneous and continuous measurements of phasic and mean left circumflex coronary artery diameter, phasic and mean aortic root pressure, left ventricular (LV) pressure, LV *dP/dt*, and phasic and mean left circumflex coronary blood flow. Methoxamine increases coronary diameters only initially and transiently and then induces striking sustained reductions in coronary diameters at a markedly elevated aortic pressure. (From Vatner *et al.*, 1980; reproduced with permission from the *Journal of Clinical Investigation.*)

1971; Mark *et al.*, 1972). In contrast, most studies conducted on *in vitro* coronary arteries suggest that isoproterenol stimulates β_1-adrenergic vascular receptors (Baron *et al.*, 1972; Drew and Levy, 1972; Johannson, 1973; De La Lande *et al.*, 1974). Since the latter investigations examined primarily large coronary arteries, whereas the studies in open-chest anesthetized animals assessed primarily effects on smaller, coronary resistance vessels, it is conceivable that both points of view are correct.

The specific question as to whether stimulation of β_1- or of β_2-adrenergic receptors results in dilatation of large coronary arteries was recently addressed (Vatner *et al.*, 1982). In the intact, conscious dog, it was found that iso-proterenol induced a substantial dilatation of large coronary arteries (Fig. 10.5). To determine whether the mechanism responsible was mediated by β_1 or β_2 receptors, the effects of isoproterenol were examined both in the presence and in the absence of selective β_1-adrenergic receptor blockade with atenolol. After the use of such a drug, isoproterenol induced similar reductions in arterial pressure (indicating that peripheral vascular β_2-adrenergic receptors were intact), but it failed to increase LV dP/dt significantly (indicating that myocardial β_1-adrenergic receptors were essentially blocked). Under these conditions, isoproterenol still dilated both small and large coronary arteries, but to a significantly lesser degree (Fig. 10.5). The conclusion reached was that the residual dilatation by isoproterenol after β_1 adrenergic receptor blockade was mediated by β_2-adrenergic receptors. These studies indicate that β_1, as well as β_2, mechanisms can dilate large and small coronary arteries. Furthermore, pirbuterol, a

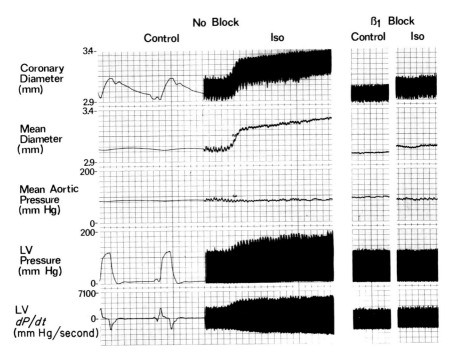

FIG. 10.5. Effects of isoproterenol (Iso) before (left panel) and after (right panel) selective β_1-adrenergic blockade with atenolol on recordings of phasic and mean coronary diameter, mean aortic pressure, left ventricular (LV) pressure, and LV dP/dt. Isoproterenol before β_1 blockade decreases mean aortic pressure and increases LV dP/dt and coronary diameter. After β_1 blockade, it still reduces mean aortic pressure, but it fails to increase LV dP/dt and increases coronary diameter to a lesser extent. (From Vatner and Macho, 1981; reproduced with permission from *Basic Research in Cardiology*.)

selective β_2-adrenergic agonist (Moore *et al.*, 1978), and prenalterol, a selective β_1-adrenergic agonist (Manders *et al.*, 1980), both dilate large, as well as resistance, coronary arteries (Vatner *et al.*, 1982).

Thus, isoproterenol induces dilatation of large coronary arteries, not only by stimulation of vascular β_2-adrenergic receptors, but possibly also by direct stimulation of vascular β_1-adrenergic receptors. However, the above results also are consistent with the hypothesis that regulation of large, as well as small, coronary arteries occurs secondarily to changes in myocardial metabolism (Macho *et al.*, 1981). To discern between primary vascular responses and effects secondary to changes in myocardial metabolism, coronary sinus O_2 content and A-V O_2 content difference were measured by Vatner *et al.* (1982). As observed previously by Klocke *et al.* (1965) in anesthetized open-chest dogs, studies in conscious animals indicated that isoproterenol, either in the presence or absence of selective β_1-adrenergic receptor blockade, induced an increase in coronary blood flow, an elevation of coronary sinus O_2 content, and a reduction in A-V O_2 difference. Similar responses were observed with pirbuterol, whereas prenalterol increased A-V O_2 difference only slightly. It is therefore likely that additional β_2-adrenergic receptor stimulation, induced by isoproterenol or pirbuterol, activated β_2-vascular receptors and caused dilatation of small and large coronary vessels independently of changes in myocardial O_2 demand. By contrast, the component of large vessel coronary dilatation due to direct β_1-adrenergic receptor stimulation or secondary to β_1-adrenergic receptor stimulation of myocardial resistance vessels was not of sufficient magnitude to elicit augmented coronary blood flow, increased coronary sinus O_2 content, and narrowing of the A-V O_2 difference.

The cellular mechanisms by which large coronary vessels are regulated in response to interventions that alter myocardial metabolic demands is not clear. One possibility is that adjacent myocardial cells liberate metabolites (e.g., adenosine) with increases in myocardial metabolic requirements and that these substances then dilate coronary vessels. However, there are no data to support such a conjecture. Another possibility is that large coronary arteries dilate in response to increases in coronary blood flow. In support of the concept is recent work demonstrating dilation of large coronary arteries following brief periods of myocardial ischemia and reactive hyperemia. When the reactive hyperemia is prevented by constricting the coronary artery, no dilation of the large coronary artery is observed. Regardless of the mechanism, it is important to recognize that the caliber of large coronary arteries varies in response to interventions that alter myocardial metabolic demands. For example, simply increasing either heart rate or ventricular afterload leads to substantial dilatation of the large coronary arteries (Macho *et al.*, 1981). In this regard, vasodilator responses of resistance coronary vessels to increases in myocardial metabolic demand have been recognized for some time (Berne and Rubio, 1979). It has only recently been recognized, however, that large coronary arteries also respond to interventions that increase myocardial metabolic demand.

3. Pharmacologic Therapy for Coronary Vasospasm

The classic agent utilized as therapy in angina pectoris has been nitroglycerin. This drug is exceedingly effective regardless of whether the symptom complex is due to myocardial ischemia on the basis of imbalance between O_2 demand and supply or to inadequate blood flow resulting from vasospasm of the coronary arteries. The studies by Winbury et al. (1969) demonstrated an important feature of nitroglycerin, i.e., its action on large coronary arteries is much more prolonged than its effect on resistance coronary vessels. It is this action of the drug that may be particularly important in the relief of coronary vasospasm. Recently such a possibility has been studied in intact, conscious animals (Vatner et al., 1980; Macho and Vatner, 1981; Hintze and Vatner, 1982) using a single dose of nitroglycerin. It was found that there was a peak dilatation of large coronary arteries within 2–5 minutes after administration, with maintenance of the vasodilator effect for the subsequent hour. In contrast, the dilatation of resistance coronary vessels subsided within 1 minute and, in fact, was rapidly replaced by a period of coronary vasoconstriction. The resulting reduction in total coronary blood flow was most likely secondary to lowered metabolic requirements due to decreases in preload and afterload, other important features of the action of nitroglycerin. The prolonged dilatation of large coronary arteries, however, is most important and is probably the basis for the beneficial mechanism of action of the drug in the presence of coronary vasospasm.

Newer agents that are particularly useful in the treatment of coronary vasospasm are the calcium channel antagonists (Fleckenstein, 1977; Henry, 1980). These drugs, characterized by such prototypes as verapamil, diltiazem, and nifedipine, are potent relaxants of vascular smooth muscle (Fleckenstein, 1977; Henry, 1980). A number of investigators have demonstrated that these agents relieve spasm or dilate large coronary arteries in patients with coronary artery disease studied in the cardiac catherization laboratory. More systematic investigations in intact, conscious animals have shown that a single dose of nifedipine can induce marked dilatation of large coronary arteries for up to 1 hour (Hintze and Vatner, 1983; Vatner and Hintze, 1982). Although the response seems similar to that elicited by nitroglycerin, it is important to recognize that the calcium channel antagonists are not as specific for large coronary arteries as is nitroglycerin. While both induce prolonged relaxation of the large coronary arteries, the calcium channel antagonists also cause substantial and sustained dilatation of resistance vessels, whereas nitroglycerin induces only transient dilatation of the arterioles. Due to their action on coronary resistance vessels, the calcium channel antagonists profoundly increase coronary sinus oxygen content when administered to either people or animals with a normal coronary circulation. Even when given to patients with myocardial ischemia, they still elicit coronary vasodilatation in the ischemic myocardium (Lichtlen, 1975). In contrast, nitroglycerin, while having only a transient dilator effect on peripheral resistance vessels, does induce prolonged relaxation of veins, which, in turn, reduces preload, a reaction not found with currently utilized calcium channel antagonists. However, it is the strong vasodilating effect on large arteries that both types of agents possess that can be crucial in alleviating coronary vasospasm, a phenomenon frequently noted in such vessels.

E. PATHOPHYSIOLOGY, PATHOGENESIS, AND PATHOLOGY OF BLOOD CIRCULATION

By G. ROSS AND M. C. FISHBEIN

Although many disorders may involve the coronary arteries, the etiologic agent in over 99% of patients with clinically significant coronary artery disease is atherosclerosis. This section deals with those characteristics of the atherosclerotic plaque that are particularly relevant to the pathogenesis of ischemic syndromes of the heart.

1. PATHOPHYSIOLOGY AND PATHOGENESIS OF BLOOD CIRCULATION

a. Hemodynamic effects of coronary artery stenosis: The effects of atherosclerosis on the pressure–flow relationship of human coronary arteries has been studied *in vitro* by Logan (1975), who reported that in normal coronary arteries, the resistance was low, only 0.005 mm Hg/ml/minute for a 6 cm length of 3.2 mm diameter artery perfused at 100 mm Hg. This is only 0.5% of the total vascular resistance of the bed normally perfused by such an artery. Stenoses have a relatively small influence on resistance until the minimal cross-sectional area (A_{min}) of the partially obstructed segment is less than 80% of the area of a nonstenosed vessel (A_o); this is graphically shown in Fig. 10.6, in which stenosis resistance (pressure drop across the stenosis/flow) is plotted against percent stenosis $(A_o - A_{min})100/A_o$. In a 3 mm diameter artery with an 80% reduction in cross-sectional area, a further diameter reduction of less than 0.1 mm markedly increases flow resistance. The sudden rise in resistance beyond a critical degree of stenosis is due mainly to the finite entrance length of the stenotic lesion and to the development of local turbulence. The down-

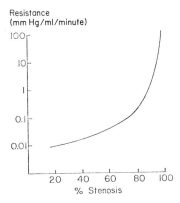

Resistance
(mm Hg/ml/minute)

FIG. 10.6. Effect of stenosis on coronary arterial resistance. Note sharp increase in resistance when stenosis exceeds 80%.

stream resistance (Rd) importantly influences the effect of a stenosis. When it is high, the reduction of cross-sectional area by 67–86% has little effect on flow, but when it is low (equivalent to maximal arteriolar dilatation), a 67% stenosis reduces flow by approximately 30% and an 86% stenosis, by 70% (Logan, 1975). These observations provide an explanation for angina of effort since stenoses which have no effect when myocardial oxygen demands are low (high downstream resistance) become flow limiting when they are high (low downstream resistance).

  b. *Effects of stenosis on myocardial blood flow distribution:* Studies with radioactive microspheres have shown that severe coronary artery stenoses limit flow to the subendocardium more than to the subepicardium. Under resting normal conditions, subendocardial flow in dogs is approximately 20% greater than subepicardial flow, even though the subendocardium receives almost no blood flow during systole. This suggests that the subendocardial arterioles are more dilated and have a lower "vasodilator reserve" than subepicardial vessels. Thus, maximal arteriolar dilatation will be achieved earlier, and flow will become pressure dependent sooner in the subendocardium than in the subepicardium. The pressure driving flow through the myocardium is opposed by the tissue pressure, which normally is low during diastole and negligible when compared to coronary artery pressure. However, when severe stenosis is present, the arterial pressure beyond the stenosis may be only a few mm Hg above ventricular diastolic pressure so that now tissue pressure assumes much greater importance as a factor opposing flow, especially in the subendocardium where it is highest. Thus, when coronary artery stenosis is present and the subendocardial arterioles are maximally dilated, subendocardial flow will be lower than subepicardial flow. This, in part, explains the increased vulnerability of the subendocardium to ischemic injury.

c. Coronary artery vasomotion: Angiographic studies of normal human arteries have shown spontaneous decreases in diameter of coronary arteries of up to 20% (Gensini *et al.*, 1971); nitroglycerin (Feldman *et al.*, 1978) may increase and ergonovine (Heupler *et al.*, 1978) may decrease diameter by 30%. There appears to be a diurnal variation in coronary arterial tone, with minimal arterial diameter occurring in the early morning (Yasue *et al.*, 1979). In some patients with variant angina, much greater changes in diameter, at times sufficient to occlude the lumen, may occur. The term "coronary spasm" is appropriate in the case of such individuals.

The factors that influence normal human coronary arterial tone are largely unknown. Effects of several exogenous stimuli have been studied, but their relevance to spontaneous fluctuation of vasomotor tone remains uncertain. Brown (1981) has reported the responses to epinephrine after β-adrenergic blockade in patients with variant angina and in those with ordinary exertional angina. The "normal" vessels of both patient groups demonstrated similar changes, consisting of a 13–16% reduction in lumen area despite a substantial rise of arterial pressure. However, the stenotic lesions of patients with variant angina showed a significantly greater reduction in area (70%) than did the comparably severe stenotic lesions of patients with exertional angina. Sublingual nitroglycerin appeared to dilate the normal and stenotic segments equally in all groups. Brown (1981) also reported the results of isometric handgrip in two groups of patients, one with angina only at rest and the other with angina only on exertion, with none in the study having variant angina. This sympathetic stimulus constricted the large coronary arteries of both groups of patients. In those with rest angina, the stenotic segments constricted more strongly than did the normal segments (33% versus 14%). These observations indicate that some stenoses are not fixed lesions, as was once thought, but may show greater responses to stimuli than nonstenosed segments. The experiments also demonstrated that the larger coronary arteries of man have an α-adrenergic mechanism that can be activated by exogenous agonists and by reflexly induced sympathetic neural discharge. Reflex coronary arterial constriction has also been shown in response to immersion of the forearm in ice-cold water (Raizner *et al.*, 1978).

A number of nonadrenergic stimuli likewise produce coronary artery constriction. Studies (e.g., Heupler *et al.*, 1978) have shown that ergonovine constricts normal coronary arteries by up to 20% and that the coronary arteries of patients with variant angina are more sensitive to this drug and respond with greater constriction. Intravenous cholinomimetic agents, such as methacholine, also produce coronary artery constriction (Endo *et al.*, 1976), but it is possible that the constriction resulted from reflex sympathetic activation.

d. Effect of wall thickness: Coronary arteries of man differ from those of other species in that the intima progressively thickens with age (Neufeld and

Schneeweiss, 1981). In many individuals, the degree of thickening is also increased by the development of atheroma (see Section E-2, below). MacAlpin (1980) has pointed out that wall thickening may considerably magnify the effects of vascular smooth muscle contraction, with the result that even minor degrees of shortening, well within the range shown to occur angiographically in normal coronary arteries, could conceivably cause complete occlusion of the lumen. In the case of thickening sufficient to produce a "critical" stenosis of 80%, it can be calculated that a shortening of the circumferential muscle of only 6% would cause such a response, provided the stenosis was pliable. Unfortunately, little is known about the pliability of atheromatous lesions, the effects of "tethering" of coronary arteries by surrounding tissues, the orientation of smooth muscle fibers, and the characteristics of the collagenous and other connective tissue elements of the media. All are important factors that may influence the effect of contraction on the arterial lumen. Hence, they need to be explored further before the significance of wall thickening on luminal constriction can be fully assessed.

  e. *In vitro studies of human coronary arterial reactivity:* Ring segments of human coronary arteries obtained from patients undergoing cardiac transplantation show marked spontaneous rhythmic contractions (Golenhofen, 1978; Ross *et al.*, 1980). These responses are myogenic and occur after degeneration of the intramural nerves. They are highly dependent on the external calcium concentration and are abolished by "calcium channel blocking agents." Slight depolarization of the smooth muscle cells by low concentrations of potassium chloride induces rhythmic contractions in quiescent preparations or increases the frequency and magnitude of the contractions of spontaneously active arteries. Agonists, such as norepinephrine, ergonovine, and histamine, have similar effects.

  f. *Arachidonic acid metabolites and coronary arterial tone:* Human coronary arteries continuously synthesize prostaglandin I_2 (PGI_2), an arachidonic acid metabolite which is a potent vasodilator and inhibitor of platelet aggregation. Another arachidonic acid metabolite, thromboxane A_2, is produced by platelets and is released when platelets are activated by contact with collagen. It is a powerful vasoconstrictor and enhances platelet aggregation. The possibility has been raised that coronary artery tone depends on the relative concentrations of PGI_2 and thromboxane A_2 in the arterial wall (Moncada and Vane, 1979), but direct evidence for this hypothesis is lacking.

  g. *Pathophysiology of variant angina:* It has been convincingly shown that variant angina is caused by contraction of coronary arterial smooth muscle, which, in the majority of patients, begins in the vicinity of an atherosclerotic lesion. One explanation for such a response is that it is caused by normal vasomotion exaggerated by increased wall thickness. Another is that the coronary arteries in patients with atherosclerosis are hypersensitive to adrenergic stimuli. However, α-adrenergic blocking drugs often are unsuccessful in preventing attacks. Even cardiac denervation produced by autotransplantation may not control coronary spasm, although it does abolish pain (Clark *et al.*, 1977).

Still another explanation for contraction of coronary arterial smooth muscle is that it is mediated by the action of platelets. In support of this possibility is the finding that when a mechanical constriction device is placed on a coronary artery of the dog *in vivo*, spontaneous cyclic fluctuations of flow occur, which appear to be due to intermittent plugging of the narrowed area by platelet aggregates. Both plugging and the cyclic flow changes can be prevented by pretreatment with inhibitors of platelet aggregation, such as aspirin and indomethacin (Folts *et al.*, 1982). As mentioned above, platelet aggregates release thromboxane A_2, which could cause coronary artery constriction unless its effects were opposed by a vasodilator, such as PGI_2 produced in the coronary artery wall. Atheromatous lesions contain hydroperoxyarachidonic acid, a substance that inhibits PGI_2 formation, and this might explain the increased frequency with which variant angina is associated with coronary constriction beginning at or near an atheromatous plaque. Although the platelet hypothesis is plausible, there is no direct evidence to support it. Recent measurements of the thromboxane A_2 metabolite, thromboxane B_2, in patients with recurrent variant angina has failed to show a causative link. Interestingly, thromboxane B_2 plasma levels are elevated in these individuals, but the time course suggests that coronary artery constriction leads to thromboxane A_2 production, rather than being initiated by it (Robertson *et al.*, 1981).

Finally, it is necessary to point out that the spontaneous phasic contractions of human coronary arteries *in vitro* sometimes show a periodicity similar to that observed during variant anginal attacks. It is possible, therefore, that all human coronary arteries have a latent capacity to contract rhythmically, but that this tendency normally is suppressed, whereas in patients with variant angina, it is activated, either by excitatory stimuli or by removal of inhibitory factors. The abnormality may thus be considered to be a hyperresponsiveness of smooth muscle, not to any particular stimulus, but possibly to all constrictor agents. This would help explain the multiplicity of stimuli that can induce variant angina in susceptible patients, e.g., ergonovine, epinephrine, propranolol, methacholine, and alkalosis. Moreover, it would account for the failure of any mode of therapy to influence the condition unless the final common path—the rise in intracellular calcium—is prevented, as for example, by nitrates or calcium channel blocking agents.

Unfortunately, none of the above hypotheses explains why, in a given patient, different segments of the coronary arterial tree constrict at different times or why long periods of freedom from variant angina may occur.

Fig. 10.7. Histologic sections of serial slices of left anterior descending coronary artery from a patient who died from acute transmural myocardial infarction. Note that in the first section (left), there is an occlusive luminal thrombus, apparently separate and independent from the adjacent, circumferential atheromatous plaque. The next section (right), however, shows an area of plaque rupture (arrow), with cholesterol clefts and atheromatous material admixed with the luminal thrombus. This case illustrates the focal nature of changes in atheromatous plaques and the potential for missing lesions due to sampling problems. Trichrome stain. ×10.

2. PATHOLOGY OF BLOOD CIRCULATION

In clinically significant coronary artery disease, the responsible lesion is almost always the complicated atherosclerotic plaque: an altered intimal fibrous lesion in which there are varying amounts of intracellular and extracellular lipid, smooth muscle cell proliferation, hemorrhage, calcification, cell necrosis, thrombosis, and inflammation. The process usually extends into the media, causing atrophy and thinning and is accompanied by adventitial fibrosis and inflammation (Fig. 10.7). (For a general discussion of the histologic changes in atherosclerosis, see Section E-2, Chapter 1.)

a. Role of thrombosis: For many years, "coronary thrombosis" was the phrase used to denote an acute lesion superimposed on an atherosclerotic plaque responsible for acute myocardial infarction. This was because a thrombus often was found occluding the vessel providing flow to the region of the damaged myocardium. The relationship of thrombosis to acute myocardial infarction has been a subject of great controversy, with the reported frequency of coronary thrombosis associated with myocardial infarction varying from 54 to 97% (Chandler *et al.*, 1974; Ridolfi and Hutchins, 1977). With the advent of aggressive diagnostic and therapeutic approaches to patients with acute myocardial infarction, the role of thrombosis has been clarified. DeWood *et al.*

(1980) reported that thrombi were present in the coronary arteries of 88% of patients undergoing coronary artery bypass surgery during the acute phase of myocardial infarction. Ganz *et al.* (1981) attempted rapid intracoronary infusion of streptokinase in 20 patients with evolving myocardial infarction catheterized within 3 hours of the onset of chest pain and reported that vascular patency was reestablished in 19 of 20 patients, findings which strongly suggested that a thrombus was present at the site of occlusion. Others have noted similarly good results with thrombolytic therapy (Rentrop *et al.*, 1981; Reduto *et al.*, 1981; Markis *et al.*, 1981). Such consistent findings provide strong evidence that coronary thrombosis is present early in most patients with acute myocardial infarction. The preliminary evidence suggesting clinical improvement after thrombolysis (Reduto *et al.*, 1981) also supports the view that thrombosis has a primary rather than secondary role in causing the myocardial injury. If, as the above studies indicate, myocardial infarction is due to thrombosis at sites of atherosclerotic plaques, it would be of great importance to know what alterations occurring in these regions precipitate the acute thrombosis. (For further discussion of the role of thrombosis in myocardial infarction, see Section E-2b, Chapter 1.)

  b. Mechanism of thrombosis: Recent experimental studies indicate that endothelial integrity is necessary to prevent platelet aggregation and subsequent thrombosis. When endothelial injury occurs, several factors favor platelet aggregation and thrombus formation: (1) platelets are exposed to subendothelial collagen, a potent stimulus to ADP release and platelet aggregation; (2) tissue thromboplastin is released, activating the extrinsic pathway of the coagulation system; (3) PGI_2 (produced by normal endothelial cells), which inhibits platelet aggregation, is locally depleted; (4) plasminogen activator (also formed by endothelial cells), which is important in prevention of intravascular clotting, is also depleted; (5) the adherent activated platelets release serotonin (a vasoconstrictor), platelet factor 3 which participates in the intrinsic pathway of the coagulation system, platelet factor 4 which has an antiheparin effect, and thromboxane A_2 which induces platelet aggregation and vasoconstriction; and (6) if conditions favor progression, thrombin is generated and a larger more stable thrombus develops (Robbins and Cotran, 1979). Local changes in the coronary arteries that could precipitate the above sequence of events include: (1) hemorrhage or rupture of the underlying plaque, a common finding in arteries with acute thrombosis (Chandler *et al.*, 1974; Ridolfi and Hutchins, 1977) (Fig. 10.6); (2) spasm or other hemodynamic events which could cause endothelial injury (Maseri *et al.*, 1978); and (3) abnormalities in platelet function, in prostaglandin metabolism, or in both, and/or in the coagulation system. All of these factors have been demonstrated in patients with ischemic heart disease and in experimental models of myocardial ischemia (Haft, 1979; Sobel *et al.*, 1981; Smitherman *et al.*, 1981). It is of interest that thrombosis superim-

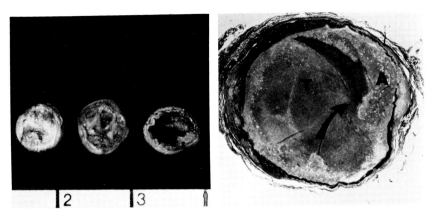

Fig. 10.8. Circumflex coronary artery–saphenous vein graft removed 5 years after implantation following reevaluation of the patient for recurrent angina pectoris. Note the slices of vein graft (left) showing complete luminal occlusion by a mixture of dark thrombotic material and lighter atheromatous material. (Scale below this portion of figure indicates centimeters.) Histologically (right), there is disruption of the fibrous cap (fc) of the atheroma (arrow), with rupture of the atheroma (A) and its contents merging with an occlusive thrombus (T) in the lumen of the graft. Trichrome stain. ×8.

posed on plaque rupture also occurs at sites of atheroma formation in implanted saphenous vein–coronary artery bypass grafts (Walts et al., 1982) (Fig. 10.8). It is possible that such grafts could provide a human model in which the progression of atherosclerotic plaques, plaque rupture, and subsequent thrombosis could be studied prospectively. (For further discussion of the role of endothelium in thrombosis, see Section C-6, Chapter 1.)

c. Topography of coronary artery atherosclerosis: In most cases, significant coronary atherosclerosis (greater than 75–80% luminal narrowing) is located angiographically in the proximal segments of coronary arteries, especially in the left coronary system where severe narrowing occurs in the first 3–4 cm of the anterior descending and circumflex arteries. Although proximal lesions are also noted in the right coronary artery, significant atheromata are just as common in the middle third of the vessel, particularly at the acute margin of the heart where the artery curves around to the posterior surface.

However, at autopsy, clinically significant coronary artery disease usually is found to be diffuse, severely narrowing multiple vessels, and often more marked than estimated angiographically (Vlodaver et al., 1973); distinct patterns frequently exist. Significant disease affects the left anterior descending artery twice as often as any other vessel, with an acute occlusion proving fatal more frequently than with circumflex or right coronary occlusion (Schuster et al., 1981). Left main coronary artery involvement, the most lethal form of coronary disease, is less frequent and is present in less than 10% of patients with clinically significant coronary artery disease. Atherosclerosis rarely is limited to this vessel, and severe disease of the other three major vessels can usually be expected in association with it.

Perhaps one of the most intriguing aspects of atherosclerotic coronary artery disease is the lack of correlation between postmortem and clinical findings. At autopsy, patients who die suddenly and with no clinical history of coronary artery disease tend to have similar degrees and patterns of narrowing in their coronary arteries as do individuals with typical clinical pictures of ischemic syndromes of the heart (Roberts and Jones, 1979).

F. LYMPHATIC SYSTEM

By A. J. MILLER

1. HISTORY

In 1653, Rudbeck recorded that he saw lymphatics leaving the heart to enter lymph glands. In 1692, Nuck visualized lymphatics by filling them with mercury, a technique later perfected by Hunter and Hunter (cited by Cruikshank, 1786), Hewson (1774), and Cruikshank (1786). In 1787, Mascagni published his beautiful illustration of the surface lymphatics of the heart (Fig. 10.9). Subsequently, the lymphatic anatomy of the mammalian heart was fairly extensively studied, with the works of Aagaard (1924) and Patek (1939) being of particular value. (For general consideration of lymphatic system, see Sections B and C, Chapter 4.)

2. ANATOMY OF LYMPHATIC SYSTEM

a. Atria, ventricles, and heart valves: The lymphatics in the atria are relatively sparse. The ventricles have an extensive lymphatic system, consisting of a subendocardial plexus, which lies parallel to the heart surface and which drains into a subepicardial plexus via myocardial lymphatics (Fig. 10.10). The subepicardial plexus, which also lies in a plane parallel to the heart surface, then drains into valved collecting channels on the surface of the heart (Aagaard, 1924; Patek, 1939). Lymphatics are present in the atrioventricular valves of

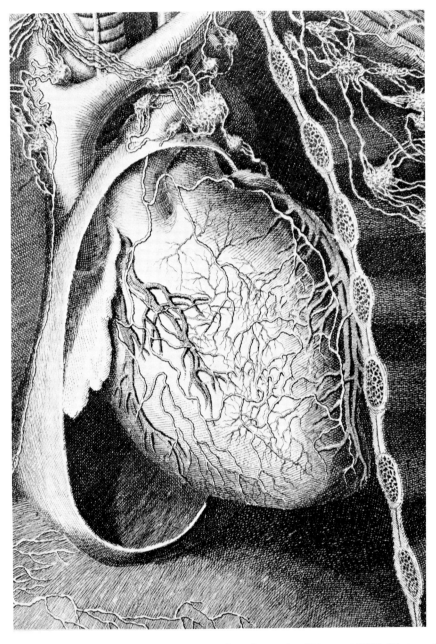

Fig. 10.9. Surface lymphatics of human heart from classic atlas of Mascagni (1787). (Photograph from book in collection of Northwestern University Medical School Library.)

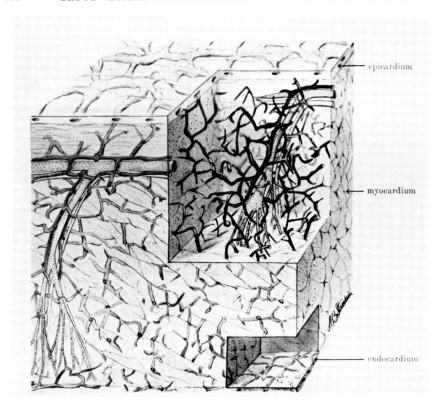

FIG. 10.10. A schematic drawing of a section of the ventricular wall of the dog heart showing the relationships of the lymphatics (black), arteries (white), and veins (stippled). (From Patek, 1939; reproduced with permission from the *American Journal of Anatomy*.)

mammalian hearts, but it is doubtful whether they are present in the semilunar valves (Miller *et al.*, 1961).

b. Conduction system: Lymphatics drain the areas of the SA and A-V nodes, bundle of His, and bundle branches (Eliska and Eliskova, 1976, 1980). However, the frequently observed "space" around the A-V node and the bundle of His has not been found to communicate with the lymphatic system. The excellent work of Aagaard and Hall (Aagaard, 1924) differentiated the fibers of the Purkinje system from ventricular lymphatics.

c. Lymphatics draining aorta and coronary arteries: Walls of arteries are metabolically active, and it is not surprising that the larger vessels have an abundant lymphatic drainage, as noted in the aorta (Lee, 1922) and the coronary arteries (Johnson, 1969).

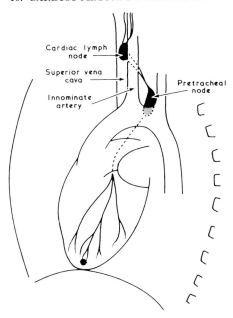

FIG. 10.11. Sketch of principal lymphatic drainage pathway in dog, as seen from the left lateral surgical approach. (From Miller *et al.*, 1963; reproduced with permission from the *British Heart Journal.*)

d. Mediastinal lymphatic drainage pathways from the heart: The collecting lymphatics on the surface of the heart join to form the major cardiac lymphatic trunks: a principal one and a lesser one. The *principal lymphatic trunk* forms on the left ventricle, ascends under the left atrial appendage, passes beneath the main pulmonary artery, and enters a node between the arch of the aorta and the trachea. From this pretracheal node, one or more vessels arise and pass to the right and upward, to enter the cardiac lymph node, which lies between the superior vena cava and the innominate artery (Fig. 10.11). From the cardiac lymph node, lymphatics ascend to enter the right lymphatic duct.

The *lesser lymphatic trunk* ascends from the right ventricle, passes near the right atrial appendage, and travels toward the base of the pulmonary artery. It may enter the principal lymphatic trunk, or it may ascend independently to empty near the upper reaches of the thoracic duct (Feola *et al.*, 1977).

3. PHYSIOLOGY OF LYMPHATIC SYSTEM

Though the heart is a uniquely functioning organ, there is evidence to indicate that its lymphatic system acts in a manner comparable to that of lymphatics elsewhere in the body (see Section C, Chapter 4).

TABLE 10.I

AVERAGE VALUES OF LYMPH FLOW AND COMPOSITION IN 15 DOGS[a]

Dog no.	Weight (kg)	Anesthetic[c]	No. of ½-hour collection periods	Blood pressure (mm Hg)	Respiratory rate/minute	Arterial hematocrit (%)	Volume lymph collected during ½-hour period (ml)	Total protein (g/100 ml)		Chloride (mEq/liter)		Sodium (mEq/liter)		Potassium (mEq/liter)	
								Lymph	Plasma	Lymph	Plasma	Lymph	Plasma	Lymph	Plasma
1	16	Pentobarb.	6	125/98	30		1.2	4.22	4.88			155.5		3.50	
2	18	Pentobarb.	3	135/118	25		1.9	3.90	5.88						
3	18.4	Pentobarb.	8	152/92	23		0.6	4.00	5.60	128.0	115.7	154.5	146.3	5.26	5.93
4	15.2	Pentobarb. and ether	1		15		1.5	5.37		119.7					
5	17.4	Pentobarb. and ether	6	157/89	18	40	2.8	2.51	4.76	120.7	125.0	152.2	156.8	5.04	4.91
6[d]	17	Pentobarb. and ether	6		15		0.9	5.17	6.51	122.2	112.9	153.6	150.9	4.62	4.66
7	12.3	Pentobarb. and ether	8				0.9			120.5	113.5	152.2	149.2	5.61	5.12
8	23	Pentobarb. and ether	6	171/122		52	1.3	3.69	5.53	114.5	110.5	152.6	152.8	5.70	5.77
9	17.7	Pentobarb.	12	164/119	22		1.3	4.39	6.60	116.9	113.8	143.7	144.0	3.64	3.84
10	13.6	Pentobarb. and ether	9	147/109	48	40	3.4	3.59	5.10	117.2	116.6	151.2	146.4	4.31	4.37
11	21.1	Pentobarb.	12	168/109	23	35	1.6[e]	3.74	5.46	122.1	117.4	149.2	146.1	3.21	3.84
12	16.6	Pentobarb.	12	151/102	23	37	1.7[e]	3.67	5.70	130.0	124.6	150.9	149.6	5.05	5.40
13	24	Pentobarb.	6	155/103	23	34	1.0	3.84	5.30	120.2	115.1	153.7	146.3	4.36	4.38
14	20.5	Pentobarb. and ether					f	4.76		108.7		147.2		6.14	
15	25	Pentobarb. and ether					f		5.92	122.3	120.2				
Means[b]							1.6	3.92	5.63	122.2	117.0	151.2	148.3	4.65	4.77
SEM								±0.07	±0.07	±0.7	±0.6	±0.4	±0.5	±0.29	±0.03

[a] From Miller et al. (1964b).
[b] Means and standard error of the means were calculated using the N corresponding to the number of individual samples determined for each parameter.
[c] Pentobarb., pentobarbital.
[d] This dog received intravenous Ringer's solution at rate of 20 drops/minute; all other dogs, 10 drops/minute.
[e] Data on lymph flow taken before period of transitory anoxia.
[f] Data omitted because lymph flow was collected for less than ½ hour.

a. Flow and composition: Studies in the anesthetized dog have established an average cardiac lymph flow of somewhat over 3.0 ml/hour (Miller *et al.*, 1964a). The amount increases after ischemia, with resistance loads placed on the left ventricle, with drugs which enhance myocardial contractility, after hypoxia, and after ventricular fibrillation (Miller, 1982). The lymph flow also rises markedly when there is interference with myocardial venous outflow. In such a situation, a pericardial effusion results because the ability of the lymphatics to clear the interstitial space is overwhelmed (Miller *et al.*, 1971).

The composition of cardiac lymph reflects that of the blood, including all the proteins and electrolytes (Table 10.I). The lymph chloride levels are slightly higher than in the blood. Red and white blood cells are present, the former increasing markedly after hypoxia.

4. PATHOLOGY OF LYMPHATIC SYSTEM

a. Cardiac lymphatics and infection: Experimental interference with cardiac lymph flow in the dog predisposes to valvular and myocardial infections and inflammation (Miller *et al.*, 1964b). Similarly, following injury to cardiac lymph-obstructed dogs, myocardial inflammatory reactions are altered and protracted.

b. Myocardial reparative processes: In dogs with coronary lymphatic obstruction, areas of myocardial injury produced by injection of autologous blood into the left ventricular myocardium are associated with altered healing processes and greater scarring than occur in control animals (Kline *et al.*, 1963). Also, the rate of healing and the associated histologic changes that follow myocardial infarction produced by ligation of a coronary artery branch are different in the cardiac lymph-obstructed dog. The resultant scar is larger, the healing process around the coronary artery ligature is altered, and inflammatory reactions are prolonged (Kline *et al.*, 1964).

c. Role of lymphatics in diseases of the heart and blood vessels: Little is known about the relationship between cardiac lymphatic circulation and diseases of the heart, other than some suggestive experimental results. In this regard, it has already been noted that, in lymph-obstructed dogs, there is a predisposition to infection and inflammation of the heart valves and myocardium and healing processes are altered. Moreover, studies on chronically cardiac lymph-obstructed dogs have revealed ventricular endocardial thickening, with an increase in fibrous and elastic tissue (Miller *et al.*, 1960, 1963). Some of these investigations remain controversial.

In man, congenital lymphangiectasis has been reported in association with endocardial fibroelastosis, and A-V block has been noted in the presence of

TABLE 10.II

DISEASE PROCESSES IN WHICH INVOLVEMENT OF THE CORONARY LYMPHATICS MAY BE
IMPORTANT IN THE PATHOGENESIS

A. Endocardial fibroelastosis
   1. Congenital, including congenital lymphangiectasia
   2. Secondary to other pathologic process
B. Endomyocardial fibrosis (Uganda)
C. Endocardial fibrosis
   1. Secondary to chronic myocardial disease, such as chronic myocarditis (e.g., Chagas' disease)
   2. Secondary to adhesive or constrictive pericarditis
D. Pericardial effusion
   1. Inflammatory, infectious
   2. Inflammatory, noninfectious (e.g., irradiation)
   3. Secondary to myocardial congestion and edema
E. Congestive heart failure of various etiologies
F. Acute myocardial insults
   1. Healing of acute myocardial infarction
   2. Healing of areas of myocardial trauma (e.g., surgery)
   3. Inflammatory (e.g., acute viral myocarditis)
G. Myocarditis, chronic
   1. Inflammatory
   2. Noninflammatory
H. Cardiomyopathies
   1. Alcohol
   2. Other causes, including heavy metals
I. Rheumatic fever and rheumatic heart disease
   1. Exacerbation of inflammation; predisposition to infection
   2. Progressive fibrosis; Aschoff body
J. Endocarditis, infective
K. Constrictive pericarditis
   1. Infectious
   2. Noninfectious (e.g., posttraumatic)
L. Myxomatous disease of atrioventricular valves

lymphangitis (Rossi, 1965). It has been suggested that interference with lymph drainage across the walls of larger arteries may be a factor in the pathogenesis of cardiac diseases, including atherosclerosis (Sims, 1979). The Aschoff nodule has been described as forming in inflamed lymphatic tissue, and it has been known for years that rheumatic fever involves the lymphatics. Some of the increased vascularity found on the atrial side of heart valves in this condition may be due to lymphatics. The chronic recurrent aspects of rheumatic fever or rheumatoid valvular involvement may also be related to interference with lymph drainage, and chronic myocarditis may smolder for a similar reason. In such syndromes as Kawasaki's disease, there may be lymphatic inflammation. In Chagas' myocarditis, lymphatic involvement may explain the thin-walled subendocardial vessels seen in this disorder, as well as endocardial thickening. (For a list of some of the situations in which a significant element in the pathogenesis might be interference with cardiac lymph flow, see Table 10.II.)

5. FUTURE AVENUES FOR INVESTIGATION

The lymphatics in the heart have been relatively ignored, partly because of technical difficulties in studying them in mammalian hearts and partly because at postmortem they are not readily identified. Direct injection techniques are laborious and *in vivo* indirect injection methods are often unsatisfactory. Studies of the composition and rates of flow of cardiac lymph present the formidable problem of lymphatic cannulation, and hence few investigators have been interested in studying the subject. Nevertheless, despite all the obstacles, a body of information has become available, and astute workers will look to cardiac lymph for important answers to tissue metabolic processes. It would be particularly exciting if a safe, simple radiopaque substance were to become available that could be injected into tissue and selectively be picked up by lymphatics.

REFERENCES*

Aagaard, O. C. (1924). "Les Vaisseaux Lymphatique du Coeur Chez L'Homme et Chez Quelques Mammifères." Levin & Munksgaard, Copenhagen. (F)

Adam, K. R., Boyles, S., and Scholfield, P. C. (1970). Cardio-selective beta adrenoceptor blockade and the coronary circulation. *Br. J. Pharmacol.* **40**, 534–536. (D)

Armour, J. A., and Klassen, G. A. (1981). Epicardial coronary venous pressure. *Can. J. Physiol. Pharmacol.* **59**, 1250–1259. (B)

Baron, G. D., Speden, R. N., and Bohr, D. F. (1972). β-Adrenergic receptors in coronary and skeletal muscle arteries. *Am. J. Physiol.* **223**, 878–881. (D)

Bellamy, R. F. (1978). Diastolic coronary artery pressure–flow relations in the dog. *Circ. Res.* **43**, 92–101. (C)

Bellamy, R. F., Lowensohn, H. S., Ehrlich, W., and Baer, R. W. (1980). Effect of coronary sinus occlusion on coronary pressure–flow relations. *Am. J. Physiol.* **239**, H57–H64. (B)

Berne, R. M., and Rubio, R. (1979). Coronary circulation. *In* "Handbook of Physiology" (R. M. Berne, N. Sperelakis, and S. R. Geiger, eds.), 2nd ed., Sect. 2, Vol. I, pp. 873–952. Am. Physiol. Soc., Bethesda, Maryland. (B, D)

Berne, R. M., DeGeest, H., and Levy, M. N. (1965). Influence of the cardiac nerves on coronary resistance. *Am. J. Physiol.* **208**, 763–769. (C)

Brown, B. G. (1981). Coronary vasospasm. *Arch. Intern. Med.* **141**, 716–722. (E)

Chandler, A. B., Chapman, I., Erhardt, L. R., Roberts, W. C., Schwartz, C. J., Sinarius, D., Spain, D. M., Sherry, S., Ness, P. M., and Simon, T. L. (1974). Coronary thrombosis in myocardial infarction. Report of a workshop on the role of coronary thrombosis in the pathogenesis of acute myocardial infarction. *Am. J. Cardiol.* **34**, 823–833. (E)

Christensen, G. C., and Campeti, F. L. (1959). Anatomic and functional studies of the coronary circulation in the dog and pig. *Am. J. Vet. Res.* **20**, 18–26. (B)

*In the reference list, the capital letter in parentheses at the end of each reference indicates the section in which it is cited.

Clark, D. A., Quint, R. A., and Mitchell, R. L. (1977). Coronary artery spasm. Medical management, surgical denervation and autotransplantation. *J. Thorac. Cardiovasc. Surg.* **73**, 332–339. (E)

Cruikshank, W. (1786). "The Anatomy of the Absorbing Vessels of the Human Body." G. Nicol, London. (F)

De La Lande, I. S., Harvey, J. A., and Holt, S. (1974). Response of the rabbit coronary arteries to autonomic agents. *Blood Vessels* **11**, 319–337. (D)

DeWood, M. A., Spores, J., Notske, R., Mouser, L. T., Burroughs, R., Golden, M. S., and Lang, H. T. (1980). Prevalence of total coronary occlusion during the early hours of transmural myocardial infarction. *N. Engl. J. Med.* **303**, 897–902. (E)

Downey, H. F., Bashour, F. A., Stephens, A. J., Kechejian, S. J., and Underwood, R. H. (1974). Transmural gradient of retrograde collateral blood flow in acutely ischemic canine myocardium. *Circ. Res.* **35**, 365–371. (B)

Drew, G. M., and Levy, G. P. (1972). Characterization of the coronary vascular β-adrenoceptor in the pig. *Br. J. Pharmacol.* **46**, 348–350. (D)

Eckstein, R. W. (1957). Effect of exercise and coronary narrowing on coronary collateral circulation. *Circ. Res.* **5**, 230–235. (B)

Eliska, O., and Eliskova, M. (1976). Lymph drainage of sinu-atrial node in man and dog *Acta Anat* **3**, 418–428. (F)

Eliska, O., and Eliskova, M. (1980). Lymphatic drainage of the ventricular conduction system in man and in the dog. *Acta Anat.* **107**, 205–213. (F)

Ellis, A. K., and Klocke, F. J. (1980). Effects of preload on the transmural distribution of perfusion and pressure–flow relationships in the canine coronary vascular bed. *Circ. Res.* **46**, 68–77. (C)

Endo, M., Hirosawa, K., Kaneko, N., Hase, K., Inoue, Y., and Konno, S. (1976). Prinzmetal's variant angina: Coronary arteriogram and left ventriculogram during angina attack induced by methacholine. *N. Engl. J. Med.* **294**, 252–255. (E)

Feigl, E. O. (1967). Sympathetic control of coronary circulation. *Circ. Res.* **20**, 262–271. (C)

Feigl, E. O. (1969). Parasympathetic control of coronary blood flow in dogs. *Circ. Res.* **25**, 509–519. (C)

Feigl, E. O. (1975). Reflex parasympathetic coronary vasodilation elicited from cardiac receptors in the dog. *Circ. Res.* **37**, 175–182. (C)

Feldman, R. L., Pepine, C. J., Curry, R. C., and Contin, R. (1978). Coronary artery responses to graded doses of nitroglycerin. *Circulation* **58**, Suppl. II, II-25. (E)

Feola, M., Merklin, R., Sung, C., and Brockman, S. K. (1977). The terminal pathway of the lymphatic system of the human heart. *Ann. Thorac. Surg.* **24**, 531–536. (F)

Fleckenstein, A. (1977). Specific pharmacology of calcium in myocardium, cardiac pacemakers, and vascular smooth muscle. *Annu. Rev. Pharmacol. Toxicol.* **17**, 149–166. (D)

Folts, J. D., Gallagher, K., and Rowe, G. G. (1982). Blood flow reductions in stenosed canine coronary arteries: Vasospasm or platelet aggregation. *Circulation* **65**, 248–254. (E)

Forbes, M. S., Rennels, M. L., and Nelson, E. (1977). Ultrastructure of pericytes in mouse heart. *Am. J. Anat.* **149**, 47–69. (B)

Fulton, W. F. M. (1965). "The Coronary Arteries; Arteriography, Microanatomy, and Pathogenesis of Obliterative Coronary Artery Disease." Thomas, Springfield, Illinois. (B)

Ganz, W., Buchbinder, N., Marcus, H., Mondkar, A., Maddahi, J., Charuzi, Y., O'Connor, L., Shell, W., Fishbein, M. C., Kass, R., Miyamoto, A., and Swan, H. J. C. (1981). Intracoronary thrombolysis in evolving myocardial infarction. *Am. Heart J.* **101**, 4–13. (E)

Gensini, G. G., Kelly, H. E., Dacosta, B. C. B., and Huntington, P. P. (1971). Quantitative angiography: The measurement of coronary vasomobility in the intact animal and man. *Chest* **60**, 522–530. (E)

Gerova, M., Bonta, E., and Gero, J. (1979). Sympathetic control of major coronary artery diameter in the dog. *Circ. Res.* **44**, 459–467. (C)

Golenhofen, K. (1978). Activation mechanisms in smooth muscle of human coronary arteries and their selective inhibition. *Naunyn-Schmiedeberg's Arch. Pharmacol.* **302**, Suppl. R36. (E)

Gregg, D. E. (1950). "Coronary Circulation in Health and Disease." Lea & Febiger, Philadelphia, Pennsylvania. (B)

Gregg, D. E., and Fisher, L. C. (1963). Blood supply to the heart. *In* "Handbook of Physiology" (W. F. Hamilton, ed.), Sect. 2, Vol. II, pp. 1517–1584. Am. Physiol. Soc., Washington, D.C. (B)

Gwirtz, P. A., and Stone, H. L. (1981). Coronary blood flow and myocardial oxygen consumption after α-adrenergic blockade during submaximal exercise. *J. Pharmacol. Exp. Ther.* **217**, 92–98. (C)

Hackett, J. G., Abboud, F. M., Mark, A. L., Schmid, P. G., and Heistad, D. D. (1972). Coronary vascular response to stimulation of chemoreceptors and baroreceptors. Evidence for reflex activation of vagal cholinergic innervation. *Circ. Res.* **31**, 8–17. (C)

Haft, J. I. (1979). Role of blood platelets in coronary artery disease. *Am. J. Cardiol.* **43**, 1197–1206. (E)

Halpern, M. H. (1954). Arterial supply to the nodal tissue in the dog heart. *Circulation* **9**, 547–554. (B)

Hardie, E. L., Jones, S. B., Euler, D. E., Fishman, D. L., and Randall, W. C. (1981). Sinoatrial node artery distribution and its relation to hierarchy of cardiac automaticity. *Am. J. Physiol.* **241**, H45–H53. (B)

Hellerstein, H. K., and Orbison, J. L. (1951). Anatomic variations of the orifice of the human coronary sinus. *Circulation* **3**, 514–523. (B)

Henry, P. D. (1980). Comparative pharmacology of calcium antagonists: Nifedipine, verapamil and diltiazem. *Am. J. Cardiol.* **46**, 1047–1058. (D)

Heupler, F. A., Proudfit, W. L., Razavi, M., Shirley, E. K., Greenstreet, R., and Sheldon, W. C. (1978). Ergonovine maleate provocative test for coronary artery spasm. *Am. J. Cardiol.* **41**, 631–640. (E)

Hewson, W. (1774). "The Lymphatic System in the Human Subject, and in Other Animals." J. Johnson, London. (F)

Heyndrickx, G. R., Muylaert, P., and Pannier, J. L. (1982). α-Adrenergic control of oxygen delivery to the myocardium during exercise in conscious dogs. *Am. J. Physiol.* **242**, H805–H809. (C)

Hintze, T. H., and Vatner, S. F. (1983). Comparison of effects of nifedipine and nitroglycerin on large and small coronary arteries and cardiac function in conscious dogs. Symposium on calcium channel blocking agents. *Circ. Res.* **52**, Suppl. I, 139–146. (D)

Hunter, J., and Hunter, H. cited by Cruikshank (1786). (F)

James, T. N. (1960). The arteries of the free ventricular walls in man. *Anat. Rec.* **136**, 371–384. (B)

James, T. N. (1961). "Anatomy of the Coronary Arteries." Harper (Hoeber), New York. (B)

James, T. N. (1977). Small arteries of the heart. *Circulation* **56**, 2. (B)

James, T. N., Isobe, J. H., and Urthaler, F. (1975). Analysis of components in a hypertensive cardiovascular chemoreflex. *Circulation* **52**, 179–192. (B)

James, T. N., Urthaler, F., and Hageman, G. R. (1980). Reflex heart block. Baroreflex, chemoreflex, and bronchopulmonary reflex causes. *Am. J. Cardiol.* **45**, 1182–1188. (B)

Johannson, B. (1973). The β-adrenoceptors in the smooth muscle of pig coronary arteries. *Eur. J. Pharmacol.* **24**, 218–224. (D)

Johnson, R. A. (1969). Lymphatics of the blood vessels. *Lymphology* **2**, 44–56. (F)

Kelly, K. O., and Feigl, E. O. (1978). Segmental α-receptor-mediated vasoconstriction in the canine coronary circulation. *Circ. Res.* **43**, 908–917. (C, D)

Kline, I. K., Miller, A. J., and Katz, L. N. (1963). Cardiac lymph flow impairment and myocardial fibrosis. *Arch. Pathol.* **76**, 424–433. (F)

Kline, I. K., Miller, A. J., Pick, R., and Katz, L. N. (1964). The effects of chronic impairment of cardiac lymph flow on myocardial reactions after coronary artery ligation in dogs. *Am. Heart J.* **68**, 515–523. (F)

Klocke, F. J., Kaiser, G. A., Ross, J., Jr., and Braunwald, E. (1965). An intrinsic adrenergic vasodilator mechanism in the coronary vascular bed of the dog. *Circ. Res.* **16**, 376–382. (D)

Lee, F. C. (1922). On the lymphatic vessels in the wall of the thoracic aorta of the cat. *Anat. Rec.* **23**, 343–349. (F)

Licata, R. (1962). Pulmonary circulation. *In* "Blood Vessels and Lymphatics" (D. I. Abramson, ed.), pp. 258–261. Academic Press, New York. (A)

Lichtlen, P. (1975). Coronary and left ventricular dynamics under nifedipine in comparison to nitrates, β blocking agents and dipyradamole. *In* "Second International Adalat Symposium" (W. Lochner, W. Braasch, and G. Kroneberg, eds.), pp. 212–224. Springer-Verlag, Berlin and New York. (D)

Logan, S. E. (1975). On the fluid mechanics of human coronary artery stenosis. *IEEE Trans. Biomed. Eng.* **BME-22**, 327–334. (E)

MacAlpin, R. N. (1980). Contribution of dynamic vascular wall thickening in luminal narrowing during coronary arterial constriction. *Circulation* **61**, 296–301. (E)

McAlpine, W. A. (1975). "Heart and Coronary Arteries," pp. 133–150. Springer-Verlag, Berlin and New York. (B)

McRaven, D. R., Mark, A. L., Abboud, F. M., and Mayer, H. E. (1971). Responses of coronary vessels to adrenergic stimuli. *J. Clin. Invest.* **50**, 773–778. (D)

Macho, P., and Vatner, S. F. (1981). Effects of nitroglycerin and nitroprusside on large and small coronary vessels in conscious dogs. *Circulation* **64**, 1101–1107. (D)

Macho, P., Hintze, T. H., and Vatner, S. F. (1981). Regulation of large coronary arteries by increases in myocardial metabolic demands in conscious dogs. *Circ. Res.* **49**, 594–599. (D)

Manders, W. T., Vatner, S. F., and Braunwald, E. (1980). Cardio-selective β-adrenergic stimulation with prenalterol in the conscious dog. *J. Pharmacol. Exp. Ther.* **215**, 266–270. (D)

Mark, A. L., Abboud, F. M., Schmid, P. G., Heistad, D. D., and Mayer, H. E. (1972). Differences in direct effects of adrenergic stimuli on coronary, cutaneous, and muscular vessels. *J. Clin. Invest.* **51**, 279–287. (D)

Markis, J. E., Malagold, M., Parker, J. A., Silverman, K. J., Barry, W. H., Als, A. V., Paulin, S., Grossman, W., and Braunwald, E. (1981). Myocardial salvage after intracoronary thrombolysis with streptokinase in acute myocardial infarction. *N. Engl. J. Med.* **305**, 777–782. (E)

Mascagni, P. (1787). "Vasorum lymphaticorum corporis humani historia et iconographia." P. Carli, Sensi, Italy. (F)

Maseri, A., L'Abbate, A., Baroldi, G., Chierchia, S., Marzilli, M., Ballestra, A. M., Severi, S., Parodi, O., Biagini, A., Distante, A., and Pesola, A. (1978). Coronary vasospasm as a possible cause of myocardial infarction. *N. Engl. J. Med.* **299**, 1271–1277. (E)

Mautz, F. R., and Gregg, D. E. (1937). The dynamics of collateral circulation following chronic occlusion of the coronary arteries. *Proc. Soc. Exp. Biol. Med.* **36**, 797–801. (B)

Miller, A. J. (1982). "The Lymphatics of the Heart." Raven Press, New York. (F)

Miller, A. J., Pick, R., and Katz, L. N. (1960). Ventricular endomyocardial pathology produced by chronic cardiac lymphatic obstruction in the dog. *Circ. Res.* **8**, 941–947. (F)

Miller, A. J., Pick, R., and Katz, L. N. (1961). Lymphatics of the mitral valve of the dog. *Circ. Res.* **9**, 1005–1009. (F)

Miller, A. J., Pick, R., and Katz, L. N. (1963). Ventricular endomyocardial changes after impairment of cardiac lymph flow in dogs. *Br. Heart J.* **25**, 182–190. (F)

Miller, A. J., Ellis, A., and Katz, L. N. (1964a). Cardiac lymph: Flow rates and composition in dogs. *Am. J. Physiol.* **206**, 63–66. (F)

Miller, A. J., Pick, R., Kline, I. K., and Katz, L. N. (1964b). The susceptibility of dogs with chronic impairment of cardiac lymph flow to staphylococcal valvular endocarditis. *Circulation* **30**, 417–442. (F)

Miller, A. J., Pick, R., and Katz, L. N. (1971). The production of acute pericardial effusion: The effects of various degrees of interference with venous blood and lymph drainage from the heart muscle in the dog. *Am. J. Cardiol.* **28**, 463–466. (F)

Mohrman, D. E., and Feigl, E. O. (1978). Competition between sympathetic vasoconstriction and metabolic vasodilation in the canine coronary circulation. *Circ. Res.* **42**, 79–86. (C, D)

Moncada, S., and Vane, J. R. (1979). Arachidonic acid metabolites and the interaction between platelets and the blood-vessel walls. *N. Engl. J. Med.* **300**, 1142–1147. (E)

Moore, P. F., Constantine, J. W., and Barth, W. E. (1978). Pirbuterol, a selective β₂-adrenergic bronchodilator. *J. Pharmacol. Exp. Ther.* **207**, 410–418. (D)

Mudge, G. H., Jr., Grossman, W., Mills, R. M., Jr., Lesch, M., and Braunwald, E. (1976). Reflex increase in coronary vascular resistance in patients with ischemic heart disease. *N. Engl. J. Med.* **295**, 1333–1337. (C)

Murray, P. A., and Vatner, S. F. (1979). Alpha adrenoceptor attenuation of coronary vascular response to severe exercise in the conscious dog. *Circ. Res.* **45**, 654–660. (C)

Murray, P. A., and Vatner, S. F. (1981a). Carotid sinus baroreceptor control of right coronary circulation in normal, hypertrophied and failing right ventricles of conscious dogs. *Circ. Res.* **49**, 1339–1349. (C)

Murray, P. A., and Vatner, S. F. (1981b). α-Adrenergic constriction and decrease in right coronary flow in response to carotid chemoreflex activation in conscious dogs. *Circulation* **65**, 441 (abstr.). (C)

Netter, F. H. (1969). "Heart," Vol. 5 of the Ciba Collection of Medical Illustrations, pp. 16–17. Ciba, Summit, New Jersey. (B)

Neufeld, H. N., and Schneeweiss, A. (1981). Etiology and prevention of coronary obstructive lesions in infancy and childhood. *Clin. Cardiol.* **4**, 217–222. (E)

Nuck, A. (1692). "Adenographia curiosa et uteri foemenei anatome nova. Appendix: De inventis novis epistola anatomica." J. Luchtmans, Lugd. Bat. (F)

O'Rahilly, R. (1971). The timing and sequence of events in human cardiogenesis. *Acta Anat.* **79**, 70–75. (A)

Palade, G. E. (1961). Blood capillaries of the heart and other organs. *Circulation* **24**, 368–384. (B)

Parratt, J. R. (1980). Effects of adrenergic activators and inhibitors on the coronary circulation. *Handb. Exp. Pharmacol.* 735–822. (D)

Patek, P. R. (1939). The morphology of the lymphatics of the mammalian heart. *Am. J. Anat.* **64**, 203–219. (F)

Patten, B. M. (1968). The development of the heart. *In* "Pathology of the Heart and Blood Vessels" (S. E. Gould, ed.), 3rd ed., pp. 20–90. Thomas, Springfield, Illinois. (A)

Provenza, D. V., and Scherlis, S. (1959). Coronary circulation in dog's heart. Demonstration of muscle sphincters in capillaries. *Circ. Res.* **7**, 318–324. (B)

Raizner, A. E., Ishimoro, T., Chahine, R. A., and Nasiruddin, J. (1978). Coronary artery spasm induced by the cold pressor test. *Am. J. Cardiol.* **41**, 358. (E)

Randall, W. C., Wehrmacher, W. H., and Jones, S. B. (1981). Hierarchy of supraventricular pacemakers. *J. Thorac. Cardiovasc. Surg.* **82**, 797–800. (B)

Reduto, L. A., Smalling, R. W., and Freund, G. C. (1981). Intracoronary infusion of streptokinase in patients with acute myocardial infarction: Effects of reperfusion on left ventricular performance. *Am. J. Cardiol.* **48**, 403–409. (E)

Rentrop, K. P., Blanke, H., Karsh, K. R., Kaiser, H., Kostering, H., and Leitz, K. (1981). Selective intracoronary thrombolysis in acute myocardial infarction and unstable angina pectoris. *Circulation* **63**, 307–317. (E)

Ridolfi, R. L., and Hutchins, G. M. (1977). The relationship between coronary artery lesions and myocardial infarcts: Ulceration of atherosclerotic plaques precipitating coronary thrombosis. *Am. Heart J.* **93**, 468–486. (E)

Robbins, S. L., and Cotran, R. S. (1979). "Pathologic Basis of Disease." Saunders, Philadelphia, Pennsylvania. (E)

Roberts, W. C., and Jones, A. A. (1979). Quantitation of coronary arterial narrowing at necropsy in sudden coronary death. *Am. J. Cardiol.* **44**, 39–45. (E)

Robertson, R. M., Robertson, D., Roberts, L. J., Maas, R. L., Fitzgerald, G. A., Friesinger, G. C., and Oates, J. A. (1981). Thromboxane A_2 in vasotonic angina pectoris: Evidence from direct measurements and inhibitor trials. *N. Engl. J. Med.* **306**, 998–1003. (E)

Ross, G., and Jorgensen, C. R. (1970). Effects of cardio-selective β adrenergic blocking agent on the heart and coronary circulation. *Cardiovasc. Res.* **4**, 148–153. (D)

Ross, G., Stinson, E., Schroeder, J., and Ginsburg, R. (1980). Spontaneous phasic activity of isolated human coronary arteries. *Cardiovasc. Res.* **14**, 613–618. (E)

Rossi, L. (1965). Case of cardiac lymphangitis with atrioventricular block. *Br. Med. J.* **2**, 32–33. (F)

Rudbeck, O. (1653). "Nova Exercitato Anatomia Exhibens Ductus Hepatico Aquosos et Vasa Glandularum Serosa." Arosiae, Upsala, Sweden. (F)

Schaper, W. (1971). "The Collateral Circulation of the Heart." Am. Elsevier, Amsterdam. (B)

Schaper, W., Schaper, J., Xhonneux, R., and Vandesteene, R. (1969). The morphology of the intercoronary artery anastomoses in chronic coronary artery occlusion. *Cardiovasc. Res.* **3**, 315–323. (B)

Schaper, W., Flameng, W., Snoeckx, L., and Jagerneau, A. (1971). Der Einfluss korperlichen Trainings auf den Kollateralkreislauf des Herzens. *Verh. Dtsch. Ges. Kreislaufforsch.* **37**, 112–121. (B)

Scheel, K. W., and Ingram, L. A. (1981). Relationship between coronary collateral growth in the dog and ischemic bed size. *Basic Res. Cardiol.* **76**, 305–312. (B)

Scheel, K. W., Ingram, L. A., and Wilson, J. L. (1981). Effects of exercise on the coronary and collateral vasculature of beagles with and without coronary occlusion. *Circ. Res.* **48**, 523–530. (B)

Schuster, E. H., Griffith, L. S., and Bulkley, B. H. (1981). Preponderance of acute proximal left anterior descending coronary arterial lesions in fatal myocardial infarction: A clinicopathologic study. *Am. J. Cardiol.* **47**, 1189–1196. (E)

Sims, F. H. (1979). The arterial wall in malignant disease. *Atherosclerosis* **32**, 445–450. (F)

Smitherman, T. C., Milam, M., Woo, J., Willerson, J. T., and Frenkel, E. P. (1981). Elevated β-thromboglobulin in peripheral venous blood of patients with acute myocardial ischemia: Direct evidence for enhanced platelet reactivity *in vivo*. *Am. J. Cardiol.* **48**, 395–402. (E)

Sobel, M., Salzman, E. W., Davies, G. C., Handin, R. I., Sweeney, J., Ploetz, J., and Kurland, G. (1981). Circulating platelet products in unstable angina pectoris. *Circulation* **63**, 300–306. (E)

Sobin, S. S., Frasher, W. G., and Tremer, H. M. (1962). Vasa vasorum of the pulmonary artery of the rabbit. *Circ. Res.* **11**, 257–263. (B)

Truex, R. C., and Angulo, A. W. (1952). Comparative study of the arterial and venous systems of the ventricular myocardium with special references to the coronary sinus. *Anat. Rec.* **113**, 467–491. (B)

Truex, R. C., and Schwartz, M. J. (1951). Venous system of the myocardium with special reference to the conduction system. *Circulation* **4**, 881–889. (B)

Vatner, S. F., and Hintze, T. H. (1982). Effects of a calcium channel antagonist on large and small coronary vessels in conscious dogs. *Circulation* **66**, 579–588. (D)

Vatner, S. F., and Macho, P. (1981). Regulation of large coronary vessels by adrenergic mechanisms in conscious dogs. *Basic Res. Cardiol.* **76**, 408–517. (D)

Vatner, S. F., and McRitchie, R. L. (1975). Interaction of the chemoreflex and the pulmonary inflation reflex in the regulation of coronary circulation in conscious dogs. *Circ. Res.* **37**, 664–673. (C)

Vatner, S. F., Higgins, C. B., and Braunwald, E. (1974). Effects of norepinephrine on coronary circulation and left ventricular dynamics in the conscious dog. *Circ. Res.* **34**, 812–823. (C, D)

Vatner, S. F., Pagani, M., Manders, W. T., and Pasipoularides, A. D. (1980). α-Adrenergic vasoconstriction and nitroglycerin vasodilation of large coronary arteries in the conscious dog. *J. Clin. Invest.* **65**, 5–14. (D)

Vatner, S. F., Hintze, T. H., and Macho, P. (1982). Regulation of large coronary arteries by β-adrenergic mechanisms in the conscious dog. *Circ. Res.* **51**, 56–66. (D)

Vlodaver, Z., Frech, R., Van Tassel, R. A., and Edwards, J. E. (1973). Correlation of the ante-mortem coronary arteriogram and the postmortem specimen. *Circulation* **47**, 162–169. (E)

Walmsley, R., and Watson, H. (1978). "Clinical Anatomy of the Heart." Churchill-Livingstone, Edinburgh and London. (A)

Walts, A. E., Fishbein, M. C., Sustaita, H., and Matloff, J. K. (1982). Ruptured atheromatous plaques in saphenous vein coronary artery bypass grafts: A mechanism of acute, thrombotic, late graft occlusion. *Circulation* **65**, 197–201. (E)

Wiggers, C. J. (1945). The functional consequences of coronary occlusion. *Ann. Intern. Med.* **23**, 158. (B)

Winbury, M. M., Howe, B. B., and Hefner, M. A. (1969). Effects of nitrates and other coronary dilators on large and small coronary vessels: An hypothesis for the mechanism of action of nitrates. *J. Pharmacol. Exp. Ther.* **168**, 70–95. (D)

Yasue, H., Omote, S., Takizawa, A., Nagao, M., Miwa, K., and Tanaka, S. (1979). Circadian variation of exercise capacity in patients with Prinzmetal's variant angina: Role of exercise-induced coronary arterial spasm. *Circulation* **59**, 938–947. (E)

Zuberbuhler, R. C., and Bohr, D. R. (1965). Responses of coronary smooth muscle to cate-cholamines. *Circ. Res.* **16**, 431–440. (B, D)

Chapter 11
Cardiopulmonary System: Lungs

A. EMBRYOLOGY OF BLOOD CIRCULATION
By C. E. CORLISS

1. ESTABLISHMENT OF THE PULMONARY CIRCULATION

a. Early development: Human lung buds appear in the fourth week as a bifurcation of the distal tracheal primordium. At the same time, capillary nets approach the lung primordia from two different sources: the aortic sac and the dorsal aorta itself (Congdon, 1922). Together, these form a profuse vascular network over the foregut that soon separates into individual pulmonary and esophageal components.

The aortic sac contribution coalesceses into the right and left pulmonary arteries, and, with sprouts from the aorta joining them at the 5 mm stage, they comprise the symmetric sixth aortic arches. However, by the seventh week, the right aortic segment degenerates, whereas the left retains its link with the parent vessel via the ductus arteriosus. The latter structure persists throughout gestation, providing a low resistance bypass for pulmonary blood to the systemic circulation, "the exercising channel for the right ventricle" (Patten, in Corliss, 1976, p. 440).

b. Later development: During the last months of gestation, there is a surprising amount of pulmonary development and a concomitant increase in pulmonary vasculature. Consequently, the abrupt change to air breathing at birth is accomplished with a minimum of functional disturbance (Patten, in Corliss, 1976, p. 445).

c. Histogenesis: Four periods of lung development have been described (Hislop and Reid, 1978; cited in Inselman and Mellins, 1981). During the third or canalicular period (16–24 weeks), cartilage, muscle, and other tissues appear, along with proliferation of a rich vascular supply that brings capillaries into a functional position for gaseous exchange. At this stage, the fetal pulmonary arteries contain a great deal of elastic tissue and are twice as thick-walled as their adult counterparts (Inselman and Mellins, 1981). They seem also to be in a state of vasoconstriction. Similarly, smaller vessels are equipped with spirally arranged muscle fibers (Licata, 1962).

d. Segmental supply: From the seventh week on, the pulmonary arteries parallel bronchial branching and, by the sixteenth week, all preacinar arterial branching is complete (Inselman and Mellins, 1981). After the seventeenth week, there is gradual invasion of the acinar substance itself, and by the twenty-sixth week, actual respiratory saccules appear (Perelman *et al.*, 1981).

e. Pulmonary veins: These vessels have a dual origin. The cardiac primodium of the pulmonary veins arises as an endothelial sprout from the medial superior wall of the left atrium. An extracardiac portion begins in an angioblastic plexus, which appears at the 3–5 mm stage, and by 6 mm joins the cardiac segment just caudal to the bifurcation of the trachea (Neill, 1956). The pulmonary veins enter the lung along the interseptal boundaries and return blood to the left atrium via two pairs of vessels. These branches of the original single vein are incorporated by the expanding atrial wall during cardiac development.

f. Bronchial arteries: These vessels originate as ventral visceral segmental arteries off the aorta and intercostal arteries. They may be induced by the maturing tissues and "propinquity" of the developing bronchus (Boyden, 1970). At the level of the main bronchi, bronchial arteries anastomose with their counterparts across the midline.

B. ANATOMY OF BLOOD CIRCULATION

By K. H. Albertine and N. C. Staub

The function of the lung is to exchange oxygen and carbon dioxide for the benefit of all the cells of the body. This organ receives blood from both pulmonary and bronchial circulations, each of which demonstrates substantial differences in architecture, distribution, and function.

1. Pulmonary Arteries

a. Arrangement: The pulmonary arteries follow the airways from the lobar bronchi (hilum) to the respiratory bronchioles (Miller, 1947b; von Hayek, 1960;

Blood Vessels and Lymphatics in Organ Systems
Copyright © 1984 by Academic Press, Inc.

Nagaishi, 1972). This anatomic arrangement emphasizes the basic function of the lung, which is to match perfusion with ventilation for efficient gas exchange.

b. Structure: The pulmonary arterial tree consists of extrapulmonary and intrapulmonary portions. The extrapulmonary vessels include the pulmonary trunk and the main pulmonary arteries. Histologically, the extrapulmonary vessels are of the elastic type, but the elastic laminae are interrupted and fragmented (Wagenvoort and Wagenvoort, 1979). The walls of the extrapulmonary arteries are heavily innervated with diffuse, undifferentiated receptors to which numerous, but largely unsubstantiated, cardiopulmonary reflexes have been attributed.

The intrapulmonary vessels are continuous with the pulmonary arteries at the hilum. They are also elastic vessels for about half the length of the bronchial tree, with the elastic laminae being arranged regularly. Branches of the elastic arteries separate at wider angles than do systemic arteries.

At vessel diameters between 1000 and 500 μm (in the regions of the bronchioles), a transition from elastic to muscular arteries occurs (Hislop and Reid, 1978). The muscular arteries (Fig. 11.1) distribute blood to different regions by active vasomotion. The pulmonary arterioles ($<$ 100 μm in diameter) are the terminal divisions of the pulmonary artery, and it is here that the tunica media disappears.

Elliott and Reid (1965) claimed that, in addition to the regular branching of the pulmonary arteries with the bronchi, there are numerous other tiny vessels that directly supply respiratory tissue. These are said to arise along the entire length of the pulmonary artery and presumably function as an auxiliary blood supply.

c. Function: The pulmonary arteries are thin-walled compared with comparable systemic vessels. This reflects the fact that the pulmonary arterial pressure after birth is only about one-fifth that in the aorta (about 19 cm H_2O at midchest level in man). Since pulmonary blood flow equals systemic blood flow, the calculated pulmonary vascular resistance is correspondingly low. Because pulmonary arterial pressures are so low, the pressure surrounding the pulmonary vessels (perivascular interstitial pressure) is of great significance. For this reason, the pulmonary circulation is also divided into two functional components. Some arteries are exposed to external pressures less than alveolar pressure ("extraalveolar vessels") (West, 1977), whereas others, located within the thin alveolar walls, are exposed to an external pressure that is similar to alveolar pressure and hence are referred to as "alveolar vessels." The pulmonary arteries down to the arterioles are considered to be mainly extraalveolar.

2. PULMONARY VEINS

a. Arrangement: The pulmonary venous system is not as well described as the arterial system. It begins by the convergence of the pulmonary capillary networks in the connective tissue septa at the periphery of the lung lobules; thus the veins are distant from the airways (Miller, 1947b).

b. Structure: This is somewhat different from that of the arteries. The venous wall is made up of an endothelial layer and a connective tissue adventitia, between which is located a thin, poorly developed media (Fig. 11.2). Smooth muscle cells, seen first in venules >20 μm in diameter (Rhodin, 1978), are scattered among collagen and elastin fibers.

c. Function: Because the pulmonary veins are more numerous, thinner walled, and of greater diameter than corresponding arteries, they function as a reservoir of blood for the filling of the left atrium and ventricle. For example, when right ventricular stroke volume decreases suddenly, left ventricular filling is maintained for 2 to 3 heart beats. The pulmonary veins have distensibility characteristics similar to systemic veins, that is, they hold a relatively large volume of blood at a low distending pressure. With further small increases in volume, they become relatively stiff. They offer much less resistance to flow than do the pulmonary arteries (Bhattacharya and Staub, 1980). Pulmonary venous flow has both a forward and retrograde pulsation. The forward pulsation is due to the thrust of right ventricular stroke output and the retrograde pulsation, to the systolic contraction of the left atrium.

3. PULMONARY MICROCIRCULATION

a. Arrangement: The pulmonary microcirculation is composed of arterioles (<100 μm in diameter) that branch into a complex anastomotic network of capillaries; the latter are distributed through the three-dimensional structure of the pulmonary alveoli. The capillaries eventually converge into venules (<100 μm in diameter). The main function of the lung, which is gas exchange, occurs within the microcirculation.

The alveolar capillaries are arranged in networks (Fung and Sobin, 1969). Figure 11.3 shows that these vessels take on the pattern of irregular vascular rings or hexagons. In contrast to the capillaries in other organs, the alveolar capillary network contains a relatively large quantity of blood and occupies most of the volume of the alveolar walls.

b. Structure: Knowledge of the anatomy of the blood–air barrier has been enhanced by the use of the transmission electron microscope (Weibel, 1963).

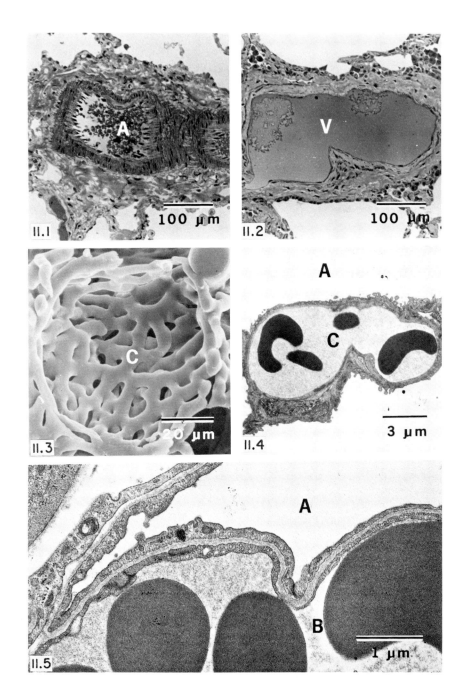

Structurally, the alveolar capillary barrier is composed of three layers: capillary endothelial cells, interstitial space, and alveolar epithelial cells (Figs. 11.4 and 11.5). The alveolar capillaries have a continuous endothelium that overlays a continuous basal lamina. The interstitial space is composed of collagen and elastin fibers and cells, predominantly fibroblasts. The pericapillary interstitium is continuous with that surrounding the airways and the blood vessels. Alveolar epithelial cells (type I and type II), which line the air spaces, have their own basal lamina.

   c. *Function:* The design of the pulmonary microcirculation is to spread the blood out in an extremely thin film so as to promote gaseous exchange quickly, while at the same time maintaining a very low resistance to a high volume of blood flow. The alveolar capillaries in man are said to have a maximum capacity of about 200 ml and a surface area of approximately 100 m², although at rest, their functional volume is only about 75 ml, i.e., equal to the stroke volume of the right ventricle. These vessels have an elliptical cross section, with major and minor axes of about 15 and 6 μm, respectively (Weibel, 1963); thus they are generally wider than systemic capillaries. Pulmonary capillaries in the dog and cat have a length averaging 500–600 μm and pass over several alveoli before becoming venules. The overall length of the capillary network and its functional volume mean that blood remains within the gas exchange capillaries for approximately 1 cardiac cycle, which at rest is about 0.75 sec. Because erythrocytes in the alveolar wall capillaries are no more than about 1.6–1.8 μm from the gas space, this time is more than adequate for near equilibration of oxygen and carbon dioxide exchange.
   Recently, the pressure profile in the pulmonary microcirculation has been measured. Nearly half of the total passive resistance to pulmonary blood flow is located within the alveolar wall capillaries (Bhattacharya and Staub, 1980).

4. BRONCHIAL CIRCULATION

   The bronchial circulation receives only 1% of the output of the left ventricle and carries well-oxygenated systemic blood at high pressure. It serves as the

FIG. 11.1. Small muscular pulmonary artery (A) in a 47-year-old man. Smooth muscle cells are circularly arranged. ×140.
   FIG. 11.2. Small pulmonary vein (V) in a 52-year-old woman. Wall of vein is thin compared to that of an artery. ×130.
   FIG. 11.3. Scanning electron micrograph of a latex cast of the pulmonary microcirculatio,. in a rat. Capillaries (C) form an anastomotic mesh in walls of the single alveolus shown. ×700. (From Guntheroth *et al.*, 1982; reproduced with permission of the American Physiological Society.)
   FIG. 11.4. Transmission electron micrograph of an alveolus (A) and an alveolar capillary (C) in a sheep. Most of the alveolar wall is capillary volume. ×4100.
   FIG. 11.5. Transmission electron micrograph of the blood (B)–air (A) barrier in a sheep. Gas exchange is made easier by a thin diffusion barrier. ×17,640.

vasa vasorum of the support structures of the lung, namely, the airways, the large blood vessels, and the connective tissue septa.

a. Arteries: Bronchial arteries in the human and in the dog vary in number and origin, although three vessels usually are described (Nagaishi, 1972). In sheep, on the other hand, the bronchial circulation arises from a single bronchoesophageal artery (Nakahara *et al.*, 1979). Histologically the bronchial arteries are similar to muscular arteries in the systemic circulation.

Once the bronchial arteries enter the lung, they travel in the peribronchial connective tissue, supplying nutrients (other than oxygen) to the airways down to the terminal bronchioles and the walls of the pulmonary arteries. The matter of the blood supply to the pleura has been controversial for decades. In those species with a thick visceral pleura (man and sheep), the bronchial arteries are its principal nutrient supply (Miller, 1947b; Albertine *et al.*, 1982b), whereas in those animals with thin pleura (dog and cat), the pleural surface blood supply is mainly via the pulmonary circulation.

b. Capillaries and veins: The bronchial arteries break up into capillary networks that supply the entire length of the airways and the walls of the arteries and then empty into nearby pulmonary venules. Most of the bronchial venous blood returns to the left atrium and is part of the so-called venous admixture component of pulmonary blood flow. In the region of the trachea, carina, and main stem bronchi, there are several bronchial veins that carry blood from this region and adjacent pleura through the azygos system of veins to the right atrium.

c. Bronchopulmonary anastomoses: Connections between the bronchial and pulmonary circulations seem to exist, but their functional importance in normal lungs is believed to be negligible (Wagenvoort and Wagenvoort, 1979). Since the lung does not usually die when either the bronchial arteries or the pulmonary artery is ligated, it appears that either system is capable of supplying sufficient nutrient to the other to maintain tissue viability.

C. PHYSIOLOGY OF BLOOD CIRCULATION*

By S. S. SOBIN

1. GENERAL CONSIDERATIONS

The interrelationship between structure and function in biologic systems is nowhere more elegantly seen than in the pulmonary circulation. The complex-

*The author wishes to acknowledge the critical review of this section by Dr. Herta M. Tremer.

ity of the interrelationship, however, is not simple, for it consists of air–tissue and blood–tissue interfaces; in addition, blood is a two-phase system containing deformable particles and a viscosity that changes with tube dimension and hematocrit. All of this is contained within a highly complex array of branching airways and blood vessels and a tight network of distensible alveolar capillaries arranged in sheets. There is no unitary arrangement of inflow and outflow vessels to individual alveoli so that capillary path lengths often span a number of alveoli. Blood flow through these vessels is influenced by intraluminal pressure gradients, rhythmic pressure changes of the pleural space (transmitted differently to alveolar and extraalveolar vessels), gravitational pressure gradients, and airway pressures. The pulmonary vascular system accepts blood flows that vary from basal values to up to six times this amount. The arrangement of vascular tubes and terminal alveolar microvascular sheets appears receptive and passive, with little apparent control by neural or humoral mechanisms under normal conditions.

Both the vascular and the respiratory systems converge at the alveolar level, and it is at the interalveolar wall that the mechanical behavior of the respiratory membrane and the microvascular interalveolar sheet interact.

2. THE ORGANIZATION OF THE PULMONARY CIRCULATION

a. Arterial tree: The exact number of generations of the human pulmonary arterial tree from its origin at the pulmonary valve to its termination in the interalveolar capillary sheets is not known. There are a number of problems in determination of such branching sequences; one concerns the technique of generation counting, proximal to distal, using the method of Weibel and Gomez (1962) made on airway casts. Such counting was based on the dichotomous branching pattern of the human airway, uncommon in the asymmetric human pulmonary arterial tree. Singal *et al.* (1973) made casts of the human pulmonary arterial tree and used a branch-counting method called "ordering," beginning at the most distal vessel of 0.1 mm diameter (order 1) and counting proximally. When two branches of the same size (or order), such as order 1, were joined together, the conjoined branch was assigned order 2; when two branches of differing orders were joined, the conjoined vessel retained the number of the larger branch, e.g., if branch 6 joined branch 11, the combined branch was given order 11. The latter system of counting proximally from the periphery was formulated by Strahler (1957) for physical geography and is now the preferred method of ordering branching systems. It avoids the common problem of counting in an asymmetric branching system. Using these methods, Singal *et al.* counted 17 generations in the human pulmonary arterial tree.

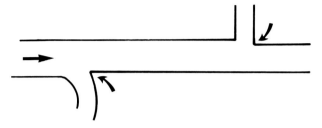

Fig. 11.6. Diagram of arterial branching and cylindricity between branches. The straight arrow within the artery points in the direction of flow. The curved arrows beyond the branches point to the arterial wall offset immediately beyond each branch. Narrowing of vessels below each branch site is accomplished by such steps.

b. Geometry and arterial branch sites: Arterial branching in the systemic and pulmonary beds classically has been described as arborization, with tapering between branches so that the arterial segments between branches have been described as truncated cones (Knisely, 1965). Recently Sobin and Tremer (1980) described step decreases in vessel diameter immediately distal to branches in both the pulmonary and systemic arterial vessels (Fig. 11.6). Thus, vessels diminish in dimension by a decrease in diameter at the distal border of the branch vessel, and cylindricity, not taper, is the rule for intervening segments. The physiologic consequences of such geometry have not been investigated but should be related to wave reflection, volume–flow distribution, viscosity, and flow resistance.

c. Microvascular bed of the interalveolar wall (sheet-flow model): Weibel and Gomez (1962) modeled the capillary network in the postmortem human lung as a hexagonal network of interconnected, truncated, cylindrical tubes of finite length and width ($1/d = 1.3$). Subsequently, Sobin and Fung (1967), Fung and Sobin (1969), and Sobin *et al.* (1970) described a sheet configuration for the network and showed that the density of the latter could be expressed as a ratio between the interalveolar microvascular space and its circumscribing tissue volume. This was termed a vascular space–tissue ratio (VSTR) (Figs. 11.7 and 11.8). For various laboratory mammals and man, geometric considerations demonstrated that a microvascular bed composed of an assembly of tubes would be 1.3 times wider than long and therefore would not be compatible with a tube blood flow model. Therefore, the microvasculature of the interalveolar wall was represented instead as a sheetlike endothelium-lined space, the opposing walls of which are held apart (or together) by endothelium-lined posts. Thus the interalveolar walls contain interalveolar microvascular sheets and blood flow can be considered sheet-flow, the characteristics of this arrangement being described below.

It is worthwhile to clarify the relationship between the alveoli and the

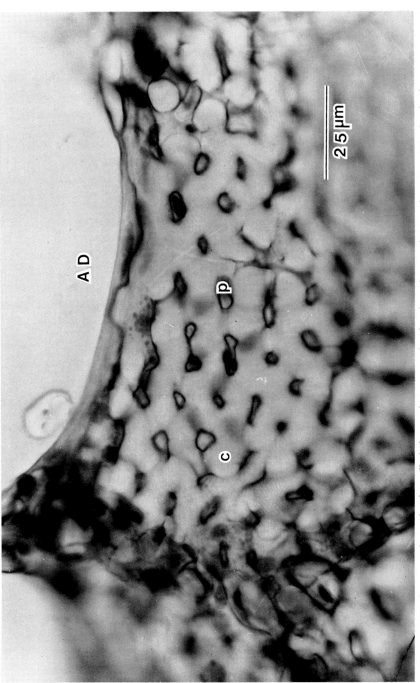

FIG. 11.7. En face view of interalveolar wall of silicone elastomer perfused cat lung showing sheet configuration. AD, alveolar duct; c, a segment of the continuous silicone filled sheet; p, post, the avascular portion of the sheet. The dark ringlike structure delineating the posts is the basement membrane stained with cresyl violet.

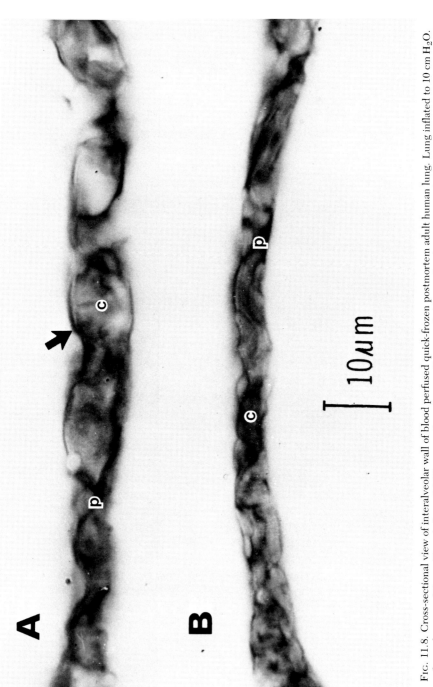

FIG. 11.8. Cross-sectional view of interalveolar wall of blood perfused quick-frozen postmortem adult human lung. Lung inflated to 10 cm H_2O. Lung frozen by pour of Freon-22 cooled to −156°C over surface of vertically suspended lung. The photomicrographs show change in sheet width with change in hydrostatic pressure of blood column. (A) 25 cm H_2O transmural pressure. (B) 5 cm H_2O transmural pressure. There is photomicrographic distortion of the actual width due to the thickness of the histologic section. Arrow shows convexity of sheet at post mortem. Symbols as in Fig. 11.7.

interalveolar microvascular bed. Alveoli result from partitioning or subdividing of air spaces postnatally by the ingrowth of connective tissue, on which capillary bilayers grow; the latter structures, in turn, are converted to a single capillary layer in the interalveolar wall. Alveoli exist because there are capillary sheets, with their connective tissue and alveolar cell coverings, and when the latter disappear, as in disease, alveoli also disappear and proper gas exchange is disturbed.

   d. *Topology of pulmonary microvascular bed:* True Poiseuillean capillaries are absent in the sheet-flow model of the interalveolar wall, and the red cell path through the microvascular sheet is, by convention, called a capillary path. The distance between an arteriole and a venule is thus a capillary path length and has physical meaning. Initially the alveolar circulation was illustrated with an arteriolar supply and venular drainage for each alveolus (Miller, 1947b); for such a model, the average capillary path length would be the average distance between arterioles and venules. However, Staub and Schultz (1968) infused oil into the pulmonary artery of the blood-filled pulmonary circulation of the rabbit, cat, and dog and described the pulmonary capillary length, the distance between oil-filled 60 μm-sized pulmonary arterioles and blood filled venules, as extending over a number of alveoli, with an approximate length of 400–600 μm. Miyamoto and Moll (1971) studied rapidly frozen lungs and estimated the capillary length at half or less than that found by Staub and Schultz (1968). More recently, Sobin *et al.* (1980) examined silicone elastomer injected preparations and reported that the microvascular bed of the lung could be topologically represented by an unique cluster distribution of arterioles and venules, best described as islands of pulmonary arterioles surrounded by oceans of pulmonary venules (Fig. 11.9). Stereological measurements of these maps revealed that the average capillary length in cat lung was 556 ± 286 (SD) μm at 10 cm H_2O transpulmonary pressure. The average capillary length (L) is an important determinant of pulmonary blood flow; for given arterial and venous pressures, the flow rate is inversely proportional to L^2, the square of the average length. The transit time (time available for diffusion and oxygenation) varies directly with L^2, and the regional differences in blood flow depend on L.

3. ELASTICITY OF THE PULMONARY VASCULATURE

   a. *Interalveolar microvascular sheet:* The thesis that capillaries are rigid, which arose from Burton's (1959) observations that those in the systemic circulation are rigid, was for some time also applied to pulmonary capillaries. However, Glazier *et al.* (1969) demonstrated by quick-freezing the blood-filled

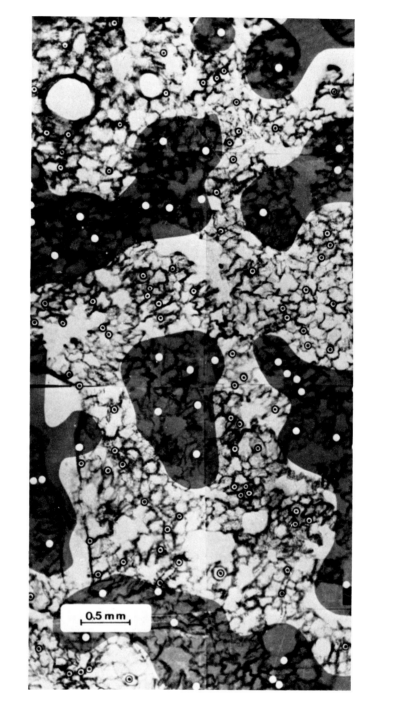

0.5 mm

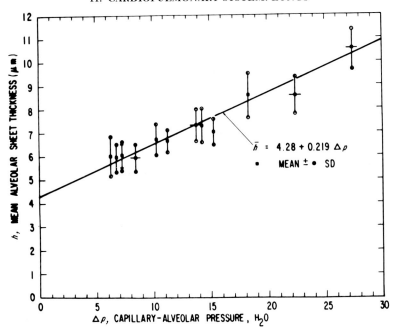

$$\bar{h} = 4.28 + 0.219\ \Delta p$$

• MEAN ± • SD

FIG. 11.10. Plot of thickness of microvascular sheet of silicone perfused cat lung for different values of transmural pressure of the microvascular sheet. Perfused *in situ* from pulmonary artery. (From Sobin *et al.*, 1972; reproduced with permission from American Heart Association.)

dog lung that the interalveolar wall capillaries are distensible. Subsequently, Sobin *et al.* (1972) used silicone microvascular casting methods and found that the pulmonary microvascular sheet is linearly distensible over the normal range of physiologic pressures. Figure 11.10 shows that sheet thickness is a function of sheet pressure in the statically loaded silicone prefused cat lung. The equation for the regression line is

$$\bar{h} = h^o + \alpha(p_c - p_A) \tag{C.1}$$

where $\bar{h}$ is the mean sheet width or thickness; h^o is the extrapolated sheet thickness at $p_c - p_A = 0$; p_c is the capillary sheet pressure; p_A is the alveolar pressure; α is the compliance coefficient which defines the elastic behavior of

---

FIG. 11.9. Map made from montage of individual photomicrographs of thick histologic section of cat lung. Branches of pulmonary vessels from 15 to 100 μm diameter are individually identified. The domains of the pulmonary arteries are made darker by overlay of the area with blue transparency film. Individual alveoli are clearly seen as well as branching vessels in the plane of the photomicrograph. The pulmonary artery domains are discontinuous, and the pulmonary venous areas are continuous: the arterial areas are islands immersed in an ocean of pulmonary veins. (From Sobin *et al.*, 1980; reproduced with permission of Academic Press.)

the sheet. Since in the capillary bed, $p_c - p_A = \Delta p$ (transmural pressure), $h = h^o + \alpha(\Delta p)$. The compliance coefficient α for the cat is 0.219 μm/cm H_2O pressure. The α values for dog and human lung are approximately 0.1 μm/cm H_2O (Fig. 11.8). Fung et al. (1966) demonstrated that the rigidity of systemic capillaries is due to the mechanical properties of the surrounding tissue. These authors termed systemic capillaries "tunnels in a gel" and the properties of the capillaries were found to be those of the gel. However, pulmonary capillary sheets are devoid of a significant gel, the basis for the elasticity in this tissue being attributed to the collagen and elastin in the interalveolar walls (Sobin et al., 1972; Rosenquist et al., 1973). The microvascular sheet is quite compliant and can accommodate increased blood volume with very little increase in pressure; the membranes that limit the space also are compliant and provide the alveolar planar elasticity necessary for changing lung volumes (Fung and Sobin, 1972a).

b. *Compliance of juxtaalveolar vessels:* These noncapillary vessels have been studied using silicone elastomer microvascular casting methods (Sobin et al., 1978). The juxtaalveolar vessels are circular, less than 50 μm in diameter and not classified, since they are a mixed population of arterioles and venules. The mechanical compliance of these vessels is similar to that of the capillary sheet: $\alpha = 0.274$ μm/cm H_2O, a finding which could be anticipated since they do not possess normal vascular smooth muscle cells (Sobin et al., 1977).

c. *Elasticity of pulmonary arteries and veins:* The elastic properties of the main pulmonary artery and its major branches have been well characterized; however, information on the elastic properties of large pulmonary veins appears lacking (for a critical review, see Milnor, 1972). The elasticity of small pulmonary arteries and veins (100 to 1600 μm diameter) has been studied by Yen et al. (1980) and Yen and Foppiano (1981), respectively, in the excised cat lung, using radiographic methods of considerable accuracy. The compliance of small arteries was found to be one-fourth to one-half that of the capillary sheet and varied among vessels of different dimensions. Similar studies on the small pulmonary veins showed a range of compliance not remarkably different from that of small arteries [see Caro et al. (1978) and Dobrin (1983) for detailed reviews of vascular elasticity].

4. PULMONARY BLOOD FLOW

Pulmonary gas exchange in various physiologic states requires a vascular bed that can accommodate cardiac outputs of up to six times resting values. To accomplish this with minimal loading of the heart requires a pulmonary vascular system that imposes minimal resistance with increasing blood flow. In a series of papers, Cournand and colleagues (see Cournand, 1958) showed that pulmonary blood flow increases with exercise without a concomitant rise in resistance, at least with moderate increases in cardiac output; in fact, computed resistance falls, indicating that the cross-sectional area of the microvascular bed is enlarged. However, the precise mechanism of this behavior is still unclear.

   a. Vascular sluicing or waterfall: The variable vascular resistor first described by Knowlton and Starling (1912) for the isolated heart–lung preparation was a "fingerstall" or a segment of readily collapsible tubing, attached at both ends to glass tubing; with all enclosed in an outside glass tube, it was possible to apply different pressures and thus to impose variable resistances on the collapsible segment. This has been since called a Starling resistor. With such a concept, Bannister and Torrance (1960) demonstrated in the cat lung that pulmonary capillaries could act as a pressure (and therefore flow) regulating system, with alveolar pressure closing, and pulmonary arterial and venous pressures opening, these vessels. Thus, transmural capillary pressure participates in a hydraulic regulatory system to control local blood flow. Since regulation of flow in hydraulic channels by valve or flood-gate is accomplished by a sluice, Bannister and Torrance called this effect "sluicing." Their studies were followed by the suggestion of Permutt *et al.* (1962) to utilize the analogy to a "waterfall" because of the following reasoning and experimental proof: (1) the pulmonary vascular system has a constant arterial pressure; (2) the pulmonary capillaries are collapsible tubes inside an alveolar pressure chamber; and (c) resistance in the postcapillary vessels is minimal. Under such circumstances, when venous pressure is less than alveolar pressure and the end capillary pressure is not less than alveolar pressure, the latter determines the lower end of the pulmonary vascular pressure gradient; this, in turn, dictates flow. Furthermore, with venous pressure below alveolar pressure, blood flow between capillaries and veins is like a waterfall, since waterfall flow is not determined by the height of the fall of the waterfall but by the pressure gradient between the source and the point immediately before the waterfall.

   b. Gravitational effect: The sluice or waterfall effect and the relationship among pulmonary artery pressure (p_a), alveolar pressure (p_A), and pulmonary venous pressure (p_v) have been further refined into a zonal distribution of pulmonary blood flow (Q_p) by West *et al.* (1964), as noted below.

$$\text{Zone } 1 = p_a < p_A > p_v \text{ and } Q_p = 0 \tag{C.2}$$
$$\text{Zone } 2 = p_a > p_A > p_v \text{ and } Q_p \approx (p_a - p_A) \tag{C.3}$$
$$\text{Zone } 3 = p_a > p_A < p_v \text{ and } Q_p \approx (p_a - p_v) \tag{C.4}$$

In normal upright man, all three zones may exist. There is an intravascular hydrostatic gravitational effect down the lung. Pulmonary artery mean pressure in normal standing man is usually measured at 15 mm Hg. Thus it will perfuse the lung only 20 cm above the pulmonary artery. Since alveolar pressure is generally equal throughout the lung and equal to ambient pressure, the vascular pressure at the top of the lung may be less than alveolar pressure and hence the pulmonary capillary sheets will be collapsed. With physical activity, pressure rises sufficiently to perfuse the apex of the lung and zone 1 is transformed into zone 2. Pulmonary venous pressure remains below alveolar pressure until the hydrostatic level is below that of the left atrium, at which time hydrostatic gravitational pressure leads to increased pulmonary venous pressure for zone 3. Glazier *et al.* (1969) validated the zonal distribution concept anatomically in the quick-frozen lung of the dog.

   c. Sheet-flow: The concept of sheet-flow in the interalveolar wall not only provides the means by which to analyze various features in the sheet model, but it also is consistent with experimental laboratory and clinical data. The microvascular sheet is elastic and can readily accommodate increased blood volume with minimal increase in transmural pressure. Theoretically (Fung and Sobin, 1969) and experimentally (Fig. 11.8), it can be demonstrated that with increased vascular or transmural pressure, the sheet becomes convex between posts. Thus microvascular distensibility is seen with locally elevated pressure and results in greater flow with diminished downstream resistance. The term, compliance coefficient, discussed above, characterizes the elastic behavior in micrometers change of sheet thickness per cm H_2O pressure change across the sheet wall (Δp). As noted, it is 0.219 for the cat, and 0.1 for man, and it is linear over the physiologic range of pressure.

   Sheet-flow theory is consistent with the Starling resistor model, explains in more formal mathematical terms zonal distribution of blood flow, and demonstrates the merging of zones 2 and 3 under dynamic conditions of flow. Sheet-flow is accurately described by

$$Q = (1/C) \, (h_a{}^4 - h_v{}^4) \tag{C.5}$$

where C includes various physical characteristics of the sheet and of the blood (Fung and Sobin, 1972a,b). This demonstrates the responsiveness of flow to arterial pressure as the primary determinant of pulmonary blood flow.

   The question of recruitment versus distensibility as a mechanism of increased flow with decreased resistance does not concern the Starling resistor mechanism within zone 1, but it does deal with local hemodynamics within interalveolar walls or sheets. Both the experimental findings of patchy alveolar blood flow (Warrell *et al.*, 1972), as well as the computer simulation studies (West *et al.*, 1975), indicate capillary recruitment. However, these observations are compatible with sheet-flow when considered from the point of view of stochastic flow (Fung, 1973) and a nonuniform distribution of red cells in a

nonuniform velocity field (Yen and Fung, 1978). Sheet distensibility is sufficient to explain many of the phenomena associated with increased flow.

d. Influence of surface forces: The demonstration of surfactant in the alveolar lining was critically dependent on specialized electron microscopic
methods. Until recently, alveoli were considered spheres or structures consisting of curved walls. With the development of quick-freezing methods to preserve lung morphology (Staub and Storey, 1962) and intravascular fixation of
the air-filled lung (Gil *et al.*, 1979), the interalveolar walls have been demonstrated to be essentially linear. In zone 1, with quick-freezing techniques,
"corner" vessels are found to be open despite collapsed mural capillaries or
sheets (Glazier *et al.*, 1969; Warrell *et al.*, 1972). On the basis of the results
obtained with intravascular fixation of the air-filled lung, Gil *et al.* (1979) have
suggested that these corner vessels are probably bundles of capillaries altered
by septal wall pleating at sites where three septa meet. This phenomenon is not
seen in saline-filled lungs or in ones fixed by intratracheal infusion, in which
surfactant is diluted or destroyed and surface forces are therefore markedly
increased.

e. Pulsatile blood flow: In 1955, Lee and DuBois postulated that the
pulsatile outflow of the right ventricle is demonstrated as pulsatility in the
pulmonary capillaries as well. By recording instantaneous pulmonary capillary
blood flow by the nitrous oxide method in a whole body plethysmograph, they
were able to demonstrate pulsatility of blood flow in pulmonary capillaries.
Such findings accentuate the importance of the systolic component of pulsatile
flow to perfuse zonal regions of the lung usually considered not to be perfused
when mean pulmonary artery pressure alone is used.

f. Transmural pressure of pulmonary arteries and veins: This is the difference between intraluminal pressure and pressure at the exterior wall surface. For the lung, it is more appropriate to consider an effective transmural
pressure due to the fact that, with lung expansion, there are differences in the
response of intraalveolar intraparenchymal vessels and extraalveolar intraparenchymal vessels. The latter have a connective tissue sheath with a perivascular space continuous with the pleural space. With airway expansion, the
extraalveolar vessels enlarge, whereas the intraalveolar vessels are compressed.
Therefore, a statement of the forces operative in transmural pressure must
include consideration of circumferential and longitudinal tethering, interdependence of the mechanical forces between lung parenchyma and blood vessels
contained within it, alveolar pressure, tension on septal or alveolar membrane,
the perivascular space, etc. (For discussion of these points, see Culver and
Butler, 1980; Lai-Fook, 1979; Lai-Fook and Hyatt, 1979.)

5. Pressure Gradient in the Pulmonary Vascular Bed

The pressure gradient down the pulmonary vascular bed and the distribution of peripheral vascular resistance have been repeatedly investigated using indirect methods. Recently, however, Bhattacharya and Staub (1980) directly measured the subpleural microvascular pressures in the isolated dog lung and determined that 46% of the peripheral vascular resistance was in the microcirculatory bed. Overholser *et al.* (1982) modeled the microvascular system of the lung based on averaged anatomic and physiologic data from the literature using mathematical formulations appropriate for each model. The derived values showed that 45–50% of the flow resistance is located in the alveolar microvasculature for both sheet and tube network models. (For background review, see Milnor, 1972.)

6. Humoral Control Mechanisms

It is important to distinguish between blood-borne biochemical agents that can affect the pulmonary circulation and those that are important in its physiologic control. The distinction in meaning is precise although experimental differentiation and proof are often difficult to obtain. The problem is compounded by the many animal species studied and the different types of experimental preparations. The latter include: isolated lungs perfused under constant volume flow so that changes in pulmonary artery pressure are an index of arterial constriction or dilatation; test strips, helices, or rings of vessels in which changes in tension or force are measured; anesthetized animals with extensive surgery and complex instrumentation; and unanesthetized animals and man with minimal instrumentation. Each preparation presents special and unique problems, often not understood at the time of study. For example, strips of vessels make receptor sites available for chemical agents via incised exposed surfaces, whereas the normal pathway for physiologic vasoactive agents is by way of the endothelial surface (Furchgott and Zawadski, 1980). Extensive surgical preparations and anesthetic agents have individual problems related to altered physiologic states, including regional vascular tone, reflex sensitivity, etc. Substances active in isolated strips and rings often are significantly less or even differently active in organ perfusion or in intact unanesthetized preparations. For example, ATP is a powerful dilator in the intact animal, whereas it constricts strips of the rabbit aorta. Bergofsky (1980) recently tabulated a variety of chemical agents studied in isolated organs or blood-perfused lungs and listed their effective doses, most of which appear too large to be considered as physiologic regulators.

Discrepancies are notable in the distribution of receptor sites and drug effects. Su *et al.* (1978) examined segments of pulmonary arteries from the main pulmonary artery to 200 μm diameter by fluorescent glyoxylic acid methods and found a dense adrenergic plexus at the adventitial–medial junction that paralleled the neuronal uptake of tritiated norepinephrine. Arteries larger than 0.6 mm diameter responded strongly to nerve stimulation, whereas those below 0.6 mm did so poorly, if at all. However, the latter vessels did respond to serotonin, histamine, and potassium chloride. Su *et al.* (1978) suggested the possibility that sympathetic control of vascular tone could be restricted to relatively large arteries and speculated that "in the pulmonary vasculature, through which the output of the right heart must pass to the left, an extensive arterial constriction accompanying sympathetic discharge, resulting in major changes in blood flow, is undesirable." Thus, vascularly active materials may not play a role in physiologic control, but when employed as pharmacologic agents, they may modulate physiologic action.

Therefore, at present it may be useful to consider the pulmonary circulation to be independent of extraneous humoral physiologic control mechanisms, except possibly under stressful conditions.

7. Hypoxia

a. Acute hypoxia: The role of respiratory gases, especially low oxygen tension, in the control of the pulmonary circulation has interested physiologists since the late 1800's, with little agreement existing regarding the experimental results and their interpretation. In 1946, von Euler and Liljestrand clearly demonstrated that in the chlorolose-anesthetized cat with a closed thorax and breathing spontaneously, 10% oxygen in nitrogen produced a prompt rise in pulmonary artery pressure without a concomitant elevation in left atrial pressure. They suggested that regional hypoxia with attendant local arterial constriction could redistribute blood into better-oxygenated alveoli, an intuitively powerful physiologic insight. Von Euler and Liljestrand also observed relative insensitivity of the pulmonary vascular bed to vasoactive agents, baroreceptor reflexes, and sympathetic stimulation. Their clearly defined hypoxic response led to extensive investigations by others in animals and man (for reviews, see Fishman, 1961; Daly and Hebb, 1966).

The response to acute hypoxia is consistent in most mammals examined, occurring promptly and disappearing soon after removal of the stimulus. An acute rise in pulmonary artery blood pressure is noted with 14–15% oxygen at sea level, although the usual experimental challenge is 10% oxygen or ambient air at 0.5 atm pressure. Very low oxygen tension, as used in some experiments, is acutely damaging to pulmonary capillaries, as revealed by electron microscopy. A major problem in such preparations is separation of hypoxia from tissue acidosis as the effective physiologic stimulus.

The mechanisms producing constriction of small arteries and arterioles in response to hypoxia are unresolved, since investigation of pulmonary and extrapulmonary baroreceptor and chemoreceptor reflexes and of a large number of vasoactive agents has not demonstrated consistent controlling factors. Blockade of receptors and smooth muscle contraction and of the prostaglandin cascade has often produced contradictory results and an inability to confirm isolated organ data in intact animals. It may be that hypoxia is a direct vasoconstrictor of pulmonary vascular smooth muscle, in contrast to its vasodilator effect on systemic vascular smooth muscle (Fishman, 1980).

   b. *Chronic hypoxia:* This state is found in man living at high altitude and has been reported to be associated with pulmonary hypertension (Arias-Stella and Saldaña, 1962). It has been established that low pO_2 is critical to the hypoxic response and that if continued, hypoxia could lead to persistent or chronic hypertension. Changes similar to those in man are found in all mammals except in those whose natural habitat is at a high altitude (llama, alpaca, and mountain viscacha), probably indicating a genetic adaptation to lowered oxygen tension.

   The ease with which chronic hypoxia can be produced by hypobaria has fostered many experimental studies in the rat using a standardized experimental exposure of 0.5 atm pressure. The pulmonary arterioles of the normal adult land mammal are thin-walled, with only occasional vascular smooth muscle cells. This is in marked contrast to the thick-walled systemic arterioles. In chronic pulmonary hypertension, the thin-walled arterioles are transformed into thick-walled vessels by muscularization. The source of the vascular smooth muscle has been variously attributed to activation of the relatively dormant smooth muscle, outward growth of vascular smooth muscle from proximal small arteries, pericyte transformation to vascular smooth muscle, and transformation of a specialized cell, the interstitial cell, into vascular smooth muscle (Wagenvoort and Wagenvoort, 1977; Reid, 1979; Sobin *et al.*, 1977; Rabinovitch *et al.*, 1981). Recent studies of this phenomenon in the rat at 0.5 atm pressure have shown that by 24 hours after initiation of hypoxia, fibroblasts accumulate in the abluminal arteriolar wall, and, during the course of an additional 24–96 hours, these fibroblasts differentiate or transform into typical vascular smooth muscle (Sobin *et al.*, 1983). Despite prolonged hypobaric hypoxic exposure (9 months in the rat and years in man), return to sea level leads to partial regression of both pulmonary hypertension and its typical pathologic morphology.

   As in studies of the acute hypoxic response in animals, chronic hypoxia with persistent pulmonary hypertension has produced widely divergent data with reference to hematologic responses, severity of pulmonary hypertension, growth curves, etc., but it is possible that species, age, sex specificity, or subtle differences in experimental conditions cause some of these differences in response. At present, there are no convincing data to indicate that chronic pulmonary hypertension results from the same mechanisms that produce acute pulmonary hypertension, or that chronic pulmonary hypertension is a modification of the mechanisms that produce the acute response.

8. NEURAL CONTROL MECHANISMS

Although the pulmonary parenchyma and airway systems are richly supplied with nerve tissue, the neural supply to pulmonary vasculature is variously described as modest to sparse, depending upon the species examined. In an exhaustive study of the human lung, combining both light and electron microscopy, Nagaishi (1972) described the vasomotor nerves to the pulmonary arteries and veins as extremely sparse, with fewer nerves to veins than to arteries. He did not find postganglionic nerve fibers in vessels less than 100 μm internal diameter (arterioles, capillaries, or venules). However, Rhodin (1978) did describe nerve filaments in arterioles and venules in the cat, even in the precapillary sphincter area. In the rat, such structures are rarely seen in the distal microvasculature. Hebb (1969) studied various nonhuman animal lungs by histochemical methods and clearly demonstrated adrenergic and cholinergic fibers to the pulmonary vasculature, with remarkable differences being noted among species and in different portions of the vascular bed. A major innervation was seen in larger vessels. However, this worker emphasized that the demonstration of such fibers did not establish their function.

Electrical stimulation of autonomic fibers to the pulmonary vasculature has not produced consistent results, but the consensus is that the major effect of such stimulation is to increase the stiffness of the large pulmonary arteries, with relatively small changes occurring in the resistance vessels.

With regard to reflex control of the pulmonary circulation, von Euler and Liljestrand (1946) reported an absence of significant direct stimulation and reflex control. The reflex effects on the pulmonary circulation have since been examined by various modalities, including electrical stimulation from the hypothalamic integrative areas, general systemic and central ischemia, and carotid baroreceptor and chemoreceptor stimuli. Such studies again have involved various species and experimental preparations, with nonuniform results being observed. The principal conclusion drawn from these investigations is that the various neural mechanisms may serve to modulate pulmonary vascular phenomena but do not appear to play a definitive role in physiologic control.

9. PERMEABILITY

The normal permeability of the pulmonary capillary bed to respiratory gases is a primary function of the lung. The moment-to-moment bidirectional gas exchange in the lung depends upon a normally dry capillary membrane. However, all tissues require either free or bound water for their biologic processes. The very thin respiratory membrane (respiratory epithelium and capillary endothelium separated by a fused basal lamina) is the basis for minimal diffusion distance with a harmonic mean thickness of 0.6 μm. Elastin and collagen form the elastic and structural framework for alveoli, and these connective tissues pass within the thickened basal lamina between the epithelium and endothelium and produce a potential interstitial space. The region between segments of the capillary network (posts) provide a still larger interstitial space. Thus, water and solutes are normally present, and the permeability characteristics of the endothelial barrier become very important in the physiologic economy of the lung.

The features of tissues that generally control the movement of fluids and their dissolved substances across the endothelium also apply to the pulmonary capillary bed: the characteristics of endothelial junctions, the biophysical properties of endothelial cells, and the biophysical properties of the subendothelial interstitium. The intercellular junctions of the nonfenestrated pulmonary endothelium are tight junctions, with occasional limited intermembrane spaces (Schneeberger, 1978). Inasmuch as the theoretical cellular "pores" necessary for the passage of solvent and solutes have not been seen by electron microscopy, it is likely that the intercellular junction areas are the site of fluid movement. They act as a sieve for dissolved substances relative to size. Gas molecules pass most readily through the endothelial cells, governed by their properties and by the water-impermeable cell membrane. Water moves through intercellular junctions in accordance with the physical principles governing fluid movement, as expressed by Starling (1896) and formulated by Landis (1928; see review by Landis and Pappenheimer, 1963).

A third component of fluid to be considered is the dissolved substances, which vary in size from crystalloids to proteins, and in chemical composition, primarily polarity. Small crystalloid molecules pass freely in aqueous channels over a fairly wide range of size; as the materials become increasingly polar, they are more capable of cellular passage (Chinard, 1980). Large molecular substances in the normal lung are believed to be transported by a pinocytotic mechanism that is more properly called "cytopempsis." This is a process of vesicular transport by which vesicles containing large molecules are pinched off from the plasmalemma and shuttled across the endothelial cell, carrying materials through the endothelial cell and discharging the contained substances at the basal laminal surface. In the process they are thought to form channels and fenestrae (Simionescu, 1979).

10. OTHER PARAMETERS

Other important considerations of the pulmonary circulation include pulmonary blood volume and its distribution, local vasoactive substance production and inactivation, and local enzyme storage and release. These and other functions have come under intensive investigation and frequent updating and review.

D. PHARMACOLOGY OF BLOOD CIRCULATION*

By C. N. GILLIS AND R. J. ALTIERE

Several recent reviews have focused on pharmacologic aspects of the pulmonary circulation, including autonomic innervation and regulation (Downing and Lee, 1980), effects of vasoactive substances (Bergofsky, 1980; Su and Bevan, 1976) and hypoxia (Fishman, 1976), the influence of mechanical factors on the pulmonary microvasculature (Culver and Butler, 1980), metabolic functions of the pulmonary circulation (Gillis and Pitt, 1982), and fetal pulmonary circulation (Rudolph, 1979). This section reviews briefly major pharmacologic characteristics of the adult pulmonary circulation.

1. AUTONOMIC CONTROL

a. Adrenergic actions: Sympathetic nerve stimulation (SNS) evokes varied responses of both pulmonary arteries and veins. Thus, in large pulmonary arteries (PA), SNS may decrease distensibility without altering vascular resistance (PVR), as noted by Ingram *et al.* (1968), although others (Kadowitz and Hyman, 1973; Hakim and Dawson, 1979) reported increased PVR with no change in vascular compliance. Norepinephrine (NE), in contrast, invariably increases PVR, an effect that is inhibited by adrenergic blocking agents. Since phentolamine, bretylium, or 6-hydroxydopamine blocks responses to SNS, it is likely that SNS releases NE, which, in turn, acts at α-receptors in the vasculature (Kadowitz *et al.*, 1973). Release of NE from rabbit lung vasculature has been confirmed directly (Tong *et al.*, 1978).

Isoproterenol causes β-adrenergic receptor-mediated vascular relaxation (hereafter termed β-relaxation) *in vitro* (Su and Bevan, 1976) and in the intact lung of several species, including man. Sympathetic nerve stimulation also produces β-relaxation of pulmonary vessels, suggesting that the net response to SNS reflects a balance between α- and β-receptor-mediated effects. This is

*Work from the authors' laboratory supported by Grants HLBI 13315 and HLBI 23245 from the National Institutes of Health.

consistent with the fact that propranolol increases constrictor tone, whereas α-receptor blockade enhances vasodilatory responses to NE and epinephrine (EPI) (Hyman et al., 1981). In general, however, there is considerable species variation in the net pulmonary vascular response to SNS requiring caution in extrapolating to man the results of experiments with laboratory animals.

Adrenergic receptor characteristics of small intrapulmonary arteries and veins in vitro also have been studied. Norepinephrine contracts these vessels from dog, sheep, swine, and human lungs, although there is considerable species variation (Joiner et al., 1975). Greater decreases in contractile responses to NE or transmural nerve stimulation than to 5-hydroxytryptamine or potassium are observed in progressively distal intrapulmonary arteries and veins of the rabbit (Su et al., 1978), a finding that may reflect, in part, persistence of β-relaxation in smaller intrapulmonary vessels (Altiere et al., 1981). Again, however, species variation is evident, for isolated intrapulmonary vessels of other species do not show this phenomenon (Greenberg et al., 1981).

Analysis of adrenergic receptor subtypes in pulmonary vessels has received little attention. β_1- and β_2-adrenergic receptor subtypes, both mediating relaxation, exist in rat isolated main PA (O'Donnell and Wanstall, 1981). Similar studies with other species or with intrapulmonary vessels are lacking, although β_2-receptors may mediate vasodilatation in the perfused cat lung (Hyman et al., 1981). Whereas the rabbit main PA has both pre- and postsynaptic α-adrenergic receptors (Starke et al., 1974), no evidence exists for α-receptor subtypes in intrapulmonary blood vessels. In the dog, postsynaptic α-receptors in intrapulmonary veins may differ from those in intrapulmonary arteries (Greenberg et al., 1981). Clearly, kinetic analyses of adrenergic receptors in pulmonary vascular smooth muscle are needed, since these might reveal other points of similarity or difference. Such studies might form the basis for design of pulmonary specific vasodilators, a drug category that does not now exist but that would be of great value in treatment of pulmonary hypertension.

Another neglected but important area of vascular adrenergic pharmacology concerns mechanisms that terminate effects of neuronally released NE. Inhibition of neuronal uptake with cocaine potentiates contractile response to SNS in main pulmonary artery (Bevan, 1962) and to NE in intrapulmonary vessels (Altiere et al., 1981). Hydrocortisone, which inhibits extraneuronal uptake, also potentiates NE contractile responses in canine pulmonary vessels (Greenberg et al., 1981), but apparently not in rabbit pulmonary vessels (Altiere et al., 1981).

b. Cholinergic actions: There is little agreement about the role of the parasympathetic nervous system in regulating tone of the pulmonary vascular bed (Bergofsky, 1979). Responses of intrapulmonary arteries and veins of several species to acetylcholine (ACh) are variable (Joiner et al., 1975). Isolated segments of rat, cat, and rabbit PA contract in response to ACh, whereas vagal stimulation dilates and ACh either constricts or dilates the perfused cat lung, depending on the dose used and the level of preexisting tone (Barer and Thompson, 1973; Nandiwada *et al.*, 1980). All of these responses are blocked by atropine, indicating mediation by muscarinic receptors.

2. VASOACTIVE SUBSTANCES

a. Biogenic amines: 5-Hydroxytryptamine (5-HT) is a potent pulmonary vasoconstrictor in many species, including man (Freeman *et al.*, 1981; Gruetter *et al.*, 1981). Responses to it are blocked by methysergide, suggesting involvement of D-type serotonin receptors. The physiologic role for 5-HT in regulating pulmonary vascular tone is uncertain. It apparently is not involved in hypoxic pulmonary vasoconstriction (Fishman, 1980), presynaptic regulation of NE release (Freeman *et al.*, 1981), or vasoconstriction of sensitized pulmonary vessels (Kong and Stephens, 1981).

Histamine has a dual action on pulmonary vasculature. In several species, including man, it contracts vessels via H-1 receptors and dilates them via H-2 receptors (Tucker *et al.*, 1975; Boe *et al.*, 1980). The effect depends, in part, on the intrinsic tone of the vessel. Physiologically, histamine may attenuate hypoxic pulmonary vasoconstriction (Martin *et al.*, 1978), may participate in the anaphylactic response of the lung vascular bed (Kong and Stephens, 1981), and may increase vascular permeability (Brigham and Owen, 1975).

b. Prostanoids: There are species, as well as regional, variations in vascular responses to prostaglandins (Altura and Chand, 1981). Among the cyclooxygenase products studied, PGA_1, PGE_1, and PGI_2 generally relax pulmonary vascular smooth muscle; most of the other prostanoids either lack effect or cause vasoconstriction (Hyman *et al.*, 1978). A bolus injection of arachidonic acid (AA) causes pulmonary vasoconstriction; however, slow infusion of this substance into cat lung results in vasodilatation (Hyman *et al.*, 1980). Inhibitors of cyclooxygenase (e.g., indomethacin or meclofenamate) prevent both of these responses, indicating that vasoactive prostanoids are synthesized *de novo* from administered AA. Administration of cyclooxygenase inhibitors also increases PVR in dog lung, suggesting that a vasodilator prostanoid may normally maintain pulmonary vascular dilatation (Kadowitz *et al.*, 1975a). Since PGI_2 is the only known cyclooxygenase product that causes vasodilatation in adult cat pulmonary vasculature (Hyman and Kadowitz, 1979), it is the most likely prostanoid to be involved in maintaining low vascular tone (Kadowitz *et al.*, 1981a). Rabbit PA, in contrast, contracts in response to AA, a difference that may reflect the predominant production of thromboxane A_2 rather than PGI_2 in this species (Salzman *et al.*, 1980).

Knowledge of the precise physiologic role of prostanoids in regulating pulmonary vascular tone is incomplete, and mechanisms by which they affect pulmonary vessels only now are being studied and appear to involve altered calcium and magnesium fluxes (Greenberg, 1981). The relatively few investigations concerning effects of lipoxygenase derivatives (SRS-A) and leukotrienes on pulmonary vasculature are inconclusive (Burka and Eyre, 1977; Hanna *et al.*, 1981). However, since leukotrienes are increasingly implicated in pulmonary pathophysiology, the possibility that inhibition of cyclooxygenase enzyme may "redirect" arachidonic acid through the lipoxygenase pathway to produce leukotrienes should prove to be a fruitful field of research in the future.

c. Peptides: Angiotensin II (A-II) contracts pulmonary arterial smooth muscle, an effect that is antagonized by saralasin, a specific A-II antagonist (McMurtry *et al.*, 1976). Sensitivity to A-II is species dependent. The ferret lung is unresponsive to A-II, whereas vessels of cat, rat, and dog lungs contract to varying degrees (Barer *et al.*, 1977).

Bradykinin (BK) causes profound vasodilatation in the fetal circulation (Assali *et al.*, 1971), whereas in the adult lung, it may constrict (Collier, 1970) or dilate vessels (Howard *et al.*, 1975), depending on the species. In the human lung, BK reverses pulmonary hypertension but also can cause vasoconstriction in the normotensive pulmonary circulation.

d. Vasodilators: Other substances capable of dilating pulmonary vessels are the phosphodiesterase inhibitors, theophylline and papaverine, as well as hydralazine, diazoxide, and α-adrenergic receptor antagonists (Bergofsky, 1980). Sodium nitroprusside, a potent systemic vasodilator, also has the same effect on the pulmonary vascular bed in the dog, especially when tone is high (Kadowitz *et al.*, 1981b). Possible clinical relevance of these findings was suggested by Bixler *et al.* (1981), who showed that nitroprusside reverses dopamine- or EPI-induced pulmonary vasoconstriction in dogs.

There currently is considerable interest in the effects of calcium channel blockers on the pulmonary circulation. Verapamil and nifedipine decrease elevated PVR induced by infusion of prostaglandin endoperoxide (Lippton *et al.*, 1981). Interestingly, this effect is more prominent in the pulmonary than in the systemic circulation, suggesting that calcium antagonists may be useful in the treatment of pulmonary hypertension. However, it should be noted that these agents act only when influx of extracellular calcium is involved in the contractile process under study. Thus, verapamil has different effects on 5-HT-, phenylephrine-, and potassium chloride-induced contractions of rabbit pulmonary artery (Rodger, 1979) and does not lower PVR in rat lung when tone is induced by angiotensin or $PGF_{2\alpha}$ (McMurtry *et al.*, 1976).

A recent development in pulmonary pharmacology involves the possibility of a nonadrenergic, noncholinergic (NANC) nervous system in the lung, which has been studied primarily in relation to airway smooth muscle; its association with the pulmonary vasculature also has been proposed (Richardson, 1981). Identification of the NANC neurotransmitter and the physiologic relevance of the system await further investigation.

3. Vascular Sites of Action

The longitudinal distribution of pulmonary vascular responsiveness has important mechanistic and clinical significance. Information of this type is provided by use of perfused lobe preparations or of small intrapulmonary arteries and veins in vitro (Kadowitz et al., 1975b). Recently, techniques involving bolus injections of a low viscosity solution (Grimm et al., 1978), or outflow occlusion during forward and retrograde perfusion (Hakim et al., 1979) have been used to study the distribution of vascular resistance in perfused dog lung. Both approaches give comparable data indicating that histamine causes primarily venoconstriction; NE has equipotent constrictor action on both arterial and venous segments; and $PGF_{2\alpha}$, 5-HT, hypoxia, and SNS affect primarily upstream (presumably arterial) segments. Dawson et al. (1981) have developed a double occlusion (arterial and venous) method to determine capillary pressures and sites of vascular resistance and compliance changes in perfused lung lobes. This technique may ultimately yield answers to the difficult questions concerning sites of altered resistance in the disease-injured lung.

4. Role of Endothelium

The ability of the pulmonary vascular bed to remove vasoactive substances (including NE, 5-HT, BK, and prostaglandins of the E and F series) from the pulmonary circulation has attracted the attention of many groups (see Gillis and Pitt, 1982). The primary sites of removal are believed to be endothelial cells lining the vascular lumen, especially those of the microvasculature. The possibility that endothelial cell removal processes influence pulmonary vascular responses to 5-HT was studied by Rickaby et al. (1980) using the outflow occlusion technique. These authors observed that pulmonary removal of the amine did not affect the site of 5-HT action (primarily arterial vasoconstriction). However, arterial responses to 5-HT during retrograde perfusion were reduced because of prior "upstream" removal of 5-HT by capillary endothelium.

Removal processes also may be important in certain clinical situations. Naeije et al. (1982) reported that PGE_1 is an effective vasodilator in patients with pulmonary hypertension. When infused at a rate slow enough to permit removal and inactivation by capillary endothelium, PGE_1 produces pulmonary but not systemic vasodilatation. This work supports the hypothesis of a primarily arterial site of action for PGE_1 in the lung (Hakim et al., 1979). Pulmonary uptake processes also may be responsible in part for the selective action of several calcium channel blocking agents on the pulmonary circulation (Lippton *et al.*, 1981), although direct evidence for this mechanism is lacking. Further study of sites of drug action and influence of pulmonary removal processes may provide valuable insights into therapeutic control of the pulmonary circulation.

Endothelial removal processes also have been shown to be important in terminating the actions of neuronally released NE in pulmonary vessels (Rorie and Tyce, 1981). Probably the large synaptic gap between sympathetic nerve terminals and smooth muscle effector cells in pulmonary artery diminishes the effectiveness of neuronal uptake of NE (Verity, 1971), and endothelial cell uptake of neuronally released NE assumes added significance. Thus, pulmonary endothelial cells apparently can take up NE both from the luminal and abluminal surfaces.

An intriguing area of research concerns modification of pulmonary vascular responses *in vitro* by physical removal of the endothelium. Furchgott and Zawadzki (1980) found that intact endothelial cells are required for the normal ACh-induced relaxation of the aorta. Extending this theme, Chand and Altura (1981) reported that removal of endothelium from isolated pulmonary arteries abolished ACh- or BK-induced relaxations but not those induced by a variety of other agents. In the absence of endothelium, BK and ACh caused contraction. Such mechanisms may explain the equivocal effects of ACh and BK on pulmonary vessels reported previously and also may be involved in the pathogenesis of some pulmonary vascular disorders.

5. FUTURE AVENUES FOR INVESTIGATION

Development of selective pulmonary vasodilators for treatment of pulmonary hypertensive disorders remains an elusive but important goal. More detailed knowledge of mechanisms and sites of action of pharmacologic agents and of the interrelationship between pulmonary metabolic functions and vascular responsiveness is needed for understanding the processes that normally regulate the pulmonary vascular bed.

E. PATHOPHYSIOLOGY, PATHOGENESIS, AND PATHOLOGY OF BLOOD CIRCULATION

By J. U. BALIS AND U. DESAI

Recent reviews have dealt in detail with the pathology of pulmonary blood vessels in various disease states (Wagenvoort and Wagenvoort, 1979; Heath

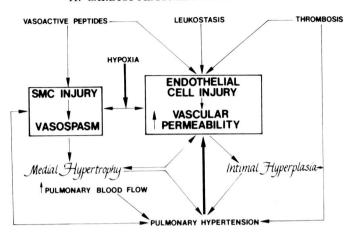

FIG. 11.11. Development and progression of hypertensive pulmonary artery disease: Interrelationships of major etiologic and pathogenetic factors operating in pulmonary arteries and arterioles.

and Smith, 1979). Causes of pathologic changes in various segments of the pulmonary vasculature include elevated pulmonary artery pressure, hypoxia, thrombosis, increased pulmonary blood flow, and inflammatory mediators. These etiologic factors often operate in combination with one another to trigger a broad range of adaptive and pathologic responses.

1. PATHOPHYSIOLOGY AND PATHOGENESIS OF BLOOD CIRCULATION

The development and progression of pulmonary artery disease are mainly dependent upon the severity and duration of structural and functional alterations in the intima and media of muscular arteries and arterioles. Endothelial damage holds a central position in the pathogenesis of vascular lesions (Mason and Balis, 1980). Injury to endothelium and underlying smooth muscle cells (SMCs) is believed to initiate a sequence of events leading to intimal hyperplasia and medial hypertrophy. A proposed interrelationship of major pathophysiologic factors (hypoxia, hypertension, vasoactive peptides) and pathogenetic factors (vasospasm, leukostasis, platelet-fibrin thrombi) operating in these vessels is shown schematically in Fig. 11.11. Certain aspects of vascular response that appear to play a basic role in the pathophysiology and pathogenesis of pulmonary vascular lesions are discussed below.

a. Vasospasm and endothelial injury: Segmental arterial "spasm," a term denoting severe vasoconstriction as a pathophysiologic event, is readily pro-

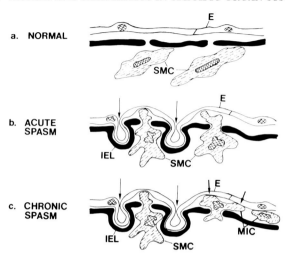

FIG. 11.12. Effects of acute and chronic spasm on small pulmonary arteries. E, endothelium; IEL, internal elastic lamina; SMC, medial smooth muscle cell; MIC, myointimal cell; long arrows, sites of endothelial injury; short arrows, myoendothelial junctions.

duced with vasoactive peptides (Joris and Majno, 1981). It results in endothelial damage caused by the tight folding of the internal elastic lamina following contraction of medial SMCs (Fig. 11.12). This type of endothelial injury leads to the development of interendothelial gaps when the artery relaxes, and this leads to increased permeability at sites of previous endothelial pinching. In addition, other experimental studies have shown that spasm-associated endothelial damage triggers formation of thrombi (Gertz, 1979). These considerations are relevant with respect to the pathogenesis of hypoxic pulmonary hypertension. Recent investigations have clearly demonstrated that severe pulmonary vascular constriction, with medial and intimal alterations similar to those seen in vasospasm, are characteristically produced by prolonged hypoxia (Smith and Heath, 1977; Dingemans and Wagenvoort, 1978). These changes were found to be independent of the mode of fixation and included folding of the internal elastic lamina, with formation of endothelial gaps containing cytoplasmic processes of contracted medial SMCs. The processes developed in places where the cells were not separated by connective tissue elements, and they often bulged through openings of the internal elastic lamina into subendothelial blisters of overlying endothelial cells. Similar structural interrelationships between contracted medial SMCs and injured endothelium have been described in small arteries and arterioles following acute endotoxemia (Balis *et al.*, 1974), which is known to induce transient increases in pulmonary vascular resistance. These findings support the concept (schematically presented in Fig. 11.12c) that vasoconstriction promotes intimal "migration" of medial SMCs and the development of myoendothelial junctions that are known to exist in pulmonary venules, arterioles, and precapillary sphincter areas (Rhodin, 1978). Such a concept is consistent with tracer studies, using horseradish peroxidase as an ultrastructural marker, that indicate that in fluid-overload pulmonary edema, the anatomic pathway of fluid leakage into the wall is at the myoendothelial junctions (Yoneda, 1980).

    b. Vasoactive peptides and leukostasis: The mammalian lung takes an active role in activation, formation, release, and inactivation of a myriad of polypeptides which, in turn, are capable of inducing either dilatation or contraction of vessels and aggregation of leukocytes and platelets, leading to intravascular stasis. The anaphylotoxins and chemotactic factors released during activation of complement may result in contraction of endothelial cells, with formations of subendothelial blebs and interendothelial gaps. Similar changes occurring in association with increased vascular permeability are observed in pulmonary vessels following exposure of the lung to chemotactic fragments of the complement fraction C5 (Desai *et al.*, 1979). Various lines of evidence indicate that intravascular activation of complement results in leukostasis, with release of various proteolytic enzymes and superoxide radicals by activated polymorphonuclear leukocytes (PMNs) (Craddock *et al.*, 1977; O'Flaherty *et al.*, 1978; Bass *et al.*, 1978). A great potential exists for damage to pulmonary vasculature induced by chemotactic factors, since resident lung cells synthesize and release various complement components and complement-derived chemotactic factors for PMNs and monocytes (Kreutzer *et al.*, 1981). In addition, Ala-Gly-Ser-Glu and Val-Gly-Ser-Glu, tetrapeptides, which are chemotactic for eosinophils, also are formed and released in the lung (Lewis and Austen, 1977). It is reasonable to conclude, therefore, that chemotactic factors play an important role in the pathophysiology of pulmonary vascular disease via margination of activated leukocytes, thus leading to either intravascular aggregation and stagnation or transvascular migration of leukocytes.

    Other vasoactive peptides have been isolated from mammalian lungs. A vasodilator peptide similar to VIP (vasoactive intestinal peptide) relaxes vascular smooth muscle and is immunohistochemically localized in mast cells, nerve cell bodies, and nerve fibers in bronchial walls and pulmonary vessels. These peptides may play an active part in pulmonary edema (Said *et al.*, 1980). Substance P, on the other hand, contracts smooth muscle and has been found in nerve fibers in the lung. Its pathophysiologic role in pulmonary vascular disease remains to be studied.

    c. Vascular injury and thrombosis: Platelet activating factor (PAF) is released from leukocytes, including IgE-sensitized basophils, in response to stimulation by C3a, C5a, neutrophil cationic proteins, and IgE-induced anaphylactic reactions. PAF induces aggregation of platelets, leading to sequestration of platelets within the pulmonary vasculature (Pinckard *et al.*, 1977). Increasing quantities of PAF result in release of 5-[^{14}C]hydroxytryptamine and adenine nucleotides and ultimately cause cytolysis of platelets. Vasoactive amines thus released from platelets have been implicated in deposition of immune complexes in blood vessels, ultimately leading to vasculitis and tissue injury (Henson and Pinkard, 1977). In addition, upon activation, platelets release various vascular permeability factors, including cationic proteins from α granules. The latter also contain platelet growth factor (PGF), a potent mitogen which recruits quiescent cells into S phase of cell cycle for further growth. Aggregation and adherence of platelets, followed by endothelial injury, inevitably expose arterial smooth muscle cells to PGF. In animals made thrombocytopenic, there is a markedly reduced rate of intimal proliferation (Friedman *et al.*, 1977). It is likely, therefore, that intimal smooth muscle cell proliferative responses to platelet-derived mitogen play an important role in the pathogenesis of intimal hyperplasia.

2. PATHOLOGY OF BLOOD CIRCULATION

a. Age-associated vascular changes: Such alterations of both pulmonary arteries and veins include widespread intimal thickening composed predominently of acellular connective tissue components, in association with elongated cells showing features of both SMCs and fibroblasts (myointimal cells or myofibroblasts). In arteries, intimal fibrosis is more cellular and the myointimal cells have more pronounced features of SMCs than in veins (Smith and Heath, 1980).

b. Elastic arteries: Intimal fibrosis and accelerated formation of atheromatous plaques are consistently found in all forms of pulmonary hypertension, regardless of the age group studied. Large fibrotic intimal plaques or bands and web-like structures spreading out into the arterial lumen are generally due to organization of emboli originating in the lower limbs.

c. Muscular pulmonary arteries and arterioles: These vessels, which play a key role in regulating arterial blood flow, develop characteristic medial and intimal changes in various forms of hypertensive pulmonary artery disease. In hypoxic pulmonary hypertension, early structural alterations include increased thickness of the media of muscular arteries and muscularization of the arterioles. In general, thickening of the muscular coat of the pulmonary arteries results from either vasoconstriction or "hypertrophy" of the media. Vasoconstriction of muscular arteries is a characteristic response to hypoxia (Fishman, 1976), and it is structurally manifested by increased thickness of the media, nuclear identations and cytoplasmic processes of medial SMCs, and crenation of elastic laminae (Dingemans and Wagenvoort, 1978). Medial hypertrophy is the most common alteration in all forms of pulmonary hypertension, and it represents the outcome of hypertrophy and, to a less extent, of hyperplasia of SMCs in association with increase accumulation of acellular connective tissue elements (Meyrick and Reid, 1979, 1982). Muscularization of the arterioles involves development of new muscle in the normally nonmuscular regions of the arteriolar walls, and this change occurs in hypoxic and hyperkinetic forms of pulmonary hypertension (Smith and Heath, 1977; Dingemans and Wagenvoort, 1978; Meyrick and Reid, 1979, 1980, 1982).

Characteristic patterns of intimal fibrosis have been described in various forms of hypertensive pulmonary vascular disease. In hypoxic pulmonary hypertension, bundles of longitudinally arranged myointimal cells are usually observed within the thickened intima. Similar changes in muscular arteries are also found in association with pulmonary venous hypertension. By contrast, concentric laminar intimal fibrosis, "plexogenic pulmonary arteriopathy," is observed in primary pulmonary hypertension and in congenital heart disease with shunt (Wagenvoort and Wagenvoort, 1979). The plexogenic pulmonary arteriopathy is characterized by proliferation of circularly oriented myointimal cells which, together with collagen fibers, form a concentric, onionlike arrangement that narrows and in some instances causes occlusion of the lumen. In more severe cases of pulmonary hypertension, the muscular arteries develop additional lesions of the whole wall, including fibrinoid necrosis, arteritis, and plexogenic pulmonary arteriopathy of the intima. The sequence of vascular alterations, from fibrinoid necrosis and arteritis to plexiform lesions, has been determined in experimental pulmonary hypertension, using creation of a shunt between systemic and pulmonary circulations as a model (Saldaña et al., 1968). However, it is necessary to point out that this type of pathologic change has not been described in association with hypoxic pulmonary hypertension.

d. Microcirculation: Endothelial damage with fluid leakage occurs in a diversity of edematous and exudative conditions of the lung (Pietra, 1978; Schneeberger, 1979; Hurley, 1982). Lesions of the endothelium, in the form of swelling, cytoplasmic vacuolation, alterations of tight junctions, formation of subendothelial blisters, and endothelial disruption or fragmentation, have been observed in various forms of acute lung injury, especially following endotoxic and hemorrhagic shock, oxygen toxicity, and other conditions leading to "shock lung." These abnormalities are often associated with leukostasis and platelet aggregates.

e. Veins: Fibrosis of the adventitia and hypertrophy of the media of veins are usually found in mitral stenosis and other conditions associated with pulmonary venous hypertension. These changes are often accompanied by some intimal fibrosis.

3. FUTURE AVENUES FOR INVESTIGATION

At the present time there is considerable need to provide accurate quantitative data of structural alterations in animal and in tissue culture models, elucidating alterations of endothelium, SMCs, and myoendothelial junctions in response to various mediators and drugs. Significant information is still lacking with respect to the morphogenesis and quantitation of vascular lesions induced by hydroxyl and oxygen radicals, arachidonic acid metabolites (especially leukotrienes), and vasoactive peptides (including those of neuroendocrine origin). Studies attempting to define receptors for various mediators may provide a powerful tool for investigating the pathophysiology of the pulmonary vasculature and for developing modalities for prevention and treatment of vascular lesions.

F. LYMPHATIC SYSTEM

By K. H. ALBERTINE AND N. C. STAUB

1. INTRODUCTION

The lungs have an extensive lymphatic system, maintaining lung fluid homeostasis and playing a role in respiratory defense mechanisms. From a hemodynamic standpoint, the lungs are different from other organs because they are perfused by the entire output of the right ventricle. There is a very large microvascular surface across which fluid filtration occurs. The pulmonary microvascular pressure, however, is considerably lower than systemic microvascular pressure, which helps reduce the total filtration rate. A unique feature of the lungs is that the airways and alveolar gas exchange surfaces may be exposed to inhaled toxins and particulate matter. Although most of these materials are transported up the airways by mucociliary mechanisms, some are removed by the lymphatics. (For general considerations of the functions of the lymphatics, see Section C, Chapter 4.)

2. STRUCTURE OF PULMONARY LYMPHATICS

Lung lymph is formed from interstitial fluid, enters the lymphatic capillaries, and is drained away from the lungs by the collecting vessels (Figs. 11.13 and 11.14).

Blood Vessels and Lymphatics in Organ Systems
Copyright © 1984 by Academic Press, Inc.

a. Lymphatic capillaries: Pulmonary lymphatic capillaries, which are formed by a continuous endothelium resting over an interrupted basal lamina (Lauweryns and Baert, 1977; Leak, 1977; Albertine *et al.*, 1982b), do not differ from their counterpart in other body regions.

The pathways for interstitial fluid transport into the capillaries include intercellular channels and intracellular routes. The intercellular channels appear to be the principal pathway for fluid and solute movement (Lauweryns and Baert, 1977; Leak, 1977). Cell contacts range from end-to-end abutment to complex interdigitation. Specialized junctional complexes are not present between all cell contacts, and gaps are seen occasionally.

Intracellular transport of fluid can occur directly through cells by diffusion. Solute movement may also take place by way of pinocytotic vesicles that are present in lymphatic endothelial cells as well as in blood vessel endothelium. The quantitative importance of the vesicular pathway is unclear and probably is insignificant.

b. Collecting lymphatics: These vessels, which contain valves, function as conduits for the transport of lymph centrally. They are lined by continuous endothelium and variable numbers of smooth muscle cells. The basal lamina of most of these vessels is fragmented.

According to Lauweryns (1971), the valves in human pulmonary lymphatics are monocuspid funnels. Other investigators, who studied valves of lymphatics from various body regions and species, disagree (Kampmeier, 1928; Albertine *et al.*, 1982a). The latter maintain that the general morphologic plan is the bicuspid valve. Opposed to Lauweryns' interpretation is the fact that monocuspid valves lack structural reinforcement to prevent inversion, whereas bicuspid valves are reinforced by buttresses. These mesenteric-like folds anchor the cusps to the vessel wall, strengthen the valves, and help them resist inversion (Albertine *et al.*, 1982a).

3. ORGANIZATION OF THE PULMONARY LYMPHATIC SYSTEM

Historically the lymphatics of the lung have been subdivided into two principal groups (Miller, 1947a; von Hayek, 1960; Nagaishi, 1972): a superficial plexus and a deep plexus. By far, the lithographs prepared by Miller (1900) are the most impressive depictions of the human pulmonary lymphatic system available. Drawings of comparable quality for the ruminants have been prepared by Baum (1912).

a. Superficial plexus: This drainage system is located in the connective tissue of the visceral pleura, although not developed to the same extent in all species. In animals with a thin visceral pleura (dog and cat), there is a relatively

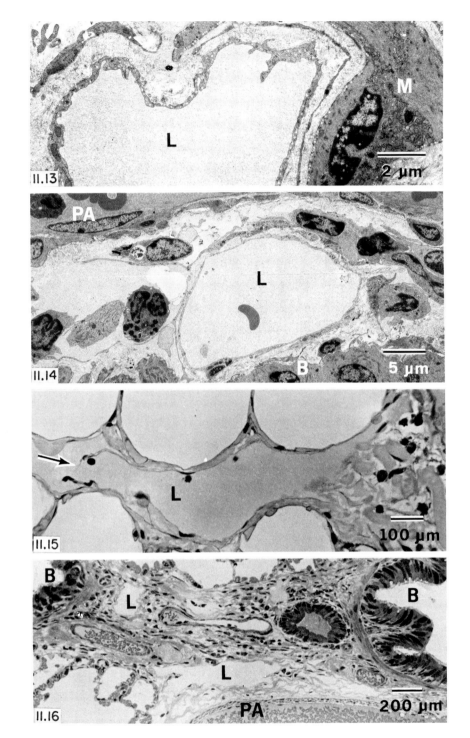

II.13

M

L

2 µm

II.14

PA

L

B

5 µm

II.15

L

100 µm

II.16

B

L

B

L

PA

200 µm

sparse system of lymphatics, whereas in those animals with a thick pleura (human, sheep, cow), the superficial plexus is more prominent. The extensive network of lymphatic capillaries and collecting lymphatics of the visceral pleura can be appreciated when these vessels are filled with opaque substances (Miller, 1947a; Lauweryns, 1971); they ramify in a complex fashion throughout this structure.

In the sheep lung, the distribution and density of pleural lymphatics have been quantified (Albertine et al., 1982b). Collecting vessels with a diameter > 100 μm have a density of 1/cm on all lobe surfaces except the costal surface of the cranial (upper) and middle lobes, where the density is < 1/cm. Smaller vessels are evenly distributed over all lobe surfaces. No comparable data in man are available.

b. Deep plexus: Peribronchovascular lymphatic capillaries and collecting lymphatics are distributed around the airways to the level of the respiratory bronchioles and alongside branches of the pulmonary arteries and veins (Figs. 11.15 and 11.16) (Miller, 1947a; von Hayek, 1960; Lauweryns and Baert, 1977; Nagaishi, 1972; Leak, 1977). There are no lymphatics in the alveolar walls (interalveolar septa). Surprisingly, no quantitative data on the distribution and density of the peribronchovascular lymphatic vessels are available.

c. Lymphoid tissue: Besides the lymphatic vessels, the lungs contain scattered patches of lymphoid tissue in the wall of the entire tracheobronchial tree and, to a lesser extent, around blood vessels (Miller, 1947a; von Hayek, 1960; Nagaishi, 1972; Bienenstock et al., 1973a,b). Apparently these aggregates develop in response to antigenic stimulation, for the lungs at birth and those of germ-free animals have little or no such tissue (Bienenstock et al., 1973a,b).

d. Drainage patterns: Lymph from human, dog, and cat lungs is carried by the deep and superficial lymphatic plexi toward the hilum. In those species endowed with long pulmonary ligaments (sheep, goat, cow), many pleural lymphatic vessels on the caudal lobes course through these structures to enter the mediastinal lymph nodes (Baum, 1912).

Fig. 11.13. Transmission electron micrograph of a sheep lung lymphatic (L). A smooth muscle cell (M) partially surrounds the vessel wall, which otherwise is composed of a continuous endothelium and fragmented basal lamina. ×6300.

Fig. 11.14. Low-power transmission electron micrograph of a peribronchovascular lymphatic vessel in a sheep lung. The luminal contents of the lymphatic (L) are palely stained compared to those of the pulmonary artery (PA). B, bronchiole. ×2200.

Fig. 11.15. Interlobular septal lymphatic (L) in a sheep lung. This lymphatic has a valve (arrow) in its lumen. ×88.

Fig. 11.16. Peribronchovascular lymphatics in a sheep lung. Two lymphatic vessels (L) are seen alongside a pulmonary artery (PA) and bronchioles (B). ×43.

Taking advantage of such a drainage pattern in the sheep, Staub *et al.* (1975) developed the lung lymph fistula preparation, whereby the efferent duct of the caudal mediastinal lymph node is cannulated. The preparation makes it possible to collect predominantly lung lymph in acute and chronic experiments designed to study the physiology and pathophysiology of lung fluid and solute exchange (Staub, 1980).

Debate persists on whether the superficial and deep systems are independent. This question arose because of the apparent orientation of valves in the interlobular septal lymphatics. According to Miller (1947a), the flow is predominantly from deep to superficial. Others have expressed the opposite point of view (Lauweryns, 1971). Lung lymph probably flows in both directions, although in sheep, it is predominantly toward the pleural surface.

In the human, pulmonary lymph flows to extrapulmonary lymph nodes located around the primary bronchi and trachea (Miller, 1947a; von Hayek, 1960; Nagaishi, 1972). In ruminants, it also enters mediastinal nodes, from which it follows an anastomotic network of paratracheal, mediastinal, and subdiaphragmatic channels (Baum, 1912; Dyon, 1973).

Ultimately lung lymph goes either to the right lymphatic duct (right bronchomediastinal duct) or to the thoracic duct. There has been much argument about which pathway is generally taken. In this regard, it must be emphasized that the patterns of pulmonary lymph drainage are complex and variable, making prediction of percentage drainage by the right bronchomediastinal duct and thoracic duct tenuous (Warren and Drinker, 1942; Dyon, 1973).

4. FUNCTION OF THE PULMONARY LYMPHATIC SYSTEM

The pulmonary lymphatics drain the interstitial space of the lung, which consists of alveolar and extraalveolar components. The alveolar interstitial compartment is exposed to alveolar pressure, whereas the remaining loose interstitial (peribronchovascular) compartment behaves as though it is protected from the full effect of alveolar pressure. Gee and Havill (1978) have shown that the lymph capillaries and the extraalveolar interstitial connective tissue spaces are arranged in parallel with each other and in series with the alveolar wall interstitium. Most importantly, they noted that interstitial fluid entering the lymphatics comes directly from microvascular filtrate. Thus, in pulmonary edema, interstitial fluid represents the contribution of microvascular filtrate at an earlier time than that of the lung lymph.

The pulmonary lymphatic system not only functions to maintain fluid and solute balance for the lung (Staub, 1980), but it also plays a significant role in pulmonary defense mechanisms (Nagaishi, 1972; Leak, 1977). An important contribution to both antibody-mediated and cell-mediated immune responses is apparently made by the tracheobronchial lymphoid tissue. The lymphatic vessels are also involved in metastasis of lung cancer and in the development of some respiratory diseases.

REFERENCES*

Albertine, K. H., Fox, L. M., and O'Morchoe, C. C. C. (1982a). The morphology of canine lymphatic valves. *Anat. Rec.* **202**, 453–461. (F)

Albertine, K. H., Wiener-Kronish, J. P., Roos, P. J., and Staub, N. C. (1982b). Structure, blood supply and lymphatic vessels of the sheep's visceral pleura. *Am. J. Anat.* **165**, 277–294. (B, F)

Altiere, R. J., Douglas, J. S., and Gillis, C. N. (1983). Norepinephrine responses in rabbit pulmonary arteries and veins *in vitro*. *J. Pharmacol. Exp. Ther.* **224**, 579–589. (D)

Altura, B. M., and Chand, N. (1981). Differential effects of prostaglandins on canine intrapulmonary arteries and veins. *Br. J. Pharmacol.* **73**, 819–827. (D)

Arias-Stella, J., and Saldaña, M. (1962). The muscular pulmonary arteries in people native to high altitude. *Med. Thorac.* **19**, 292–301. (C)

Assali, N. S., Johnson, G. H., Brinkman, C. R., III, and Huntsman, D. J. (1971). Effects of bradykinin on the fetal circulation. *Am. J. Physiol.* **221**, 1375–1382. (D)

Balis, J. U., Gerber, L. I., Rappaport, E. S., and Neville, W. E. (1974). Mechanisms of blood-vascular reactions of the primate lung to acute endotoxemia. *Exp. Mol. Pathol.* **21**, 123–137. (E)

Bannister, J., and Torrance, R. W. (1960). The effects of tracheal pressure upon flow: Pressure relations in the vascular bed of isolated lungs. *Q. J. Exp. Physiol. Cogn. Med. Sci.* **45**, 352–367. (C)

Barer, G. R., and Thompson, B. (1973). Acetylcholine and the pulmonary circulation. *J. Physiol. (London)* **230**, 47P–48P. (D)

Barer, G. R., Mohammed, F. H., and Suggett, A. J. (1977). Angiotensin, hypoxia, verapamil and pulmonary vessels. *J. Physiol. (London)* **270**, 43P–44P. (D)

Bass, D. A., DeChatalet, L. R., and McCall, C. E. (1978). Independent stimulation of motility and the oxidative metabolic burst of human polymorphonuclear leukocytes. *J. Immunol.* **121**, 172–178. (E)

Baum, H. (1912). "Das lymphgefassystem des Rindes." Hirschwald, Berlin. (F)

Bergofsky, E. H. (1979). Active control of the normal pulmonary circulation. *Lung Biology. Health Dis.* **14**, 233–277. (D)

Bergofsky, E. H. (1980). Humoral control of the pulmonary circulation. *Annu. Rev. Physiol.* **42**, 221–233. (C, D)

Bevan, J. A. (1962). Some characteristics of the isolated sympathetic nerve–pulmonary artery preparation of the rabbit. *J. Pharmacol. Exp. Ther.* **137**, 213–218. (D)

Bhattacharya, J., and Staub, N. C. (1980). Direct measurement of microvascular pressures in isolated, perfused dog lung. *Science* **210**, 327–328. (B, C)

Bienenstock, J., Johnston, N., and Perey, D. Y. E. (1973a). Bronchial lymphoid tissue. I. Morphologic characteristics. *Lab. Invest.* **28**, 686–692. (F)

Bienenstock, J., Johnston, N., and Perey, D. Y. E. (1973b). Bronchial lymphoid tissue. II. Functional characteristics. *Lab. Invest.* **28**, 693–698. (F)

Bixler, T. J., Gott, V. L., and Gardner, T. J. (1981). Reversal of experimental pulmonary hypertension with sodium nitroprusside. *J. Thorac. Cardiovasc. Surg.* **81**, 537–545. (D)

*In the reference list, the capital letter in parentheses at the end of each reference indicates the section in which it is cited.

Boe, J., Boe, M. A., and Simonsson, B. G. (1980). A dual action of histamine on isolated human pulmonary arteries. *Respiration* **40**, 117–122. (D)

Boyden, E. A. (1970). The developing bronchial arteries in a fetus of the twelfth week. *Am. J. Anat.* **129**, 357. (A)

Brigham, K. L., and Owen, P. J. (1975). Increased sheep lung vascular permeability caused by histamine. *Circ. Res.* **37**, 647–657. (D)

Burka, J. F., and Eyre, P. (1977). Effects of bovine SRS-A (SRS-A[bov]) on bovine respiratory tract and lung vasculature *in vitro*. *Eur. J. Pharmacol.* **44**, 169–177. (D)

Burton, A. C. (1959). The relation between pressure and flow in the pulmonary bed. *In* "Pulmonary Circulation—International Symposium, 1958" (W. R. Adams, Jr. and I. Veith, eds.), pp. 26–35. Grune & Stratton, New York. (C)

Caro, C. G., Pedley, T. J., Schroter, R. C., and Seed, W. A. (1978). "The Mechanics of the Circulation," pp. 89–90, 96. Oxford Univ. Press, London and New York. (C)

Chand, N., and Altura, B. M. (1981). Acetylcholine and bradykinin relax intrapulmonary arteries by acting on endothelial cells: Role in lung vascular diseases. *Science* **213**, 1376–1379. (D)

Chinard, F. P. (1980). The alveolar–capillary barrier: Some data and speculations. *Microvasc. Res.* **19**, 1–17. (C)

Collier, H. O. J. (1970). Kinins and ventilation of the lungs. *Handb. Exp. Pharmacol.* **25**, 409–420. (D)

Congdon, E. D. (1922). The transformation of the aortic-arch system during the development of the human embryo. *Carnegie Inst. Washington Publ.* **277**, 47. (A)

Corliss, C. E. (1976). "Patten's Human Embryology." McGraw-Hill, New York. (A)

Cournand, A. (1958). Control of the pulmonary circulation in normal man. *In* "Circulation (Proceedings of the Harvey Tercentenary Congress)" (J. McMichael, ed.), pp. 219–237. Thomas, Springfield, Illinois. (C)

Craddock, P. R., Fehr, J., Bringham, K. L., Viorenenberg, R. S., and Jacob, H. S. (1977). Complement and leukocyte mediated pulmonary dysfunction in hemodialysis. *N. Engl. J. Med.* **296**, 769–774. (E)

Culver, B. H., and Butler, J. (1980). Mechanical influences on the pulmonary microcirculation. *Annu. Rev. Physiol.* **42**, 187–198. (C, D)

Daly, I. deB., and Hebb, C. (1966). "Pulmonary and Bronchial Vascular Systems." Williams & Wilkins, Baltimore, Maryland. (C)

Dawson, C. A., Rickaby, D. A., and Linehan, J. H. (1981). Pulmonary capillary pressure in the dog lung as determined by lobar arterial and venous occlusion. *Fed. Proc., Fed. Am. Soc. Exp. Biol.* **40**, 503. (D)

Desai, U., Kreutzer, D. L., Showell, H., Arroyave, C. V., and Ward, P. A. (1979). Acute inflammatory pulmonary reaction induced by chemotactic factors. *Am. J. Pathol.* **96**, 71–83. (E)

Dingemans, K. P., and Wagenvoort, C. A. (1978). Pulmonary arteries and veins in experimental hypoxia. *Am. J. Pathol.* **93**, 353–361. (E)

Dobrin, P. B. (1983). Vascular mechanics. *In* "Handbook of Physiology" (J. T. Shepherd and F. M. Abboud, eds.), 2nd ed., Sect. 2, Vol. III. Am. Physiol. Soc., Bethesda, Maryland, 65–102. (C)

Downing, S. E., and Lee, J. C. (1980). Nervous control of the pulmonary circulation. *Annu. Rev. Physiol.* **42**, 199–210. (D)

Dyon, J. F. (1973). Contribution à l'étude du drainage des lymphatiques du poumon. Ph.D. Dissertation, Université Scientifique et Médicale de Grenoble. (F)

Elliott, F. M., and Reid, L. (1965). Some new facts about the pulmonary artery and its branching patterns. *Clin. Radiol.* **16**, 193–198. (B)

Fishman, A. P. (1961). Respiratory gases in the regulation of the pulmonary circulation. *Physiol. Rev.* **41**, 214–280. (C)

Fishman, A. P. (1976). Hypoxia on the pulmonary circulation. How and where it acts. *Circ. Res.* **38**, 221–231. (D, E)

Fishman, A. P. (1980). Vasomotor regulation of the pulmonary circulation. *Annu. Rev. Physiol.* **43**, 211–220. (C, D)

Freeman, W. K., Rorie, D. K., and Tyce, G. M. (1981). Effects of 5-hydroxytryptamine on neuroeffector junction in human pulmonary artery. *J. Appl. Physiol.: Respir., Environ. Exercise Physiol.* **51**, 693–698. (D)

Friedman, R. J., Stemerman, M. B., Moore, S., Gauldie, J., Gent, M. C., and Spaet, T. H. (1977). The effect of thrombocytopenia on experimental arteriosclerotic lesion formation in rabbits. *J. Clin. Invest.* **60**, 1191–1201. (E)

Fung, Y. C. B. (1973). Stochastic flow in capillary blood vessels. *Microvasc. Res.* **5**, 34–48. (C)

Fung, Y. C. B., and Sobin, S. S. (1969). Theory of sheet flow in lung alveoli. *J. Appl. Physiol.* **26**, 472–488. (B, C)

Fung, Y. C. B., and Sobin, S. S. (1972a). Elasticity of the pulmonary alveolar sheet. *Circ. Res.* **30**, 451–469. (C)

Fung, Y. C. B., and Sobin, S. S. (1972b). Pulmonary alveolar blood flow. *Circ. Res.* **30**, 470–490. (C)

Fung, Y. C. B., Zweifach, B. W., and Intaglietta, M. (1966). Elastic environment of the capillary bed. *Circ. Res.* **19**, 441–461. (C)

Furchgott, R. F., and Zawadski, J. V. (1980). The obligatory role of endothelial cells in the relaxation of arterial smooth muscle by acetylcholine. *Nature (London)* **288**, 373–376. (C, D)

Gee, M. H., and Havill, A. M. (1978). The relationship between pulmonary perivascular cuff fluid and lung lymph in dogs with edema. *Microvasc. Res.* **19**, 209–216. (F)

Gertz, S. D. (1979). Vascular damage and thrombosis from spasm. *N. Engl. J. Med.* **300**, 197. (E)

Gil, J., Bachofen, H., Gehr, P., and Weibel, E. R. (1979). Alveolar volume-surface area relation in air- and saline-filled lungs fixed by vascular perfusion. *J. Appl. Physiol.: Respir., Environ. Exercise Physiol.* **47**(5), 990–1001. (C)

Gillis, C. N., and Pitt, B. R. (1982). The fate of circulating amines within the pulmonary circulation. *Annu. Rev. Physiol.* **44**, 269–281. (D)

Glazier, J. B., Hughes, J. M. B., Maloney, J. E., and West, J. B. (1969). Measurements of capillary dimensions and blood volume in rapidly frozen lungs. *J. Appl. Physiol.* **26**, 65–76. (C)

Greenberg, S. (1981). Effect of prostacyclin and 9a,11a-epoxymethanoprostaglandin H_2 on calcium and magnesium fluxes and tension development in canine intralobar pulmonary arteries and veins. *J. Pharmacol. Exp. Ther.* **219**, 326–337. (D)

Greenberg, S., Kadowitz, P. J., Hyman, A., and Curro, F. A. (1981). Adrenergic mechanisms in canine intralobar pulmonary arteries and veins. *Am. J. Physiol.* **240**, H274–H1285. (D)

Grimm, D. J., Dawson, C. A., Hakim, T. S., and Linehan, J. H. (1978). Pulmonary vasomotion and the distribution of vascular resistance in a dog lung lobe. *J. Appl. Physiol.: Respir., Environ. Exercise Physiol.* **45**, 545–550. (D)

Gruetter, C. A., Ignarro, L. J., Hyman, A. L., and Kadowitz, P. J. (1981). Contractile effects of 5-hydroxytryptamine in isolated intrapulmonary arteries and veins. *Can. J. Physiol. Pharmacol.* **59**, 157–162. (D)

Guntheroth, H. G., Luchtel, D. L., and Kawabori, I. (1982). Pulmonary microcirculation: Tubules rather than sheet-and-post? *J. Appl. Physiol.* **53**, 510–515. (B)

Hakim, T. S., and Dawson, C. A. (1979). Sympathetic nerve stimulation and vascular resistance in a pump-perfused dog lung lobe. *Proc. Soc. Exp. Biol. Med.* **160**, 38–41. (D)

Hakim, T. S., Dawson, C. A., and Linehan, J. H. (1979). Hemodynamic responses of dog lung lobe to lobar venous occlusion. *J. Appl. Physiol.: Respir., Environ. Exercise Physiol.* **47**, 145–152. (D)

Hanna, C. J., Bach, M. K., Pare, P. D., and Schellenberg, R. R. (1981). Slow-reacting substances (leukotrienes) contract human airway and pulmonary vascular smooth muscle *in vitro*. *Nature (London)* **290**, 343–344. (D)

Heath, D., and Smith, P. (1979). Pulmonary vascular disease secondary to lung disease. In "Pulmonary Vascular Disease" (K. M. Moser, ed.), pp. 387–426. Dekker, New York. (E)

Hebb, C. (1969). Motor innervation of the pulmonary blood vessels of mammals. In "The Pulmonary Circulation and Interstitial Space" (A. P. Fishman and H. H. Hecht, eds.), pp. 195–222. Univ. of Chicago Press, Chicago, Illinois. (C)

Henson, P. M., and Pinckard, R. N. (1977). Platelet activating factor. Monogr. Allergy 12, 13–26. (E)

Hislop, A., and Reid, L. (1978). Normal structure and dimensions of the pulmonary arteries in the rat. J. Anat. 175, 71–84. (A, B)

Howard, P., Barer, G. R., Thompson, B., Warren, P. M., Abbott, C. J., and Mungall, I. P. F. (1975). Factors causing and reversing vasoconstriction in unventilated lung. Respir. Physiol. 24, 325–345. (D)

Hurley, J. V. (1982). Types of pulmonary microvascular injury. In "Mechanisms of Lung Microvascular Injury" (A. B. Malik and N. C. Staub, eds.), Part IV, pp. 269–289. N.Y. Acad. Sci., New York. (E)

Hyman, A. L., and Kadowitz, P. J. (1979). Pulmonary vasodilator activity of prostacyclin (PGI_2) in the cat. Circ. Res. 45, 404–409. (D)

Hyman, A. L., Spannhake, E. W., and Kadowitz, P. J. (1978). Prostaglandins and the lung. Am. Rev. Respir. Dis. 117, 111–136. (D)

Hyman, A. L., Spannhake, E. W., and Kadowitz, P. J. (1980). Divergent actions of arachidonic acid on the feline pulmonary vascular bed. Am. J. Physiol. 239, H40–H46. (D)

Hyman, A. L., Nandiwada, P., Knight, D. S., and Kadowitz, P. J. (1981). Pulmonary vasodilator responses to catecholamines and sympathetic nerve stimulation in the cat. Evidence that vascular β-2 adrenoceptors are innervated. Circ. Res. 48, 407–415. (D)

Ingram, R. H., Szidon, J. P., Skalak, R., and Fishman, A. P. (1968). Effects of sympathetic nerve stimulation on the pulmonary arterial tree of the isolated lobe perfused in situ. Circ. Res. 22, 801–815. (D)

Inselman, L. S., and Mellins, R. B. (1981). Growth and development of the lung. J. Pediatr. 98, 1. (A)

Joiner, P. D., Kadowitz, P. J., Hughes, J. P., and Hyman, A. L. (1975). NE and ACh responses of intrapulmonary vessels from dog, swine, sheep and man. Am. J. Physiol. 228, 1821–1827. (D)

Joris, I., and Majno, G. (1981). Endothelial changes induced by arterial spasm. Am. J. Pathol. 102, 346–358. (E)

Kadowitz, P. J., and Hyman, A. L. (1973). Effect of sympathetic nerve stimulation on pulmonary vascular resistance in the dog. Circ. Res. 32, 221–227. (D)

Kadowitz, P. J., Joiner, P. D., and Hyman, A. L. (1973). Differential effects of phentolamine and bretylium on pulmonary vascular responses to norepinephrine and nerve stimulation. Proc. Soc. Exp. Biol. Med. 144, 172–176. (D)

Kadowitz, P. J., Chapnick, B. M., Joiner, P. D., and Hyman, A. L. (1975a). Influence of inhibitors of prostaglandin synthesis on the canine pulmonary vascular bed. Am. J. Physiol. 229, 941–946. (D)

Kadowitz, P. J., Joiner, P. D., Hyman, A. L., and George, W. J. (1975b). Influence of prostaglandins E_1 and $F_{2\alpha}$ on pulmonary vascular resistance, isolated lobar vessels and cyclic nucleotide levels. J. Pharmacol. Exp. Ther. 192, 677–687. (D)

Kadowitz, P. J., Gruetter, C. A., Spannhake, E. W., and Hyman, A. L. (1981a). Pulmonary vascular responses to prostaglandins. Fed. Proc., Fed. Am. Soc. Exp. Biol. 40, 1991–1996. (D)

Kadowitz, P. J., Nandiwada, P., Gruetter, C. A., Ignarro, L. J., and Hyman, A. L. (1981b). Pulmonary vasodilator responses to nitroprusside and nitroglycerin in the dog. J. Clin. Invest. 67, 893–902. (D)

Kampmeier, O. F. (1928). The genetic history of the valves in the lymphatic system of man. Am. J. Anat. 40, 413–457. (F)

Knisely, M. H. (1965). Intravascular erythrocyte aggregation (blood sludge). In "Handbook of

Physiology" (W. F. Hamilton, ed.), Sect. 2, Vol. III, pp. 2249–2292. Am. Physiol. Soc., Washington, D.C. (C)

Knowlton, F. P., and Starling, E. H. (1912). The influence of variations in temperature and blood-pressure on the performance of the isolated mammalian heart. *J. Physiol. (London)* **44**, 206–219. (C)

Kong, S. K., and Stephens, N. L. (1981). Pharmacological studies of sensitized canine pulmonary blood vessels. *J. Pharmacol. Exp. Ther.* **219**, 551–557. (D)

Kreutzer, D. L., Desai, U., Douglas, W. H. J., and Blazka, M. (1981). Mechanisms of inflammation in lung tissue. *In* "Hamster Immune Responses in Infections and Oncologic Diseases" (D. L. Kreutzer, J. S. Streilein, J. Stein-Streilein, W. R. Duncan, and R. E. Billingham, eds.), pp. 221–230. Plenum, New York. (E)

Lai-Fook, S. J. (1979). A continuum mechanics analysis of pulmonary vascular interdependence in isolated dog lobes. *J. Appl. Physiol.* **46**, 419–429. (C)

Lai-Fook, S. J., and Hyatt, R. E. (1979). Effect of parenchyma and length changes on vessel pressure-diameter behavior in pig lungs. *J. Appl. Physiol.* **47**, 666–669. (C)

Landis, E. M. (1928). Microinjection studies of capillary permeability. III. The effect of lack of oxygen on the permeability of capillary wall to fluid and plasma proteins. *Am. J. Physiol.* **83**, 528–542. (C)

Landis, E. M., and Pappenheimer, J. R. (1963). Exchange of substances through the capillary walls. *In* "Handbook of Physiology" (W. F. Hamilton and P. Dow, eds.), Sect. 2, Vol. II, pp. 961–1034. Am. Physiol. Soc., Washington, D.C. (C)

Lauweryns, J. M. (1971). Stereomicroscopic funnel-like architecture of pulmonary lymphatic valves. *Lymphology* **4**, 125–132. (F)

Lauweryns, J. M., and Baert, J. H. (1977). Alveolar clearance and the role of the pulmonary lymphatics. *Am. Rev. Respir. Dis.* **115**, 625–683. (F)

Leak, L. V. (1977). Pulmonary lymphatics and their role in the removal of interstitial fluids and particulate matter. *In* "Respiratory Defense Mechanisms" (J. D. Brain, D. F. Proctor, and L. M. Reid, eds.), pp. 631–685. Dekker, New York. (F)

Lee, G. de J., and Dubois, A. B. (1955). Pulmonary capillary blood flow in man. *J. Clin. Invest.* **34**, 1380–1390. (C)

Lewis, R. A., and Austen, K. F. (1977). Nonrespiratory functions of pulmonary cells. *Fed. Proc., Fed. Am. Soc. Exp. Biol.* **36**, 2676–2683. (E)

Licata, R. (1962). Pulmonary circulation. *In* "Blood Vessels and Lymphatics" (D. I. Abramson, ed.), pp. 292–294. Academic Press, New York. (A)

Lippton, H. L., Nandiwada, P. A., Kadowitz, P. J., and Hyman, A. L. (1981). Vasodilator actions of nifedipine and verapamil in the pulmonary vascular bed. *Fed. Proc., Fed. Am. Soc. Exp. Biol.* **40**, 590. (D)

McMurtry, I. F., Davidson, A. B., Reeves, J. T., and Grover, R. F. (1976). Inhibition of hypoxic pulmonary vasoconstriction by calcium antagonists in isolated rat lungs. *Circ. Res.* **38**, 99–104. (D)

Martin, L. F., Tucker, A., Munroe, M. L., and Reeves, J. T. (1978). Lung mast cells and hypoxic pulmonary vasoconstriction in cats. *Respiration* **35**, 73–77. (D)

Mason, R. G., and Balis, J. U. (1980). Pathology of the endothelium. *In* "Pathobiology of Cell Membranes" (B. F. Trump and A. U. Arstila, eds.), Vol. 2, pp. 425–472. Academic Press, New York. (E)

Meyrick, B. O., and Reid, L. (1979). Hypoxia and incorporation of ^{3}H-thymidine by cells of the rat pulmonary arteries and alveolar wall. *Am. J. Pathol.* **96**, 51. (E)

Meyrick, B. O., and Reid, L. (1980). Ultrastructural findings in lung biopsy material from children with congenital heart defects. *Am. J. Pathol.* **101**, 527–537. (E)

Meyrick, B. O., and Reid, L. M. (1982). Crotalaria-induced pulmonary hypertension: Uptake of ^{3}H-thymidine by the cells of the pulmonary circulation and alveolar walls. *Am. J. Pathol.* **106**, 84–94. (E)

Miller, W. S. (1900). Das lungenläppchen, seine blut- und lymphgefässe. *Arch. Anat. Physiol.*, *Anat. Abt.* 3/4, 197–228. (F)

Miller, W. S. (1947a). "The Lung," 2nd ed., pp. 75–77. Thomas, Springfield, Illinois, (B, C)

Miller, W. S. (1947b). "The Lung," 2nd ed., pp. 89–118. Thomas, Springfield, Illinois. (F)

Milnor, W. R. (1972). Pulmonary hemodynamics. In "Cardiovascular Fluid Dynamics" (D. H. Bergel, ed.), Vol. 2, pp. 299–340. Academic Press, New York. (C)

Miyamoto, Y., and Moll, W. A. (1971). Measurements of dimensions pathway of red blood cells in rapidly frozen lungs *in situ*. *Respir. Physiol.* 12, 141–156. (C)

Naeije, R., Melot, C., Mols, P., and Hallemans, R. (1982). Reduction in pulmonary hypertension by prostaglandin E_1 in decompensated chronic obstructive pulmonary disease. *Am. Rev. Respir. Dis.* 125, 1–5. (D)

Nagaishi, C. (1972). "Functional Anatomy and Histology of the Lung." University Park Press, Baltimore, Maryland. (B, C, F)

Nakahara, K., Ohkuda, K., and Staub, N. C. (1979). Effect of infusing histamine into pulmonary or bronchial artery on sheep pulmonary fluid balance. *Am. Rev. Respir. Dis.* 120, 875–882. (B)

Nandiwada, P. A., Kadowitz, P. J., and Hyman, A. L. (1980). Pulmonary vasodilator resonses to vagal nerve stimulation in the cat. *Circulation* 62, Suppl. III, 133. (D)

Neill, C. A. (1956). Development of the pulmonary veins with reference to the embryology of anomalies of pulmonary return. *Pediatrics* 18, 880. (A)

O'Donnell, S. R., and Wanstall, J. C. (1981). Demonstration of both β_1- and β_2-adrenoceptors mediating relaxation of isolated ring preparations of rat pulmonary artery. *Br. J. Pharmacol.* 74, 547–552. (D)

O'Flaherty, J. T., Kreutzer, D. L., and Ward, P. A. (1978). Chemotactic factor influences on the aggregation, swelling and adhesiveness of human leukocytes. *Am. J. Pathol.* 90, 537–550. (E)

Overholser, K. A., Bhattacharya, J., and Staub, N. C. (1982). Microvascular pressures in the isolated, perfused dog lung: Comparison between theory and measurement. *Microvasc. Res.* 23, 67–76. (C)

Perelman, R. H., Engle, M. J., and Farelli, P. M. (1981). Perspectives on fetal lung development. *Lung* 159, 53. (A)

Permutt, S., Bromberger-Barnea, B., and Bane, H. N. (1962). Alveolar pressure, pulmonary venous pressure and the vascular waterfall. *Med. Thorac.* 19, 47–68. (C)

Pietra, G. G. (1978). The basis of pulmonary edema, with emphasis on ultrastructure. In "The Lung Structure, Function and Disease" (W. M. Thurlbeck and M. R. Abel, eds.), pp. 215–234. Williams & Wilkins, Baltimore, Maryland. (E)

Pinckard, R. N., Halonen, M., Palmer, J. D., Butler, C., Shaw, J. O., and Henson, P. M. (1977). Intravascular aggregation and pulmonary sequestration of platelets during IgE-induced systemic anaphylaxis in the rabbit: Abrogation of lethal anaphylactic shock by platelet. *J. Immunol.* 119, 2185–2193. (E)

Rabinovitch, M., Gamble, W. J., Miettinen, O. S., and Reid, L. (1981). Age and sex influence on pulmonary hypertension of chronic hypoxia and on recovery. *Am. J. Physiol.* 240 (*Heart Circ. Physiol.* 9), H62–H72. (C)

Reid, L. M. (1979). The pulmonary circulation: Remodeling in growth and disease. *Am. Rev. Respir. Dis.* 119, 531–546. (C)

Rhodin, J. A. G. (1978). Microscopic anatomy of the pulmonary vascular bed in the cat lung. *Microvasc. Res.* 15, 169–193. (B, C, E)

Richardson, J. B. (1981). Nonadrenergic inhibitory innervation of the lung. *Lung* 159, 315–322. (D)

Rickaby, D. A., Dawson, C. A., and Maron, M. B. (1980). Pulmonary inactivation of serotonin and site of serotonin pulmonary vasoconstriction. *J. Appl. Physiol.: Respir., Environ. Exercise Physiol.* 48, 606–612. (D)

Rodger, I. W. (1979). Effects of Ca^{2+} withdrawal and verapamil on excitation–contraction coupling in rabbit pulmonary vascular smooth muscle. *Br. J. Pharmacol.* 67, 467P. (D)

Rorie, D. K., and Tyce, G. M. (1981). The importance of endothelium in the economy of nor-epinephrine released from adrenergic nerve endings in canine pulmonary artery. *Physiologist* **24,** 74. (D)

Rosenquist, T. H., Bernick, S., Sobin, S. S., and Fung, Y. C. (1973). The structure of the pulmonary interalveolar microvascular sheet. *Microvasc. Res.* **5,** 199–212. (C)

Rudolph, A. M. (1979). Fetal and neonatal pulmonary circulation. *Annu. Rev. Physiol.* **41,** 383–395. (D)

Said, S. I., Mutt, V., and Erdos, E. G. (1980). The lung in relation to vasoactive polypeptides. *Ciba Found. Symp.* **78,** 217–237. (E)

Saldaña, M. E., Harley, R. A., Liebow, A. A., and Carrington, C. B. (1968). Experimental extreme pulmonary hypertension and vascular disease in relation to polycythemia. *Am. J. Pathol.* **52,** 935–981. (E)

Salzman, P. M., Salmon, J. A., and Moncada, S. (1980). Prostacyclin and thromboxane A_2 synthesis by rabbit pulmonary artery. *J. Pharmacol. Exp. Ther.* **215,** 240–247. (D)

Schneeberger, E. E. (1978). Structural basis for some permeability properties of the air–blood barrier. *Fed. Proc., Fed. Am. Soc. Exp. Biol.* **37,** 2471–2478. (C)

Schneeberger, E. E. (1979). Barrier function of intercellular junctions in adult and fetal lungs. *In* "Pulmonary Edema" (A. P. Fishman and E. M. Renkin, eds.), pp. 21–37. Am. Physiol. Soc., Bethesda, Maryland. (E)

Simionescu, M. (1979). Transendothelial movement of large molecules in the microvasculature. *In* "Pulmonary Edema" (A. P. Fishman and E. M. Renkin, eds.), pp. 39–52. Am. Physiol. Soc., Bethesda, Maryland. (C)

Singal, S., Henderson, R., Horsfield, K., Harding, K., and Cumming, G. (1973). Morphometry of the human pulmonary artery tree. *Circ. Res.* **33,** 190–197. (C)

Smith, P., and Heath, D. (1977). Ultrastructure of hypoxic hypertensive pulmonary vascular disease. *J. Pathol.* **121,** 93–100. (E)

Smith, P., and Heath, D. (1980). The ultrastructure of age-associated intimal fibrosis in pulmonary blood vessels. *J. Pathol.* **130,** 247–253. (E)

Sobin, S. S., and Fung, Y. C. (1967). A sheet-flow concept of the pulmonary alveolar microcirculation. *Physiologist* **10,** 308 (abstr.). (C)

Sobin, S. S., and Tremer, H. M. (1980). Cylindricity of the arterial tree in the dog and cat. *Fed. Proc., Fed. Am. Soc. Exp. Biol.* **39,** 269 (abstr.). (C)

Sobin, S. S., Tremer, H. M., and Fung, Y. C. (1970). Morphometric basis of the sheet-flow concept of the pulmonary alveolar microcirculation in the cat. *Circ. Res.* **26,** 397–414. (C)

Sobin, S. S., Fung, Y. C., Tremer, H. M., and Rosenquist, T. H. (1972). Elasticity of the pulmonary alveolar microvascular sheet in the cat. *Circ. Res.* **30,** 440–450. (C)

Sobin, S. S., Lindal, R. G., and Bernick, S. (1977). The pulmonary arteriole. *Microvasc. Res.* **14,** 227–239. (C)

Sobin, S. S., Lindal, R. G., Fung, Y. C., and Tremer, H. M. (1978). Elasticity of the smallest noncapillary pulmonary blood vessels in the cat. *Microvasc. Res.* **15,** 57–68. (C)

Sobin, S. S., Fung, Y. C., Lindal, R. G., Tremer, H. M., and Clark, L. (1980). Topology of pulmonary arterioles, capillaries, and venules in the cat. *Microvasc. Res.* **19,** 217–233. (C)

Sobin, S. S., Tremer, H. M., Hardy, J. D., and Chiodi, H. P. (1983). Changes in arteriole in acute and chronic hypoxic pulmonary hypertension in rat. *J. Appl. Physiol.: Respir. Environ. Exercise Physiol.* **55**(5), 1445–1455. (C)

Starke, K., Montel, H., Gayk, W., and Merker, R. (1974). Comparison of the effects of clonidine on pre- and postsynaptic adrenoceptors in the rabbit pulmonary artery. *Naunyn-Schmiedeberg's Arch. Pharmacol.* **285,** 133–150. (D)

Starling, E. H. (1896). On the absorption of fluids from the connective tissue spaces. *J. Physiol. (London)* **19,** 312–326. (C)

Staub, N. C. (1980). The pathogenesis of pulmonary edema. *Prog. Cardiovasc. Dis.* **23,** 53–80. (F)

Staub, N. C., and Schultz, E. L. (1968). Pulmonary capillary length in dog, cat and rabbit. *Respir. Physiol.* **5**, 371–378. (C)

Staub, N. C., and Storey, W. F. (1962). Relation between morphological and physiological events in lung studied by rapid freezing. *J. Appl. Physiol.* **17**, 381–390. (C)

Staub, N. C., Bland, R. D., Brigham, K. L., Demling, R., Erdmann, A. J., and Woolverton, W. C. (1975). Preparation of chronic lung lymph fistulas in sheep. *J. Surg. Res.* **19**, 315–320. (F)

Strahler, A. N. (1957). Quantitative analysis of watershed geomorphology. *Trans. Am. Geophys. Union* **38**, 913–920. (C)

Su, C., and Bevan, J. A. (1976). Pharmacology of pulmonary blood vessels. *Pharmacol. Ther., Part B* **2**, 275–288. (D)

Su, C., Bevan, R. D., Duckles, S. P., and Bevan, J. A. (1978). Functional studies of the small pulmonary arteries. *Microvasc. Res.* **15**, 37–44. (C, D)

Tong, E. Y., Mathe, A. A., and Tisher, P. W. (1978). Release of norepinephrine by sympathetic nerve stimulation from rabbit lungs. *Am. J. Physiol.* **235**, H803–H808. (D)

Tucker, A., Weir, E. K., Reeves, J. T., and Grover, R. F. (1975). Histamine H_1- and H_2-receptors in pulmonary and systemic vasculature of the dog. *Am. J. Physiol.* **229**, 1008–1013. (D)

Verity, M. A. (1971). Morphologic studies of vascular neuroeffector apparatus. *Physiol. Pharmacol. Vasc. Neuroeff. Syst., Proc. Symp., 1969*, pp. 2–12. (D)

von Euler, U. S. and Liljestrand, G. (1946). Observations on the pulmonary arterial blood pressure in the cat. *Acta Physiol. Scand.* **12**, 301–320. (C)

von Hayek, H. (1960). "The Human Lung," pp. 298–314. Hafner, New York. (B, F)

Wagenvoort, C. A., and Wagenvoort, N. (1977). "Pathology of Pulmonary Hypertension." Wiley, New York. (C)

Wagenvoort, C. A., and Wagenvoort, N. (1979). Pulmonary vascular bed. *In* "Pulmonary Vascular Disease" (K. M. Moser, ed.), pp. 1–109. Dekker, New York. (B, E)

Warrell, D. A., Evans, J. W., Clarke, R. O., Kingaby, G. P., and West, J. B. (1972). Pattern of filling in the pulmonary capillary bed. *J. Appl. Physiol.* **32**, 346–356. (C)

Warren, M. F., and Drinker, C. K. (1942). The flow of lymph from the lungs of the dog. *Am. J. Physiol.* **136**, 207–221. (F)

Weibel, E. R. (1963). Principles and methods for the morphometric study of the lung and other organs. *Lab. Invest.* **12**, 131–155. (B)

Weibel, E. R., and Gomez, D. M. (1962). Architecture of the human lung. *Science* **137**, 577–585. (C)

West, J. B. (1977). Blood flow. *In* "Regional Differences in the Lung" (J. B. West, ed.), pp. 85–165. Academic Press, New York. (B)

West, J. B., Dollery, C. T., and Naimark, A. (1964). Distribution of blood flow in isolated lung; Relation to vascular and alveolar pressures. *J. Appl. Physiol.* **19**, 713–724. (C)

West, J. B., Schneider, A. M., and Mitchell, M. M.(1975). Recruitment in networks of pulmonary capillaries. *J. Appl. Physiol.* **39**, 976–984. (C)

Yen, R. T., and Foppiano, L. (1981). Elasticity of small pulmonary veins in the cat. *J. Biomech. Eng.* **103**, 38–42. (C)

Yen, R. T., and Fung, Y. C. (1978). Effect of velocity distribution on red cell distribution in capillary blood vessels. *Am. J. Physiol.* **235**(2), H251–H257. (C)

Yen, R. T., Fung, Y. C., and Bingham, N. (1980). Elasticity of small pulmonary arteries in the cat. *J. Biomech. Eng.* **102**, 170–177. (C)

Yoneda, K. (1980). Anatomic pathway of fluid leakage in fluid-overload pulmonary edema in mice. *Am. J. Pathol.* **101**, 7–13. (E)

Chapter 12
Digestive System: Esophagus and Stomach

A. EMBRYOLOGY OF BLOOD CIRCULATION
By C. E. CORLISS

1. ESTABLISHMENT OF THE ESOPHOGASTRIC CIRCULATION

a. Early development: In the fourth week, the paired dorsal aortae of the human embryo send out three sets of branches to the surrounding tissues: dorsal segmental, lateral segmental, and ventral segmental arteries. The dorsal segmentals are destined for the central nervous system (see Section A, Chapter 5 and Section A, Chapter 6), the laterals, for the urogenital structures, and the ventrals (actually the first to appear), for the primitive digestive tract (Evans, 1912).

Only the dorsal segmentals are segmental. The ventrals tend to arise more frequently from the aorta, and during the first month, many of them extend ventrad to anastomose with the vascular plexus over the yolk sac (Evans, 1912). This union of embryonic with extraembryonic vessels (together with the accompanying veins) creates the vitelline arc of embryonic circulation. The paired vitelline vessels soon fuse to form the three unpaired vessels to the digestive tract—the celiac, superior mesenteric, and inferior mesenteric arteries of adult anatomy (see Section A-1a and 1b, Chapter 13).

b. Celiac artery: At about 1 month, the seventh and eighth cervical segmental arteries form the primordium of the celiac artery (trunk). This can be followed in the direction of the future stomach and liver region of the digestive complex (Evans, 1912). All three of the unpaired vessels supplying the gut are found at successively lower levels as development proceeds. This "caudal wandering" (Evans, 1912) shifts the base of the celiac caudad from its site of origin at C7 to T12, and after $6\frac{1}{2}$ weeks, it is stabilized at the adult level of L1.

Several secondary branches of the celiac artery have been identified in a 43-day-old human embryo injected by Mall (quoted by Evans, 1912). These include the inferior phrenic, left gastric with esophageal branches, and hepatic artery. Another branch present at this stage is the anterior superior pancreatico-duodenal artery. The point of anastomosis between this vessel and the

Blood Vessels and Lymphatics in Organ Systems
Copyright © 1984 by Academic Press, Inc.

anterior inferior pancreatico-duodenal (from the superior mesenteric) marks the junction between the original embryonic foregut and midgut. In the 43-day-old embryo, several small twigs are also present, approaching the esophagus directly from the aorta at T6 and T7 (Evans, 1912).

B. ANATOMY OF BLOOD CIRCULATION

By K. R. LARSEN AND F. G. MOODY

1. EXTRAMURAL VASCULAR SUPPLY

a. Arteries: The arterial supply to the esophagus and stomach is redundant. All three branches of the celiac artery supply the stomach through two concentric circles (Fig. 12.1). The one on the lesser curvature consists of the left gastric, right gastric, and common hepatic arteries (clockwise in the figure). The circle to the greater curvature includes (clockwise) the splenic, left gastroepiploic, right gastroepiploic, gastroduodenal, and common hepatic arteries. Additional arterial blood supply is available from the esophageal branches of the thoracic aorta, through the esophageal submucosal plexus, and into the left gastric artery. Also, the gastroduodenal artery can receive arterial blood via the

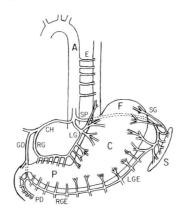

FIG. 12.1. Named arteries of the esophagus and stomach. A, aorta; C, corpus of stomach; CH, common hepatic; E, esophageal branches of aorta; F, fundus of stomach; GD, gastroduodenal; LG, left gastric; LGE, left gastroepiploic; P, pylorus of stomach; PD, pancreaticoduodenal; RG, right gastric; RGE, right gastroepiploic; S, spleen; SG, short gastric; SP, splenic.

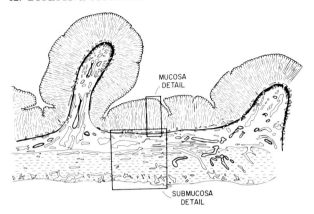

FIG. 12.2. Section of anterior corpus of dog stomach showing abundance and complexity of intramural vessels and their involvement in a rugal fold. All outlined structures within the muscularis externa (bottom layer) and submucosa (center) are arteries and veins.

pancreaticoduodenal arteries, which are supplied by the superior mesenteric artery. Branches of the splenic artery to the gastric fundus (top of Fig. 12.1) are the short gastric arteries. (For further discussion, see Wheaton *et al.*, 1981.)

b. Veins: These vessels run parallel to the arteries, with the gastric and splenic veins emptying into the portal system. The esophageal veins are tributaries of the inferior thyroid veins (top of Fig. 12.1), azygos and hemiazygos veins (middle), and short gastrics or left gastric (coronary) vein (bottom). Portosystemic shunting of venous blood can occur via the left gastric vein, esophageal submucosal plexus, and azygous veins. The subsequent engorgement of the submucosal plexus leads to esophageal varices.

2. Intramural Vascular Supply

Besides the external vessels described above, there may be as many as eight additional levels of arterial plexi or anastomosing networks, some of which, however, may be absent in parts of the esophagus and stomach. The abundance and complexity of intramural vessels and their involvement in a rugal fold are shown in Fig. 12.2, drawn from a 4 μm section through the anterior corpus of a dog stomach. Inserts show the location of schematic drawings of submucosal detail (Fig. 12.3) and mucosal detail (Fig. 12.4). The veins are not included in Fig. 12.3 because they parallel the arterial pattern. (For further discussion, see Barlow *et al.*, 1951.)

a. Subserous plexus: Immediately after piercing the serosa, the gastric arteries give rise to the subserous plexus (SE in Fig. 12.3) (Piasecki, 1980). This

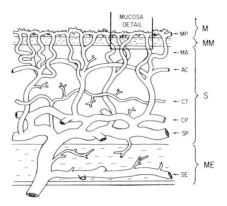

FIG. 12.3. Detail of submucosal arteries. Names are mostly from Barlow *et al.* (1951). AC, anastomosing channels of mucosal arteries; CT, connective tissue plexus; CP, cross-anastomosing channels of submucosal plexus; M, mucosa; MA, mucosal artery; ME, muscularis externa; MM, muscularis mucosae; MP, mucosal plexus; S, submucosa; SE, subserous plexus; SP, submucosal plexus.

plexus provides the origin of vessels that serve the superficial layers of the muscularis externa (ME in Fig. 12.3).

b. Submucosal plexus: The main branches of the gastric (and esophageal) vessels traverse the muscularis externa and form an anastomosing network— the submucosal plexus (SP in Fig. 12.3). The arteries in this network are about

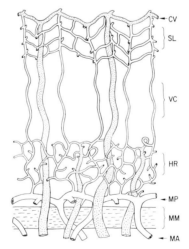

FIG. 12.4. Detail of mucosal arteries, veins, and capillaries. The arrangement of the veins is controversial, the existence of deep mucosal venous drainage being debatable. CV, collecting venule; HR, horizontal capillary ramifications; MA, mucosal artery; MM, muscularis mucosae; MP, mucosal plexus; SL, surface capillary loops; VC, vertical capillaries.

200 μm in diameter (Barlow *et al.*, 1951). They cover the entire stomach and esophagus except for the lesser curvature, where mucosal arteries are derived directly from extramural vessels. Smaller branches from the submucosal plexus perfuse the inner layers of the muscularis externa.

c. Cross-anastomosing channels: The main submucosal plexus is interconnected by loops of cross-anastomosing channels (CP in Fig. 12.3) deeper in the submucosa. These vessels are about 150 μm in diameter (Barlow *et al.*, 1951).

d. Connective tissue plexus: This vascular network (CT in Fig. 12.3) arises from the main or cross-anastomosing channels of the submucosal plexus. The extensive vessels (10–100 μm in diameter) making up the plexus fill the submucosal spaces without further communication with the mucosal arteries or the submucosal plexus.

e. Mucosal arteries: Vertical branches from the submucosal plexus (or cross-anastomosing channels) traverse the submucosa and branch into 3 or 4 mucosal arteries (MA in Figs. 12.3 and 12.4) before entering the muscularis mucosae (MM). According to Barlow *et al.* (1951), these vessels intercommunicate before piercing the muscularis mucosae (MM). The anastomosing channels (AC) of mucosal arteries are about 50 μm in diameter and form a network on the submucosal side of the muscularis mucosae (Fig. 12.3).

f. Mucosal plexus: On the mucosal side of the muscularis mucosae, the mucosal arteries again form an extensive plexus (MP in Figs. 12.3 and 12.4). Mucosal capillaries arise vertically from the horizontal plexus.

g. Capillaries: Barlow *et al.* (1951) described the main vessels of the mucosa as vertical vessels (VC in Fig. 12.4) about 20 μm in diameter, which appear to fill the substance of the mucosa. Between them are found smaller (8 μm) horizontal capillary ramifications (HR in Fig. 12.4), which provide a thick meshwork among the gastric glands. They may or may not communicate with collecting venules (CV) before reaching the surface region.

Capillary loops near the surface (SL in Fig. 12.4) form a meshwork around the gastric pits, giving a honeycomb appearance from above (Reynolds and Kardon, 1981). They are intermediate in diameter (15 μm) between the vertical capillaries and deeper horizontal ramifications (Barlow *et al.*, 1951).

h. Collecting venules: Capillary surface loops enter collecting venules (CV in Fig. 12.4) near the surface cells, and the latter vessels join other mucosal veins, descend the thickness of the mucosa, and pierce the muscularis mucosae. Barlow *et al.* (1951) reported that there was no venous plexus on the glandular side of the muscularis mucosa in man, and Guth (1981) noted a similar absence in the rat. However, Bulkley *et al.* (1970) did find both an arterial and a venous plexus on the glandular side of the muscularis mucosae in the rabbit. The differences may be due to species variations, different techniques of visualization, or both.

3. End Arteries

Piasecki (1980) reported that mucosal arteries become end arteries (with no distal arterial anastomoses) on the glandular side of the muscularis mucosae and that capillary beds within the mucosa do not communicate with similar beds of adjacent mucosal arteries. Most mucosal arteries communicate freely on the submucosal side of the muscularis mucosae and within the submucosal plexus itself. However, such connections are lacking in certain parts of the lesser curvature. Barlow *et al.* (1951) reported that gastric arteries on the lesser curvature pierce the muscularis externa and the submucosa without forming a plexus. Furthermore, Piasecki (1980) noted that the lack of a submucosal plexus is most pronounced in the distal third of the lesser curvature and the first $\frac{1}{2}$-inch of the duodenum. Such extramural "end arteries" form the anatomic basis of a "local ischemia" hypothesis for peptic and duodenal ulcers (Piasecki, 1980).

4. Arteriovenous Anastomoses

Barlow *et al.* (1951) presented histologic and glass microsphere data to demonstrate the existence of channels, as large as 140 μm in diameter, between arteries and veins of the submucosal plexus and also between mucosal arteries and their adjacent veins. Serial sections have shown that these anastomotic connections are thick-walled and are surrounded by "musculoepithelial"-type cells. On the other hand, Guth (1981) found no evidence of functioning arteriovenous shunts during *in vivo* microscopy in the gastric microcirculation of the rat. The possible physiologic role of arteriovenous anastomoses and, indeed, their very existence, in the gastric submucosa remain controversial.

5. Antral-Body Portal System

Taylor and Torrance (1975) presented evidence for the existence of an antral-body portal system. Their ^{86}Rb movement supports the hypothesis that gastrin may be transported directly to the parietal cell mass without passing through the systemic circulation. Piasecki (1980) speculated that Barlow's connective tissue plexus (CT in Fig. 12.3) may be the anatomic system for such an antral-body transport. In this regard, it is anatomically and theoretically possible for venous spasm and antral muscle contractions to work in harmony, forcing antral venous blood into the connective tissue plexus and from there to the body of the stomach.

C. PHYSIOLOGY OF BLOOD CIRCULATION

By B. L. Tepperman and E. D. Jacobson

The present section deals briefly with the physiology of the gastric circulation. For a more comprehensive discussion of the subject, the reader is referred to the reviews by Lanciault and Jacobson (1976) and Guth and Ballard (1981).

1. Circulatory Response to a Meal

The major physiologic stimulus in the alimentary tract is the introduction of food. Ingestion of food increases total blood flow to the stomach via the celiac artery and also raises mucosal blood flow (Chou *et al.*, 1976). It is not apparent as to which factors are responsible for the observed postprandial hyperemia, a response that could be mediated by either the autonomic nervous system, hormones, paracrines, or changes in the metabolic environment associated with stimulation of secretion and motility. In many cases, administration of parasympathomimetic or sympathomimetic drugs, gastrointestinal peptides, and paracrine substances can mimic the effect of ingestion of a meal, although there is no certainty about the physiologic significance of responses to large amounts of specific substances given under experimental circumstances. Therefore, in this section, blood flow is described only as it relates to physiologic functions of the stomach.

a. Relationship to acid secretion: Interventions that initiate gastric acid secretion generally also increase gastric mucosal blood flow (Jacobson *et al.*, 1966b). Agents, such as gastrin and histamine, that stimulate both gastric acid secretion and blood flow may either relax vascular smooth muscle directly, have their effect indirectly through activation of oxyntic glandular metabolic

Blood Vessels and Lymphatics in Organ Systems
Copyright © 1984 by Academic Press, Inc.

activity, or operate through both mechanisms. Conversely, agents that diminish both secretion and blood flow could do so through a primary antisecretory action on the oxyntic cell or via direct vasoconstriction. In the nonsecreting stomach, such a drug as isoproterenol raises blood flow without promoting acid secretion; however, vasodilatation may further augment secretion in the stomach by increasing the rate of delivery of secretagogue and nutrients to the oxyntic cell. Such findings suggest that mucosal blood flow can be a rate-limiting step in the process of gastric secretion, but that an increase in mucosal circulation by itself will not initiate active secretion. Finally, it has been found repeatedly that *in vitro* segments of the oxyntic mucosa of a number of species will secrete acid in response to secretagogues, indicating that blood flow per se is not essential to the initiation of acid secretion (Tepperman *et al.*, 1975). However, the maximal rates of secretion *in vitro* are low, compared with the rates measured in the stomachs of conscious animals. In summary, like many functions upon which a variety of cells depend, an adequate blood flow plays a permissive and supportive role in acid secretion.

   b. Relationship to gastric motility: Changes in blood flow within an organ are reciprocally related to alterations in vascular resistance, which, in turn, reflect active changes in the smooth muscle of resistance vessels and passive influence of transmural pressure. Vasoactive agents induce active changes in resistance vessels, whereas muscular action affects resistance passively through extravascular compression. Gastric motor activity is evident *in vitro*, whereas vasopressin decreases blood flow *in vivo* without affecting motility (Schuurkes and Charbon, 1978). Therefore, alterations in blood flow appear not to be essential for the maintenance of normal motor activity. Rather the relationship appears to be in the direction of alterations in motility sometimes affecting blood flow, but not the reverse.

   Although the effect of muscular activity on local blood flow in the stomach has not been studied as extensively as in the small intestine (Walus and Jacobson, 1981), it is likely that the relationship between gastric motor activity and blood flow may be similar to that in the gut. Distention of the canine stomach to 20 cm H_2O evokes a profound reduction in blood flow to the mucosa and submucosa of the corpus without altering circulation to the adjacent muscularis (Edlich *et al.*, 1970). Distention of an exteriorized loop of small bowel redistributes blood flow through the manipulated segment, but it does not affect total or compartmental flow within the gastric body (Chou and Grassmick, 1978). Rhythmic contractions of the stomach are accompanied by a corresponding rise and fall in instantaneous gastric arterial and venous flows (Semba *et al.*, 1970). Administration of physostigmine prompts a sustained tonic contraction and decreased flow to the whole wall of the stomach, with the reduced circulation confined to the mucosa and submucosa (Chou and Grassmick, 1978). The effects of administration of gastrointestinal hormones, such as gastrin and cholecystokinin, on gastric motility and blood flow are variable (Schuurkes and Charbon, 1978). The vascular changes induced by exogenously administered peptides may reflect the direct vasoactive properties of these compounds, as well as their ability to alter gastric motility.

    c. Relationship to regulation of mucosal integrity: The precise mechanisms involved in the development of acute and chronic gastric erosions or ulcers are unknown. Although alterations in gastric mucosal blood flow have been proposed as a causal factor in their etiology, proof is lacking that vascular changes represent events other than those that are secondary to the development of the ulcerative lesion.

    Several lines of evidence suggest that some degree of gastric mucosal ischemia may be present in patients who develop stress ulcers (Lucas *et al.*, 1971). Similarly, in experimental animals, direct observation of the gastric mucosa reveals the development of erosions at sites of apparent ischemia (Merserau and Hinchey, 1973), although increased blood flow has also been observed in the area immediately surrounding the ulcer (Skarstein *et al.*, 1979). Gastric lesions induced with a topically applied bile salt are significantly increased in size by simultaneous reduction of gastric mucosal blood flow with vasopressin (Ritchie, 1975); vasodilator agents have been observed to ameliorate the damaging actions of ulcerogenic agents (McGreevy and Moody, 1977).

    It has been noted that ulcerogenic agents directly alter gastric mucosal blood flow, with most experiments showing an increase in local circulation in response to their intraluminal administration (Ritchie, 1975; Cheung and Chang, 1977). In most cases, however, damage accompanies an augmentation in blood flow only if combined with increased back-diffusion of acid (Puurunen, 1980). The hyperemia has been thought to represent an important compensatory mechanism by which the gastric mucosa attempts to clear or neutralize excessive intramural H^+ (Mersereau and Hinchey, 1978). However, if the capacity of the gastric mucosa to dispose of back-diffusing H^+ is compromised (as occurs during ischemia induced by hemorrhage or vasopressin), ulceration results (Starlinger *et al.*, 1981). The actual cause of the increased blood flow appears to be the direct action of H^+ on the blood vessels (Bruggeman *et al.*, 1979). Alternatively, it may be that an adequate mucosal blood flow is necessary for the maintenance of a HCO_3-dependent protective system that neutralizes back-diffusing H^+ (Starlinger *et al.*, 1981).

    Against the thesis that ischemia is a significant etiologic factor in the formation of gastric erosions is the finding that a reduction in mucosal blood flow does not lead to an increase in H^+ back-diffusion (Davenport and Barr, 1973). Ulcerative lesions may appear even in the absence of both H^+ influx and ischemia (Moody and Aldrete, 1971). In some instances, gastric blood flow is found to be unchanged even 3 days after induction of gastritis in cats (Screide *et al.*, 1980). Thus, doubt must be expressed about the critical nature of the vasculature in the early stages of gastric erosions and ulcers. (For a more detailed discussion of the pathophysiology of peptic ulcer, see Section E-2a, below.)

d. Intrinsic vascular regulatory mechanisms: The ability of the intestinal vascular bed to evade continuous constriction imposed by prolonged stimulation of sympathetic nerves or infusion of catecholamines or angiotensin II has been termed "autoregulatory escape" (Shehadeh *et al.*, 1969). The recovery of flow usually occurs 1 to 2 min after the onset of escape and appears to be independent of autoregulation. A similar phenomenon is observed in the celiac artery and the gastric vascular bed (Guth and Smith, 1977). *In vivo* microscopic observations of the gastric microcirculation indicate that escape is due to relaxation of initially constricted vessels and not to a general redistribution of flow by shunting through arteriovenous anastomoses. The submucosal arterioles appear to be the segment controlling mucosal blood flow during escape, with constriction decreasing mucosal blood flow and dilatation increasing it.

Autoregulation refers to an intrinsic mechanism within a vascular bed that allows maintenance of a steady blood flow over a range of arterial pressures. Although the phenomenon is noted consistently in the mesenteric circulation, the stomach appears to be a variable autoregulator of its own blood flow. In sympathetically innervated preparations, autoregulation of gastric blood flow is not evident (Jacobson *et al.*, 1962). However, the ability of the stomach to autoregulate blood flow and oxygen uptake is significantly augmented after denervation (Holm-Rutili *et al.*, 1981). Reduction in perfusion pressure in the denervated stomach decreases vascular resistance, whereas oxygen uptake stays constant.

The exact mechanism ascribed to autoregulation is controversial. Two models have been considered to explain this function. In one, the myogenic model, an increase in vascular transmural pressure produces active arteriolar constriction and an increase in vascular resistance. In the other, the metabolic model, oxygenation of tissues rather than blood flow is regulated and a decrease in the ratio of tissue O_2 availability to O_2 demand relaxes arteriolar and/or precapillary smooth muscle. It has been suggested that both myogenic and metabolic factors may be involved in the intrinsic autoregulation of gastric blood flow and oxygenation (Granger and Kvietys, 1981). The exposure of autoregulation after sympathetic denervation may result from either a reduction in an artificially excessive sympathetic tone or an increase in gastric O_2 demand; another possibility is that it may not be present in the intact stomach of conscious animals with an intact sympathetic nerve supply.

Increased perfusion observed during secretagogue-induced secretion reflects either a direct vascular action of the stimulant on vascular muscle or the release of a vasodilator agent (Gerkens *et al.*, 1977b) associated with increased O_2 consumption during acid secretion. However, it is likely that the O_2 demand during secretion is met by increases in blood flow rather than by enhanced O_2 extraction (Cheung *et al.*, 1978). Hence, little evidence exists supporting a metabolic control system linking oxidative metabolism and blood flow.

D. PHARMACOLOGY OF BLOOD CIRCULATION

By B. L. Tepperman and E. D. Jacobson

Many of the agents described in this section are endogenous to the stomach and may influence the gastric vasculature in a physiologic manner. However, most of the information obtained about these substances is based upon experiments in which large exogenous doses were administered. Although their physiologic significance remains to be proved, nevertheless, these studies have elucidated mechanisms whereby vasoactive agents control the gastric circulation.

1. Autonomic Nervous System

a. Sympathetic stimulation and catecholamine release: Electrical stimulation of the gastric sympathetic nerves or infusion of catecholamines reduces total and mucosal blood flow (Reed *et al.*, 1971; Jacobson *et al.*, 1966b). Under *in vitro* conditions, both norepinephrine and epinephrine contract gastric vessels (Van Hee and Vanhoutte, 1978); α-adrenergic blockade inhibits such a response, whereas β blockade is ineffective (Yano *et al.*, 1981). However, β receptors appear to mediate a portion of the increase in blood flow associated with autoregulatory escape. Only inconsistent information is available regarding the actions of dopamine on the gastric circulation. It dilates the canine left gastric artery but constricts the splenic artery (Ross and Brown, 1967); it has also been reported to have no effect on gastric mucosal blood flow or tissue oxygenation (Bowen *et al.*, 1981).

b. Vagus and acetylcholine release: Vagal stimulation appears to increase gastric mucosal blood flow (Swan and Jacobson, 1967), a response that apparently is independent of its secondary effect on stimulation of gastric acid secretion (Guth and Smith, 1975), whereas vagotomy decreases mucosal blood flow (Hunter *et al.*, 1979). Administration of exogenous acetylcholine provokes both

Blood Vessels and Lymphatics in Organ Systems
Copyright © 1984 by Academic Press, Inc.

increases (Yano *et al.*, 1981) and decreases in gastric blood flow (Necheles *et al.*, 1936). However, such vascular effects are difficult to dissociate from the action of the drug on acid secretion, gastric motility, and cardiac output.

2. HORMONES

The amine precursor uptake decarboxylase-containing cells of the gastrointestinal mucosa produce and release a variety of peptides into the blood, extracellular space, and gastric lumen. Exogenous administration of some peptides affects the gastric circulation and, to complicate matters further, also strongly influences other gastric functions or alters hepatic, pancreatic, and intestinal function; such responses, in turn, may influence the circulation of the stomach indirectly. In many of the experiments from which this information derives, the doses of exogenous administration of the peptide were well beyond the physiologic range, thereby prompting responses not normally taking place.

Gastrin and its tetrapeptide analogue increase both acid secretion and gastric blood flow (Jacobson *et al.*, 1966b). The vascular response in tissue appears to be secondary to the rise in acid secretion, suggesting that vasodilatation is being mediated through an indirect metabolic effect. The structurally similar peptide, cholecystokinin (CCK), is also a potent gastric vasodilator (Guth and Smith, 1976), as is the peptide, metenkephalin (Konturek *et al.*, 1978). However, two other members of the gastrin family, motilin and somatostatin, reduce gastric blood flow secondary to inhibition of acid secretion (Konturek *et al.*, 1977, 1981).

By analogy, the structurally related members of the secretin family (secretin, vasoactive intestinal peptide, and glucagon) decrease gastric mucosal blood flow through their inhibitory effects on gastric secretion (Jacobson *et al.*, 1966b; Konturek *et al.*, 1976; Bond and Levitt, 1980). Only when these hormones are administered by intraarterial injection can one discriminate between their direct and indirect effects on the gastric circulation. In such experiments, both secretin and CCK (compounds that are unrelated structurally) produce vasodilatation in the stomach (Guth and Smith, 1976).

Another hormone that affects the gastric circulation is vasopressin. This pituitary peptide decreases gastric secretion and mucosal blood flow by its predominant vasoconstrictor action (Jacobson *et al.*, 1966b). Chronic administration of the glucocorticoid hormone-like steroid, prednisone, increases both total and mucosal blood flow and is associated with the development of gastric lesions (Zamora *et al.*, 1980). Chronic use of this drug permissively augments gastric secretory responses to gastrin, possibly by raising mucosal blood flow (Jacobson and Price, 1969).

3. PROSTANOIDS

The natural prostaglandins and their methylated analogues inhibit gastric acid secretion and reduce gastric mucosal blood flow (Jacobson, 1970; Cheung, 1980). The circulatory effect probably is a consequence of the inhibition of the secretory process and of decreased metabolic activity of the parietal cells, in view of the fact that normally prostaglandins E_1 and E_2 have been found to be potent vasodilator agents in the resting stomach (Jacobson, 1970; Cheung, 1980; Walus et al., 1980a,b). However, a decrease in resting mucosal blood flow after administration of prostaglandin E_2 has also been reported (Konturek et al., 1980a). Prostacyclin has been found both to increase blood flow in the resting gastric mucosa (Konturek et al., 1980b; Walus et al., 1980a) and to decrease it (Kauffman et al., 1979). The stable metabolite of prostacyclin, 6-ketoprostaglandin $F_{1\alpha}$, does not alter gastric blood flow (Walus et al., 1980b). When blood flow is elevated during gastric secretion or when vascular tone is raised by norepinephrine, prostacyclin becomes a more potent dilator than prostaglandin E_2 (Konturek et al., 1980b; Salvati and Whittle, 1981).

The precursor of the prostaglandin, arachidonic acid, in constant blood flow experiments increases perfusion pressure to the stomach, signifying heightened vascular resistance, whereas prostaglandin E_2 and prostacyclin decrease pressure, indicating decreased resistance (Walus et al., 1980a). Since indomethacin blocks the first response, it may be assumed that some metabolite of arachidonate (possibly thromboxane) is responsible for the gastric vasoconstriction. By itself, indomethacin decreases unstimulated gastric mucosal blood flow (Kauffman et al., 1980). Hydroperoxy products of arachidonate have some vasodilator activity, but the hydroxylipooxygenase products are inactive (Salvati and Whittle, 1981).

4. HISTAMINE

Like gastrin, histamine increases both gastric acid secretion and mucosal blood flow (Jacobson et al., 1966a); however, the circulatory response appears to combine a direct pharmacologic vasodilatation with the indirect functional dilatation secondary to stimulation of secretion (Jacobson and Chang, 1969).

It has been suggested that histamine mediates its action via two separate receptors, termed H_1 and H_2. There is, however, some controversy with respect to which receptor mediates the increase in secretion versus mucosal blood flow. Many of the differences of opinion may stem from the species of animal under investigation, the doses of agonist employed, the antagonist drugs selected for study, and the methodology utilized for measurement of gastric blood flow. Both H_1 and H_2 receptors are active in histamine-induced gastric vasodilatation in the cat (Harvey *et al.*, 1980), a species in which the interaction between histamine and its vascular receptors involves first H_1 receptor responses and then H_2 receptor responses. When small doses of H_1 and H_2 antagonists are used in man, it is difficult to find evidence for either H_1 or H_2 receptors within the gastric circulation (Knight *et al.*, 1980), whereas in other species, both are demonstrable, although they appear to be quantitatively and functionally different. In the dog, the two receptors apparently mediate vasodilatation, although H_2 receptors predominate (Charbon *et al.*, 1980). By contrast, in the rabbit H_1 receptors predominate and may mediate vasoconstriction, whereas H_2 receptors produce a dilator effect (Curwain and Turner, 1981).

5. Cyclic Nucleotides

Cyclic adenosine monophosphate (cAMP) and its dibutyryl derivative (N^6-2-O-dibutyryl-3′,5′-cyclic AMP) are vasodilator agents with regard to the gastric circulation (Mao *et al.*, 1972). Furthermore, drugs which increase the synthesis of cAMP in some tissues (glucagon, isoproterenol, prostaglandin E_1) or inhibit its metabolism (papaverine, theophylline) also dilate gastrointestinal blood vessels. The vasodilator action of dibutyryl cAMP does not appear to depend upon the gastric secretory response to the drug (Konturek *et al.*, 1980a). Papaverine significantly increases total gastric blood flow, although epithelial P_{O_2} and total gastric O_2 consumption are depressed (Bowen *et al.*, 1981). Sodium fluoride, a nonspecific stimulant of adenylate cyclase, has been shown to decrease gastric mucosal blood flow and acid secretion. Thus, the significance of cAMP as a regulator of gastric blood flow is unresolved.

6. Serotonin

It has been suggested that serotonin, a potent inhibitor of gastric acid secretion (Haverback *et al.*, 1958), exerts this action, in part, through a direct modulation of gastric blood flow; however, it still retains the ability to inhibit acid secretion even under *in vitro* conditions (Canfield and Spencer, 1981). Administration of low doses of serotonin in the rat elicits an increase in gastric mucosal blood flow, whereas higher doses produce a decrease (Yano *et al.*, 1981). This is analogous to the situation observed when the drug is administered into the mesenteric circulation (Fara, 1976).

E. PATHOPHYSIOLOGY, PATHOGENESIS, AND PATHOLOGY OF BLOOD CIRCULATION

By J. M. McGreevy and F. G. Moody

1. Esophagus

The contribution of blood vessels to esophageal diseases has not been the subject of much investigation. Of the problems commonly encountered in medicine today, only three have been examined for a vascular etiology: achalasia, atresia, and esophagitis.

a. Achalasia: This is a motor disorder of the esophagus which results in a dilated organ with poorly coordinated peristalsis. The distal esophageal sphincter does not relax in response to swallowing, which causes functional distal esophageal obstruction. Earlham (1972a) advanced the hypothesis that achalasia is due to a vascular obstruction in the developing embryonal upper alimentary canal, which may be the consequence of rotational stresses. If the resultant ischemia is temporary, only the anoxia-sensitive neural tissue succumbs, producing an esophagus deficient in myenteric plexuses. If it is permanent, the entire organ fails to develop, resulting in esophageal atresia.

b. Atresia: Experimental support for the theory of embryonal mesenteric ischemia causing congenital alimentary defects is fairly conclusive for intestinal atresia. In the chick and dog embryo, interruption of mesenteric vessels results in intestinal stenosis and atresia (Barnard, 1956; Earlham, 1972b). Although analogous experiments have not been performed for the esophagus, the hypothesis seems plausible.

c. Esophagitis: This condition may be caused by a variety of offending agents, both chemical and infectious. The easiest of these factors to quantitate is stomach acid. Esophagitis from reflux of stomach acid into the distal esophagus is a well-established entity, frequently requiring operative intervention. The distal esophageal sphincter, which normally prevents reflux of acid into the esophagus, may malfunction in response to ischemia (Earlham *et al.*, 1967).

Although the offensive nature of hydrochloric acid is mitigated by a rich mucosal blood flow in the stomach (Cheung and Chang, 1977), the mucosa of the esophagus is not as well vascularized (Harell *et al.*, 1972). This difference is

Blood Vessels and Lymphatics in Organ Systems
Copyright © 1984 by Academic Press, Inc.

thought to be due to differing functions. The esophagus is a mechanical transport conduit with an enhanced blood supply to the muscle, whereas the stomach is a secretory organ with emphasis on vascular perfusion of its mucosa. The relative paucity of blood flow to the epithelial lining of the esophagus may explain its sensitivity to stomach acid. Endoscopic biopsies of patients with reflux esophagitis show neovascularization, probably as an attempt to overcome this deficit (Geboes *et al.*, 1980). Microscopic study of the specimens demonstrates an increase in the diameter of the blood vessel-bearing papillary dermis, an increase in diameter of the blood vessels, an ingrowth of capillaries into the squamous cell layer, and large lakes of blood within the squamous cell layers only several cells from the lumen. This hyperemic response may be analogous to that which occurs in the stomach as a protective mechanism in response to back-diffusion of acid. The relationship of esophageal mucosal blood flow to intramural pH of the esophagus in regard to the production of esophagitis has not been studied.

2. STOMACH

Gastric blood vessels and particularly gastric blood flow have well-established roles as contributors to stomach pathology. For example, experimental evidence has demonstrated a relationship between mucosal ischemia and ulcerative lesions of the stomach.

a. Role of mucosal ischemia: Mucosal ischemia due to arteriolar occlusion or embolus was proposed as the mechanism for peptic ulcer disease over a century ago (Virchow, 1853). The relative constancy of the position of ulcers along the distal lesser curvature and in the first inch of the duodenum suggested to Virchow that the blood supply to these areas might be deficient. Evidence for such a possibility has been found in human stomachs in which there are a reduced number and smaller anastomoses within the submucosal plexus of the ulcer-bearing areas (Piasecki, 1974). Similarly, such regions of the stomach have arterioles supplying the mucosa that originate extramurally, traverse the muscle, and have no intramural communication with the submucosal plexus (Piasecki, 1974). It is also possible that these extramural end-arteries are subject to occlusion from sustained muscular contraction (Dai and Ogle, 1975). Of interest in this regard is the finding that the dog stomach, an organ that does not spontaneously develop ulcers, does not have a microvascular architecture similar to that found in the human stomach (Piasecki, 1975).

Attempts to demonstrate mucosal ischemia morphologically during ulcer formation have been limited by technical considerations and the lack of a good animal model. Microvascular injection studies of rat stomachs developing erosive gastritis in response to stress have demonstrated early generalized mucosal ischemia that later becomes focal in nature (Hase and Moss, 1973). Prominent submucosal channels, which develop in this ulcer model, have been interpreted both as open arteriovenous shunts (Hase and Moss, 1973) and as vascular engorgement resulting from muscular contraction (Dai and Ogle, 1975; Guth, 1972). Such postmortem injection studies, however, only give an indication of the volume of blood present in the mucosa at the time of death, with no reliable estimate of the actual blood flow.

For the above reason, other approaches for demonstrating mucosal ischemia during ulcerogenesis have been tried using various methods of measuring gastric blood flow. Early models of gastric ulceration produced by hemorrhagic or septic shock provide supportive evidence for an ischemic cause. However, even though ischemia has an established contributing role in experimental ulcer disease, it is neither the only nor the unifying etiologic factor. First, intraluminal acid must be present during low flow states for ulcers to occur (see below). Second, many studies have demonstrated hyperperfusion during the development of mucosal injury. Moreover, human trauma patients have higher gastric blood flows than normal ones; yet they develop stress ulcers (McClelland et al., 1971). Similarly, septic pigs are subject to stress ulcers despite a measured increase in gastric blood flow (Lucas et al., 1976). Animal ulcers induced by exposure to aspirin, acetic acid, bile salt, and alcohol demonstrate increased gastric blood flow during injury (Cheung et al., 1975; Screide et al., 1976). Also, enhanced blood flow has been noted in the early phases of healing experimental gastric ulcers (Skarstein et al., 1979). Of further interest is the finding that morphologic studies of healing ulcers show neovascularization of the ulcer bed within 24 hr of injury, with revascularization preceding epithelial regeneration (Kairaluoma, 1972). Increased blood flow to focal areas of the mucosa that eventually ulcerate have been demonstrated using radioactive microspheres (McGreevy and Moody, 1981), but the hyperemia may be secondary to inflammation. Increased mucosal blood flow also is protective, as demonstrated by a canine, aspirin-ulcer model, in which injury was prevented by artificial gastric hyperperfusion (McGreevy and Moody, 1977).

   b. Role of intraluminal acid: In order to understand fully the role of ischemia in gastric ulcer formation, the gastric blood flow rate must be considered relative to the amount of acid that penetrates the epithelium (Starlinger et al., 1981). In virtually every experimental model of ulcer disease, ulcers will not develop unless intraluminal acid is present (Moody and Aldrete, 1971). This is true clinically as well. Severely stressed intensive care unit patients do not develop mucosal erosions if their gastric acidity is continuously neutralized (Hastings et al., 1978), for the acid produces damage by diffusing into the mucosa (Davenport, 1966). For years the importance of acid back-diffusion was debated because experimental ulcers could form with both low and high back-diffusion rates. Also, certain studies demonstrated large amounts of back-diffusion with no damage (Moody *et al.*, 1975), suggesting that gastric mucosal perfusion neutralizes the influxing acid and prevents injury (McGreevy and Moody, 1977). If gastric blood flow is sufficiently high, large amounts of hydrogen ion back-diffusion are tolerated without damage (Cheung and Chang, 1977). Ischemia of the gastric epithelium promotes and enhances the lesion produced by any rate of acid back-diffusion (Mersereau and Hinchey, 1973; Ritchie, 1975). Thus, although intraluminal acid is necessary for mucosal ulceration to occur, it is the gastric microcirculation that determines the rate of acid back-diffusion necessary to cause damage.

c. Local regulation of gastric blood flow: With the current understanding of the protective nature of the gastric blood flow, future research might profitably concentrate on exogenous and endogenous controls of the gastric mucosal perfusion. Preliminary efforts have already defined some of the neural, humoral, and pharmacologic influences on gastric blood flow regulation (see Sections C and D, above).

With regard to neural control, it has been found that splanchnic nerve stimulation decreases gastric blood flow (Reed and Sanders, 1971) whereas vagal stimulation increases it (Reed *et al.*, 1971); vagotomy diminishes blood flow, probably through removal of acetylcholine inhibition of nonepinephrine release (Van Hee and Vanhoutte, 1978). Morphologic studies support a neural influence by demonstrating atrophy of nerve endings near arterioles after vagotomy (Kalahanis *et al.*, 1976). This procedure also produces capillary constriction, as revealed by electron microscopy (Halaris, 1971). Clinically, the latter response is important because parietal cell vagotomy, one of the most promising and new operations for peptic ulcer, decreases gastric blood flow by 50% (Seifert *et al.*, 1980).

With regard to humoral control, it has been found that many gastrointestinal hormones have an effect on the gastric blood flow (Konturek *et al.*, 1976; Guth and Smith, 1976). Whether these changes are physiologic or pharmacologic remains to be identified.

Prostaglandins, which are a family of C_{20} fatty acids occurring universally in mammalian tissues, have also been implicated in the problem of ulcer formation. They are derivatives of the phospholipids of cell membranes via arachidonic acid metabolism, and they function as local regulators of biologic processes. In many organs, prostaglandins regulate the regional microcirculation. With regard to the stomach, different prostaglandins have opposite vasoactive effects. PGE_2 and PGE_1, the principle ones, both increase gastric blood flow (Cheung, 1980; Cheung and Lowry, 1978), whereas thromboxane has the opposite effect. The balance among these naturally occurring, local regulators could determine the adequacy of mucosal perfusion. As would be expected, thromboxane infusion causes experimental ulcers (Whittle *et al.*, 1981), whereas prostaglandin E_2 prevents such lesions when given topically or intravenously (Robert *et al.*, 1979). In the future, drug therapy aimed at augmenting or diminishing certain prostaglandins in the stomach may help heal or prevent gastric mucosal damage by regulating the microcirculation (Gerkens *et al.*, 1977a). Indeed, the key to understanding mucosal disease of the stomach may lie in elucidation of prostaglandin autoregulation of gastric blood flow. (For further discussion of the gastric circulatory effects of prostaglandins, see Section D-3, above.)

F. LYMPHATIC SYSTEM

By M. A. Perry and D. N. Granger

1. Anatomy

a. Large channels: The esophagus and stomach are richly supplied with lymphatic vessels. These form plexuses in the mucosa just above the muscularis mucosa and again in the submucosal and muscularis layers. Lymph moves between the plexuses via short vessels that pass at right angles through the tissue layers. The lymph channels in the submucosa and muscularis are larger and possess valves that direct lymph to nodes located either in the neck or on both curvatures of the stomach (Yoffey and Courtice, 1970). The arrangement of the lymph vessels draining the stomach and the associated lymph nodes has recently been described for a number of species, including man (Durovicova and Munka, 1979).

b. Lymphatic capillaries: In the gastric mucosa, the lymphatic capillaries originate as blind projections among the basal portions of the gastric glands, but they do not extend as close to the epithelial surface as do the blood capillaries (Fig. 12.5). This arrangement of lymphatic and blood capillaries is similar to that observed in the colon (Kvietys *et al.*, 1981). Under pathologic conditions (gastric ulceration), the lymphatic capillaries appear to be closely adherent to the secretory epithelium of the gastric glands (Renyi-Vamos and Szinay, 1955).

2. Physiology

a. Composition of gastric lymph: The composition of gastric lymph has not been widely studied due to the difficulty in obtaining adequate samples. Brugeman (1975) collected samples of gastric lymph in the dog using micropuncture techniques and found that, at a normal venous pressure, the lymph to

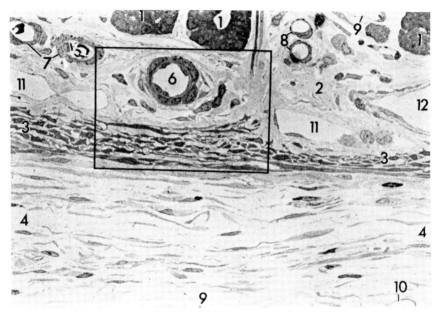

FIG. 12.5. Electron micrograph of the wall of the cat stomach. ×680. 1, Base of gastric glands proper; 2, lamina propria; 3, lamina muscularis mucosae; 4, submucosa; 6, arteriole; 7, terminal arterioles; 8, precapillary sphincters; 9, blood capillaries; 10, postcapillary venule; 11, lymphatic capillary; 12, lymphatic vessel. (From Rhodin,1974; reproduced with permission of the Oxford University Press, New York.)

plasma ratio (L/P) for total plasma proteins was 0.51. This figure decreased with increasing molecular size from 0.68 for albumin to 0.39 for fibrinogen. Such data suggested selective restriction, on the basis of molecular size, to the movement of plasma proteins across gastric capillaries. (For general considerations of lymph formation and flow, see Section C-3, Chapter 4.)

Recently, Perry et al. (1981) cannulated a lymphatic crossing the lesser curvature of the cat stomach and reported a mean flow rate of 4 μl/min. The L/P ratios recorded under control conditions were similar to those observed by Bruggeman (1975) in the dog. Elevation of gastric venous pressure in the cat caused an increase in lymph flow and a reduction in the L/P ratio for total protein from 0.50 to 0.16. The L/P ratio for each of the protein fractions studied decreased to a constant value at lymph flow rates above twice control. The minimum (filtration rate independent) L/P ratio (L/P_{min}) was used to calculate the osmotic reflection coefficient (σ_d) for each protein fraction as $\sigma_d = 1 - L/P_{min}$. Reflection coefficients ranged from 0.73 for albumin to 0.91 for fibrinogen. A plot of $1 - \sigma_d$ as a function of solute radius for each protein fraction was analyzed by the method of Renkin et al. (1977) in order to estimate the porosity of the gastric capillaries. Gastric capillaries were described by two populations of "equivalent" pores of 47 and 250 Å radii.

The concentration of smaller solute molecules in gastric lymph has also been investigated recently by Keyl *et al.* (1981) through cannulation of single lymphatic vessels on the greater curvature of the dog stomach. The results indicate that the concentration of glucose and electrolytes in gastric lymph is the same as in plasma.

Bruggeman (1975) studied the effect of dithiothreitol, a gastric mucosal barrier breaker, on the composition of gastric lymph and found no significant change from control values. However, Wood and Davenport (1982), using different techniques, found that dithiothreitol causes a reduction in the osmotic reflection coefficients of albumin and fibrinogen, consistent with an increase in vascular permeability. In this regard, it should be noted that the selectivity of the gastric capillaries for macromolecules can only be assessed by analysis of lymph at high capillary filtration rates where L/P is independent of lymph flow. Since Bruggeman collected lymph samples only at control venous pressure, it appears that further investigation of the effects of disruption of the gastric mucosal barrier on gastric capillary permeability is warranted.

b. Fluid exchange in the stomach: The lymphatics are responsible for the removal of fluid and solutes that have entered the tissues either from the plasma or by absorption across the mucosa and have not been subsequently removed either by the capillaries or as secretions. This function is essential to the maintenance of normal fluid exchange in an organ. The relationship among the factors that govern fluid exchange in the stomach is described by the Starling equation, i.e.,

$$J_v = K_f [(P_c - P_t) - \sigma_d(\pi_p - \pi_t)] \qquad \text{(F.1)}$$

where J_v is net capillary filtration rate, K_f is the capillary filtration coefficient, P_c is the capillary hydrostatic pressure, P_t is the interstitial fluid pressure, σ_d is the capillary reflection coefficient to total plasma proteins, π_p is the plasma oncotic pressure, and π_t is the interstitial fluid oncotic pressure.

Each of the factors in the Starling equation has been measured or calculated for the resting stomach. Bill (1979) calculated that lymph flow in the stomach was 0.06 ml/min/100 gm which, in the resting state, is assumed to be equal to J_v. The capillary filtration coefficient has been reported to be 0.058 ml/min/mm Hg/100 gm (Jansson *et al.*, 1970), although preliminary studies (M. Perry, personal observations) indicate a different value of 0.18 ml/min/mm Hg/100 gm, which is close to figures reported for the intestines. Interstitial fluid pressure measured by capsules is 0.53 mm Hg (Altaminaro *et al.*, 1975) and σ_d, π_p, and π_t are 0.78, 17.9, and 6.3 mm Hg, respectively (Perry *et al.*, 1981). Since all but one of the factors in the Starling equation are known, it is possible to calculate capillary pressure, which is 10.6 mm Hg (assuming $K_f = 0.058$) or 9.9 mm Hg (if $K_f = 0.18$). An interesting feature of the forces governing fluid exchange in the nonsecreting stomach is the relatively high capillary filtration coefficient, which indicates that only a very small net capillary filtration pressure is required to generate the observed lymph flow, i.e., $(P_c - P_t) - \sigma_d(\pi_p - \pi_t)$ ranges between 0.3 and 1.0 mm Hg.

One of the major functions of the stomach is the secretion of hydrochloric acid. During stimulated acid secretion, approximately 0.7 ml/min/100 gm of gastric juice is transported from the mucosal interstitium into the gastric lumen. This secretion rate represents a fluid flux that is approximately 12 times greater than the net capillary filtration rate (lymph flow) in the nonsecreting stomach. Since the fluid for gastric secretion is derived from the plasma, the capillary filtration rate must increase markedly (12-fold) when the stomach is stimulated to produce acid. Changes in capillary filtration rate of such magnitude would require dramatic changes in the forces and/or membrane parameters governing the rate of transcapillary fluid filtration.

Theoretically, as protein-free fluid is secreted into the gastric lumen, mucosal fluid volume should diminish, thereby causing tissue oncotic pressure to rise, and tissue pressure to decrease. The reduction of tissue pressure should cause a concomitant decrease in lymph flow. These changes in interstitial forces should tend to enhance capillary filtration rate, which, in turn, would serve to provide the fluid necessary for gastric secretion.

The effect of acid secretion on the forces governing gastric transcapillary fluid exchange have not been studied. Preliminary studies (M. Perry, personal observations) support the concept that changes in the interstitial forces are responsible for enhancing capillary filtration rate during gastric secretion. During pentagastrin-stimulated acid secretion, there is a 50–100% increase in K_f, whereas capillary pressure remains unchanged. The increase in K_f may result from either recruitment of additional capillaries or greater vascular permeability. In order to account for the augmentation in capillary filtration rate required during gastric secretion, the net pressure driving fluid across the gastric capillaries must increase 6-fold. The observation that capillary pressure is unaffected by pentagastrin suggests that the increased capillary filtration results from an alteration in the interstitial forces, i.e., π_t increases and/or P_t decreases.

3. FUTURE AVENUES FOR INVESTIGATION

The effects of stimulated acid secretion on lymph flow rate, capillary and interstitial hydrostatic pressures, osmotic forces, and the permeability characteristics of the capillary wall are poorly understood and require further study. Similarly, the way in which each of these parameters changes during portal hypertension to protect the stomach against edema has not been investigated. The role of gastric lymphatics in preventing the edema associated with gastritis and ulcers warrants attention.

REFERENCES*

Altaminaro, M., Requena, M., and Perez, T. (1975). Interstitial fluid pressure in canine gastric mucosa. *Am. J. Physiol.* **229,** 1414–1420. (F)

Barlow, T. E., Bentley, F. H., and Walder, D. N. (1951). Arteries, veins and arteriovenous anastomoses in the human stomach. *Surg., Gynecol. Obstet.* **93,** 657–671. (B)

Barnard, C. (1956). The genesis of intestinal atresia. *Surg. Forum* **7,** 393–396. (E)

Bill, A. (1979). Regional lymph flow in unanesthetized rabbits. *Upsala J. Med. Sci.* **84,** 129–136. (F)

Bond, J. H., and Levitt, M. D. (1980). Effect of glucagon on gastrointestinal blood flow in dogs in hypovolemic shock. *Am. J. Physiol.* **238,** G434–G439. (D)

Bowen, J. C., LeDoux, J. C., Ochsner, J. L., Jr., Ochsner, M. G., Jr., and Payne, J. G. (1981). Contrasting effects of vasodilators on oxygen tension and membrane potential of canine gastric surface epithelium. *Surgery* **90,** 41–48. (D)

Bruggeman, T. M. (1975). Plasma proteins in canine gastric lymph. *Gastroenterology* **68,** 1204–1210. (F)

Bruggeman, T. M., Wood, J. G., and Davenport, H. W. (1979). Local control of blood flow in the dog's stomach: Vasodilation caused by lack of acid back diffusion following topical application of salicylic acid. *Gastroenterology* **77,** 736–744. (C)

Bulkley, G., Goldman, H., Trencis, L., and Silen, W. (1970). Gastric microcirculatory changes in hemorrhagic shock. *Surg. Forum* **21,** 27–30. (B)

Canfield, S. P., and Spencer, J. E. (1981). The effect of 5-hydroxytryptamine on gastric acid secretion by the rat isolated stomach. *Br. J. Pharmacol.* **74,** 253P. (D)

Charbon, G. A., Brouwers, H. A., and Sala, A. (1980). Histamine H_1- and H_2-receptors in the gastrointestinal circulation. *Naunyn-Schmiedeberg's Arch. Pharmacol.* **312,** 123–129. (D)

Cheung, L. Y. (1980). Topical effects of 16,16,-dimethylprostaglandin E_2 on gastric blood flow in dogs. *Am. J. Physiol.* **238,** G514–G519. (D, E)

Cheung, L. Y., and Chang, N. (1977). The role of gastric mucosal blood flow and H+ back-diffusion in the pathogenesis of acute gastric erosions. *J. Surg. Res.* **22,** 357–361. (C, E)

Cheung, L. Y., and Lowry, S. F. (1978). Effects of intraarterial infusion of prostaglandin E_1 on gastric secretion and blood flow. *Surgery* **83,** 699–704. (E)

Cheung, L. Y., Moody, F. G., and Reese, R. S. (1975). Effect of aspirin, bile salt and ethanol on canine gastric mucosal blood flow. *Surgery* **77,** 786–792. (E)

Cheung, L. Y., Moody, F. G., Larson, K., and Lowry, S. F. (1978). Oxygen consumption during cimetidine and prostaglandin E_2 inhibition of acid secretion. *Am. J. Physiol.* **234,** E445–E450. (C)

Chou, C. C., and Grassmick, B. (1978). Motility and blood flow distribution within the wall of the gastrointestinal tract. *Am. J. Physiol.* **235,** H34–H39. (C)

Chou, C. C., Hsieh, C. P., Yu, Y. M., Kvietys, P., Yu, C. C., Pittman, R., and Dabney, J. M. (1976). Localization of mesenteric hyperemia during digestion in dogs. *Am. J. Physiol.* **230,** 583–589. (C)

Curwain, B. P., and Turner, N. C. (1981). Histamine H_1- and H_2-receptors in the gastric vasculature of the rabbit. *Eur. J. Pharmacol.* **22,** 515–519. (D)

Dai, S., and Ogle, C. W. (1975). Effects of stress and of autonomic blockers on gastric mucosal microcirculation in rats. *Eur. J. Pharmacol.* **30,** 86–92. (E)

Davenport, H. W. (1966). Fluid produced by the gastric mucosa during damage by acetic and salicylic acids. *Gastroenterology* **50,** 487–499. (E)

Davenport, H. W., and Barr, L. L.(1973). Failure of ischemia to break the dog's gastric mucosal barrier. *Gastroenterology* **65,** 619–624. (C)

Durovicova, J., and Munka, V. (1979). Lymph drainage from stomach. *Lymphol., Proc. Int. Cong., 6th, 1977* pp. 33–34. (F)

*In the reference list, the capital letter in parentheses at the end of each reference indicates the section in which it is cited.

Earlham, R. J. (1972a). A vascular cause for aganglionic bowel. *Am. J. Dig. Dis.* **17**, 255–261. (E)

Earlham, R. J. (1972b). A study of the etiology of congenital stenosis of the gut. *Ann. R. Coll. Surg. Engl.* **51**, 126–130. (E)

Earlham, R. J., Schlegel, J. F., and Ellis, F. H. (1967). Effect of ischemia of lower esophagus and esophagogastric junction on canine esophageal motor function. *J. Thorac. Cardiovasc. Surg.* **54**, 822–831. (E)

Edlich, R. F., Borner, J. W., Kuphal, J., and Wangensteen, O. H. (1970). Gastric blood flow. Its distribution during gastric distention. *Am. J. Surg.* **120**, 35–37. (C)

Evans, H. M. (1912). The development of the vascular system. *In* "Manual of Human Embryology" (F. Keibel and F. P. Mall, eds.), Vol. II, pp. 570–709. Lippincott, Philadelphia, Pennsylvania. (A)

Fara, J. W. (1976). Mesenteric vasodilator effect of 5-hydroxytryptamine: Possible enteric neuron mediation. *Arch. Int. Pharmacodyn. Ther.* **221**, 235–249. (D)

Geboes, K., Desmet, V., Vantrappen, G., and Mebis, J. (1980). Vascular changes in the esophageal mucosa. *Gastrointest. Endoscopy* **26**, 29–32. (E)

Gerkens, J. F., Flexner, C., Oates, J. A., and Shand, D. G. (1977a). Prostaglandin and histamine involvement in the gastric vasodilator action of pentagastrin. *J. Pharmacol. Exp. Ther.* **201**, 421–426. (E)

Gerkens, J. F., Shand, D. G., Flexner, C., Weis, A. S., Oates, J. A., and Data, J. L. (1977b). Effect of indomethacin and aspirin on gastric blood flow and acid secretion. *J. Pharmacol. Exp. Ther.* **203**, 646–652. (C)

Granger, D. N., and Kvietys, P. R. (1981). The splanchnic circulation: Intrinsic regulation. *Annu. Rev. Physiol.* **43**, 409–418. (C)

Guth, P. H. (1972). Gastric blood flow in restraint stress. *Am. J. Dig. Dis.* **17**, 807–813. (E)

Guth, P. H. (1981). *In vivo* microscopy of the gastric microcirculation. *In* "Measurement of Blood Flow" (D. N. Granger and G. B. Bulkley, eds.), pp. 105–119. Williams & Wilkins, Baltimore, Maryland. (B)

Guth, P. H., and Ballard, K. W. (1981). Physiology of the gastric circulation. *In* "Physiology of the Gastrointestinal Tract" (L. R. Johnson, ed.), pp. 709–731. Raven Press, New York. (C)

Guth, P. H., and Smith, E. (1975). Neural control of gastric mucosal blood flow in the rat. *Gastroenterology* **69**, 935–940. (D)

Guth, P. H., and Smith, E. (1976). The effect of gastrointestinal hormones on the gastric microcirculation. *Gastroenterology* **71**, 435–438. (D, E)

Guth, P. H., and Smith, E. (1977). Nervous regulation of the gastric microcirculation. *In* "Nerves and the Gut" (F. P. Brooks and P. W. Evers, eds.), pp. 365–373. Charles B. Slack, Inc., Thoroughfare, New Jersey. (C)

Halaris, A. E. (1971). Mucosal vasculature after truncal vagotomy. *Experientia* **27**, 78–79. (E)

Harell, G., DeNardo, G., Archibald, R., Bradley, B., and Zboralske, F. (1972). Regional distribution of feline esophageal blood flow. *Gastroenterology* **63**, 627–633. (E)

Harvey, C. A., Owen, D. A., and Shaw, K. D. (1980). Evidence for both histamine H_1- and H_2-receptors in the gastric vasculature of the cat. *Br. J. Pharmacol.* **69**, 21–27. (D)

Hase, T., and Moss, B. (1973). Microvascular changes of gastric mucosa in the development of stress ulcer in rats. *Gastroenterology* **65**, 224–234. (E)

Hastings, P. R., Skillman, J. J., Bushnell, L. S., and Silen, W. (1978). Antacid titration in the prevention of acute gastrointestinal bleeding. *N. Engl. J. Med.* **298**, 1041–1045. (E)

Haverback, B. J., Bogdanski, D., and Hogben, C. A. M. (1958). Inhibition of gastric acid secretion in the dog by the precursor of serotonin, 5-hydroxytryptophan. *Gastroenterology* **34**, 188–195.

Holm-Rutili, L., Perry, M. A., and Granger, D. N. (1981). Autoregulation of gastric blood flow and oxygen uptake. *Am. J. Physiol.* **241**, G143–G149. (C)

Hunter, G. C., Goldstone, J., Villa, R., and Way, L. W. (1979). Effect of vagotomy upon intragastric redistribution of microvascular flow. *J. Surg. Res.* **26**, 314–319. (D)

Jacobson, E. D. (1970). Comparison of prostaglandin E_1 and norepinephrine on the gastric mucosal circulation. *Proc. Soc. Exp. Biol. Med.* **133**, 516–519. (D)

Jacobson, E. D., and Chang, A. C. K. (1969). Comparison of gastrin and histamine on gastric mucosal blood flow. *Proc. Soc. Exp. Biol. Med.* **130**, 484–486. (D)

Jacobson, E. D., and Price, W. E. (1969). Effect of hydrocortisone on gastric mucosal blood flow and secretion. *Gastroenterology* **56**, 36–43. (D)

Jacobson, E. D., Scott, J. B., and Frohlich, E. D. (1962). Hemodynamics of the stomach I. Resistance-flow relationship in the gastric vascular bed. *Am. J. Dig. Dis.* **7**, 779–790. (C)

Jacobson, E. D., Eisenberg, M. M., and Swan, K. G. (1966a). Effects of histamine on gastric blood flow in conscious dogs. *Gastroenterology* **51**, 466–472. (D)

Jacobson, E. D., Linford, R. H., and Grossman, M. I. (1966b). Gastric secretion in relation to mucosal blood flow studied by a clearance technique. *J. Clin. Invest.* **45**, 1–73. (C, D)

Jansson, G., Lundgren, O., and Martinson, J. (1970). Neurohormonal control of gastric blood flow. *Gastroenterology* **58**, 424–429. (F)

Kairaluoma, M. I. (1972). Experimental gastric ulcer in Shay rat. Healing of gastric ulcers, development of their vascular supply, and effect of vagotomy on healing. *Acta Chir. Scand., Suppl.* **428**, 1–36. (E)

Kalahanis, N. G., Das Gupta, T. K., and Nyhus, L. M. (1976). Neural control of blood flow in gastric mucosa. *Am. J. Surg.* **131**, 86–90. (E)

Kauffman, G. L., Jr., Whittle, B. J. R., Aures, D., Vane, J. R., and Grossman, M. I. (1979). Effects of prostacyclin and a stable analogue 6β-PGI$_2$ on gastric acid secretion mucosal blood flow and blood pressure in conscious dogs. *Gastroenterology* **77**, 1301–1306. (D)

Kauffman, G. L., Jr., Aures, D., and Grossman, M. I. (1980). Intravenous indomethacin and aspirin reduce basal gastric mucosal blood flow in dogs. *Am. J. Physiol.* **238**, G131–G134. (D)

Keyl, M. J., Chang, A. C. K., and Dowell, R. T. (1981). Constituents of lymph from the nonsecreting stomach of the dog. *Lymphology* **14**, 118–121. (F)

Knight, S. E., McIsaac, R. L., and Rennie, C. D. (1980). The effect of histamine and histamine antagonists on gastric acid secretion and mucosal blood flow in man. *Br. J. Surg.* **67**, 266–268. (D)

Konturek, S. J., Dembinski, A., Thor, P., and Krol, R. (1976). Comparison of vasoactive intestinal peptide (VIP) and secretin on gastric secretion and mucosal blood flow. *Pfluegers Arch.* **361**, 175–181. (D, E)

Konturek, S. J., Dembinski, A., Krol, R., and Wunsch, E. (1977). Effect of 13 NLE-motilin on gastric secretion, serum gastrin level and mucosal blood flow in dogs. *J. Physiol. (London)* **264**, 665–672. (D)

Konturek, S. J., Pawlik, W., Walus, K. M., Coy, D. H., and Schally, A. F. (1978). Methionine-enkephalin stimulates gastric secretion and gastric mucosal blood flow. *Proc. Soc. Exp. Biol. Med.* **158**, 156–160. (D)

Konturek, S. J., Pawlik, W. W., Walus, K. M., and Jacobson, E. D. (1980a). Effect of dibutyryl cyclic AMP on gastric secretion and mucosal blood flow. *Hepato-Gastroenterol.* **27**, 204–207. (D)

Konturek, S. J., Robert, A., Hanchar, A. J., and Nezamis, J. E. (1980b). Comparison of prostacyclin and prostaglandin E$_2$ on gastric secretion, gastrin release and mucosal blood flow in dogs. *Dig. Dis. Sci.* **25**, 673–679. (D)

Konturek, S. J., Tasler, J., Jaworek, J., Pawlik, W., Walus, K. M., Schusdziarra, V., Meyers, C. A., Coy, D. H., and Schally, A. V. (1981). Gastrointestinal secretory, motor, circulatory and metabolic effects of prosomatostatin. *Proc. Natl. Acad. Sci. U.S.A.* **78**, 1967–1971. (D)

Kvietys, P. R., Wilborn, W. H., and Granger, D. N. (1981). Effects of net transmucosal volume flux on lymph flow in the canine colon. Structural-functional relationship. *Gastroenterology* **81**, 1080–1090. (F)

Lanciault, G., and Jacobson, E. D. (1976). The gastrointestinal circulation. *Gastroenterology* **71**, 851–873. (C)

Lucas, C. E., Sugawa, C., and Riddle, J.(1971). Natural history and surgical dilemna of stress gastric bleeding. *Arch. Surg. (Chicago)* **102**, 266–273. (C)

Lucas, C. E., Ravikant, T., and Walt, A. J. (1976). Gastritis and gastric blood flow in hyperdynamic septic pigs. *Am. J. Surg.* **131**, 73–77. (E)

McClelland, R. N., Shires, G. T., and Prager, M. (1971). Gastric secretory and splanchnic blood flow studies in man after severe trauma and hemorrhagic shock. *Am. J. Surg.* **121**, 134–142. (E)

McGreevy, J. M., and Moody, F. G. (1977). Protection of gastric mucosa against aspirin-induced erosions by enhanced blood flow. *Surg. Forum* **28**, 357–359. (C, E)

McGreevy, J. M., and Moody, F. G. (1981). Focal microcirculatory changes during the production of aspirin-induced gastric mucosal erosions. *Surgery* **81**, 337–341. (E)

Mao, C. C., Shanbour, L. L., Hodgins, D. S., and Jacobson, E. D. (1972). Adenosine 3'5'-monophosphate (cyclic AMP) and secretion in the canine stomach. *Gastroenterology* **62**, 427–438. (D)

Mersereau, W. A., and Hinchey, E. J. (1973). Effect of gastric acidity on gastric ulceration induced by hemorrhage in the rat, utilizing a gastric chamber technique. *Gastroenterology* **64**, 1130–1135. (C, E)

Mersereau, W. A., and Hinchey, E. J. (1978). Interaction of gastric blood flow, barrier breaker and hydrogen in back diffusion during ulcer formation in the rat. *Surgery* **83**, 248–251. (C)

Moody, F. G., and Aldrete, J. S. (1971). Hydrogen permeability of canine gastric secretory epithelium during formation of acute superficial erosions. *Surgery* **70**, 154–160. (C, E)

Moody, F. G., Simons, M. A., and Jackson, T. (1975). Effect of p-chloromercuribenzene sulfonate on gastric parietal and surface cell function in the dog. *Gastroenterology* **68**, 279–284. (E)

Necheles, H., Frank, R., Kaye, W., and Rosenman, E. (1936). Effect of acetylcholine in blood flow through the stomach and legs of the rat. *Am. J. Physiol.* **114**, 695–699. (D)

Perry, M. A., Crook, W. J., and Granger, D. N. (1981). Permeability of gastric capillaries to small and large molecules. *Am. J. Physiol.* **241** (*Gastrointest. Liver Physiol.* 4), G478–G486. (F)

Piasecki, C. (1974). Blood supply to the human gastroduodenal mucosa with special reference to the ulcer-bearing areas. *J. Anat.* **118**, 295–335. (E)

Piasecki, C. (1975). Observations on the submucous plexus and mucosal arteries of the dog's stomach and first part of the duodenum. *J. Anat.* **119**, 133–148. (E)

Piasecki, K. (1980). Patterns of blood supply to human gastroduodenal mucosa: A basis for local ischaemia. In "Gastro-Intestinal Mucosal Blood Flow" (L. P. Fielding, ed.), pp. 3–16. Churchill-Livingstone, Edinburgh and London. (B)

Puurunen, J. (1980). Gastric mucosal blood flow in ethanol-induced mucosal damage in the rat. *Eur. J. Pharmacol.* **16**, 275–280. (C)

Reed, J. D., and Sanders, D. J. (1971). Splanchnic nerve inhibition of gastric acid secretion and mucosal blood flow in anesthetized cats. *J. Physiol. (London)* **219**, 555–570. (E)

Reed, J. D., Sanders, D. J., and Thorpe, V. (1971). The effect of splanchnic nerve stimulation on gastric acid secretion and mucosal blood flow in the anesthetized cat. *J. Physiol. (London)* **214**, 1–13. (D, E)

Renkin, E. M., Watson, P. D., Sloop, C. H., Joiner, W. M., and Curry, F. E. (1977). Transport pathways of fluid and large molecules in microvascular endothelium of dog's paw. *Microvasc. Res.* **14**, 205–214. (F)

Renyi-Vamos, F., and Szinay, G. (1955). Lymphatic system of the stomach and its behavior in gastric ulcer. *Moden Tetkik Arama Oszt. Kozl.* **7**, 11 (cited in Rusznyak et al., p. 93). (F)

Reynolds, D. G., and Kardon, R. H. (1981). Methods of studying the splanchnic microvascular architecture. In "Measurement of Blood Flow" (D. N. Granger and G. B. Bulkley, eds.), pp. 69–88. Williams & Wilkins, Baltimore, Maryland. (B)

Rhodin, J. A. G. (1974). "Histology: A Text and Atlas," p. 547. Oxford Univ. Press, London and New York. (F)

Ritchie, W. P., Jr. (1975). Acute gastric mucosal damage induced by bile salts, acid and ischemia. Gastroenterology **68**, 699–707. (C, E)

Robert, A., Nezemis, J. E., Lancaster, C., and Hanchar, A. J. (1979). Cytoprotection by prostaglandins in rats. Gastroenterology **77**, 433–443. (E)

Ross, G., and Brown, H. W. (1967). Cardiovascular effect of dopamine in the anesthetized cat. Am. J. Physiol. **212**, 823–828. (D)

Rusznyak, I., Foldi, M., and Szabo, G., eds. (1967). "Lymphatics and Lymph Circulation." Pergamon, Oxford. (F)

Salvati, P., and Whittle, B. J. R. (1981). The vasoactive effects of some arachidonate lipoxygenase and cyclooxygenase products on the isolated perfused stomach of rabbit and rat. Br. J. Pharmacol. **73**, 2568–2578. (D)

Schuurkes, J. A. J., and Charbon, G. A. (1978). Motility and hemodynamics of the canine gastrointestinal tract. Stimulation by pentagastrin, cholecystoknin and vasopressin. Arch. Int. Pharmacodyn. Ther. **236**, 214–227. (C)

Screide, O., Skarstein, A., Varhang, J. E., and Svanes, K. (1976). Acetic acid induced gastritis in cats. J. Surg. Res. **21**, 191–200. (E)

Screide, O., Svanes, K., Varhang, J. E., and Skarstein, A. (1980). Changes in gastric mucosal morphology, capillary permeability and blood flow during the first 3 days of acute gastritis in cats. Eur. J. Surg. Res. **12**, 108–122. (C)

Seifert, J., Lenz, J., Bruckner, W., Brendel, W., and Holle, F. (1980). Are blood flow changes after selective proximal gastric vagotomy responsible for a necrosis in the gastric wall? Res. Exp. Med. **177**, 263–266. (E)

Semba, T., Fuji, K., and Fuji, Y. (1970). Influence of persistaltic contraction of the stomach on blood flow through the gastrosplenic vein. Hiroshima J. Med. Sci. **19**, 87–97. (C)

Shehadeh, Z., Price, W. E., and Jacobson, E. D. (1969). Effects of vasoactive agents on intestinal blood flow and motility in the dog. Am. J. Physiol. **216**, 386–392. (C)

Skarstein, A., Svanes, K., Varhang, J. E., and Screide, O. (1979). Blood flow distribution in the stomach of cats with acute gastric ulcer. Scand. J. Gastroenterol. **14**, 897–903. (C, E)

Starlinger, M., Schiessel, R., Hung, C. R., and Silen, W. (1981). H$^+$ back-diffusion stimulating gastric mucosal blood flow in the rabbit fundus. Surgery **89**, 232–236. (C, E)

Swan, K. G., and Jacobson, E. D. (1967). Gastric blood flow and secretion in conscious dogs. Am. J. Physiol. **212**, 891–896. (D)

Taylor, T. V., and Torrance, B. (1975). Is there an antral-body portal system in the stomach? Gut **16**, 781–784. (B)

Tepperman, B. L., Schofield, B., and Tepperman, F. S. (1975). Effect of metiamide on acid secretion from isolated kitten fundic mucosa. Can. J. Physiol. **53**, 1141–1146. (C)

Van Hee, R. H., and Vanhoutte, P. M. (1978). Cholinergic inhibition of adrenergic neurotransmission in the canine gastric artery. Gastroenterology **74**, 1266–1270. (D, E)

Virchow, R. (1853). Historiches, kritisches, und positives lehre der unterleibsaffektionem. Virchows Arch. Pathol. Anat. Physiol. **5**, 281–387, 632. (E)

Walus, K. M., and Jacobson, E. D. (1981). Relation between small intestinal motility and circulation. Am. J. Physiol. **241**, G1–G15. (C)

Walus, K. M., Gustaw, P., and Konturek, S. J. (1980a). Differential effects of prostaglandins and arachidonic acid on gastric circulation and oxygen consumption. Prostaglandins **20**, 1089–1102. (D)

Walus, K. M., Pawlik, W., and Konturek, S. J. (1980b). Prostacyclin-induced gastric mucosal vasodilation and inhibition of acid secretion in the dog. Proc. Soc. Exp. Biol. Med. **163**, 228–232. (D)

Wheaton, L. G., Sarr, M. G., Schlossberg, L., and Bulkley, G. B. (1981). Gross anatomy of the splanchnic vasculature. *In* "Measurement of Blood Flow" (D. N. Granger and G. B. Bulkley, eds.), pp. 9–45. Williams & Wilkins, Baltimore, Maryland. (B)

Whittle, B. J. R., Kauffman, G. L., and Moncada, S. (1981). Vasoconstriction with thromboxane A_2 induces ulceration of the gastric mucosa. *Nature (London)* **292**, 472–474. (E)

Wood, J. G., and Davenport, H. W. (1982). Measurement of canine gastric vascular permeability to plasma proteins in the normal and protein-losing states. *Gastroenterology* **82**, 725–733. (F)

Yano, S., Hoshino, E., and Harada, M. (1981). Effect of vasoactive drugs on gastric blood flow measured by a cross thermo couple method in rats. *Jpn. J. Pharmacol.* **31**, 117–124. (D)

Yoffey, J. M., and Courtice, F. C. (1970). "Lymphatics, Lymph and the Lymphomyeloid Complex," p. 210. Academic Press, New York. (F)

Zamora, C. S., Reddy, V. K., Frandle, K. A., and Samson, M. D. (1980). Effect of prednisone on gastric blood flow in swine. *Am. J. Vet. Res.* **41**, 885–888. (D)

Chapter 13
Digestive System: Small and Large Intestines

A. EMBRYOLOGY OF BLOOD CIRCULATION

By C. E. CORLISS

1. ESTABLISHMENT OF THE INTESTINAL CIRCULATION

Transformation of ventral intersegmental branches of the dorsal aorta into the definitive esophogastric blood supply has been described in Section A, Chapter 12. The development of the intestinal supply from the more caudal vessels is considered below.

a. Superior mesenteric artery: By 1 month, ventral intersegmental arteries, 9 through 13, have anastomosed longitudinally to produce the superior mesenteric artery, the largest of the unpaired aortic branches. Once formed, this vessel extends ventrad, encircles the midline developing gut tube, and joins the previously formed yolk sac plexus of arteries (Evans, 1912). The site of exit of the superior mesenteric artery from the aorta shifts caudad from T2 to L1 during development, reaching its definitive position in the sixth week.

A series of intestinal branches leaves inferiorly from the proximal segment of the superior mesenteric artery and extends through the mesentery to the convoluted small intestine. Distally, three larger branches, the ileocolic, right colic, and middle colic, fan out to supply the ascending and two-thirds of the transverse colon.

b. Inferior mesenteric artery: This vessel, smallest of the three unpaired digestive tract arteries, is responsible for a limited segment of gut. At about 37 days, it has its origin, apparently from the twentieth ventral segmental artery (Broman, 1907, cited by Evans, 1912), among several similarly sized aortic twigs in the region of the lower colon. It divides into the left colic, which supplies the distal one-third of the transverse colon, and the superior rectal and sigmoid arteries, which supply the lower gut. "Caudal wandering" (Evans, 1912) of the inferior mesenteric "root" is short—from T12 to L3.

There is an anastomotic overlap of the middle colic (from the superior mesenteric artery) and left colic (from the inferior mesenteric artery) on the

Blood Vessels and Lymphatics in Organ Systems
Copyright © 1984 by Academic Press, Inc.
All rights of reproduction in any form reserved.
ISBN 0-12-042520-3

distal one-third of the transverse colon, marking the junction of the original mid- and hindgut portions of the embryonic digestive tube. Drawings of Mall's 43-day-old human embryo (reproduced by Evans, 1912, p. 650) dramatize the size difference among the aortic branches of this stage, the inferior mesenteric artery being dwarfed by the relatively huge paired umbilical vessels that are located a few segments caudal to it.

The caudal-most part of the intestinal tract is supplied, not by vessels derived from ventral segmental branches, but by inferior and middle rectal arteries arising from dorsal branches.

B. ANATOMY OF BLOOD CIRCULATION

By H. J. GRANGER

1. GROSS ANATOMY

The small and large intestines are perfused by branches of the celiac, superior mesenteric, and inferior mesenteric arteries (Wheaton *et al.*, 1981). In man, the celiac artery arises from the abdominal aorta immediately below the diaphragm. Approximately 1 cm distal from its mouth, it trifurcates into the splenic, common hepatic, and left gastric branches. In turn, the common hepatic artery gives off a gastroduodenal branch that bifurcates into the superior pancreaticoduodenal artery and the right gastroepiploic artery. The superior pancreaticoduodenal artery divides into anterior and posterior branches that provide a rich vascular network supplying the duodenum.

a. Superior mesenteric artery: The distal small intestine and the proximal colon are perfused by this vessel. After originating approximately 1 cm distal to the celiac artery, the superior mesenteric trunk provides 12 or more mesenteric branches to the jejunum and ileum. In addition, it gives rise to the inferior pancreaticoduodenal, middle colic, right colic, and ileocolic branches. The inferior pancreaticoduodenal artery, the first branch of the superior mesenteric artery, anastomoses with branches of the superior pancreaticoduodenal artery, perfuses the duodenum, and provides a collateral pathway between the celiac and superior mesenteric flow streams. The middle colic artery fans out near the bowel wall into two branches that supply the transverse colon. The midascending colon is perfused by the right colic branch, which also unites with the right

Smith *et al.*, 1975). Taking into account the differential in net pressure head and filtration coefficient, the calculated fraction of intestinal transvascular fluid flux occurring in the mucosa is greater than 0.9. Thus, in the nonabsorptive state, less than 10% of intestinal lymph flow derives from the muscularis. In the absorptive state, the mechanics of transcapillary fluid transport is substantially altered (see Section C-5, below).

An interesting feature of the exchange system in the villi of the small intestine is the possible existence of countercurrent exchange between input microvessels traveling up toward the tip and output vessels coursing down toward the base of the villus (Hallback *et al.*, 1978). In theory, highly diffusible substances could leave the input vessel and shunt across to the output vessel long before the blood arrives at the villus tip; consequently, the tissue concentration of the materials would be greatest at the base and smallest at the tip of the villus. Indeed, a tissue P_{O_2} gradient does exist in the villus, P_{O_2} at the base and apex averaging 26 and 15 mm Hg, respectively (Bohlen, 1980). The concepts of countercurrent exchange and multiplication may also explain the gradient of osmolality observed in the villus (Hallback *et al.*, 1978).

3. Local Control of the Intestinal Circulation

The small intestine is capable of regulating its own perfusion in the absence of neurogenic and endocrine inputs to the organ (Johnson, 1964). Reduction of arterial perfusion pressure to the resting intestine elicits a decrease in vascular resistance that is weak to moderate, and, consequently, intrinsic flow autoregulation is less than perfect. More specifically, the degree of autoregulation is 30 to 60% in the unfed state (Norris *et al.*, 1979). Below perfusion pressures of 40 mm Hg, the intrinsic vasodilator response to local arterial hypotension reaches its limit, whereas the upper pressure limit of flow autoregulation appears to reside near 140 mm Hg. Although the degree of autoregulation of total intestinal flow is moderate, villus blood flow does not change substantially with small reductions in perfusion pressure (Lundgren and Svanvik, 1973). Hence, it would appear that the crypts and muscularis autoregulate their flows very poorly if at all. The intrinsic resistance responses reflect alterations in arteriolar caliber. The precapillary sphincters of the intestine also respond to reduction of arterial perfusion pressure, with local arterial hypotension eliciting a 2- to 4-fold increase in capillary exchange capacity in the nonabsorbing small bowel (Granger and Shepherd, 1979).

4. Metabolic and Myogenic Mechanisms of Local Vasoregulation

Local arterial hypotension and vascular occlusion are only two of a number of stresses that elicit intrinsic feedback modulation of arteriolar and precapillary sphincter tone. These local vasoregulatory phenomena can be explained on the basis of metabolic and myogenic mechanisms (Johnson, 1964). The metabolic theory envisages a linkage between microvascular tone and intestinal oxygenation; in essence, arteriolar conductance and the number of perfused capillaries are inversely related to the tissue P_{O_2}. The myogenic hypothesis is based on the proportional relationship between vascular smooth muscle tone and transmural pressure; in other words, stretch of the arteriole elicits an intrinsic vasoconstriction, whereas vasodilatation occurs in response to reduced distention. Thus, the metabolic mechanism seeks to buffer tissue oxygenation and the myogenic feedback attempts to maintain a constant intravascular pressure at the arteriolar and capillary levels.

In accord with the metabolic theory are the following observations: First, intestinal vascular conductance and capillary exchange capacity rise during local arterial hypoxemia (Shepherd, 1978). Second, elevating oxygen uptake elicits vasodilatation of intestinal arterioles and precapillary sphincters (Shepherd, 1979). Third, the degree of flow autoregulation is enhanced when the prevailing tissue P_{O_2} is lowered below normal (Granger and Norris, 1980a). These responses are consistent with a feedback system triggered by alterations in tissue oxygenation.

By contrast, the increased tone of the intestinal vessels in response to elevated venous pressure is consistent with a control system seeking to regulate transcapillary fluid filtration. Thus, in venous hypertension, the rise in arteriolar resistance returns the capillary pressure toward normal, and the derecruitment of capillaries reduces the filtration area, both responses serving to stabilize filtration rate at or near the normal level (Mortillaro and Taylor, 1976). After release of complete venous occlusion, the magnitude of reactive hyperemia in the small intestine is much less than that observed following arterial ischemia (Mortillaro and Granger, 1977), a differential response which may reflect the competitive interplay of high transmural pressure and tissue hypoxia in venous occlusion. During arterial occlusion, the metabolic and myogenic mechanisms both act to dilate the intestinal vessels. It is apparent that a complete picture of intrinsic vasoregulation cannot be presented without considering the coexistence of metabolic and myogenic feedback mechanisms (Parker and Granger, 1979).

5. Postprandial Vascular Reactions

A primary function of the intestinal circulation is the absorption of water and nutrients transported across the villus epithelium following feeding. The intestinal vascular reactions set in motion during feeding and digestion are complex (Chou and Kvietys, 1981). Upon presentation and ingestion of food, a transient intestinal hyperemia is elicited via neurogenic influences on intestinal blood vessels. This cephalic phase of intestinal hyperemia is small in magnitude (14%) and short in duration (30 min). As the chyme travels through the enteric tract, the intestinal phase of postprandial hyperemia occurs. This is large in magnitude (50 to 100%) and long in duration (several hours). The basic requirement for induction of postprandial hyperemia is the presence of the hydrolytic products of food on the surface of the villus lining. The potential mediators of postprandial hyperemia are numerous. Since digestion requires increased secretion, absorption, and motility, the oxygen demand of the intestine is elevated. Consequently, tissue P_{O_2} falls as uptake exceeds supply, and a local metabolic signal causes a reduction of arteriolar and precapillary sphincter tone. The subsequent increase in blood flow and in number of open capillaries enhances the delivery of oxygen to the cells of the intestine. Local neural reflexes and hormones may also contribute to vascular relaxation. In addition, the presence of bile in the intestinal lumen sensitizes the microvessels to vasodilator stimuli.

During rapid absorption of water by the epithelial cells, the arrival of protein-free fluid in the interstitium dilutes the interstitial protein pool and causes a lowering of interstitial oncotic pressure. At the same time, interstitial fluid pressure rises with expansion of interstitial water volume. The alterations of interstitial oncotic and hydrostatic pressures reverse the net transcapillary pressure difference to favor absorption of fluid into the bloodstream (Granger, 1981). Moreover, the metabolically mediated augmentation of capillary surface area facilitates the fluid removal from the interstitium. The intestinal lymphatics also contribute to the rapid removal of interstitial fluid during epithelial absorption (see Section F-2 below).

6. REMOTE CONTROL OF THE INTESTINAL CIRCULATION

Signals originating outside of the enteric tract also impinge on the intestinal blood vessel. In general, the remote control systems are concerned with modulating the intestinal circulation in accord with maintenance of systemic variables (i.e., blood pressure and volume) important to the survival of the whole organism. These remote vasoregulators include the sympathetic nervous system, the renin–angiotensin system, and vasopressin. Stimulation of the sympathetic nerves to the intestine elicits an initial increase in vascular resistance and a sustained reduction of capillary exchange capacity and tissue blood volume (Svanvik, 1973). Over a period of a few minutes, intestinal blood flow and vascular resistance return to normal in spite of continued nerve stimulation. During vasoconstrictor fiber stimulation, muscularis flow falls and remains low, whereas mucosal flow falls initially and returns to control or above control. The secondary dilatation of mucosal arterioles may be due to triggering of the intrinsic vasoregulator upon reducing mucosal blood flow and capillary exchange capacity (Shepherd and Granger, 1973). The resultant tissue hypoxia elicits powerful local vasodilatory signals that override the neurogenic stimulus at the arteriolar level. This "autoregulatory escape" from sympathetic stimulation does not occur at the precapillary sphincters or venules and veins. Indeed, adrenergic stimuli can produce sustained reductions of capillary exchange capacity to 50% of normal and intestinal blood volume to 60% of normal.

Angiotensin II (Pawlik *et al.*, 1975) and vasopressin (Quillen *et al.*, 1977) produce sustained increases in vascular tone in the resistance, exchange, and capacitance sections of the intestinal circulation. These vasoregulators may play a major role in recruitment of the intestinal vasculature for participation in arterial pressure and blood volume regulation. Thus, although the sympathetic vasoconstrictor response escapes over a period of time, stresses initiating the baroreceptor, chemoreceptor, and atrial receptor reflexes can produce a sustained intestinal vascular reaction via reflex control of angiotensin II and vasopressin release (McNeil *et al.*, 1970).

D. PHARMACOLOGY OF BLOOD CIRCULATION

By H. J. GRANGER

1. GENERAL CONSIDERATIONS

Knowledge of the effects of various drugs and naturally occurring compounds on resistance and exchange vessels of the intestine is accumulating rapidly. However, defining the direct vascular actions of these agents is frequently difficult because, at the same time, many of the compounds alter intestinal absorption, secretion, and/or motility. Under such conditions, vascular reactions may also be influenced by local metabolic and myogenic feedback signals secondary to a primary effect of the agent on parenchymal function. Although this obstacle may be circumvented by studies of drug action on isolated mesenteric arteries and veins, it is necessary to point out that there is substantial support for the idea that arterioles, precapillary sphincters, and venules do not necessarily mimic the behavior of their larger parent vessels. In recent years, this dilemma has been recognized, and hence more studies of intestinal vascular pharmacology have included coupled measurements of circulatory and parenchymal functions in an effort to uncover the primary action of chemical agents (Kvietys and Granger, 1982).

As discussed in Section C-6, above, stimulation of the α-receptor in the intestine causes arteriolar, precapillary sphincter, and venous constriction, whereas vascular relaxation follows administration of a β-adrenergic stimulant (Patel *et al.*, 1981). Stimulation of sympathetic nerves elicits only the α-receptor response, apparently because the β-receptors are located some distance away from the postsynaptic membrane, the loci of the α-receptors. However, preliminary blockade of both vascular α-receptors and presynaptic norepinephrine reuptake produces intestinal vasodilatation, presumably because the concentration of the transmitter now rises to levels sufficient to allow diffusion to the distant β-receptors.

2. Vascular Responses to Specific Agents

a. Action of autacoids: In general, the autacoids (i.e., bradykinin, histamine, and serotonin) cause arteriolar and precapillary sphincter dilatation in the intestine, thereby augmenting blood flow and capillary exchange capacity (Chou and Kvietys, 1981). Of the three autacoids, histamine has received the most attention. Infusion of this drug into the mesenteric artery produces a sharp temporary rise, followed by a subsequent drop of blood flow to a steady-state level above control (Pawlik *et al.*, 1977). The transient spike component is blocked by tripelennamine, a H_1-receptor antagonist. The secondary sustained component of vasodilatation is abolished by metiamide, a H_2-receptor blocker. Histamine (as well as the other autacoids) increases intestinal capillary permeability to macromolecules, the molecular radius of which is 96 Å or less; such a change is blocked by H_2, but not H_1, antagonists (Mortillaro *et al.*, 1981).

b. Action of prostaglandins: Prostaglandins (PGs) have numerous effects on the intestines, including induction of secretion and motility (Waller, 1973). In general, prostacyclin, prostaglandin D_2, and the E series are vasodilatory in nature, whereas $PGF_{2\alpha}$ and B_2 cause vasoconstriction (Chou and Kvietys, 1981). The vasodilator prostaglandins probably predominate: (1) infusion of arachidonate leads to dilatation of intestinal microvessels and (2) inhibition of cyclooxygenase causes vasoconstriction. The impact of the prostaglandins on transvascular exchange has received little attention. PGE_1 increases mucosal capillary pressure and permeability, and the resultant expansion of the interstitial space disrupts the epithelial barrier and allows loss of protein and water into the intestinal lumen (Granger *et al.*, 1979).

c. Action of miscellaneous agents: In general, the gastrointestinal hormones [pentagastrin, secretin, cholecystokinin, and vasoactive intestinal polypeptide (VIP)] cause vasodilatation of intestinal microvessels (Chou and Kvietys, 1981). A large component of their vascular action may occur via local metabolic vasoregulation, secondary to altered absorption, secretion, and motility. Acetylcholine and smooth muscle relaxants (papaverine, sodium nitroprusside, and sodium nitrate) act directly on intestinal vasculature to cause an increase in vessel caliber. As discussed in Section C-6, above, angiotensin II and vasopressin are potent vasoconstrictors in the intestine. Finally, inhibitors of the vascular $Na^+ - K^+$ pump, such as oubain and digoxin, increase the tone of intestinal microvessels (Pawlik *et al.*, 1974).

d. Action of adenyl derivatives: Such substances as ATP, ADP, AMP, and adenosine are potent vasodilators of intestinal vasculature (Chou and Kvietys, 1981). Adenosine may be a potential chemical link between intestinal metabolism and local vasoregulation. For example, the adenosine concentration in ileal interstitium rises 4-fold following a 60-sec arterial occlusion, suggesting that adenosine accumulation in the perivascular spaces may contribute to the dilatation of reactive hyperemia (Mortillaro and Mustafa, 1978). In addition, the adenosine receptor blocker, theophylline, shortens the duration of reactive hyperemia. Also, blockade of adenosine receptors on intestinal microvessels attenuates the intense flow autoregulation observed during digestion. Finally, inhibition of parenchymal reuptake of adenosine by dipyridimole causes a 20% increase in postprandial hyperemia. Thus, adenosine may be an important metabolic vasoregulator under conditions of severe metabolic stress (Granger and Norris, 1980b). By contrast, theophylline and dipyridimole do not alter vascular tone in nonabsorbing intestine. Moreover, adenosine receptor blockade does not alter flow autoregulation in the fasted state. It would appear, therefore, that adenosine does not contribute to resting vascular tone, nor do modest reductions in perfusion pressure in resting intestine cause a significant increase in tissue adenosine levels. Thus, the role of adenosine in local intestinal vasoregulation is dependent on the prevailing metabolic state and the severity of the stresses applied to the intrinsic feedback mechanisms (Granger and Norris, 1980b).

E. PATHOPHYSIOLOGY, PATHOGENESIS, AND PATHOLOGY OF BLOOD CIRCULATION

By H. J. GRANGER

1. HYPERTENSION

Chronic arterial hypertension usually is associated with normal cardiac output, elevated total peripheral resistance, increased venous filling pressure, and augmented albumin extravasation (Ulrych, 1979), with the intestinal circulation playing a role in these systemic alterations. In genetic and high-renin forms of hypertension, intestinal blood flow is elevated above control (Nyhof *et al.*, 1983), the intestinal resistance vessels contributing to the systemic vaso-constriction. Moreover, the transvascular flux of protein and water in the small intestine is doubled in one-kidney, one-clip, renal hypertension, suggesting that the small bowel may be a major site of the elevated albumin extravasation from the systemic circulation (Laine and Granger, 1981). The augmented pro-tein and water flux is due, at least in part, to increased microvascular protein permeability (Laine and Granger, 1981). Since acute intraarterial infusion of angiotensin II at pathophysiologic doses does not induce endothelial barrier opening, it would appear that the proposed angiotoxic effect of this substance cannot account for the enhanced protein permeability (Nyhof and Granger, 1982). With regard to capacitance function, the pressure–volume relationship for mesenteric veins shifts toward the pressure axis in chronic renal hyperten-sion, indicating increased vascular stiffness; thus, the intestinal vasculature may play a major role in elevating mean circulatory filling pressure in hypertensive states (Simon *et al.*, 1975). Elimination of smooth muscle tone does not remove the differences in distensibility between normotensive and hypertensive veins, suggesting that increased stiffness of noncontractile, elastic elements is the cause of decreased vein wall compliance.

Severe hypertension has a dramatic impact on the structure of intestinal blood vessels and surrounding tissue. At the microscopic level, wall hypertro-phy and necrosis are commonly observed in the intestinal arterioles (Giese, 1964a). Such a degenerative process may involve angiotensin-induced contrac-tion of arteriolar endothelium and subsequent infiltration of plasma protein into the subendothelial space (Giese, 1964b). Or, the endothelial disruption of the arterioles may be the direct result of high intraarteriolar pressure. Also in severe forms of hypertension, augmented macromolecular leakage is evident at capillary and venular levels (Surtees *et al.*, 1979). The large increase in intra-vascular pressure and/or disruption of capillary and venular endothelium prob-ably accounts for the intestinal edema and peritoneal effusions observed in hypertension related to severe renal ischemia.

2. INTESTINAL ISCHEMIA

Acute global intestinal ischemia is usually fatal, the two major causes of the condition being massive hemorrhage and occlusion of the superior mesenteric artery. The pathophysiologic consequences of bowel ischemia secondary to mucosal hypoxia include (1) enhanced transcapillary filtration, (2) interstitial fluid accumulation, and (3) net fluid movement into the lumen of the intestine (Granger *et al.*, 1981). One hour of ischemia causes a substantial increase in the permeability of intestinal capillaries to macromolecules, the resulting reduced effectiveness of the microvascular membrane in generating osmotic pressure accounting for the rapid filtration. The elevated interstitial pressure apparently facilitates mechanical disruption of the mucosal barrier. Possible mediators of increased microvascular permeability include (1) histamine, (2) bradykinin, (3) prostaglandins, (4) bacterial endotoxins, and (5) lysosomal and other enzymes released from necrotic enterocytes. With regard to the role of enzymes, xanthine oxidase may catalyze the formation of the cytotoxic superoxide free radical. Blockade of xanthine oxidase provides protection against the effects of ischemia; in addition, pretreatment with superoxide dismutase, a superoxide radical scavenging enzyme, attenuates the change in intestinal capillary permeability observed in ischemia (Granger *et al.*, 1981). Histamine may also be involved, possibly as a result of the inhibitory effects of ischemia on the activity of the inactivating enzyme, diamine oxidase (Kusche *et al.*, 1981).

While gradual occlusion of the intestinal blood supply usually elicits the development of adequate collaterals, acute occlusion causes catastrophic intestinal necrosis (Granger and Molstad, 1982).

3. Intestinal Hemorrhage

Bleeding can occur because of hemorrhoids, diverticula, or any inflammatory process that erodes the intestinal vessels. A unique cause is from angiodysplasias, i.e., clusters of small dilated arteries and capillaries which bleed into the intestinal lumen. These lesions occur most frequently in the right colon.

F.   LYMPHATIC SYSTEM

By D. N. Granger and P. R. Kvietys

1. Anatomy of Lymphatic System

a. Lymphatics of small intestine: A characteristic feature of the small intestine is the presence of large central lacteals in the villi of the mucosa. The lacteals are situated roughly 50 μm from the epithelial cells and are approximately 20 μm in diameter. These vessels unite to form a narrow-meshed plexus of lymph capillaries surrounding the bases of a few glands in the deeper lamina propria, above the muscularis mucosa. The lymphatics of the mucosa (villi) are connected with submucosal lymphatics by means of numerous anastomoses. The lymph vessels in the submucosa are considered to be true collecting vessels, with the structure of their walls resembling the walls of small mesenteric lymphatics. Within the muscularis the lymphatics are not as numer-

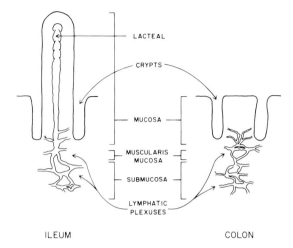

ILEUM   COLON

FIG. 13.1. Schematic representation of the mucosal–submucosal lymphatic microcirculation of the canine ileum and colon. (From Kvietys *et al.*, 1981; reproduced with permission from Elsevier Science Publishing Co.)

ous as those in the submucosa. They run both circularly and longitudinally according to the muscle layers and have abundant anastomoses forming a lymphatic network around each muscle layer. The lymphatics of the serosa are also characterized by many anastomoses and leave the intestine at its mesenteric margin to drain into the principal efferent trunks that course with the blood vessels in the mesentery (Kalima, 1981; Barrowman, 1978; Granger, 1981).

b. Lymphatics of large intestine: The lymphatic microcirculation of the large intestine is in many respects similar to that of the small intestine (see above). The major difference lies in the caliber and distribution of the mucosal lymphatics, as represented schematically in Fig. 13.1. In general, the lymphatics of the colonic mucosa are smaller in caliber and originate in the lamina propria, near the bases of glands in the lower third of the mucosa. Thus, the initial lymphatics of the mucosa are located 300–400 μm from the surface epithelium. Like their counterparts in the small intestine, the lymph vessels of the muscularis mucosa, submucosa, and muscularis are characterized by extensive anastomotic plexuses. The relative sparsity and smaller caliber of the lymphatics in the mucosal region of the colon suggest that the colonic mesenteric lymph trunks primarily drain the muscularis region (Kvietys *et al.*, 1981).

2. PHYSIOLOGY OF LYMPHATIC CIRCULATION

a. Lymph propulsion: The lymphatic system in the small intestine plays an important role in removing fluid and protein escaping from the blood circulation and in transporting absorbed fluid out of the lamina propria. The rate of lymph formation in this organ (which ranges between 0.03 and 0.07 ml/min/100 gm at normal portal pressures) is determined primarily by the rate of fluid filtration across the capillaries and the rate of fluid absorption by the mucosal membrane. The fluid pressure within the interstitium is considered a key factor, allowing the interaction among capillary, mucosal, and lymphatic fluid fluxes. A sigmoid relationship between steady-state lymph flow and interstitial fluid pressure (P_t) has been demonstrated in the cat ileum (Mortillaro and Taylor, 1976). Increases in P_t within the range of -0.10 to $+1.80$ mm Hg do not significantly alter lymph flow; however, at transmural pressures (P_t) between 3 and 5 mm Hg, lymph flow increases dramatically. Lymph flow remains relatively constant when $P_t > 5$ mm Hg. A sigmoid relation between P_t and lymph flow is expected, irrespective of whether P_t is increased due to enhanced capillary filtration or to stimulation of transmucosal fluid absorption. However, there is evidence suggesting that the relation obtained with absorption may be shifted to the right of the curve obtained by altering capillary filtration rate, possibly representing a difference in interstitial compliance (Granger *et al.*, 1980).

In order for lymphatic filling to occur, lacteal pressure must occasionally fall below mucosal interstitial fluid pressure. Intrinsic and extrinsic mechanisms exist that permit this to occur. Rhythmic contractility of lacteals has been demonstrated in rat and guinea pig small intestine, while intestinal lacteals of cat and man do not exhibit rhythmic contractions (Barrowman, 1978). The absence of an active pump (rhythmic contractions) in lacteals of some animals presupposes the existence of extrinsic influences for the propulsion of intestinal lymph.

The effects of intestinal motility on lymph flow remain uncertain. However, studies in rats indicate that drugs that enhance or reduce intestinal motility produce a corresponding change in lymph flow. Villus contraction (shortening) also appears to influence lymphatic filling in the intestine (Lee, 1971). Sustained villus shortening is invariably associated with a diminution of lacteal filling due to occlusion of the lymphatics. However, fluid movement into the central lacteals occurs during the relaxation phase of villus shortening. This observation suggests that intermittent contraction and relaxation of villi, particularly during the absorption of nutrients, may produce the occasional decrease in intralymphatic pressure necessary for lacteal filling.

b. Chylomicron absorption: One of the primary functions of intestinal lymphatics is the removal of newly synthesized chylomicrons from the mucosal interstitium. Since chylomicrons range in size from 0.1 to 0.35 μm in diameter, they are much too large to enter directly into the fenestrated capillaries of the mucosa. Consequently, the large pathway provided by the lacteal clefts allows ready access of chylomicrons to the systemic circulation. There is evidence suggesting that a major portion of the chylomicrons may enter the lacteals via pinocytotic transport. Of interest is the frequent observation that conditions associated with enhanced chylomicron production (fat or ethanol absorption) produces a greater rise in intestinal lymph flow than that observed during carbohydrate absorption (Barrowman, 1978).

c. *Fluid absorption:* Although it is well recognized that absorbed fluid is removed from the mucosal interstitium of the small intestine by both lymphatics and capillaries, the relative contribution of each system remains controversial (Granger, 1981). Estimates of the proportion of absorbed water leaving the intestine via lymph vessels range between 1 and 85%. Most of the more recent studies indicate that less than 40% of the absorbed fluid is transported by lymphatics. The very low reported values (1–5%) result from measurements of lymphatic isotopic water content, which provide an estimate of water diffusion, yet do not reflect net volume movement within the lymphatics. Several factors that affect lymph formation in the nonabsorptive state could account for some of the variability in estimates of lymph contribution during absorption. These include: portal vein pressure, motility, luminal pressure, and tonicity of the luminal fluid. When these influences are minimized or eliminated, the rate of fluid absorption becomes a major determinant of the absolute and relative amounts of absorbed volume removed by the lymphatics. At normal and high fluid absorption rates, the relative contributions of each system remain relatively constant, at 80–85% capillary removal and 15–20% lymph removal. In contrast to the small bowel, lymph flow is unaffected by fluid absorption in the colon, a finding attributed to the paucity of lymphatic drainage from the colonic mucosa (Kvietys *et al.*, 1981).

d. *Fluid secretion:* Although the small intestine is usually an absorbing organ, excess stimulation of active secretory processes (e.g., cholera) can lead to net fluid movement into the bowel lumen. Theoretically, secretion of protein-free fluid across the mucosa should result in a reduction in mucosal interstitial fluid volume and pressure. In turn, the reduction in interstitial fluid pressure should produce a concomitant decline in intestinal lymph flow. Experimental support for these predictions is provided by reports that villus lymph (lacteal) pressure and total intestinal lymph flow decrease after exposing the mucosa of the small intestine to cholera toxin and other active secretagogues (Lee and Silverberg, 1973; Granger *et al.*, 1978). Colonic lymph flow, however, is not altered by stimulation of active secretory processes.

e. *Safety factor against edema:* Any imbalance in the forces governing transcapillary fluid exchange that facilitates filtration into the interstitium will be accompanied by a concomitant increase in tissue pressure and lymph flow and a decrease in interstitial oncotic pressure. These changes tend to oppose further filtration of fluid into the interstitium and are referred to as "edema safety factors." The effective force provided by an increased lymph flow in opposing edema formation can be estimated using the change in net capillary filtration pressure (calculated by dividing lymph flow by the capillary filtration coefficient) for a given increment in capillary pressure. Results obtained using this approach indicate that the lymphatic safety factor provides as much as 4 mm Hg (30% of the total edema safety factor) in opposition to excess fluid filtration in the small intestine when capillary pressure is increased by 12 mm Hg (Mortillaro and Taylor, 1976). For a comparable increment in capillary pressure in the colon, the lymphatic safety factor is <0.5 mm Hg. The relatively small lymphatic safety factor may be explained by the normally low lymph flow (0.015 ml/min/100 gm) from the colon.

3. Pharmacology and Pathology of Lymphatic System

Lymph flow in the intestines is known to be altered by a wide variety of drugs and pathologic conditions (Table 13.I). Most of these do so by altering one or more of the forces (e.g., capillary pressure) or membrane parameters (vascular permeability) that govern transcapillary fluid exchange. Some drugs affect intestinal lymph flow by altering the rate or intensity of lymphatic con-

TABLE 13.I

Physiologic, Pharmacologic, and Pathologic Conditions Known to Alter Intestinal Lymph Flow

Increased lymph flow		Decreased lymph flow	
Conditions	Drugs and other substances	Conditions	Drugs and other substances
Portal hypertension	Bradykinin	Arterial hypotension	Vasopressin
Plasma dilution	Isoproterenol	Lymphangiectasia	Morphine
Luminal distention	Histamine	Crohn's disease	Epinephrine
Fluid absorption	Cholecystokinin		Sodium pentobarbital
Tissue compression	Secretin		Vasoactive intestinal
Inflammation	Prostaglandin E_1		polypeptide (VIP)
	Glucagon		Theophylline
	Carbachol		Cholera toxin[a]
	Bethanechol		Hypertonic glucose
	Serotonin		
	Ricinoleic acid[a]		
	Furosemide		
	Ethacrynic acid		
	Mannitol		

[a]Response to intraluminal placement.

tractility (e.g., serotonin). Other drugs alter lymph flow by virtue of their effects on transmucosal water transport, i.e., by stimulating absorption or secretion. Alterations in lymphatic permeability (Crohn's disease) and lymphatic obstruction (lymphangiectasias) account for the reduced intestinal lymph flow associated with some pathologic conditions.

4. FUTURE AVENUES FOR INVESTIGATION

Several areas of physiology, pharmacology, and pathology of intestinal lymphatics deserve further research attention. These include the role of lymph vessels in the absorptive and secretory functions of the intestine; mechanisms involved in the development of lymphangiectasia; factors governing lymphatic contractility; and the relative contribution of the mucosal, submucosal, and muscle layers to total lymph flow in the small and large bowel.

REFERENCES*

Baez, S. (1977). Skeletal muscle and gastrointestinal microvascular morphology. In "Microcirculation" (G. Kaley and B. M. Altura, eds.), Vol. 1, pp. 69–94. University Park Press, Baltimore, Maryland. (B)

Barrowman, J. A. (1978). "Physiology of the Gastrointestinal Lymphatic System." Cambridge Univ. Press, London and New York. (F)

Bohlen, H. G. (1980). Intestinal tissue P_{O_2} and microvascular responses during glucose exposure. Am. J. Physiol. 238, H164–H171. (C)

Casley-Smith, J. R., O'Donoghue, P. J., and Crocker, K. W. J. (1975). The quantitative relationship between fenestrae in jejunal capillaries and connective tissue channels: Proofs of "tunnel-capillaries." Microvasc. Res. 9, 78–100. (C)

Chou, C. C., and Kvietys, P. R. (1981). Physiological and pharmacological alterations in gastrointestinal blood flow. In "Measurement of Blood Flow" (D. N. Granger and G. B. Bulkley, eds.), pp. 477–509. Williams & Wilkins, Baltimore, Maryland. (C,D)

Clementi, F., and Palade, G. E. (1969). Intestinal capillaries. II. Structural effects of EDTA and histamine. J. Cell Biol. 42, 706–724. (B)

Eade, M. N., and Ginn, R. W. (1978). The distribution of blood flow along the small intestine of the dog. Proc. Soc. Exp. Biol. Med. 157, 390–392. (C)

Eto, T., Oniki, H., Omae, T., and Yamamoto, T. (1977). Increased capillary permeability in muscularis layer of rat intestine caused by kidney extract. Virchows Arch. B 25, 83–93. (B)

Evans, H. M. (1912). The development of the vascular system. In "Manual of Human Embryology" (F. Keibel and F. P. Mall, eds.), Vol. II, pp. 570–709. Lippincott, Philadelphia, Pennsylvania. (A)

Gallavan, R. H., Chou, C. C., Kvietys, P. R., and Sit, S. P. (1980). Regional blood flow during digestion in the conscious dog. Am. J. Physiol. 238, H220–H225. (C)

Gannon, B. J., Gore, R. W., and Rogers, P. A. W. (1981). Is there an anatomical basis for a vascular countercurrent mechanism in rabbit and human intestinal villi? Biomed. Res. 2, 235–241. (B)

*In the reference list, the capital letter in parentheses at the end of each reference indicates the section in which it is cited.

Giese, J. (1964a). Acute hypertensive vascular disease. 1. Relation between blood pressure changes and vascular lesions in different forms of acute hypertension. *Acta Pathol. Microbiol. Scand.* **62**, 481–496. (E)

Giese, J. (1964b). Acute hypertensive vascular disease. 2. Studies of vascular reaction patterns and permeability changes by means of vital microscopy and colloidal tracer technique. *Acta Pathol. Microbiol. Scand.* **62**, 497–515. (E)

Gore, R. W., and Bohlen, H. G. (1977). Microvascular pressures in rat intestinal muscle and mucosal villi. *Am. J. Physiol.* **233**, H685–H693. (C)

Granger, D. N. (1981). Intestinal microcirculation and transmucosal fluid transport. *Am. J. Physiol.* **240**, G343–G379. (C,F)

Granger, D. N., and Taylor, A. E. (1980). Permeability of intestinal capillaries to endogenous macromolecules. *Am. J. Physiol.* **238**, H457–H464. (C)

Granger, D. N., Mortillaro, N. A., and Taylor, A. E. (1978). Interactions of intestinal lymph flow and secretion. *Am. J. Physiol.* **232**, E13–E18. (F)

Granger, D. N., Schackleford, J. S., and Taylor, A. E. (1979). PGE$_1$-induced intestinal secretion: Mechanism of enhanced transmucosal protein efflux. *Am. J. Physiol.* **236**, E788–E796. (D)

Granger, D. N., Mortillaro, N. A., Kvietys, P. R., and Taylor, A. E. (1981). Regulation of interstitial fluid volume in the small bowel. *In* "Tissue Fluid Pressure and Composition" (A. R. Hargens, ed.), pp. 173–183. Williams & Wilkins, Baltimore, Maryland. (F)

Granger, D. N., Rutilli, G., and McCord, J. (1981). Superoxide radicals in feline intestinal ischemia. *Gastroenterology* **81**, 22–29. (E)

Granger, H. J., and Molstad, C. (1981). Collateral circulation in the splanchnic vasculature. *Fed. Proc., Fed. Am. Soc. Exp. Biol.* **40**, 491 (abstr.). (E)

Granger, H. J., and Norris, C. P. (1980a). Intrinsic regulation of intestinal oxygenation in the anesthetized dog. *Am. J. Physiol.* **238**, H836–H843. (C)

Granger, H. J., and Norris, C. P. (1980b). Role of adenosine in local control of intestinal circulation in the dog. *Circ. Res.* **46**, 764–770. (D)

Granger, H. J., and Nyhof, R. A. (1982). Dynamics of intestinal oxygenation: Interactions between oxygen supply and uptake. *Am. J. Physiol.* **243**, G91–G96. (C)

Granger, H. J., and Shepherd, A. P. (1979). Dynamics and control of the microcirculation. *Adv. Biomed. Eng.* **7**, 1–63. (C)

Hallback, K. A., Hulten, L., Jodal, M., Lindhagen, J., and Lundgren, O. (1978). Evidence for the existence of a countercurrent exchanger in the small intestine in man. *Gastroenterology* **74**, 683–690. (C)

Johnson, P. C. (1964). Review of previous studies and current theories of autoregulation. *Circ. Res., Suppl.* **I**, 12–19. (C)

Kalima, T. V. (1981). The structure and function of intestinal lymphatics and the influence of impaired lymph flow on the ileum in rats. *Scand. J. Gastroenterol.* **6**, Suppl. 10, 9–87. (F)

Kusche, J., Lorenz, W., Stahlknecht, C. D., Richter, H., Hesterberg, R., Schmal, A., Hinterlang, E., Weber, D., and Ohmann, C. (1981). Intestinal diamine oxidase and histamine release in rabbit mesenteric ischemia. *Gastroenterology* **80**, 980–987. (E)

Kvietys, P. R., and Granger, D. N. (1982). Vasoactive agents and splanchnic oxygen uptake. *Am. J. Physiol.* **243**, G1–G9. (D)

Kvietys, P. R., Wilborn, W. H., and Granger, D. N. (1981). Effects of net transmucosal volume flux on lymph flow in the canine colon. Structural–functional relationship. *Gastroenterology* **81**, 1080–1090. (F)

Laine, G. A., and Granger, H. J. (1981). Permeability of intestinal capillaries in chronic arterial hypertension. *Microvasc. Res.* **21**, 248. (E)

Lanciault, G., and Jacobson, E. D. (1976). The gastrointestinal circulation. *Gastroenterology* **71**, 851–873. (C)

Lee, J. S. (1971). Contraction of villi and fluid transport in dog jejunal mucosa *in vitro*. *Am. J. Physiol.* **221**, 488–495. (F)

Lee, J. S., and Silverberg, J. W. (1973). Effects of cholera toxin on fluid absorption and villus lymph pressure in dog jejunal mucosa. *Gastroenterology* **62**, 993–1000. (F)

Lundgren, O., and Svanvik, J. (1973). Mucosal hemodynamics in the small intestine of the cat during reduced perfusion pressure. *Acta Physiol. Scand.* **88**, 551–563. (C)

McNeil, J. R., Stark, R. D., and Greenway, C. V. (1970). Intestinal vasoconstriction after hemorrhage: Roles of vasopressin and angiotensin. *Am. J. Physiol.* **219**, 1342–1360. (C)

Maxwell, L. C., Shepherd, A. P., Riedel, G. L., and Morris, M. D. (1981). Effect of microsphere size on apparent intramural distribution of intestinal blood flow. *Am. J. Physiol.* **241**, H408–H414. (C)

Mortillaro, N. A., and Granger, H. J. (1977). Reactive hyperemia and oxygen extraction in the feline small intestine. *Circ. Res.* **41**(6), 859–865. (C)

Mortillaro, N. A., and Mustafa, S. J. (1978). Possible role of adenosine in the development of intestinal postocclusion reactive hyperemia. *Fed. Proc., Fed. Am. Soc. Exp. Biol.* **37**, 874 (abstr.). (D)

Mortillaro, N. A., and Taylor, A. E. (1976). Interaction of capillary and tissue forces in the cat small intestine. *Circ. Res.* **39**, 348–358. (C,F)

Mortillaro, N. A., Granger, D. N., Kvietys, P. R., Rutili, G., and Taylor, A. E. (1981). Effects of histamine and histamine antagonists on intestinal capillary permeability. *Am. J. Physiol.* **240**, G381–G386. (D)

Norris, C. P., Barnes, G. E., Smith, E. E., and Granger, H. J. (1979). Autoregulation of superior mesenteric flow in fasted and fed dogs. *Am. J. Physiol.* **237**, H174–H177. (C)

Nyhof, R. A., and Granger, H. J. (1982). The acute effects of angiotensin II on canine intestinal vascular permeability. *Physiologist* **25**, 232. (E)

Nyhof, R. A., Laine, G. A., Meininger, G. A., and Granger, H. J. (1983). The splanchnic circulation in hypertension. *Fed. Proc., Fed. Am. Soc. Exp. Biol.* **42**, 1690–1693. (E)

Parker, R. E., and Granger, D. N. (1979). Effect of graded arterial occlusion on ileal blood flow distribution. *Proc. Soc. Exp. Biol. Med.* **162**, 146–149. (C)

Patel, P., Bose, D., and Greenway, C. (1981). Effects of prazosin and phenoxybenzamine on α- and β-receptor-mediated responses in intestinal resistance and capacitance vessels. *J. Cardiovasc. Pharmacol.* **3**, 1050–1059. (D)

Pawlik, W., Shepherd, A. P., Mailman, D., and Jacobson, E. D. (1974). Effects of ouabain on intestinal oxygen consumption. *Gastroenterology* **67**, 100–106. (D)

Pawlik, W., Shepherd, A. P., and Jacobson, E. D. (1975). Effects of vasoactive agents on intestinal oxygen consumption and blood flow in dogs. *J. Clin. Invest.* **56**, 484–490. (C)

Pawlik, W., Tague, L. L., Tepperman, B. L., Miller, T. A., and Jacobson, E. D. (1977). Histamine H_1- and H_2-receptor vasodilation of canine intestinal circulation. *Am. J. Physiol.* **233**, E219–E224. (D)

Perry, M. A., and Granger, D. N. (1981). Permeability of intestinal capillaries to small molecules. *Am. J. Physiol.* **241**, G24–G30. (C)

Quillen, E. W., Granger, D. N., and Taylor, A. E. (1977). Effects of arginine vasopressin on capillary filtration in the cat ileum. *Gastroenterology* **73**, 1290–1295. (C)

Richardson, P. D. I., and Granger, D. N. (1981). Capillary filtration coefficient as a measure of perfused capillary density. *In* "Measurement of Blood Flow" (D. N. Granger and G. B. Bulkley, eds.), pp. 321–336. Williams & Wilkins, Baltimore, Maryland. (C)

Rothe, C. F., Bennett, T. D., and Johns, B. L. (1980). Linearity of the vascular pressure–volume relationship of the canine intestine. *Circ. Res.* **47**, 551–558. (C)

Shepherd, A. P. (1978). Intestinal oxygen consumption and ^{86}Rb extraction during arterial hypoxia. *Am. J. Physiol.* **234**, E248–E251. (C)

Shepherd, A. P. (1979). Intestinal capillary blood flow during metabolic hyperemia. *Am. J. Physiol.* **237**, E548–E554. (C)

Shepherd, A. P., and Granger, H. J. (1973). Autoregulatory escape in the gut: A systems analysis. *Gastroenterology* **65**, 77–91. (C)

Simionescu, N., Simionescu, M., and Palade, G. E. (1972). Permeability of intestinal capillaries. *J. Cell Biol.* **53**, 365–392. (B)

Simon, G., Pamnani, M. B., Dunkel, J. F., and Overbeck, H. W. (1975). Mesenteric hemodynamics in early experimental renal hypertension in dogs. *Circ. Res.* **36**, 791–798. (E)

Surtees, V. M., Ham, K. N., and Tange, J. D. (1979). Visceral oedema and increased vascular permeability in early experimental hypertension. *Pathology* **11**, 663–670. (E)

Svanvik, J. (1973). The effect of reduced perfusion pressure and regional sympathetic vasoconstrictor activation on the rate of absorption of ^{85}Kr from the small intestine of the cat. *Acta Physiol. Scand.* **89**, 239–248. (C)

Ulrych, M. (1979). Pathogenesis of essential hypertension. *Angiology* **30**, 104–116. (E)

Waller, S. L. (1973). Prostaglandins and the gastrointestinal tract. *Gut* **14**, 402–417. (D)

Wheaton, L. G., Sarr, M. G., Schlossberg, L., and Bulkley, G. B. (1981). Gross anatomy of the splanchnic vasculature. *In* "Measurement of Blood Flow" (D. N. Granger and G. B. Bulkley, eds.), pp. 9–45. Williams & Wilkins, Baltimore, Maryland. (B)

Chapter 14
Digestive System: Liver

A. EMBRYOLOGY OF BLOOD CIRCULATION
By C. E. CORLISS

Evans, in 1912, reviewed in detail the available literature on development of hepatoportal circulation, and more recently Dickson (1957) and Severn (1972) have added pertinent information on the subject.

1. HEPATOPORTAL SYSTEM CIRCUITS

Three sets of venous channels are involved in the hepatoportal system: (1) paired vitelline veins from the yolk sac plexus, (2) intrahepatic sinusoids and veins developing *in situ*, and (3) umbilical veins returning chorionic blood to the heart. All three groups appear early in the fourth embryonic week and assume a definitive pattern by the sixth week (Fig. 14.1A and C).

2. EARLY ANGIOGENESIS

a. Vascular plexuses: In the fourth week, the vitelline plexus is an embryonic extension of the early yolk sac vessels (see Section A-1a, Chapter 10). Concurrently, intrahepatic vessels develop as hepatic cords invade endothelium-lined spaces in the septum transversum. The classic concept that sinusoids result from "tapping" vitelline and umbilical vessels is challenged by Severn (1972), who claims that they form *in situ*. Umbilical veins originate early as allantoic vessels in the body stalk and, by the fourth week, return placental blood from the chorion.

b. "Symmetric" stage: This 1-month stage is nearly bilaterally symmetric (see Fig. 14.1B). Both umbilical veins reach the sinus venosus directly, but soon lose this connection as the ductus venosus becomes a midline shunt, conducting umbilical blood efficiently to the heart via symmetric hepatic veins and the vena cava (Fig. 14.1A and D). According to Dickson (1957), the symmetric posthepatic veins have a fourth, subdiaphragmatic, anastomosis.

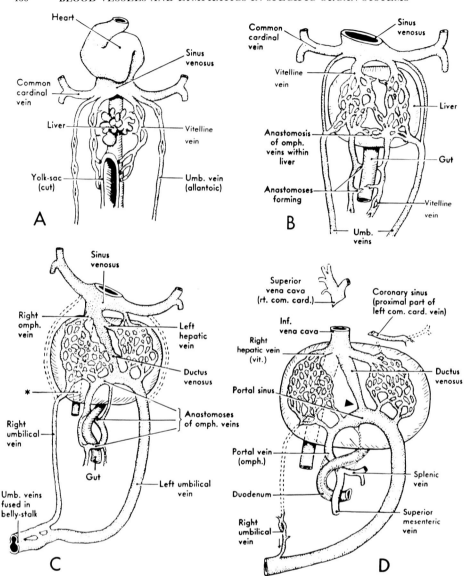

FIG. 14.1. Diagrams showing development of hepatic portal circulation and route of placental blood through the liver. (A) Embryo of the fourth week. (B) Embryo of the fifth week. (C) Embryo of the sixth week. (D) Embryo of the eighth week and older. Key: Asterisk in (C) indicates hepatic part of the inferior vena cava; arrowhead in (D) indicates site of sphincter. (From Corliss, 1976; reproduced with permission of McGraw-Hill Book Co.)

3. DIFFERENTIATION OF HEPATOPORTAL SYSTEM

At the fifth week, a final pattern emerges and, by the sixth week, it is almost complete (Fig. 14.1B and C). The paired posthepatic vitelline veins, after complex anastomoses around the gut and selective degeneration, become a single, short portal vein. Splenic and superior mesenteric veins join the latter distally (Fig. 14.1B–D). The newly formed portal sinus connects the right half of the subhepatic anastomosis to the portal circulation by the sixth week (Fig. 14.1D).

The final route, then, for umbilical blood returning via the hepatic portal system includes the following vessels: (1) left umbilical vein, (2) left subhepatic anastomosis, (3) ductus venosus, (4) right subdiaphragmatic anastomosis, and (5) common hepatic vein to the inferior vena cava (Fig. 14.1D). Blood that returns via the portal vein is directed into the liver by right and left portal branches and thence into hepatic plexuses or the ductus venosus. The amount that passes into the latter is controlled by a sphincter near the junction of the ductus with the portal sinus (site indicated by arrowhead in Fig. 14.1D). When closed, this sphincter deflects the flow of deoxygenated portal blood through the capillary beds of the liver.

B. ANATOMY OF BLOOD CIRCULATION

By P. K. CHAUDHURI*; A. M. RAPPAPORT**

The liver is a highly vascular organ with a unique arrangement of blood supply, different from any other abdominal viscera. Blood is carried to it by two entirely separate channels: the hepatic artery and the portal vein. From the liver, blood returns to the systemic venous system through hepatic veins.

1. GROSS ANATOMY OF EXTRAHEPATIC BLOOD VESSELS†

a. Hepatic artery: Arterial blood supply to the liver is through the common hepatic artery. Although major variations in the origin of this vessel fre-

*Author of Sections B-1 and B-2.
**Author of Sections B-3 to B-7, inclusive.
†By P. K. Chaudhuri.

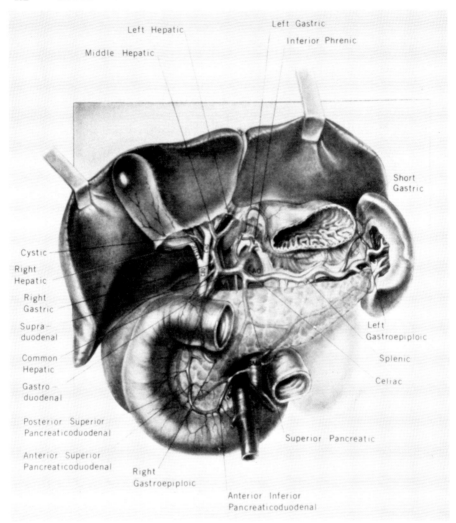

FIG. 14.2. Arterial blood supply to the liver and biliary system. (From Orloff, 1981; reproduced with permission from W. B. Saunders Co.)

quently occur (Michels, 1966), in most instances it arises as a branch of the celiac axis (Fig. 14.2). From its origin, it courses to the right, along the superior margin of the pancreas, to the hepatoduodenal ligament. Here it lies medial to the common bile duct and anterior to the portal vein. As it courses superiorly toward the porta hepatis, it usually gives off the right gastric branch and then the gastroduodenal artery. The common hepatic artery generally divides into a right and left hepatic artery. The middle hepatic artery, which supplies the caudate and quadrate lobes, usually comes off from either the left or right hepatic branch; at times, however, it may arise from the common hepatic trunk as a third terminal branch (Elias and Petty, 1952; Michels, 1966). The cystic artery is usually a branch of the right hepatic artery; less commonly, it may arise from the common hepatic artery and, very infrequently, from the left hepatic artery. Prior to, and following its termination as the right and left hepatic artery, the common hepatic artery may have a variable relationship with the right and left hepatic, common, and cystic ducts.

b. *Anomalies of hepatic artery:* The classically described origin and distribution of the hepatic artery is seen in 55% of cases (Michels, 1966; Suzuki *et al.*, 1971), whereas in 25%, the left hepatic artery may arise from the left gastric artery and in approximately half of this group, a branch of the left gastric artery may take the place of the left hepatic artery. In 12% of the cases, the right hepatic artery may originate from the superior mesenteric artery. In 5%, a smaller right hepatic artery may arise from the superior mesenteric artery, behind the head of the pancreas, in addition to the standard right hepatic artery originating as a branch of the common hepatic trunk. Following its origin, the anomalous right hepatic artery passes upward, and in its course, it lies to the right and posterior to the portal vein. In the lesser omentum it is found behind the common bile duct, gaining entry to the liver through the porta hepatis, posterior to the common hepatic and right hepatic ducts (Warwick and Williams, 1973). Rarely, the arterial blood supply to the liver may arise either entirely from the left gastric artery, from a branch of the superior mesenteric artery, or from a branch of the abdominal aorta. Figure 14.3 illustrates some of the more common anomalies.

c. *Collateral circulation:* There are numerous collateral vessels connecting the hepatic artery with other systemic arteries (Michels, 1966). Although the majority of them are celiac in origin and branches of the hepatic artery proper (e.g., right gastric artery, gastroduodenal artery), a number also are outside this system. Of the latter, the following are important and may constitute the major arterial blood supply to the liver in case of ligation of the hepatic artery, just below its bifurcation, or of its main branches: (1) vessels from the inferior and superior phrenic arteries communicating with branches of the hepatic artery in the diaphragm; (2) vessels in the falciform ligament and ligamentum teres via the rumulus appendicem ensiformen of the internal mammary artery, uniting with branches of the left hepatic artery; (3) arteries in the left and right triangular and coronary ligaments; (4) intercostal arteries and arteries of the posterior abdominal wall.

d. *Portal vein:* The portal venous system carries blood from all of the abdominal viscera (except the anal canal, spleen, pancreas, and biliary tracts) to the liver (Fig. 14.4). The portal vein itself is devoid of valves (Warwick and

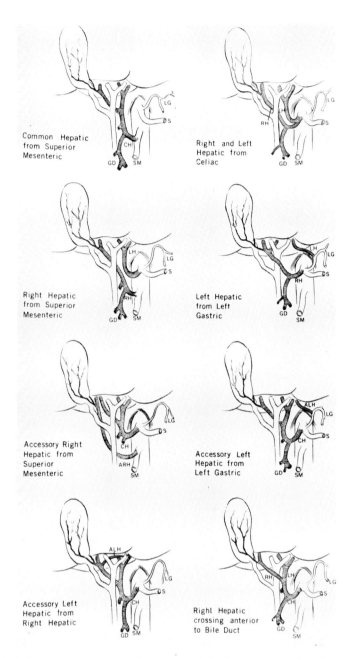

FIG. 14.3. Common variations in the anatomy of the hepatic artery, CH, Common hepatic artery; RH, right hepatic artery; LH, left hepatic artery; ARH, accessory right hepatic artery; S, splenic artery; SM, superior mesenteric artery; GD, gastroduodenal artery; ALH, accessory left hepatic artery; LG, left gastric artery. (From Orloff, 1981; reproduced with permission from W. B. Saunders Co.)

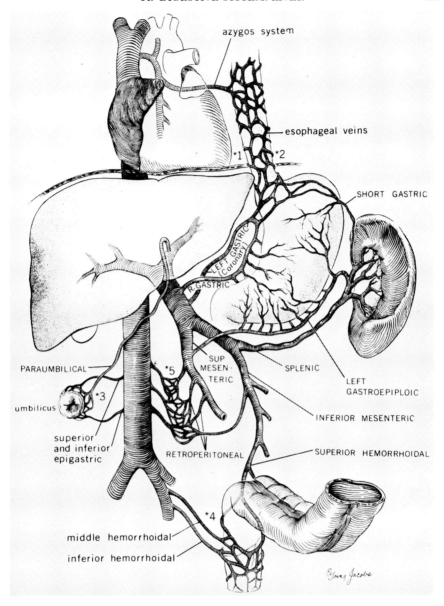

FIG. 14.4. Portal venous system. Numbers denote sites of portal-systemic anastomoses which may act as collateral communications. (From Orloff, 1981; reproduced with permission from W. B. Saunders Co.)

Williams, 1973). It is 6 to 8 cm long and starts behind the neck of the pancreas, at the level of the second lumbar vertebra, by union of the splenic and superior mesenteric veins. Following its origin, the portal vein passes upward and slightly to the right, first behind the pancreas and the first portion of the duodenum and then behind the common bile duct and hepatic artery. In this position, it lies anterior to the inferior vena cava. Near the porta hepatis, it divides into the left and right branches, which then accompany the corresponding branches of the hepatic artery into the liver substance through the porta hepatis.

The right branch of the portal vein, which is shorter and wider than the left branch, passes to the right lobe of the liver. The left branch supplies the left lobe and gives off branches to the quadrate and caudate lobes, usually before entry into liver substance.

Although at the level of the porta hepatis the relationship of branches of the portal vein to hepatic arteries and ducts may be variable, the vein usually lies posterior to the other structures. The left branch of the portal vein is joined by paraumbilical veins and the obliterated left umbilical vein (ligamentum teres). The tributaries of the portal vein consist of (1) superior mesenteric vein, (2) splenic vein, (3) coronary veins (left and right gastric veins), (4) pancreaticoduodenal veins, (5) pancreatic veins, (6) paraumbilical veins (usually draining into the left branch of the portal vein), and (7) cystic veins (generally draining into the right branch of the portal vein).

The length and direction of the portal vein usually are constant, although the location and number of its tributaries may vary. Rarely there may be a cavernomatous transformation which may be congenital, although an acquired type cannot be completely ruled out (Simonds, 1936).

Under normal conditions, communications between the portal and systemic venous systems are insignificant. However, in cirrhosis of the liver or portal venous thrombosis, these collateral vessels may become very large. Although many are found, only the following pathways assume any significant role (Orloff, 1981): (1) coronary vein to azygos veins through vessels in the fundus of the stomach and esophagus, and left gastroepiploic and splenic veins to the azygos vein through the esophageal vein and veins in the fundus of the stomach; (2) superior hemorrhoidal vein to the middle and inferior hemorrhoidal veins through the wall of the rectum; (3) paraumbilical vein to the superior and inferior epigastric veins; and (4) tributaries of the superior mesenteric vein to the inferior vena cava through the retroperitoneal veins.

e. Hepatic veins: These vessels carry blood from the liver to the systemic circulation, their number and position being quite variable (Healey, 1954). They commence in the intralobular veins, which collect blood from the sinusoid and then join to form sublobular veins. The latter, in turn, form the

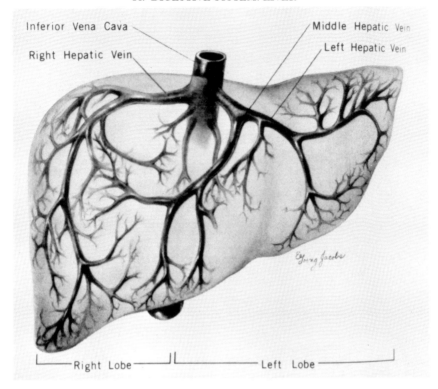

Inferior Vena Cava

Right Hepatic Vein

Middle Hepatic Vein

Left Hepatic Vein

Right Lobe   Left Lobe

FIG. 14.5. Hepatic systemic venous system. (From Orloff, 1981; reproduced with permission from W. B. Saunders Co.)

hepatic veins (Fig. 14.5). As these vessels emerge from the posterior surface of the liver, they drain directly into the inferior vena cava, usually at the point where the latter lies in the groove on the posterior surface of the liver.

The hepatic veins, depending on their position, generally are arranged into two groups: upper and lower. The upper group usually consists of three large vessels, the right branch draining blood from the right lobe; the left branch, from the left lobe; and the middle (which is located slightly lower than the other two), from the caudate lobe. Occasionally, the middle and left branches may join to form a common vessel before opening into the inferior vena cava. The tributaries in the lower group vary from three to five in number. They usually are small in size and originate from the right and caudate lobes. All the hepatic veins are short, thin-walled and, like portal veins, are devoid of valves (Warwick and Williams, 1973).

2. Gross Anatomy of Intrahepatic Blood Vessels*

Although the liver classically is divided by anatomists into left and right lobes by attachment of the falciform ligament, a true division of the lobes is different, being based on blood supply. In accord with such a view, the lobar fissure divides the liver into true left and right lobes, the division corresponding to a line drawn through the liver from the gallbladder fossa anteriorly to the fossa of the inferior vena cava posteriorly. In turn, the right lobe can be divided, based on distribution of vessels and biliary ducts, into anterior and posterior segments. This division, however, is somewhat imperfect and is located in the midplane of the right lobe of the liver. The distribution of the intrahepatic veins does not follow the segmental distribution of the hepatic artery and portal vein.

a. Intrahepatic distribution of the hepatic artery: The main arterial supply for the right lobe derives from the right hepatic artery. Just before entering the parenchyma of the liver, this vessel divides into anterior and posterior segmental branches, each of which then divides into superior and inferior branches. The caudate and quadrate lobes are supplied by the middle hepatic artery, which may originate from either the left or the right hepatic artery. Occasionally a middle hepatic artery may arise as a terminal branch of the main hepatic artery. The left lobe is supplied by the left hepatic artery, which, following its origin, ascends upward and posteriorly in the left lobe and divides into a medial and a lateral segmental artery. Like the right side, each segmental vessel again divides into a superior and an inferior subsegmental branch.

b. Intrahepatic distribution of portal vein: The right branch of the portal vein supplies the right lobe of the liver. The branches of this vessel follow closely the pattern of distribution of the hepatic artery. From its origin, the right branch of the portal vein divides into an anterior and a posterior segmental branch. Each of the branches then divides into superior and inferior subsegmental branches.

The distribution of the left portal vein is somewhat different from that of the artery. The vessels to the medial segment arise from the left portal vein near its entrance into parenchyma and are two to four in number. Usually two of these branches drain the superior portion, and the other two drain the inferior portion of the medial segment. The lateral segment of the left lobe drains via an anterior and a posterior branch of the portal vein. The portal venous branch draining the caudate process arises from the right main trunk, whereas that draining the caudate lobe arises from the left main trunk.

*By P. K. Chaudhuri.

c. Intrahepatic distribution of the hepatic veins: Unlike the hepatic artery and the portal vein, the hepatic veins are not segmentally distributed. These vessels lie in intersegmental planes and drain adjacent segments. The right hepatic vein is found in the intersegmental fissure of the right lobe and drains the entire posterior segment and the superior portion of the anterior segment of this lobe. The left hepatic vein lies in the upper portion of the segmental fissure and drains the entire lateral segment and the superior portion of the medial segment of the left lobe. The middle hepatic vein lies in the interlobar fissure and drains the inferior portion of the anterior segment of the right lobe and the inferior portion of the medial segment. The veins in the caudate lobe usually drain directly into the inferior vena cava by one or two branches.

3. General Considerations of Physioanatomy of Microcirculation**

The anatomy of the microcirculatory vessels in the liver can best be described as the framework for the parenchymal units they supply. These units are, in increasing order of magnitude: (a) the simple acinus, (b) the complex acinus, and (c) the acinar agglomerate (Rappaport *et al.*, 1954). Each of the acini harbors a correspondingly named microvascular unit that consists of afferent and efferent vascular branches and sinusoids.

The simple acinar microvascular unit (SAMU) is formed by the terminal arterial and portal branches, peribiliary arterial plexus, sinusoidal glomus, and terminal hepatic venules (Fig. 14.6). Afferent and efferent vessels interdigitate, mostly at right angles.

The complex acinar microvascular unit (CAMU) is composed of three SAMUs, the afferent vessels of which originate by a trichotomy of the preterminal arterial and portal branches, to form the axis of the CAMU.

The agglomerate microvascular unit (AMU) is composed of at least three CAMUs, the preterminal afferent vessels of which originate from the same axial portal or arterial branch; these supply all the microcirculatory units contained within an acinar agglomerate.

4. Structure of the Simple Acinar Microvascular Unit (SAMU)**

a. Terminal hepatic arterioles (THA): These vessels have an average diameter of 25 μm and a single layer of smooth muscle cells, but no elastica interna; the endothelium has a basement membrane. Projections from the endothelium

** By A. M. Rappaport.

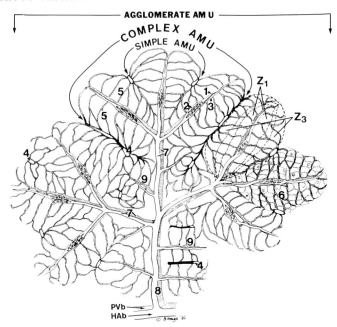

FIG. 14.6. Three microcirculatory units of liver, supplying simple acini, complex acini, and acinar agglomerates (diagrammatic). 1, Terminal hepatic arteriole (THA, white); the THA forms a plexus (2). 3, Terminal portal venule (TPV, cross hatched). 4, Terminal hepatic venule (ThV). It interdigitates with the supplying vessels and collects blood from sinusoids of several microcirculatory units. 5, Sinusoidal glomus, consisting of wide capillary loops separated by single or double cell plates and arranged tridimensionally to form simple liver acinus (6). 7, Preterminal hepatic arteriole and portal venule supply three microcirculatory units. 8, The axial arterial and portal branch, each a stem vessel supplying blood to all microcirculatory units within an acinar agglomerate. 9, Tiny microvascular units sustaining a small acinus (acinulus). HAb, Hepatic arterial branch; PVb, portal venous branch; Z_1, periportal microcirculatory zone; Z_3, microcirculatory periphery.

into the tunica media form myoendothelial junctions that facilitate the transmission of stimuli from vasoactive substances circulating in the blood. There are also unmyelinated nerve fibers in the arteriolar wall. Thus, the arteriole contains all of the elements necessary to change the diameter of the vessel and to regulate blood flow. The arterioles break up into arterial capillaries, 10 μm in diameter, which are of the same structure as other systemic arterial capillaries. They are endothelial tubes with a basement membrane, and some have a precapillary smooth muscle sphincter with closely apposed unmyelinated nerve fibers. Sphincters are commonly noted in the afferents of the peribiliary arterial plexus, that is densely woven around the terminal bile ducts (Fig. 14.7). Some of the efferent capillaries and arterioles of the plexus join the sinusoids at their periportal (zone 1) portion; others empty into the terminal portal venules (TPV) and increase the *vis a tergo* in the venules. No arteriole empties into the terminal hepatic venules (ThV) (McCuskey, 1966; Nakata and Kanbe, 1966;

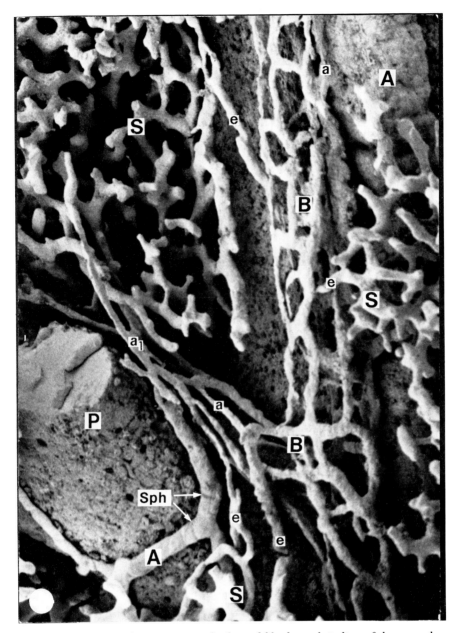

FIG. 14.7. Scanning electron micrograph of casted blood vessels in liver of rhesus monkey. Peribiliary plexus (B) receives its blood from arterial branches (A) via afferent arterioles (a), and it supplies the sinusoids (S) through efferent arterioles (e). Note the constrictions in the arterial cast caused by the circular smooth muscle fibers. Sph, Grooves indicating arteriolar sphincters; a_1, arterioles bypassing the plexus and emptying directly into sinusoids; P, portal venous branch. Methyl methacrylate cast. × 135. (Courtesy of Dr. T. Murakami, Department of Anatomy, Okayama University, Japan.)

Olds and Stafford, 1930). Cast studies have shown that threads of methyl methacrylate run "through all the lobules" (Kardon and Kessel, 1980), and it has been assumed that they are arterioles connecting directly with ThVs. However, this is not sufficient proof for the presence of "translobular" arterioles, for the tissue elements of the walls of such vessels would first have to be demonstrated histologically. *In vivo* observations in the transilluminated mammalian liver reveal selective arterialization and dilatation of transacinar sinusoids receiving arterial blood (Rappaport, 1979). The peribiliary plexus plays an important function in the secretion and resorption of duct bile, in view of fact that it mediates a countercurrent exchange of ions between blood and bile (Rappaport, 1980). The anatomic arrangement of this plexus is similar to that of the glomeruli in the kidney (Andrews, 1955). There are species differences in the distribution of the terminal hepatic arteriole. In the monkey arterial blood appears to reach the sinusoids only after passing through the peribiliary plexus (Grisham and Nopanitaya, 1981; Murakami *et al.*, 1974).

b. Terminal portal venule (TPV): In the liver, this vessel is an endothelial tube of 47 μm in average width and 720 μm in length. It is surrounded by scant connective tissue and limiting plates of parenchymal cells, with no smooth muscle fibers in its wall. There are openings in the wall that lead either directly into sinusoids or into inlet venules that supply a group of sinusoids. The tip of the TPV may divide into a number of inlet venules. The latter have large endothelial cells at their origin from the TPV. Their intracellular contractile protein permits these cells either to bulge into the lumen or to flatten out and thus, respectively, impede or facilitate the passage of erythrocytes into the sinusoids. Terminal portal twigs branch from preterminal portal venules to supply small glomi of sinusoids inside acinuli (Rappaport, 1963). An acinulus, a tiny acinus, possesses vessels that are not part of a terminal trichotomy of preterminal artery or portal vein but, instead, originate as single lateral twigs from either of them. The acinuli fill the gaps between the simple acini and complete the mantle of parenchyma around the portal space.

c. Sinusoids: These are large capillaries of the parenchyma, remnants of the vitelline venous system, and they radiate from the ThVs connecting them with the THA and TPV. The length of a sinusoid averages 332 μm, which corresponds to one-half the width of an acinus. The width of the sinusoidal lumen is 9–12 μm but it can enlarge to 30 μm. The sinusoids of the acinar zone 1 have a smaller volume fraction and larger surface to volume ratio than those in zone 3 (Miller *et al.*, 1979). The wall of the sinusoids consists of a lining of endothelial cells interspersed with Kupffer cells (Fig. 14.8). The endothelial cells have a long attenuated portion and a short, thicker part that harbors the small cell nucleus. The cytoplasm is electrolucent and contains granular endoplasmic reticulum and micropinocytotic vesicles, some of them bristle coated. The endothelium is interlocked at cell junctions; it has no basal membrane, and

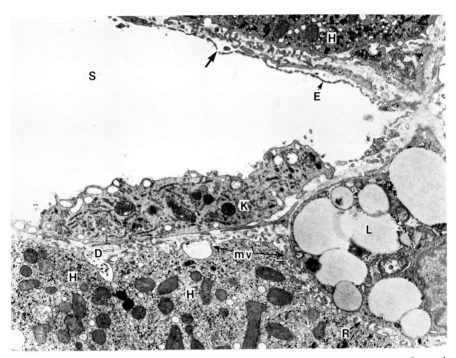

FIG. 14.8. Transmission electron micrograph of normal human liver. Seen are portions of several hepatocytes (H); a Kupffer cell (K); processes of an endothelial cell (E) with fenstrations (arrow); and lipocyte (Ito cell) (L), bordering a sinusoid (S). Note that microvilli (mv) of hepatocytes project into space of Disse (D) and perisinusoidal recess (R). Lead citrate stain. × 7,175. (Courtesy Dr. M. J. Phillips, Department of Pathology, Sick Children's Hospital, University of Toronto.)

the attenuated regions of the cytoplasm have "fenestrae," i.e., holes 100 nm wide. These "sieve plates" (Wisse, 1977) permit a free communication between the sinusoidal lumen and the perisinusoidal space of Disse into which the hepatocytes dip their microvilli (Fig. 14.8). The fenestrae in the sieve plates appear to change diameter (Fraser, 1978) and filter out molecules of large size. Because of the fenestrae, the intrasinusoidal hydrostatic pressure is equal to that in the Disse space, an arrangement that keeps both spaces open. The sinusoidal endothelial tube is also supported by adjacent modified mesenchymal cells, the fat-storing lipocytes (Ito cells) (Fig. 14.8), which can develop hemopoietic or fibrogenic activity. There are also scant reticular fibers in the perisinusoidal spaces supporting the endothelial tube. The Kupffer cells are, according to newer investigations (Wisse, 1977) not modified endothelial cells, but macrophages derived from monocytes. Frequently they are suspended within the sinusoidal lumen by pseudofilia that lock into microvilli. Their large nucleus resembles that of macrophages, and their cytoplasm contains wormlike tubules, i.e., a reservoir of cell membranes for rapid phagocytic activity. Within the cytoplasm there are few mitochondria, many granular endoplasmic reticula, a minor Golgi apparatus, lysosomes rich in acid phosphatase, phagosomes, and dense bodies, the secondary lysosomes; also present are pinocytotic vesicles and hemosiderin from erythrocytes that have been destroyed. The cytoplasm of the Kupffer cell of the rat gives a positive peroxidase reaction, similar to that seen in monocytes. The Kupffer cells, as well as the sinusoidal endothelium, are part of the reticuloendothelial system.

d. Terminal hepatic venule (ThV): This vessel is a relatively wide endo-thelial tube (average of 60 μm in width and 1 to 1.5 mm in length) into which the sinusoids empty singly or via outlet venules (Deysach channels); the latter collect blood from several sinusoids and convey it into the ThV. The ThV is a drainage center for several adjacent SAMU, into which they pour their blood. *In vivo* observation of blood flow in the transilluminated rodent liver provides evidence of intermittent emptying of single sinusoids into the hepatic venule (Rappaport, 1972). As the venules interdigitate with the afferent terminal ves-sels, the rapid transfer of hepatic blood into the systemic veins is ensured. The ThVs are situated at the periphery of the microcirculatory network, which contradicts their classic term "central veins." Cells surrounding the ThV are most sensitive to ischemia, congestion, nutritional deficiencies, and toxic damage (Rappaport, 1963). In the rat, the endothelial cells of the ThV have fenestrae but rest on a basement membrane. The hepatic veins have no valves, and their wall is surrounded by scant reticular fibers and contains discontinuous primitive myocytes (Sasse and Staubesand, 1976) and elastic fibers. In the larger venous branches, there is a continuous muscle coat, often double layered with fibroblasts between them. The ThV has no limiting plate, which indicates that the surrounding tissue is not bound together but belongs to different individual acini. The most remote portion of the acinar zone 3 apposes the venule. The ThVs come together to form the preterminal hepatic venules draining the CAMU.

5. Structure of the Complex Acinar Microvascular Unit (CAMU)**

The CAMU consists of three simple microvascular units, together with acinular units sustaining the mantle of parenchyma surrounding the pretermi-nal axial supplying vessels and draining ducts.

a. Preterminal hepatic arteriole: This vessel has an average width of 43 μm, and its wall has, besides the endothelial lining, a double ring of smooth

**By A. M. Rappaport.

smooth muscle in these vessels being affected predominantly by local mechanisms. The latter adjust arterial flow to compensate for changes in portal flow
(the arterial buffer response). The sympathetic nerves also modify arterial flow.

Important physiologic aspects of the hepatic circulation involve stabilization
of hepatic blood flow, portal pressure, and blood content to allow control of the
blood reservoir provided by the liver. Changes in hepatic venous compliance
play a major role in overall control of cardiac output. These basic features are
discussed in more detail below (for further discussion, see Greenway, 1983a;
Greenway and Stark, 1971; Lautt, 1981a; Richardson and Withrington, 1981).

2. PORTAL VENOUS CIRCUIT

a. Portal flow: Blood flow in the portal vein is determined by arteriolar
resistances in the gastrointestinal and splenic circulations. Although a few studies have demonstrated preferential distribution of blood flow from certain areas
of the gastrointestinal tract to specific lobes of the liver, it is likely that in
conscious mobile animals, portal blood is well mixed. This is important, since a
hepatotrophic factor is found in pancreatic blood, and lobes supplied with this
factor show enlargement, whereas those deprived of it demonstrate atrophy.
Although presinusoidal sphincters cause portal flow in individual sinusoids to
be intermittent when observed in transilluminated livers, microsphere studies
show overall distribution of portal blood to the parenchyma to be uniform
throughout the liver.

b. Hepatic venous resistance: Although portal inflow is determined by pre-
hepatic splanchnic organs, the resistance to flow in the portal circuit within the
liver determines portal pressure (at rest 7–10 mm Hg). It has been assumed
that this resistance is presinusoidal, hence the term, portal resistance. However, recent work in the cat and dog has shown that pressure in lobar hepatic
veins approaches portal pressure, even when great care is taken to avoid recording a wedged pressure (Fig. 14.11). This suggests that the major site of
resistance in the portal circuit is in the lobar hepatic veins (Fig. 14.10). The
hepatic venous resistance should be calculated as portal minus caval pressure
divided by total hepatic flow. Occlusion of the hepatic artery reduces portal
pressure by 2–3 mm Hg and total hepatic flow by 20–33%. Thus, venous
resistance remains unchanged.

The functions of hepatic venous resistance involve control of both portal
pressure and transsinusoidal fluid movements. No myogenic or metabolic intrinsic regulation has been demonstrated, and the major controlling mechanism
appears to be sympathetic innervation mediated through α-receptors. Although hepatic nerve stimulation raises portal pressure, more generalized sympathetic activity, e.g., splanchnic nerve stimulation, reduces portal flow by
vasoconstriction in the gastrointestinal and splenic beds. Hepatic venous re-

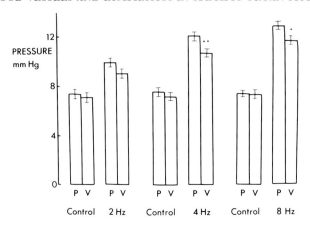

FIG. 14.11. Simultaneous unwedged pressures in portal (P) and hepatic (V) veins before and during stimulation of hepatic nerves (means ± SE; n = 10 cats). (From Greenway, 1981a; reproduced with permission from Raven Press.)

sistance then tends to maintain portal and sinusoidal pressure constant. This reduces passive effects of changes in portal flow on splanchnic blood volume.

It is important to note that in the dog, but not in other species, hepatic venous resistance is extremely sensitive to both exogenous and endogenous histamine. Endogenous histamine may be released in the liver of the dog, not only by drugs, endotoxin, and anaphylaxis, but also by extensive surgery and perfusion through extracorporeal circuits. The action of histamine is to elevate portal and sinusoidal pressures, causing congestion and transsinusoidal filtration (outflow block). The physiologic function of such a mechanism is unknown, but it can be a confounding factor in experiments on the canine hepatic circulation.

c. Hepatic blood volume: The walls of the portal and hepatic venules contain smooth muscle that modifies wall compliance and hence controls the blood content of the liver. Hepatic venous compliance, defined as total hepatic blood volume divided by portal pressure, is 2–4 ml/mm Hg/100 gm liver in cats and dogs and can decrease to less than 1 ml/mm Hg/100 gm during sympathetic stimulation (Greenway, 1983a).

For the blood reservoir to be effectively controlled by the central nervous system, it must be relatively insensitive to passive factors. Although isolated manipulation of hepatic vascular parameters, such as portal flow, results in passive changes in hepatic blood volume, the integrated splanchnic circulation minimizes passive changes resulting from a variety of external forces. Maintenance of portal pressure at 7–10 mm Hg reduces the effects of small changes in caval pressure and portal flow. Increased hepatic venous resistance, which occurs during splanchnic nerve activity, also reduces passive effects of flow (see Section D-2b, above). The hepatic arterial buffer response (see Section C-3, below) stabilizes total hepatic flow and hence portal pressure. Since the liver is entirely enclosed within the peritoneal cavity, increases in intraperitoneal pressure cause corresponding alterations in hepatic venous pressure by compression of the larger hepatic veins. Thus, transmural pressure in the sinusoids and portal circuit and hence the hepatic blood content are stabilized during intraperitoneal pressure changes, which can be as large as 200 mm Hg during vigorous muscular activity. It seems reasonable to suggest that an important factor in the evolution of the splanchnic circulation is stabilization of blood content in the face of changes in flow and external pressures. On such a basis, the major factor controlling splanchnic blood volume then becomes the compliance of the venular walls, determined by sympathetic innervation mediated through α-receptors (Greenway, 1983a).

d. Transsinusoidal fluid movements: The sinusoids are freely permeable to protein, although there is some barrier to diffusion of protein from sinusoid to lymph under resting conditions. As sinusoidal pressure is increased, filtration of fluid rises and protein passes more readily into the lymphatics. If the latter are occluded or if the filtration rate exceeds their capacity, fluid exudes across the liver capsule into the peritoneal cavity. However, large increases (15–20 mm Hg) in transmural sinusoidal pressure are unphysiologic and cause disruption of the parenchyma and cessation of bile flow. In cirrhosis, for example, accumulation of large amounts of fluid in the peritoneal cavity raises intraperitoneal pressure, reduces sinusoidal transmural pressure, and limits filtration. The fluid entering the peritoneal cavity during high sinusoidal pressures in acute experiments has a high protein content. The origin of the relatively low-protein ascites seen under chronic conditions in alcoholic cirrhosis is not clear (for review of transsinusoidal fluid movements, see Greenway, 1981b).

3. HEPATIC ARTERIAL CIRCUIT

The hepatic arterioles terminate mainly at the portal ends of the sinusoids, although some enter part way along them. Transillumination studies show vasomotion in arteriolar flow into individual sinusoids, but overall hepatic arterial flow is uniformly distributed to the hepatic parenchyma and also supplies the gallbladder and bile ducts. Hepatic arteriolar resistance is high compared with hepatic venous resistance. It usually is calculated as arterial minus caval pressure divided by arterial flow. However, if sinusoidal pressure is close to portal pressure, it is more accurate to calculate it as arterial minus portal pressure divided by arterial flow. Unlike hepatic venous resistance, arteriolar resistance can be modified by local intrinsic mechanisms, one of which is the hepatic arterial buffer response. Hepatic arteriolar resistance varies directly with portal flow; thus, hepatic arterial flow increases when portal flow decreases and decreases when portal flow increases. This type of response serves to stabilize total hepatic flow and portal and sinusoidal pressures, but since the capacity of the hepatic arterial circuit to dilate is limited (the maximal flow being 2–3 times resting flow at constant arterial pressure), it can compensate only for moderate changes in portal flow. The mechanism responsible for this response is not clear, although various myogenic and metabolic theories have been suggested (Lautt, 1981b).

The smooth muscle of the hepatic arterioles can be acted upon by vasoactive substances in both arterial and portal blood, but it is unlikely that vasoactive metabolites produced by the parenchymal cells play any role in this regard unless they are carried back to the smooth muscle by the lymphatics. Changes in hepatic metabolic activity produced by SKF 525A or by 2,4-dinitrophenol have been shown by Lautt (1980a) not to elicit corresponding changes in hepatic arterial resistance. He therefore has concluded that hepatic arterial flow is not dependent on hepatic metabolism but tends to maintain total hepatic blood flow constant. Substances in portal blood can influence hepatic arterial resistance, but, in conscious dogs, hepatic arterial flow remains unchanged during digestion (Gallavan *et al.*, 1980). Gastrointestinal hormones in pharmacologic doses have a marked action on the hepatic circulation (Richardson and Withrington, 1981), but it seems unlikely that they exert important effects at physiologic levels.

4. NERVOUS CONTROL OF HEPATIC VASCULAR BED

a. Hepatic nerve stimulation: Parasympathetic nerves from the vagus supply the hepatic parenchyma, gallbladder, and bile ducts but not the blood vessels. Stimulation of such nerves has no obvious effect on overall flows, pressures, or volume in the liver, but local microvascular effects and metabolic responses may occur.

The sympathetic nerves exert major effects on all parts of the hepatic vascular bed (Fig. 14.12). Responses are dependent on frequency of nerve stimulation over the range 0.5 to 15 Hz and are mediated only by α-receptors. Changes produced by nerve stimulation mediated by β-receptors are seen only after administration of nonselective α-blockers. It seems likely that they involve increased release and diffusion of transmitter after presynaptic α_2-receptor block, as in the intestine. Stimulation of the hepatic nerves increases venous resistance, thereby raising portal pressure. Normally, however, an elevation in hepatic venous resistance would prevent a decrease in portal pressure rather

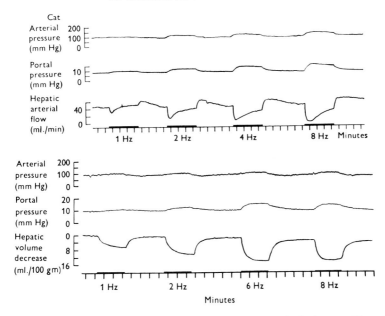

FIG. 14.12. Effects of hepatic nerve stimulation on hepatic vascular bed in cats. (From Green-way and Oshiro, 1972; reproduced with permission from the *British Physiology Soc.*)

than produce an increase (see Section C-2b, above). Stimulation of hepatic nerves causes hepatic arterial resistance to be increased. In cats, but not in dogs, this vasoconstriction is not maintained, and autoregulatory escape occurs (Fig. 14.12), a response seen in a variety of vascular beds and species. It involves relaxation of the same vessels originally constricted, and it may be due to accumulation of an unknown vasodilator substance or to failure of the smooth muscle to maintain its contracted state (Greenway, 1983b). It is not known whether autoregulatory escape occurs in human hepatic arterioles. The smooth muscle controlling hepatic venous compliance is markedly contracted by sympathetic nerve stimulation, and, as a result, up to 50% of the hepatic blood volume can be expelled from the liver (Fig. 14.12). This response is maintained for as long as the hepatic nerves continue to be stimulated. Other reactions to hepatic nerve stimulation have been less extensively studied. Transient constriction of presinusoidal sphincters may alter diffusion parameters, and trans-sinusoidal filtration may be reduced.

Thus, the hepatic nerves have major effects on the hepatic circulation. It is not as yet clear whether specific nerve fibers innervate the different parts of the hepatic vascular bed, permitting vessel responses to be elicited independently. The mechanisms and conditions under which the central nervous system increases hepatic sympathetic nerve activity remain largely unknown. Under resting conditions in anesthetized animals, there is little sympathetic tone present in arteriolar and venous resistance vessels but considerable tone in the venules, thus controlling hepatic blood volume (Greenway, 1983a). (For a review of the effect of hepatic nerve activity on the liver, see Lautt, 1980b.)

  b. *Central nervous control of hepatic circulation:* In isolated perfusion studies of the carotid sinuses in dogs, reducing carotid sinus pressure increases hepatic arterial and venous resistances and decreases hepatic blood volume, whereas raising carotid sinus pressure produces opposite effects. It is clear that in the isolated perfused preparation, the arterial baroreflex affects the hepatic circulation, but the normal closed-loop function of the carotid sinus has not been studied. In cats, the baroreceptor reflex influences hepatic arteriolar resistance but has no effect on hepatic blood volume. It has generally proved difficult to elicit reflex effects on venous compliance in anesthetized cats even though arterial pressure control by the baroreflex is very effective. It seems likely that postural reflex mechanisms are not as well developed in cats as in dogs (Greenway, 1983a). In general, cardiopulmonary receptors appear to inhibit sympathetic tone in the venous system, including the hepatic venous bed. Control of the hepatic vascular bed by high centers in the central nervous system has not yet been studied. (For a review of the reflex effects of the central nervous system on hepatic veins, see Rothe, 1984.)

5. Future Avenues for Investigation

  The most important area requiring further studies is the confirmation in conscious humans of the many ideas developed from animal studies. Since in man the spleen is insignificant as a blood reservoir, the liver may be even more important in this regard than in animals. Changes in hepatic flow and blood volume in exercise, various postures, after mild hemorrhage, and in many pathologic situations are almost completely unknown. Computed tomography and ultrasonic scanning appear to be potential noninvasive approaches to measurement of liver volume in man.

  Subjects requiring further work are confirmation that sinusoidal pressure is closer to portal pressure than to caval pressure and that venous resistance can be controlled independently of the smooth muscle that alters venous compliance. Intrinsic myogenic and metabolic hepatic arteriolar mechanisms, the mechanism of the arterial buffer response, and the mechanism and significance of autoregulatory escape are incompletely resolved. Finally, reflex and central nervous control mechanisms have hardly begun to be elucidated. It is clear that much remains to be done.

D. PHARMACOLOGY OF BLOOD CIRCULATION

By C. V. Greenway

Pharmacologic aspects of the hepatic blood vessels have received even less attention than physiologic problems. The hepatic circulation is altered to some degree, directly or indirectly, by many drugs, but this section is concerned only with direct drug actions on the hepatic vessels. Agents which alter arteriolar resistance in the gastrointestinal tract and spleen also change portal flow (see Section D, Chapter 13). Hence, they may have indirect effects on hepatic arterial flow through the arterial buffer response, or their direct actions may mask this intrinsic mechanism. Three areas of study will be discussed: (1) drug receptors in the hepatic blood vessels, (2) drug effects on hepatic blood flow in relation to drug metabolism and pharmacokinetics, and (3) drug effects on hepatic blood volume in relation to control of cardiac output and treatment of cardiac failure. (For reviews of some aspects of the effects of drugs on hepatic blood vessels, see Richardson and Withrington, 1981; Greenway, 1981c.)

1. Drug Receptors in Hepatic Blood Vessels

a. Adrenergic receptors: The presence of innervated α-receptors in all smooth muscle of the hepatic blood vessels and of noninnervated β-receptors in the hepatic arterioles has already been discussed in Section C, above. The effects of α-receptor agonists generally resemble those of sympathetic nerve stimulation. Nonselective α-receptor blockers, such as phenoxybenzamine and phentolamine, abolish these actions, but the selective α_1-receptor blocker, prazosin, appears to have little effect on hepatic blood volume responses to sympathetic nerve stimulation. Epinephrine increases portal flow due to a β-receptor-mediated vasodilatation in the intestine, but its effects on hepatic arterial flow are minimal and variable, due to a more or less balanced effect on α- and β-receptors in the hepatic artery. This drug also causes a marked decrease in hepatic venous compliance and hepatic blood volume in spite of the increased blood flow. After administration of α-blockers, epinephrine dilates the hepatic artery, whereas following propranolol, constriction occurs. Isoproterenol has little effect on hepatic venous resistance, reduces hepatic arterial resistance, and causes an unexpected indirect decrease in venous compliance (the mechanism of which is not yet known). The mobilization of blood

485

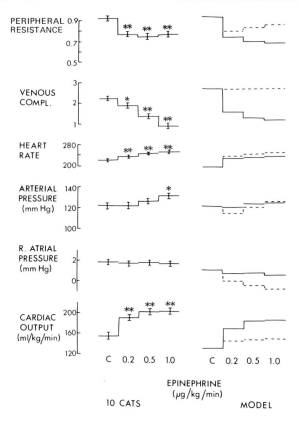

FIG. 14.13. Effects of epinephrine infusions in 10 cats (left) and predictions by computer model (right). Hepatic venous compliance (COMPL.) for cats in milliliters per millimeter Hg, 100 gm liver; data for model of total splanchnic bed in milliliters per millimeter Hg; body weight in kilograms. Dotted lines show predicted responses if epinephrine did not reduce splanchnic venous compliance; cardiac preload falls and increase in cardiac output is markedly attenuated. Single asterisk denotes $p < .05$; double asterisk denotes $p < .01$. (For details of model, see Greenway, 1981c.)

from the hepatic bed plays an important role in the increased cardiac output noted during epinephrine and isoproterenol infusions (Fig. 14.13).

Infusions of acetylcholine, but not vagal stimulation, relax the hepatic arterioles and inhibit responses to sympathetic nerve stimulation; this suggests the presence of noninnervated presynaptic cholinergic receptors on the sympathetic terminals. Infusions of dopamine have complex effects on the hepatic artery, mediated through α-receptors and specific dopamine receptors that relax the smooth muscle and are blocked by haloperidol. The overall response to dopamine is increased total hepatic and portal flows, decreased hepatic arterial flow, and reduced hepatic blood volume. Direct actions of adrenergic drugs are modified when given intravenously by changes in arterial pressure and other cardiovascular parameters. An overall synthesis of the effects of adrenergic drugs on the circulation has been presented by Greenway (1981c).

  b. *Autacoids:* Bradykinin causes hepatic arterial vasodilatation, with no effect on hepatic venous resistance. Serotonin has weak and variable actions. Histamine causes hepatic arterial vasodilatation in all species and markedly increased venous resistance in the dog but not in other species. In the dog, it raises hepatic venous resistance without affecting smooth muscle controlling venous compliance, in contrast to norepinephrine which contracts all the venous smooth muscle. Both drugs therefore raise portal pressure. However, norepinephrine decreases hepatic blood volume, whereas histamine markedly increases it, causing outflow block. This confirms the view that venous compliance and venous resistance are controlled by pharmacologically different smooth muscle. Since the liver in the dog is rich in endogenous histamine, any mechanism producing release of this substance causes hepatic congestion. The pharmacologic effects of histamine are reportedly blocked by antihistamines. Cimetidine, the H_2-receptor blocking agent widely used in the treatment of peptic ulcers, reduces hepatic blood flow in man by 25–33% (Feely *et al.*, 1981). The effects of H_2-receptor agonists and antagonists require further study. The prostaglandins, PGE and probably PGI, cause hepatic arterial vasodilatation, which may be offset by the arterial buffer response to increased portal flow. PGA_2 produces vasoconstriction. In summary, autacoids, with the exception of histamine in dogs, do not have pronounced effects on the hepatic circulation. (For a review of the actions of autacoids, see Richardson and Withrington, 1981.)

  c. *Peptides:* Angiotensin produces more marked vasoconstriction of both arteriolar and venous smooth muscle in the splanchnic and hepatic beds than it does in skeletal muscle; however, the importance of endogenous release of renin as a mediator of hepatic vascular responses under physiologic conditions has not yet been clarified. Vasopressin causes a marked reduction in portal flow due to intestinal and splenic arteriolar vasoconstriction, but its direct effects on the hepatic blood vessels appear to be weak. Peptides from the gastrointestinal tract generally cause hepatic arterial vasodilatation, but the doses required probably exceed the amounts produced endogenously.

2. DRUG METABOLISM AND LIVER BLOOD FLOW

Hepatic blood flow is a major factor controlling the clearance of certain drugs, such as β-blockers, lidocaine, verapamil, opiate analgesics, and tricyclic antidepressants. Clearance of these substances has been shown to change with age, posture, and in various diseases, but in only few instances has this been proved to be due to alterations in hepatic blood flow. Questions of hepatic flow effects on drug metabolism and of the latter on hepatic flow have not been studied extensively. Cimetidine reduces hepatic flow in man, as well as decreasing clearance of such drugs as propranolol. However, it also appears to interfere with microsomal enzyme activity and to produce a whole range of clinically significant drug interactions. The role of changes in hepatic flow in these interactions has not been clarified. When a variety of agents that alter hepatic microsomal enzyme activities (chlordiazepoxide, antipyrine, phenytoin, and barbiturates) is given to rats, either no change or a rise in hepatic flow, proportional to an increase in liver weight, occurs (McDevitt *et al.*, 1977). Such data support the physiologic concepts described earlier, namely, that changes in hepatic metabolism do not alter hepatic blood flow and constancy of hepatic flow is important for normal clearance of many drugs and endogenous substances. However, a much wider variety of both acute and chronic studies is required before any definitive conclusions can be reached.

3. Drug Effects on Hepatic Blood Volume

Arteriolar vasodilators used in the treatment of heart failure have variable effects on hepatic venous compliance. Nitroglycerin reduces basal venous tone but does not alter sympathetically mediated decreases in compliance. Nitroprusside, prazosin, and diazoxide have little effect, whereas hydralazine decreases hepatic venous compliance. In heart failure, drugs that produce splanchnic venoconstriction in addition to afterload reduction (dopamine and hydralazine) cause less reduction in pulmonary congestion than do those which produce only afterload reduction (nitroprusside and prazosin). Such actions have been integrated into the overall cardiovascular effects of these drugs on cardiac output in normal people and in patients with heart failure (Greenway, 1981c).

4. Future Avenues for Investigation

It is clear that investigation of the pharmacology of the hepatic vessels has just begun, largely due to technical difficulties (see Section C-1, above). In animals and man, acute and chronic effects of many commonly used therapeutic agents are unknown. Relationships among drug delivery, drug uptake by the parenchyma, and drug metabolism are largely obscure. The role of the splanchnic venous system in control of cardiac output, in heart failure, and in drug-induced postural hypotension requires much further study.

E. PATHOPHYSIOLOGY, PATHOGENESIS, AND PATHOLOGY OF BLOOD CIRCULATION

By S. Glagov

1. General Considerations

Three principal pathogenetic mechanisms cause abnormalities of the hepatic circulation: (1) diseases that primarily affect the lobular parenchymal cells and sinusoids or the intrahepatic bile ducts, causing obstruction or diversion of hepatic blood flow; (2) disturbances of hepatic perfusion due to primary abnormalities of the afferent portal or arterial vessels, either outside the liver or within the hepatic portal fields; and (3) primary obstructions to outflow at the level of the hepatic veins. Interference with hepatic circulation by any of these means may occur rapidly or gradually and may be reversible or irreversible. Although the portal vein furnishes approximately 70% of hepatic flow and about 50% of the oxygen supply, sudden portal vein occlusion is compatible with survival in man and primates, whereas collateral channels may be inadequate to sustain hepatic viability in other mammals (Child, 1954; Elias and Sherrick, 1969). On the other hand, obstruction of the hepatic artery, a vessel that normally accounts for only 25% of the hepatic flow, may result in hepatic infarction due to the relative paucity of arterial collateral channels in both primates as well as subprimate species. Sudden occlusion of the efferent hepatic veins results in marked engorgement of the liver, with necrosis occurring in the vicinity of the central veins. Acute or subacute forms of primary liver cell degeneration and necrosis also may interfere with flow, depending upon the extent of hepatic injury.

Regardless of the nature of the initiating process, chronic forms of hepatocellular injury or hepatic vascular disease result in the replacement of the liver parenchyma by fibrous tissue and in modifications of hepatic architecture. Both of these changes interfere with hepatic blood flow, causing further replacement of parenchyma by fibrous tissue and disruption of hepatic lobule. Eventually the normal perfusion of blood from afferent vessels to efferent veins by way of the lobular or acinar sinusoids is severely modified, and the liver cell mass is progressively reduced. As a consequence, flow across the hepatic circulatory bed is obstructed and hepatic function deteriorates, leading to the principal complications of chronic hepatic disease, i.e., portal hypertension and hepatic parenchymal failure. The sequence and relative severity of these complications depend on the nature of the pathogenic process, as discussed below.

2. PORTAL HYPERTENSION

Portal venous pressure is normally about 7 mm Hg, while sinusoidal pressure and hepatic venous pressure are slightly lower. Obstruction of the hepatic circulation at the level of the hepatic veins by fibrous tissue compression in and about the portal fields or by perisinusoidal fibrosis may raise the portal vein or hepatic wedge pressure up to 20 mm Hg. Under these conditions, portal blood flow may be reduced to 10–15% of normal, and portal hypertension may be said to exist. This abnormality may be classified according to the initial site of obstruction to flow. Portal pressure elevations due to hepatic vein obstructions usually are called postsinusoidal, whereas conditions affecting the portal veins directly are referred to as presinusoidal and may be either intrahepatic or extrahepatic in location. The portal vein is devoid of valves so that elevation of portal venous pressure due to increased resistance to flow is readily transmitted to the contributing mesenteric and splenic channels. This gives rise to rapid engorgement and distention of the splenic and mesenteric veins. In addition, abdominal splanchnic circulation may actually be increased, thus adding a hyperdynamic component to elevate portal pressure (Witte *et al.*, 1974). Splenomegaly and intestinal edema develop and the elevated pressure contributes, along with other metabolic consequences of liver disease, to the formation of ascites (see Section F-3a,b, below). Major extrahepatic portal–systemic anastomoses normally present between the gastric coronary veins and the diaphragmatic esophageal veins, in the region of the lower esophagus and the cardia of the stomach, and between the superior hemorrhoidal and inferior hemorrhoidal veins in the region of the anus, enlarge, become tortuous, and may form varices when portal hypertension results in diversion of portal blood flow into them. As a result of the increase in diameter and the elevated pressure, tension in the vein walls is markedly increased, and the walls of both the portal veins and the varices tend to become fibrotic. Rupture of the varices may occur, most commonly in the submucosa of the lower esophagus. Disruption at this site may be potentiated by erosions of the overlying mucosa due to exposure to mechanical stresses and the chemical environment of the contents. Because luminal pressure and wall tension in varices are elevated, and because these distended vessels are continually exposed to the esophageal lumen, bleeding from esophageal varices is not easily stemmed and thrombi do not form readily. Exsanguinating hemorrhage is therefore a frequent terminal complication of ruptured esophageal varices.

3. Vascular Pathology in Cirrhosis

Cirrhosis of the liver results from chronic, usually irreversible, destructive processes, and regardless of the cause, is characterized by progressive reduction of functional hepatic parenchyma, extensive fibrosis, parenchymal regeneration, and the replacement of the normal lobular architecture by regenerative nodules consisting of islands of parenchyma surrounded by fibrous tissue. The normal anatomic and functional relationships among portal fields, central veins, and liver cells are largely absent and the usual lobular or acinar organization is obscured. This architectural alteration is generally referred to as "reconstruction" and is a major morphologic basis for abnormalities of hepatic blood flow.

Depending on etiology and pathogenesis, the fibrosing process may be diffuse and uniform, involving nearly all lobules, or patchy, involving only some lobules. A typical example of diffuse fibrosis occurs as a result of chronic ingestion of alcohol. In this case, diffuse fibrotic obliteration of the central veins is a prominent early feature (Edmonson et al., 1963). In addition, collagen fibers, which normally form a fine supporting framework for the hepatic sinusoids, become increasingly abundant (Orrego et al., 1979). With obstruction of many central lobular venous channels, blood is diverted to those vessels that are not yet obliterated. Such connections form between and across lobules, bypassing the parenchyma by way of fibrous bands or septa. Progression of the sclerosing process causes compression and reduced compliance of the sinusoids and increasing centrizonal venous obstruction. Thus, blood is increasingly shunted away from the liver cell plates and through ectatic channels that are imbedded in the fibrous tissue septa. Fibrous bridges between central zones and between central zones and portal fields become increasingly prominent. Eventually, the islands of hepatic parenchyma that remain are of lobular or sublobular size. This type of cirrhosis is, therefore, termed micronodular.

Patchy hepatocellular necrosis occurs typically in viral hepatitis. In the overwhelming majority of cases, the infection consists of a single clinical episode with restitution of the parenchyma and preservation of the normal lobular architecture. However, in some instances, for reasons that are not clear, regeneration is delayed or inadequate and necrosis of liver cells is complicated by the formation of bridges of collapsed sinusoids between central veins and between central veins and portal fields. When this occurs (Boyer and Klatskin, 1970), the likelihood for formation of persistent bridging scars and shunts appears to be greatly increased, and the disease may pass into a chronic destructive phase with the formation of nodules. Because the process is patchy, the zones of collapse may be extensive, and large areas with normal lobular architecture may be present adjacent to the broad bands of fibrous tissue. Portal triads may still be discernible in these islands of normal tissue. Because hepatocellular nodules frequently encompass several lobules, cirrhosis developing in this manner is characterized as macronodular. In chronic forms of hepatitis, collapse and fibrosis with fibrous bridge formation may advance inter-mittently. At first, hepatic destruction appears to progress mainly by erosion of the periportal parenchyma, with replacement by fibrous tissue. Eventually, fibrous tissue compresses the intrahepatic portal vessels with the formation of septa. This morphologic reconstructive pattern is largely macronodular.

Hepatic fibrosis also may take the form of progressive, uniform fibrotic enlargement of the portal fields, associated with atrophy of periportal liver cells. Such changes occur in hemochromatosis. Patchy inflammation with fibro-sis of the portal fields associated with necrosis of intrahepatic bile ducts occurs in primary biliary cirrhosis. In such a case, morphologic reconstruction with formation of nodules may eventually take place, but portal hypertension may supervene early, before cirrhosis is manifest. This occurs providing portal fibro-sis is extensive and causes compression and obstruction of the portal veins within the portal fields. The nature of the process and the rate at which it develops may determine whether nodular reorganization occurs before severe hepatic insufficiency intervenes.

Regardless of the pathogenesis, flow is reduced in the presence of cirrhosis, despite the formation of direct anastomoses between portal veins and hepatic veins. Except for instances of marked, uniform, and progressive portal fibrosis, the severity of the resulting portal hypertension is more closely related to the degree of nodular transformation than to the relative quantity of fibrous tissue. The relative proportion of hepatic flow supplied by the hepatic artery is in-creased, and arteriovenous shunting may account for some of the portal pres-sure elevation. The walls of the larger portal vein branches and the portal vein itself become thickened (Hou and McFadzean, 1965) in relation to the degree and duration of the associated portal hypertension. Thrombi may occur within the dilated portal venous system, but thrombotic occlusion is unusual in the absence of complicating infection or dehydration. Central veins and hepatic lobular veins that remain patent are difficult to distinguish from other dilated portal-sinusoidal channels, for they can no longer be related to lobular or acinar landmarks.

4. PRIMARY PORTAL VEIN OBSTRUCTION

Although disease-producing agents may gain access to hepatic parenchyma by way of the portal vein, primary portal venous abnormalities that result in hepatic disease and obstructions to portal flow include two major categories: thrombosis and obstruction by parasites. Thrombosis of the portal vein may complicate cirrhosis, abdominal inflammatory processes such as pancreatitis or peritonitis, involvement of the vein by intraabdominal neoplasms with or with-out hepatic metastases, severe dehydration and cachexia, trauma, or manipula-tion of the portal vein during abdominal surgery. Acute obstructive thrombosis may be catastrophic. Gradual compression leading to portal vein obstruction is, however, compatible with survival, eventual organization of the thrombus, and restoration of portal flow if the cause is removed. Obstruction may result from infestation by *Schistosoma mansoni* or *S. japonicum* (schistosomiasis). The parasites penetrate into the mesenteric and portal veins, with their eggs passing into the distal portal venous channels and lodging in the terminal portal vein ramifications. Here a local reaction produces granulomas and marked fibrosis, with compression and destruction of portal veins. Large fibrous bands may bridge adjacent portal fields, but many portal fields may be spared (Alves *et al.*, 1971; Kamel *et al.*, 1978). In general, hepatic perfusion is adequate to permit hepatocellular function, but portal hypertension may be extreme, with development of marked splenomegaly and ascites. Since the fibrous bridges connect mainly recognizable large portal fields and many normal lobules are preserved, generalized reconstruction of hepatic parenchyma into nodules is not a prominent feature. Other parasitic infestations of the liver, such as amebiasis, begin in the intestine, gain access to portal circulation, and may be carried to the liver. However, except for hemorrhages that may occur when hepatic necrosis induced by proteolytic enzymes liberated by the organisms disrupt vessels, chronic abnormalities of the hepatic circulation do not occur. Congenital hepatic fibrosis (Kerr *et al.*, 1961) is essentially a vascular malformation in which abnormal intrahepatic portal vein connections result in early severe portal hypertension.

5. ABNORMALITIES OF HEPATIC ARTERIES

Primary arterial occlusions due to thrombosis or surgical ligation occur but are unusual. Generalized diseases that affect the hepatic arteries include arteritis, embolization, amyloidosis, and hypertension. Arteritis may result in focal hepatic ischemic necrosis and may be a significant clinical aspect of generalized polyarteritis nodosa (Cowen *et al.*, 1977). Involvement of the main hepatic artery can result in hepatic infarction. Amyloid deposition in artery walls may narrow their lumina and compromise the active contractility of vessels, but seldom results in complete thrombosis. Necrotizing lesions of intrahepatic small arteries and arterioles have been noted in malignant hypertension, occasionally with consequences to hepatic blood flow similar to those observed in cases of arteritis, but vascular sclerosis associated with chronic forms of systemic hypertension have not been shown to create specific circulatory difficulties in the liver. The large hepatic artery branches may also be sites of aneurysm formation. These abnormalities are usually true aneurysms that form at the inner or acute angle of branch points in the hilus of the liver and may be large and saccular. Rupture may occur with hemorrhage into the liver or into the hilus, but most aneurysms remain undetected. Intrahepatic vascular malformations, such as arteriovenous fistulae (Foley *et al.*, 1971), may result in elevations of portal pressure. Abdominal trauma and liver biopsy may lead to the formation of similar lesions (Okuda *et al.*, 1977). Mycotic aneurysms due to embolization of infected thrombi to the hepatic artery may complicate infective endocarditis. Hepatic neoplasms are supplied mainly by the hepatic artery, and associated elevations of portal pressure are usually attributable to underlying abnormalities, such as cirrhosis rather than to shunting within or about the tumor.

6. INTERFERENCE WITH HEPATIC VENOUS OUTFLOW

Reductions in venous outflow may occur as a result of interference at any level of the hepatic outflow tract. Acute or subacute forms of hepatic vein obstruction result in the Budd-Chiari syndrome. Regardless of the cause or level of venous obstruction, corresponding alterations in the liver consist of marked dilatation of the centrilobular veins and sinusoids, often with disruption and necrosis of centrizonal hepatic cells, at times with the formation of central vein thrombi. Hepatic vein thrombosis, particularly of the larger vessels, may follow abdominal trauma or invasion of the major hepatic veins or the inferior vena cava by neoplasms or blood dyscrasias, e.g., sickle cell anemia or polycythemia vera. During a sickle cell crisis, obstruction of central veins results in disturbances of liver function and clinical findings similar to those noted in other forms of acute or subacute hepatic vein occlusion. However, flow may be restored, with variable residual damage, when the crisis terminates. On the other hand, repeated crises lead to the development of prominent collars of dense fibrous tissue about centrilobular veins that tend to narrow the venous lumina and may restrict venous outflow. A peculiar form of chronic obstruction results from the formation of a web or fibrous band in the inferior vena caval segment between the ostia of the hepatic veins and the right atrium. The origin of this formation is obscure. If it is actually the residue of an organized thrombus, the nature of the original thombogenic episode is seldom evident (Schaffner *et al.*, 1967; Takeuchi *et al.*, 1971).

Oral contraceptives (Hoyumpa *et al.*, 1971) and other therapeutic agents (Clain *et al.*, 1967) have been incriminated as causes of hepatic vein thrombosis, but there is also an idiopathic form of hepatic venous thrombosis in which there is no history of trauma and no known hematologic disorder or exposure to drugs. Patchy obstruction of the veins by thrombi, often with partial organization, is noted in all of these instances. Accompanying dilatation of portal filled lymphatics may also be seen. The consequences with respect to hepatic function are variable and depend on the extent of involvement. Acute extensive hepatic vein obstruction is often fatal, whereas chronic forms may respond to therapy. Veno-occlusive disease is a term often applied to the consequences of obstruction of the centrilobular veins, usually with sparing of the larger veins (Stuart and Bras, 1957). The effects are similar to those noted with widespread hepatic vein thrombosis, but the condition is not clearly initiated by thrombus formation. It is associated with exposure to substances that may be toxic to central vein endothelium, including pyrrolizidine alkaloids (McLean, 1970), chemotherapeutic agents (Griner et al., 1976), and the contraceptive hormones (Alpert, 1976).

Other forms of interference with hepatic venous outflow (Dunn et al., 1973) occur with passive congestion due to cardiac failure and chronic lung disease with cor pulmonale. In these cases, hepatic vein thickening may be marked, especially of the larger sublobular veins. Acute passive congestion results in centrizonal sinusoidal dilatation and may progress, when severe, to centrilobular hepatocellular necrosis. Constrictive pericarditis may produce changes similar to those seen with acute or chronic passive congestion. When left ventricular output failure is present in conjunction with right heart failure, oxygen supply to the centrilobular cells is further reduced. Shock, i.e., acute systemic vascular collapse, regardless of the initiating mechanism, also results mainly in necrosis of the vulnerable centrizonal parenchyma (Birgens et al., 1978). Chronic cardiac output failure may be followed by portal field enlargement and fibrosis and, in the presence of chronic passive congestion, may lead to central vein fibrosis and portocentral bridging. Nodular reconstruction of sufficient degree to be characterized as cirrhosis seldom occurs, even in severe cases of chronic heart failure. However, in instances of widespread fibrosis associated with long-standing tricuspid valve regurgitation or constrictive pericarditis, regenerative nodules may form, and many of the circulatory consequences of cirrhosis may be present (Luna et al., 1968).

7. ABNORMALITIES AFFECTING PRIMARILY SINUSOIDAL CHANNELS

Sinusoidal lumina may be obstructed by fibrin thrombi under conditions that result in disseminated intravascular coagulation. Septic states and widespread neoplastic disease are the most frequent causes. Hepatic function may be moderately compromised and associated with hepatic necrosis and cholestasis. Permanent interference with hepatic flow does not occur if the underlying causal state is controlled and lysis of the thrombi is accomplished. Sinusoidal walls may be involved in processes that affect the microcirculation and the perivascular matrix elsewhere. Amyloidosis, particularly the forms associated with chronic inflammatory disease, results in the deposition of amyloid in the perisinusoidal region between the liver cells and the Kupffer cells. Congestion and interference with diffusion lead eventually to atrophy of the hepatocellular cords, but portal hypertension is unusual, since the sinusoids, though narrowed, generally remain open. Metabolic abnormalities, which result in parenchymal cell storage, may involve the liver. Hepatocytes may be the major site of abnormal accumulations, as in the glycogenoses, fatty liver, and α_1-Antitrypsin deficiency. Compression of sinusoids sufficient to cause portal hypertension is unusual despite marked increases in cell volume. α_1-Antitrypsin deficiency may, however, be associated with the development of cirrhosis, in which case portal hypertension may occur. The lipidoses, Gaucher's disease and Niemann-Pick disease, generally involve the Kupffer cells and the perisinusoidal macrophages. These disorders may compress the sinusoidal lumina, occasionally resulting in portal hypertension. Similarly, metastatic neoplasms cause hepatic functional abnormalities associated with liver cell replacement or with bile duct compression if the process is widespread and occupies over 50% of the liver volume. In such cases, portal angiomas are common and often multiple in affected livers. These lesions may obstruct hepatic flow sufficient to cause portal hypertension. Small capillary or cavernous angiomas are usually asymptomatic and often undergo involution and sclerosis with obliteration of the channels, whereas large angiomas may hemorrhage if disrupted by trauma. Peliosis hepatis is a condition in which sinusoidal dilatation is marked, and small cystic blood-filled spaces form in sites where liver cell plates have become atrophic. The condition is associated with the therapeutic administration of anabolic steroids and also occurs in the presence of long-standing severe debilitating disease. No explanation is available for the formation of these blood-filled spaces. They may be the result, at least in part, of inadequate tensile support by the perisinusoidal collagen fibers, due to metabolic alterations in matrix composition.

F. LYMPHATIC SYSTEM

By M. H. WITTE AND C. L. WITTE

1. ANATOMY OF LYMPHATIC SYSTEM

The liver's lymphatic apparatus, consisting of lymph, lymphocytes, lymphatic channels, and lymph nodes, is a vital regulator of the hepatocellular microenvironment. Electron microscopy has firmly validated the existence of a perisinusoidal interstitial space (of Disse), and markers, such as ferritin parti-

Blood Vessels and Lymphatics in Organ Systems
Copyright © 1984 by Academic Press, Inc.

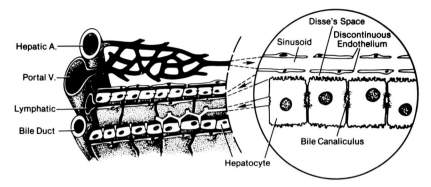

Fig. 14.14. Schematic outline of cellular and lobular architecture of the liver, demonstrating the interrelationship of blood, bile, and lymph. Despite perfusion by both arterial and venous blood, hydrostatic pressure in the hepatic sinusoid is extremely low, and an incomplete basement membrane allows protein to escape almost freely into the perisinusoidal space of Disse (see Fig. 14.15). Ultimately, this filtrate reaches lymphatics in the periphery of the lobule where the vast bulk runs, like bile, countercurrent to blood flow.

cles that readily pass from plasma into the perisinusoidal space, have documented the highly porous nature of the hepatic sinusoid. Distinct lymphatic capillaries first appear in the periphery of the liver lobule, and a rich lymphatic plexus is interspersed in the interlobular connective tissue among hepatic arterioles, portal venules, and bile ductules. The bulk of hepatic lymph drains countercurrent to blood flow, exiting through hilar lymph nodes, to empty into multiple efferent lymph trunks paralleling the portal vein (Fig. 14.14). Eventually this lymph returns to the bloodstream via the thoracic duct, a shared central "lymphshed" conveying nearly all visceral and peripheral lymph formed below the diaphragm. A small volume of hepatic lymph also ascends in channels running along hepatic veins, through substernal and intrathoracic nodes, to reach the right lymph duct. Hepatic lymphatics from the liver capsule course over the surface of the gallbladder and within the hepatoduodenal ligament; ordinarily they are barely visible but become especially prominent in various disease states. For example, in patients with hepatic cirrhosis or chronic congestive heart failure, these tenuous, thin-walled structures distend and thicken remarkably, assuming at times the size of nearby blood vessels.

2. PHYSIOLOGY OF LYMPHATIC SYSTEM

a. Hydraulic principles: The partition of extracellular fluid between blood and tissues is governed by the balance of hydrostatic and oncotic pressure gradients across capillary membranes, and the small surplus of tissue fluid formed (net capillary filtrate) returns to the bloodstream via lymphatics. While the hepatic microcirculation, in general, conforms to these hydrodynamic principles, hepatic lymph dynamics are in several respects unique. These distinctive features are traceable to the dual afferent blood supply (arterial and venous) and the unusual sinusoidal circulation lacking a capillary basement membrane.

Intermingled with reticuloendothelial (Kupffer) cells, the endothelium of sinusoids is punctuated by many large gaps (>1000 Å) that allow nearly free exchange of protein and other macromolecules between plasma and the spaces of Disse. Therefore, in striking contrast to other extracellular fluid subcompartments where endothelial apertures are much narrower (e.g., aqueous humor, cerebrospinal fluid, and peripheral lymph), the protein concentration of hepatic lymph, a fluid derived primarily from perisinusoidal filtrate, approximates that of plasma. The proteins arise almost entirely from capillary leakage rather than from *de novo* synthesis by liver cells, as amply demonstrated by timed recovery in lymph of a variety of intravenously administered protein tracers (vital colored dyes and radiotagged albumin) and high molecular weight dextrans. Serial bloodstream disappearance curves and simultaneous lymph appearance curves from different organs have further established the unique permeability of hepatic capillaries to these macromolecules.

b. Protein flux: In both anesthetized and conscious animals, hepatic lymph forms continously (30–100 μl/kg/hr), contributing 15–20% of total body lymph production. The ratio of hepatic lymph protein to that of plasma is 0.85 to 0.95, with a slightly lower fraction of high molecular weight globulins and lipoproteins indicative of mild restrictive diffusion or sieving. This slight sieving probably reflects a minor contribution of lymph from intrahepatic sites other than sinusoids, that is, peribiliary capillaries of low permeability. The nonsinusoidal fraction, however, rapidly disappears after even a minimal rise in sinusoidal pressure and subsequent increase in hepatic lymph flow. Under this circumstance, plasma entering and hepatic lymph exiting the liver are virtually identical in protein content (Fig. 14.15).

The lymphatic network of the liver, therefore, serves a crucial "safety valve" function for the hepatic interstitium, and, not surprisingly, shifts in transcapillary fluid and protein exchange and in the adaptability of the lymph system play a key role in certain complications of liver disease. Because escape of plasma protein from the sinusoid into the spaces of Disse is almost unrestricted, oncotic (colloid osmotic) pressure normally exerts only a minor role in regulating sinusoidal filtration (Greenway and Stark, 1971). Instead, the extremely low intravascular pressure (4–7 mm Hg) acts to limit fluid extravasation. Nonetheless, hydraulic balance is particularly precarious, and even a minimal rise in resistance to hepatic venous outflow causes a precipitous rise in sinusoidal filtration with an outpouring of lymph from periportal lymphatics and the thoracic duct. Not only does fluid extravasation promptly increase, but

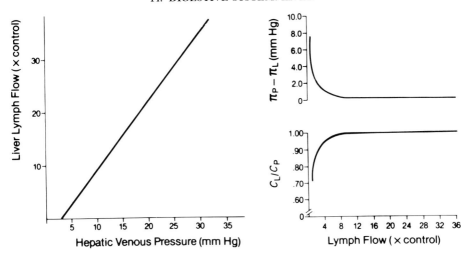

FIG. 14.15. Relationship of venous pressure to lymph flow in the liver and of accelerated hepatic lymph flow to hepatic lymph/plasma protein ratio (C_L /C_P), on the one hand, and "effective" oncotic gradient ($\Pi_P - \Pi_L$), on the other. Note the extreme sensitivity of the hepatic sinusoid to increased hydrostatic pressure, resulting in an outpouring of hepatic lymph (left). The extraordinary permeability of the sinusoid to protein is exemplified by the virtually identical protein content of plasma entering and lymph exiting the liver during hepatic congestion, accounting for an oncotic gradient of essentially 0 mm Hg (right). (Modified from Granger *et al.*, 1979; reproduced with permission from *Gastroenterology*.)

with unrestricted protein movement, leakage of protein increases proportionately (Fig. 14.15). This phenomenon, where solute flux is closely linked to water flux and hydrostatic pressure, is termed "convective transport" or simply "convection." In contrast, in less permeable capillary beds where intercellular junctions are tight (e.g., brain and extremity) or fenestrae are smaller (kidney and intestine), comparable elevations in hydrostatic pressure increase water flux far out of proportion to protein transport. The resultant dilution of interstitial fluid effectively lowers tissue oncotic pressure, widens the plasma–tissue oncotic gradient, and thereby plays a key role in forestalling continued filtration and edema formation. Moreover, the marked disparity between water and solute flux in response to long-standing elevation in portal hydrostatic pressure suggests that protein movement in the intestine, in contrast to the liver, proceeds primarily by an independent permeative mechanism such as diffusion or vesicular transport (pinocytosis) (Witte *et al.*, 1981).

3. PATHOPHYSIOLOGY OF LYMPHATIC SYSTEM

a. Congestion: In keeping with hepatic sinusoidal kinetics, disorders associated with liver congestion (e.g., hepatic venous outflow block) are characterized by an outpouring of hepatic and thoracic duct lymph high in protein content. Moreover, when formation of hepatic lymph exceeds the limited transport capacity of efferent lymphatics, excess fluid weeps off Glisson's capsule into the free peritoneal cavity to form ascites. This intraabdominal fluid characteristically is rich in protein although slightly lower than its source, hepatic lymph, due to osmotic equilibration with surrounding splanchnic viscera. Long-standing congestive heart failure, Budd-Chiari syndrome, and its experimental counterpart, constriction of the inferior vena cava above the diaphragm, are all notable examples of this phenomenon.

b. *Cirrhosis:* Ascites is also a common complication of cirrhosis of the liver, which is characterized by sinusoidal hypertension and increased hepatic lymph production. There is, however, considerable doubt that the cirrhotic liver is the primary source of ascites. Concomitant elevation of splanchnic portal venous pressure and generally low protein intestinal and thoracic duct lymph and ascitic fluid point rather to the congested digestive tract as the origin of the ascitic fluid (Witte *et al.*, 1980). This view is supported by experimental simulation of low protein intestinal and thoracic duct lymph and ascitic fluid in dogs with marked extrahepatic portal hypertension but without liver congestion (Witte *et al.*, 1969). Further, the multinodular fibrotic liver bears only a superficial resemblance to the turgid liver of hepatic venous outflow block, and the densely thickened capsule of the cirrhotic liver seems a substantial mechanical barrier to egress of hepatic tissue fluid (Witte, 1979). Indeed, while lymph is rarely, if ever, seen to leak from the surface of the cirrhotic liver spontaneously, intense edema and oozing of fluid from the engorged mesentery and bowel are a common finding at laparotomy in patients with portal hypertension accompanying hepatic cirrhosis (Wantz and Payne, 1961). Paradoxically, the cirrhotic liver, in contradistinction to the congested liver, also demonstrates evidence of restrictive diffusion (Witte *et al.*, 1981). This abnormality probably reflects the deposition of a continuous sinusoidal endothelial lining and progressive collagen deposition in the spaces of Disse, developments likely to impede free movement of macromolecules from plasma to lymph. A further consequence of the highly permeable nature of the hepatic microvasculature is that whenever the liver is injured, intrahepatic edema that accumulates, whether an outgrowth of "static" (sluggish) or "dynamic" (accelerated) insufficiency of lymph flow, is invariably protein rich. When this edema is unremitting over many months or years, diffuse parenchymal fibrosis ensues.

c. *Bile–lymph kinetics:* The two extracellular fluid circulations of the liver—lymphatic and biliary—communicate through tight endothelial junctions and interrelate in the regurgitation of biliary constituents during cholestasis, transport of immune proteins and cells, and possibly also in the formation of bile. During acute biliary tract obstruction when intrabiliary pressure is high, protein-bound Chicago blue dye, barium sulfate, or ethiodol injected into the biliary duct colors or opacifies hepatic lymphatics promptly (Mallet-Guy *et al.*, 1965). On the other hand, diffusible water-soluble dyes readily leak into the interstitium and gain access directly to the bloodstream. Similarly, within a few hours after acute ligation of the common bile duct, intravenously injected Bromsulphalein (BSP) in unconjugated (free) form appears in high concentration in thoracic duct lymph derived from the liver. This lipid-soluble substance appears to enter lymphatics through "solubility pathways" selectively, whereas water-soluble BSP metabolite bound to glutathione gains entry directly into blood (Witte *et al.*, 1968). As obstruction persists, biliary radicles become more leaky and then lipid-soluble material regurgitates into blood as well as lymph. Biliary lymphatic communications also may be a site of transfer of proteins such as IgA from lymph into bile or vice versa (Bradley, 1979). Interestingly, patients with hepatic cirrhosis may exhibit an extraordinarily high flow of watery bile (bilorrhea), suggesting bulk transfer of perisinusoidal capillary filtrate directly into biliary canaliculi (Dumont, 1982).

4. CONCLUSION

Hepatic lymph dynamics play a crucial role in liver homeostasis, facilitating and stabilizing hepatocellular metabolism. The pathomechanism(s) of edema, inflammation, fibrosis, repair, regeneration, malignant spread, and bile formation ultimately take place in the extracellular circulation of tissue fluid and lymph, the intrahepatic "plasma" that surrounds and bathes all liver cells.

REFERENCES*

Alpert, L. I. (1976). Veno-occlusive disease of the liver associated with oral contraceptives: Case report and review of the literature. *Hum. Pathol.* **7**, 709–718. (E)

Alves, C. A. P., Alves, A. R., Abreu, W. N., and Andrado, Z. A. (1971). Hepatic artery hypertrophy and sinusoidal hypertension in advanced schistosomiasis. *Gastroenterology* **72**, 126–128. (E)

Andrews, W. H. H. (1955). Excretory function of the liver, a reassessment. *Lancet* **2**, 166–169. (B3,B7)

Birgens, H. S., Henriksen, J., Matzen, P., and Poulsen, H. (1978). The shock liver. *Acta Med. Scand.* **204**, 417–421. (E)

Boyer, J. L., and Klatskin, G. (1970). Pattern of necrosis in acute viral hepatitis. Prognostic value of bridging (subacute hepatic necrosis). *N. Engl. J. Med.* **283**, 1063–1071. (E)

Bradley, S. (1979). Hepatic lymph formation. *In* "Problems in Liver Diseases" (C. S. Davidson, ed.), pp. 53–65. Stratton International Medical, New York. (F)

Child, C. G., III (1954). "The Hepatic Circulation and Portal Hypertension." Saunders, Philadelphia, Pennsylvania. (E)

*In the reference list, the capital letter in parentheses at the end of each reference indicates the section in which it is cited.

Clain, D., Freston, J., Kreel, L., and Sherlock, S. (1967). Clinical diagnosis of the Budd-Chiari syndrome. *Am. J. Med.* **43**, 544–554. (E)

Corliss, C. E. (1976). The circulatory system. In "Patten's Human Embryology," pp. 388–452. McGraw-Hill, New York. (A)

Cowen, R. E., Mallinson, C. N., Thomas, G. E., and Thomson, A. D. (1977). Polyarteritis of the liver: A report of two cases. *Postgrad. Med. J.* **53**, 89–93. (E)

Dickson, A. D. (1957). Development of the ductus venosus in man and in the goat. *J. Anat.* **91**, 358–368. (A)

Dumont, A. E. (1982). Significance of excess drainage of bile in patients with chronic hepatic disease. *Surg., Gynecol. Obstet.* **154**, 209–213. (F)

Dunn, G. D., Hayes, P., Breen, K. J., and Schenker, S. (1973). The liver in congestive heart failure: A review. *Am. J. Med. Sci.* **265**, 174–189. (E)

Edmonson, H. A., Peters, R. L., Reynolds, T. D., and Kuzma, O. T. (1963). Sclerosing hyaline necrosis of the liver in the chronic alcoholic. *Ann. Intern. Med.* **59**, 646–673. (E)

Elias, H., and Petty, D. (1952). Gross anatomy of the blood vessels and ducts within the human liver. *Am. J. Anat.* **90**, 59–111. (B)

Elias, H., and Sherrick, J. C. (1969). "Morphology of the Liver," Chapter 2, pp. 75–103. Academic Press, New York. (E)

Evans, H. M. (1912). The development of the vascular system. In "Manual of Human Embryology" (F. Keibel and F. P. Mall, eds.), Vol. II, pp. 570–709. Lippincott, Philadelphia, Pennsylvania. (A)

Feely, J., Wilkinson, G. R., and Wood, A. J. J. (1981). Reduction of liver blood flow and propranolol metabolism by cimetidine. *N. Engl. J. Med.* **304**, 692–695. (D)

Foley, W. J., Turcotte, J. G., Hoskins, P. A., Brant, R. L., and Ause, R. G. (1971). Intrahepatic arteriovenous fistulas between the hepatic artery and portal vein. *Ann. Surg.* **174**, 849–865. (E)

Fraser, R. (1978). Thoughts on the liver sieve. *Bull. Kupffer Cell Found.* **1**, 46. (B4)

Gallavan, R. H., Chou, C. C., Kvietys, P. R., and Sit, S. P. (1980). Regional blood flow during digestion in the conscious dog. *Am. J. Physiol.* **238**, H220–H225. (C)

Granger, D. N., Miller, T., Allen, R., Parker, R. E., Parker, J. C., and Taylor, A. E. (1979). Permselectivity of cat blood–lymph barrier to endogenous macromolecules. *Gastroenterology* **77**, 103–109. (F)

Greenway, C. V. (1981a). Hepatic plethysmography. In "Hepatic Circulation in Health and Disease" (W. W. Lautt, ed.), pp. 41–56. Raven Press, New York. (C)

Greenway, C. V. (1981b). Hepatic fluid exchange. In "Hepatic Circulation in Health and Disease" (W. W. Lautt, ed.), pp. 153–168. Raven Press, New York. (C)

Greenway, C. V. (1981c). Mechanisms and quantitative assessment of drug effects on cardiac output using a new model of the circulation. *Pharmacol. Rev.* **33**, 213–251. (D)

Greenway, C. V. (1983a). Role of splanchnic venous system in overall cardiovascular homeostasis. *Fed. Proc., Fed. Am. Soc. Exp. Biol.* **42**, 1678–1684. (C)

Greenway, C. V. (1983b). Autoregulatory escape in arteriolar resistance vessels. In "Smooth Muscle Contraction" (N. Stephens, ed.). Dekker, New York (in press). (C)

Greenway, C. V., and Oshiro, G. (1972). Comparison of the effects of hepatic nerve stimulation on arterial flow, distribution of arterial and portal flows and blood content in the livers of anaesthetized cats and dogs. *J. Physiol. (London)* **227**, 487–501. (C)

Greenway, C. V., and Stark, R. D. (1971). Hepatic vascular bed. *Physiol. Rev.* **51**, 23–65. (C,F)

Griner, P. F., Elbadawi, A., and Packman, C. H. (1976). Veno-occlusive disease of the liver after chemotherapy of acute leukemia. *Ann. Intern. Med.* **85**, 578–582. (E)

Grisham, J. W., and Nopanitaya, W. (1981). Scanning electron microscopy of casts of hepatic microvessels: Review of methods and results. In "Hepatic Circulation in Health and Disease" (W. W. Lautt, ed.), pp. 97–98. Raven Press, New York. (B4)

Healey, J. E. (1954). Clinical anatomic aspect of radical hepatic surgery. *J. Int. Coll. Surg.* **12,** 542–549. (B)

Hou, P. C., and McFadzean, A. J. S. (1965). Thrombosis and intimal thickening in the portal system in cirrhosis of the liver. *J. Pathol. Bacteriol.* **89,** 473–480. (E)

Hoyumpa, A. M., Jr., Schiff, L., and Helfman, E. L. (1971). Budd-Chiari syndrome in women taking oral contraceptives. *Am. J. Med.* **50,** 137–140. (E)

Kamel, I. A. A., Fiwi, A. M., Cheever, A. W., Mosimann, J. E., and Danners, R. (1978). Schistosomiasis mansoni and N haematosis infections in Egypt. IV. Hepatic lesions. *Am. J. Trop. Med. Hyg.* **27,** 931–938. (E)

Kardon, R. H., and Kessel, R. G. (1980). Three-dimensional organization of the hepatic microcirculation in the rodent as observed by scanning electron microscopy of corrosion casts. *Gastroenterology* **79,** 72–81. (B4)

Kerr, D. N. S., Harrison, C. V., Sherlock, S., and Walker, R. M. (1961). Congenital hepatic fibrosis. *Q. J. Med.* **30,** 91–117. (E)

Lautt, W. W. (1980a). Control of hepatic arterial blood flow: Independence from liver metabolic activity. *Am. J. Physiol.* **239,** H559–H564. (C)

Lautt, W. W. (1980b). Hepatic nerves: A review of their functions and effects. *Can. J. Physiol. Pharmacol.* **58,** 105–123. (C)

Lautt, W. W., ed. (1981a). "Hepatic Circulation in Health and Disease." Raven Press, New York. (C)

Lautt, W. W. (1981b). Role and control of the hepatic artery. *In* "Hepatic Circulation in Health and Disease" (W. W. Lautt, ed.), pp. 203–226. Raven Press, New York. (C)

Luna, A., Meister, H. P., and Szanto, P. B. (1968). Esophageal varices in the absence of cirrhosis. Incidence and characteristics of congestive heart failure and neoplasm of the liver. *Am. J. Clin. Pathol.* **49,** 710–717. (E)

McCuskey, R. S. (1966). A dynamic and static study of hepatic arterioles and hepatic sphincters. *Am. J. Anat.* **119,** 455–477. (B)

McDevitt, D. G., Nies, A. S., and Wilkinson, G. R. (1977). Influence of phenobarbital on factors responsible for hepatic clearance of indocyanine green in the rat. *Biochem. Pharmacol.* **26,** 1247–1250. (D)

McLean, E. K. (1970). The toxic actions of pyrrolizidine (Senecio) alkaloids. *Pharmacol. Rev.* **22,** 429–483. (E)

Mallet-Guy, P., Michoulier, J., and Baev, S. (1965). Experimental studies on liver lymph flow: Conditions affecting bilio-lymphatic transfer. *In* "The Biliary System" (W. Taylor, ed.), pp. 69–78. F. A. Davis Co., Philadelphia, Pennsylvania. (F)

Michels, N. A. (1966). Newer anatomy of liver and its variant blood supply and collateral circulation. *Am. J. Surg.* **112,** 337–347. (B1)

Miller, D. L., Zarolli, C. S., and Gumucio, J. J. (1979). Quantitative morphology of sinusoids of the hepatic acinus. *Gastroenterology* **76,** 965–969. (B4)

Murakami, T., Itoshima, T., and Shimaka, Y. (1974). Peribiliary portal system in the monkey liver as evidenced by the injection replica scanning electron microscope method. *Arch. Histol. Jpn.* **37,** 245–260. (B4,B5)

Nakata, K., and Kanbe, A. (1966). The terminal distribution of the hepatic artery and its relationship to the development of focal liver necrosis following interruption of the portal blood supply. *Acta Pathol. Jpn.* **16,** 313–321. (B4)

Okuda, K., Musha, H., Yamasaki, T., Jinnouchi, S., Nagasaki, Y., Kubo, Y., Shimokawa, Y., Nakayama, T., Kojiro, M., Sakamoto, K., and Nadashima, T. (1977). Angiographic demonstration of intrahepatic arterio-portal anastomoses in hepatocellular carcinoma. *Radiology* **122,** 53–58. (E)

Olds, J. M., and Stafford, E. S. (1930). On the manner of anastomosis of the hepatic and portal circulation. *Bull. Johns Hopkins Hosp.* **47,** 176–185. (B4)

Orloff, M. J. (1981). The liver. In "Davis-Christopher's Textbook of Surgery" (D. C. Sabiston, ed.), pp. 1131–1194. Saunders, Philadelphia, Pennsylvania. (B1)

Orrego, H., Medline, A., Blendis. L. M., Rankin, J. G., and Kreaden, D. A. (1979). Collagenisation of the Disse's space in alcoholic liver disease. Gut 20, 673–679. (E)

Rappaport, A. M. (1963). Pathophysiology of the liver acinus. In "The Liver" (C. Rouiller, ed.), Vol. 1, pp. 265–328. Academic Press, New York. (B4,B5)

Rappaport, A. M. (1972). "Normal Microcirculation of the Mammalian Liver," 16 mm color film, sound, 20 min duration. Division of Instructional Media Services, Faculty of Medicine, University of Toronto, Toronto, Ontario, Canada. (B4)

Rappaport, A. M. (1973). The microcirculatory hepatic unit. Microvasc. Res. 6, 212–228. (B7)

Rappaport, A. M. (1976). The microcirculatory acinar concept of normal and pathological hepatic structure. Beitr. Pathol. 157, 215–243. (B7)

Rappaport, A. M. (1979). "The Pathologic Microcirculation of the Mammalian Liver," 16 mm color film, sound, 25 min duration. Division of Instructional Media Services, Faculty of Medicine, University of Toronto, Ontario, Canada. (B4)

Rappaport, A. M. (1980). Hepatic blood flow: Morphological aspects and physiologic regulation. Int. Rev. Physiol. 21, 1–63. (B4)

Rappaport, A. M. (1982). Physio-anatomic considerations. In "Diseases of the Liver" (L. Schiff, ed.), 5th ed., Chapter I, pp. 9–11. Lippincott, Philadelphia, Pennsylvania. (B7)

Rappaport, A. M., Borowy, Z. J., Loughheed, W. M., and Lotto, W. N. (1954). Subdivision of hexagonal liver lobules into a structural and functional unit; role in hepatic physiology and pathology. Anat Rec. 119, 11–34. (B3)

Richardson, P. D. I., and Withrington, P. G. (1981). Liver blood flow. Gastroenterology 81, 159–173, 356–375. (C,D)

Rothe, C. F. (1984). Reflex control of the venous system. Physiol. Rev. (in press). (C)

Sasse, D., and Staubesand, J. (1976). Electron microscopic evidence for a muscle layer in the wall of the ThV in the rat. Cell Tissue Res. 165, 391–396. (B4,B6)

Schaffner, F., Gradboys, H. L., Safran, A. P., Baron, M. G., and Aufses, A. H., Jr. (1967). Budd-Chiari syndrome caused by a web in the inferior vena cava. Am. J. Med. 42, 838–843. (E)

Severn, C. B. (1972). A morphological study of the development of the human liver. II. Establishment of liver parenchyma, extrahepatic ducts and associated venous channels. Am. J. Anat. 133, 85–108. (A)

Simonds, J. P. (1936). Chronic occlusion of the portal vein. Arch. Surg. (Chicago) 33, 397–424. (B1)

Stuart, K. L., and Bras, G. (1957). Veno-occlusive disease of the liver. Q. J. Med. 26, 291–315. (E)

Suzuki, T., Nakayasu, A., and Kawabe, K. (1971). Surgical significance of anatomic variation of hepatic artery. Am. J. Surg. 122, 505–512. (B1)

Takeuchi, J., Takada, A., Hasumura, Y., Matsuda, Y., and Ikegami, F. (1971). Budd-Chiari syndrome associated with obstruction of the inferior vena cava. Am. J. Med. 51, 11–20. (E)

Wantz, G. E., and Payne, M. A. (1961). Experience with portacaval shunt for portal hypertension. N. Engl. J. Med. 265, 721–728. (F)

Warwick, R., and Williams, P. L., eds. (1973). "Gray's Anatomy," 35th Br. ed. Saunders, Philadelphia, Pennsylvania. (B1)

Wisse, E. (1977). Ultrastructure and function of Kupffer cells and other sinusoidal cells in the liver. Med. Chir. Dig. 6, 409–418. (B4)

Witte, C. L., Chung, Y. C., Witte, M. H., Sterle, O. F., and Cole, W. R. (1969). Observations on the origin of ascites from experimental extrahepatic portal congestion. Ann. Surg. 170, 1002–1015. (F)

Witte, C. L., Witte, M. H., Bair, G., Mobley, W. P., and Morton, D. (1974). Experimental study of hyperdynamic vs stagnant mesenteric blood flow in portal hypertension. Ann. Surg. 179, 304–310. (E)

Witte, C. L., Witte, M. H., and Dumont, A. E. (1980). Lymph imbalance in the genesis and perpetuation of the ascites syndrome in hepatic cirrhosis. *Gastroenterology* **78**, 1059–1068. (F)

Witte, M. H. (1979). Ascitic, thy lymph runneth over. *Gastroenterology* **76**, 1066–1068. (F)

Witte, M. H., Dumont, A. E., Levine, N., and Cole, W. R. (1968). Patterns of distribution of sulfobromophthalein in lymph and blood during obstruction to bile flow. *Am. J. Surg.* **115**, 69–74. (F)

Witte, M. H., Witte, C. L., and Dumont, A. E. (1981). Estimated net transcapillary water and protein flux in the liver and intestine of patients with portal hypertension from hepatic cirrhosis. *Gastroenterology* **80**, 265–272. (F)

Chapter 15
Genitourinary System: Kidney

A. EMBRYOLOGY OF BLOOD CIRCULATION
By C. E. Corliss

1. Establishment of the Renal Circulation

a. Early development: The human embryonic kidney begins late in the third week, differentiating from the paraxial mesoderm at somite levels 7–14. The transitory segmented pronephric tubules merge with mesonephric elements forming caudal to them in a nephrogenic band extending through the somite 26 level (L2). Beyond this point the nephrogenic cord is unsegmented, and during the fifth week, it is induced by the ureteric bud (from the mesonephric duct at L4) to differentiate into metanephric secretory primordia. The mesonephros functions for a time as an embryonic kidney, supplied by about 30 lateral "segmental" arteries that branch off the aorta early in the fifth week (Felix, 1912). These vessels are not truly segmental, and many disappear along with the degeneration of cephalic pronephric and mesonephric elements. The number of arteries actually increase in the caudalmost regions, and from the sixteenth to the third lumbar segments, all persist (Felix, 1912).

Distal parts of the caudal arteries anastomose to form the rete arteriosum urogenitale in the angle bounded by the gonad, mesonephros, and metanephros (Felix, 1912). In embryos of 7 weeks, this internal network of mesonephric-derived vessels connects to the renal arteries entering the hilus, with the result that internal metanephric vessels join mesonephric (renal) arteries and, through them, connect to the dorsal aorta (Felix, 1912).

b. Later development: Positional changes of the kidney occurring during development alter the organ's vascular relationships. As the caudal body gradually straightens, the kidney "ascends" relatively (Gruenwald, 1943). At 6 weeks, it lies in the pelvic cavity; at 7 weeks, it is caudal to the root of the umbilical artery; and at 9 weeks, it rests at its definitive level (L1). Thus, this organ has "climbed" up along the persisting mesonephric arteries "as if on a ladder" (Felix, 1912). At its destination, it may have several attached vessels, one of which is usually identifiable by 8 weeks as the permanent renal artery.

506

The renal artery gives off anterior and posterior branches in the hilus that, in turn, supply the arcuate arteries along the corticomedullary junction. From the arcuate vessels, the interlobular arteries branch and send afferent twigs to individual glomeruli.

Glomerular and tubular development in the kidney proceeds in a centrifugal manner from the hilus. As newer elements appear distally, older proximal glomeruli in the corticomedullary junctional area degenerate, leaving their afferent arterioles unattached. These vessels, together with the glomerular efferents from the region, comprise the arteriolae rectae spuriae that, with their accompanying veins, become the vasa rectae of the medullary tissue. The vasa rectae act as the "lesser" circulation of the kidney, whereas, the "greater" is composed of the branches of the cortical interlobular arterioles with their pre- and postglomerular branches (Licata, 1962). Existence of additional medullary vessels (arteriolae rectae verae), arising directly from arcuate and interlobular arteries, is now considered questionable (Ham and Cormack, 1979).

c. Renal venous circulation: Two sets of veins drain the early mesonephric kidney: the posterior cardinal (in a posterolateral position) and the subcardinal (in a ventromedial position). Both anastomose at the caudal margin of the organ. As the metanephric kidney develops, a renal vein appears on each side to drain into the subcardinal anastomosis. Only the right portion of the subcardinal contribution survives, being incorporated into the renal segment of the definitive inferior vena cava. This shift of blood to the right via the newly formed inferior vena cava allows the right renal vein to empty directly into the adjacent vena cava, whereas the longer left renal vein must cross the midline aorta to reach the same vessel. Like the arteries, the veins have anterior and posterior branches in the renal pelvis and, deeper in the organ, their smaller ramifications follow the general pattern of corresponding arteries.

B. ANATOMY OF BLOOD CIRCULATION

By R. Beeuwkes III

1. Introduction

The renal circulation incorporates distinct capillary beds connected in series. The first of these, the vascular tufts of glomeruli, are specialized for the formation of a large volume of nearly protein-free plasma ultrafiltrate. The

ISBN 0-12-042520-3

glomerular capillaries reconverge into efferent vessels that then divide to form the complex peritubular capillary network of the cortex or the characteristic vasa recta of the medulla. Cortical peritubular capillaries are specialized for removal of large volumes of water and solute from the cortical interstitium, whereas the medullary vessels are involved in the urine concentrating mechanism. Because studies of the renal vasculature have been pursued continuously for 300 years, it might be supposed that there is little current research interest in this area. On the contrary, both the intrinsic complexity of the structures of the kidney and the stimulus provided by new physiologic insights have provoked much recent activity. Thus this section can serve only as an introduction to the subject. More complete information can be found in recent reviews (Fourman and Moffat, 1971; Graves, 1971; Kriz *et al.*, 1976; Beeuwkes, 1980a,b; Beeuwkes *et al.*, 1981).

2. MAJOR VESSELS

a. Arteries: Each kidney is usually supplied by a single renal artery. In humans, this typically divides just outside the organ into anterior and posterior main branches (Graves, 1971). The anterior vessel supplies the anterior surface, the entire apex, and the lower pole (Fig. 15.1). The posterior vessel supplies the upper and middle segments of the posterior surface. In about 25% of normal individuals, "accessory" renal arteries may extend directly from the aorta to the lower pole. Because there are no anastomoses between renal arterial branches, each is the sole blood supply to a particular region of the organ. Interlobar arteries arise from main branch divisions and extend toward the renal cortex between medullary pyramids. From these vessels, arcuate arteries extend across the bases of the pyramids. Interlobular vessels, which arise at relatively sharp angles from the arcuates, extend toward the kidney surface. Afferent arterioles run from interlobular arteries to the glomeruli (Fig. 15.2). Such a pattern and nomenclature are essentially identical in all mammalian species.

b. Veins: The peritubular capillary network arising from glomerular efferent vessels makes venous connections at every cortical level. In humans, superficial veins run parallel to the capsule before descending near interlobular arteries. The midcortex is drained by interlobular veins. In the inner cortex, these vessels receive additional return from venous channels of the medullary vascular bundles. Anastomoses are found between venous trunks at the arcuate and interlobar levels. Thus, the venous system, unlike the arterial, can maintain flow despite the obstruction of a major branch. Convergence to the single renal vein occurs within the hilus of the kidney and follows no clearly defined segmental plan.

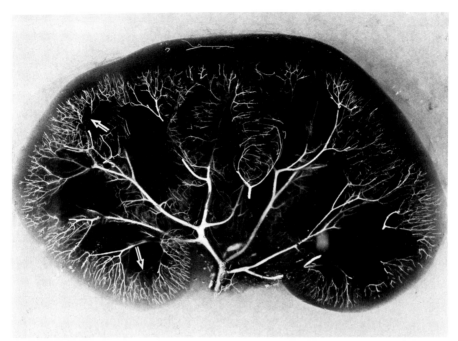

FIG. 15.1. The distribution of the anterior main branch of the renal artery of a human kidney, as revealed by injection of silicone rubber followed by tissue clearing. Divisions of this main branch lead to interlobar arteries that pass between the renal pyramids (dark areas). Arcuate arteries (arrows) branch from interlobar vessels and give rise to small interlobular vessels which extend toward the surface of the organ. [Modified from Beeuwkes and Rosen (1982); reproduced with permission from J. B. Lippincott, Philadelphia.]

3. MICROCIRCULATION

a. Afferent arterioles: These vessels contain only one or two layers of smooth muscle and lack intimal and adventitial lamina (Kriz *et al.*, 1976). Near the glomerulus, medial muscle cells become modified into the renin-secreting granular cells of the "juxtaglomerular apparatus." Other components of this apparatus are the macula densa of the distal tubule and the lacis cells of the glomerular mesangium at the vascular pole. The association of tubular macula densa cells and arteriolar granular cells has suggested a feedback mechanism, linking distal fluid composition to arteriolar function or renin secretion (Wright and Briggs, 1979). However, serial section studies have shown that granular cells are not in direct contact with the macula densa and are often located at sites far removed from the glomerular vascular pole. Nevertheless, distal tubular basement membranes away from the macula densa region sometimes do contact basement membranes of arterioles (Gorgas, 1978), allowing a distal influence on arteriolar function. The macula densa may exert its effect on

Fig. 15.2. Interlobular arteries, glomerular vascular tufts, and partially filled postglomerular vessels shown by silicone rubber injection in a cleared human kidney. The kidney surface is at the top of the picture. × 10.

contractile elements of the glomerular mesangium (see Section B-3b, below). Whatever the site of the feedback mechanism, it is clear that afferent arterioles constitute an endocrine organ of great importance to systemic circulatory regulation.

b. Glomerular structure: Each human kidney contains about 10^6 glomeruli. Since the total rate of formation of ultrafiltrate is approximately 60 ml/min, the amount filtered by each glomerulus is about 6×10^{-8} liter/min. This rate

requires marked specialization to achieve high water and small solute permeability while preventing the passage of high molecular weight solutes such as
serum albumin (molecular radius 36 Å), a need resulting in a unique vascular
structure. Studies by injection or reconstruction show that many channels of
capillary dimension course around the tuft and then converge abruptly within
its center to form the efferent vessel. Several distinct channels, each with a
complete endothelial cell lining, may be found within a single perimeter of
basement membrane. They are separated by a syncytial structure of mesangial
cells extending to the vascular pole of the glomerulus. Thus, the glomerular
vascular pathways cannot be regarded as independent capillaries. Mesangial
cells contain contractile elements, and hence hormonal influences may alter the
shape or surface area of the glomerular pathways.

   c. *Glomerular filtration barrier:* The structural and physiologic properties
of the filtration barrier have been recently reviewed by Brenner *et al.* (1981).
Endothelial cells lining the glomerular vascular channels are fenestrated with
1000 Å openings, suggesting that they probably do not have permselective
characteristics. The basement membrane has a dense central region (lamina
densa) bounded by less dense areas (lamina rara). Outside the basement membrane, but within the urinary space, each vascular pathway is covered with
epithelial cells of unique appearance ("podocytes"). These structures give rise
to foot processes or "pedicels" that rest upon the basement membrane. Adjacent pedicels are joined by a slit diaphragm, with openings too small to allow
free albumin passage. Tracer studies have shown that the slit pore can restrict
ferritin (radius 61 Å) but not horseradish peroxidase (radius 30 Å). Myeloperox-
idase (radius 44 Å) has been found in the space between basement membrane
and slit pores.
   Although the above studies suggest a mechanical barrier, an electrical component is also involved. When kidneys are fixed under free flow conditions,
albumin is restricted at a strongly electronegative region, associated with the
presence of sialoglycoproteins in the basement membrane. Since albumin is
negatively charged, electrostatic repulsion could constitute an effective filtration barrier. Electrically neutral dextran of the same size as albumin is filtered
20 times faster than albumin, whereas negatively charged dextran is nearly
identical to albumin in its filtration properties. Thus, glomerular permselectivity results from a combination of a physical size restriction and an electrical
barrier.

   d. *Cortical postglomerular vessels:* Many glomerular efferent vessels consist of simple endothelial tubes without muscular elements; hence, the common term "efferent arteriole" should be employed with caution. True efferent
arterioles are most prevalant in the outer and inner regions of the cortex. The
pattern of efferent vessels and of peritubular capillary networks is very different

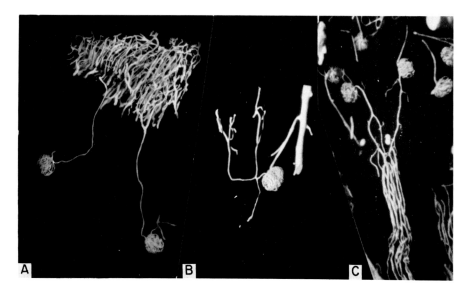

FIG. 15.3. Glomeruli and efferent vascular structures from canine kidneys as shown by silicone injection. (A) Two glomeruli from the region near the kidney surface. These have efferent vessels of the muscular type which extend nearly 1 mm toward the kidney surface before dividing to form a dense peritubular capillary network. (B) A glomerulus from midcortex, with a large afferent vessel that derives from an interlobular artery to the right. The efferent vessel is very short and supplies a simple peritubular network in a medullary ray. (C) Inner cortical glomeruli having large efferent vessels that extend downward and divide to form bundles of vasa recta. [Modified from Beeuwkes (1971); reproduced with permission from the *American Journal of Physiology*.]

in each cortical region (Fig. 15.3). Some efferent capillary vessels of the inner cortex divide to form the vascular bundles of the outer medulla (see Section B-3f, below), whereas all others perfuse only cortical regions. Near interlobular arteries, peritubular capillaries form complex networks without definable orientation. Between interlobular vessels, a long-meshed vascular pattern is found that corresponds to the straight tubular segments located in the "medullary rays." The endothelium of cortical capillaries is highly fenestrated, the resulting high permeability enabling them to take up fluid at a rate equal to one-half of the cortical mass every minute. Because glomerular efferent vessels are small in diameter, they cause a major drop in hydraulic pressure between the glomerular and peritubular components of the renal circulation. The resulting low hydraulic pressure and the high intracapillary oncotic pressure due to water loss in the glomeruli lead to a strong reabsorptive balance of Starling forces. Venous connections are made at every level within the cortex. Capillaries in the subscapular cortex drain into superficial veins, whereas those in the midcortex join with veins along the interlobular axes. Vessels within the medullary rays may drain laterally through the convoluted networks located on either side or vertically along the ray axis, depending on local pressure gradients. The pressure relationships remain unknown.

e. Vascular–tubular relations: Many textbook diagrams show entire nephrons perfused by an efferent vessel arising from the parent glomerulus, thus making each nephron an independent vascular–tubular unit. Such a unit, if it exists, could balance filtration and reabsorption through Starling forces, and interruption of the blood flow to or from a single glomerulus would have consequences for only a single nephron. However, even the most cursory examination of renal vascular pattern shows that peritubular networks formed by individual glomerular efferent vessels rarely extend more than 1 or 2 mm, whereas the nephron arising from the same glomerulus may extend 4 cm. Thus, it is an anatomic necessity for the Henle loop of a superficial nephron to be perfused by blood vessels arising from glomeruli lying deeper within the cortex. The relationships between vessels and nephrons have been directly investigated in both canine and human kidneys (Beeuwkes and Bonventre, 1975; Beeuwkes, 1980b), and it has been conclusively shown that there is no fixed association between any nephron segment and the efferent vessel arising from the same glomerulus. Convoluted tubule segments, which usually extend above the parent glomerulus, are generally perfused by efferent networks arising from more superficial glomeruli. Henle loops and collecting ducts lie within long-meshed efferent ray capillary networks. Since the medullary rays are common pathways for many nephrons, interruption of blood flow from a single glomerulus, the efferent vessel of which supplies a short ray segment, could lead to ischemic injury to loops from many nephrons. Also, because efferent vessels supplying quite different structures exist at single cortical levels, studies of renal blood flow with glomerular flow indicators such as microspheres give no information about distribution of postglomerular peritubular capillary perfusion. The current prevalence of erroneous diagrams is perplexing since a great number of the facts were known many years ago. The correct relationships are shown in Fig. 15.4. More complete diagrams are available in the literature (Beeuwkes and Bonventre, 1975).

f. Medullary vasculature: The outer and inner zones of the renal medulla are defined by characteristic tubular and vascular structures. The inner border of the outer zone is defined by the transition of thin to thick ascending limbs of Henle loops. Thus, the outer medulla includes the region of overlap between thick ascending and thin descending limbs. This region includes bundles of straight vessels ("vasa recta"), formed by the division of inner cortical or "juxtamedullary" glomeruli. Vessels extend from the cores of these bundles to supply the inner medulla. Descending vessels diverging from the surface of the bundles supply a dense capillary network that characterizes the inner strip of the outer medulla. Although inner medullary capillary networks are relatively

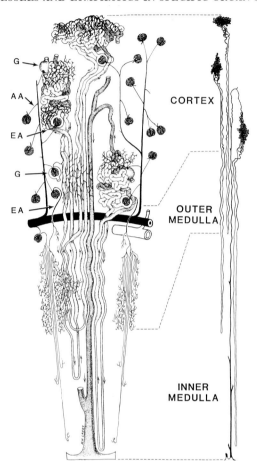

FIG. 15.4. Simplified diagram showing some features of the vascular and tubular organization of the kidney. The vascular structures have been simplified and the veins omitted. To indicate the alterations in scale, the three nephrons of the simplified diagram are shown in their correct proportions to the right. The major zones, cortex, outer medulla, and inner medulla are indicated by dashed lines. Afferent arterioles (AA) arise from interlobular arteries and connect to glomeruli (G). Efferent arterioles (EA) divide to form peritubular capillaries which surround convoluted tubules and other nephron segments. [Modified from Beeuwkes and Bonventre (1975); reproduced with permission from the *American Journal of Physiology*.]

sparse, stereologic studies show that the capillary volume fraction of the inner medulla is, in fact, twice that of the cortex (Pfaller and Rittinger, 1977), and the rate of blood flow in this region is comparable to that of the brain. Blood returns from the inner medulla in the inner ascending venous vasa recta of the vascular bundles, in countercurrent apposition to the descending flow. These venous pathways are both larger and more numerous than the descending vessels. The outer medullary capillary network drains through veins rising between the outer parts of the bundles. These, together with the venous channels of the bundles, make connection with the large veins at the lower interlobular or arcuate level. Ultrastructural studies show the presence of smooth muscle along the inner cortical efferent vessels, extending along descending vasa recta for about one-third of the bundle length. The endothelium of these descending vessels, whether in the outer or inner medulla, is continuous until capillary branching occurs. The ascending vessels are fenestrated. In rodents, many thin descending limbs of Henle loops lie in the peripheral regions of medullary vascular bundles (Kriz *et al.*, 1976). In the human, the medulla is much less rigorously organized. For example, the junction between its inner and outer border is not sharply defined, and there is no evidence for the presence of loop segments within the vascular bundles (Beeuwkes, 1980b). In general, the close apposition of descending and ascending vessels should facilitate countercurrent exchange processes and preserve medullary solute content despite substantial rates of blood flow. An association between descending thin limbs of short Henle loops and ascending vessels could facilitate urea recycling. Such exchange processes involving the medullary vasculature are essential features of recent models of the urine concentrating system (Jamison, 1981).

C. PHYSIOLOGY OF BLOOD CIRCULATION

By J. P. GILMORE

Although the kidneys comprise only 0.5% of body weight, they receive 20 to 25% of the left ventricular output. This is homeostatically appropriate since one of their functions is to clear plasma of nonvolatile end products of metabolism, such as creatinine, and substances such as drugs that enter the circulation from an exogenous source. Also, the kidneys are importantly involved in the generation of hormones and in salt, water, and acid–base homeostasis. The present section deals with the physiologic mechanisms involved in the control of renal blood flow (RBF) and glomerular filtration rate (GFR).

1. AUTOREGULATION

a. Definition: Autoregulation can be defined as the ability of an organ to maintain relatively constant blood flow despite variations in perfusion pressure.

It is a phenomenon that particularly characterizes the renal vasculature, and it occurs even in the isolated denervated organ. In the intact and innervated kidney, blood flow autoregulation is accompanied by similar control of GFR and glomerular capillary pressure; through autoregulation, filtration fraction (FF) remains constant. Together, these observations support the idea that autoregulation is mediated via alterations in preglomerular (afferent) arteriolar resistance.

Although RBF autoregulation was described more than 45 years ago (Hartmann *et al.*, 1936), there is still no agreement as to the mechanism(s) involved; however, a number of hypotheses have been put forward in the intervening years. These include the metabolic, cell separation (Pappenheimer and Kinter, 1956), tissue pressure (Hinshaw *et al.*, 1959), distal tubule–glomerular feedback (DT-GFB) (Thurau and Schnermann, 1965), and myogenic (Bayliss, 1902) concepts.

b. Metabolic hypothesis: This view has been proposed at one time or another to explain blood flow control, in whole or in part, in all organs. It essentially states that local circulation is directly related to the metabolic activity of the organ. However, in the case of the kidneys, such a mechanism appears to play little, if any, role in controlling RBF and GFR. For example, when renal perfusion pressure is decreased, GFR (and thus filtered sodium) initially decreases, and, in turn, the absolute amount of reabsorbed sodium diminishes. Since the major metabolic activity of the kidneys is the result of sodium reabsorption, the production of metabolic end products should become less, leading to an increase in renal vascular resistance. However, with a decrease in renal perfusion pressure, renal vascular resistance falls.

c. Cell separation or plasma skimming hypothesis: On the basis of this concept, altering renal perfusion pressure affects the extent of red cell axial streaming. Thus, when renal perfusion pressure is increased, there is a greater tendency for red cells to migrate toward the center of the flow stream, leaving primarily plasma in the outer portion of the flow stream. Under such circumstances, as blood flows through the interlobar, arcuate, and then interlobular arteries, the deeper nephrons are perfused with blood having a low hematocrit, whereas the blood passing through the cortex, which receives the greatest circulation, has a high hematocrit and hence provides an increased resistance to flow. The net effect is that, despite an elevation of perfusion pressure, total RBF stays relatively constant. The major evidence against the cell separation or plasma skimming hypothesis is the observation that autoregulation is observed even when the isolated kidney is perfused with a cell-free solution (Waugh and Shanks, 1960).

d. Tissue pressure hypothesis: Basically this view states that extravascular pressure varies directly with renal arterial perfusion pressure. Thus, when renal perfusion pressure is elevated, GFR also rises, thereby increasing the amount of fluid in the interstitial spaces. This type of response in a rigid, encapsulated organ such as the kidney is believed to compress the resistance vessels, (afferent arterioles), thereby preventing RBF from increasing despite the rise in perfusion pressure. Support for this idea comes from studies reporting that renal tissue pressure varies directly with renal artery perfusion pressure (Hinshaw *et al.*, 1959); however, others have found no such consistent relationship (Waugh and Shanks, 1960; Gilmore, 1964). In addition, the decapsulated kidney exhibits excellent autoregulation of RBF (Gilmore, 1964).

e. Distal tubule–glomerular feedback hypothesis: The anatomic relationship of the macula densa to the renin-secreting cells of the afferent arteriole has suggested the possibility that the renin–angiotensin system may play a role in the autoregulatory response of the kidney. Support for this view has come from studies in which it was shown that microperfusion of the distal tubule (and presumably the macula densa) with solutions having an elevated NaCl concentration caused collapse of the proximal tubule of the same nephron, apparently as a result of afferent arteriolar constriction and decreased glomerular capillary pressure (Thurau and Schnermann, 1965). Furthermore, proximal tubular collapse was seldom observed in renin-depleted kidneys in response to NaCl perfusion. These observations led to the following suggestion: When renal perfusion pressure is elevated, GFR is increased, thus raising sodium delivery to the macula densa. Consequently, renin is released, which causes intrarenal formation of angiotensin II, this response, in turn, producing afferent arteriolar constriction and relative maintenance of RBF. However, the idea that the renin–angiotensin system is involved in autoregulation has been discounted, since this phenomenon occurs even in the renin-depleted kidney (Potkay and Gilmore, 1973) and in the presence of saralasin, an angiotensin antagonist (see below). In addition, the hypothesis requires that when renal perfusion pressure is lowered, renin release decreases; however, just the opposite occurs (Goldblatt, 1947). Also against the importance of the distal tubule–glomerular feedback theory is the observation that autoregulation can occur when the flow of filtrate to the distal tubule is interrupted (Maddox *et al.*, 1974; Knox *et al.*, 1974).

f. Myogenic hypothesis: In 1902, Bayliss clearly described the myogenic response of blood vessels to alterations in intravascular pressure. He noted that increasing and decreasing intravascular pressure produced constriction and relaxation, respectively. The myogenic hypothesis basically states that there is a direct relationship between renal vascular transmural pressure and afferent

A

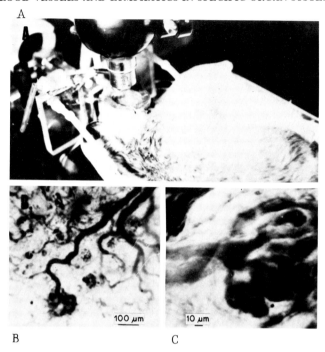

B   C

FIG. 15.5. Hamster with renal transplant prepared for microscopic observation. (A) Renal transplant tissue showing glomeruli; (B) and (C) sequentially higher magnifications of renal transplant showing a single glomerulus. [From Gilmore *et al.* (1980); reproduced with permission of American Heart Association, Inc.]

arteriolar resistance, the latter changing as a result of alterations in vascular smooth muscle tone. In the past, evidence for this hypothesis was indirect, as for example, the finding that renal autoregulation is abolished by papaverine, a smooth muscle relaxant (Thurau and Kramer, 1959; Gilmore, 1964), and that elevating ureteral pressure decreases vascular transmural pressure but, at the same time, reduces renal vascular resistance (Gilmore, 1964). More recent experiments appear to provide direct evidence for the myogenic hypothesis (Gilmore *et al.*, 1980). For example, under anesthesia, renal tissue from neonatal hamsters was placed in a sealed chamber in the cheek pouch of an adult hamster. After time was allowed for the transplanted tissue to establish its circulation with the recipient's cheek pouch membrane circulation ($\sim$ 10 days), the animal was again anesthetized and a light rod inserted into the mouth in order to transilluminate the chamber and renal tissue for direct visualization (Fig. 15.5). A positive or negative pressure was then applied to the chamber while observing changes in the diameter of pre- and postglomerular vessels, using an image shearer. When a positive pressure was applied (decrease in transmural pressure), the diameter of the preglomerular vessels increased;

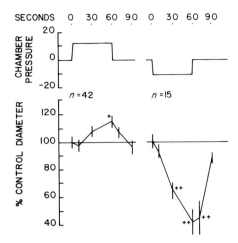

FIG. 15.6. Mean responses of afferent arterioles to changes in chamber pressure. Control diameter, 14.86 ± 0.66 μm; (+) $p < .05$; (++) $p < .01$. [From Gilmore *et al.* (1980); reproduced with permission of American Heart Association, Inc.]

when a negative pressure was applied (increase in transmural pressure), the diameter decreased (Fig. 15.6). When papaverine was added to the chamber, a positive pressure reduced the diameter and a negative pressure raised the diameter, i.e., the vessels responded passively. Neither saralasin nor indomethacin altered the response of preglomerular vessels to alterations in transmural pressure. In contrast to the reactions of preglomerular vessels, the postglomerular vessels (efferent arterioles) responded passively to alterations in transmural pressure. Failure to demonstrate an autoregulatory response in the efferent arterioles was consistent with results obtained from whole kidneys in which the calculated efferent arteriolar resistance did not change over the autoregulatory range of arterial perfusion pressure (Renkin and Gilmore, 1973).

2. NEURAL CONTROL

There is no evidence that the renal nerves exert a tonic influence on renal function under normal, unstressed conditions. Indeed, the denervated transplanted kidney still exhibits normal function (Bricker *et al.*, 1956). However, stimulation of the renal nerves in experimental animals has been reported as altering renal function significantly, producing a decrease in both RBF and GFR. The major vascular action is on the afferent arterioles. Nevertheless, it should be kept in mind that renal nerve stimulation also causes renin release, which, through the intrarenal generation of angiotensin II, could contribute to the observed vascular changes. The responses to renal nerve stimulation can be blocked by appropriate antagonists.

The control of both RBF and GFR has representation in the brain. Smith (1951) showed that fright can reduce RBF in man by 50%, with little change in GFR. Direct stimulation of the cerebral cortex (Hoff *et al.*, 1963) and widespread areas of the diencephalon (Folkow and Rubinstein, 1966; Takeuchi *et al.*, 1962) produces renal vasoconstriction. Interventions that reflexly cause renal vasodilation do so through inhibition of vasoconstrictor tone.

3. INTRARENAL HORMONES

a. Angiotensin II (AII): This is a potent renal vasoconstrictor, appearing to act on both the afferent and efferent arterioles. Hall *et al.* (1979) found that the intrarenal infusion of converting-enzyme inhibitor into dogs maintained on a normal diet increased RBF, with little influence on GFR; thus FF decreased. This type of experiment supports the idea that under normal conditions, sufficient AII is generated by the kidneys directly to modulate its blood flow. However, the anesthesia and surgical manipulations required for the study could also have led to increased release of renin. Although the renin–angiotensin system probably has little or no effect on modulating renal vascular tone under normal conditions, it undoubtedly plays a role under such conditions as volume and/or salt depletion (see also Section D-5a, below).

b. Prostaglandins: The kidney, particularly the medulla, is a rich source of prostaglandin synthetase, a group of enzymes involved in the synthesis of prostaglandins (PG) from arachidonic acid. When injected into the renal artery, a number of PGs will decrease vascular resistance (Gerber *et al.*, 1978), with little effect on GFR (Baylis *et al.*, 1976), thus indicating an effect on both afferent and efferent arterioles. When the Pg synthetase inhibitor, indomethacin, is administered to the conscious animal, RBF changes little (Swain *et al.*, 1975); in addition, it does not interfere with the autoregulatory capacity of the kidney (Venuto *et al.*, 1975). (For the role of renal PG, particularly as it relates to interactions with other renal hormones, see Section D-2, below.)

c. Kallikrein–kinin: Infusion of kallidin into the renal artery produces an increase in RBF and no consistent change in GFR. Thus, FF decreases, suggesting a vasodilatation of both the afferent and efferent arterioles (Webster and Gilmore, 1964). Although kallidin may be involved in the control of salt and water excretion, there is no good evidence that it modulates renal vascular resistance.

D. PHARMACOLOGY OF BLOOD CIRCULATION

By A. J. Lonigro

1. General Considerations

As a first approximation, pharmacologic agents that affect the renal circulation may be described as vasoconstrictor or vasodilator. Closer scrutiny reveals, however, that this classification is flawed, for a single substance may *increase* or *decrease* renal blood flow or exhibit reduced potency dependent upon dose, frequency of administration, the state of sodium balance, and the presence or absence of other substances of endogenous or exogenous origin. Moreover, perhaps even of greater moment than total renal blood flow rising or falling, are the intrarenal circulatory events that occur consequent to application of a vasoactive agent, e.g., redistribution of cortical blood flow to nephrons that are either "salt-losing" or "salt-conserving" and alterations in medullary blood flow, which may affect the osmolar gradient and, thereby, the movement of water across membranes. In this section, agents will be discussed that affect renal blood flow and/or its intrarenal distribution; moreover, whenever possible, hemodynamic responses will be defined in terms of those excretory, metabolic, or endocrine events that might influence the response. The renal prostaglandins, which are not only potent vasoactive substances but also influence the renal responses to a large number of other vasoactive agents, will be presented in detail.

2. Renal Prostaglandins

a. Historical perspective: The intricacies of arachidonic acid metabolism were prefigured by the work of Kurzrok and Lieb (1930) who observed either contraction or relaxation of the uterus upon instillation of seminal fluid. It was von Euler (1935) who identified the active principle of seminal fluid as a lipid soluble acid and named it "prostaglandin." The isolation of crystalline prostaglandins (Bergström and Sjövall, 1957), followed in quick succession by their structural definition (Bergström *et al.*, 1962) and biosynthesis (van Dorp *et al.*, 1964; Bergström *et al.*, 1964), permitted the extraordinary advances in this field. The presence of vasodepressor prostaglandins within the kidney was first described by Lee *et al.* in 1965.

Blood Vessels and Lymphatics in Organ Systems
Copyright © 1984 by Academic Press, Inc.

b. Biosynthesis: It is important to recognize that prostaglandins are not stored; therefore, any increase in their efflux from an organ or a tissue suggests *de novo* synthesis. Although several 20-carbon unsaturated fatty acids can serve as substrate for their synthesis, the dominant fatty acid is arachidonic acid (5,8,11,14-eicosatetraenoic acid) which, when acted upon by prostaglandin-synthesizing enzymes, gives rise to the bisenoic series of prostaglandins, i.e., prostaglandins E_2 (PGE_2), $F_{2\alpha}$ ($PGF_{2\alpha}$), D_2 (PGD_2), and I_2 (PGI_2). Release of arachidonic acid from tissue phospholipids, considered the rate-limiting step in the prostaglandin biosynthetic pathway, is evoked primarily by increased phospholipase A_2 activity, although in some tissues, phospholipase C, phosphorylase, and cholesterol esterase have been implicated as the enzymes responsible for providing the arachidonic acid. Phospholipase activity is inhibited by glucocorticoids, retarding thereby the synthesis of prostaglandins, as well as thromboxanes and leukotrienes (see below). The capacity of steroids to inhibit phospholipase activity correlates well with the antiinflammatory properties of the drugs.

Once released, arachidonic acid may be metabolized via the cyclooxygenase pathway to form the unstable cyclic endoperoxides, PGG_2 and PGH_2, which serve as common precursors for the synthesis of both prostaglandins and thromboxanes. The end product formed seems to be largely dependent on the enzymes present in a particular tissue. Isomerases lead to the production of PGE_2 and PGD_2, whereas reductases promote $PGF_{2\alpha}$ formation. Tissues rich in prostacyclin synthetase, such as blood vessels, synthesize large amounts of PGI_2. When the cyclic endoperoxides are acted upon by thromboxane synthetase, a nonprostaglandin product, thromboxane A_2, is formed. Thromboxane synthetase is found primarily in platelets and its activation has been implicated in the process of platelet aggregation. The kidneys possess the capacity to synthesize all of the aforementioned products of the cyclooxygenase pathway. (For reviews dealing with the ever-increasing complexities of arachidonic acid metabolism, see McGiff, 1981; Levenson *et al.*, 1982.)

The enzyme cyclooxygenase is susceptible to blockade with aspirin and other nonsteroidal antiinflammatory agents, such as indomethacin, the degree of susceptibility appearing to vary greatly from tissue to tissue. However, when cyclooxygenase is blocked, synthesis of all prostaglandins, including PGI_2 and the thromboxanes, is markedly reduced.

In the circulation, prostaglandins E_2, $F_{2\alpha}$, and D_2 exhibit a relatively short biologic half-life, being degraded either within the organ of their origin or upon passage through the pulmonary circulation by the sequential actions of a 15-hydroxyprostaglandin dehydrogenase and a $\Delta^{13,14}$-reductase. PGI_2, which is not degraded within the lungs, has an appreciably longer circulating half-life. In part, PGI_2 is degraded nonenzymatically to form 6-keto-$PGE_{1\alpha}$; estimates of concentrations of the latter compound are frequently used as an index of PGI_2

activity. Thromboxane A_2 (TXA_2) degrades rapidly to thromboxane B_2 (TXB_2), which is stable and is often used as an index of TXA_2 activity.

Recently described is the lipoxygenase pathway for arachidonic acid metabolism that rivals in complexity the cyclooxygenase pathway (Samuelsson *et al.*, 1980). Lipoxygenase transforms arachidonic acid into hydroxyfatty acids and a group of highly biologically active compounds, the leukotrienes. Several of the latter compounds appear to be active constituents of slow-reacting substance of anaphylaxis (SRS-A). Whether the products of the lipoxygenase pathway have important actions in the renal or general circulations awaits clarification.

    c. Renal vascular responses to prostaglandins: When infused into the renal artery, PGE_2, PGD_2, and PGI_2 generally produce vasodilatation, whereas $PGF_{2\alpha}$ and 6-keto-$PGF_{1\alpha}$ are without effect. The renal vasculature of the rat, however, responds differently from that of other species. In the Krebs perfused rat kidney, PGE_1, PGE_2, and PGA_2 produce vasoconstriction, as does PGE_2 in the anesthetized rat, although PGI_2 is usually vasodilator in its action (Baer and McGiff, 1979). Whether the vasoconstrictor effect of PGE_1, PGE_2, and PGA_2 in the rat kidney is due to a direct vascular action or is mediated by activation of other constrictor stimuli is unknown. No matter what the response to exogenously administered prostaglandins, physiologic interpretations or implications derived from such studies must be viewed cautiously, for no artificial route of administration reproduces precisely the effects evoked by the endogenous release of prostaglandins within the kidney, in terms of sites of action or sequence of vascular elements affected.

    d. Renal prostaglandins as local hormones: Initially, PGA_2 of renal origin was considered to be the prime candidate for subserving the role of renal antihypertensive hormone, for, of the prostaglandins originally identified within the kidney, this was the only one not degraded upon passage through the pulmonary circulation. Hence, it could have had access to, and affected the tone of, systemic resistance vessels. Evidence has accumulated, however, to suggest that PGA_2 is not enzymatically produced by the kidney but rather represents an artifact resulting from the dehydration of PGE_2. The latter realization, coupled with the recognition that PGE_2 and $PGF_{2\alpha}$ are metabolized upon passage through the pulmonary circulation, has led to the concept of renal prostaglandins functioning as local hormones, i.e., their site of action is at or near their site of synthesis. This hypothesis has been strengthened by the findings that augmentation of renal prostaglandin synthesis, produced by either mechanical reduction of renal blood flow (McGiff *et al.*, 1970a) or by administration of pressor agents (McGiff *et al.*, 1970b, 1972), is associated with a reduction in renal vasoconstriction and an increase in renal blood flow. On the other hand, inhibition of prostaglandin synthesis eliminates the increase in renal blood flow noted during administration of pressor agents (Aiken and Vane, 1973) and reduces renal blood flow in anesthetized dogs when control levels of prostaglandins in renal venous blood are elevated (Lonigro et al., 1973). Moreover, influences that tend to suppress synthesis of prostaglandins are associated with redistribution of intrarenal blood flow to outer cortical nephrons, whereas enhanced rates of renal prostaglandin synthesis favor redistribution of flow to juxtamedullary areas (Itskovitz et al., 1974).

Although it is generally agreed that renal prostaglandins protect the kidney against excess vasoconstriction, a more comprehensive definition of their role in the regulation of renal blood flow must include considerations relative to other products of arachidonic acid metabolism, to possible interactions with nervous and hormonal systems, and to as yet poorly defined factors. The latter include oxygen tension, pH, and/or P_{CO_2}, which have been reported to influence renal prostaglandins (Lonigro et al., 1982).

The relationship between renal prostaglandins and the excretion of salt and water is unclear. When either arachidonic acid or vasodilator prostaglandins are infused into a renal artery, the excretion of sodium and water generally increases. However, in the unanesthetized animal, sodium excretion has been reported to be either enhanced, impaired, or not affected by inhibitors of cyclooxygenase, whereas in the anesthetized state, these inhibitors produce antinatriuresis and antidiuresis (Levenson et al., 1982). A possible relationship between intrarenal prostaglandins and antidiuretic hormone (ADH) also may be important, for prostaglandins appear to oppose maximal water conservation by inhibiting the effects of ADH on collecting ducts (Anderson et al., 1976).

A prostaglandin component of renin release has been inferred from the report that arachidonic acid increases and indomethacin decreases plasma renin activity in the rabbit (Larsson et al., 1974). In man, nonsteroidal antiinflammatory agents consistently lower plasma renin activity and impair the stimulation of renin release associated with sodium restriction, furosemide infusion, upright posture, cirrhosis with ascites, and Bartter's syndrome. It is unknown which prostaglandin serves the prostaglandin-dependent mechanism regulating renin release, although evidence to date supports prostacyclin or a closely related compound (McGiff, 1981). (For further information regarding the role of prostaglandins, angiotensin, and kallikrein–kinin, see Section C-3, above.)

e. Extrarenal role for renal prostaglandins: The concept of prostaglandins functioning solely as local hormones has been amended by the advent of PGI_2, a highly vasoactive compound of endothelial origin not degraded upon passage through the lungs and, hence, capable of serving as a circulating hormone. Whether PGI_2 serves as the elusive antihypertensive hormone of renal origin remains controversial. There are, however, other considerations that must be taken into account before discarding the concept that renal prostaglandins affect distant organs. For example, $PGF_{2\alpha}$ is produced in rather large amounts by the kidney and, to a large extent, enters the venous side of the circulation, to be degraded ultimately within the lungs. Since it is thought to be devoid of renal hemodynamic and excretory effects, any significant biologic action must occur between its site of synthesis within the kidneys and its site of degradation within the lungs. In view of the observations that $PGF_{2\alpha}$ constricts the pulmonary vasculature, that its site of action is proximal to its site of degradation, and that the pulmonary responses are potentiated by both acidosis and hypoxia (Lonigro and Dawson, 1975), a possible role for it as a participant in the control of pulmonary blood flow must be evaluated. Recently both $PGF_{2\alpha}$ and PGE_2, administered to achieve concentrations less than 100 pg/ml in the pulmonary artery, have been found to produce a redistribution of pulmonary blood flow away from areas of hypoxic ventilation to those which are ventilated with oxygen (Sprague et al., 1982). Thus, prostaglandins of renal origin may eventually be found to participate in the regulation of distant vascular resistances, i.e., PGI_2 affecting the systemic circulation and PGE_2 and/or $PGF_{2\alpha}$, the pulmonary circulation.

3. DOPAMINE

a. Historical perspective: Dopamine, the immediate precursor of norepinephrine, differs from other endogenous catecholamines by the lack of a side-chain hydroxyl group. Its synthesis was reported independently in 1910 by Mannich and Jacobsohn and by Barger and Ewins. In the same year, Barger and Dale noted that dopamine was pressor in the spinal cat, although less potent than either epinephrine or norepinephrine. Holtz and Credner (1942) were the first to recognize that dopamine was unusual, compared with other catecholamines, in that at low doses, it decreased the blood pressure of the guinea pig and rabbit, whereas at high doses, pressor responses occurred. Subsequently, similar changes were observed in the anesthetized cat and dog and in unanesthetized man (Goldberg, 1972). Because the pressor effects of dopamine are reversed by α-adrenergic blocking agents and its depressor effects are unaffected by atropine, antihistamines, or β-adrenergic blocking agents, Goldberg (1975) and Goldberg et al. (1978) have advanced the hypothesis that this substance not only affects α- and β-receptors but also operates via a specific dopamine receptor. Such a construct makes the diverse cardiovascular effects of this compound more comprehensible.

b. Responses of the renal vasculature to dopamine: The renal vascular response to dopamine is dose dependent; viz. vasodilatation occurs at low doses and vasoconstriction at high doses, with biphasic responses being observed at intermediate levels of the agent. The renal vasoconstrictor response to dopamine is antagonized by phentolamine and phenoxybenzamine, suggesting that it is an α-adrenergic-mediated action. After administration of phenoxybenzamine, vasoconstriction is eliminated and vasodilatation occurs with all doses. In contrast, the renal vasodilator effects of dopamine are not antagonized by propranolol, atropine, or antihistamines; nor are they eliminated by prior treatment with reserpine or monoamine oxidase inhibitors. Instead, they are selectively attenuated by butyrophenones, phenothiazines, bulbocapnine, metoclopramide, and sulpiride (Goldberg *et al.*, 1978). Tachyphylaxis to the renal vasodilating properties of dopamine have not been observed. These observations lend strong support to the concept of the existence of a unique dopamine receptor.

c. Clinical implications: Because of its capacity to increase myocardial contractility and cardiac output by direct stimulation of β-adrenergic receptors and to increase renal blood flow, presumably through stimulation of specific dopamine receptors, dopamine has been a useful drug in the treatment of shock. It has been reported to increase cardiac output, blood pressure, and urine flow in patients unresponsive to other agents (Goldberg, 1974). Adverse effects associated with the use of dopamine have been related primarily to the development of ventricular arrhythmias, hypotension, and angina pectoris.

4. OTHER RENAL VASODILATORS

a. Bradykinin: This vasoactive peptide constricts some vascular beds and dilates others. When infused into a renal artery, renal blood flow increases in a dose-related manner and intrarenal prostaglandin synthesis is enhanced. Indeed renal prostaglandins have been thought to mediate, at least in part, the renal vasodilator response to bradykinin. Such a possibility has not been borne out, however, by studies in which bradykinin has been infused into a renal artery both prior to and after inhibition of prostaglandin synthesis by indomethacin. The results of such experiments demonstrate that indomethacin enhances the renal vasodilator response to bradykinin, although the efflux of both PGE_2 and $PGF_{2\alpha}$ is reduced virtually to zero and remains so even after administration of bradykinin (Lonigro *et al.*, 1978).

b. Furosemide: The loop diuretic, furosemide, in therapeutic doses, increases total renal blood flow and redistributes intrarenal flow from outer cortical to midcortical zones (Stein *et al.*, 1972). In addition, it increases concentrations of prostaglandin E in renal venous blood (Williamson *et al.*, 1975). Pretreatment with the cyclooxygenase inhibitor, indomethacin, reduces not only the renal hemodynamic response to furosemide but also its capacity to release renin; in contrast, it has no effect on the natriuretic effect of the diuretic (Bailie *et al.*, 1976). Thus, the renal vascular responses to furosemide appear to be mediated by prostaglandins, whereas its natriuretic response is not.

c. Hydralazine: The antihypertensive drug, hydralazine, not only lowers blood pressure but also produces an increase in renal blood flow. The mechanism of action on vascular smooth muscle appears to be a direct one and not mediated by prostaglandins or dopamine (Khayyal *et al.*, 1981).

d. Other agents: Acetylcholine is a potent renal vasodilator. Because of its safety and rapid metabolism, its use has been recommended for renal angiographic studies to improve visualization of smaller arteries. The antihypertensive agents, diazoxide and sodium nitroprusside, produce increases in renal blood flow, although they are not as potent as either acetylcholine or prostaglandins. In general, the administration of agents that augment renal blood flow is associated with increased salt and water excretion.

5. OTHER RENAL VASOCONSTRICTORS

a. Angiotensin II: The administration of this potent vasoconstrictor octapeptide produces a dose-related decrease in renal blood flow and a redistribution of intrarenal blood flow to inner cortical zones (Carrière and Friborg, 1969). The reduction in renal blood flow is not sustained, i.e., despite continued infusion of the pressor hormone, blood flow recovers, this occurring simultaneously with the appearance of PGE_2 in renal venous blood (McGiff *et al.*, 1970b). Pretreatment with indomethacin exaggerates the renal vasoconstrictor response to angiotensin II and eliminates the recovery of renal blood flow with continued infusion of the peptide (Aiken and Vane, 1973). Thus, the response of the renal vasculature to angiotensin is thought to be modified by intrarenal cyclooxygenase activity.

b. Norepinephrine: This catecholamine produces dose-related reductions in renal blood flow which can be inhibited by α-receptor blocking agents. The response of renal blood flow to infusion of norepinephrine is similar to that observed with angiotensin II, in that the reduction is not sustained, a response apparently related to enhanced intrarenal prostaglandin synthesis (McGiff *et al.*, 1972).

c. Other agents: Several diuretic agents, such as hydrochlorothiazide, produce minor reductions in renal blood flow, the clinical significance of which is small under most conditions, but it may be of importance in patients with compromised renal function. Other agents that cause renal vasoconstriction are those that one way or another affect adrenergic mechanisms, such as α-receptor agonists or indirect-acting sympathomimetic amines.

E. PATHOPHYSIOLOGY, PATHOGENESIS, AND PATHOLOGY OF BLOOD CIRCULATION

By M. M. Schwartz

The kidney is a highly vascularized organ, susceptible to all the diseases that affect blood vessels in general. In most of these conditions, renal involvement is seen as part of a systemic process, but some are characteristic of the kidney.

1. Renal Artery Stenosis and Aneurysm

Renal artery stenosis is one of the recognized causes of systemic hypertension (Goldblatt, 1947). The most frequent basis of stenosis is atherosclerosis with the offending lesion usually being short in length and located near the origin of the renal artery at the aorta. This condition occurs most frequently in older males, and may be bilateral. Another cause of stenosis is fibrous and fibromuscular hyperplasia. Microscopically, this lesion is characterized by degenerative changes with myxomatous fibrous tissue distorting the muscle cells. Angiographically, these lesions appear as a "string of beads." This pathology occurs most often in young women.

Renal artery aneurysms are infrequent, usually caused by atherosclerosis or medial necrosis. In most cases they are found in the main renal artery or its branches but in about a sixth of cases they occur intrarenally. They may be located distal to an area of stenosis. Aneurysms that are not calcified frequently rupture.

2. Renal Vein Thrombosis

In the past, renal vein thrombosis was considered to be a primary cause of the nephrotic syndrome, but in the last 10 years, it has been shown to be secondary to the nephrotic syndrome of diverse etiologies (Llach *et al.*, 1980).

a. Experimental renal vein thrombosis: Proteinuria in the human has been associated with conditions that elevate renal venous pressure, particularly renal vein thrombosis. However, in this regard, simply constricting the renal veins of experimental animals does not cause nephrotic range proteinuria (Fisher *et al.*, 1968).

b. Clinical observations: The temporal relationship between renal vein thrombosis and the nephrotic syndrome has failed to support a causal relationship. For example, in an autopsy series (McCarthy *et al.*, 1963) and in surgical experience (Jackson and Thomas, 1970), renal vein thrombosis has frequently been found to occur without the nephrotic syndrome. In a prospective study, Llach *et al.* (1980) reported that renal vein thrombosis developed after the onset of the nephrotic syndrome and that it was most frequently associated with membranous glomerulonephropathy. However, it also occurred in patients with other primary glomerular lesions.

   c. Pathogenetic mechanism: Individuals with the nephrotic syndrome are in a hypercoagulable state (Thomson *et al.*, 1974), and they develop renal vein thrombosis, pulmonary embolism, and other thromboembolic phenomena at a greater than normal rate (Llach *et al.*, 1980). Of interest in this regard is the finding that such patients have an increase in a number of clotting factors, including platelets, fibrinogen, factor V, factor VIII, and factor X. Moreover, Kauffmann *et al.* (1978) have reported decreased levels of antithrombin III in eight of nine individuals with thromboembolic disease in a group of 48 nephrotic patients. Although low levels of antithrombin III may be related to urinary loss of serum proteins, the mechanism underlying elevation of other clotting factors in the nephrotic syndrome is unknown.

3. DIABETES MELLITUS

   Clinical involvement of the kidneys in diabetes mellitus is an ominous prognostic sign. In a study of 112 juvenile-onset diabetics, Kussman *et al.* (1976) found that the mean duration of disease was 17.3 ± 6.0 years, and that severe renal failure developed 4 years after the onset of proteinuria, followed by death within 6 months.

   a. Relationship between clinical manifestations of diabetes mellitus and diabetic renal disease: When diabetes mellitus can be excluded on clinical grounds, the diagnostic histologic glomerular lesions are rarely seen (Heptinstall, 1974). Using ultrastructural morphometry, Østerby (1975) has shown that juvenile diabetics with normal kidneys at the onset of diabetic renal involvement develop detectable thickening of the glomerular basement membrane and mesangial expansion within 2 years and definite morphologic changes in 4 years. In addition, allografted kidneys from nonrelated nondiabetic donors form similar pathologic alterations after 2 years (Mauer *et al.*, 1976a). In animals made diabetic by the β-cell toxin, streptozocin, or by pancreatic ablation, similar changes occur coincident with the onset of the diabetic state (Foglia *et al.*, 1950). Finally, glomerular pathology is prevented in the rat by islet transplantation or by careful control of diet and insulin replacement (Mauer *et al.*, 1981). Thus, it is well established that the renal pathology seen in diabetes mellitus results from the absence of insulin and the consequent metabolic abnormalities of the diabetic state.

b. Biochemical abnormalities of diabetic glomerular basement membrane: The accumulation of basement-like material has been inferred from thickening of the glomerular basement membrane and from expansion of the mesangial areas observed in kidneys from diabetic patients. Although a basic biochemical abnormality in the development of the diabetic glomerular basement membranes has not been identified, qualitative differences in composition have been reported in diabetic humans (Kefalides, 1974; Mauer *et al.*, 1976b; Miller and Michael, 1976; Spiro, 1976) and in diabetic rats (Brownlee and Spiro, 1979; Spiro and Spiro, 1971).

c. Glomerular pathophysiology: The diabetic kidney is enlarged, with a corresponding increase in size of glomeruli, aneurysmal dilatation of glomerular capillaries, and enlargement of tubules. Recently, functional studies have shown that in the diabetic patient, the glomerular filtration rate rises at the onset of overt disease and persists for many years (Hirose *et al.*, 1980). The possibility that hyperperfusion may be the ultimate cause of glomerular obsolescence in diabetes mellitus has not been explored, but the same mechanism has been postulated for the progression of glomerular disease in the five-sixth nephrectomy model (Hostetter *et al.*, 1981) and in advanced glomerular scarring of diverse etiologies in humans. The complex relationship among pathology, pathophysiology, and biochemistry of diabetic renal disease remains unsettled. Recently, islet cell transplantation has been found to prevent the development of diabetic glomerulosclerosis in diabetic rats (Weil *et al.*, 1976). Whether such a procedure will cause regression of the renal lesion remains to be determined.

4. DISORDERS OF GLOMERULAR FILTRATION

Glomerular filtration is driven by hydrostatic pressure in the glomerular capillaries and is a function of the filtration characteristics of the capillary wall. Landis and Pappenheimer (1963) focused on the size of particles excluded from the glomerular filtrate and concluded that the filtration barrier was formed by "pores" of the requisite size in the glomerular capillaries to permit the formation of a protein-free ultrafiltrate. This theory was unchallenged until morphologic studies failed to demonstrate "pores" in the glomerular capillaries.

a. Role of glomerular charge in permselectivity: The glomerular epithelial cells are covered with a heavy anionic coat that is rich in sialic acid (Jones, 1969). It has been proposed that this coat forms part of the filtration barrier by electrostatically repulsing negatively charged serum proteins (Mohos and Skoza, 1969). In proteinuric states, the glomerular anionic coat is reduced, an observation that suggests that a deficient charge-coat causes loss of permselectivity (Michael *et al.*, 1970).

b. Role of molecular charge: The part played by the molecular charge in determining glomerular permselectivity was first explored morphologically by Rennke *et al.* (1975) using cationic, anionic, and native ferritins. They found that cationized ferritin penetrated the glomerular basement membrane to a greater extent than did the native and anionic ferritins, and they concluded that fixed charges present in the glomerular basement membrane make a major contribution to its barrier function. Brenner *et al.* (1978) used clearance techniques to demonstrate that for any given molecular size, the filtration of negatively charged macromolecules is restricted to a greater extent than are neutral and positively charged molecules. Thus, it appears that glomerular permselectivity is determined by the interaction of glomerular factors with the size and charge of the filtered molecule.

Although the loss of stainable glomerular polyanion correlates with decreased permselectivity in a number of diseases, the pathogenesis of such changes remains in question. Further research should demonstrate whether reduced glomerular charge is etiologically related to proteinuria or whether both are epiphenomena of a more proximate glomerular injury.

5. Glomerular Factors in Localization of Circulating Immune Complexes

Although circulating immune complexes may theoretically localize and initiate inflammatory damage anywhere in the vascular system, the glomerulus frequently is the exclusive site of clinically significant immune complex-mediated vascular damage. Initial studies emphasized the characteristics of the immune complexes as determinants of glomerular localization of immune complexes (Germuth, 1953), but recent evidence suggests that idiosyncracies of glomerular circulation and structure also participate in the pathogenesis of immune complex disease. In addition, the glomerular capillaries have a number of functional attributes that have been implicated in immune complex localization (see below).

a. Mesangial localization of circulating immune complexes: Circulating immune complexes and particulate material may localize in the glomerular mesangium (McClusky *et al.*, 1962), thus suggesting that this structure participates in the clearance of macromolecules from the circulation.

b. Specific glomerular receptors for immune reactants: The glomerulus contains specific receptors for the C3b fragment of complement and the Fc receptor of the immunoglobulin molecule. The C3b receptor resides on the glomerular epithelial cells (Gelfand *et al.*, 1975), and the Fc receptor is present on the surface of a mesangial cell. Since Schreiner *et al.* (1981) have shown that there is a resident mesangial population of Ia-positive bone marrow-derived monocytes, it is likely that the Fc receptor is related to this cell population. Thus, immunoglobulin and C3b may localize in the glomerulus as a result of interaction with specific receptors.

c. *Role of immune complex charge:* Border *et al.* (1982) postulated that variations in charge could enhance the localization of immune complexes in the glomerulus. Using chemically modified anionic and cationic bovine serum albumin, the authors found that cationic bovin serum albumin uniformly induced subepithelial immune complex formation, whereas anionic bovine serum albumin was predominantly deposited in the mesangium.

d. *Role of in situ immune complex formation in glomerular injury:* Couser and Salant (1980) reviewed the evidence favoring *in situ* glomerular immune complex formation and suggested three mechanisms by which antigen localized to the glomerulus could subsequently react with circulating antibodies: (1) mesangial localization of macromolecules or immune complexes (Mauer *et al.*, 1973), (2) nonimmunologic binding of nonglomerular antigens to glomerular elements (Izui *et al.*, 1976; Golbus and Wilson, 1979), and (3) reactions of circulating antibody and the recruitment of the complement system.

The above data challenge the classically held view that immune complex injury most frequently results from glomerular deposition of circulating immune complexes. Alternative mechanisms have gained support from experimental models, but it still remains to be demonstrated that a clinically significant number of patients develop glomerular injury mediated by such mechanisms. In addition, the relationship between deposits at the site of injury to readily demonstrable circulating complexes is unknown.

F. LYMPHATIC SYSTEM

By C. C. C. O'MORCHOE

1. ANATOMY OF LYMPHATIC SYSTEM

a. *General:* The kidneys contain a rich lymphatic system, one that conforms not only to the high plasma flow that they receive but also to the fenestrae of their blood capillaries. Both features are conducive to the escape of

Blood Vessels and Lymphatics in Organ Systems
Copyright © 1984 by Academic Press, Inc.

protein from the plasma and so require lymphatics to remove it. The vessels that lie within the kidney are lymph capillaries and early collecting channels. The extrarenal collecting lymphatics are visible to the naked eye and drain to the paraaortic group of nodes beside the renal pedicle. Most of our knowledge of the system stems from laboratory animals and so cannot be applied to man directly, but differences in lymphatics among mammalian kidneys appear to be quite minor.

b. Intrarenal lymphatics: The kidney, like other organs, contains a deep and a superficial plexus. Although the two drain by different routes, they interconnect to provide alternative outlets if either is interrupted (Holmes *et al.*, 1977a). The superficial component of the system lies close beneath the capsule among the outer segments of the nephrons. The early collecting vessels of this plexus pass outward through the capsule of the kidney and join upon its surface as they converge toward the poles. The early collecting vessels of the deep plexus run with the arteries and veins from which they receive their names. Thus interlobular, arcuate, and interlobar lymphatics are present within the kidney. Connections between both systems occur in the outer cortex, among their initial tributaries, and by occasional interlobular lymphatics that perforate the capsule and contribute to the capsular system (Holmes *et al.*, 1977a).

The pattern so far described is now accepted, although two issues remain in dispute. One is the extent of lymphatics in the lobules, the other is lymph drainage from the medulla. Early studies with tracers appeared to show lymph networks within the renal lobules, but dye can travel so easily through tissue spaces that the findings must remain in doubt. Pierce (1944), in a well controlled and careful study, failed to find evidence in support of their existence. More recently Kriz and Dieterich (1970), using electron microscopy, also failed to detect them. In contrast, O'Morchoe and Albertine (1980) found intralobular lymphatics in close association with tubules and with renal corpuscles. The lymphatics were then seen to pass out of the lobules, alongside the afferent arterioles, to enter the interlobular lymph vessels. These observations, originally reported in the dog, have been extended to the rat. The cause of the disparities rests largely with the techniques used, for intralobular lymphatics are sparse and elusive and therefore unlikely to be found in random tissue sections. Preliminary detection by their luminal contents, as seen with light microscopy, and subsequent confirmation by their ultrastructure provide the most complete results (O'Morchoe and Albertine, 1980).

Similar disagreement surrounds medullary lymphatics, especially since 1949 when Rawson described them in a carcinomatous human kidney. Careful studies, however, have failed to reconfirm their presence (Kriz and Dieterich, 1970; Albertine and O'Morchoe, 1980a), and now it may be accepted that the medulla does not contain lymphatics. However, functional evidence exists for a medullary component to hilar lymph (see Section F-2c, below). It might there-fore seem that morphologic and physiologic evidence are in conflict, but this may not be so if interstitial fluid ascends in the medulla before reaching lymphatics in the juxtamedullary region.

Since the pattern and distribution of intrarenal lymphatics have largely been resolved, attempts have now been made to examine their extent and relate it to the flow of lymph (O'Morchoe and Albertine, 1980). Recent ad-vances in morphometry have made such a study possible. Estimates of lympha-tic surface area and total renal lymph flow suggest that renal lymph is formed at about 1 μl/min/cm^2 of lymphatic endothelium (O'Morchoe et al., 1982).

c. *Extrarenal lymphatics:* Prenodal collecting vessels emerge from two as-pects of the kidney, those that comprise the capsular system and those of the deeper (hilar) system that form part of the renal pedicle. The efferent vessels of the capsular network leave from the renal poles to run through neighboring fat and reach the paraaortic nodes. Sometimes a vessel runs toward the hilum and enters the renal pedicle. A variable number of hilar lymphatic vessels—be-tween one and ten—divide and anastomose in their course toward the nodes. The lumbar lymphatic trunk brings lymph out of the nodes and travels to the cisterna chyli which it enters.

d. *Structure:* The structure of renal lymphatics is like that seen throughout the body (see Section B-2b and 3b, Chapter 4). Lymphatics inside the lobules are characteristic of capillaries. They stand out from the peritubular blood capillaries that display a prominent basal lamina and multiple fenestrae. In-terlobular lymphatics, though devoid of valves, exhibit an increase in suben-dothelial collagen. Thus they appear as a transition between collecting and capillary vessels. The earliest valves appear in arcuate vessels, although their walls continue to be thin. In both arcuate and interlobar lymphatics, the basal lamina is discontinuous and the subendothelium contains mainly connective tissue. The structure of these channels suggests they are engaged in lymph formation and collection. The walls of the extrarenal vessels vary considerably in thickness. Those in the rat possess more muscle and visibly contract, where-as those in the dog contain less muscle and do not show rhythmic contractions.

2. Physiology of Lymphatic System

a. *Formation and composition of lymph:* The method of lymph formation is in doubt. The concept of tissue pressure that is subatmospheric makes difficult an understanding of the passive interplay of forces. However, the kidney may be different, for evidence exists that tissue pressure there may well be positive. If true then renal lymph is formed along a hydrostatic pressure gradient—a state more easily explained than when the gradient is in the opposite direction. The transport pathway taken by macromolecules is also still in doubt but probably involves both intercellular channels and intracellular vesicles (Albertine and O'Morchoe, 1980b).

A twofold source of interstitial fluid, from which the lymph is formed, must be considered. One is the plasma filtrate from the peritubular capillaries; the other is reabsorption by the nephric tubules. The relative proportions of these sources, as they participate in renal lymph formation, has not yet been worked out, but both indeed contribute. The evidence is both structural and functional. In tissue sections, lymphatic vessels can be seen in close relationship to blood capillaries and to the reabsorbing segments of the nephron. The composition of lymph confirms these findings. The protein content of lymph must be derived from plasma transudate, for tubular fluid is almost protein free. However, were blood to be the only source of lymph, the levels of most substances would correlate in plasma and in lymph, but this is not the case. Many, including electrolytes and glucose, show concentration differences between lymph and plasma. For example, the concentration of glucose is higher in capsular lymph than in renal venous plasma. Conversely, the level in hilar lymph is about 85% that of the plasma (O'Morchoe et al., 1975). The reason, it is suggested, lies with tubular reabsorption. Fluid absorbed by the early proximal tubule, since it is rich in glucose, may influence capsular lymph because the outermost tubules are chiefly proximal. The remaining reabsorbed fluid is without glucose and it, in turn, may influence hilar lymph. If such is the case, then tubular reabsorbate contributes to both hilar and capsular lymph.

Electrolyte studies also imply that reabsorbed fluid contributes to hilar lymph, with a component derived from the medulla. Such a conclusion comes from the observation that sodium and chloride concentrations are higher in lymph than those predicted by the Gibbs-Donnan phenomenon (O'Morchoe et al., 1970). The difference has been ascribed to the electrolyte gradient in the mammalian renal medulla. The rationale is that the thick ascending limb of Henle reabsorbs sodium chloride without an equivalent volume of water. The electrolyte-rich tissue fluid so formed traverses the interstitium to reach the lymph vessels that occupy the juxtamedullary region. Support for this concept is derived from furosemide studies (O'Morchoe et al., 1970). Furosemide inhibits the electrolyte pump in the thick ascending limb of Henle and thereby reduces the medullary gradient. At the same time, the electrolyte difference between the plasma and the lymph is abolished or reduced. Thus a loss of the medullary gradient leads to a decrease in the electrolyte content of the lymph.

Lymph, it would therefore seem, comes from both plasma and the tubular fluid, although the proportions of each have yet to be determined. It is clear, nonetheless, that the bulk of reabsorbed fluid does not leave the kidney in the lymph. In the anesthetized dog, for example, the production of lymph within the kidney is less than 1% of the fluid reabsorbed by the tubules (O'Morchoe et al., 1982); the remainder reenters the plasma.

b. Flow: Estimates of renal lymph flow have been indirect, since direct measurement by simple cannulation has not been possible. Various indirect techniques have been devised, and these produce an average value of 0.30 ml/min/100 gm kidney in the anesthetized dog. In this preparation, renal lymph forms almost 40% of the thoracic duct lymph, but it contributes a smaller proportion when the animal is active because of increased flow from limbs and other regions. Renal lymph can also be increased by various lymphagogues, by certain diuretics such as mannitol, and by mechanical procedures that raise the peritubular capillary blood pressure. The means by which lymph flows in renal lymphatic vessels is similar to that in other regions of the body. It depends upon compression of the vessels and on the valves that they contain. The compression may be externally applied or come from contraction of the muscle in the walls.

c. Function: Nephrologists have asked for many years whether lymph has a special function in the kidney. The answer is not yet clear, but all the evidence suggests that it does not. This does not mean that lymph is unimportant physiologically, for its primary role in any organ is to clear macromolecules and excess fluid from the interstitium. This role, though general, must have particular importance for the kidney where fluctuations in capillary blood pressure can lead to major alterations in tissue fluid volume. The presence of a capsule, by limiting expansion of the kidney, intensifies the chance of renal damage when excess tissue fluid is not relieved by adequate lymph drainage. An unanswered question is whether lymph plays any role in urinary concentration. The absence of lymph vessels from the medulla renders this unlikely; yet interstitial fluid may serve to drain extravasated proteins, or even glycoproteins and the water bound to them, from the medulla of the kidney. No evidence on this point is yet available.

Solutions to questions on the role of renal lymph might be expected to evolve from studies on occlusion of the system. In this regard, experiments suggest that lymphedematous kidneys secrete a dilute urine. However, ligation of lymphatics without damage to the nerves is technically not possible, and even partial denervation provokes an increased flow of dilute urine. Thus the effects of renal lymph obstruction have yet to be distinguished from those of renal denervation.

d. Hemodynamic disturbances: If lymph does not have a special role in normal renal function, does it attain another role in altered renal function? Can it, for instance, serve as a diversionary system if the normal flow of urine is impeded? Studies in support of this conclusion confirm that the flow of renal lymph goes up when the ureter is occluded (Heney *et al.*, 1971). Also, when tracers are injected retrogradely into the ureter, they appear in hilar lymph. Both types of observation are interpreted to mean that when the ureter is occluded, urine escapes into the lymph, and so the damage caused by increased renal pressure is diminished. Attractive as this may seem, there is evidence against it. Not the least is the fact that hydronephrosis does develop when the ureter is obstructed. Additional evidence is gained by simultaneous analysis of renal hilar lymph and urine trapped within the pelvis of the kidney. Were urine to mix with lymph, even in small amounts, the composition of lymph would change in the direction of the urine. However, this proves not to be the case, despite wide differences in the content of both fluids (Heney *et al.*, 1971; Holmes *et al.*, 1977b). If lymph does not protect the kidney in this way, why then is there an increased flow of lymph when urine flow is hindered? The answer is thought to lie in altered hemodynamics. Obstruction of the ureter raises urinary pressure and distends the renal pelvis in the sinus of the kidney. The structures in the sinus are compressed, causing congestion of the veins and so an increase in the permeability of intrarenal blood capillaries. The increased tissue fluid which is formed, in turn, enhances renal lymph formation. There is evidence to support this concept. Within 2 or 3 days of urinary obstruction, an increased vascularity of the renal surface and a dilatation of the capsular lymph vessels can be seen, both indicating an impediment to drainage through the hilum (Holmes *et al.*, 1977b). A second line of evidence is that the changes in renal lymph that follow urinary obstruction are mimicked by the changes that occur when renal venous pressure is experimentally raised (Heney *et al.*, 1971).

One common hemodynamic change that alters renal lymph is a reduction in blood pressure. An expected drop in the production of lymph ensues, but there is little alteration in its content unless some tissue damage is produced. When the latter occurs, enzymes, such as lactate dehydrogenase, appear in increasing concentrations in the lymph.

3. PATHOLOGY OF LYMPHATIC SYSTEM

Little is known about the role lymphatics play in the pathology of the kidney. This applies to the manner in which lymphatics may affect diseases of the kidney and, conversely, how renal pathology may affect lymphatics. There is some evidence to suggest that obstruction of the lymphatics predisposes the kidney to infection. Although the data are not conclusive, the concept is not improbable. The urinary system may be affected when the mechanism of lymph flow is deranged and the valves become incompetent. Since flow cannot proceed, the vessels become dilated and lymph accumulates in dependent regions. When this occurs, lymphatics frequently erode through epithelial linings, and if the urinary system is affected, chyle may seep into the urine (chyluria).

Lymphatics in the kidney, like their counterparts elsewhere, can rapidly regenerate or show new growth. If a foreign body is placed within the kidney, an impressive growth of new lymphatics surrounds the affected region. Whether dilatation or new growth occurs when the kidneys are diseased is not yet known, but prominent lymphatics are sometimes seen in canine kidneys with advanced nephritis. One other condition associated with the kidney and its lymphatics is a lymphocele. It follows renal transplantation when lymph accumulates from the cut ends of the lymphatics. Why it occurs in only a minority of cases is not known, but when it does, it may reflect a poor regenerative capacity.

REFERENCES*

Aiken, J. W., and Vane, J. R. (1973). Intrarenal prostaglandin release attenuates the renal vaso-constrictor activity of angiotensin. *J. Pharmacol. Exp. Ther.* **184**, 678–687. (D)

Albertine, K. H., and O'Morchoe, C. C. C. (1980a). An integrated light and electron microscopic study on the existence of intramedullary lymphatics in the dog kidney. *Lymphology* **13**, 100–106. (F)

Albertine, K. H., and O'Morchoe, C. C. C. (1980b). Renal lymphatic ultrastructure and trans-lymphatic transport. *Microvasc. Res.* **19**, 338–351. (F)

Anderson, R. J., Berl, T., McDonald, K. M., and Schrier, R. W. (1976). Prostaglandins: Effects on blood pressure, renal blood flow, sodium and water excretion. *Kidney Int.* **10**, 205–215. (D)

Baer, P. G., and McGiff, J. C. (1979). Comparison of effects of prostaglandins E_2 and I_2 on rat renal vascular resistance. *Eur. J. Pharmacol.* **54**, 359–363. (D)

Bailie, M. D., Crosslan, K., and Hook, J. B. (1976). Natriuretic effect of furosemide after inhibition of prostaglandin synthetase. *J. Pharmacol. Exp. Ther.* **199**, 469–476. (D)

Barger, G., and Dale, H. H. (1910). Chemical structure and sympathomimetic actions of amines. *J. Physiol. (London)* **41**, 18–59. (D)

Barger, G., and Ewins, A. J. (1910). Some phenolic derivatives of β-phenylethylamine. *Q. J. Chem. Soc.* **97**, 2253–2261. (D)

Baylis, C., Deen, W. M., Myers, B. D., and Brenner, B. M. (1976). Effects of some vasodilator drugs on transcapillary fluid exchange in renal cortex. *Am. J. Physiol.* **230**, 1148–1158. (C)

Bayliss, W. M. (1902). On the local reactions of the arterial wall to changes of internal pressure. *J. Physiol. (London)* **4**, 220–231. (C)

Beeuwkes, R. (1971). Efferent vascular patterns and early vascular-tubular relations in the dog kidney. *Am. J. Physiol.* **221**, 1361–1374. (B)

Beeuwkes, R. (1980a). The vascular organization of the kidney. *Annu. Rev. Physiol.* **42**, 531–542. (B)

Beeuwkes, R. (1980b). Vascular-tubular relationships in the human kidney. In "Renal Pathophysiology—Recent Advances" (A. Leaf, G. Giebisch, L. Bolis, and S. Gorini, eds.), pp. 155–163. Raven Press, New York. (B)

Beeuwkes, R., and Bonventre, J. V. (1975). Tubular organization and vascular tubular relations in the dog kidney. *Am. J. Physiol.* **229**, 695–713. (B)

Beeuwkes, R., and Rosen, S. (1982). The structure of the human kidney. In "Nephrology" (W. Flamenbaum and R. Hamburger, eds.), pp. 6–27. Lippincott, Philadelphia, Pennsylvania. (B)

Beeuwkes, R., Ichikawa, I., and Brenner, B. (1981). The renal circulations. In "The Kidney" (B. Brenner and F. Rector, eds.), Vol. 1, pp. 249–283. Saunders, Philadelphia, Pennsylvania. (B)

*In the reference list, the capital letter in parentheses at the end of each reference indicates the section in which it is cited.

Bergström, S., and Sjövall, J. (1957). The isolation of prostaglandin. *Acta Chem. Scand.* 11, 1086. (D)

Bergström, S., Ryhage, R., Samuelsson, B., and Sjövall, J. (1962). The structure of prostaglandin E, F_1 and F_2. *Acta Chem. Scand.* 16, 501–502. (D)

Bergström, S., Danielsson, H., Klenberg, D., and Samuelsson, B. (1964). The enzymatic conversion of essential fatty acids into prostaglandins. *J. Biol. Chem.* 239, PC4006–PC4008. (D)

Border, W. A., Ward, H. J., Kamil, E. S., and Cohen, A. H. (1982). Induction of membranous nephropathy in rabbits by administration of an exogenous cationic antigen. Demonstration of pathogenic role for electrical charge. *J. Clin. Invest.* 69, 451–461. (E)

Brenner, B., Ichikawa, I., and Deen, W. (1981). Glomerular filtration. *In* "The Kidney" (B. Brenner and F. Rector, eds.), Vol. 1, pp. 289–327. Saunders, Philadelphia, Pennsylvania. (B)

Brenner, B. M., Hostetter, T. H., and Humes, H. D. (1978). Glomerular permselectivity: Barrier function based on discrimination of molecular size and charge. *Am. J. Physiol.* 234, F455–F460. (E)

Bricker, N. S., Guild, W. R., Reardan, J. B., and Merrill, J. P. (1956). Studies on the functional capacity of a denervated hemotransplanted kidney in an identical twin with parallel observations in the donor. *J. Clin. Invest.* 35, 1364–1380. (C)

Brownlee, M., and Spiro, R. G. (1979). Glomerular basement membrane metabolism in the diabetic rat. *Diabetes* 28, 121–125. (E)

Carrière, S., and Friborg, J. (1969). Intrarenal blood flow and PAH extraction during angiotensin infusion. *Am. J. Physiol.* 217, 1708–1715. (D)

Couser, W. G., and Salant, D. J. (1980). *In situ* immune complex formation and glomerular injury. *Kidney Int.* 17, 1–13. (E)

Felix, W. (1912). The development of the urogenital organs. *In* "Manual of Human Embryology" (F. Keibel and F. P. Mall, eds.), Vol. II, pp. 752–979. Lippincott, Philadelphia, Pennsylvania. (A)

Fisher, E. R., Sharkey, D., Pardo, V., and Vuzeuski, V. (1968). Experimental renal vein constriction: Its relation to renal lesions observed in human renal vein thrombosis and the nephrotic syndrome. *Lab. Invest.* 18, 689–699. (E)

Foglia, V. G., Mancini, R. E., and Cardeza, A. F. (1950). Glomerular lesions in the diabetic rat. *Arch. Pathol. Lab. Med.* 50, 75–83. (E)

Folkow, B., and Rubinstein, E. H. (1966). Cardiovascular effects of acute and chronic stimulations of the hypothalamic defence area in the rat. *Acta Physiol. Scand.* 68, 48–57. (C)

Fourman, J., and Moffat, D. B. (1971). "The Blood Vessels of the Kidney." Blackwell, Oxford. (B)

Gelfand, M. C., Frank, M. M., and Green, I. (1975). A receptor for the third component of complement in the human renal glomerulus. *J. Exp. Med.* 142, 1029–1034. (E)

Gerber, J. G., Branch, R. A., Nies, A. S., Gerkens, J. F., Shand, D. G., Hollifield, J., and Oates, J. A. (1978). Prostaglandins and renin release. II. Assessment of renin secretion following infusion of PGI_2, E_2 and D_2 into the renal artery of anesthetized dogs. *Prostaglandins* 15, 81. (C)

Germuth, F. G., Jr. (1953). A comparative histologic and immunologic study in rabbits of induced hypersensitivity of the serum sickness type. *J. Exp. Med.* 97, 257–282. (E)

Gilmore, J. P. (1964). Influence of tissue pressure on renal blood flow autoregulation. *Am. J. Physiol.* 206, 707–713. (C)

Gilmore, J. P., Cornish, K. G., Rogers, S. D., and Joyner, W. L. (1980). Direct evidence for myogenic autoregulation of the renal microcirculation in the hamster. *Circ. Res.* 47, 226–230. (C)

Golbus, D. M., and Wilson, C. B. (1979). Experimental glomerulonephritis induced by *in situ* formation of immune complexes in glomerular capillary wall. *Kidney Int.* 16, 257–282. (E)

Goldberg, L. I. (1972). Cardiovascular and renal actions of dopamine: Potential clinical applications. *Pharmacol. Rev.* 24, 1–29. (D)

Goldberg, L. I. (1974). Dopamine—Clinical uses of an endogenous catecholamine. *N. Engl. J. Med.* **291**, 707–710. (D)

Goldberg, L. I. (1975). The dopamine vascular receptor. *Biochem. Pharmacol.* **24**, 651–653. (D)

Goldberg, L. I., Kohli, J. D., Kotake, A. N., and Volkman, P. H. (1978). Characteristics of the vascular dopamine receptor: Comparison with other receptors. *Fed. Proc., Fed. Am. Soc. Exp. Biol.* **37**, 2396–2402. (D)

Goldblatt, H. (1947). The renal origin of hypertension. *Physiol. Rev.* **27**, 120–165. (C)

Gorgas, K. (1978). Structure and innervation of the juxtaglomerular apparatus of the rat. *Adv. Anat.* **54**, 1–84. (B)

Graves, F. T. (1971). "The Arterial Anatomy of the Kidney." Williams & Wilkins, Baltimore, Maryland. (B)

Gruenwald, P. (1943). The normal changes in the position of the embryonic kidney. *Anat. Rec.* **85**, 163. (A)

Hall, J. E., Coleman, T. G., Guyton, A. C., Balfe, J. W., and Salgado, H. C. (1979). Intrarenal role of angiotensin II and [des-Asp[1]] angiotensin II. *Am. J. Physiol.* **236**, F252–F259. (C)

Ham, A. W., and Cormack, D. H. (1979). "Histology," 8th ed., pp. 774–775. Lippincott, Philadelphia, Pennsylvania. (A)

Hartmann, H., Orskov, S. L., and Rein, H. (1936). Die Gefässreaktionen der Niere im Verlaufe allgemeiner Kreislauf-Regulationsvorgänge. *Pfluegers Arch. Gesamte Physiol. Menschen Tiere* **238**, 239–250. (C)

Heney, N. M., O'Morchoe, P. J., and O'Morchoe, C. C. C. (1971). The renal lymphatic system during obstructed urine flow. *J. Urol.* **106**, 455–462. (F)

Heptinstall, R. H. (1974). Diabetes mellitus and gout. *In* "The Pathology of the Kidney," 2nd ed., pp. 929–962. Little, Brown, Boston, Massachusetts. (E)

Hinshaw, L. B., Day, S. B., and Carlson, C. H. (1959). Tissue pressure as a causal factor in the autoregulation of blood flow in the isolated perfused kidney. *Am. J. Physiol.* **197**, 309–312. (C)

Hirose, K., Tsuchida, H., Østerby, R., and Gundersen, H. J. G. (1980). A strong correlation between glomerular filtration rate and filtration surface in diabetic kidney hyperfunction. *Lab. Invest.* **43**, 434–437. (E)

Hoff, E. C., Kell, J. F., Jr., and Carroll, M. N., Jr. (1963). Effects of cortical stimulation and lesions on cardiovascular function. *Physiol. Rev.* **43**, 68–114. (C)

Holmes, M. J., O'Morchoe, P. J., and O'Morchoe, C. C. C. (1977a). Morphology of the intrarenal lymphatic system. Capsular and hilar communications. *Am. J. Anat.* **149**, 333–352. (F)

Holmes, M. J., O'Morchoe, P. J., and O'Morchoe, C. C. C. (1977b). The role of renal lymph in hydronephrosis. *Invest. Urol.* **15**, 215–219. (F)

Holtz, P., and Credner, K. (1942). Die enzymatische Entstehung von Oxytyramin in Organismus und die physiologische Bedeutung der Dopadecarboxylase. *Naunyn-Schmiedeberg's Arch. Exp. Pathol. Pharmakol.* **200**, 356–388. (D)

Hostetter, T. H., Olson, J. L., Rennke, H. G., Venkatachalam, M. A., and Brenner, B. M. (1981). Hyperfiltration in remnant nephrons: A potentially adverse response to renal ablation. *Am. J. Physiol.* **241**, F85–F93. (E)

Itskovitz, H. D., Terragno, N. A., and McGiff, J. C. (1974). Effect of a renal prostaglandin on distribution of blood flow in the isolated canine kidney. *Fed. Proc., Fed. Am. Soc. Exp. Biol.* **34**, 770–776. (D)

Izui, S., Lambert, P.-H., and Miescher, P. A. (1976). *In vitro* demonstration of a particular affinity of glomerular basement membrane and collagen for DNA: A possible basis for local formation of DNA–anti-DNA complexes in systemic lupus erythematosus. *J. Exp. Med.* **144**, 428–443. (E)

Jackson, B. T., and Thomas, M. L. (1970). Postthrombotic inferior vena caval obstruction. A review of 24 patients. *Br. Med. J.* **1**, 18–22. (E)

Jamison, R. L. (1981). Urine concentration and dilution. *In* "The Kidney" (B. Brenner and F. Rector, eds.), Vol. 1, pp. 495–550. Saunders, Philadelphia, Pennsylvania. (B)

Jones, D. B. (1969). Mucosubstances of the glomerulus. *Lab. Invest.* **21**, 119–125. (E)

Kauffmann, R. H., Veltkamp, J. J., Van Tilburgh, N. H., and Van Es, L. A. (1978). Acquired antithrombin III deficiency and thrombosis in the nephrotic syndrome. *Am. J. Med.* **65**, 607–613. (E)

Kefalides, N. A. (1974). Biochemical properties of human glomerular basement membrane in normal and diabetic kidneys. *J. Clin. Invest.* **53**, 403–407. (E)

Khayyal, M., Gross, F., and Kreye, V. A. W. (1981). Studies on the direct vasodilator effect of hydralazine in the isolated rabbit renal artery. *J. Pharmacol. Exp. Ther.* **216**, 390–394. (D)

Knox, F. G., Ott, C., Cuche, J. L., Gasser, J., and Haas, J. (1974). Autoregulation of single nephron filtration rate in the presence and absence of flow to the macula densa. *Circ. Res.* **34**, 836–842. (C)

Kriz, W., and Dieterich, H. J. (1970). Das lymphgefasssystem der niere bei einigen saeugetieren. Licht-und elektronenmikroskopische untersuchungen. *Z. Anat. Entwicklungs Gesch.* **131**, 111–147. (F)

Kriz, W., Barrett, J. M., and Peter, S. (1976). The renal vasculature: Anatomical–functional aspects. *Int. Rev. Physiol. Ser.* **12**, 1–22. (B)

Kurzrok, R., and Lieb, C. C. (1930). Biochemical studies of human semen. II. The action of semen on the human uterus. *Proc. Soc. Exp. Biol. Med.* **28**, 268–272. (D)

Kussman, M. J., Goldstein, H. H., and Gleason, R. E. (1976). The clinical course of diabetic nephropathy. *JAMA, J. Am. Med. Assoc.* **236**, 1861–1863. (E)

Landis, E. M., and Pappenheimer, J. R. (1963). Exchange of substances through capillary walls. *In* "Handbook of Physiology" (W. F. Hamilton, ed.), Sect. 2, Vol. II, pp. 961–1034. Am. Physiol. Soc., Washington, D.C. (E)

Larsson, C., Weber, P., and Änggard, E. (1974). Arachidonic acid increases and indomethacin decreases plasma renin activity in the rabbit. *Eur. J. Pharmacol.* **28**, 391–394. (D)

Lee, J. B., Covino, B. G., Takman, B. H., and Smith, E. R. (1965). Renomedullary vasodepressor substance, medullin: Isolation, chemical characterization and physiological properties. *Circ. Res.* **17**, 57–77. (D)

Levenson, D. J., Simmons, C. E., Jr., and Brenner, B. M. (1982). Arachidonic acid metabolism, prostaglandins and the kidney. *Am. J. Med.* **72**, 354–374. (D)

Licata, R. (1962). Renal circulation. *In* "Blood Vessels and Lymphatics" (D. I. Abramson, ed.), pp. 388–390. Academic Press, New York. (A)

Llach, F., Papper, S. G., and Massry, S. G. (1980). The clinical spectrum of renal vein thrombosis: Acute and chronic. *Am. J. Med.* **69**, 819–827. (E)

Lonigro, A. J., and Dawson, C. A. (1975). Vascular responses to prostaglandin $F_{2\alpha}$ in isolated cat lungs. *Circ. Res.* **36**, 706–712. (D)

Lonigro, A. J., Itskovitz, H. D., Crowshaw, K., and McGiff, J. C. (1973). Dependency of renal blood flow on synthesis of prostaglandin E_2. *Circ. Res.* **32**, 712–717. (D)

Lonigro, A. J., Hagemann, M. H., Stephenson, A. H., and Fry, C. L. (1978). Inhibition of prostaglandin synthesis by indomethacin augments the renal vasodilator response to bradykinin in the anesthetized dog. *Circ. Res.* **43**, 447–455. (D)

Lonigro, A. J., Brash, D. W., Stephenson, A. H., Heitmann, L. J., and Sprague, R. S. (1982). Effect of ventilatory rate on renal venous PGE_2 and $PGF_{2\alpha}$ efflux in anesthetized dogs. *Am. J. Physiol.* **242**, F38–F45. (D)

McCarthy, L. J., Titus, J. L., and Daugherty, G. W. (1963). Bilateral renal vein thrombosis and the nephrotic syndrome in adults. *Ann. Intern. Med.* **58**, 837–857. (E)

McClusky, R. T., Benacerraf, B., and Miller, F. (1962). Passive acute glomerulonephritis induced by antigen–antibody complexes solubilized in hapten excess. *Proc. Soc. Exp. Biol. Med.* **111**, 764–768. (E)

McGiff, J. C. (1981). Prostaglandins, prostacyclin, and thromboxanes. *Annu. Rev. Pharmacol. Toxicol.* **21**, 479–509. (D)

McGiff, J. C., Crowshaw, K., Terragno, N. A., Lonigro, A. J., Strand, J. C., Williamson, M. A.,

Lee, J. B., and Ng, K. K. F. (1970a). Prostaglandin-like substances appearing in canine renal venous blood during renal ischemia: Their partial characterization by pharmacologic and chromatographic procedures. *Circ. Res.* **27**, 765–782. (D)

McGiff, J. C., Crowshaw, K., Terragno, N. A., and Lonigro, A. J. (1970b). Release of prostaglandin-like substances into renal venous blood in response to angiotensin II. *Circ. Res.* **26, 27**, Suppl. I, I-121–I-130. (D)

McGiff, J. C., Crowshaw, K., Terragno, N. A., Malik, K. U., and Lonigro, A. J. (1972). Differential effect of noradrenaline and renal nerve stimulation on vascular resistance in the dog kidney and the release of a prostaglandin E-like substance. *Clin. Sci.* **42**, 223–233. (D)

Maddox, D. A., Troy, J. L., and Brenner, B. M. (1974). Autoregulation of filtration rate in the absence of macula densa–glomerulus feedback. *Am. J. Physiol.* **227**, 123–131. (C)

Mannich, C., and Jacobsohn, W. (1910). Über Oxyphenylalkylamine und Dioxyphenylalkylamine. *Ber. Dsch. Chem. Ges.* **43**, 189–197. (D)

Mauer, S. M., Sutherland, D. E. R., Howard, R. J., Fish, A. J., Najarian, J. S., and Michael, A. F. (1973). The glomerular mesangium. III. Acute mesangial injury: A new model of glomerulonephritis. *J. Exp. Med.* **137**, 553–570. (E)

Mauer, S. M., Barbosa, J., Vernier, R. L., Kjellstrand, C. M., Buselmeier, T. J., Simmons, R. L., Najarian, J. S., and Goetz, F. C. (1976a). Development of diabetic vascular lesions in normal kidneys transplanted into patients with diabetes mellitus. *N. Engl. J. Med.* **295**, 916–920. (E)

Mauer, S. M., Steffes, M. W., Michael, A. F., and Brown, D. M. (1976b). Studies of diabetic nephropathy in animals and man. *Diabetes* **25**, 850–857. (E)

Mauer, S. M., Steffes, M. W., and Brown, D. M. (1981). The kidney in diabetes. *Am. J. Med.* **70**, 603–612. (E)

Michael, A. F., Blau, E., and Vernier, R. L. (1970). Glomerular polyanion. Alteration in amino nucleoside nephrosis. *Lab. Invest.* **23**, 649–657. (E)

Miller, K., and Michael, A. F. (1976). Immunopathology of renal extracellular membranes in diabetes mellitus. Specificity of tubular basement-membrane immunofluorescence. *Diabetes* **25**, 701–708. (E)

Mohos, S. C., and Skoza, L. (1969). Glomerular sialoprotein. *Science* **164**, 1519–1521. (E)

O'Morchoe, C. C. C., and Albertine, K. H. (1980). The renal cortical lymphatic system in dogs with unimpeded lymph and urine flow. *Anat. Rec.* **198**, 427–438. (F)

O'Morchoe, C. C. C., O'Morchoe, P. J., and Heney, N. M. (1970). Renal hilar lymph. Effects of diuresis on flow and composition in dogs. *Circ. Res.* **26**, 469–479. (F)

O'Morchoe, C. C. C., O'Morchoe, P. J., and Donati, E. J. (1975). Comparison of hilar and capsular renal lymph. *Am. J. Physiol.* **229**, 416–421. (F)

O'Morchoe, C. C. C., Albertine, K. H., and O'Morchoe, P. J. (1982). The rate of translymphatic endothelial fluid movement in the canine kidney. *Microvasc. Res.* **23**, 180–187. (F)

Østerby, R. (1975). Early phases in the development of diabetic glomerulopathy. *Acta Med. Scand.* **574**, Suppl., 1–80. (E)

Pappenheimer, J. R., and Kinter, W. B. (1956). Hematocrit ratio of blood within mammalian kidney and its significance for renal hemodynamics. *Am. J. Physiol.* **185**, 377–390. (C)

Pfaller, V. W., and Rittinger, M. (1977). Quantitative morphologie der Niere. *Mikroskopie* **33**, 74–79. (B)

Pierce, E. C. (1944). Renal lymphatics. *Anat. Rec.* **90**, 315–335. (F)

Potkay, S., and Gilmore, J. P. (1973). Autoregulation of glomerular filtration in renin-depleted dogs. *Proc. Soc. Exp. Biol. Med.* **143**, 508–513. (C)

Rawson, A. J. (1949). Distribution of the lymphatics of the human kidney as shown in a case of carcinomatous permeation. *Arch. Pathol.* **47**, 283–292. (F)

Renkin, E. M., and Gilmore, J. P. (1973). Glomerular filtration. *In* "Handbook of Physiology" (J. Orloff and P. W. Berliner, eds.), Sect. 8, pp. 185–248. Williams & Wilkins Co., Baltimore, Maryland. (C)

Rennke, H. G., Cotran, R. S., and Venkatachalam, M. A. (1975). Role of molecular charge in glomerular permeability. Trace studies with cationized ferritins. *J. Cell Biol.* **67**, 638–646. (E)

Samuelsson, B., Borgeat, P., Hammarström, S., and Murphy, R. C. (1980). Leukotrienes: A new group of biologically active compounds. *Adv. Prostaglandin Thromboxane Res.* **6**, 1–18. (D)

Schreiner, G. F., Kiely, J.-M., Cotran, R. S., and Uranue, E. R. (1981). Characterization of resident glomerular cells in the rat expression Ia determinants and manifesting genetically restricted interaction with lymphocytes. *J. Clin. Invest.* **68**, 920–931. (E)

Smith, H. W. (1951). "The Kidney. Structure and Function in Health and Disease." Oxford Univ. Press, London and New York. (C)

Spiro, R. G. (1976). Investigations into the biochemical basis of diabetic basement membrane alterations. *Diabetes* **25**, 909–913. (E)

Spiro, R. G., and Spiro, M. J. (1971). Effect of diabetes on the biosynthesis of the renal glomerular basement membrane. Studies on the glucosyltransferase. *Diabetes* **20**, 641–648. (E)

Sprague, R. S., Stephenson, A. H., Heitmann, L. J., and Lonigro, A. J. (1982). Effects of prostaglandins on the pulmonary circulation in the presence of unilateral alveolar hypoxia. *Fed. Proc., Fed. Am. Soc. Exp. Biol.* **41**, 1685 (abstr). (D)

Stein, J. H., Mauk, R. C., Boonjarern, S., and Ferris, T. F. (1972). Differences in the effect of furosemide and chlorothiazide on the distribution of renal cortical blood flow in the dog. *J. Lab. Clin. Med.* **79**, 995–1003. (D)

Swain, J. A., Heyndrickx, G. R., Boettcher, D. H., and Vatner, S. F. (1975). Prostaglandin control of renal circulation in the unanesthetized dog and baboon. *Am. J. Physiol.* **229**, 826–830. (C)

Takeuchi, J., Yagi, S., Takeda, T., Uchida, E., Inoue, G., and Ueda, H. (1962). Experimental studies on the nervous control of the renal circulation—Changes in renal circulation induced by electrical stimulation of the diencephalon and effect of several drugs on the circulatory responses. *Jpn. Heart J.* **3**, 57–72. (C)

Thomson, C., Forbes, C. D., Prentice, C. R. M., and Kennedy, A. C. (1974). Changes in blood coagulation and fibrinolysis in the nephrotic syndrome. *Q. J. Med.* **43**, 399–407. (E)

Thurau, K., and Kramer, K. (1959). Weitere Untershuchungen zur myogenen natur der autoregulation des Nierenkreislaufes. *Pfluegers Arch. Gesamte Physiol. Menschen Tiere* **269**, 77–93. (C)

Thurau, K., and Schnermann, J. (1965). The sodium concentration in the macula densa cells as a regulating factor for the glomerular filtrate. *Klin. Wochenschr.* **43**, 410–413. (C)

van Dorp, D. A., Beerthuis, R. K., Nugteren, D. H., and Vonkeman, H. (1964). The biosynthesis of prostaglandins. *Biochim. Biophys. Acta* **90**, 204–207. (D)

Venuto, R. C., O'Dorisio, T., Ferris, T. F., and Stein, J. H. (1975). Prostaglandins and renal function. II. The effect of prostaglandin inhibition on autoregulation of blood flow in the intact kidney of the dog. *Prostaglandins* **9**, 817–828. (C)

von Euler, U. S. (1935). Über die spezifische blutdrucksenkende Substanz des menschlichen Prostata-und Samenblasensekretes. *Klin. Wochenschr.* **14**, 1182–1183. (D)

Waugh, W. H., and Shanks, R. G. (1960). Cause of genuine autoregulation of the renal circulation. *Circ. Res.* **8**, 871–888. (C)

Webster, M. E., and Gilmore, J. P. (1964). Influences of kallidin-10 on renal function. *Am. J. Physiol.* **206**, 714–718. (C)

Weil, R., III, Nozawa, M., Koss, M., Weber, C., Reemtsma, K., and McIntosh, R. M. (1976). The kidney in streptozotocin diabetic rats. Morphological ultrastructure and function studies. *Arch. Pathol. Lab. Med.* **100**, 37–49. (E)

Williamson, H. E., Bourland, W. A., Marchand, G. R., Farley, D. B., and Van Orden, D. E. (1975). Furosemide induced release of prostaglandin E to increase renal blood flow. *Proc. Soc. Exp. Biol. Med.* **150**, 104–106. (D)

Wright, F., and Briggs, J. (1979). Feedback control of glomerular blood flow, pressure and filtration rate. *Physiol. Rev.* **59**, 958–1006. (B)

Chapter 16
Genitourinary System: Male Reproductive Organs and Lower Urinary Tract

A. BLOOD VESSELS AND LYMPHATICS OF THE TESTIS

By A. R. Pavlakis and R. J. Krane

1. Anatomy of Blood Circulation

The human testis derives its blood supply from the testicular (internal spermatic), deferential, and cremasteric (external spermatic) arteries.

a. Testicular artery: This vessel arises from the abdominal aorta in the lumboiliac region as embryonic development of the testis takes place at the dorsal abdominal wall in the gonadal ridge. At this stage, the components of the vascular pedicle of the testis enter the organ from the posterior aspect, an orientation which is preserved throughout its descent into the scrotum. The manner in which the testicular artery descends caudally from its aortic origin reflects the positional change of the testis subsequent to the time it acquires its blood supply (Patten, 1968). Duplication of the testicular artery is rare, occurring in approximately 1% and always on the left side (Anson and Kurth, 1955). The vessel enters the inguinal canal at the internal ring and forms one constituent of the spermatic cord, being bound to the other components by loose connective tissue. In eutherian mammals with scrotal testes, the testicular artery develops convolutions as it courses through the inguinal canal, which, in association with the venous pampiniform plexus, form a "vascular cone" (Setchell, 1970). However, in man, the vessel passes down the testis in a relatively straight fashion and is, in this respect, almost unique among mammals (Harrison, 1949a). The testicular artery in the vicinity of the testis varies in diameter from 0.5 to 1 mm and either remains undivided or gives off up to four branches at a variable distance from the organ. Harrison and Barclay (1948) reported that the most common pattern is division into a medial and a lateral branch, the former being the larger. However, in a study of 78 autopsy testes, Kormano and Suoranta (1971a) observed that the artery remained undivided in 56% of cases, whereas two branches entered the testis in 31%, and three or more branches in 13%.

A distinct arterial trunk to the epididymis originates from the main stem of the testicular artery or its divisions close to the testis (McMillan, 1954); however, Harrison and McGregor (1957) described one case of abnormally high branching of both testicular arteries in the abdomen, forming well-defined epididymal arteries. In their classical study, Harrison and Barclay (1948) noted the arterial branches to the epididymis to be very convoluted and observed a relatively poor arterial supply to the cauda, as compared with the caput epididymis.

The single or divided testicular artery reaches the posterior border of the testis medial to the epididymis (Harrison, 1949b), a relationship that is of surgical significance at the time of epididymectomy. It then pierces the tunica albuginea and divides into several branches (centripetal arteries) that converge toward the rete testis (Hundeiker and Keller, 1963). Many of the major branches of the centripetal arteries bend back upon themselves before reaching the mediastinum, thus running in an apposite direction, and it is for this reason that they are called the centrifugal vessels. Both centripetal and centrifugal arteries, located in the connective tissue septa and the parenchyma, further divide and terminate in the intertubular arterioles that run between the seminiferous tubules (Kormano and Suoranta, 1971b).

   b. Deferential and cremasteric arteries: The testis and epididymis also receive blood from the deferential artery, which derives from the umbilical artery and accompanies the vas through the inguinal canal. This vessel invariably communicates with the testicular artery, either directly or most commonly via the epididymal branch of the latter, at the lower testicular pole (Harrison, 1949b). Finally, the cremasteric artery, a branch of the inferior epigastric, also contributes to the testicular blood supply by anastomosing with the other two vessels. However, this pattern is not always present, as reported by Harrison (1949b), who was only able to fill the cremasteric artery by injecting the deferential in 17 of 24 cases. The contribution of collateral circulation via the cremasteric, external pudendal, and posterior scrotal arteries was demonstrated by Neuhof and Mencher (1940). These workers followed 24 patients with complete severance of the spermatic cord and observed that gangrene of the testis never occurred, although some degree of testicular atrophy developed in 80% of the cases.

   c. Arteriolar and capillary beds: Study of the microvascular organization of the human testis was made possible by the development of delicate microangiographic and histochemical techniques (Kormano, 1969). The branches of the terminal intertubular arterioles form a capillary network within triangular columns of interstitial tissue that connects with vessels of adjacent columns by branches forming a rope-ladder-like pattern (Kormano and Suoranta, 1971b). This morphologic organization allows for close contact of the Leydig cells with the capillaries. Kormano and Suoranta (1971b) noted the seminiferous tubules to be devoid of blood vessels and also observed a sparse vascular supply to the rete testis, suggestive of low metabolic activity of the latter structure.

d. Venular bed: The postcapillary venules converge to form collecting venules and deep veins that, similar to the arteries, assume a centrifugal and centripetal configuration, draining the peripheral part and the parenchyma of the testis, respectively (Kormano and Suoranta, 1971b). The two groups of veins, also connected by large short-circuiting anastomotic channels, eventually coalesce around the rete testis to form a complex network of vessels; ultimately they emerge from the organ as the pampiniform plexus (Harrison and Barclay, 1948). Although the presence of arteriovenous anastomoses has been suspected, based on the recovery from the venous circulation of microspheres injected into the testicular artery (Setchell, 1970), Kormano and Suoranta (1971b) failed to confirm the existence of such communications by microangiography.

e. Venous bed: The pampiniform plexus near the testis is subdivided into two groups, on the basis of their relation to the vas: a larger anterior group surrounding the testicular artery and a smaller posterior group. The numerous convoluted veins of the pampiniform plexus coalesce until they become one large vessel in the upper part of the inguinal canal. This is the internal spermatic vein that, together with the deferential and external spermatic veins, constitutes the primary or deep venous system of the testis. The deferential and external spermatic veins, which also drain into the pampiniform plexus, anastomose with the internal spermatic vein at the level of the external inguinal ring (Olson and Stone, 1949). The secondary or superficial venous system consists of the superficial and deep inferior epigastric veins, the superficial internal circumflex vein, and the scrotal tributaries of the superficial and deep external and internal pudendal veins (Lich *et al.*, 1978). The superficial and deep venous systems communicate freely with each other, as well as with the plexuses of the contralateral side. The adequacy of the secondary system in draining the intrascrotal contents and preserving normal testicular function has been clearly demonstrated following ligation of the internal spermatic vein for the treatment of varicocele. Finally, the right internal spermatic vein enters the inferior vena cava obliquely, whereas the left drains into the left renal vein at a perpendicular angle. The veins of the body and cauda epididymis drain into a prominent venous channel coursing along the posterior border of the organ, the vena marginalis epididymis, which eventually flows into the pampiniform plexus. The veins of the caput epididymis may drain either directly into the pampiniform plexus or into the vena marginalis (McMillan, 1954).

2. Physiology of Blood Circulation

Although the anatomic vascular supply to the testis has been investigated in detail, interest in the assessment of testicular blood flow in the human has developed only recently, following similar measurements on animals. Calculation of testicular flow by venous occlusion plethysmography on six patients with prostatic carcinoma about to undergo orchiectomy revealed a value of 9 ml/100 gm/min (Petterson *et al.*, 1973), which is in agreement with flow rates observed in testes of other mammalian species (Setchell *et al.*, 1966). In a study using radioactive xenon-133, Fritjofsson *et al.* (1969) noted the testicular flow rate to range from 3.2 to 38.5 ml/100 gm/min on the right side and from 1.6 to 12.4 ml/100 gm/min on the left side and speculated that the decreased flow to the left testis was a reflection of poorer venous drainage secondary to the anatomic characteristics of the internal spermatic vein.

The intimate association of the testicular artery with a plexus of veins surrounding it appears to provide a means for cooling arterial blood before it is delivered to the testes. The possibility of such a vascular mechanism for maintaining testicular temperature by counter exchange has been investigated by Dahl and Herrick (1959), who observed a temperature gradient within the testicular artery of the dog, with an average decrease of 3°C during periods of blood flow through this vessel. The presence of such a mechanism, as well as the ramification of the branches of the artery on the testicular surface before penetrating the parenchyma, and the thermoregulatory function of the scrotum result in a difference of 2 to 4°C between testicular and rectal temperature in the human (Agger, 1971).

3. LYMPHATIC SYSTEM

The lymphatic capillaries form an extensive network within the intertubular space, in close association with the seminiferous tubules, but apparently not penetrating them. According to Ewing (1978), obstruction of the spermatic cord lymphatic ducts results in interstitial dilatation, but with no such effect on the tubules. This preferential fluid accumulation suggests that the extracellular space of the interstitium is drained via lymphatics, whereas the seminiferous tubules are not. There still exists some controversy over whether actual lymphatic capillaries or just "lymphatic spaces" are present in the interstitial tissue (Setchell, 1970). At any rate, testicular lymph drains via lymphatic vessels in the fibrous septa into a well-developed plexus of collecting ducts in the tunica albuginea and then into some draining trunks in the mediastinum testis (Yoffey and Courtice, 1970). (For a detailed description of lymphatic sinusoids and spaces in the testis, see Section C-2b, Chapter 4.)

Four to eight collecting lymphatic channels emerge ultimately from the mediastinum testis and ascend along the spermatic cord superficial to the pampiniform plexus, anastomosing freely with each other at both their proximal and distal ends (Sayegh *et al.*, 1966). The lymphatic drainage of the testis to the area of its embryologic origin has been demonstrated by Jamieson and Dobson (1910) by injection of Prussian blue into the testes of full-term fetuses. Their original anatomic observations have been subsequently confirmed by means of pedal and especially direct testicular funicular lymphangiography (Busch and Sayegh, 1963; Chiappa *et al.*, 1966). The lymphatic trunks follow the spermatic vessels cephalad, and at the point where the vessels cross the ureter they fan out medially to the lumbar glands, which are the only barrier between the testes and the thoracic duct. The lymphatics from the right testis drain into nodes lateral, medial, and anterior to the vena cava and anterior to the aorta, whereas those from the left testis terminate in nodes located lateral and anterior to the aorta. In addition to direct drainage into primary glands, there are also three groups of secondary nodes that receive efferent vessels from the primary glands (Jamieson and Dobson, 1910). These consist of the primary glands of the same and opposite side; glands behind and between the two great trunks, below and above the level of the renal veins; and a chain along the outer side of each common iliac artery. By means of lymphangiography with direct cannulation of the tunicular lymphatics, Busch and Sayegh (1963) demonstrated that the nodes draining the testes extend from the level of the eleventh thoracic to the fourth lumbar vertebra and noted the greatest concentration of these structures around the renal pedicles.

Based on reports from anatomic and radiologic literature and their own experience from 283 patients with testicular germinal tumors subjected to retroperitoneal node dissection, Ray *et al.* (1974) summarized the primary lymphatic drainage as follows: Lymphatics from the right testis drain into the interaortocaval, precaval, preaortic, paracaval, right common iliac, and right external iliac nodes, in that order, with subsequent drainage to the paraaortic, left common iliac, and left external iliac nodes. The primary lymphatic drainage of the left testis is to the paraaortic, preaortic, left common iliac, and left external iliac nodes, in that order, with subsequent drainage to the interaortocaval, precaval, paracaval, right common iliac, and right external iliac nodes. The predictability of the lymphatic drainage of the testes has had a great influence on the surgical therapy for testicular cancer. Unlike other tumor models properly performed in cases of testicular tumor, retroperitoneal lymphadenectomies, in continuity with the internal spermatic vessels, can result in potential cure even in the face of nodal metastases.

B. BLOOD VESSELS AND LYMPHATICS OF THE
PROSTATE GLAND

By J. S. WHEELER, JR., AND R. J. KRANE

The prostate gland functions by the production and storage of secretions that, along with those of the seminal vesicles, constitute over 90% of the volume of the seminal fluid. By virtue of the fact that both the urethra and the paired ejaculatory ducts course through the substance of the prostate, the anatomic integrity of the latter is essential for the maintenance of normal micturition as well as ejaculation and fertility.

1. ANATOMY OF BLOOD CIRCULATION

a. Main arterial tree: The major blood supply to the prostate derives from the inferior vesical artery, a branch of the internal iliac artery. This vessel courses medially over the fascia of the levator ani muscles to the base of the bladder where multiple branches supply the bladder, seminal vesicles, distal ureter, and prostate. The middle hemorrhoidal and internal pudendal arteries send off small branches to the apex of the prostate; a vessel to the distal vas deferens also supplies the veru montanum. No single artery to the prostate has been demonstrated. However, the term, prostatic artery, is commonly used to designate the four or five branches arising from various points along the inferior vesical artery that traverse in parallel to the prostate (Flocks, 1937; Clegg, 1955). The "prostatic" vessel divides into two divisions upon reaching the prostate gland, a deep penetrating urethral group and a more superficial capsular group, as described by Flocks (1937) (Fig. 16.1). This worker studied the intraprostatic arterial anatomy and described urethral and capsular groupings of vessels that were consistent with the clinical observations of bleeding patterns noted during transurethral resections. The urethral group courses directly to the bladder neck, supplying the median lobe, bladder neck, and prostatic urethra. The vessels then continue distally from the bladder neck beneath the urethral surface and along the lateral lobes, terminating at the level of the veru montanum. The capsular group runs posterolaterally and branches both ventrally and dorsally to supply the prostatic capsule. The arterial distribution in

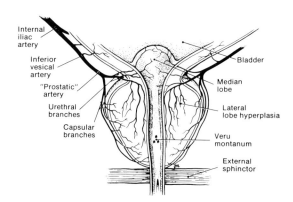

FIG. 16.1. Arterial blood supply to the prostate.

the prostate can be divided into three zones: a deep urethral zone, superficial capsular zone, and an intermediate zone supplied by vessels freely anastomosing between vessels supplying both major groups. Clegg (1956) confirmed Flocks' anatomic arrangement and further demonstrated that some urethral branches enter the bladder neck anteriorly and posteriorly as well as laterally.

b. Alterations in vascular architecture: In the infant prostate with immature periurethral glands, the urethral group of vessels is not prominent. With growth and development, the urethral arteries increase in importance as the normal enlargement of the periurethral glands proceeds. Median and lateral lobe hyperplasia results in further prominence of these vessels. Urethral branches supplying a hypertrophic median lobe enter the bladder neck at the cleft between the median and lateral lobes. If the bladder neck is viewed cystoscopically as the face of a clock, it is noted that branches supplying hypertrophic lateral lobes enter this region between seven and eleven o'clock on the right side and between one and five o'clock on the left side. In the non-hypertrophic postpubertal prostate, the capsular group of arteries is usually the predominant blood supply (Flocks, 1937). This rather constant vascular anatomic relationship is important to the urologist performing transurethral resection. After bladder neck resection and coagulation, removal of the remaining distal lateral lobe tissue is relatively bloodless because proximal control of the predominant urethral branches already has been accomplished.

c. Venous bed: The major venous drainage of the prostate follows a regional pattern. Laterally, intrinsic prostatic veins course to the capsular veins that drain into the lateral prostatic plexus. Anteroinferiorly, the drainage is to the anterior prostatic plexus, whereas the vessels in the region in proximity to the ejaculatory ducts drain to the veins of the vas deferens (Clegg, 1956). In the puboprostatic space, prostatic veins join with the dorsal vein of the penis to form Santorini's plexus. These veins become large and tortuous with age and frequently give rise to phleboliths that may be noted on pelvic radiographs. Santorini's plexus empties into both the pudendal and inferior vesical veins, depending on valvular arrangement and blood flow, and eventually into the internal iliac vein (Lowsley, 1915). Venous drainage may also occur through the hemorrhoidal veins, a finding that accounts for the rare metastatic deposit of prostatic cancer in the liver (Lich et al., 1978). In addition to the primary venous drainage patterns described above, there is communication between the tortuous prostatic plexus and the sacral vein plexus of Batson. Batson's plexus invests the sacrum, lumbar spine, and the iliac wings and may account for the diffuse bony metastases often seen early in the course of prostatic cancer (Batson, 1940).

2. LYMPHATIC SYSTEM

a. Lymphatic channels: The subject of lymphatic drainage of the prostate gland has been associated with much controversy. Originally it was theorized that fine lymphatic capillaries arise in a network around each prostatic acinus and join larger vessels radiating to peripheral connective tissue, to form periprostatic lymphatic plexuses and eventually to drain into major nodal chains. Attempts to confirm this concept have encountered difficulty, especially in demonstrating the intrinsic prostatic lymphatics. Smith (1966) injected vital dyes into the prostate but found only capsular lymphatics that immediately drained into regional lymph nodes. He concluded that there are no intraprostatic lymphatics and that prostatic cancer spreads to the lymph nodes only as a late phenomenon, when the disease has already penetrated the prostatic capsule.

Additional experimental data against intraprostatic lymphatics have been provided by electron microscopic studies of the perineural space, a region that was thought to contain lymphatic channels involved in the dissemination of prostatic cancer (Rodin *et al.*, 1967). However, this approach failed to reveal any endothelial cells, from which it was deduced that there were no lymphatics present. Injection of radiopaque contrast material by McCullough (1975) also failed to demonstrate intraprostatic lymphatics; instead, only fine capsular lymphatics were observed, thus supporting the earlier findings of Smith. This apparent lack of demonstrable intraprostatic lymphatics, along with the high incidence of occult slow-growing prostatic cancer, has raised the question as to whether the prostate is an immunologically privileged site, where antigenic tumors are not subjected to normal immune surveillance (Gittes and Mc-Cullough, 1974).

The above speculation has been refuted in rat experiments in which prostate glands failed to demonstrate prolonged allograft survival, a response presumably due to normal immune surveillance, as proposed by Neaves and Billingham (1979). Furthermore, these investigators contradicted prior electron microscopic studies by demonstrating the presence of intraprostatic lymphatics in a murine prostate. Other evidence in support of the existence of intraprostatic lymphatics was presented by Connolly *et al.* (1968) who injected India ink intraprostatically and claimed to have found lymph vessels within the gland. Also, prostatic injection of radioactive colloid has been reported to be followed by direct uptake of the agent by regional lymph nodes (Menan *et al.*, 1977). When technetium-radiolabeled antimony trisulfide was similarly introduced into the prostate, lymph node drainage was again apparent (Whitmore *et al.*, 1980). The use of Ethiodol intraprostatic injection has also been advocated to delineate intraprostatic anatomy and its lymphatic drainage (Raghavaiah, 1979; Raghavaiah and Jordon, 1979). Refinements in intraprostatic injection techniques may further clarify the intra- and periprostatic lymphatic patterns.

b. Lymph nodes: The pelvic lymph nodes provide primary lymphatic drainage for the prostate. Included in this group are the external iliac, obturator, hypogastric, and common iliac lymph nodes. The external iliac nodes are subdivided into three chains: the external chain, which lies lateral to the external iliac artery; the middle chain, which is anterior to the iliac vein; and the internal chain, which is found below the external iliac vein along the pelvic side wall (Herman *et al.*, 1963). The obturator nodes are located in the obturator fossa, lying ventral to the obturator nerve. Herman *et al.* have suggested that these nodes are actually part of the internal chain of external iliac nodes, although this opinion is not universally accepted. The hypogastric nodes are the most posterior chain and are occasionally referred to as lateral sacral or presciatic nodes since they lie near the second and third lateral sacral foramena. The common iliac nodes are found along the common iliac artery and the sacral promontory, where they are sometimes referred to as presacral nodes.

c. Summary: Intraprostatic lymphatics drain directly to the periprostatic lymphatics. Major lymphatic channels then course from the periprostatic lymphatic plexuses, following closely the prostatic vascular supply to the major pelvic lymph node chains. Although these lymphatic channels anastomose freely and lymphatic crossover has been described, ipsilateral drainage appears to be the rule. Discrepancies concerning lymphatic drainage appear to revolve around controversies in nomenclature rather than anatomic location.

C. BLOOD VESSELS AND LYMPHATICS OF THE
PENIS

By I. GOLDSTEIN AND R. J. KRANE

Although for most organs blood and lymphatic circulations are well documented and well understood, for the penis, there are several aspects of the circulation which, even today, remain poorly defined. The main questions that persist concern the exact anatomy and mechanisms for the generation and maintenance of penile erection. For example, what is the anatomy of the penile arteriovenous shunts? What is the role of the venous effluents in erection? What is the neuropharmacologic regulatory mechanism of penile blood flow? What is the effect of disease on the circulation of the penis? As a result of this lack of understanding and an increased clinical interest, anatomic, physiologic, and neuropharmacologic research on the penile circulation has increased dramatically. This section reviews both the accepted and the more recent advances in the literature. A summary of the current theories of erection and penile blood flow regulation also is presented.

1. ANATOMY OF BLOOD CIRCULATION
(FIGS. 16.2 AND 16.3)

a. Internal iliac artery: The common iliac arteries originate at the abdominal aorta, which ends on the left side of the body, at the fourth lumbar vertebra. Five centimeters on the right and 4 cm on the left, downward and laterally, the common iliac arteries branch into the paired internal and external iliac arteries. The internal iliac artery, also known as the hypogastric artery, is 4 cm long and takes origin overlying the sacroiliac joint, opposite the lumbosacral intervertebral disc. As it descends to the area of the greater sciatic foramen, the internal iliac artery divides into an anterior and posterior trunk. The posterior trunk passes backward toward the foramen and gives off the superior gluteal, as well as the lateral sacral and iliolumbar, arteries. The anterior trunk passes toward the ischial spine and gives origin to the superior vesical, inferior vesical, middle

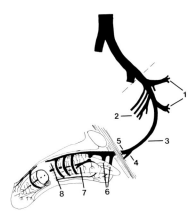

FIG. 16.2. Schematic representation of hypogastric cavernous arterial bed. The dashed line represents the origin of the internal iliac artery. Distal to the origin are drawn the superior and inferior gluteal arteries passing posteriorly (1); the vesical, obturator, and deferential arteries passing anteriorly (2); the pudendal artery continuing toward the pelvis (3); the perineal artery branching just prior to the urogenital diaphragm (4); the penile artery piercing the urogenital diaphragm (5); the bulbar artery and spongiosal artery penetrating the corpus spongiosum (6); the cavernosal artery penetrating the corpus cavernosum giving off the helical arteries (7); and the dorsal artery, beneath Buck's fascia, giving off circumflex branches before ending in short branches in the glans penis (8).

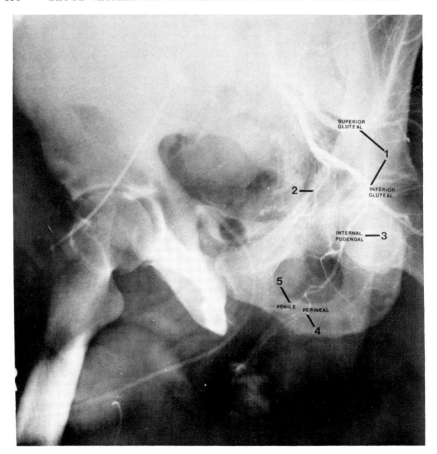

FIG. 16.3. Selective arteriogram of the internal iliac artery. Radiograph obtained from a 51-year-old man with suspected vasculogenic impotence, demonstrating the major branches of the internal iliac and internal pudendal arteries: superior and inferior gluteal arteries (1); vesical artery (2); internal pudendal artery (3); perineal artery (4); and penile artery (5). No contrast media identified beyond the origin of the penile artery, even on delayed films.

rectal, obturator, inferior gluteal, and internal pudendal arteries (Davies and Davies, 1962).

   b. Internal pudendal artery: The entire blood supply of the penis, except for some minor scrotal and epigastric arterial branches, is derived from the internal pudendal artery, the origin of which is somewhat variable. In almost three-quarters of the cases, this vessel arises from the lowest division of the ischiopudendal trunk of the internal iliac artery. Under such circumstances, origin is at the level of the sciatic notch. In the remainder of cases, the vessel derives from the terminal bouquet of the internal iliac artery (14% of cases) or from a terminal trifurcation of this vessel (11% of cases). In the latter two situations, the level of origin is lower in the pelvis than when the artery arises at the ischiopudendal trunk (Genestie, 1978).

The location of the internal pudendal artery can be divided into four arbitrary divisions: the pelvic, gluteal, ischiorectal, and perineal segments. In the pelvic portion, the direction of the artery is toward the ischial spine, and the vessel may be very short when its origin is low. In the gluteal segment, the artery direction changes; it begins at the lower part of the greater sciatic foramen, crosses the back of the tip of the ischial spine, and enters the perineum through the lesser sciatic foramen. In the ischiorectal segment, the vessel passes in the pudendal (Alcock's) canal in the lateral wall of the ischiorectal fossa, and, as it approaches the ischiopubic ramus, it enters the perineal segment. In the latter, the course of the internal pudendal artery is horizontal (Genestie, 1980). The perineal artery takes origin in the beginning of the perineal segment, and distal to it, all blood flow in the internal pudendal artery ends in the penis. As such, this portion of vessel has frequently been termed the penile artery, a name not accepted by all (Wagner, 1981) but used in this presentation. As the penile artery pierces the urogenital diaphragm at the level of the symphysis pubis, it divides into four terminal branches: the artery of the penile bulb (bulbar artery), the spongiosal (urethral) artery, the cavernosal (deep) artery, and the dorsal artery. These vessels have varying patterns in degree of cross communication of blood flow (Forster, 1903).

The internal pudendal artery also demonstrates variations in anatomic architecture. For example, one in ten patients studied by Genestie (1978) using radiographic techniques showed some type of anomaly of this vessel. In the absence of the internal pudendal artery, an accessory vessel was identified. The origin of the so-called accessory internal pudendal artery was located either at the internal iliac, at the internal pudendal, or at a branch of either vessel. The course of the accessory artery was found to be different from that of the internal pudendal artery, projecting either above the obturator foramen, along the ischiopubic ramus to the pubis, or under the pubic arch. Beyond any of these points, it gave origin to the terminal branches of the penis.

c. Bulbar artery: The artery to the bulb of the corpus spongiosum is the first branch of the penile artery. It is short and runs downward, penetrating the bulb 2 to 3 mm beyond its origin. Once in the bulb, the vessel divides into numerous small branches that anastomose with the spongiosal (urethral) artery (Newman and Northrup, 1981). On pudendal arteriograms, the bulbar artery is easily identified because of simultaneous opacification of a cone-shaped area of bulbar parenchyma.

d. Spongiosal (urethral) artery: Approximately 5 to 8 mm distal to the bulbar branch, the spongiosal artery takes origin from the penile artery. This vessel penetrates the tunica of the corpus spongiosum just proximal to the area where the two corpora cavernosa meet in the midline. The spongiosal vessel runs in a longitudinal direction, passing anteriorly within the spongiosal tissue. It breaks up into short branches along the way, supplying the corpus spongiosum and urethra and finally the glans penis. To some extent, the spongiosal artery communicates with vessels in the cavernous bodies.

e. *Dorsal artery:* Distal to the origin of the bulb and urethral arteries, the paired penile arteries continue deep to Buck's fascia, just outside the corpora cavernosa, as the dorsal arteries. The paired vessels course on either side of the unpaired dorsal vein in a relatively straight path until they terminate in multiple short, helical branches in the glans penis. Along the way, four to five circumflex branches take origin. These pass around the corpus cavernosum and penetrate both it and the corpus spongiosum. Since the dorsal artery provides the main blood flow to the glans penis, its state has particular relevance to the success of numerous plastic reconstructive procedures on the distal glans penis. In addition, it provides blood flow to the skin and superficial tissues of the penis. The latter function is shared to a small extent with the external pudendal artery (Newman and Northrup, 1981).

f. *Cavernosal artery:* The paired cavernosal arteries share a common origin with the dorsal artery, both being branches of the penile artery outside of the tunica of the corpora cavernosa. Proximal to the point where the two cavernous bodies merge, the cavernosal artery pierces the tunica albuginea. It first gives off a small branch that passes retrograde and supplies the proximal portion of the cavernous body attached to the pubic arch. The main branch of the cavernosal artery passes distally within the cavernous body, at a location much closer to the septum than previously appreciated. In the past, most anatomy textbooks placed this vessel in the middle of the cavernous bodies (Newman and Reiss, 1982). In their longitudinal course, the cavernosal artery trunks supply: (a) nutrient vessels to the tunica and trabecular tissue, (b) vessels that anastomose with the contralateral cavernous body, (c) vessels that pass ventrally to the corpus spongiosum, and (d) vessels (helical arteries) that, under appropriate stimulation, act to shunt blood to the trabecular spaces of the cavernous body.

g. *Helical arteries:* These unique vessels were evaluated first by Müller in 1835 and then by numerous other investigators, including Vastarini-Cresi (1902), Clara (1939), Rotter and Schürmann (1950), and Conti (1952, 1953). Their name reflects the tortuous nature of these vessels in the flaccid state. Each helical artery divides, in an arbor-like pattern, into several subdivisions. From the latter, small end-arteries take origin. These open into either the trabecular spaces of the cavernous tissue or into arteriovenous shunts that pass into the cavernous venous system without passing through any capillary bed. The inner diameter of the arteriovenous anastomoses has been approximated by microsphere injection at 100 μm (Newman *et al.*, 1964). There are about three helical arteries per centimeter in the proximal half of the cavernous body, slightly more than the number found in the distal half (Newman and Northrup, 1981). This may explain the common complaint of loss of penile rigidity that develops in the distal portion before it does in the proximal portion. In addition, most of the intracavernous shunts (helical arteries) are located in the peripheral rather than the central portion of the cavernous body. Such an arrangement may provide the shortest means of shunting arterial blood to the venous system in the flaccid state (Wagner, 1981).

Wagner *et al.* (1982) reported an alternative pathway for helical arterial blood flow by demonstrating in 47 cadaver penises the existence of "shunt" arteries connecting the helical arteries with arteries of the glans and corpus spongiosum. In the flaccid state, these arterial shunts help divert blood away from the trabecular spaces. In the erect state, closure of these shunts by neuropharmacologic stimulation or by mechanical compression helps keep blood within the cavernous body.

h. Deep veins of the penis: It is postulated that cavernous venous effluent is collected into short postcavernous veins from either the cavernous spaces or from the cavernous interstitium. This effluent then drains into large vessels, the deep veins of the penis, which arise at the proximal end of each corpus cavernosum. The deep veins eventually communicate with the prostatic venous plexus (Newman and Northrup, 1981). Of note is the fact that no muscular deep veins have ever been identified in the corpus cavernosum (Khodos, 1962). It is of further interest that several authors were not able to identify postcavernous veins and therefore could not demonstrate a venous system within the cavernous tissue (Goldstein and Fitzpatrick, 1980; Wagner, 1981).

i. Deep dorsal vein: In the corpus spongiosum, numerous small veins drain the venous effluent. Urethral veins are divided into anterior ones, which perforate the tunica of the corpus spongiosum, and posterior ones, which drain toward the spongiosal bulb. The anterior perforating veins join trunks that pass around the corpus cavernosum (circumflex veins), superficial to the tunica, and back to Buck's fascia. This venous complex eventually ends in the deep dorsal vein, the large unpaired vessel on the dorsum, traveling longitudinally in a sulcus between the two corpora cavernosa and surrounded by the paired dorsal arteries. The deep dorsal vein receives venous output from numerous vessels in the glans (retrocoronal plexus) and from one additional source, the emissary veins. The latter group of vessels pierces the tunica albuginea of the corpus cavernosum to open directly into the deep dorsal vein. The emissary veins collect venous outflow from the peripheral portion of the cavernous body.

j. Superficial dorsal vein: Drainage from the skin and subcutaneous structures passes by way of the superficial dorsal vein. The latter structure is superficial to Buck's fascia and travels in the same longitudinal direction as the deep dorsal vein. In 70% of cases, it drains into the left great saphenous vein. Other drainage sources include the right great saphenous vein, left femoral vein, and epigastric veins. Anastomotic channels may exist between the superficial dorsal and deep dorsal veins in the coronal sulcus and at the pelvic bone. Through such a pathway, superficial venous flow may eventually drain into the prostatic venous plexus.

2. PHYSIOLOGY OF BLOOD CIRCULATION

a. Regulatory mechanism: The existence of a regulatory mechanism to direct blood flow either into the efferent venous system (flaccidity) or into the trabecular spaces (erection) has been studied by several investigators. Ercolani (1869) first described intravascular entities that compromised the vessel lumen. Von Ebner (1900) identified similar intraluminal intrusions that have since been called penile pads, cushions, or polsters. Kiss (1921) analyzed the histology of involved vessels and their intrusions and found an outermost endothelial lining, a subendothelial layer, a tunica elastica layer surrounding the intrusion, a longitudinally oriented smooth muscle layer (the intrusion), and a medial and adventitial layer of the vessel wall. Rotter and Schürmann (1950) found similar intrusions in virtually all the branches of the arterial and venous tree of the penis. These investigators attributed the regulatory nature of penile blood flow to such intraluminal entities, proposing that when the polsters contracted, penile vascular resistance was raised, whereas when they relaxed, vascular resistance was lowered. Several authors have recently doubted the role of these protrusions in blood flow regulation. Attention has been drawn to the fact that similar structures have been identified in most of the muscular arteries throughout the body, even in places where no regulatory mechanism is known to exist. A second opposing point is found in the reports of Benson *et al.* (1980) and Newman and Tchertkoff (1980), who could not identify the pads in the penile vessels of young males and of the newborn, respectively. It is possible polsters are early prominent manifestations of atherosclerosis, associated with aging (Ruzbarsky and Michal, 1977).

Recent neuropharmacologic studies support the existence of short adrenergic neurons on the helical arteries, postsynaptic to the cholinergic pelvic nerves. The neurotransmitter is possibly a β-adrenergic agent, such as norepinephrine (Siroky and Krane, 1979). An agent producing relaxation of the trabecular smooth muscle, vasoactive intestinal polypeptide (VIP) (Willis *et al.*, 1981; Polak *et al.*, 1981), has been found in high concentration in the corpus cavernosum. The net effect of either a β-adrenergic hormone or a VIP vessel relaxant would be an increased blood flow to the cavernous body, thus resulting in an accumulation of blood under pressure (erection).

b. Current theories of penile erection: When the penis is flaccid, the helical arteries remain tortuous, and intracorporeal blood flow is preferentially directed away from corporal trabecular spaces. This may be via shunt arteries, sending the blood into the corpus spongiosum or glans penis.

Erection may be considered in two phases: generation and maintenance. The generation phase is accomplished by neuropharmacologic stimulation of the helical arteries and trabecular smooth muscle, causing vasodilatation and relaxation, as well as possibly neuropharmacologic stimulation of the corporal shunt arteries and arteriovenous shunts, producing vasoconstriction. The net effect is accumulation of blood in the trabecular spaces. The maintenance phase is accomplished by closing corporal venous outflow so that blood accumulates in the trabecular spaces under pressure, or this phase may simply represent a secondary effect of increased blood entering the trabecular spaces. Emissary and circumflex veins draining the corpora may become mechanically passively obstructed as the pressure in the tunica increases (35 mm Hg), according to Wagner (1981).

Detumescence first begins as a result of the loss of helical artery vasodilatation or loss of shunt artery and arteriovenous shunt vasoconstriction. As inflow to the trabecular spaces decreases, cavernous pressure falls. At some critical point, passive venous occlusion is overcome and detumescence is quickly completed.

c. Regulatory function of veins: What, if any, is the regulatory role of the penile veins (deep and superficial dorsal veins) during tumescence and detumescence? Fitzpatrick (1980) performed cavernosography in normal subjects following erection and noted corporal venous blood passing to the deep dorsal vein via the circumflex and emissary veins. During erection, venous drainage by this route was mechanically impeded, largely because of the 90° insertion of the circumflex veins into the deep dorsal vein. The circumflex veins, on the other hand, provided an alternate route for cavernous blood to drain during detumescence. Fitzpatrick and Cooper (1975) identified venous valves in the deep dorsal vein just distal to the suspensory ligament, the role of which in the generation of an erection remains controversial. Infusion cavernosography, another means to visualize the venous anatomy during erection, was used by Ebbehoj *et al.* (1980) to demonstrate that during the flaccid state in normal subjects, venous outflow is reduced, and at 35 mm Hg pressure in the cavernous bodies, it is minimal or actually stops. These investigators postulated that outlet regulation is essential to the maintenance of the hydraulic action of an erection.

3. LYMPHATIC SYSTEM

a. Glans: Numerous anastomosing lymphatic vessels can be found in a network within the tissue of the glans penis. The drainage from this plexus begins ventrally but passes on the sides of the penis by lymphatic collecting vessels to an area under Buck's fascia, on the dorsum next to the deep dorsal vein. The effluent then flows to a lymphatic plexus at the root of the penis. Drainage continues into superficial lymph nodes, deep inguinal nodes within the femoral canal, and external iliac lymph nodes. The lymphatics from the glans also form numerous anastomoses with those from the skin and urethra (Davies and Davies, 1962; Hollinshead, 1971).

b. Corpora: Lymphatics draining the corpora may enter the pelvis, either with the urethra or with the pudendal vessels. The primary nodal site of the distal network is ascribed to the external iliac and deep subinguinal lymph nodes. More proximally in the urethra, especially the membraneous and prostatic regions, lymphatic drainage is to the internal iliac lymph nodes. There are multiple anastomotic channels between lymphatics in the corpora and those in the glans and the skin of the penis, especially in the distal portion. Distal lymphatic drainage tends to parallel drainage of the glans (Nesselrod, 1936).

c. Skin: Lymphatics draining the skin and prepuce pass superficially to Buck's fascia, to follow the course of the superficial dorsal vein, and then laterally with the external pudendal vein. The primary lymph node drainage of this pathway is the superficial inguinal lymph nodes. As noted previously, there may be anastomosing lymphatics, especially to the ones draining along the deep dorsal vein.

D. BLOOD VESSELS AND LYMPHATICS OF THE

URETER

By R. M. WEISS

1. ANATOMY OF BLOOD CIRCULATION

a. Arteries: The ureter is supplied by a variable number of segmental arteries that arise from the aorta and from a variety of its branches or sub-

All rights of reproduction in any form reserved.
ISBN 0-12-042520-3

branches (Daniel and Shackman, 1952), including the renal, gonadal, adrenal, common iliac, internal iliac, external iliac, superior vesical, inferior vesical, vesiculodeferential, uterine, obturator, gluteal, vaginal, and middle hemorrhoidal arteries. These segmental arteries divide into ascending and descending branches that run in the adventitial coat of the ureter. The descending branches of proximally located segmental arteries anastomose with the ascending branches of more distally located segmental vessels, thus forming longitudinally running vascular channels. These anastomosing arteries may give off secondary branches that form plexuses on the surface of the ureter. From the latter arise small tributaries that pierce the muscular coat and form more delicate plexuses in the submucosal layer. A less frequent pattern is that in which the segmental arteries break into branches that anastomose with each other and form adventitial plexuses without a longitudinally running vascular channel (Shafik, 1972). DeSousa (1966) suggested that there are two types of perforating vessels: (1) relatively large caliber arteries that run directly to the submucosal plexus and give off few branches to the muscle layer and (2) a larger number of smaller caliber vessels that appear to supply mainly the muscle elements.

b. Microvasculature and nervous innervation: Hoyes *et al.* (1976) and Hoyes and Barber (1981) identified three ultrastructurally distinct types of arterioles in the ureters: vessels of 50–100 μm diameter with a media consisting of two layers of muscle cells (arterioles); vessels of 30 μm in diameter containing one layer of muscle cells (terminal arterioles); and vessels of 10–15 μm in diameter with muscle cells in the media being peripherally attenuated (metarterioles).

Hoyes *et al.* (1976) also described two types of nerve plexuses with intercommunications, associated with the microvasculature: a perivascular plexus, located between the adventitia and the media and observed in all vessels, and a paravascular plexus, found on the outer aspect of the adventitia and present only on larger terminal arterioles. The unmyelinated nerves forming these networks exhibited three types of terminal varicosities: adrenergic terminals containing small dense-cored vesicles, cholinergic terminals possessing small clear vesicles, and sensory terminals with large dense-cored vesicles. Adrenergic terminals predominated in each vessel type, and nerve density was greater in terminal arterioles than in arterioles or metarterioles. The latter finding led to the suggestion that the terminal arterioles may play an important role in regulating blood flow through the capillary networks (Hoyes *et al.*, 1976).

c. Vascular response to ischemia: Experimental ureterolysis has been shown to decrease ureteral blood supply temporarily (Sirca *et al.*, 1978), and a similar relative ischemia has been observed proximal to a site of ureteral transection and subsequent reanastomosis (Saidi *et al.*, 1973). Small vessels begin to cross the anastomosis within 1 week, and by 2 weeks, large dilated vessels are observed. In the obstructed rabbit ureter, there appears to be no change in the number of vessels, but there is an increase in the diameter and tortuosity of those already present.*

2. Lymphatic System

Ureteral lymph vessels begin in submucous, intramuscular, and adventitial plexuses that communicate with each other (Williams and Warwick, 1980). Collecting vessels from the proximal ureter may join lymphatics of the kidney; the latter follow the course of the renal vein, to end in the lateral aortic nodes, or they may drain directly into the lateral aortic nodes near the origin of the gonadal arteries. Lymphatics from the midureter drain into the common iliac nodes and those from the lower portion of the ureter may end in the common, external, or internal iliac glands.

E. BLOOD VESSELS AND LYMPHATICS OF THE URINARY BLADDER AND URETHRA

By R. M. Weiss

1. Anatomy of Blood Circulation of Bladder

a. Arteries: The internal iliac or hypogastric artery, which arises at the bifurcation of the common iliac artery, divides into two major branches: anterior and posterior divisions. The posterior division gives off three branches, the iliolumbar, lateral sacral, and superior gluteal arteries. In the male, the anterior division divides into seven branches: the superior vesical, vesiculodeferential, inferior vesical, middle rectal, obturator, internal pudendal, and inferior gluteal arteries (Braithwaite, 1952). In the female, there are two additional branches, the uterine and vaginal arteries. Although all nine branches of the anterior division have been reported to supply the bladder, the only constant

*The author is not aware of pertinent data in the literature on the physiology, pharmacology, or pathology of the blood vessels of the ureters.

sources are the superior vesical, inferior vesical, and vesiculodeferential arteries (Braithwaite, 1952), with the first two vessels providing most of the circulation (Williams and Warwick, 1980).

The superior vesical arteries supply the largest area of the bladder and are comprised of one to four vessels (Braithwaite, 1952; Shehata, 1976). These commonly arise from the umbilical artery in the fetus (which obliterates following birth) but may rarely take origin from the vesiculodeferential, uterine, or obturator arteries (Shehata, 1976). The superior vesical arteries frequently anastomose with each other and also with the vesiculodeferential and inferior vesical vessels. There are also communications between the branches of the superior vesical and deep epigastric arteries, which may provide important collateral circulation following bilateral ligation of the internal iliac artery.

The vesiculodeferential arteries are primarily distributed on the posterior aspect of the bladder (Braithwaite, 1952) and, in the male, frequently provide the main blood supply to the trigone (Shehata, 1976). Branches of the uterine artery generally supply the trigone in females (Shehata, 1976).

The inferior vesical artery has a variable origin, either directly from the internal iliac artery or from one of its branches. This has led Braithwaite (1952) to suggest that the various descriptions of vesical branches arising from the inferior gluteal, internal pudendal, and obturator arteries resulted from the variable origin of the inferior vesical arteries. The latter vessels supply the inferolateral segments of the anterior surface of the bladder and bladder neck and possibly the trigone. They frequently anastomose with the superior vesical and vesiculodeferential vessels. In the female, branches of the vaginal artery may supply the area that usually receives its circulation from the inferior vesical artery in the male.

b. Microvasculature: Sarma (1981a) used microangiographic techniques to study the blood supply to the bladder and noted three distinct components: extramural, intramural, and submucous, with free intercommunications among them. The extramural plexus is located on the outer surface of the detrusor and is most prominent over the posterolateral walls of the bladder and bladder neck. A broad midline zone of the trigone and of the anterior and posterior bladder walls is relatively avascular. Small tributaries from the extramural plexus pierce the muscular layer to communicate with the intramural and submucous plexuses. Occasional large vessels appear to supply areas of the mucosa. The intramural plexus is situated within the muscular wall and frequently extends into the submucous region. Occasionally, branches reach the mucosa. This plexus is prominent, not only in the posterolateral walls, but also around the bladder neck and adjoining anterolateral surfaces. Finally, the submucous plexus is situated adjacent to the inner surface of the intramural plexus.

c. Variations in blood supply: Sarma (1981a) noted considerable variation in the distribution of the blood supply to the bladder mucosa. There are large avascular or sparsely supplied regions of mucosa that seem to receive nourishment by simple diffusion. On the other hand, other areas, mainly on the posterolateral and anterolateral walls, are more richly supplied. Sarma (1981b) has suggested that these regional variations in mucosal blood supply might account for observed distributions of bladder tumors.

d. Veins: Shehata (1979) studied the venous drainage of the bladder and noted that it originates by the formation of three intercommunicating plexuses: submucous, intramuscular, and external. In the male, the external or vesical plexus lies on the outer surface of the muscular coat of the bladder and is densest around the bladder neck, where it is continuous with the prostatic plexus, and at the bladder base, where it extends around the seminal vesicles and the terminal ends of the ureter and vas deferens. In the female, the vesical plexus surrounds the proximal urethra and the bladder neck.

In close association with the vesical plexus and sharing in the venous drainage of the bladder is a venous plexus that is located behind the lower part of the symphysis pubis and the inferior pubic ligaments. It receives the deep dorsal vein of the penis or the clitoris in addition to draining the anterior surface of the bladder. The different plexuses drain into the internal iliac (hypogastric) vein via two to five vesical veins. The latter are commonly joined by the obturator, prostatic, and vaginal veins (Shehata, 1979). In some cases, a common trunk of united obturator and vesical veins drains into the external iliac vein (Shehata, 1979). Furthermore, occasionally an inferior vesical vein accompanies the inferior vesical artery when the latter arises from the obturator artery (Shehata, 1979). This is in contradistinction to the usual finding that the vesical veins do not accompany the vesical arteries.

2. Physiology of Blood Circulation of Bladder

a. Vascular responses to bladder distention: Lapides *et al.* (1968) and Mehrotra (1953) postulated that bladder distention causes diminished blood flow through the bladder wall, with a resultant breakdown of the natural host defense mechanisms of the bladder and a greater susceptibility to infection. Such a relationship between bladder blood flow and intravesical distention has been documented, using light microscopy (Mehrotra, 1953), radioisotopic washout techniques (Dunn, 1974; Finkbeiner and Lapides, 1974), photoelectric pulse wave detection methods (Nanninga, 1976), and radioisotope-labeled microsphere techniques (Nemeth *et al.*, 1977). With radioisotope-labeled microspheres, Nemeth *et al.* (1977) noted a decrease in canine bladder

blood flow with vesical distention from 2.80 to 2.06 ml/min. These data correlated with previous studies in the dog showing a 25 to 30% reduction in blood
flow following 2 hr of bladder distention (Finkbeiner and Lapides, 1974). Furthermore, Nemeth *et al.* reported that bladder distention elicited a decrease in
mucosa to muscularis blood flow ratio, which they interpreted to mean that the
relative decrease in mucosal blood flow was compatible with a reduced host
resistance to infection. They did not observe any significant difference in the
ratio of trigone to dome blood flow with bladder distention. However, Finkbeiner and Lapides (1974) were unable to document such redistribution of
blood flow with distention.

   b. Vascular responses to carcinogenesis: Hemodynamic changes have also
been observed during carcinogenesis. Lurie *et al.* (1979) have shown an increase in bladder vascular volume during this state induced by N-butyl-N-(4-
hydroxybutyl)nitrosamine (BBN) in the rat urinary bladder. The early increase
in vascular volume appeared to be due to either neovascularity secondary to the
effect of BBN directly on the vasculature, indirectly through inflammation, or
both. Later, increases in vascular volume correlated with growth of tumors.
Lurie *et al.* (1979) have suggested that such changes in vascular volume represent a response to secretion of an angiogenesis factor from malignant cells
(Folkman and Cotran, 1976).

3. Pharmacology of Blood Circulation
of Bladder

   There have only been a few studies of the responses of the bladder vasculature to pharmacologic agents. Matsumura *et al.* (1968) classified a group of
drugs on the basis of their effects on the canine urinary bladder and its vasculature.

   a. Agents causing both bladder contraction and vesical vasoconstriction:
Such drugs include: (1) the hypertensive polypeptides, angiotensin II, Lys-
vasopressin, and oxytocin; (2) the ergot alkaloid, ergotamine; (3) an α-adrenergic agonist, phenylephrine; (4) a biogenic amine, 5-hydroxytryptamine
(serotonin); (5) the competitive ganglionic blocking agent, tetraethylammonium
(TEA); and (6) the anticholinesterase, physostigmine. Serotonin produces a
variable vascular response, but its pressor component is greater than its depressor action.

   b. Agents causing bladder contraction and vesical vasodilatation: Such
drugs include: (1) the muscarinic and nicotinic cholinergic agonists, acetyl-

choline, bethanechol (Urecholine), carbamylcholine (carbachol), tetramethyl-ammonium (TMA), lobeline, nicotine, and 1,1-dimethyl-4-phenylpiperazinium (DMPP); (2) the anticholinesterase, neostigmine; (3) the hypotensive polypep-tides, bradykinin, eledoisin, and kallikrein; (4) a biogenic amine, histamine; (5) a nucleotide, adenosine triphosphatate (ATP); (6) morphine; and (7) potassium chloride (KCl).

c. Agents causing other combinations of changes: Another group consist-ing of a variety of sympathomimetics, i.e., norepinephrine, epinephrine, dopamine, and tyramine, produce bladder relaxation and constriction of the vesical vascular bed. The β-adrenergic agonist, isoproterenol, causes both blad-der relaxation and vesical vasodilatation. The α-adrenergic agonists, ephedrine and methoxamine, elicit vesical vasoconstriction without affecting bladder con-tractility. Finally, such drugs as aminophylline; dipyridamole; hydralazine; ni-troglycerin; papaverin; adenosine, adenosine mono- and diphosphate (AMP and ADP); uridine mono-, di-, and triphosphate (UMP, UDP and UTP); pro-caine; and tetrodotoxin dilate the vesical vascular bed without affecting bladder contractility.

d. Comparison of actions of drugs on vesical vasculature with those on other vascular beds: In this regard, Matsumura *et al.* (1968) and Hashimoto and Kumakura (1965) reported the following differences: (1) UMP produces vesical vasodilatation, whereas it constricts the coronary, renal, mesenteric, and femoral arteries; (2) the ganglionic stimulants, nicotine, lobeline, and DMPP, cause vesical vasodilatation but constrict mesenteric and renal arteries; (3) adenosine, AMP, dipyridamole, procaine, and morphine dilate the vesical arteries but constrict the renal artery; (4) TEA constricts the vesical artery but dilates the femoral artery; and (5) norepinephrine, epinephrine, and serotonin constrict the vesical artery but dilate the coronary vessels.

e. Vascular effects of prostaglandins: Young *et al.* (1979) have shown that PGE_1, PGE_2, PGA_1, PGA_2, and $PGF_{1\alpha}$ cause dilatation of vesical vasculature, whereas $PGF_{2\alpha}$ produces vesical arteriolar constriction. These workers also confirmed the arteriolar constrictor effect of norepinephrine and serotonin and the vasodilatory response to histamine. Pretreatment with PGE_1, but not with histamine, diminishes the vasoconstrictor responses to $PGF_{2\alpha}$ and nor-epinephrine. The latter, but not $PGF_{2\alpha}$ or serotonin, has been shown to con-strict the vesical veins (Young *et al.*, 1979), whereas histamine, PGE_1, PGE_2, and PGA_1 all produce significant venular dilatation.

4. LYMPHATIC CIRCULATION OF THE BLADDER

Lymph vessels from the bladder take origin in three plexuses: mucous, intramuscular, and extramural. Nearly all of the collecting vessels arising in these sites terminate in the external iliac nodes (Williams and Warwick, 1980). Lymphatics from the anterior and posterior surfaces of the bladder drain along the course of the umbilical artery to terminate in the external iliac nodes (Parker, 1936). Occasional vessels may enter the internal and common iliac nodal chains (Williams and Warwick, 1980). Lymphatics from the trigonal region pass around the lower end of the ureters to the external iliac nodes and between the superior and inferior vesical arteries as they run in the pelvic fascia to the internal iliac chain.

Using intraepithelial injections of iron–dextran and India ink, Milroy and Cockett (1973) demonstrated lymphatics in all layers of the canine bladder, including the endothelium. These vessels may play a role in removing substances that traverse the bladder urothelium. In the dog, tetracycline, nitrofurantoin, and thio-TEPA have been shown to pass from the bladder lumen into the lymphatics and ultimately into the systemic systems (Milroy *et al.*, 1974).

5. Blood Vessels and Lymphatics of the Urethra

a. Urethral vasculature: The posterior urethra is supplied by the internal pudendal artery, which is a terminal branch of the anterior division of the internal iliac artery. The bulbous urethra receives its blood circulation from the artery of the urethral bulb, which is also a branch of the internal pudendal artery.

b. Urethral lymphatics: Lymphatics draining the prostatic and membranous urethra in the male and the entire urethra in the female mainly pass to the internal iliac nodes, although a few vessels may terminate in the external iliac lymph nodes (Williams and Warwick, 1980). Lymph vessels from the penile urethra accompany lymphatics from the glans penis and terminate in the deep inguinal and external iliac nodes.

References*

Agger, P. (1971). Scrotal and testicular temperature: The relation to sperm count before and after operation for varicocele. *Fertil. Steril.* **22,** 286–297. (A)
Anson, B. J., and Kurth, E. L. (1955). Common variations in the renal blood supply. *Surg., Gynecol. Obstet.* **100,** 156–162. (A)
Batson, O. V. (1940). The function of the vertebral veins and their role in the spread of metastases. *Ann. Surg.* **112,** 138–149. (B)
Benson, G. S., McConnell, J., Lipshultz, L. I., Corrière, J. N., Jr., and Wood, J. (1980). Neuromorphology and neuropharmacology of the human penis. *J. Clin. Invest.* **65,** 506–513. (C)

*In the reference list, the capital letter in parentheses at the end of each reference indicates the section in which it is cited.

Braithwaite, J. L. (1952). The arterial supply of the male urinary bladder. *Br. J. Urol.* **24**, 64–71. (E)

Busch, F. M., and Sayegh, E. S. (1963). Roentgenographic visualization of human testicular lymphatics: A preliminary report. *J. Urol.* **89**, 106–110. (A)

Chiappa, S., Uslenghi, C., Bonadonna, G., Marano, P., and Ravasi, G. (1966). Combined testicular and foot lymphangiography in testicular carcinoma. *Surg., Gynecol. Obstet.* **123**, 10–14. (A)

Clara, M. (1939). "Die Arteri-venose Anastomosen." Barth, Leipzig. (C)

Clegg, E. J. (1955). The arterial supply of the human prostate and seminal vesicles. *J. Anat.* **89**, 209–216. (B)

Clegg, E. J. (1956). The vascular arrangements within the human prostate gland. *Br. J. Urol.* **28**, 428–435. (B)

Connolly, J. G., Thomson, A., Jewett, M. A. S., Hartman, N., and Webber, M. (1968). Introprostatic lymphatics. *Invest. Urol.* **5**, 371–378. (B)

Conti, G. (1952). L'érection du pénis humain et ses bases morphologicovasculaires. *Acta Anat.* **14**, 217–262. (C)

Conti, G. (1953). Über das workommen von sperrvorrichtungen in arterien mit speziellei berucksichtigung der "Gestielten Polster." *Acta Anat.* **18**, 234–255. (C)

Dahl, E. V., and Herrick, J. F. (1959). A vascular mechanism for maintaining testicular temperature by counter-current exchange. *Surg., Gynecol. Obstet.* **108**, 697–705. (A)

Daniel, O., and Shackman, R. (1952). The blood supply of the human ureter in relation to ureterocolic anastomoses. *Br. J. Urol.* **24**, 334–343. (D)

Davies, D. V., and Davies, F. (1962). "Gray's Anatomy," 33rd ed., pp. 927–938, 1528–1530. Longmans, Green, New York. (C)

deSousa, L. A. (1966). Microangiographic aspects of the ureter. *J. Urol.* **95**, 179–183. (D)

Dunn, M. (1974). A study of the bladder blood flow during distension in rabbits. *Br. J. Urol.* **46**, 67–72. (E)

Ebbehoj, J., Uhrenholdt, A., and Wagner, G. (1980). Infusion caverosography in the human in the unstimulated and stimulated situations and its diagnostic value. In "Vasculogenic Impotence" (A. W. Zorgniotti and G. Rossi, eds.), pp. 185–190. Thomas, Springfield, Illinois. (C)

Ercolani, G. B. (1869). "Dei tessuti e degli organi erettili." Gamberini e Parmeggiani, Bologna. (C)

Ewing, L. L. (1978). Physiology of male reproduction. In "Campbell's Urology." 4th ed. (H. J. Harrison, R. F. Gittes, A. D. Perlmutter, T. A. Stamey, and P. C. Walsh, eds.) Saunders, Philadelphia, Pennsylvania. (A)

Finkbeiner, A., and Lapides, J. (1974). Effect of distension on blood flow in dog's urinary bladder. *Invest. Urol.* **12**, 210–212. (E)

Fitzpatrick, T. (1980). The venous drainage of the corpus cavernosum and spongiosum. In "Vasculogenic Impotence" (A. W. Zorgniotti and G. Rossi, eds.), pp. 181–184. Thomas, Springfield, Illinois. (C)

Fitzpatrick, T. J., and Cooper, J. F. (1975). A cavernosogram study of the valvular competence of the human deep dorsal vein. *J. Urol.* **113**, 497–500. (C)

Flocks, R. H. (1937). The arterial distribution within the prostate gland: Its role in transurethral prostatic resection. *J. Urol.* **37**, 524–549. (B)

Folkman, J., and Cotran, R. (1976). Relation of vascular proliferation to tumor growth. *Int. Rev. Exp. Pathol.* **16**, 207–248. (E)

Forster, A. (1903). Beitrage zur anatomie der ausseren mannlichen geschlechtosorgane des menschen. *Z. Morphol.* **6**, 435–501. (C)

Fritjofsson, A., Persson, J. E., and Petterson, S. (1969). Testicular blood flow in man measured with Xenon-133. *Scand. J. Urol. Nephrol.* **3**, 276–280. (A)

Genestie, J.-F. (1978). Pudendal angiography. In "Vasculogenic Impotence" (A. W. Zorgniotti and G. Rossi, eds.), pp. 125–142. Thomas, Springfield, Illinois. (C)

Gittes, R. F., and McCullough, D. L. (1974). Occult carcinoma of the prostate: An oversight of immune surveillance—A working hypothesis. *J. Urol.* **112**, 241–244. (B)

Goldstein, A. N. B., and Fitzpatrick, T. J. (1981). Architecture of the corpora cavernosa and a possible intrinsic mechanism of erection. *Annu. Meet. Am. Urol. Assoc. 76th, May 10–14, Boston, Mass.*, p. 644. (C)

Harrison, R. G. (1949a). The comparative anatomy of the blood supply of the mammalian testis. *Proc. Zool. Soc. London* **119**, 325–344. (A)

Harrison, R. G. (1949b). The distribution of vasal and cremasteric arteries to the testis and their functional importance. *J. Anat.* **83**, 267–282. (A)

Harrison, R. G., and Barclay, A. E. (1948). The distribution of the testicular artery (internal spermatic artery) to the human testis. *Br. J. Urol.* **20**, 57–66. (A)

Harrison, R. G., and McGregor, G. A. (1957). Anomalous origin and branching of the testicular arteries. *Anat. Rec.* **129**, 401–405. (A)

Hashimoto, K., and Kumakura, S. (1965). The pharmacological features of the coronary, renal, mesenteric and femoral arteries. *Jpn. J. Physiol.* **15**, 540–551. (E)

Herman, P. G., Benninghoff, D. L., Nelson, J. H., and Mellins, H. Z. (1963). Roentgen anatomy of the ilio-pelvic-aortic lymphatic system. *Radiology* **80**, 182–193. (B)

Hollinshead, W. H. (1971). "Anatomy for Surgeons," Vol. 2, pp. 838–850. Harper, New York. (C)

Hoyes, A. D., and Barber, P. (1981). Quantitative ultrastructural studies on arteriolar innervation in the rat ureter. *Microvasc. Res.* **21**, 165–174. (D)

Hoyes, A. D., Bourne, R., and Martin, B. G. H. (1976). Ureteric vascular and muscle coat innervation in the rat. A quantitative ultrastructural study. *Invest. Urol.* **14**, 38–43. (D)

Hundeiker, M., and Keller, L. (1963). Die Gefässarchitektur des menschlichen. Hodens. *Morphol. Jahrb.* **105**, 26–73. (A)

Jamieson, J. K., and Dobson, J. F. (1910). The lymphatics of the testicle. *Lancet* **1**, 493–495. (A)

Khodos, A. B. (1962). The vascular supply of the penis. *Arkh. Anat. Gistol. Embriol.* **43**, 52–57. (C)

Kiss, F. (1921). Anatomisch-histologishe untersuchungen über die erekition. *Z. Anat. Entwicklungsgesch.* **61**, 455–521. (C)

Kormano, M. (1969). Demonstration of microvascular structures using microangiographic and histochemic techniques. *Invest. Radiol.* **4**, 391–395. (A)

Kormano, M., and Suoranta, H. (1971a). An angiographic study of the arterial pattern of the human testis. *Anat. Anz.* **128**, 69–76. (A)

Kormano, M., and Suoranta, H. (1971b). Microvascular organization of the adult human testis. *Anat. Rec.* **170**, 31–40. (A)

Lapides, J., Costella, R. T., Jr., Zierdt, D. K., and Stone, T. E. (1968). Primary cause and treatment of recurrent urinary infection in women: Preliminary report. *J. Urol.* **100**, 552–555. (E)

Lich, R., Howerton, L. W., and Amin, M. (1978). Anatomy and surgical approach to the urogenital tract in the male. *In* "Campbell's Urology" (H. J. Harrison, R. F. Gittes, A. D. Perlmutter, T. A. Stamey, and P. C. Walsh, eds.), 4th ed., pp. 1–33. Saunders, Philadelphia, Pennsylvania. (A,B)

Lowsley, O. S. (1915). The gross anatomy of the human prostate gland and contiguous structures. *Surg., Gynecol. Obstet.* **20**, 183–192. (B)

Lurie, A. G., Rippey, R. M., Conran, P. B., Tatematsu, M., and Ito, N. (1979). A technique for multiparametric analysis of hemodynamic changes in rat urinary bladder during carcinogenesis by N-butyl-N-(4-hydroxy-butyl)nitrosamine. *Gann* **70**, 717–718. (E)

McCullough, D. L. (1975). Experimental lymphangiography. *Invest. Urol.* **13**, 211–219. (B)

McMillan, E. W. (1954). The blood supply of the epididymis in man. *Br. J. Urol.* **26**, 60–71. (A)

Matsumura, S., Taira, N., and Hashimoto, K. (1968). The pharmacological behavior of the urinary bladder and its vasculature of the dog. *Tohoku J. Exp. Med.* **96**, 247–258. (E)

Mehrotra, R. M. L. (1953). An experimental study of the vesical circulation during distension and cystitis. *J. Pathol. Bacteriol.* **66**, 79–88. (E)

Menan, M., Merron, S., Strauss, H. W., and Catalona, W. J. (1977). Demonstration of the existence of canine prostatic lymphatics by radioisotope technique. *J. Urol.* **118**, 274–277. (B)

Milroy, E. J. G., and Cockett, A. T. K. (1973). Lymphatic system of canine bladder. An anatomic study. *Urology* **2**, 375–377. (E)

Milroy, E. J. G., Cockett, A. T. K., and Roberts, A. P. (1974). The bladder lymphatics. A study of drug transport. *Invest. Urol.* **12**, 69–73. (E)

Müller, J. (1835). Entdeckung der bei der erektion des mannlichen gliedes wirksamen arterien bei dem menschen und em thieren. *Arch. Anat., Phyiol. Wiss. Med.* **8**, p. 202. (C)

Nanninga, J. B. (1976). Effect of distension on vesical blood flow. *Surg. Forum* **27**, 598–599. (E)

Neaves, W. B., and Billingham, R. E. (1979). The lymphatic drainage of the rat prostate and its status as an immunologically privileged site. *Transplantation* **27**, 127–132. (B)

Nemeth, C. J., Khan, R. M., Kirchner, P., and Adams, R. (1977). Changes in canine bladder perfusion with distension. *Invest. Urol.* **15**, 149–150. (E)

Nesselrod, J. P. (1936). An anatomic restudy of the pelvic lymphatics. *Ann. Surg.* **104**, 905–916. (C)

Neuhof, H., and Mencher, W. H. (1940). The viability of the testis following complete severance of the spermatic cord. *Surgery* **8**, 672–685. (A)

Newman, H. F., and Northrup, J. D. (1981). Mechanism of human penile erection: An overview. *Urology* **17**, 399–408. (C)

Newman, H. F., and Reiss, H. (1982). Method for exposure of cavernous artery. *Urology* **19**, 61–62. (C)

Newman, H. F., and Tchertkoff, V. (1980). Penile vascular cushions and erection. *Invest. Urol.* **18**, 43–45. (C)

Newman, H. F., Northrup, J. D., and Devlin, J. (1964). Mechanism of human penile erection. *Invest. Urol.* **1**, 350–353. (C)

Olson, R. O., and Stone, E. P. (1949). Varicocele. Symptomatologic and surgical concepts. *N. Engl. J. Med.* **240**, 877–880. (A)

Parker, A. E. (1936). The lymph collectors from the urinary bladder. *Anat. Rec.* **65**, 443–460. (E)

Patten, B. M. (1968). "Human Embryology," 3rd ed., p. 515. McGraw-Hill, New York. (A)

Petterson, S., Soderholm, B., Persson, J. E., Eriksson, S., and Fritjofsson, A. (1973). Testicular blood flow in man measured with venous occlusion plethysmography and xenon-133. *Scand. J. Urol. Nephrol.* **7**, 115–119. (A)

Polak, J. M., Gu, J., Nina, S., and Bloom, S. R. (1981). VIP-ergic nerves in the penis. *Lancet* **2**, 217–219. (C)

Raghavaiah, N. V. (1979). Prostatography. *J. Urol.* **121**, 174–177. (B)

Raghavaiah, N. V., and Jordan, W. P. (1979). Prostatic lymphagraphy. *J. Urol.* **121**, 178–181. (B)

Ray, B., Hajdu, S. I., and Whitmore, W. F., Jr. (1974). Distribution of retroperitoneal lymph node metastases in testicular germinal tumors. *Cancer* **33**, 340–348. (A)

Rodin, A. E., Larson, D. L., and Roberts, D. K. (1967). Nature of the perineural space invaded by prostatic carcinoma. *Cancer* **20**, 1772–1779. (B)

Rotter, W., and Schürmann, R. (1950). Die blutegefässe des menschlichen penis. Beitrag zur orthologie und pathologi der regulationssyteme des peripheren kreislaufs (arterio-venöse anastomosen sperrarterien und drosselvenen). *Virchows Arch. Pathol. Anat. Physiol.* **318**, 352–393. (C)

Ruzbarsky, V., and Michal, V. (1977). Morphological changes in the arterial bed of the penis with aging: Relationship to the pathogenesis of impotence. *Invest. Urol.* **15**, 194–199. (C)

Saidi, F., Osmond, J. D., III, and Hendren, W. H. (1973). Microangiographic study in experimentally produced megaureters in rabbits. *J. Pediatr. Surg.* **8**, 117–123. (D)

Sarma, K. P. (1981a). Microangiography of the bladder in health. *Br. J. Urol.* **53**, 237–240. (E)

Sarma, K. P. (1981b). Genesis of papillary tumors: Histological and microangiographic study. *Br. J. Urol.* **53**, 228–236. (E)

Sayegh, E., Brooks, T., Sacher, E., and Busch, F. (1966). Lymphangiography of the retro-peritoneal lymph nodes through the inguinal route. *J. Urol.* **95**, 102–107. (A)

Setchell, B. P. (1970). Testicular blood supply, lymphatic drainage and secretion of fluid. *In* "The Testis" (A. D. Johnson, W. R. Gomes, and N. L. Van Demark, eds.), Vol. 1, pp. 101–239. Academic Press, New York. (A)

Setchell, B. P., Waites, G. M. H., and Thorburn, G. D. (1966). Blood flow in the testis of the conscious ram measured with Krypton-85. *Circ. Res.* **18**, 755–765. (A)

Shafik, A. (1972). A study of the arterial pattern of the normal ureter. *J. Urol.* **107**, 720–722. (D)

Shehata, R. (1976). The arterial supply of the urinary bladder. *Acta Anat.* **96**, 128–134. (E)

Shehata, R. (1979). Venous drainage of the urinary bladder. *Acta Anat.* **105**, 61–64. (E)

Sirca, A., Dekleva, A., and Kordas, I. (1978). Vascularization of the ureter after experimental ureoterolysis in rabbits. *Invest. Urol.* **15**, 422–424. (D)

Siroky, M. B., and Krane, R. J. (1979). Physiology of male sexual function. *In* "Clinical Neuro-urology" (R. J. Krane and M. B. Siroky, eds.), pp. 45–62. Little, Brown, Boston, Massachu-setts. (C)

Smith, M. J. V. (1966). The lymphatics of the prostate. *Invest. Urol.* **3**, 439–444. (B)

Vastarini-Cresi, G. (1902). Communicazioni dirette tra le arterie e el vene (anastomosi-arterio-venose) nei mammiferi. *Monit. Zool. Ital.* **13**, 136–144. (C)

von Ebner, V. (1900). Über klappenartige vorrichtungen in der arterien der Schwellkorper. *Anat. Anz.* **18**, 79–81. (C)

Wagner, G. (1981). Erection: Anatomy. *In* "Impotence" (G. Wagner, and R. Green, eds.), pp. 7–24. Plenum, New York. (C)

Wagner, G., Willis, E. A., Bro-Rasmussen, F., and Nelson, M. H. (1982). New theory on the mechanism of erection involving hitherto undescribed vessels. *Lancet* **1**, 416–418. (C)

Whitmore, W. F., Blute, R. D., Kaplan, W. D., and Gittes, R. I. (1980). Radiocolloid scintigraphic mapping of the lymphatic drainage of the prostate. *J. Urol.* **124**, 62–67. (B)

Williams, P. L., and Warwick, R., eds. (1980). "Gray's Anatomy," 36th ed. Saunders, Philadelphia, Pennsylvania. (D,E)

Willis, E., Ottensen, B., Wagner, G., Sundler, F., and Fahrenkrug, J. (1981). Vasoactive intestinal polypeptide (VIP) as a possible neurotransmitter involved in penile erection. *Acta Physiol. Scand.* **113**, 545–547. (C)

Yoffey, J. M., and Courtice, F. L. (1970). "Lymphatics, Lymph and the Lymphomyeloid Com-plex," p. 251. Academic Press, New York. (A)

Young, W. F., Dey, R. D., and Echt, R. (1979). Comparisons of prostaglandin vasoactive effects and interactions in the *in vivo* microcirculation of the rat urinary bladder. *Microvasc. Res.* **17**, 1–11. (E)

Chapter 17
Genitourinary System: Female Reproductive Organs

A. BLOOD VESSELS AND LYMPHATICS OF THE OVARY

By K. E. Clark

The primary function of the ovaries is to produce oocytes and steroids. This process is regulated by the hypothalamus and the pituitary luteinizing hormones (LH) and follicle stimulating hormone (FSH). Raised levels of FSH stimulate thecal and follicular cells to produce estrogen that, in turn, causes the pituitary gland to secrete LH. The latter is important in inducing ovulation, following which it stimulates follicular cells to produce progesterone. Luteinizing hormone and FSH help to complete the reproductive cycle by preparing the uterus for implantation (see Section B, below) and, by feeding back to the hypothalamus, to regulate the further formation of these hormones.

1. ANATOMY OF BLOOD CIRCULATION

a. Blood vessels: The ovary receives its blood supply from the ovarian arteries that arise from the anterior aspect of the abdominal aorta, just below the renal arteries. These vessels run up the broad ligament, breaking up over the ovary into terminal branches that enter the organ through the hilus along its anterior border. Divisions of the ovarian arteries also perfuse a portion of the fallopian tubes and anastomose with the uterine artery. The ovarian veins arise from the ovary and form a plexus in the broad ligament. This gives rise to a single ovarian vein from each ovary, the right ovarian vein draining directly into the inferior vena cava, and the left, into the left renal vein.

b. Lymphatic vessels: The lymphatic system of the ovary originates with blind ending "fingerlike" thin-walled capillaries which form a rich plexus draining through the medulla via numerous large lymphatic channels that contain valves. In general, the latter vessels parallel the blood vascular supply. After leaving the ovary, they drain to the middle lumbar lymph nodes, the purified lymph eventually being returned to the thoracic duct and the venous system.

c. Nervous innervation: The plexus of nerves that follows the ovarian vessels contains afferent fibers from the tenth thoracic nerve and efferents, both sympathetic and parasympathetic (thoracolumbar and vagal fibers). The adrenergic fibers may be secretory, as well as vasomotor, in function, since some nerve endings have been found to be in no way related to the blood vessels. They form wreaths around normal and atretic follicles, giving off many small branches that have been traced up to, but not through, the membrana granulosa.

2. PHYSIOLOGY AND PHARMACOLOGY OF BLOOD CIRCULATION

Early studies of corpus luteum regression focused on the vascular actions of $PGF_{2\alpha}$. This prostaglandin was known to be elevated at the time of luteal regression and was shown to be a vasoconstrictor of the ovarian circulation. However, Pang and Behrman (1979) demonstrated that at the time of luteal regression, no changes in ovarian blood flow occurred. In contrast, LH has been shown to cause ovarian hyperemia, which appears to be mediated by prostaglandins, since its action can be blocked by indomethacin (Lee and Novy, 1978).

Only limited studies have been performed to evaluate the pharmacology of the ovarian circulation. The vasculature responds to angiotensin II and norepinephrine with vasoconstriction and to β-receptor stimulation (Fenoteral) with vasodilatation (Phernetton and Rankin, 1978; Varza *et al.*, 1979). Although the role of prostaglandins in regulating ovarian blood flow has not been established, indomethacin does increase vascular resistance (vasoconstriction) in this vascular bed.

B. BLOOD VESSELS AND LYMPHATICS OF THE UTERUS

By K. E. CLARK

The primary function of the uterus is to provide an environment for the fertilized ovum to develop into first a blastocyst, then an embryo, and, finally, a fetus. Pregnancy creates a unique cardiovascular challenge for the maternal organism. A new vascular bed develops, thus imposing greatly increased demands upon the maternal circulation. Indeed, maternal cardiac output is raised about 30% in order to perfuse the placenta adequately and provide critical nutrients and oxygen for the fetus. As the end of gestation approaches, the previously quiescent uterus undergoes increased contractile activity, and finally parturition occurs.

Because of the cyclic hormonal fluctuations during the reproductive period, the thickness of the endometrium of the uterus changes continuously. In primates, this layer undergoes three distinct phases: menstrual, proliferative (follicular), and secretory (luteal). Each phase of the ovarian cycle causes specific vascular changes.

1. ANATOMY OF BLOOD VESSELS AND LYMPHATICS

a. Blood vessels: The uterus receives its blood supply from the ovarian, vaginal, and uterine arteries. The uterine artery arises from the internal iliac artery and then enters the base of the broad ligament, to follow along the side of the uterus. Just proximal to the level of the supravaginal portion of the cervix, it divides, with the smaller branch, the cervicovaginal artery, going to the cervix and upper portion of the vagina where it anastomoses with the vaginal artery. The uterine artery finally terminates into three branches: tubal, fundal, and ovarian, the latter anastomosing with the ovarian artery. The anastomosis between the ovarian and vaginal arteries probably plays an important role in providing adequate arterial perfusion during pregnancy. The small veins of the uterus give rise to a plexus that accompanies the uterine artery laterally and terminates in the internal iliac vein.

b. Lymphatic vessels: The endometrium (subserosal) is abundantly supplied with lymphatic vessels that are largely confined to the base of the uterus. The lymphatic drainage from the body of the uterus ends in the hypogastric nodes, or if the lymph vessels follow ovarian arterial channels, they terminate in the lumbar nodes. Lymph vessels from the fundus join ovarian channels and drain into lumbar nodes, whereas cervical lymphatics end in either the external or internal iliac nodes.

c. Nervous innervation: The uterus receives both sympathetic and parasympathetic innervation. The sympathetic nervous system enters the pelvis through the hypogastric plexus, which arises from the aortic plexus. Stimulation of the sympathetic fibers causes myometrial contraction and also vasoconstriction. The parasympathetic system arises from the pelvic nerve, which consists of a few fibers derived from the second, third, and fourth sacral nerves. Stimulation of the parasympathetic nerves inhibits myometrial contraction and also causes vasodilatation.

2. PHYSIOLOGY OF BLOOD VESSELS

a. Menstrual cycle: The vasculature of the uterus undergoes well-defined changes during the menstrual cycle (Ramsey and Donner, 1980). The branches of the uterine arteries penetrate the uterine wall to the middle of the myometrium where they subdivide into arcuate arteries. The latter give rise to radial arteries, which come off at a 90° angle and proceed to the myometrial–endometrial junction where, upon entering the endometrium, they are called endometrial spiral arteries. Branches of the radial arteries, the basal arteries, begin at the myometrial–endometrial junction. The endometrial spiral arteries are hormone-sensitive, whereas basal arteries do not respond to steroids (Ramsey and Donner, 1980). In contrast to primate studies, both the myometrial and endometrial vessels of sheep appear sensitive to the vascular effects of estrogen (Makowski, 1977).

b. Circulatory responses to estrogen and progesterone: Although estrogen and progesterone have numerous effects on the uterus and cervix, the following discussion is limited to their vascular actions. Estrogen stimulates the growth of both the endometrium and endometrial arteries, in primates the spiral arteries responding to a greater extent than the endometrium. Estrogen also produces profound vasodilatation of the uterine vasculature (Makowski, 1977), the mechanism of which is still unclear. The effect is not a direct action, since vasodilatation does not begin to occur for a period of 45 min, with a maximum response being noted at 90 to 120 min. It is apparent from pharmacologic studies that, at least in the sheep, estrogen-induced vasodilatation is not mediated by acetylcholine, histamine, or prostaglandins. Since the response can be prevented by protein synthesis inhibition (Makowski, 1977) and since it is associated with a time delay of 45 to 60 min, it is currently thought to be mediated by either a peptide or a compound produced by a newly formed or activated enzyme.

During the luteal phase of the ovarian cycle, progesterone levels increase dramatically. As a result, the proliferative endometrium is converted to secretory endometrium and then to decidua. Progesterone also inhibits uterine contractile activity and, in the presence of estrogen, it continues to support endometrial and arterial growth. It has been reported to have little direct effect on uterine blood flow, but it has been shown to decrease estrogen-induced augmentation in circulation by approximately 25%. The mechanism of such a response is currently unknown. Upon progesterone withdrawal, the endometrium and, subsequently, endometrial arteries (spiral arteries in primates) undergo degeneration. Moreover, the spiral arteries, which normally manifest periodic contractions throughout the ovarian cycle and pregnancy, now show much more intense contractions, the response appearing to produce ischemia.

c. Mechanisms involved in alterations in uteroplacental blood flow: Pregnancy is associated with dramatic increases in uteroplacental blood flow. In the late stage, the vascular response amounts to 20% of the maternal cardiac output, with approximately 80% of the uterine blood flow going to the placenta. Such a large increase is thought to be the result of both an accelerated vessel growth and vasodilatation of the vasculature. Although the blood vessels can still dilate in response to estrogen in late pregnancy, the magnitude of the response is decreased, suggesting that the vasculature is already almost maximally dilated. In addition, the vessels in the myometrium and the endometrium dilate to a much greater extent than those in the placenta (Makowski, 1977). The agent or agents responsible for uterine and placental vasodilatation have not been clearly identified. Numerous studies have suggested a role for estrogens, vasoactive peptides, and prostaglandins; other pharmacologic investigations have ruled out acetylcholine, dopamine, and histamine.

d. Autoregulation: Several studies have dealt with the ability of the uterus to autoregulate. Investigations by Greiss (1966) in sheep during the last 30 days of gestation demonstrate a linear pressure–flow relationship, with no indication of autoregulation. A similar lack of autoregulatory response also exists in the case of the placenta but not in the myometrium of the pregnant sheep and in the nonpregnant uterus (Makowski, 1977). In addition to arterial blood pressure, uterine contractile activity can significantly modify uterine blood flow. Increased uterine tone and contractile activity cause mechanical compression of the uterine vasculature and thus decreased uterine blood flow.

3. PHARMACOLOGY OF BLOOD VESSELS

As described in Section B-2b, above, the estrogens are potent vasodilators of the pregnant and nonpregnant uterine vasculature (Makowski, 1977). Direct administration of norepinephrine or stimulation of sympathetic nerves leads to uterine arterial vasoconstriction. Circulating angiotensin II also produces vasoconstriction, but in pregnancy both uterine and systemic vascular responses are depressed. The latter response appears to be due to increased circulating angiotensin II, which decreases the sensitivity or subsequent number of angiotensin receptors. Serotonin and dopamine cause vasoconstriction in both pregnant and nonpregnant sheep following direct administration into the uterine vasculature. In contrast, in nonpregnant sheep, the polypeptides, bradykinin, and vasoactive intestinal polypeptide (VIP) are potent uterine vasodilators (Clark *et al.*, 1981a); in pregnant sheep they either have no effect (VIP) or produce vasoconstriction (bradykinin). In recent years, the uterine vascular effects of prostaglandins have been extensively investigated (Clark *et al.*, 1981b, 1982; Clark and Brody, 1982). From these studies, it is clear that only prostacyclin (PGI_2) and PGD_2 are potent vasodilators of the pregnant and nonpregnant uterine vasculature. PGE_2 acts as a weak uterine dilator in nonpregnant sheep and decreases uterine blood flow in pregnant sheep.

4. Pathophysiology of Blood Vessels

Although there are numerous diseases that affect uterine function, pregnancy-induced hypertension (preeclampsia) is the one most clearly associated with a reduction in uterine blood flow in the late stage. Recent studies have clearly indicated that in this disorder the uteroplacental vasculature has a reduced ability to produce prostacyclin. Presently it is not known what causes such a defect, but decreased production of a uterine vasodilator such as prostacyclin could clearly cause the clinical symptoms that occur in preeclampsia. Further research into this possible mechanism is required.

C. BLOOD CIRCULATION OF THE PLACENTA

By E. M. Ramsey

1. Embryology

In the development of the placenta and its vasculature, there are three fundamental differences between this organ and the intrinsic organs of the developing embryo. In the first place, the placenta is at no time a part of the embryonic body. From earliest preimplantation stages, the trophoblastic wall of the blastocyst, which will give rise to the placenta, is distinct from the inner cell mass that forms the embryo, and the two pursue entirely different developmental paths. Second, the placenta does not accompany the fetus into its postnatal life but is cast off at parturition. Finally, the embryo itself contributes only half of the material that enters into the formation of the placenta; the other half is the mother's contribution. This is particularly true of the vasculature.

All rights of reproduction in any form reserved.
ISBN 0-12-042520-3

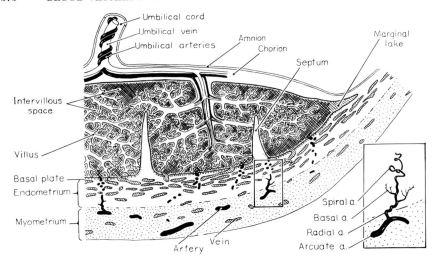

Fig. 17.1. Diagrammatic representation of a portion of the primate (hemochorial) placenta attached to the uterine wall. The insert shows, at higher magnification, the components of a single arterial stem. From E. M. Ramsey, *in* "Respiratory Problems in the Premature Infant," Report of 15th M & R Pediatrics Research Conference, 1955. (Courtesy of M & R Laboratories.)

Therefore, in the development of the vascular system of the placenta, the two components must be considered separately.

a. Maternal vasculature: This is an outgrowth of the progravid uterine vascular network, which must be traced first. In doing so, it is necessary to revert to the time when the mother herself was an embryo. In the sixth week of development, the Müllerian (paramesonephric) ducts arise as bilateral inrollings of coelomic epithelium at the level of the urogenital ridge. The ducts grow downward toward the pelvis, and in the ninth week, their lower segments fuse, forming the body of the uterus. At the same time, mesoderm forms around the ducts and, with subsequent growth, becomes the myometrium and endometrium. Vascular channels arise within the mesoderm at an early stage and gradually proliferate and interlace to form the network characteristic of the mature uterus (Fig. 17.1).

The spiral arteries of the endometrium have a unique property. Other arteries of the body, once formed, remain constant except when altered by trauma or pathology. The endometrial spiral arteries, on the other hand, react to cyclic hormonal stimuli and vary on a recurring schedule from one menstrual cycle to the next and on a different but fixed schedule in each pregnancy.

b. Fetal vasculature: The fetal blood vessels, which occupy the villi and follow the ramifications of the villous tree, originate in the second week of pregnancy as angioblasts in the mesodermal cores of the villi and in the chorion. By progressive coalescence, the independent units form a continuous system that connects with the vessels of the body stalk. This, in turn, forms the umbilical cord and its vessels: two arteries carrying venous blood from the fetus to the placenta and a single vein returning oxygenated blood to the fetus. Branches of the umbilical vessels course in all directions through the chorionic plate.

The absorptive units of the fetal vasculature, which run in the chorionic villi, lie in the terminal, twig-size fronds. The precapillary bed is intricately ramified and connects with two types of richly anastomosing capillary complexes: a superficial network and a paravascular system that provides extravillous shunts. The latter act as safety valves to prevent overloading of the villous circulation or local stoppage should fibrosis or pathologic developments occur (Bøe, 1953). The terminal capillaries are located deep in the villous core at first, but approach ever closer to the trophoblastic border until, at term, only the most delicate mantle of syncytium separates them from maternal blood in the intervillous space (Aladjem, 1970). Occasional microscopic breaks in this covering permit passage of a few fetal blood cells into the maternal circulation.

2. ANATOMY

a. Gross and microscopic anatomy of vasculature of nonpregnant uterus: Figures 17.2A and B illustrate the changes that humoral stimuli evoke in the endometrial arteries of the nonpregnant monkey uterus with each menstrual cycle. It may be noted that reproduction in the rhesus monkey and in man is so similar (both anatomically and physiologically) that the monkey may be used as a model for primate reproduction in general. This is a matter of importance, since many experimental procedures necessary for analysis of reproductive phenomena are inappropriate for use in human patients.

Histologically, the myometrial blood vessels of the nonpregnant uterus present the pattern characteristic of arteries and veins supplying smooth muscle elsewhere. In functional activity, the local contractility of the radial arteries is highly significant. This function appears to be humorally controlled (see Section B-2b, above, and Fig. 17.2C). The current work of Zuspan *et al.* (1981) highlights the occurrence of catecholamine fluorescent nerves in the walls of myometrial vessels.

The basal and spiral endometrial arteries differ markedly from one another. The former are characterized by paucity of elastic fibrils and predominance of a myoepithelioid type of muscle cell, displaying abundant cytoplasm and large nuclei (Okkels and Engle, 1938). They are not responsive to hormonal stimulation and remain intact at the time of menstruation. The spiral arteries, on the other hand, have a well-developed muscle coat, containing a rich network of elastic fibrils which extends into the precapillary area of the subepithelial zone.

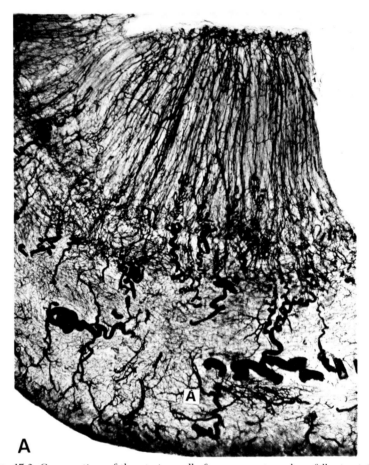

A

Fig. 17.2. Cross sections of the uterine wall of nonpregnant monkeys following injection of India ink and starch by G. W. Bartelmez. (A) Twelfth day of menstrual cycle. Minimal coiling of spiral artery restricted to lower one-fourth of endometrium. (B) Seventeenth day of cycle. Extensive coiling and growth toward endometrial surface. (C) Eleventh day of cycle. Marked constriction of radial artery at myometrial–endometrial junction (arrow). Monkeys B338, B309, and B323, respectively (see pp. 581 and 582). (Courtesy of the Carnegie Institution of Washington.)

As the menstrual cycle advances and the spiral arteries lengthen, the elastic and muscle constituents progress distally. In the composition of their walls, the spiral arteries so closely resemble the radial arteries that they are currently considered as prolongations of them, a similarity heightened by the intermittent contractility that they also display. In contractile capacity, the individual arteries are independent units, and constriction is not necessarily synchronous. The vasoconstrictive activity persists during pregnancy (for further discussion of the vasculature of the nonpregnant uterus, see Section B-2a and 2b, above).

FIG. 17.2 (Continued)

b. Gross and microscopic vascular adaptations in pregnant uterus: Within a matter of hours after implantation of a fertilized ovum, invading trophoblast opens up the walls of regional capillaries, permitting maternal blood to seep into lacunar spaces within the trophoblastic shell (Fig. 17.3A). During the succeeding week, in consequence of progressive growth of spiral arteries and increasing invasion of the endometrium by trophoblast, the tips of maternal arteries are opened. Blood entering the lacunae under arterial pressure expands them and promotes formation of intercommunications. The trophoblastic columns intervening between the lacunae differentiate into characteristic chorionic villi, with a mesodermal core and a border of cytotrophoblast and syncytiotrophoblast (Fig. 17.3B). The latter forms the outer covering of the villi, lining the lacunae and coming into contact with the maternal blood. The lacunae thus become intervillous spaces or, in the aggregate, "the intervillous space" of the placenta. When slow seepage of blood into the lacunae from low

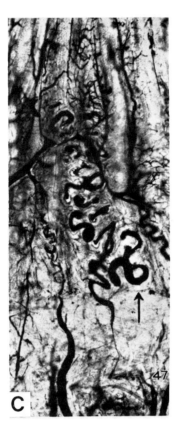

FIG. 17.2 (Continued)

pressure capillaries is replaced by active arterial flow, true placental circulation is established.

At the same time, the changes in endometrial spiral arteries that were initiated during the menstrual cycle continue. During the first third of pregnancy, there is progressive growth of the arteries, combined with decreasing width of the endometrium, the latter occasioned by trophoblastic erosion and by pressure of the overlying comceptus. In order that the vessels can be accommodated within the endometrium, not only increased coiling of the arteries but

FIG. 17.3. (A) Section through an implanted monkey ovum 10 days after fertilization. Arrow indicates inner cell mass. There is direct communication between maternal capillaries and trophoblastic lacunae, with maternal blood elements seen in the latter. Carnegie specimen C-524; section 3-4-10. (B) Section of the wall of the blastocyst of a 16-day human embryo showing the three layers of an early chorionic villus. i.v.s., intervillous space; Mes., mesodermal core; Sync., syncytiotrophoblast; Cyto., cytotrophoblast. Carnegie specimen 7802; section 26-8-5. [Both (A) and (B), courtesy of the Carnegie Laboratories of Embryology, University of California, Davis.]

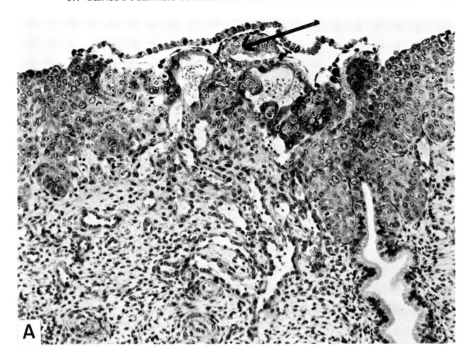

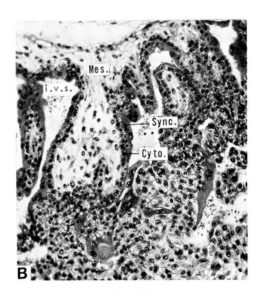

FIG. 17.3

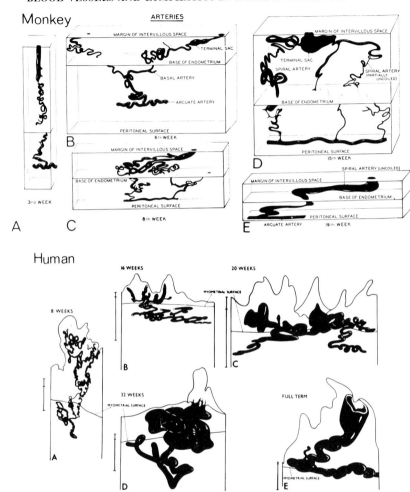

FIG. 17.4. Diagrammatic representations of the course and configuration of monkey and human uteroplacental arteries based on three-dimensional models constructed from serial sections. Comparable stages of pregnancy are shown in the two species. (Courtesy of the Carnegie Institution of Washington.)

also back and forth looping becomes necessary. At mid-term, when growth of the component parts of the uterus is superseded by simple stretching as the effective force in uterine enlargement (the epoc named "conversion" by Reynolds, 1947), coils of the spiral arteries are paid out, and their subsequent course is a gently undulating one (Fig. 17.4). A terminal dilatation of the arteries, the "terminal sac," occurring just before connection of the vessels with the intervillous space, is a striking feature in both rhesus monkey and man. The mechanism of formation of the dilatation and its role in the hemodynamics of placental circulation are discussed below.

In general, in nonhuman primates, the base of the placenta does not expand after the original implantation, completed by the end of the first trimester. In man, however, there is significant expansion, so that new arteries are tapped throughout pregnancy (Gruenwald, 1972). Thus at the fourth month, there are 100 to 500 communicating arteries and at term, some 180 to 300 such vessels (Boyd, 1956). In contrast, in the bidiscoid placenta of rhesus, there are only 8–12 arteries per disc from the time of maturity until parturition. The distribution of maternal vessels tapped by the trophoblast is haphazard, reflecting their distribution in the progravid endometrium. Characteristic pregnancy changes occur only in the endometrial arteries connecting directly with the intervillous space. The remaining vessels, which are concerned with mural circulation, remain small and unchanged.

The alterations in endometrial veins during pregnancy are essentially passive, namely, stretching and dilatation. External compression by the endometrial stroma effects reduction in the number of vessels as the placenta matures. Patent channels remain, however, in all regions of the placental base. Some of those draining the central portion are slender and run an oblique course, especially in the human; those in the monkey are wider and often run directly from the intervillous space to the muscularis. In all primates, more prominent veins frequently drain the periphery of the placenta, the so-called "marginal sinus" (Spanner, 1935). These account for about one-third of the total drainage. Such vessels are not to be confused with the dilated veins that form a wreath within the endometrium at the margin of the placenta and are not connected with the organ. Presently, the marginal sinus itself is regarded as a downward extension of the subchorial lake. It is often discontinuous and may fluctuate from time to time. The alternate designation "marginal lake" or "lakes" is preferred by many authors.

The microscopic vascular adaptations in the uterus during pregnancy are manifestations of one of the trophoblast's activities. They consist first, of tapping of endometrial vessels and second, of cytotrophoblastic entry into the lumina and walls of the spiral arteries. The channel of entry of the artery into the intervillous space may initially be completely or partially blocked (Fig. 17.5A). Proximally the trophoblast progresses toward the myometrium in "candle drip" fashion. The process decreases as the muscularis is approached. In man it reaches the inner third of the myometrium around the twelfth week and thereafter disappears. The process is of shorter duration in other primates and halts at the myometrial–endometrial junction. Originally regarded as a proliferation of the endothelial lining of the arteries, the invading cells have been investigated by sex chromatin studies (Harris et al., 1917) and electron microscopy (Panigel, 1969; DeWolf et al., 1973) and their trophoblastic nature established.

Following the twelfth week, trophoblastic invasion of the arterial wall proper commences, progressing proximally and with the same distribution as the intraarterial invasion. The cytotrophoblast replaces the elastic and muscle

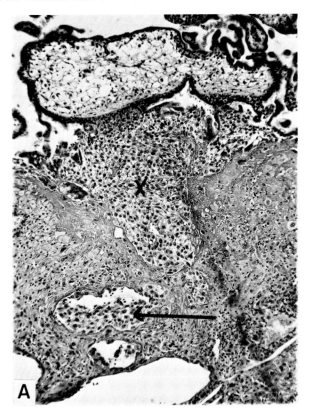

FIG. 17.5. (A) Section of the site of attachment of the placenta in a 12-week human pregnancy. Cytotrophoblast is penetrating the lumen of an endometrial spiral artery, occluding its orifice at the point of entry to the intervillous space (X) and "dripping" into proximal portions of its lumen (arrow). University of Virginia No. 32-2. (Courtesy of the University of Virginia, Department of Obstetrics and Gynecology.)

tissues of the wall. In man alone, there is also migration of trophoblastic giant cells into the endometrial stroma, especially perivascularly. This process, in turn, is followed by disappearance of the trophoblast in the arterial walls by a mechanism not yet understood. The trophoblast is replaced by fibrosis that lasts to term (Fig. 17.5B).

The activity of the trophoblast has been extensively studied over the past 20 years by an international, interdisciplinary team (Brosens *et al.*, 1967; Pijnenborg *et al.*, 1981a,b; DeWolf *et al.*, 1973), which has established that the process is "physiologic," i.e., entirely normal and indeed necessary. The effect upon the arterial walls is a weakening, and loss of elasticity and contractility, though the final fibrosis prevents danger of actual rupture under normal systemic pressure. The force of the latter does produce dilatation of the endo-

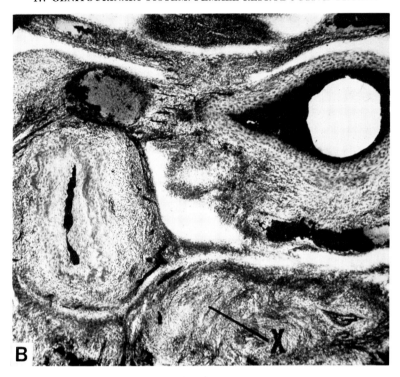

FIG. 17.5. (B) Section of the uterine wall in the area of placental attachment in a monkey on the 102nd day of pregnancy. Segments of two spiral arteries are seen. In the one on the right, the wall is replaced by trophoblast, in that on the left, by fibrosis. At X a segment of the radial artery is occluded by vasoconstriction. Carnegie specimen C-679; section B125. (Courtesy of the Carnegie Laboratories of Embryology, University of California, Davis.)

metrial portion of the arteries and, in particular, creation of the terminal sac. The dilatation is necessary to slow the circulatory flow, reduce the afferent pressure, and allow for the handling of greatly increased volumes of blood. For, in the nonpregnant human, the spiral arteries transport only a few milliliters of blood per minute, as compared with the state of placental maturity, when they convey some 600 ml/min (Robertson *et al.*, 1975). On the other hand, at the point of the artery's entrance to the intervillous space, pressure is reduced to only about 12 mm Hg (Moll and Freese, 1971).

   c. Vasculature of mature maternal placenta: In Fig. 17.1 are presented in schematic form the components of the mature primate placenta and their relationships as described above. Also to be noted is the manner in which the mature placenta is irregularly subdivided into a variable number of maternal cotyledons or lobes by slender septa of connective tissue in which endometrial and trophoblastic elements mingle. In the human, entering arteries often rise within these septa to orifices located along the sides. The septa run from the base of the placenta almost to the chorionic plate, a few actually connecting with it. Communication via the subchorial lake and through perforations in the septa prevents the cotyledons from being entirely self-contained circulatory units (Boyd, 1956; Ramsey, 1954). The maternal cotyledons (lobes) are best observed on the maternal surface of the delivered placenta, in contradistinction to the fetal cotyledons described below.

Of interest is the question of the capacity of the intervillous space. On the basis of highly sophisticated morphometry, Aherne and Dunnil (1966) determined that the volume of the space in normal human placentas at term is 144 ml and the villous surface is 11.00 m².

d. Gross and microscopic anatomy of fetal vasculature: Differentiation of the primitive trophoblastic columns into chorionic villi is an achievement of the first 2 weeks after implantation. Subsequent growth and development of the villi result in the attainment of a final form, which may be likened to that of a deciduous tree rooted in the chorionic plate, with trunk, limbs, branches, and twigs hanging into the intervillous space. Many of the terminal twigs adhere to the endometrium or to the sides of the septa as anchoring villi (Fig. 17.6).

The chorionic villi are grouped into fetal cotyledons. There are some 200 in the mature human placenta and 8–12 per disc in the rhesus placenta (Boyd and Hamilton, 1970). The unfortunate use of the term cotyledon for two entirely separate structures necessitates careful differentiation. The maternal cotyledon (better referred to as "lobe") is the portion of the placenta lying between two septa, and the fetal cotyledon is that portion of the placenta related to the ramifications of a single stem villus. Several cotyledons may be embraced in a single maternal lobe (although only one is shown in the drawing in Fig. 17.1), whereas fetal cotyledons frequently overlap between lobes. Maternal arteries enter the areas of fetal cotyledons in the ratio of 1:1 in all primates so far studied.

The chorionic villi are, for the most part, closely packed together in the fetal cotyledons, except that at the points of inflow of maternal arterial blood, the free villi surrounding the orifices of entry are swept aside. Villous population in these central areas of cotyledons is appreciably less dense than that at the periphery, creating sites of markedly diverse chemical concentrations (Freese, 1968; Schuhmann, 1981). Reduced villous population is also seen in the sub-chorial lake and marginal sinus regions. (For further discussion of anatomy of fetal vasculature, see Section C-1b, above.)

3. Physiology

a. Maternal blood flow: Students of placental hemodynamics in the early years of this century were confronted by a dilemma: If orifices of entry to and

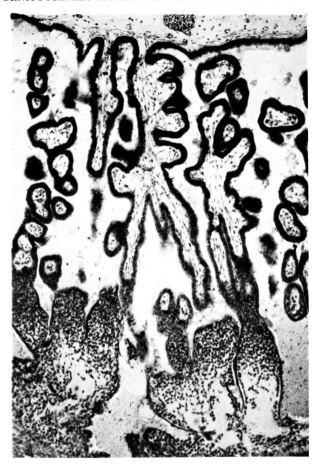

FIG. 17.6. Section of the blastocyst wall of a 29-day monkey embryo showing arborization of chorionic villi and differentiation of free and anchoring types. Fetal capillaries are seen as delicate strands in the mesodermal cores of the villi. At the top, a larger capillary appears in the chorionic plate. Carnegie specimen C-477; section 89. (Courtesy of the Carnegie Laboratories of Embryology, University of California, Davis.)

exit from the intervillous space are located indiscriminately along the base of the placenta, why does not inflowing maternal blood short-cut directly to the nearest venous exit, without circulating through the intervillous space? Various explanatory theories were formulated. Spanner (1935) denied the presence of venous exits except at the placental margin. He maintained that arterial blood rises to the subchorial lake between septal dividers and passes laterally to the "marginal sinus," where drainage takes place. This is the "overflow filling" concept. Bumm (1893) accepted basal exits but postulated upward coiling of arteries through the septa to orifices at the septal summits, with blood flowing back to venous exits, bathing the chorionic villi en route.

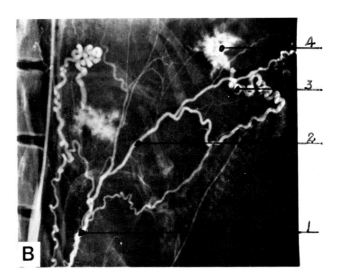

FIG. 17.7. (A) Section of the entire placenta attached to the uterine wall in a monkey on day 123 of pregnancy. A "spurt" of arterial blood into the intervillous space is seen on the left. Venous drainage from a different vascular field is seen on the right. Carnegie specimen C-750; section 87. (Courtesy of the Carnegie Laboratories of Embryology, University of California, Davis.) (B) Radiogram of a portion of a monkey uterus on day 107 of pregnancy. The full course of a representative uteroplacental artery is seen. 1, Arcuate artery; 2, radial artery; 3, spiral artery; 4, "spurt." (From Ramsey *et al.*, 1979, reproduced with permission of the *American Journal of Obstetrics and Gynecology.*)

Recent studies combining modern histologic and injection techniques have supplanted these older theories. Work, carried out by many independent investigators (Daron, 1936, 1937; Bartelmez, 1957; Borell *et al.*, 1958; Ramsey, 1949, 1954), has shown that arterial and venous orifices are indeed distributed throughout the basal plate as well as at the placental margin. Anatomic preparations and *in vivo* radioangiographic studies in both the rhesus monkey and man (Borell *et al.*, 1958; Donner *et al.*, 1963; Ramsey and Donner, 1980) demonstrate characteristic "spurts" or fountainlike "jets" entering the intervillous space (Figs. 17.7A and B). The "Physiologic Concept" (Ramsey and Donner, 1980) envisages the *vis a tergo* of maternal blood pressure as the controlling factor preventing short cutting, rather than morphologic arrangements. This pressure, though slight at the entry to the intervillous space, is sufficiently higher than the 5 mm Hg prevailing within the space that arterial blood is driven well up toward the chorion before lateral dispersion occurs. The incoming blood crowds the blood already present in the intervillous space toward the basal plate, where it enters orifices of exit connecting with maternal veins.

The pressure differentials are enhanced by the intermittent myometrial contractions that occur throughout pregnancy (Braxton–Hicks contractions). When maximal, the contractions completely prevent arterial inflow and venous drainage. Less powerful contractions decrease but do not eliminate flow. Radioangiography shows that the pool of blood in the intervillous space is maintained throughout uterine contractions. This negates the old idea that contractions "squeeze the placenta like a sponge." In fact, quite the opposite is the case. Blood is trapped in the intervillous space, a useful provision assuring that exchange continues throughout the contraction phase.

Constriction of uterine arteries during myometrial contractions is to be distinguished from the intrinsic vasoconstriction of individual arteries noted in the nonpregnant uterus. This capacity persists into pregnancy. As in the menstrual cycle, the constrictions are not synchronous, vessels functioning quite independently of one another.

With the progressing elimination of elastic and muscle tissue from the walls of the uteroplacental arteries, as the result of trophoblastic invasion, the endometrial spiral arteries become less able to constrict. The myometrial arteries, which remain intact, therefore become the effective site of vasoactivity. Radioangiography has shown that each constriction tends to last for some 6–12 sec (Ramsey *et al.*, 1979).

b. Fetal blood flow: The head of fetal blood pressure dominates the flow of fetal blood through the vessels within the chorionic villi in the same way as the *vis a tergo* of maternal systemic pressure controls the circulation in the intervillous space. Radioangiography has demonstrated that uterine contractions do not affect the fetal vessels, even those in the umbilical cord, presumably because of the cushioning effect of the blood pool in the intervillous space (Ramsey *et al.*, 1966).

c. Exchange: In 1927 Grosser formulated a morphologic classification of the placentas of various animals, based on the number of tissue layers separating the fetal and maternal bloodstreams. The diversity from species to species is probably a reflection of variations in the "invasive potential" of the trophoblast, though this vague explanation requires clarification. Suffice it to say that the placentas of primates (like those of lagomorphs, rodents, some bats, etc.) are of the hemochorial type, i.e., the trophoblast destroys all the maternal tissues that intervene between maternal and fetal blood: the uterine surface epithelium, the endometrial stroma, and the walls of the endometrial arteries. Thus the maternal blood circulates freely around the chorionic villi and is separated from the fetal blood only by three fetal tissue layers: the trophoblastic covering of the villi, the mesodermal cores of the villi, and the endothelium of the fetal capillaries.

In lower animals, the mechanism of circulation within the placenta is by "countercurrent flow." Thus the blood, which in these animals is contained in completely closed vessels (no amorphous intervillous space), flows in opposite directions. In this way, reduced blood makes its first contact with maternal blood in the venous state and gradually comes in contact with increasingly pure arterial maternal blood (Mossman, 1926). It has been shown that maximum crossing of a permeable membrane takes place under such circumstances (Noer, 1946), a finding of wide occurrence in biologic systems (e.g., the gills of fish). It seems unlikely that such countercurrent flow takes place in the hemochorial placenta, in which fetal vessels within the chorionic villi are immersed in a maternal pool of blood that has no restraining walls to direct flow. Furthermore, the differing chemical constitution of the maternal blood in various regions of the intervillous space renders the system highly irregular. It is presently considered that the mechanism of circulation is a "multivillous flow," as described by Bartels *et al.* (1962).

References*

Aherne, W., and Dunnill, M. S. (1966). Quantitative aspects of placental structure. *J. Pathol. Bacteriol.* **91**, 123–139. (C)

Aladjem, S. (1970). Studies in placental circulation: Vascular area of the terminal villus in normal and abnormal pregnancies. *Am. J. Obstet. Gynecol.* **107**, 88–92. (C)

Bartelmez, G. W. (1957). The phases of the menstrual cycle and their interpretation in terms of the pregnancy cycle. *Am. J. Obstet. Gynecol.* **74**, 931–955. (C)

Bartels, H., Moll, W., and Metcalfe, J. (1962). Physiology of gas exchange in the human placenta. *Am. J. Obstet. Gynecol.* **84**, 1714–1730. (C)

Bøe, F. (1953). Studies on the vascularization of the human placenta. *Acta Obstet. Gynecol. Scand.* 5 Suppl. **32**, 1–92. (C)

Borell, U., Fernström, I., and Westman, A. (1958). Eine arteriographische Studie des Plazentakreislaufs. *Geburtshilfe Frauenheilkd.* **18**, 1–9. (C)

Boyd, J. D. (1956). Morphology and physiology of the uteroplacental circulation. *In* "Gestation" (C. A. Villee, ed.), pp. 132–194. Macy, New York. (C)

Boyd, J. D., and Hamilton, W. J. (1970). "The Human Placenta." Heffer, Cambridge, England. (C)

*In the reference list, the capital letter in parentheses at the end of each reference indicates the section in which it is cited.

Brosens, I. A., Robertson, W. B., and Dixon, G. (1967). The physiological response of the vessels of the placental bed to normal pregnancy. *J. Pathol. Bacteriol.* **93**, 569–679. (C)

Bumm, E. (1893). Über die Entwicklung des mütterlichen Blutkreislaufes in der menschlichen Placenta. *Arch. Gynaekol.* **43**, 181–195. (C)

Clark, K. E., and Brody, M. J. (1982). Prostaglandins and uterine blood flow. *In* "Prostaglandins: Organ and Tissue Specific Actions" (S. Greenburg, P. J. Kadowitz, and T. Burks, eds.), pp. 107–130. Dekker, New York. (B)

Clark, K. E., Stys, S. J., Mills, E. G., and Seeds, A. E. (1981a). Effects of vasoactive polypeptides on the uterine vasculature. *Am. J. Obstet. Gynecol.* **139**, 182–190. (B)

Clark, K. E., Austin, J. E., and Stys, S. J. (1981b). Effect of bisenoic prostaglandins on uterine blood flow in nonpregnant sheep. *Prostaglandins* **22**, 333–348. (B)

Clark, K. E., Austin, J. E., and Seeds, A. E. (1982). Effect of bisenoic prostaglandins and arachidonic acid on the uterine vasculature of pregnant sheep. *Am. J. Obstet. Gynecol.* **142**, 261–268. (B)

Daron, G. H. (1936). The arterial pattern of the tunica mucosa of the uterus in Macacus rhesus. *Am. J. Anat.* **58**, 349–419. (C)

Daron, G. H. (1937). The veins of the endometrium (*Macacus rhesus*) as a source of the menstrual blood. *Anat. Rec.* **67** (Suppl.) p. 13. (C)

DeWolf, F., DeWolf-Peeters, C., and Brosens, I. (1973). Ultrastructure of the spiral arteries in the human placental bed at the end of normal pregnancy. *Am. J. Obstet. Gynecol.* **117**, 833–848. (C)

Donner, M. W., Ramsey, E. M., and Corner, G. W., Jr. (1963). Maternal circulation in the placenta of the rhesus monkey: A radioangiographic study. *Am. J. Roentgenol., Radium Ther. Nucl. Med.* **90**, 638–649. (C)

Freese, U. E. (1968). The uteroplacental vascular relationship in the human. *Am. J. Obstet. Gynecol.* **101**, 8–16. (C)

Greiss, F. C. (1966). Pressure–flow relationship in the gravid uterine vascular bed. *Am. J. Obstet. Gynecol.* **96**, 41–47. (B)

Grosser, O. (1927). Frühentwincklung, Eihautbildung und Placentation des Menschen und der Säugetiere. *In* "Deutsche Frauenheilkunde, Geburtshilfe, Gynäkologie und Nachbargebiete in Einzeldarstellungen" (Begrundet E. von Opitz, ed.), Vol. 5. Bergman, München. (C)

Gruenwald, P. (1972). Expansion of placental site and maternal blood supply of primate placentas. *Anat. Rec.* **173**, 189–204. (C)

Harris, J. W. S., Houston, M. L., and Ramsey, E. M. (1971). Intravascular trophoblast in human, baboon and monkey uteri. *Anat. Rec.* **169**, 334. (C)

Lee, W., and Novy, M. J. (1978). Effects of luteinizing hormone and indomethacin on blood flow and steroidogenesis in the rabbit ovary. *Biol. Reprod.* **17**, 799–807. (A)

Makowski, E. L. (1977). Vascular physiology. *In* "Biology of the Uterus" (R. M. Wynn, ed.), pp. 77–99. Plenum, New York. (B)

Moll, W., and Freese, U. E. (1971). Hämodynamik des intervillösen Raumes der Primatenplazenta. *Perinat. Med. (Stuttgart)* **3**, 680–687. (C)

Mossman, H. W. (1926). The rabbit placenta and the problem of placental transmission. *Am. J. Anat.* **37**, 433–497. (C)

Noer, R. (1946). A study of the effect of flow direction on the placental transmission, using artificial placentas. *Anat. Rec.* **96**, 383–389. (C)

Okkels, H., and Engle, E. T. (1938). Studies on the finer structure of the uterine blood vessels of the Macacus monkey. *Acta Pathol. Microbiol. Scand.* **15**, 150–168. (C)

Pang, C. V., and Behrman, H. R. (1979). Relationship of luteal blood flow and corpus luteum function in pseudopregnant rats. *Am. J. Physiol.* **237**, 30–34. (A)

Panigel, M. (1969). Structure et ultrastructure comparée de la membrane placentaire chez certains primates non humains. *Bull. Assoc. Anat.* **145**, 319–337. (C)

Phernetton, T. M., and Rankin, J. H. G. (1978). The effect of several vasoactive drugs on ovarian blood flow in the near-term sheep. *Proc. Soc. Exp. Biol. Med.* **158,** 105–108. (A)

Pijnenborg, R., Robertson, W. B., Brosens, I., and Dixon, G. (1981a). Trophoblast invasion and the establishment of haemochorial placentation in man and laboratory animals. *Placenta* **2,** 71–91. (C)

Pijnenborg, R., Bland, J. M., Robertson, W. B., Dixon, G., and Brosens, I. (1981b). The pattern of interstitial trophoblastic invasion of the myometrium in early human pregnancy. *Placenta* **2,** 303–316. (C)

Ramsey, E. M. (1949). The vascular pattern of the endometrium of the pregnant rhesus monkey (*Macaca mulatta*). *Carnegie Contrib. Embryol.* **33,** 113–147. (C)

Ramsey, E. M. (1954). Venous drainage of the placenta of the rhesus monkey (*Macaca mulatta*). *Carnegie Contrib. Embryol.* **35,** 151–173. (C)

Ramsey, E. M., and Donner, M. W. (1980). "Placental Vasculature and Circulation." Thieme, Stuttgart. (B,C)

Ramsey, E. M., Martin, C. B., Jr., McGaughey, H. S., Jr., Kaiser, I. H., and Donner, M. W. (1966). Venous drainage of the placenta in rhesus monkeys: Radiographic studies. *Am. J. Obstet. Gynecol.* **95,** 948–955. (C)

Ramsey, E. M., Chez, R. A., and Doppman, J. L. (1979). Radioangiographic measurements of the internal diameters of the uteroplacental arteries in rhesus monkeys. *Am. J. Obstet. Gynecol.* **135,** 247–251. (C)

Reynolds, S. R. M. (1947). Relation of maternal blood flow within the uterus to change in shape and size of the conceptus during pregnancy. Physiological basis of uterine accommodation. *Am. J. Physiol.* **148,** 77–85. (C)

Robertson, W. B., Brosens, I., and Dixon, G. (1975). Uteroplacental vascular pathology. *Eur. J. Obstet., Gynecol. Reprod. Biol.* **5**(½), 47–65. (C)

Schuhmann, R. (1981). Plazenton: Begriff, Entstehung, funktionelle Anatomie. *In* "Die Plazenta des Menschen" (V. Becker, T. H. Schiebler, and F. Kubli, eds.), pp. 197–207. Thieme, Stuttgart. (C)

Spanner, R. (1935). Mütterlicher und kindlicher Kreislauf des menschlichen Placenta und seine Strohmbahnen. *Z. Anat. Entwicklungsges.* **105,** 163–242. (C)

Varza, B., Zsolnai, B., and Bernard, A. (1979). Stimulation of the alpha- and beta-adrenergic receptors in human ovarian vasculature in vitro. *Gynecol. Obstet. Invest.* **10,** 31–87. (A)

Zuspan, F. P., O'Shaughnessy, R. W., Vinsel, J., and Zuspan, M. (1981). Adrenergic innervation of uterine vasculature in human term pregnancy. *Am. J. Obstet. Gynecol.* **139,** 678–680. (C)

Chapter 18
Integumentary System: Skin and Adipose Tissue

A. EMBRYOLOGY OF CUTANEOUS BLOOD
CIRCULATION

By K. Christensen

At the beginning of the third month of human development, the epidermis is only 2-cell layers' thick and the dermis is a "gelatinous bed of proliferating mesenchymal cells" (Ryan, 1973). At this time, the skin is devoid of blood vessels. During the third month, accumulating cells having a positive alkaline phosphatase reaction appear in portions of the cutaneous mesenchyme. These are the first to produce endothelium of the blood vessels, although also giving rise to a variety of other cell types. Interestingly, the alkaline phosphatase reaction is greatest where the early epidermis proliferates and later forms hair follicles or nail beds of the fingers and toes.

In both the third and fourth months, arterial and venous channels mature in the subcutaneous connective tissues, and blood begins to circulate through them after the superficial vessels in the dermis have made connections with deeper vessels.

At the fourth month, a few vessels develop muscular and adventitial layers, but before this takes place, budding and anastomoses are probably unrestricted and occur wherever there is stimulation of growth and organization. Endothelium continues to proliferate close to developing sweat glands and hair roots, producing a mesh of capillaries.

At the end of the fourth month, proliferative activity results in the development of a superficial vascular plexus close to the epidermis and a deeper one about the glomerular apparatus of the sweat glands and around the terminal papillae of hair follicles. The superficial plexus is connected to the deeper plexus by vessels alongside the hair follicles, and projections from the deep plexus join the subcutaneous vascular plexus.

At the fifth month, new vessel formation occurs, largely by budding from preexisting channels, but some development and differentiation from the mesenchymal cells still occur. At this time, arteries and veins, as well as arterioles and venules, are clearly organized into patterns, some of which remain in adult life. In fact, every vessel in the adult skin was at one time part of a capillary network. The capillaries in the embryo destined for preservation are those that

Blood Vessels and Lymphatics in Organ Systems
Copyright © 1984 by Academic Press, Inc.

show greatest differentiation into arteries and veins. As the skin becomes thick and its surface increases, the number of vessels also rise. The bulk of endothelial proliferation continues to be subepidermal and periadnexal. When the hair follicles and sweat glands become separated by greater distances and the length of the former exceeds that of the sweat glands, the blood vessels to them become removed from those supplying the epidermis. This results in some reorganization of the vascular pattern. Networks of capillaries are no longer restricted to levels connected by perifollicular vessels. Instead, sites of endothelial proliferation are now found subepidermally: (1) around sweat glands and sebaceous glands, (2) around hair roots, and (3) especially in the deeper dermis and subcutaneous connective tissues.

Development and differentiation of the microvasculature in the skin continues for some time after birth. One reason for this is the rapid increase in surface area in the first few weeks of postnatal development. The redness of the newborn skin is due to a dense network of capillaries in the superficial dermis. Ryan (1973), in a series of diagrams, illustrated the progressive changes in the newborn cutaneous vasculature. These include new growth of some vessels and the absorption of others, thus bringing about the adult pattern. The latter is developed with vascular loops in the dermal papillae interconnected by a horizontal vascular plexus. The same blood vessels that supplied the network originally now supply the loops, and those that drained the vascular network carry blood away from the loops. The rate and extent of proliferation of all tissues in the skin slow down when the vascular patterns become fixed.

B. ANATOMY OF CUTANEOUS BLOOD

CIRCULATION

By J. E. Bernstein and A. L. Lorincz

1. Gross Morphologic Pattern

a. General architecture: Spalteholz (1893) attempted the first detailed description of the anatomy of the cutaneous blood circulatory system, and his elegant, simplified, summarizing diagrams have tended too dogmatically to be erroneously interpreted as a substitute for the much more complex and variable real nature of cutaneous vascular patterns in different sites and kinds of skin. In general, there are large-meshed plexuses of anastomosing arteries in the fascia and at the dermal subcuticular junction and a finer-meshed network of smaller arterioles in the subpapillary region. The large-meshed fascial network supplies the subcutaneous tissue and only rarely achieves major significance with regard to the circulation of the skin. From the muscular arteries forming the ill-defined subdermal plexus, arterioles arise that enter the dermis perpendicularly to the skin surface and branch horizontally as they ascend toward the epidermis, the whole dermis being riddled by a rich continuous meshwork of anastomosing arterioles and their associated still finer capillary networks.

   b. Architecture of vascular tree in the dermis: The distribution of blood vessels to the superficial dermis often occurs in a "candelabra" pattern (Ryan, 1976). The ascending arterioles divide in a treelike fashion and send their terminal branches as far as the most superficial dermis, where they form the subpapillary plexus from which capillary loops arise and course through the dermal papillae to nourish the epidermis. Generally, each dermal papilla is served by one capillary loop. The papillary loops drain into a horizontal sub-papillary venular plexus, which, in turn, drains into a profuse network of larger venules throughout the dermis and then into the larger hypodermal venous network and the efferent veins that accompany the deep arterial plexus.

   c. Variations in cutaneous capillary bed: In various parts of the body, the skin shows significant differences in morphology and number of blood vessels supplying the papillary dermis and epidermis (Moretti, 1968). In areas where the latter structure is thin and the rete ridges are poorly formed, capillary loops are sparse. In such sites as the palms, with well-developed rete ridges, the capillary loops are numerous and form complex candelabras. The number of capillary loops is limited by the number of papillae, the latter being about 60–70 mm^2 (Figs. 18.1 and 18.2) (Moretti, 1968). Within the capillary loop, the diameter of the lumen of the vessel varies from 5 to 7.5 μm in the descending extrapapillary part (Braverman and Yen, 1977), with the intrapapillary portion being the narrowest part (3.5–6.0 μm). It is within the bend in the loop that resistance to flow for the least flexible blood cells (the neutrophils) is greatest.

   d. Typical findings in cutaneous vascular bed: Several basic aspects of the gross morphologic pattern of the cutaneous circulation are as follows: (1) The richness of cutaneous vascularization far exceeds the vascular supply required for supporting the metabolic needs of skin tissues, this vascular abundance playing a major role especially in temperature regulation (see Section C-3, below) and in blood pressure control. (2) The major vessels have a rather serpentine course, which allows considerable stretching of the skin without interfering with its circulatory supply. (3) The venous vessels of the various

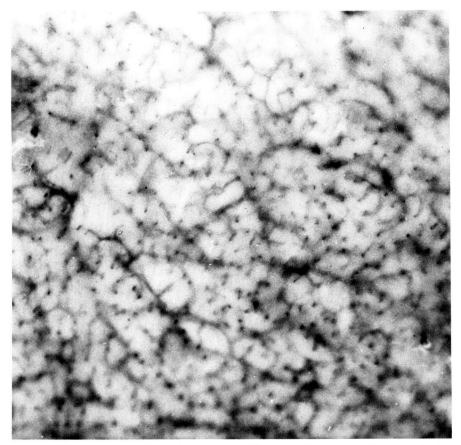

FIG. 18.1. Low power microscopic appearance of the subpapillary and papillary microcirculation of normal human forearm skin *in vivo* as viewed from the skin surface after stripping away of the stratum corneum with cellophane adhesive tape.

portions of the cutaneous vascular networks are always larger in diameter than their corresponding thicker-walled arterial elements. (4) The general vascular schema does not follow a pattern of progressive splitting from arteries to arterioles, to capillaries, and their degressive joining into venules and veins. Rather the pattern is that of a meshwork at all levels, with incredibly complex interconnection within the macro- and micronetworks (Saunders, 1961).

  e. Architecture of terminal bed: Zweifach (1949) called attention to the basic pattern of the cutaneous microvascular network at the level of its finest terminal meshes. Metarterioles or preferential arcading channels emerge from terminal arterioles and loop back into venules accompanying the latter. Side

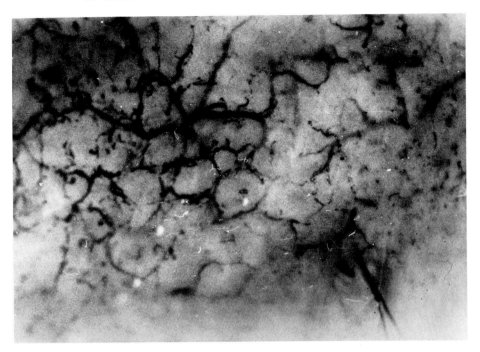

FIG. 18.2. Enlarged view of the superficial microcirculation of normal human forearm skin *in vivo* by capillary microscopy under conditions as in Fig. 18.1.

branches from the metarteriolar portion of this arcading loop, which are controlled by precapillary sphincters, form an interconnected finer meshwork of true capillaries, across which vascular respiratory, nutrient, and metabolic exchange takes place. Blood flow through the true capillary bed can be varied and its distribution therein shifted by various factors that regulate and alter the state of contraction of the smooth muscle elements in the arterioles, metarterioles, precapillary sphincters, and muscular venules, to balance both general body needs for thermal and blood pressure regulation and local tissue metabolic requirements. Moreover, arteriovenous anastomoses are often found directly between the terminal arterioles and corresponding muscular venules. Contraction of the metarterioles, therefore, can shunt blood via these anastomoses directly into the muscular venules and thus restrict flow around the preferential channel loops and their capillary submeshworks.

f. Glomus bodies: In acral areas, especially in digital skin, there are numerous more complex, highly muscularized arteriovenous anastomoses, the glomus bodies. Flow through these specialized shunts is controlled by vasomotor innervation, and when they are opened, marked increases in blood flow can take place through them, mainly for purposes of temperature regulation (see Section C-3, below). These structures also function as stopcocks that close if blood pressure does not exceed a critical value (Moretti, 1968). There are from one to several hundred glomus bodies per mm² of digital skin. With age these bodies tend to involute.

g. Blood supply to appendages of the skin: The hair follicles and sweat glands have an arteriolar blood supply of their own (Montagna and Ellis, 1961). The hair follicle receives its nutrition from an anastomosing network of capillaries that surrounds the shaft and from a separate though interconnected richer system for its bulbar end. The upper perifollicular vasculature is connected with the subpapillary plexus, whereas the deeper portions come directly from hypodermal vessels. Vascular supply to the hair follicle is directly related to the size of this structure, with longer hairs, whiskers, and terminal hairs displaying a dense vascular network, whereas small vellus hairs are less generously supplied.

The coil of the eccrine gland is surrounded by a dense network of capillaries (Montagna and Ellis, 1961). In the case of smaller sweat glands, such vessels usually arise from one arteriole, but for the larger ones, there may be several arterioles involved. The capillaries surrounding the duct are in close apposition to it, and numerous intercapillary anastomoses can be observed between these vessels and those of the subpapillary plexus.

2. MICROSCOPIC AND ELECTRON MICROSCOPIC STRUCTURE

a. General characteristics: The histologic structure of cutaneous vessels is similar to that of the vascular system elsewhere in the body (Rhodin, 1968). The small arteries of the fascial plexuses and the arterioles of the dermis possess three layers: (1) an intima composed of endothelial cells and an internal elastic lamina, (2) a media containing muscle cells in a discontinuous layer (in the upper dermis) or in one or more layers (in the lower dermis and subcutaneous tissue), and (3) an adventitia of connective tissue. The capillaries of the dermis consist essentially of a single layer of endothelial cells separated from an incomplete layer of pericytes (see Section B-2d, below) by a well-defined basement membrane (see Section B-2c, below). Walls of dermal veins are usually thinner than those of arteries and less clearly divided into the three main layers. Large veins and venules are similar to arteries but are distinguishable by the presence of valves (Moretti, 1968).

b. Endothelial cells: Although these structures appear flat by light microscopy, on electron microscopic examination, they are much more complex (see Section C, Chapter 1). In undamaged skin, the intercellular space is very narrow (200 Å wide), and toward the lumen, cells may be so tightly applied that it is completely obliterated. Large fenestrations, found both between and within endothelial cells in many internal organs, are not usually present in the skin. In fact, large gaps between endothelial cells may indicate a pathologic process involving the skin. The luminal surface of the endothelial cell often possesses numerous microvilli, which are especially prominent in dermal veins and venules (Higgins and Eady, 1981).

Endothelial cells are actively functioning structures containing abundant Golgi apparatus and pinocytotic vesicles. Lysosomes are common, although not as numerous as in the dedifferentiated endothelium of Kaposi's sarcoma (Sterry et al., 1979). Weibel-Palade bodies are less frequently observed in cutaneous endothelial cells than in those of other organs (Higgins and Eady, 1981). Endothelial cells in human skin possess a large number of microfilaments similar to actin and myosin of muscle (Cooke and Fay, 1972). The function of these structures is unclear, but they may represent contractile elements (Hibbs et al., 1958). (For more detailed discussion of endothelial cells, see Sections A and C, Chapter 1.)

c. Basement membrane: This structure is well developed in cutaneous vessels, being relatively homogeneous in appearance on the arterial side and multilaminated on the venous side. Yen and Braverman (1976) and Braverman and Yen (1977) have demonstrated that the homogeneous-appearing basement membrane changes to multilaminate at the extrapapillary portion of the descending limb of the papillary loop. However, this has been disputed by Higgins and Eady (1981), who observed multilaminate basement membranes at the peak of the loop. The multilaminate looser structures possibly facilitate protein transport on the venous side, whereas the homogeneous arterial basement membranes are more likely resistant to the higher intraluminal pressure of an arterial capillary and thus may prevent excessive leakage.

d. Pericytes: These are cells of disputed origin that lie outside the basement membrane and surround, and possibly support, the endothelium. Like smooth muscle cells, pericytes contain fine cytoplasmic filaments, but, in contrast to them, the filaments are sparse and do not form dense bodies (Kuhn and Rosai, 1969). The role of the filaments is much debated. The small vessels of the papillary dermis have few pericytes, but the large vessels of the lower dermis and subcutaneous tissue are well supplied with them. Although the function of the pericytes is as controversial as their origin, it is likely that such cells help provide a supporting structure for the larger arteries exposed to a high pressure.

C. PHYSIOLOGY OF CUTANEOUS BLOOD CIRCULATION

By D. I. ABRAMSON

1. GENERAL CONSIDERATIONS

In addition to its nutritive role, the cutaneous circulation has several other important functions, foremost among which are the preservation of a steady body temperature and the delivery of materials essential for cutaneous wound healing. Moreover, it participates in the maintenance of adequate circulating blood volume and in the control of arterial blood pressure.

a. Methods of study: Because of its location, the cutaneous circulation in the limbs is readily studied by various experimental methods, including calorimetry, venous occlusion plethysmography, thermometry, radioisotopic clearance, cutaneous microscopy, thermography microscopic television system, and laser Doppler methods. A detailed discussion of the first six procedures can be found in a monograph by Abramson (1967), and descriptions of the last two methods are available in papers by Fagrell *et al.* (1977) and Holloway and Watkins (1977), respectively. Recently three semiquantitative means of investigating cutaneous circulation, thermal, optical, and electrical impedance techniques, have been reviewed by Rolfe (1979). The thermal approach primarily measures the nutrient component of total skin blood flow, whereas both the optical and electrical impedance methods provide an indication of circulation in deeper vessels.

b. Resting limb blood flow: Blood circulation in the hand and especially the fingers (both of which possess a preponderance of skin if nonvascular tissues and bone are excluded) varies markedly, fluctuations in rate of blood flow being present from moment to moment due to periodic rhythmic discharges of nervous impulses over sympathetic vasoconstrictor nerves. In addition, ambient temperature significantly influences cutaneous circulation (see Section C-3, below), as do circulating hormones (see Section C-2b, below). In contrast, the metabolic needs of the local tissues, which are low, only minimally affect blood flow. In this regard, it has been estimated that in the finger, only 0.8 ml of blood per minute per 100 ml of digit skin tissue is sufficient to satisfy oxygen requirements (Burton, 1961); still, in a hot climate, circulation may increase a hundredfold over such a figure. Even under physiologic conditions, flow may be 20 or 30 times greater than that necessary for local nutritive needs. Much of the excess circulation is shunted through cutaneous arteriovenous anastomoses

602

(see Section B-1f, above) and, by circumventing the exchange vessels, serves no function in the process of tissue metabolism. However, it does play a significant role in heat dissipation.

In the total hand, blood flow is not as high as in the fingers, varying from less than 1 ml/min/100 ml of limb volume to 20 or 30 ml or more. For example, with the limb immersed in a water bath at 32°C (physiologic conditions), the local circulation is 9.3 ml/min/100 ml limb volume (Abramson and Fierst, 1942); at a bath temperature of 45°C (under conditions producing almost maximal cutaneous vasodilatation), the readings rise to 22.0 ml or higher (Abramson et al., 1941). At 11–14°C, the rate of blood flow falls to approximately 2 ml, due to intense vasoconstriction elicited by the cold. An even greater fall (to 0.2 ml) occurs when the whole body (including the hands) is exposed to cold.

2. REGULATION OF CUTANEOUS BLOOD FLOW

a. Neural control: Despite its great potential volume capacity, the cutaneous circulation in the limbs is so finely and delicately regulated that it faithfully follows a pattern that permits the nutritional requirements of the organism as a whole to be adequately satisfied. At times, this may be accomplished only if the metabolic needs of the skin are temporarily sacrificed in the process. In the maintenance of a proper blood distribution, the sympathetic nervous system plays the predominant role, with local autoregulatory mechanisms having a negligible function. In fact, the autonomous control of the cutaneous circulation is poorly developed, as compared with the vascular mechanisms existing in the deeper tissues of the limbs, where the rate of blood flow is dependent primarily upon the local metabolic requirement of the various structures.

However, even in the skin, all anatomic regions of the vascular tree are not equally affected by sympathetic nervous system control, since selective regulation exists. For example, the cutaneous vascular bed in the limbs is under much stronger vasomotor influence than is that of other parts of the body. This is most apparent in the fingers and toes and becomes progressively less marked as the proximal regions of the extremities are approached (Blair et al., 1960).

In accord with the above statements is the finding that adrenergic terminal perivascular end organs are present in large numbers in the cutaneous vessels of the limbs. Ready identification of these structures has been facilitated by means of histochemical procedures (Norberg and Hamberger, 1964; Lindvall and Björklund, 1972) and electron microscopic techniques (Goodman, 1972).

In early studies on the nervous innervation of arteriovenous anastomoses (AVAs), only cholinesterase-positive sympathetic nerve endings were reported (Hurley and Mescon, 1956). However, recently Waris et al. (1980), using the glyoxylic acid fluorescence histochemical method, were able to identify adrenergic (probably vasoconstrictor) innervation of these structures in the sub-cutaneous fascia of the trunk skin of the rat. The study also suggested that the AVAs are regulated not only by neural but also by local mechanisms. In the rabbit ear, they likewise appear to be innervated by both adrenergic and cho-linergic nerves, with the former being denser (Iijima and Tagawa, 1976). In general, the AVA demonstrates a denser innervation than its artery of origin, despite the fact that the nervous network is an extension of the sympathetic ground plexus found around cutaneous arteries and arterioles (Waris et al., 1980). The anatomic finding of adrenergic vasoconstrictor innervation of AVAs supports early physiologic evidence (Grant, 1930; Clark and Clark, 1934) that these vessels are under strong sympathetic control.

b. Hormonal control: A number of systemic vasoconstrictor hormones modify the rate of blood flow in the skin. Among these is circulating epi-nephrine, which causes extensive constriction of vessels, thus enhancing the response to sympathetic nerve discharge (Celander, 1954) (see Section C-2a, above). The process responsible for the reaction involves the liberation of the hormone from the adrenal gland, the distribution of this agent via the blood stream, and local activation of α-receptor endings in the cutaneous vessel wall.

Another hormone having a similar influence, but to a much lesser degree, is circulating norepinephrine. Its minimal role is partly due to the fact that the main source of supply of this agent is primarily overflow from synaptic clefts following adrenergic sympathetic activity, with an additional even smaller com-ponent originating in the adrenal medulla. However, under abnormal condi-tions, such as in the presence of pheochromocytoma, the quantity of nor-epinephrine in the bloodstream may be great enough to produce intense end organ responses in the cutaneous circulation. As in the case of epinephrine, vasoconstriction occurs through stimulation of α-receptors in the blood vessel wall.

Recent studies support the view that renin has a local vasoconstrictor action on cutaneous vessels (Swales, 1980). This agent interacts with an α-globulin substrate derived from the liver, to form angiotensin I. The latter is then hydrolyzed in the lung to angiotensin II (angiotonin), which constricts re-sistance vessels by acting upon vascular smooth muscle. In this regard, it has been reported that the arterial wall, as well as other tissues, contains enzymes that can split the substrate molecule *in vitro* to yield angiotensin II (Ganten et al., 1976; Hackenthal et al., 1978). However, no conclusive evidence is avail-able to indicate that a similar reaction occurs *in vivo*. Angiotensin II may also have an effect upon the central nervous system, giving rise to increased sym-pathetic outflow (Peach, 1977).

Daly and Duff (1960) reported an average reduction of 45% in hand blood flow following intraarterial infusion of angiotensin II into the upper limb and concluded that this drug had a direct effect on vessels. The response persisted for several minutes and then gradually disappeared. The decreased blood flow was associated with pallor of the skin of the forearm, this helping to localize the change to the cutaneous vessels.

Besides vasoconstrictor hormones, there are several endogenous vasodilator amines that produce strong peripheral relaxation of previously contracted vascular smooth muscle. One of these is histamine, which, in concentrations low enough not to affect the basal tone of veins, is capable of causing pronounced relaxation of these vessels if administered during sympathetic nerve stimulation. The reaction parallels a corresponding reduction in evoked release of epinephrine. A dilator response of even greater magnitude is elicited when acetylcholine is substituted for histamine. Such reactions suggest that these agents act to prevent release of adrenergic substance from the nerve terminals and/or directly inhibit vascular muscle contractility.

Another hormone that counteracts adrenergic neurotransmission (as by electrical stimulation of sympathetic nerve endings) is serotonin. It appears to exert its effect by inhibiting the quantity of released adrenergic transmitter (McGrath, 1977). However, when smooh muscle tone is increased by direct application of norepinephrine, contraction is either unaffected or augmented by serotonin. This hormone, which circulates in association with blood platelets and is released during their aggregation, may play a role in the maintenance of local tone (Jarrott et al., 1975).

c. Local control: Although the sympathetic nervous system plays the major role in the control of cutaneous blood flow in the limbs, the small precapillary sphincters do demonstrate some autoregulatory function. As a result, vasomotor constrictor impulses can operate on a pronounced level of basal tone, a combination of both reducing cutaneous blood flow to a minimal level. That basal tone exists is supported by the finding that temporary sympathetic denervation does not elicit maximal cutaneous vasodilatation. For example, topical application of heat to a limb in which vasomotor impulses have been blocked produces a further increase in local circulation over that which follows sympathetic denervation alone.

Many other local influences can alter blood flow through the cutaneous vascular tree. Among these are osmolarity, oxygen concentration, and pH. At least in the case of the AVAs, such factors exert their action directly on the vascular smooth muscle cells or on the epithelioid cells themselves (Waris *et al.*, 1980).

3. THERMOREGULATION

In view of the high resting blood flow observed in the distal portions of the limbs under physiologic conditions (see Section C-1b, above), it is clear that the vascular tree in these sites is overperfused relative to the stable metabolic needs of the skin and its appendages. In great part, such a discrepancy reflects the function of the extremities in the process of thermoregulation. Because the limbs are generally exposed to and influenced by environmental temperature and because of the rich sympathetic nervous system innervation of their cutaneous vessels, they are especially adapted to play a significant role in this regard.

a. Vascular responses to elevations of ambient temperature: Inasmuch as the exposed segments of skin under resting conditions are generally much cooler than internal structures, a regulatory mechanism exists for making fine adjustments of body temperature. This is accomplished by increasing cutaneous blood flow, thus permitting the skin to come more nearly into temperature equilibrium with the remainder of the body. The cutaneous AVAs are especially involved in such a response, since, when dilated, they are responsible for flooding of the superficial veins with blood. As a result, heat is lost to the skin and then to the environment. The process is accelerated by the low peripheral resistance existing in the latter vessels (as compared with deep venae comites), thus permitting prolongation of blood cooling. Inasmuch as there is simultaneous vasodilatation of arterioles, blood flow through the nutritive circulation is also increased, a response that allows for an augmented delivery of ions and water to the sweat glands. The combined factors of evaporation of perspiration and heat loss by radiation, conduction, and convection ensure an efficient and significant elimination of heat from the body, provided the environment is at a lower than body temperature and humidity is not excessive.

Of importance in the degree of efficiency of the mechanisms discussed above is the fact that the AVAs and other vessels of the microcirculation can change in lumen size from that of complete closure to one of maximal dilatation by only small changes in sympathetic discharge (Celander and Folkow, 1953). The anatomic finding of a very rich sympathetic innervation of such vessels (see Section C-2a, above) may help explain this sensitive response to autonomic nervous system control.

b. Vascular responses to elevations of ambient temperature during physical exercise: When physical exertion is carried out in a high environmental temperature, there is competition between skin and active muscles for the available blood flow (Rowell, 1977). In this regard, Johnson and Park (1979) have demonstrated that during exercise, there is a rise in the level of internal temperature at which circulation in the skin begins to respond to thermoregulatory demands. The type of change resembles that observed with assumption of an upright position (Roberts and Wenger, 1980; Johnson and Park, 1981). The explanation for the rise in the threshold is not clear, although it is possible that it represents a true hypothalamic "set-point" shift, as described by Hammel *et al.* (1963). In any event, the change contributes to the elevated internal temperature noted during bouts of severe exercise.

c. Vascular responses to low ambient temperatures: When the body is exposed to a cold environment, flow through the cutaneous AVAs, arterioles, and capillaries practically ceases due to a marked rise in vasomotor tone and resulting strong vasoconstriction. Most of the blood is then shunted to the interior of the body and to deeper tissues of the limbs. In the case of the latter sites, it returns mainly by venae comites situated close to the main arteries, thus favoring local heat exchange, with warming of the local venous blood. Under conditions of minimal cutaneous circulation and resultant conservation of body heat, skin and subcutaneous adipose tissue are able to serve as an efficient heat insulator.

d. Remote vasomotor responses to low ambient temperatures: It is generally accepted that hypothermia produces distant, as well as local, vasomotor responses in the skin of the limbs. Major *et al.* (1981), using forelimbs of dogs isolated from the trunk except for the main arterial and venous blood supply, found that cutaneous vessels, especially veins, play an important role in heat conservation. They noted that decreased systemic temperature elicits constriction (i.e., increased vascular resistance and decreased limb weight) of cutaneous veins, a conclusion previously reached by Haddy *et al.* (1957). The vasoconstrictor response caused by systemic cooling can be overridden by active vasodilatation elicited by local cooling of the limb (Major *et al.*, 1981) (see Section C-3e, below).

e. Cold vasodilatation: If exposure of the limbs to cold is prolonged, the AVAs and, to a much lesser extent, the arteriolar and capillary beds open intermittently, thus flooding the cutaneous vascular tree with blood and transiently warming the tissues (cold vasodilatation). As a result, overcooling is prevented, a response that is especially important in the case of the fingers, which are most likely to suffer from the deleterious effects of cold because of their position and shape. Another factor that contributes to maintenance of viability is the reduction in metabolic needs of the tissues exposed to cold, due to the accompanying fall in tissue temperature. Therefore, even a minimal increase in blood flow through the arteriolar bed is sufficient to satisfy nutritional requirements under such conditions, at least for short intervals.

The mechanisms involved in the production of cold vasodilatation have not been clearly defined. Among the proposed theories is the elicitation of axon reflexes involving axon collaterals from unmyelinated nociceptive fibers that innervate neighboring arterioles. Stimulation of these structures may induce long-lasting vasodilatation. A possible associated mechanism may be local release of vasodilator substances. Cold suppression of vascular smooth muscle activity (Gaskell and Hoepper, 1966) has also been suggested to play a role. Another view is that the local cold-induced dilatation observed in the innervated limb results from antagonism to neurally mediated vasoconstrictor tone (Major *et al.*, 1981), possibly through the anesthetic effect of cold on sympathetic nerves. (For a discussion of the mechanisms involved in flushing reactions, see Section D-3, below.)

4. Role of the Cutaneous Circulation as a Blood Reservoir

An anatomic study of the vascularity of the fingers readily reveals the presence of a highly developed and extensive circulatory bed in these sites. The possibility has therefore been offered that, besides their role in thermoregulation (see Section C-3, above), the numerous vascular plexuses, capillary and venular beds, arteriovenous anastomoses, and large veins in the skin of the digits and the hand may be an important reservoir when there is need for shunting blood to inactive tissues. Such a mechanism may also contribute to the maintenance of a stable level of systemic blood pressure.

D. PHARMACOLOGY OF CUTANEOUS BLOOD CIRCULATION

By J. E. Bernstein and A. L. Lorincz

1. General Considerations

Many pharmacologically active agents can influence the cutaneous circulation, but their roles in vascular physiology in many cases remain speculative. Included among these local vasoactive compounds are histamine, bradykinin, prostaglandins, dopamine, substance P, angiotensin, acetylcholine, catecholamines, and adenine nucleotides and nucleosides. Some of these chemical mediators have been linked with important pathologic changes in the skin, such as urticaria and inflammation (see Sections E-2a and 2d, respectively, below). In addition to these naturally occurring, pharmacologically active agents, a number of exogenously derived vasoactive substances can be introduced into the skin through topical application or local injection. (For vascular responses to such agents, see Section D-5, below.)

2. AXON REFLEX VASODILATATION

a. Causative mechanism: When human skin is traumatized, an area around the site of injury becomes erythematous. This phenomenon, known as "flare" or axon reflex vasodilatation, is part of the erythema in many inflammatory processes. It is dependent upon intact sensory innervation to the area and can be reproduced by antidromic stimulation of cutaneous sensory nerves (Langley, 1923). The erythema results from a greater than normal quantity of blood within those small subpapillary, very superficial cutaneous blood vessels normally responsible for skin color, rather than to local increase in the rate of blood flow through the vascular bed.

b. Chemical mediators: Although it has been suspected for many years that axon reflex vasodilatation is mediated by the release of vasoactive humoral substances from sensory nerves, the identity of the chemical mediator of this important vascular response remains unknown. In this regard, large numbers of vasoactive agents have been implicated, including histamine (Beck, 1965; Brody, 1966), bradykinin (Fox and Hilton, 1958), prostaglandins (Bergström *et al.*, 1965), dopamine (Bell *et al.*, 1975), angiotensin (Godfraind, 1970), and adenine nucleotides and nucleosides (Folkow, 1949; Haddy and Scott, 1968). Recently, substance P (SP), the undecapeptide putative neurotransmitter released from the terminals of primary sensory neurons in the spinal cord, has also been suggested as a possible mediator of the vasodilator response (Burnstock, 1977).

For centuries, the seeds and membranes of certain species of plants of the nightshade family, notably capsicum, have been known to possess a substance that, when applied to skin or mucous membrane, initially produces intense erythema, pain, and inflammation. The active principle responsible for this reaction is capsaicin (*trans*-8-methyl-*N*-vanillyl-6-nonenamide). Jansco and Jansco-Gabor (1959) and Jansco *et al.* (1967), while studying "neurogenic inflammation," demonstrated that capsaicin causes the skin of man and animals eventually to become insensitive to various types of chemical pain agents in a manner similar to that which follows denervation. Such results suggested that capsaicin can render a subpopulation of cutaneous nerves nonreactive to chemical pain stimuli. In this regard, Jansco *et al.* (1977) have reported that systemic pretreatment of adult rats with capsaicin makes them temporarily insensitive to painful chemical sensory irritants and that neonatal systemic capsaicin therapy results in a completely irreversible impairment of the function of chemosensitive neurons. It has also been demonstrated that the erythema produced by antidromic stimulation of sensory nerves could be inhibited by capsaicin (Jansco *et al.*, 1967). Such responses have led to the belief that capsaicin interferes with the production, storage, or release of one or more vasoactive "neurohumors" released under the influence of chemical or antidromic stimulation of chemosensitive primary sensory neurons.

An early proposed explanation for axon reflex vasodilatation was that it results directly from the action of acetylcholine released at vascular endings. However, since atropine does not abolish either axon reflex or nerve stimulation vasodilatation of skin vessels, acetylcholine can almost certainly be excluded as the causative neurohumoral agent (Lorincz and Pearson, 1959).

Recent evidence supporting the view that substance P is a transmitter of primary sensory neurons and the demonstration by Jessell *et al.* (1979) that parenterally administered capsaicin results in a loss of SP from the spinal cord terminals of primary sensory neurons have suggested that SP might be the vasoactive "neurohumor" responsible for axon reflex vasodilatation. Such a possibility is further strengthened by the fact that this substance is capable of releasing histamine from mast cells and is itself an extremely potent vasodilator (Johnson and Erdos, 1973; Burcher *et al.*, 1977).

Perhaps the strongest evidence for the role of SP in axon reflex vasodilatation comes from recent studies involving the application of topical capsaicin to human skin (Swift *et al.*, 1979; Bernstein *et al.*, 1981a,b). In all these investigations, the drug was applied topically to the forearm skin of normal volunteers and a variety of chemical agents capable of producing a flare were injected intradermally at the border between capsaicin-treated and untreated skin. These included histamine, bradykinin, prostaglandin E_2, and papain. Intradermal injection of such agents evoked flare, wheal, and itching in normal skin, but only wheal and itching in the areas treated with topically applied capsaicin. Despite the inhibition of axon reflex vasodilatation by capsaicin, pinprick, touch, and temperature sensation remained intact in treated areas. From such results, it was concluded that whereas the capsaicin-treated areas remained responsive to mechanical injury, axon reflex vasodilatation secondary to trauma was abolished. The inhibition by capsaicin of the flare induced either by antidromic stimulation of sensory nerves or by chemical mediators strongly suggests a role for SP as a chemical mediator of axon reflex vasodilatation, even though there is no direct evidence that capsaicin applied topically depletes SP from local sensory nerve terminals. Further support for this view comes from studies on flare elicited by the intradermal injections of SP, adenosine triphosphate (ATP), histamine, or compound 48/80 versus saline control (Jorizzo *et al.*, 1981). Substance P, ATP, and histamine are potent inducers of axon reflex vasodilatation, and both SP and ATP demonstrate tachyphylaxis and cross-tachyphylaxis with histamine. The erythema from SP is not blocked by a constricting band, suggesting a neurogenic mechanism. However, it has not been possible to distinguish between a direct effect by SP and one mediated through histamine released from dermal mast cells (Jorizzo *et al.*, 1981).

3. FLUSHING

a. Occurrence and characteristics: Episodic flushing of the face and neck (the "flush" or "blush" areas) may be associated with a variety of agents, such as physical exertion, emotional disturbances, defecation, excessively high environmental temperatures, ingestion of alcohol, and certain foods. Similarly, a number of hormonal or enzymatic alterations, such as those occurring during menopause or in the carcinoid syndrome, may produce clinically identical-appearing flushing.

Cutaneous flushing is a manifestation of a generalized hemodynamic change in the vasomotor system. In this regard, observation of the peritoneum during abdominal laparotomy has revealed hyperemia of abdominal vasculature during facial flushing. Facial flushing ranges in color from pink to orange, to bright red, and though it is most noticeable on the face and neck, it not infrequently involves the shoulders, chest, and arms. Occasionally, the back and abdominal area are also affected. Although there are a variety of causes of flushing, recent evidence (Leslie and Pyke, 1978; Lightman and Jacobs, 1979; Bernstein and Soltani, 1980) indicates that most, though not all, types of facial flushing are mediated, in part, through an endogenous opionergic pathway.

b. Chemical mediators: Although it has been recognized for many years that facial flushing in rosacea and carcinoid syndrome and during menopause may be aggravated by the ingestion of alcoholic beverages, the chemical mediators involved in such reactions have remained elusive. Recently, reports regarding alcohol-induced facial flushing in diabetics on chlorpropamide (Pyke and Leslie, 1978) and alcohol-induced facial flushing in a large series of Orientals (Wolf, 1972) have heightened the interest in the pharmacologic mechanisms responsible for such a response. The vascular component of rosacea, characterized by erythema and telangiectasia, is seen most frequently in women aged 20 to 50 years old. Generally this reaction not only occurs earlier than the acneiform component, but it often remains the only aspect of rosacea experienced. Of interest is the finding that the affected skin shows normal vasomotor responses to epinephrine, norepinephrine, acetylcholine, and histamine (Borrie, 1955).

Chlorpropamide-alcohol flushing (CPAF) was first described by Leslie and Pyke (1978) as an intense facial flushing observed after alcohol intake in non-insulin-dependent diabetics on chlorpropamide. While subsequent attempts to utilize this reaction as a genetic marker for this subset of diabetic patients (Pyke and Leslie, 1978) have not been successful, nonetheless CPAF has been documented as a real and not uncommon cutaneous vascular reaction. The response resembles that in rosacea and during the menopause (Clayden *et al.*, 1974) and in the carcinoid syndrome (Roberts *et al.*, 1979). All three types of flushing can be induced by ethanol, they share identical anatomic distribution (the "flush" or "blush" area), and they have a similar course.

Alcohol has been demonstrated to have significant opiate-like effects (Davis and Walsh, 1970; Blum *et al.*, 1977; Schultz *et al.*, 1980), and in man, specific cross-tolerance and cross-dependence have been demonstrated between alcohol and morphine. The fact that some of the opiatelike effects of alcohol (narcosis, withdrawal convulsions, and tremor) are antagonized by the specific opiate antagonist, naloxone (Sorensen and Mattison, 1978; McKenzie, 1979; Jeffcoate *et al.*, 1979; Jeffreys *et al.*, 1980) suggested to Leslie *et al.* (1979) that CPAF might partly be mediated by endogenous opioid peptides. In this regard, they were able to demonstrate that the facial flushing and wheezing observed in patients on chlorpropamide after drinking alcohol were induced by an enkephalin analogue, both responses being prevented by naloxone.

The demonstration that CPAF and menopausal flushing are mediated, at least in part, by endogenous opioid peptides and that the responses can be inhibited by naloxone suggests that alcohol-induced facial flushing in rosacea (AIRF) might also be enkephalin- and/or endorphin-mediated. A recent study (Bernstein and Soltani, 1982) attempted to evaluate this hypothesis by administering a small dose of naloxone hydrochloride subcutaneously, in a double-blind paradigm, to nondiabetic patients with erythematotelangiectatic rosacea 10 min prior to the ingestion of 6% ethanol. It was found that this completely inhibited facial flushing, as measured visually and by recording skin temperature with cholesteric liquid crystal discs. Moreover, mean forehead skin temperature increased by 1.1°C during flushing after saline injection alone but rose only 0.4°C after naloxone administration.

It should not be concluded, however, that all cutaneous flush reactions, let alone all those elicited by alcohol, are physiologically and pharmacologically similar. For example, the alcohol-induced flushing observed in Orientals (Wolf, 1972), which may be accompanied by elevated acetaldehyde levels, shows no increase in endogenous opiate secretion and is unaffected by pretreatment with naloxone (Bernstein *et al.*, 1982). (For a review of the mechanisms involved in flushing reactions, see Wilkin, 1981.)

4. SWEATING

a. Role of vasodilatation: The 2 to 4 million eccrine sweat glands distributed over the entire human body surface are capable of producing a large volume of hypotonic sweat solution containing approximately similar concentrations of inorganic salts (mainly NaCl) and organic substances (mainly urea, but also other substances such as lactic and amino acids). Through their production of sweat, these structures have an important function in temperature homeostasis of the body (see Section C-3a, above). The secretory coils of the sweat glands are innervated by sympathetic nerve endings, and they have long been thought to produce their secretory product in response to acetyl-choline released from the sympathetic nerve endings. However, the localization of vasoactive intestinal polypeptide (VIP) and kallikreins in cholinergic exocrine glands, along with the demonstration of the vasodilator and secretory functions of the peptides, has suggested new and more complicated, possibly additional, regulatory mechanisms for cholinergic exocrine secretion. Morphologic and functional data support a complex interaction among acetyl-choline, VIP, and the kallikrein–kinin systems in the control of vasodilatation and secretion in exocrine glands. Such results demonstrate that activation of postganglionic nerve terminals to cholinergic exocrine glands causes concomitant release from the same nerve endings of both acetylcholine and VIP (Lundberg et al., 1980). Acetylcholine mainly produces secretion through its muscarinic actions, which can be blocked by atropine, whereas VIP causes chiefly vasodilatation of the dense periglandular capillary network (Lundberg et al., 1980). Through such an action, VIP may cause an increase in capillary permeability and, by its effect on venules, a rise in intracapillary pressure. The result of these two vascular phenomena may be an efflux of fluid from the blood to the eccrine gland to permit or supplement the secretory process.

The increase in sweating induced by exposure of subjects to elevated environmental temperatures is accompanied by a rise in kinin-like vasodilator activity in human eccrine sweat (Fox and Hilton, 1958). This response is similar to that produced by VIP and is thought, along with VIP, to contribute to heat-induced vasodilatation, with a resulting promotion of sweating. Finally, although a regulatory role in eccrine gland secretion has not yet been attributed to SP, demonstration of SP-positive nerve fibers around sweat glands (Hokfelt et al., 1977), coupled with its potent vasodilator activity, suggests that release of SP from distal terminals of primary afferent fibers might influence sweat production in a manner similar to that initiated by VIP and kinins.

5. Influence of Exogenous Chemical Agents

a. *Corticosteroids:* Topically applied corticosteroids constitute a treatment of choice for a variety of inflammatory disorders of the skin, particularly eczemas and psoriasis. Besides the antiinflammatory actions of topical steroids, it is generally accepted that these medications also have potent vasoconstrictive effects on the cutaneous microvascular system. In fact, a standard method for comparing the "therapeutic activity" of topical steroid preparations is to assay their vasoconstrictive powers (Stoughton, 1972).

E. PATHOPHYSIOLOGY AND PATHOLOGY OF CUTANEOUS BLOOD CIRCULATION

By J. E. Bernstein and A. L. Lorincz

1. General Considerations

The integrity of the skin is dependent upon its blood supply, and the consequences of vascular injury are determined by the size and type of vessel involved. Cutaneous blood flow can be compromised in a number of ways, each characterized by a fairly typical pathologic sequence and reaction. Mild injury to the capillaries and venules may lead to an increase in permeability of their walls with edema formation. Gross deprivation of the vascular supply to an area of skin, producing ischemia, may be followed by focal necrosis or even frank gangrene. Changes may occur in the arteries, veins, or small vessels in response to a variety of noxious stimuli, resulting in erythema, a state in which there is an increase in blood within those small blood vessels of the skin responsible for the hemoglobin-based components of skin color. Inflammation and thrombosis of small arterioles may not cause sufficient obstruction to produce necrosis but, instead, may elicit combinations of erythema, papules, nodules, purpura, and blisters. There may be functional vascular disorders, in which the vessels appear normal but exhibit impaired or altered function. Abnormalities in growth of the vascular system are common and may be genetically or congenitally determined, or the conditions may be acquired later in life. Proliferative states, which are most often benign but occasionally malignant, may also involve cutaneous vessels.

2. Pathophysiology and Pathology

a. Urticaria and angioedema: Although not often thought of as vascular responses, these reactions are based upon minor injury to small capillaries and venules of the dermis and subcutis. Subsequent to venular constriction, capillary dilatation occurs, together with a local increase in the permeability of the vascular walls; this causes extravasation of fluid into the dermis or subcutaneous tissues. Secondarily, eosinophils tend to accumulate in the area. Whatever the provoking cause, the changes in the vessels are largely due to the action of early phase chemical mediators of inflammation derived from mast cells, of which histamine is best known. In urticarial lesions, degranulation of mast cell can frequently be demonstrated histologically in skin biopsy specimens obtained

614

from involved sites. Other substances, however, may also play a role, either directly or indirectly, in the mediation of urticaria, including, bradykinin (Regali and Barabé, 1980), prostaglandins (Goldyne, 1975), and complement (Ballow *et al.*, 1975). The time required for resolution of the wheal is variable and depends upon the interaction of the chemical mediators and their inhibitors. For example, development of a wheal secondary to intradermal histamine injection is characteristically rapid in development and in regression (from 3 to 30 min). However, a more severe or continuing immune complex-mediated injury, associated with inflammation of small blood vessels (vasculitis), may result in a wheal that lasts for several days.

b. Ischemia: This state is a consequence of a sufficiently impaired arterial blood supply to produce clinical changes. The end stage of severe ischemia is gangrene or death of tissue. Hypoxia refers specifically to a deficiency of tissue oxygenation, which can sometimes occur even though blood flow may be adequate. The epidermis is relatively resistant to both hypoxia and ischemia and can survive over 24 hr with no blood supply; in comparison, necrosis of subcutaneous fat or smooth muscle occurs after only 2 hr of such deprivation (Willms-Kretschmer and Majno, 1969). Focal necrosis of the skin caused by ischemia first manifests itself as an area of painful erythematous swelling that progresses rapidly to become purpuric and finally to loss of viability (with the development of vesicles or pustules). If the blood supply is grossly impaired, full-thickness skin necrosis may result, with large ulcerations and gangrene. Although there are a considerable number of vascular disorders that can lead to ischemia and gangrene, the most common ones are arteriosclerosis, hypertension, and diabetes mellitus. Regardless of the causative factor, the skin of the extremities, especially of the legs and feet, is most often involved.

c. Vasospasm: There are a variety of functional disorders that are responsible for transient or even permanent impairment of local vascular blood flow without demonstrating pathologic alterations in vessel structure except possibly in the final stage. Included among these are a number of circulatory conditions associated with altered or abnormal responses to cold, such as Raynaud's disease, Raynaud's phenomenon, acrocyanosis, and livedo reticularis. In Raynaud's disease or Raynaud's syndrome, exposure to cold, which normally produces mild to moderate vasoconstriction, may elicit such severe constriction that local perfusion blood pressure is reduced below the point of critical closing pressure, thereby cutting off local blood flow (Edwards, 1954). Classical theories of excessive sympathetic system activity or increased sensitivity to catecholamines in Raynaud's disease or Raynaud's phenomenon are currently being challenged by the view that affected vessels may have lost their capacity to dilate (Downey *et al.*, 1971). Acrocyanosis, a vasospastic disturbance of the smaller arterioles of the skin, often is associated with endocrine disorders and anxiety states. The characteristic cyanosis is due to the associated dilatation of the subpapillary venous plexus, at the same time that the cutaneous arterioles are in spasm. Elevated blood viscosity also has been implicated in this disorder (Copeman and Ryan, 1971).

Livedo reticularis is a reticular cyanotic discoloration of the skin characterized by a reddish-blue mottled appearance. Onset is usually first related to cold exposure but often progresses to a permanent discoloration. This condition has a great diversity of causes, but in all cases the net-like color change appears to be due to dilatation of, and flow stagnation within, the capillaries and small venules in the dark areas, resulting from an interference in the cutaneous blood supply (Champion, 1965).

A non-cold-related functional disorder of the cutaneous vasculature is white dermatographism. This consists of the exaggerated development of a white line response 10–15 sec following light to moderate mechanical stroking of the skin, which lasts for several minutes. Whereas some workers believe that the reaction results from the effect of increased intradermal pressure which follows edema (Davis and Lawler, 1958), other suggest that it follows from local release of vasoconstrictor agents in the upper dermis (Ramsey, 1969).

d. Inflammation (vasculitis): Vasculitis is an inflammation of the walls of the blood vessels, sometimes leading to necrosis of their dependent tissue. Many different factors can trigger or be involved in causing cutaneous vasculitis, including the deposition of immune complexes, bacterial and viral infections, and drug reactions. In general, an injured vessel exhibits increased permeability (leakiness), retarded blood flow, fibrin deposition and platelet aggregation (clotting), proliferation, and necrosis. Various types of cutaneous vasculitis may variably involve capillaries, venules, or arterioles. In response to inflammation, capillaries proliferate (Demis and Galakatos, 1967) and develop thickened walls. Necrosis of the vessel wall may occur as an initial event if the injury is severe or may result from prolonged ischemia. Erythrocytes often escape from the injured vessel into the surrounding tissues. An infiltrate of disintegrating neutrophils is a prominent feature of many types of vasculitis (leukocytoclasia). The cells can be observed in the lumen, the vessel wall, and surrounding tissues. An infiltrate of mononuclear cells also is often observed in surrounding tissues, presumably participating in the repair response to the injury. Granuloma formation is frequently present in deeper lesions involving fat.

e. Venous stasis: When the venous drainage from the skin is impaired, pigmentation, purpura, edema, and leg ulceration may result. The causes of impaired venous outflow include primary and secondary varicosities and the postphlebitic syndrome. Raised venous pressure (venous hypertension) elevates intracapillary pressure, producing elongation and tortuosity of the papillary vessels. Such changes, in turn, increase the fragility of these vessels and may result in edema, blood leaking through the damaged vessel walls, or thrombosis. Raised venous pressure also can cause dilatation of the subpapillary venular plexus. In addition, stasis reduces the quantity of nutrients available for the repair of wounds. As a result, minor injuries or infections may develop into chronic stasis ulcers. Such alterations are most pronounced in the legs due to gravitational forces. With advancing age, elongation and tortuosity also develop in dermal vessels of the legs (Marinov and Tzvetkova, 1977).

f. Alterations in growth: Changes in growth and malformations of the cutaneous vascular system are common, consisting of numerous histologic types, including hamartomas and hyperplasias, as well as benign and malignant neoplasms. It is useful to consider several classes of cutaneous vascular growths and malformations that are illustrative of the diversity of the group.

Capillary hemangiomas are vascular malformations composed of tightly packed, thin-walled capillaries that are arranged in lobules. These red-purple, polypoid lesions may be present at birth or appear within the early weeks of life and usually involve both dermis and subcutaneous tissues. Most capillary hemangiomas involute spontaneously during childhood (Walsh and Tompkins, 1956). Cavernous hemangiomas display violaceous spongy tumor-like aggregates of dilated venular blood vessels in a mostly subcutaneous fibrous stroma. They show considerably less tendency for spontaneous involution.

Senile or cherry angiomas are small, red papules that appear commonly in middle-aged or elderly persons. The lesions are similar histologically to the capillary hemangioma of infancy, but they have fewer vascular spaces and more abundant fibrous stroma. The nevus flammeus (port-wine stain) is a pink to purplish macular lesion, composed of mature dilated capillaries arranged loosely in the dermis. The condition is usually present at birth and shows little tendency to regress.

Granuloma pyogenicum is a localized red to purple pedunculated nodule or papule which is composed of proliferating newly formed capillaries in a loose myxoid stroma. Lesions often occur at sites of injury. Although the name suggests an infectious cause, evidence for such an etiology is lacking.

Kaposi's sarcoma is usually a slowly evolving malignant vascular neoplasm of multifocal origin affecting predominantly males. It presents as multiple vascular nodules or plaques in the skin and other organs, particularly the gastrointestinal tract and the lymphoreticular system. The lesions display an accumulation of spindle cells, hemosiderosis, and the presence of vascular slits. Although the etiology of Kaposi's sarcoma is presently unknown, recent evidence has demonstrated a significant association with concomitant viral disease (Friedman-Kien, 1981) and immunosuppression (Douglass *et al.*, 1980).

F. CUTANEOUS LYMPHATIC SYSTEM*

By A. E. Taylor and D. J. Martin

The lymphatic system of skin and subcutaneous tissue is very extensive, and it removes fluid and protein that normally escapes from the circulation into the tissues.

1. Anatomy of Lymphatic System

The dynamics of lymph drainage has been extensively studied in animal paw preparations, in which the prepopliteal paw lymphatics appear to drain mostly skin and subcutaneous tissue. No lymphatic vessels are present in the epidermis. A superficial lymphatic plexus is found within the corium, which extends to the outer two-thirds of this structure, and another deeper one is located in the portion adjacent to the subcutaneous tissue. The superficial plexus possesses no valves, but such structures are present in the deeper plexus occupying the inner dermis and subcutaneous tissue layers. Although the entire lymphatic system of the skin is well developed, specific areas, such as the fingers, the palm, the sole, and the scrotum, appear to contain an even greater number of lymphatics. Structural differences in the cutaneous lymphatics appear in different types of skin. In regions where the skin is thick and tight, the lymphatics are uniform in shape, but in areas where it is thin, the lymphatics are variable in form. The reason for this difference is unknown, but it may simply reflect variations in external pressures applied to the lymphatic walls.

The lymphatics that originate in the interior of the corium join anatomically with the subcutaneous lymphatics and form Jossifow's cutaneous lymphatic plexuses. Although the capillaries in the skin are of the continuous type, a small amount of plasma proteins still leaks into the interstitium. However, the anatomic arrangement of the skin's lymphatics ensures that such materials are returned to the blood. In addition, the lymphatics provide a means of removing excessive capillary filtrate (Yoffey and Courtice, 1970; Rusznyak et al., 1967).

2. Lymph Flow and Protein Content

a. Lymph flow: The lymphatic system draining the dog's hindpaw consists of approximately three or four large collecting lymphatics that drain to the

*J. C. Parker assisted in the preparation of this section.

Blood Vessels and Lymphatics in Organ Systems
Copyright © 1984 by Academic Press, Inc.

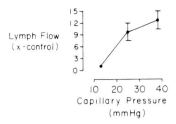

FIG. 18.3. Plot of lymph flow times control as a function of capillary pressure. Control lymph flow in these studies was 9×10^{-4} ml/min. (Modified from Chen *et al.*, 1976, reproduced with permission from the American Heart Association.)

popliteal node. The total resting lymph flow from them is about 3–5 µl/min, draining from approximately 24 gm of tissue. This can be translated into 1 ml/hr/100 gm of lymph flow from the skin and subcutaneous regions, a figure that is similar to values obtained from other organs such as the lung. Lymph flow may increase 12 to 13 times normal when venous pressure is elevated (Fig. 18.3). When plasma proteins are suddenly decreased in the circulation, paw lymph flow has been observed to rise 60- to 70-fold (Gibson, 1974).

Lymph may be propelled through the lymphatic system by at least three different mechanisms (Nicoll and Taylor, 1977): (1) The lymphatics are considered to have an intrinsic ability to contract, and since they have valves, lymph will move away from the tissue spaces with each contraction. However, although the initial lymphatics in the wing of the bat have such an ability (Nicoll and Webb, 1955), in other species, this capability is found in only the larger, more muscular lymphatics (Hall *et al.*, 1965). (2) When the surrounding tissues contract or move, they compress the lymphatics and thus expel the contents away from the tissue spaces (Casley-Smith, 1977). (3) Arterial pulsations provide a pumping mechanism that increases lymph flow in neighboring lymphatic channels. However, the response may also be due to a greater capillary filtration occurring during pulsatile flow (Parsons and McMasters, 1938). Respiration may likewise play a role in lymph flow.

b. Protein content of lymph: The total protein content of paw lymph is approximately 2 gm% (plasma = 6.5 gm%) at normal lymph flows, and it decreases to 10% of plasma values when venous pressure is elevated to 35 mm Hg (Chen *et al.*, 1976; Rutili *et al.*, 1982). The latter response indicates that capillary permeability in the skin and subcutaneous tissue is relatively low. A simple pore model demonstrating pores of 47 and 200 Å describes the kinetics of protein turnover in the paw tissue (Taylor *et al.*, 1982; Garlick and Renkin, 1970; Taylor and Granger, 1984).

Measurements of the protein concentration in human leg lymph are comparable to dog paw data; however, lymph collected from human subjects may not exclusively represent skin and subcutaneous tissue because of contamination by drainage from muscle tissue (Olszewski *et al.*, 1977).

3. Physiologic Functions of the Lymphatics

Capillary filtrate ($J_{V,C}$) is a function of the differences in hydrostatic and colloid osmotic pressure acting across the capillary wall. The relationship can be formally written in the following fashion:

$$J_{V,C} = K_{F,C}\left[(P_C - P_T) - \sigma_d(\pi_P - \pi_T)\right] \qquad \text{(F.1)}$$

where $K_{F,C}$ is the hydraulic conductance of the capillary wall, P_C and P_T refer to capillary and interstitial hydrostatic pressures, respectively, π_P and π_T refer to the colloid osmotic pressure of proteins in plasma and tissues, respectively, and σ_d refers to the osmotic reflection coefficient, a measure of the capillary wall selectivity. σ_d is approximately 0.9 for skin and subcutaneous tissue and $K_{F,C}$ is about 0.01 ml/min/100 gm/mm Hg (Rutili *et al.*, 1982).

When capillary pressure is elevated, P_T increases, π_T decreases, and lymph flow rises. Obviously, the P_T increase and the π_T decrease oppose the tendency for capillaries to filter fluid. However, lymph flow also carries away a portion of the filtered fluid and allows more capillary filtrate to occur without tissues swelling (Taylor *et al.*, 1973; Taylor, 1981). Figure 18.4 indicates how important the ability of the lymphatics to remove filtrate is to overall fluid balance in dog paw tissues. Normally, the net force acting across the capillary wall ΔP [sum of all forces on the right side of Eq. (F.1)] is only 0.5 mm Hg, with the result that only a small amount of lymph forms. If P_C is elevated to 25 mm Hg, P_T increases to 0, π_L falls to 1 mm Hg, ΔP is now 5 mm Hg, and a 10-fold increase in lymph flow carries away this increased filtrate, i.e.,

$$\text{Lymph flow} = K_{F,C}\,(\Delta P) \qquad \text{(F.2)}$$

The lower panel of Fig. 18.4 indicates what would occur if lymph flow did not rise: tissue pressure would have to increase to 4.5 mm Hg and π_T would have to decrease to 1 mm Hg, resulting in an imbalance in forces, ΔP equal to 0.5 mm Hg. But, different amounts of fluid enter the tissues in the two dissimilar situations. With a 10-fold augmentation in lymph flow, the interstitial volume rises by only 1.7 ml/100 gm, whereas when lymph flow remains unchanged, it increases by 11 ml/100 gm. Such an alteration in interstitial fluid represents

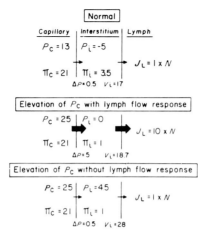

FIG. 18.4. Role of lymph flow in opposing the formation of edema. P_C is the capillary hydrostatic pressure; π_C is the capillary colloid osmotic pressure; P_i is the interstitial fluid pressure; π_i is the interstitial colloid osmotic pressure; ΔP represents the transcapillary pressure drop or imbalance in Starling forces $(P_C - P_i) - (\pi_C - \pi_i)$; J_L represents lymph flow; and V_i is the assumed interstitial volume. (From Chen *et al.*, 1976, reproduced with permission from the American Heart Association.)

almost a doubling of its volume. It can therefore be concluded that the lymphatics are important determinants of tissue volume in subcutaneous tissue and skin (Taylor *et al.*, 1973; Taylor and Gibson, 1975; Guyton *et al.*, 1975; Granger, 1979).

Lymph flow in subcutaneous tissue and skin rises with histamine infusion, burns, and endotoxin (Taylor and Granger, 1984), all of which are known to raise vascular permeability in the capillary bed. However, the increased lymph flow under these conditions does not represent a greater ability of the lymphatics to remove fluid inasmuch as the lymphatic safety factor is not large. Instead, the response can be attributed to the simultaneous increase in $K_{F,C}$. Since fluid flow through capillary walls rises as a function of r^4, then the augmentation in lymph flow following an increase in permeability is small relative to the $K_{F,C}$ increase; as a result, edema ensues. The major consequence observed when the skin lymphatics are overwhelmed or physically blocked is accumulation of a large amount of fluid and protein in the potentially large subcutaneous tissue spaces. Chronically, this leads to subcutaneous fatty acid deposits and fibrosis. An example of this is elephantiasis associated with filarial infections, in which the lymphatic system is blocked through a neighboring inflammatory reaction (Yoffey and Courtice, 1970).

G. BLOOD CIRCULATION OF ADIPOSE TISSUE

By B. Linde

1. General Considerations

Blood circulation in adipose tissue plays a central role in the storage and mobilization of lipids. In addition to supplying the tissue with absorbed substrates following meals, the circulation in adipose tissue influences metabolism by delivering hormones regulating lipolysis, as well as by supplying carrier proteins necessary for the transport of the free fatty acids from their site of production. Moreover, adipose tissue circulation has an important role in regulating overall cardiovascular and temperature homeostasis of the body. Finally, knowledge regarding the adipose tissue vascular bed has clinical implications, since many drugs are administered subcutaneously, their mobilization from the depot being influenced by local blood flow, transvascular fluid movements, and vascular permeability.

A significant advance in research on adipose tissue circulation took place when quantitative studies became possible through the development of *in situ* animal preparations with intact blood vessels and nerves. Among these, the canine inguinal subcutaneous adipose tissue preparation is the one most extensively investigated. [For information on earlier studies on adipose tissue circulation, see Ballard (1977), Rosell and Linde (1978), and Rosell and Belfrage (1979).]

2. Anatomy of Blood Circulation

For many years white adipose tissue was considered to be an isolating layer, without metabolic activity and with a sparse vascularity. In contrast, it is now known to possess a rich vascular supply, with the fat cells surrounded by a network of capillaries, each adipocyte being in contact with at least one capillary. For metabolic purposes, the vascular bed of adipose tissue is even richer than that of skeletal muscle. From studies on mesenteric fat, it has been reported that the adipocytes are located closer to the venous than the arterial side of the circulation, in a location where the permeability is known to be higher (Fig. 18.5). This site may be of importance for the transport of products of lipolysis from the fat cells into the circulation.

Although it has been known that adrenergic nerves control lipolysis, direct morphologic support for this well-established metabolic fact has been lacking

Blood Vessels and Lymphatics in Organ Systems
Copyright © 1984 by Academic Press, Inc.

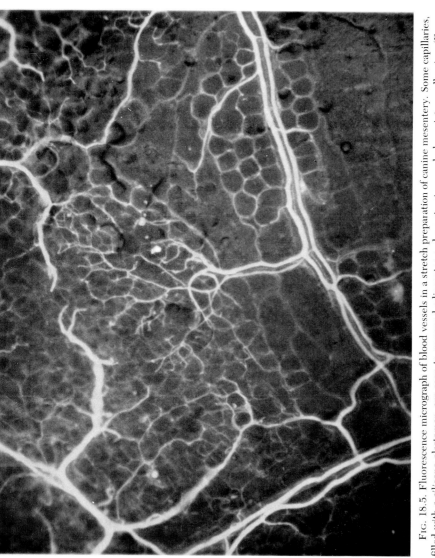

Fig. 18.5. Fluorescence micrograph of blood vessels in a stretch preparation of canine mesentery. Some capillaries, filled with a silicone elastomer, are seen to surround adipocytes and empty at several places into small veins. (From Ballard *et al.*, 1974, reproduced with permission from *Microvascular Research*.)

until recently. Adrenergic nerves have now been shown to be in close proximity not only to blood vessels, but also to fat cells in the adipose tissue.

3. PHYSIOLOGY OF BLOOD CIRCULATION

a. Vascular dimensions: Basal adipose tissue blood flow (Table 18.I) varies little between species, but does vary considerably with regard to the region of the body, fat cell size, and nutritional and neurohumoral states. Thus, omental and mesenteric fat shows higher flow rates than does subcutaneous fat, but there also are regional differences within the subcutaneous layer, as for example, a higher blood flow in the subcutaneous fat in the abdominal wall than in the extremities. In most species, blood flow per unit weight is considerably lower in adipose tissue containing large fat cells than in that with small ones. Hence, differences in fat cell size may partly explain regional differences in adipose tissue blood flow, as well as the fact that in obese persons, adipose tissue blood flow is lower than in thin subjects. Furthermore, blood flow following fasting is higher than in the fed state (see Section G3b, below).

The maximal blood flow rates in adipose tissue (25–30 ml/100 gm/min) are

TABLE 18.I

HEMODYNAMIC DATA FROM CANINE SUBCUTANEOUS ADIPOSE TISSUE DURING BASAL CONDITIONS[a]

	Ref.[b]	Value
Blood flow (ml/min/100 gm)	1	8.5 (3–19)
Capillary filtration coefficient (ml/min/100 gm/mm Hg)	2	0.027 (0.020–0.035)
PS product (ml/min/100 gm)	3	
[^{14}C]sucrose		1.5
[^{3}H]polyethylene glycol$_{800\text{-}1000}$		1
Rubidium-86	4	2.5 ± 0.5
Osmotic reflection coefficient	5	
NaCl		0.0227 ± 0.002
Sucrose		0.0333 ± 0.003
Raffinose		0.0340 ± 0.0002
Ficoll$_{70}$		0.9110 ± 0.060
Isovolumetric capillary pressure (mm Hg)	6	9.4 ± 3.2
Blood volume (ml/100 gm)	7	7.3 ± 2.7
Interstitial space, [^{14}C]sucrose (ml/100 gm)	8	10.6 (4.1–18.8)
Interstitial colloid osmotic pressure (mm Hg)	9	10.6 ± 0.8

[a]Mean values and 1 standard deviation or ranges are given.

[b]Key to references: (*1*) Ngai *et al.*, 1966; (*2*) Öberg and Rosell, 1967; (*3*) Linde *et al.*, 1974; (*4*) Linde and Gainer, 1974; (*5*) Ballard and Perl, 1978; (*6*) Rosell *et al.*, 1974; (*7*) Linde, 1976; (*8*) Linde and Chisolm, 1975; (*9*) Johnsen, 1974.

lower than those found in skeletal muscle during maximal vasodilatation (50–70 ml/100 gm/min), reflecting the lower flow resistance of the latter tissue.

The vascular exchange section in adipose tissue resembles that of skeletal muscle in that the capillaries are of the continuous type. However, vascular permeability seems to be higher in the former, since the capillary filtration coefficient in adipose tissue is greater in spite of a lower capillary surface area, as shown by both histologic and functional studies. Determinations of osmotic reflection coefficients in canine subcutaneous adipose tissue (Ballard and Perl, 1978) support the view that pores of the vascular membrane in adipose tissue are larger than those in skeletal muscle. The higher colloid osmotic pressure in its interstitial space is consistent with this possibility. The blood content of canine adipose tissue varies from 2 to 10 ml/100 gm and the interstitial space comprises approximately 10% of the wet weight.

b. *Local control:* This mechanism is similar to that of other tissues, consisting of myogenic, metabolic, and local nervous control. Autoregulation has been found in canine, as well as in human, subcutaneous adipose tissue, but it is poorly developed, compared with that of most other tissues. In the subcutaneous tissues of the forearm, blood flow is constant over the pressure range −20 to +25 mm Hg from the arterial level. Such local control is achieved by myogenic and metabolic mechanisms.

When vascular transmural pressure is increased above approximately 25 mm Hg, a vasoconstrictor response is elicited by a local venoarteriolar reflex mechanism, triggered by the elevated venous pressure (Henriksen, 1977), a similar regulation being found in cutaneous and skeletal muscle tissues. The vasoconstrictor mechanism appears to be of importance for the control of mean capillary pressure under conditions of elevated venous pressure.

Reactive hyperemia is found in both canine inguinal adipose tissue and human subcutaneous tissue from the thigh and calf. In the latter region the response is related to the length of the ischemic period (Nielsen and Sejrsen, 1972), suggesting that metabolic components play a predominant role. However, unlike in cutaneous and skeletal muscle tissues, the response in the subcutaneous tissue is not altered by the prostaglandin synthesis inhibitor indomethacin, indicating that prostaglandins are of little importance in maintaining reactive hyperemia in this tissue (Carlsson *et al.*, 1983). Arterial occlusion of short duration is also followed by hyperemia, indicating that myogenic factors may contribute to the vasodilatation.

Adipose tissue blood flow is known to be elevated during various conditions involving increased mobilization of fat from adipose tissue. Thus, functional hyperemia is found during infusions of catecholamines and glucagon, diabetic and fasting states, and muscular exercise. Two different mechanisms could mediate such a response: One is activation of vascular adrenoceptors by neurally released or blood-borne catecholamines, and the other is the local release of metabolites in connection with lipolysis.

Adenosine, which has been shown to be a physiologic regulator of blood flow in several vascular beds, is present in venous blood draining canine adipose tissue, and its rate of production is enhanced by sympathetic nerve stimulation (Fredholm, 1976). Its role in blood flow regulation in canine adipose tissue was investigated recently by Sollevi and Fredholm (1981) who showed that vascular conductance is linearly related to the venous plasma concentration of this hormone, regardless of whether the concentration is altered by exogenous adenosine or by drugs that inhibit adenosine inactivation. Blood flow rose by approximately 50% with a twofold increase in the plasma adenosine concentration. Theophylline, which antagonizes the action of adenosine in several tissues, reduced the vasodilatation caused by parenteral administration of adenosine and also decreased basal adipose tissue blood flow. Such effects were seen at therapeutic plasma concentrations of theophylline, which are lower than those required for phosphodiesterase inhibition. These results suggest that adenosine is one factor that modulates canine adipose tissue blood flow under physiologic conditions. Other factors, such as oxygen tension, hydrogen ions, and prostaglandins, also have been discussed in this connection. The relative importance of adrenergic and metabolic mechanisms for functional hyperemia in adipose tissue is not presently known.

The release of free fatty acids (FFA) from adipose tissue may be limited by blood flow, since a carrier protein is required for their transport. In fact, model experiments in the dog indicate that, even under basal blood flow conditions, this carrier function may be limiting for the transport of FFA (Bülow and Madsen, 1981). Therefore, as the lipolytic rate increases, the tissue may require more carriers if FFA are not to be reesterified or accumulate within the tissue. In this regard, accumulation of FFA has been shown to cause inhibition of lipolysis (Fredholm, 1978). Thus, blood flow per se might regulate lipid mobilization by causing variations in the supply of protein to the adipose tissue.

The effects of temperature changes illustrate the importance of metabolic factors in flow regulation. When the temperature of adipose tissue is reduced, resting blood flow is decreased and the vasoconstrictor effects of sympathetic stimulation are enhanced. These responses may be of importance with regard to the heat-insulating function of adipose tissue. Furthermore, during cooling the output of FFA is reduced although lipolysis is unaffected, suggesting increased reesterification of FFA. This is known to be a heat-generating process and may also operate during adaptation to cold.

c. Central nervous control: Although knowledge of the activation of the efferent link of the sympathetic pathways to adipose tissue is extensive, little is known regarding the central nervous control of the circulation and its relation to lipid mobilization. Recently, Parker *et al.* (1979) found experimental evidence in the dog that blood flow in subcutaneous tissues is regulated by influences from the hypothalamus. In fact, it was possible to evoke circulatory

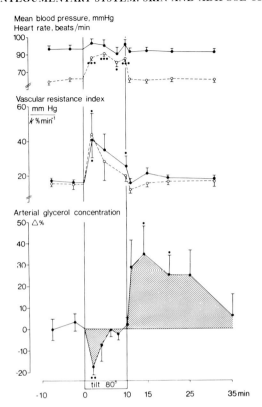

FIG. 18.6. Effect of tilting 80° with horizontal plane for 10 min on mean blood pressure (●———●) and heart rate (○- - - - -○), vascular resistance index in adipose tissue (●———●) and skeletal muscle (○- - - - -○), and arterial glycerol concentration in 8 healthy individuals. Values are means ± SE. ***, $p < .001$. **, $p < .01$. *, $p < .05$. (From Linde and Hjemdahl, 1982, reproduced with permission from *American Journal of Physiology.*)

effects without metabolic activation by precise electrode placement in this region. Such findings suggest that there are specific sympathetic neurons connected to the adipocytes that control lipolysis and others to the vasculature that regulate blood flow.

Attempts to activate blood flow and metabolism of subcutaneous adipose tissue reflexly in anesthetized dogs have not been successful. For example, occlusion of the carotid arteries has failed to stimulate the sympathetic nerves to dog adipose tissue. On the other hand, tilting humans to activate baroreceptors results in altered circulation and metabolism in adipose tissue (Linde and Hjemdahl, 1982; Skagen and Bonde-Petersen, 1982) (Fig. 18.6). The difference between the responses in dogs and humans might be explained by species variations or by regional differences in the reflex activation of the sympathetic nerves to adipose tissue. The animal experiments were performed on abdominal adipose tissue, whereas those on humans were limited to adipose tissue in the extremities. Since baroreceptors are involved in cardiovascular homeostasis, the results in man indicate that adipose tissue contributes to this mechanism during alterations of body position. The vascular responses may follow both an increase of vascular resistance and mobilization of blood from the capacitance vessels. However, the absorption of fluid from the interstitial space in response to sympathetic activity, which occurs in skeletal muscle, for example, is not noted in adipose tissue.

d. Peripheral sympathetic control: In most species, activation of sympathetic nerves is the most important stimulus for short-term regulation of lipolysis. However, there are regional differences in the sensitivity to adrenergic stimuli. For example, the mesenteric fat responds little, if at all. The effect of adrenergic stimuli on the fat cells is mediated by β-adrenoceptors and, in most species, norepinephrine is a more potent stimulus in this regard than epinephrine. Thus, there are predominantly β_1-adrenoceptors on the adipocytes of most animal species, but in dogs and humans, β_2-adrenoceptors also seem to be present. Isolated fat cells from humans also contain α_2-adrenoceptors that may reduce lipolysis, apparently by inhibiting cyclic AMP production.

Electrical activation of sympathetic nerves to canine and omental adipose tissue causes a frequency-related α-receptor-mediated vasoconstriction that, upon prolonged stimulation, fades and may even be converted into vasodilatation. Upon removal of the electrical current, there is a period of hyperemia, during which the outflow of lipolytic products, previously trapped within the tissues, is enhanced. After blockade of α-adrenoceptors, nerve stimulation now induces an immediate vasodilatation, thus revealing the presence of vascular β-adrenoceptors. In addition, the precapillary sphincters and capacitance vessels in adipose tissue also manifest vasoconstriction and vasoconstrictor escape.

Unlike the response in other tissues investigated by the capillary filtration coefficient (CFC) technique (Cobbold *et al.*, 1963), adipose tissue responds to sympathetic nerve activation with an increased filtration capacity. This occurs in spite of vasoconstriction and concomitant reduction of the capillary surface area (Fig. 18.7). In fact, the increase in CFC is even larger than that produced by maximal vasodilatation. These results indicate that vascular permeability in adipose tissue is enhanced by sympathetic activity, a possibility that has also been considered utilizing other techniques. Studies with both diffusion techniques determining permeability-surface area (PS) products for different-sized molecules, as well as measurement of the isovolumetric capillary pressure, have supported the view that there is increased permeability during sympathetic nerve activation. Such a change does not seem to be limited to the subcutaneous adipose tissue but is also present in omental adipose tissue during stimulation of the sympathetic nerves.

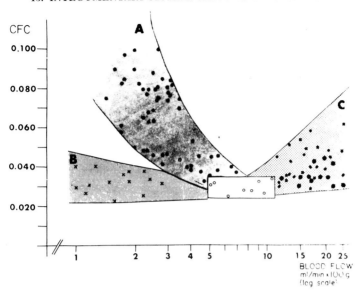

FIG. 18.7. Relation between blood flow (ml/min/100 gm) and CFC (ml/min/100 gm/mm Hg) in eight subcutaneous adipose tissue preparations during various experimental procedures. Open circles represent resting values. Values enclosed in area A represent sympathetic stimulation with no receptor blockade or, in some cases, following β-receptor blockade. Values enclosed in area B represent mechanical restriction of blood flow. Values enclosed in area C represent poststimulatory hyperemia (small dots) or during sympathetic stimulation following α-receptor blockade (large dots). (From Öberg and Rosell, 1967, reproduced with permission from *Acta Physiologica Scandinavica*.)

Present data indicate that the increase in permeability is caused by activation of α-adrenoceptors. The mechanism could involve contraction of endothelial cells, which may increase vascular permeability by enlarging interendothelial gaps, as has been suggested for histamine-like compounds (see Section E, Chapter 11). Another possibility is that vesicular transport is augmented. As a result of increased capillary permeability it is possible that greater numbers of albumin-carrier molecules reach the extravascular space, thereby facilitating the transport of FFA into the blood. Thus, increased permeability of exchange vessels may constitute a link between lipid mobilization from the adipocytes and the transport of FFA into the blood circulation. Quantitative data on FFA and albumin concentrations in the interstitial space are not presently available, but it would be of importance to evaluate the metabolic consequences of changes in permeability.

e. Circulating catecholamine control: Detailed pharmacologic studies of the adipose tissue of dogs have shown that the vascular adrenoceptors causing vasodilatation in this structure are of the β₁-subtype. Thus, in canine adipose tissue, circulating norepinephrine, which predominantly activates β₁-adrenoceptors, may induce vasodilatation, whereas epinephrine, being more β_2-selective, causes vasoconstriction via its α-stimulating effect. The current view, therefore, is that the vasculature of adipose tissue is supplied with α- and β_1-adrenoceptors.

In a study on humans, Hjemdahl and Linde (1983) found that norepinephrine at low, physiologically important plasma concentrations induced a dose-related vasoconstriction in abdominal and thigh adipose tissue and in the calf (made up predominantly of skeletal muscle), whereas epinephrine caused a dose-dependent vasodilatation in the three sites. These responses suggest the presence of vascular β_2-adrenoceptors in human adipose tissue, which is contrary to the situation in other species. Arterial concentrations of epinephrine required to obtain vascular effects in humans were in the order of 1 to 2 nM, values well within the physiologic range.

In view of the lipolytic potency of exogenous catecholamines, it appears that, in canine adipose tissue, high levels of circulating norepinephrine may be important in the regulation of lipolysis in certain situations, although, for the short-term control, the sympathetic nerves seem to have a greater role. In the dog, epinephrine is less effective than norepinephrine as a lipolytic hormone and therefore probably plays a less important function in the physiologic regulation of lipolysis in this species. In humans, however, the reverse has been found to be true (Hjemdahl and Linde, 1983). In fact, arterial glycerol concentrations increased at plasma concentrations of epinephrine of 1 to 2 nM, findings which are in agreement with those of Galster et al. (1981). In humans, both epinephrine and norepinephrine appear to function as circulating hormones that induce vascular effects and lipolysis in adipose tissue at physiologic plasma levels.

f. Other hormonal influences: Neurotensin, one of the new gastrointestinal peptides (Rosell and Rökaeus, 1981), also influences adipose tissue circulation. Neurotensin-like immunoreactivity is released from the gut and enters the blood following ingestion of fat, its main physiologic actions appearing to be located in the gastrointestinal region. However, experiments in canine adipose tissue have shown that neurotensin may also influence the circulation in this tissue, causing sustained vasoconstriction. The effect on adipose tissue vascular resistance appears to be indirect, since there is a latent period following infusion before vasoconstriction is apparent. Recently, these findings have been confirmed in humans, in whom intravenous infusion of neurotensin induced vasoconstriction in abdominal subcutaneous tissue (Linde et al., 1982). An interesting observation was that this type of response was found to be pronounced in lean subjects, whereas those with a high body fat content responded poorly or not at all. Since the uptake of FFA in peripheral organs is related to the inflow (plasma flow × arterial FFA concentration), these results indicate that neurotensin may play an endocrine role by affecting the regional deposition of fat postprandially, a possibility that requires further investigation. As already mentioned, it has been found that adipose tissue blood flow increases following fasting, the release of vasodilating agents having been proposed as a possible mechanism. However, low levels of circulating neurotensin might be another factor contributing to such a response.

4. PATHOLOGY OF BLOOD CIRCULATION

Adipose tissue appears to be unusually sensitive to hemorrhagic shock. For example, in the dog, local blood flow is reduced considerably more in this tissue than in most other structures following standardized bleeding. Blockade of α-adrenoceptors prevents the signs of vascular damage after bleeding. Under conditions of high sympathetic tone, impaired circulation leads to hypoxia, low ATP levels, and the accumulation of FFA within the tissue. Such a reaction pattern does not appear to change following denervation. This suggests that circulating catecholamines, rather than activation of sympathetic nerves, are involved in the development of irreversible changes in adipose tissue vasculature. Since epinephrine is a vasoconstrictor in canine adipose tissue, the high levels of this hormone found in connection with bleeding might contribute to the impaired circulation in adipose tissue under such conditions.

5. FUTURE AVENUES FOR INVESTIGATION

Knowledge of the regulation of adipose tissue circulation is still incomplete. Although there is considerable information from animal studies regarding separate control systems, little is presently known about how these mechanisms respond in different situations to create an integrated response in the intact organism. For example, the relative importance of the neurally released and the circulating catecholamines and the mechanism through which the latter modulate neurogenic responses under different conditions have not been determined. Nor have the role of the local metabolic control systems in adipose tissue circulation (as for example, adenosine) and the manner in which the local and remote control systems interact been elucidated. The possibility that adipose tissue blood flow is linked to some undetermined metabolic alterations, as, for example, changes associated with muscular exercise, food intake, deprivation, etc., should be tested experimentally. Furthermore, the possibility should be investigated as to whether substances released into the blood following food intake (for example, gut peptides) modulate the uptake of metabolic substrates in peripheral organs, perhaps by directing them to skeletal muscle instead of adipose tissue.

Although much remains to be elucidated concerning the regulation of adipose tissue circulation in normal individuals, even less is known regarding the function of the regulatory systems under different pathophysiologic conditions. The question of control during autonomic dysfunction, hyperthyroidism, and hypothyroidism is well worth studying, in view of the altered sensitivity to catecholamines present under these conditions.

References*

Abramson, D. I. (1967). "Circulation in the Extremities," pp. 49–113. Academic Press, New York. (C)

Abramson, D. I., and Fierst, S. M. (1942). Resting blood flow and peripheral vascular responses in hypertensive subjects. Am. Heart J. 23, 89–101. (C)

Abramson, D. I., Katzenstein, K. H., and Ferris, E. B., Jr. (1941). Observations on reactive hyperemia in various portions of the extremities. Am. Heart J. 22, 329–341. (C)

Ballard, K. W. (1977). Functional characteristics of the microcirculation in white adipose tissue. Microvasc. Res. 16, 1–18. (G)

Ballard, K. W., and Perl, W. (1978). Osmotic reflection coefficients of canine subcutaneous adipose tissue endothelium. Microvasc. Res. 16, 224–236. (G)

Ballard, K. W., Malmfors, T., and Rosell, S. (1974). Adrenergic innervation and vascular patterns in canine adipose tissue. Microvasc. Res. 8, 164–171. (G)

Ballow, M., Ward, G. W., Jr., and Gershiven, M. E. (1975). C1 - Bypass complement–activation pathway in patients with chronic urticaria and angioedema. Lancet 2, 248–250. (E)

Beck, L. (1965). Histamine as the potential mediator of axon reflex vasodilation. Fed. Proc., Fed. Am. Soc. Exp. Biol. 24, 1298–1310. (D)

Bell, C., Conway, E. L., Lang, W. J., and Padany, R. (1975). Vascular dopamine receptors in the canine hindlimb. Br. J. Pharmacol. 55, 167–172. (D)

Bergström, S., Carlson, L. A., Ekelund, I., and Oro, L. (1965). Cardiovascular and metabolic response to infusions of prostaglandin E and to simultaneous infusions of noradrenaline and prostaglandin E in man. Acta Physiol. Scand. 64, 332–339. (D)

Bernstein, J. E., and Soltani, K. (1980). Naloxone inhibition of alcohol-induced rosacea flushing. Clin. Res. 28, 718A. (D)

Bernstein, J. E., and Soltani, K. (1982). Alcohol-induced rosacea flushing blocked by naloxone. Br. J. Dermatol. 107, 59–62. (D)

Bernstein, J. E., Swift, R. M., Soltani, K., and Lorincz, A. L. (1981a). Inhibition of axon-reflex vasodilatation with capsaicin. J. Invest. Dermatol. 76, 394–395. (D)

Bernstein, J. E., Hamill, J. R., and Soltani, K. (1981b). Bradykinin, substance P, prostaglandin E_2 and papain induced flare suppressed by capsaicin. Clin. Res. 29(4), 787A. (D)

Bernstein, J. E., Levine, L. E., Gruszka, R., Robertson, G. L., and Rothstein, J. (1982). Biochemical aspects of alcohol-induced flushing in Orientals. Clin. Res. 30(2), 249A. (D)

Blair, D. A., Glover, W. E., and Roddie, I. C. (1960). Vasomotor fibres to skin in the upper arm, calf and thigh. J. Physiol. (London) 153, 232–238. (C)

Blum, K., Hamilton, M. G., and Wallace, J. E. (1977). Alcohol and opiates: A review of common neurochemical and behavioral mechanisms. In "Alcohol and Opiates" (K. Blum, ed.), pp. 203–206. Academic Press, New York. (D)

Borrie, P. (1955). The state of the blood vessels of the face in rosacea. II. Br. J. Dermatol. 67, 73–75. (D)

*In the reference list, the capital letter in parentheses at the end of each reference indicates the section in which it is cited.

Braverman, I. M., and Yen, A. (1977). Ultrastructure of the human dermal microcirculation. II. The capillary loops of the dermal papillae. *J. Invest. Dermatol.* **68**, 44–52. (B)

Brody, M. J. (1966). Neurohumoral mediation of axon reflex vasodilatation. *Fed. Proc., Fed. Am. Soc. Exp. Biol.* **25**, 1583–1592. (D)

Bülow, J., and Madsen, J. (1981). Influence of blood flow on fatty acid mobilization from lipolytically active adipose tissue. *Pfluegers. Arch.* **390**, 169–174. (G)

Burcher, E., Atterhog, J. H., Pernow, B., and Rosell, S. (1977). Cardiovascular effects of substance P: Effects on the heart and regional blood flow in the dog. *In* "Substance P" (U. S. von Euler, ed.), pp. 261–268. Raven Press, New York. (D)

Burnstock, G. (1977). Autonomic neuroeffector junctions-reflex vasodilatation of the skin. *J. Invest. Dermatol.* **69**, 47–57. (D)

Burton, A. C. (1961). Special features of the circulation of the skin. *Adv. Biol. Skin* **2**, 117–122. (C)

Carlsson, I., Linde, B., and Wennmalm, Å. (1983). Arachidonic acid metabolism and regulation of blood flow: Effect of indomethacin on cutaneous and subcutaneous reactive hyperemia in humans. *Clin. Physiol.* **3**, 365–373. (G)

Casley-Smith, J. R. (1977). The functioning of the lymphatic system under normal and pathological conditions: Its dependence on the fine structures and permeabilities of the vessels. *In* "Progress in Lymphology" (A. Ruttiman, ed.,), pp. 348–359. Thieme, Stuttgart. (F)

Celander, O. (1954). The range of control exercised by the "sympathico-adrenal system": A quantitative study on blood vessels and other smooth-muscle effectors in the cat. *Acta Physiol. Scand.* **32**, Suppl. 116, 1–132. (C)

Celander, O., and Folkow, B. (1953). A comparison of the sympathetic vasomotor fibre control of the vessels within the skin and the muscle. *Acta Physiol. Scand.* **29**, 244–250. (C)

Champion, R. H. (1965). Livedo reticularis. A review. *Br. J. Dermatol.* **77**, 167–170. (E)

Chen, H. I., Granger, H. J., and Taylor, A. E. (1976). Interactions of capillary, interstitial and lymphatic forces in the canine hindpaw. *Circ. Res.* **38**, 245–254. (F)

Clark, E. R., and Clark, E. L. (1934). Observations on living arteriovenous anastomoses as seen in transparent chambers introduced into the rabbit's ear. *Am. J. Anat.* **54**, 229–286. (C)

Clayden, J. R., Bell, J. W., and Pollard, P. (1974). Menopausal flushing: Double-blind trial of a nonhormonal medication. *Br. Med. J.* **1**, 409–412. (D)

Cobbold, A., Folkow, B., Kjellmer, I., and Mellander, S. (1963). Nervous and local chemical control of precapillary sphincters in skeletal muscle as measured by changes in filtration coefficient. *Acta Physiol. Scand.* **57**, 180–192. (G)

Cooke, P. H., and Fay, F. S. (1972). Correlations between fibre length, ultrastructure and length–tension relationship of mammalian smooth muscle. *J. Cell Biol.* **32**, 105–116. (B)

Copeman, P. W. M., and Ryan, T. J. (1971). Cutaneous angiitis. Patterns of rashes explained by: (1) Flow properties of blood, (2) anatomical disposition of vessels. *Br. J. Dermatol.* **85**, 205–214. (E)

Daly, J. J., and Duff, R. S. (1960). Direct effects of angiotonin on peripheral vessels of subjects with normal and raised blood pressures. *Clin. Sci.* **19**, 457–463. (C)

Davis, M. J., and Lawler, J. C. (1958). Observations on the delayed blanch phenomenon in atopic subjects. *J. Invest. Dermatol.* **30**, 127–132. (E)

Davis, V. E., and Walsh, M. J. (1970). Alcohol, amines, and alkaloids, a possible biochemical basis for alcohol addiction. *Science* **167**, 1005–1007. (D)

Demis, D. J., and Galakatos, E. (1967). Healing proliferation: A new microcirculation concept. *Angiology* **18**, 498–527. (E)

Douglass, M. C., Gruz, C., and Pelachyk, J. M. (1980). Kaposi's sarcoma and immunosuppression. *Curr. Concepts Skin Disord.* **1**, 7–16. (E)

Downey, J. A., Leroy, E. C., Miller, J. M., III, and Darling, R. C. (1971). Thermoregulation and Raynaud's phenomenon. *Clin. Sci.* **40**, 211–219. (E)

Edwards, E. A. (1954). Varieties of digital ischemia and their management. *N. Engl. J. Med.* **250**, 709–716. (E)

Fagrell, B., Fronek, A., and Intaglietta, M. (1977). A microscope-television system for studying flow velocity in human skin capillaries. *Am. J. Physiol.* **233**, H318–H321. (C)

Folkow, B. (1949). The vasodilatation action of adenosine triphosphate. *Acta Physiol. Scand.* **17**, 311–316. (D)

Fox, R. J., and Hilton, S. M. (1958). Bradykinin formation in human skin as a factor in heat vasodilatation. *J. Physiol. (London)* **142**, 219–232. (D)

Fredholm, B. B. (1976). Release of adenosine-like material from isolated perfused dog adipose tissue following sympathetic nerve stimulation and its inhibition by adrenergic α-receptor blockade. *Acta Physiol. Scand.* **96**, 422–430. (G)

Fredholm, B. B. (1978). Local regulation of lipolysis in adipose tissue by fatty acids, prostaglandins and adenosine. *Med. Biol.* **56**, 249–261. (G)

Friedman-Kien, A. E. (1981). Disseminated Kaposi's sarcoma syndrome in young homosexual men. *J. Am. Acad. Dermatol.* **5**, 468–471. (E)

Galster, A. D., Clutter, W. E., Cryer, P. E., Collins, J. A., and Bier, D. M. (1981). Epinephrine plasma thresholds for lipolytic effects in man. *J. Clin. Invest.* **67**, 1729–1738. (G)

Ganten, D., Schelling, P., Vecsei, P., and Ganten, U. (1976). Iso-renin of extrarenal origin: "The tissue angiotensinogenase systems." *Am. J. Med.* **60**, 760–772. (C)

Garlick, D. G., and Renkin, E. M. (1970). Transport of large molecules from plasma to interstitial fluid and lymph in dogs. *Am. J. Physiol.* **219**, 1595–1605. (F)

Gaskell, P., and Hoepper, D. L. (1966). The relative influence of nervous control and of local warming on arteriolar muscle during indirect vasodilatation. *Can. J. Physiol. Pharmacol.* **45**, 89–91. (C)

Gibson, W. H. (1974). Dynamics of lymph flow, tissue pressure and protein exchange in subcutaneous connective tissues. Ph.D. Dissertation, University of Mississippi, Columbia. (F)

Godfraind, T. (1970). "Fundamentals of Biochemical Pharmacology" (Z. M. Braeg, ed.). Pergamon, Oxford. (D)

Goldyne, M. D. (1975). Prostaglandins and cutaneous inflammation. *J. Invest. Dermatol.* **64**, 377–385. (E)

Goodman, T. F. (1972). Fine structure of the cells of the Suquet-Hoyer canal. *J. Invest. Dermatol.* **59**, 363–369. (C)

Granger, H. J. (1979). Role of the interstitial matrix and lymphatic pump in regulation of transcapillary fluid balance. *Microvasc. Res.* **18**, 209–216. (F)

Grant, R. T. (1930). Observations on direct communications between arteries and veins in rabbit's ear. *Heart* **15**, 281–303. (C)

Guyton, A. C., Taylor, A. E., and Granger, H. J. (1975). "Circulatory Physiology," Vol. II, pp. 18–22. Saunders, Philadelphia, Pennsylvania. (F)

Hackenthal, E., Hackenthal, R., and Hilgenfeldt, U. (1978). Isorenin, pseudorenin, cathepsin D and renin: A comparative enzymatic study of angiotensin-forming enzymes. *Biochim. Biophys. Acta* **522**, 574–588. (C)

Haddy, F. J., and Scott, J. B. (1968). Metabolically linked vasoactive chemicals in local regulation of blood flow. *Physiol. Rev.* **48**, 688–707. (D)

Haddy, F. J., Fleishman, M., and Scott, J. B., Jr. (1957). Effect of change in air temperature upon systemic small and large vessel resistance. *Circ. Res.* **5**, 58–63. (C)

Hall, J. G., Morris, B., and Woolley, G. (1965). Intrinsic rhythmic propulsion of lymph in the unanesthetized sheep. *J. Physiol. (London)* **180**, 349–366. (F)

Hammel, H. T., Jackson, D. C., Stolwijk, J. A. J., Hardy, J. D., and Stromme, S. B. (1963). Temperature regulation by hypothalamic proportional control with an adjustable set point. *J. Appl. Physiol.* **18**, 1146–1154. (C)

Henriksen, O. (1977). Local sympathetic reflex mechanism in regulation of blood flow in human subcutaneous adipose tissue. *Acta Physiol. Scand., Suppl.* **450**. (G)

Hibbs, R. G., Burch, G. E., and Phillips, J. E. (1958). The fine structure of the small blood vessels of normal human dermis and subcutis. *Am. Heart. J.* **56**, 662–670. (B)

Higgins, J. C., and Eady, R. A. J. (1981). Human dermal microvasculature. I. Its segmental differentiation. Light and electron microscopic study. *Br. J. Dermatol.* **104**, 117–129. (B)

Hjemdahl, P., and Linde, B. (1983). Influence of circulating norepinephrine and epinephrine on adipose tissue vascular resistance and lipolysis in humans. *Am. J. Physiol.* **245**, H447–H452. (G)

Hökfelt, T., Johansson, O., Kellerth, J. O., Ljungdahl, A., Nilsson, G., Nygards, A., and Pernow, P. (1977). Immunohistochemical distribution of substance P. *In* "Substance P" (U. S. von Euler, ed.), pp. 117–145. Raven Press, New York. (D)

Holloway, G. A., Jr., and Watkins, D. W. (1977). Laser Doppler measurement of cutaneous blood flow. *J. Invest. Dermatol.* **69**, 306–309. (C)

Hurley, H. J., Jr., and Mescon, H. (1956). Cholinergic innervation of the digital arteriovenous anastomoses of human skin: A histological localization of cholinesterase. *J. Appl. Physiol.* **9**, 82–84. (C)

Iijima, T., and Tagawa, T. (1976). Adrenergic and cholinergic innervation of the arteriovenous anastomosis in the rabbit's ear. *Anat. Rec.* **185**, 373–379. (C)

Jansco, G., Kiraly, E., and Jansco-Gabor, A. (1977). Pharmacologically induced selective degeneration of chemosensitive primary sensory neurons. *Nature (London)* **270**, 741–743. (D)

Jansco, N., and Jansco-Gabor, A. (1959). Dauerausschaltung der chemischen schmerzempfindlinchkeit durch capsaicin. *Naunyn-Schmiedebergs Arch. Exp. Pathol. Pharmako.* **236**, 142–145. (D)

Jansco, N., Jansco-Gabor, A., and Szolcsanyi, J. (1967). Direct evidence for neurogenic inflammation and its prevention by denervation and by pretreatment with capsaicin. *Br. J. Pharmacol.* **31**, 138–151. (D)

Jarrott, B., McQueen, A., Graf, L., and Louis, W. J. (1975). Serotonin levels in vascular tissue and the effects of a serotonin synthesis inhibitor on blood pressure in hypertensive rats. *Clin. Exp. Pharmacol. Physiol., Suppl.* **2**, 201–205. (C)

Jeffcoate, W. J., Cullen, M. H., Herbert, M., and Hastings, A. G. (1979). Prevention of effects of alcohol intoxication by naloxone. *Lancet* **2**, 1157–1159. (D)

Jeffreys, D. B., Flanagan, R. J., and Volans, G. N. (1980). Reversal of ethanol-induced coma with naloxone. *Lancet* **1**, 308–309. (D)

Jessell, T. M., Tsunoo, A., Kanazawa, I., and Otsuka, M. (1979). Substance P: Depletion in the dorsal horn of rat spinal cord after section of the peripheral processes of primary sensory neurons. *Brain Res.* **168**, 247–259. (D)

Johnsen, H. M. (1974). Measurement of colloid osmotic pressure of interstitial fluid. *Acta Physiol. Scand.* **91**, 142–144. (G)

Johnson, A. R., and Erdos, E. G. (1973). Release of histamine from mast cells by vasoactive peptides. *Proc. Soc. Exp. Biol. Med.* **142**, 1252–1256. (D)

Johnson, J. M., and Park, M. K. (1979). Reflex control of skin blood flow by skin temperature: Role of core temperature. *J. Appl. Physiol.* **47**, 1188–1193. (C)

Johnson, J. M., and Park, M. K. (1981). Effect of upright exercise on threshold for cutaneous vasodilation and sweating. *J. Appl. Physiol.* **50**, 814–818. (C)

Jorizzo, J. L., Coutts, A. A., Greaves, M. W., and Burnstock, G. (1981). A comparison of substance P and adenosine triphosphate as mediators of cutaneous inflammation. *J. Invest. Dermatol.* **76**, 315 (abstr.). (D)

Kuhn, C., and Rosai, J. (1969). Tumors arising from pericytes. *Arch. Pathol.* **88**, 653–663. (B)

Langley, J. N. (1923). Antidromic action. *J. Physiol. (London)* **57**, 428–446. (D)

Leslie, R. D. G., and Pyke D. A. (1978). Chlorpropamide-alcohol flushing: A dominantly inherited trait associated with diabetes. *Br. Med. J.* **2**, 1519–1521. (D)

Leslie, R. D. G., Pyke, D. A., and Stubbs, W. A. (1979). Sensitivity to enkephalin as a cause of non-insulin-dependent diabetes. *Lancet* **1**, 341–343. (D)

Lightman, S. L., and Jacobs, H. S. (1979). Naloxone: Nonsteroidal treatment for postmenopausal flushing. *Lancet* **2**, 1071. (D)

Linde, B. (1976). Effect of sympathetic nerve stimulation on net transvascular movement of fluid in canine adipose tissue. *Acta Physiol. Scand.* **97**, 166–174. (G)

Linde, B., and Chisolm, G. (1975). The interstitial space of adipose tissue as determined by single injection and equilibration techniques. *Acta Physiol. Scand.* **95**, 383–390. (G)

Linde, B., and Gainer, J. L. (1974). Disappearance of [133]Xenon and [131]Iodide and extraction of ^{86}Rb in subcutaneous adipose tissue during sympathetic nerve stimulation. *Acta Physiol. Scand.* **91**, 172–179. (G)

Linde, B., and Hjemdahl, P. (1982). Effect of tilting on adipose tissue vascular resistance and sympathetic activity in humans. *Am. J. Physiol.* **242**, H161–H167. (G)

Linde, B., Chisolm, G., and Rosell, S. (1974). The influence of sympathetic activity and histamine on the blood tissue exchange of solutes in canine adipose tissue. *Acta Physiol. Scand.* **92**, 145–155. (G)

Linde, B., Rosell, S., and Rökaeus, Å. (1982). Blood flow in human adipose tissue after infusion of (Gln[4])-neurotensin. *Acta Physiol. Scand.* **115**, 311–315. (G)

Lindvall, O., and Björklund, A. (1972). The glyoxylic acid fluorescence histochemical method: A detailed account of the methodology for the visualization of central catecholamine neurons. *Histochemistry* **39**, 97–127. (C)

Lorincz, A. L., and Pearson, R. W. (1959). Studies on axon reflex vasodilatation and cholinergic urticaria. *J. Invest. Dermatol.* **32**, 429–435. (D)

Lundberg, J. M., Anggard, A., Fahrenkrug, J., Hökfelt, T., and Mutt, V. (1980). Vasoactive intestinal polypeptide in cholinergic neurons of exocrine glands: Functional significance of coexisting transmitters for vasodilatation and secretion. *Proc. Natl. Acad. Sci. U.S.A.* **77**, 1651–1655. (D)

McGrath, M. A. (1977). 5-Hydroxytryptamine and neurotransmitter release in canine blood vessels: Inhibition by low and augmentation by high concentrations. *Circ. Res.* **41**, 428–435. (C)

McKenzie, A. I. (1979). Naloxone in alcohol intoxication. *Lancet* **1**, 733–734. (D)

Major, T. C., Schwinghamer, J. M., and Winston, S. (1981). Cutaneous and skeletal muscle vascular response to hypothermia. *Am. J. Physiol.* **240** (*Heart Circ. Physiol.* 9), H868–H873. (C)

Marinov, G., and Tzvetkova, T. Z. (1977). Age related differentiation in the local peculiarities of the terminal vascular bed of the lower limb skin. *Verh. Anat. Ges.* **71**, 689–695. (E)

Montagna, W., and Ellis, R. A., eds. (1961). "Advances in Biology of Skin," Vol. II. Pergamon, Oxford. (B)

Moretti, G. (1968). The blood vessels of the skin. *In* "Handbuch der Haut und Geschlechtskrankheiten" (O. Gans and G. K. Steigleder, eds.), Vol.1/1, pp. 491–623. Springer-Verlag, Berlin and New York. (B)

Ngai, S. H., Rosell, S., and Wallenberg, L. R. (1966). Nervous regulation of blood flow in the subcutaneous adipose tissue in dogs. *Acta Physiol. Scand.* **68**, 397–403. (G)

Nicoll, P. A., and Taylor, A. E. (1977). Lymph formation and flow. *Annu. Rev. Physiol.* **39**, 73–95. (F)

Nicoll, P. A., and Webb, R. L. (1955). Vascular patterns and active vasomotion as determinants of flow through minute vessels. *Angiology* **6**, 291–310. (F)

Nielsen, S. L., and Sejrsen, P. (1972). Reactive hyperemia in subcutaneous adipose tissue in man. *Acta Physiol. Scand.* **85**, 71–77. (G)

Norberg, K.-A., and Hamberger, B. (1964). The sympathetic adrenergic neuron: Some characteristics revealed by histochemical studies on the intraneuronal distributor of the transmitter. *Acta Physiol. Scand., Suppl.* **238**, 1–42. (C)

Öberg, B., and Rosell, S. (1967). Sympathetic control of consecutive vascular sections in canine subcutaneous adipose tissue. *Acta Physiol. Scand.* **71**, 47–56. (G)

Olszewski, W. L., Engeset, A., and Sokolowski, J. (1977). Lymph flow and protein in the normal male leg during lying, getting up, and walking. *Lymphology* **10**, 178–183. (F)

Parker, R. E., Hall, R. E., Marr, H. B., and Skinner, N. S., Jr. (1979). Vascular responses in canine subcutaneous adipose tissue to hypothalamic stimulation. *Am. J. Physiol.* **273**, H386–H391. (G)

Parsons, R. J., and McMasters, P. D. (1938). The effect of the pulse upon the formation and flow of lymph. *J. Exp. Med.* **68**, 353–376. (F)

Peach, M. J. (1977). Renin-angiotensin system: Biochemistry and mechanisms of action. *Physiol. Rev.* **57**, 313–370. (C)

Pyke, D. A., and Leslie, R. D. G. (1978). Chlorpropamide-alcohol flushing: A definition of its relation to non-insulin-dependent diabetes. *Br. Med. J.* **2**, 1521–1522. (D)

Ramsey, C. (1969). Vascular changes accompanying white dermatographism and delayed blanch in atopic dermatitis. *Br. J. Dermatol.* **81**, 37–42. (E)

Regali, D., and Barabé, J. (1980). Pharmacology of bradykinin and related kinins. *Pharmacol. Rev.* **32**, 1–46. (E)

Rhodin, J. A. G. (1968). Ultrastructure of mammalian venous capillaries, venules, and small collecting veins. *J. Ultrastruct. Res.* **25**, 452–500. (B)

Roberts, L. J., Marney, S. R., and Oates, J. A. (1979). Blockade of the flush associated with metastatic gastric carcinoid by combination histamine H_1 and H_2 receptor antagonists. *N. Engl. J. Med.* **300**, 236–238. (D)

Roberts, M. F., and Wenger, C. B. (1980). Control of skin blood flow during exercise by thermal reflexes and baroreflexes. *J. Appl. Physiol.: Respir., Environ. Exercise Physiol* **48**, 717–723. (C)

Rolfe, P. (1979). Theoretic aspects of skin blood flow estimation using thermal, optical, and electrical impedance techniques. *Birth Defects, Orig. Artic. Ser.* **15**(4), 135–147. (C)

Rosell, S., and Belfrage, E. (1979). Blood circulation in adipose tissue. *Physiol. Rev.* **59**, 1078–1104. (G)

Rosell, S., and Linde, B. (1978). "Adipose Tissue. Peripheral Circulation" (P. C. Johnson, ed.), pp. 315–336. Wiley, New York. (G)

Rosell, S., and Rökaeus, Å. (1981). Actions and possible hormonal functions of circulating neurotensin. *Clin. Physiol.* **1**, 3–20. (G)

Rosell, S., Intaglietta, M., and Chisolm, G. M. (1974). Adrenergic influence on isovolumetric capillary pressure in canine adipose tissue. *Am. J. Physiol.* **227**, 692–696. (G)

Rowell, L. B. (1977). Competition between skin and muscle for blood flow during exercise. *In* "Problems with Temperature Regulation during Exercise" (E. R. Nadel, ed.), pp. 49–76. Academic Press, New York. (C)

Rusznyak, I., Foldi, M., and Szabo, G. (1967). "Lymphatics and Lymph Circulation: Physiology and Patholgogy," 2nd ed. (in English), pp. 162–163. Pergamon, Oxford. (F)

Rutili, G., Granger, D. N., Parker, J. C., Mortillaro, N. A., and Taylor, A. E. (1982). Analysis of lymphatic protein data. IV. Comparison of the different methods used to estimate reflection coefficients and permeability-surface area products. *Microvasc. Res.* **23**, 347–360. (F)

Ryan, T. J. (1973). Structure, pattern and shape of the blood vessels of the skin. *Physiol. Pathophysiol. Skin* **2**, 577–654. (A)

Ryan, T. J. (1976). The blood vessels of the skin. *J. Invest. Dermatol.* **67**, 110–118. (B)

Saunders, R. L. de C. H. (1961). X-ray projection microscopy of the skin. *Adv. Biol. Skin* **2**, 38–56. (B)

Schultz, R., Wuster, M., Duka, T., and Herz, A. (1980). Acute and chronic ethanol treatment changes endorphin levels in brain and pituitary. *Psychopharmacology (Berlin)* **68**, 221–227. (D)

Skagen, K., and Bonde-Petersen, F. (1982). Regulation of subcutaneous blood flow during head-up tilt (45°) in normals. *Acta Physiol. Scand.* **114**, 31–35. (G)

Sollevi, A., and Fredholm, B. B. (1981). Role of adenosine in adipose tissue circulation. *Acta Physiol. Scand.* **112**, 293–298. (G)

Sorensen, S. C., and Mattison, K. W. (1978). Naloxone as an antagonist in severe alcohol intoxication. *Lancet* **2**, 688–689. (D)

Spalteholz, W. (1893). Die Verteilung der Blutgefasse in der haut. *Arch. Anat. Physiol., Anat. Abt.* **1**, 54. (B)

Sterry, W., Steigleder, G. K., and Bordeax, E. (1979). Kaposi's sarcoma: Venous capillary haemangioblastoma. *Arch. Dermatol. Res.* **266**, 253–268. (B)

Stoughton, R. B. (1972). Bioassay system for formulations of topically applied glucocorticoids. *Arch. Dermatol.* **106**, 825–827. (D)

Swales, J. D. (1980). Local vascular renin activity as a factor in circulatory control. *Cardiovasc. Rev. Rep.* **1**, 309–315. (C)

Swift, R. M., Bernstein, J. E., Soltani, K., and Lorincz, A. L. (1979). Inhibition of axon reflex vasodilatation in human skin by topically applied capsaicin. *Clin. Res.* **27**, 245A. (D)

Taylor, A. E. (1981). Capillary fluid filtration: Starling forces and lymph flow. *Circ. Res.* **39**, 557–575. (F)

Taylor, A. E., and Gibson, H. (1975). Concentrating ability of lymphatic vessels. *Lymphology* **8**, 43–49. (F)

Taylor, A. E., and Granger, D. N. (1984). Exchange of macromolecules across the circulation. *In* "Handbook of Physiology" (E. M. Renkin and C. C. Michel, eds.), Chapter 11. Am. Physiol. Soc., Washington, D.C. (in press). (F)

Taylor, A. E., Gibson, W. H., Granger, H. J., and Guyton, A. C. (1973). The interaction between intracapillary and tissue forces in the overall regulation of interstitial fluid volume. *Lymphology* **6**, 192–208. (F)

Taylor, A. E., Perry, M. A., and Shin, D. W. (1982). Calculation of the effective pore radii in dog hindpaw capillaries using lymph endogenous proteins. *Microvasc. Res.* **23**, 276. (F)

Walsh, T. S., Jr., and Tompkins, U. N. (1956). Some observations on strawberry nevus of infancy. *Cancer* **9**, 869–904. (E)

Waris, T., Kyösola, K., and Partanen, S. (1980). The adrenergic innervation of arteriovenous anastomoses in the subcutaneous fascia of rat skin. *Scand. J. Plast. Reconstr. Surg.* **14**, 215–220. (C)

Wilkin, J. K. (1981). Flushing reactions: Consequences and mechanisms. *Ann. Intern. Med.* **15**, 468–476. (D)

Willms-Kretschmer, K., and Majno, G. (1969). Ischaemia of the skin. *Am. J. Pathol.* **54**, 327–343. (E)

Wolf, P. (1972). Ethnic differences in alcohol sensitivity. *Science* **125**, 449–451. (D)

Yen, A., and Braverman, I. M. (1976). Ultrastructure of the human dermal microcirculation: The horizontal plexus of the papillary dermis. *J. Invest. Dermatol.* **66**, 131–142. (B)

Yoffey, J. M., and Courtice, F. C. (1970). "Lymphatics, Lymph and the Lymphomyeloid Complex," pp. 314–315. Academic Press, New York. (F)

Zweifach, B. W. (1949). Basic mechanisms in peripheral vascular homeostasis. *Conf. Factors Regul. Blood Pressure, Trans.* **3**. (B)

Chapter 19
Locomotor System: Skeletal Muscle

A. EMBRYOLOGY OF BLOOD CIRCULATION

By K. Christensen

None of the main blood vessels to skeletal muscle arises as a single trunk in the embryo, but instead, along the prospective course of each channel, a capillary network is laid down. Then, by the selection and enlargement of definitive paths in the vascular bed, the larger arteries and veins are defined. The following discussion is limited to the development of the vasculature in skeletal muscle of the lower limbs of man.

According to Senior (1919, 1924), four stages can be identified in the development of the lower limb arteries, as shown by his well-known chart (see Warwick and Williams, 1973, p. 162). In the first, the primary arterial trunk, which arises from the dorsal aspect of the umbilical artery and occupies the central portion of the lower limb bud, is the axis or ischiadic artery. In the same stage, the femoral artery takes origin from the common iliac artery. In the second stage, both these vessels have grown with the lengthening of the limb bud, each ending in a simple network of blood vessels in the expanded primordial foot.

In the third stage, the extensions from the femoral artery include the popliteal artery and both anterior and posterior tibial arteries, each of which ends in the vascular network in the embryonic foot. The ischiadic artery at this stage shows degenerative changes both in the thigh and in the leg. The remaining distal part below the knee becomes the peroneal artery. In the fourth stage, the femoral artery in the thigh gives off the profunda femoris artery, the branches of which join with the remains of the ischiadic artery at a level above the knee. No further changes are found in the knee and leg, but medial and lateral plantar arteries are noted in the foot. [For details of larger vascular branches given off by the major arteries to the muscles of the leg, see Warwick and Williams, (1973).]

In most embryologic descriptions of skeletal muscles and their vascularity, very little is presented about the type and number of terminal branches to a given muscle. A recent paper describes the electron microscopic structure of developing intramuscular blood vessels (Minguetti and Mair, 1979). Sections of muscles from the thigh were studied from human fetuses, 9 weeks to 9 months in development. At 9 weeks, the blood vessels had the appearance of capillar-

639

ies, but by 16 weeks larger blood vessels exhibited the structure of arteries and veins.

It is always of interest to associate vascular development in a given tissue or organ with that of the tissue supplied. Pertinent to such a subject is a recent paper (Curtis and Zalin, 1981) on the transition from proliferating single myoblasts to their fusion and formation of differentiated multinucleated skeletal muscle fibers. It was found that epinephrine and isoproterenol provoked primary myoblasts to initiate precocious cellular fusion.

B. ANATOMY OF BLOOD CIRCULATION

By D. I. ABRAMSON

The vascular tree in skeletal muscle has an extraordinarily well-developed capability to cope with marked variations in local metabolic needs, this function being clearly exhibited in the microcirculation.

1. GROSS ANATOMY

The gross arterial pattern in skeletal muscle varies in different locations relative to individual functional requirements. The definitive work on this subject was reported in the early part of the twentieth century, and very few significant contributions have appeared in the intervening years.

a. Extrinsic vascular pattern: According to the studies of Campbell and Pennefather (1919) and of Mathes and Nahai (1981), the arterial tree in skeletal muscles in the limbs and elsewhere can be divided into a number of categories. The classification proposed by the latter workers reflects clinical application to reconstructive surgery and is made up of five subgroups. Type I consists of a single vascular pedicle entering the muscle, as in the case of the gastrocnemius, rectus femoris, and tensor fascia lata (Fig. 19.1). Type II is characterized by the entrance of one or more large vascular pedicles into the muscle in close proximity either to its origin or to its insertion, together with small vascular pedicles penetrating the muscle belly (Fig. 19.1). Such a vascular arrangement is found in the adductor digiti minimi, abductor hallucis, gracilis, biceps femoris, and flexor digitorum brevis, among others. In type III, there are two large vascular

All rights of reproduction in any form reserved.
ISBN 0-12-042520-3

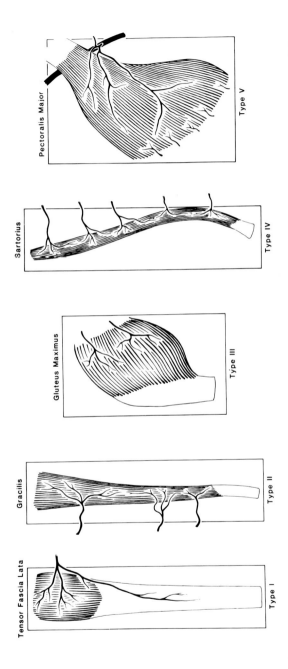

FIG. 19.1. Patterns of vascular anatomy of muscle. Type I, one vascular pedicle; type II, dominant pedicle (s) plus minor pedicles; type III, two dominant pedicles; type IV, segmental vascular pedicles; type V, dominant pedicle plus secondary segmental pedicles. (From Mathes and Nahai, 1981, reproduced with permission from Williams and Wilkins, Publishers.)

pedicles, each arising from a different regional artery, as noted in the rectus abdominus, serratus anterior, and gluteus maximus (Fig. 19.1). The local vascular tree in type IV consists of multiple vascular pedicles that enter the muscle between its origin and insertion, all being of equal size and segmental distribution. This group includes the extensor digitorum longus, extensor hallucis longus, flexor digitorum longus, tibialis anterior, and sartorius, among others (Fig. 19.1). Type V is typified by the presence of a single large vascular pedicle located close to the insertion of the muscle, together with segmental pedicles entering it close to its origin. Such a vascular arrangement is found in the pectoralis major and latissimus dorsi muscles (Fig. 19.1).

   b. Intrinsic arterial pattern: The arterial channels penetrate the muscle substance first by traversing intramuscular septi and then by forming various vascular patterns. The latter networks have been divided into several main categories, based on the injection studies of Blomfield (1945). In one, longitudinal anastomoses arise from a number of separate extrinsic arteries and enter the muscle throughout its full length, as occurs in the soleus and peroneus longus. In another, a longitudinal pattern of vascular arcades originates from a single nutrient vessel and enters the upper end of the muscle, as typified by the vascular arrangement in the gastrocnemius. In a third type, a radiating pattern of collateral channels arises from a single extrinsic artery and then enters the midportion of the muscle, as in the case of the biceps brachii. In a fourth type, an elaborate series of anastomosing arcades or loops traverses the length of the muscle, formed by a sequence of entering extrinsic vessels. From the arcades arise smaller loops that also contribute to the formation of a complex network of anastomosing vascular arcades. Such a vascular pattern is found in the tibialis anterior, extensor digitorum longus, and long extensors of the leg. In the last category, the vascular arrangement consists of an open rectangular pattern of communications with sparse anastomotic connections, as found in the extensor hallicus longus.

   Once in the muscle mass, arteries promptly branch freely to form many anastomoses comprising the primary arterial tree. From the resulting meshwork originate smaller arteries that, in turn, also interconnect to produce a secondary set of arteries. Branches of the latter are the source for the formation of the arteriolar network.

2. Microscopic and Electron Microscopic Anatomy

   With the introduction of the open cremaster muscle preparation of the rat, it has been possible to study muscle circulation at high magnifications using proper translumination (Baez, 1968; Samaji et al., 1970). The subsequent use of the thin tenuissimus muscle preparation in the cat (Brånemark and Eriksson, 1971; Eriksson and Myrhage, 1972; Ericson et al., 1973) has further advanced knowledge of the field of muscle microvascular morphology.

a. *Characteristics of the microcirculation:* Studies performed on the cat tenuissimus muscle (Eriksson and Myrhage, 1972) reveal that feeding arterioles (110 μm in diameter) divide into central arterioles (70 μm in diameter), from which arise transverse arterioles (70 μm in diameter). The latter give rise to 2–5 generations of terminal arterioles (8–15 μm in diameter), which, in turn, terminate in the capillary network. Capillaries demonstrate a wide range of vessel dimensions, with a mean of 5.5 μm in width and 100 μm in length. Consistently, tributaries from two or more terminal arterioles supply a specific capillary section. Each muscle fiber is usually surrounded by 4 or 5 capillaries.

The slow twitch oxidative-rich (type I) fibers of red muscle have a more abundant capillary network than do phosphorylase-rich (type II) fibers of white muscle (Myrhage, 1977). Most of the capillaries run parallel to the long axis of the muscle fiber, and many form 1–3 intercapillary anastomoses before reaching venules. Dichotomous branchings and subsequent mergings of a single capillary are frequent findings in the densely interconnected network (Eriksson and Myrhage, 1972). In warm-blooded species, the capillary bed in skeletal muscle is potentially very great. For example, in the gastrocnemius of the horse, it has been calculated that there is a maximum of 1350 vessels per mm^2 in transverse section of muscle (Krogh, 1929). The number of capillaries surrounding each muscle fiber appear to be directly related to the activity of the oxidative enzyme in that fiber (Romanul, 1965) and independent of tissue dimensions. The mechanisms involved in the growth of new capillaries in muscle have not been elucidated, although one plausible explanation is that capillary sprouting results from increased muscle blood flow due to elevated oxygen consumption, together with a moderate tissue hypoxia present either constantly or intermittently.

Collecting venules (10–25 μm in diameter) drain the capillary bed and join with other similar vessels to form transverse venules (40–60 μm in diameter). The latter empty in central venules (90 μm in diameter) that, in turn, contribute to the formation of the main draining venules (140 μm in diameter).

b. *Species differences in the microcirculation:* The morphology of the microcirculation in skeletal muscle varies in different animals and man. In the rat, channels course along the connective tissue planes of the muscle and then divide into metarterioles that cross at right angles to the muscle fibers, to terminate abruptly in a series of capillaries or of preferential vessels (Zweifach and Metz, 1955). Capillaries, which are found 40 to 50 μm apart, run parallel to

and between the muscle fibers, anastomosing with each other to form a surrounding meshwork. The preferential vessels (thoroughfare channels) are most
commonly observed along the edge of the muscle. For the most part, these
metarterioles break up into 8–10 capillaries; however, occasionally they begin
to receive capillaries in their course, thus becoming postcapillary venules
(Baez, 1977). Except for the latter type of vascular arrangement, no arteriovenous anastomoses or short connections between arterial and venous
pathways are found in the cremaster muscle of the rat (Samaji et al., 1970;
Baez, 1973).

Measurements by Smaji et al. (1970) of the vascular elements give values of
615 ± 194 μm for capillary lengths (from origin to termination in the primary
venule), with cross connections in the same plane occurring at 210 ± 85 μm
intervals. The internal diameter of the arteriolar end of the capillary is 5.5 ±
1.1 μm, with the venular end being 6.1 ± 1.4 μm. Calculated values for
capillary density have been found to be 1300/mm² and capillary surface area,
244 cm²/cm³ muscle.

In the cat tenuissimus preparation, there are a centrally located artery
(about 70 μm in diameter) and a vein (about 90 μm in diameter) running
together in the tissue proper. From them, transverse arterioles and venules
arise, to form "end-arterioles" and "end-venules" (25–50 μm in internal diameter). No anastomoses are observed between these vessels, and only occasional
arteriovenous anastomoses are present. Endothelial capillaries measure 1015
μm in length, with cross connections occurring at 200 μm intervals. The diameter of capillaries measures 4.7, 5.3, and 5.9 μm at the beginning, middle, and
end, respectively (Eriksson and Myrhage, 1972). No precapillary sphincters are
present (Ericson et al., 1973), although end-arterioles in the vascular bed may
serve the same purpose (Eriksson and Lisander, 1972).

By means of cytochemical techniques, Ericson et al. (1973) were able to
quantitate the number of capillaries in close contact with each type of cat
muscle fibers, based on the presence of ATPase activity (Henneman and Olson,
1965); they found 3.5 for A fibers, 3.6 for B fibers, and 3.8 for C fibers.

In the rabbit, there appear to be differences in the morphology of the
microcirculation of white muscle (fast contracting) as compared with that of red
muscle (slow contracting) (Lee, 1958). In the latter, sac-like dilatations are
found at branching points in arterioles, capillaries, and venules. However,
these structures do not occur in true transverse connections between capillaries. Where capillaries and venules converge, the dilatations are large. Another
characteristic of the microcirculation in red muscle is tortuosity of the vessels.
This is more marked in the capillaries than in the arterioles. Capillary meshes
are irregular in shape, most being formed by branches from the same capillary.
On the other hand, in white muscle, saclike dilatations are not found at sites of
branching points of arterioles and capillaries, and the contour of these channels
is uniform. Moreover, their course usually is straight and smooth, with the size

of their branches diminishing uniformly. Terminal components of an arteriole or branches of a capillary are usually set widely apart. This is different from the arrangement noted in red muscle, in which branches often are quite close together.

In man, quantitative measurement of cellular and extracellular components of skeletal muscle capillaries has been attempted by a number of workers. Endothelium has been found to be 1880 ± 120 Å in width, and the lumen, 3.09 ± 0.21 μm in width. Capillary basement membrane widths differ markedly, the figure ranging from 670 to 6000 Å (Zacks *et al.*, 1962; Fuchs, 1964; Bencosme *et al.*, 1966). Since there is a relationship between basement membrane width and lumen size, such variation in basement membrane widths may be due to naturally occurring differences in capillary size. This may result from inadvertent sampling of distended and constricted vessels. Of interest is the finding that, in the capillaries of skeletal muscle in the leg, the areas occupied by the basement membrane increase in both absolute and relative dimensions from proximal to distal sites (Vracko, 1970b). No explanation has been offered for this finding in the leg, a change not present elsewhere in the lower limb or in other capillary beds of skeletal muscle.

3. Vascular Responses to Physiologic and External Factors

a. Effect of aging on the microcirculation: Certain alterations occur in the microcirculation of skeletal muscle with aging. For example, the basement membrane is 10–15% thicker in 60-year-old human subjects than in 40-year-old subjects (Holm *et al.*, 1972). The changes are essentially segmental in character and similar to those of diabetes (Kilo *et al.*, 1972) (see Section E-2d, below). In their training experiments on rats subjected to prolonged swimming, Adolfsson *et al.* (1981) reported that, although there was a resulting increase in capillary density in both young and adult animals, the response was proportionately less in the older group.

b. Effect of training on the microcirculation: Endurance training has been reported to produce increases in density of capillaries in the exercised skeletal muscle of the rat (Petrén *et al.*, 1937) and guinea pig (Mai *et al.*, 1970), the augmentation in the rat being 45% (Petrén *et al.*, 1937). Muscles not involved in the specific exercise are not affected by training. Moreover, the vascular alterations are not associated with any increase in mean area of muscle fibers (Adolfsson *et al.*, 1981). However, other workers have not confirmed such findings (Parízková *et al.*, 1972; Müller, 1976).

In man the published data concerning training also are conflicting. For example, Andersen and Henriksson (1977) and Ingjer and Brodal (1978) have noted increased capillary density in skeletal muscle following endurance training, whereas Hermansen and Wachtlová (1971) have not. According to the latter workers, endurance exercise in human subjects enlarges muscle fiber diameter, as well as increasing the number of capillaries, the newly developed vessels running parallel to the hypertrophied fibers. However, these investigators did not find any absolute change in the number of capillaries per mm^2 when compared to muscle mass. In any event, it can be stated that endurance training in human subjects constitutes a powerful stimulus for capillary proliferation in skeletal muscle (Andersen and Henriksson, 1977), as is also the case in a variety of animals. At present a discussion of the factors responsible for the development of new capillaries in response to training can only be speculative.

c. Effect of repeated electrical stimulation of muscle on the microcirculation: It has been found that both the number of capillaries and the oxidative capacity of "fast" muscles increase when an intact motor nerve is intermittently stimulated for more than 4 days (Pette *et al.*, 1973; Brown *et al.*, 1976). In contrast to muscle exercise, such a measure does not cause fiber hypertrophy but does elicit increased capillary density.

After 14 days of electrical stimulation, the growth of new capillaries takes the form of sprouts emanating from preexisting capillaries, the proportion of sprouts to preexisting vessels being about 40% (Myrhage, 1977). Chronic muscle stimulation of fast contracting muscle results in widening of the capillaries (especially those close to the collecting venules), as well as increased capillary density.

d. Effect of hypoxia: A state of chronic hypoxia in Andean guinea pigs has been reported to cause a rise in aerobic skeletal muscle metabolism, associated with an increase in the number of muscle capillaries (Valdivia, 1958). In accord with such findings is the investigation of Cassin *et al.* (1971), in which striated muscles of chronically hypoxic rats was studied with a histochemical method for capillary visualization. After 6 weeks of exposure to a simulated high altitude, the workers found elevated values for capillary density, as compared with the results obtained at sea level.

However, the above findings have not been substantiated by others. For example, Sillau and Banchero (1979) and Sillau *et al.* (1980a), also using modern histochemical methods for capillary visualization, noted that the changes in capillarity with growth in Andean guinea pigs and in guinea pigs exposed to severe environmental hypoxia were similar to those observed at sea level. Moreover, Sillau and Banchero (1977) claimed that, in rats, the differences noted between hypoxic and normoxic animals were exclusively a reflection of differences in body weight and not attributable to the hypoxic stimulus. These authors found no evidence of increases in capillarity as a result of 3 and 6 weeks of hypoxia in growing rats exposed to a simulated altitude equivalent to 5100 m.

e. Effect of cold: Acclimatization of small animals to cold is associated with an increase in skeletal muscle capillarity (Heroux and St. Pierre, 1957). Since such a response is also noted in animals with smaller weights (due to abnormal environmental conditions), the possibility has been proposed that the increased capillary density could have resulted from the smaller fiber cross-sectional area and not necessarily from an increase in capillary supply. However, the work of Sillau *et al.* (1980b) has demonstrated that the skeletal muscle of cold-acclimatized guinea pigs is more richly vascularized than that of control animals, the increased capillarity being absolute and independent of the changes in fiber cross-sectional area.

C. PHYSIOLOGY OF BLOOD CIRCULATION

By H. V. Sparks, Jr., and L. P. Thompson

The control of skeletal muscle blood vessels is affected by two major factors: local metabolism and autonomic nerves, mechanisms that sometimes work in direct opposition. The net effect on muscle blood flow depends on the relative strength of the two factors. In this section, they will be discussed separately, but an attempt will also be made to indicate their interactions.

1. General Considerations

The largest alterations in skeletal muscle blood flow occur during exercise. The mechanisms responsible for this increase in local circulation have been under study since Gaskell proposed in 1877 that vasodilator metabolites might be the responsible agents. In some species, it appears that initiation of exercise hyperemia may be the result of activation of sympathetic cholinergic fibers to skeletal muscle resistance vessels. However, in man, the available evidence does not support the idea that this is an important mechanism. Although skeletal muscle vasodilatation in anticipation of exercise is observed in humans, this is probably the result of increased circulating catecholamines activating vascular smooth muscle β-adrenergic receptors (see Section C-3a). However, all such remote neural and humoral mechanisms are relatively unimportant compared to local factors.

2. LOCAL CONTROL

Studies over the past few years have identified several vasodilator mechanisms that may be involved in exercise hyperemia, the relative importance of each depending on (1) the exercise pattern, i.e., twitch versus tetanic contractions; (2) the duration of the exercise bout; (3) the relationship between oxygen supply and consumption; and (4) the muscle fiber type. The various vasodilator mechanisms are discussed below.

a. Arterial wall P_{O_2}: There are two viewpoints on the sensitivity of arteriolar vascular smooth muscle to lowered P_{O_2}. Duling and Pittman (1975) have presented evidence indicating that vascular smooth muscle P_{O_2} must fall into the range of the critical P_{O_2} for oxidative phosphorylation (<1 mm Hg) before vascular relaxation is observed. However, evidence obtained by other workers suggests that the sensitivity of vascular smooth muscle to oxygen may depend upon phenomena with a much higher critical P_{O_2} (Gellai and Detar, 1974; Coburn, 1977). Reduced vessel wall P_{O_2} may cause relaxation of vascular smooth muscle by the release of a vasodilator prostaglandin (Kalsner, 1976) or adenosine within the wall (Van Harn *et al.*, 1977). Decreased arteriolar wall P_{O_2} undoubtedly occurs when flow is restricted during prolonged exercise, and it may result in vasodilatation in such a situation. However, that it develops during the hyperemic response to a brief tetanic contraction (Mohrman *et al.*, 1973) or during free flow twitch exercise (Gorzynski and Duling, 1978) is unlikely.

b. Osmolarity: Increased tissue osmolarity occurs during exercise as a result of splitting of high-energy phosphate compounds and substrates, a response that ultimately develops in the vicinity of the vascular muscle. Such a change probably contributes to the initiation of vasodilatation during both free flow and ischemic exercise (Sparks, 1980a). In the cat, it probably helps to sustain exercise vasodilatation in low, but not in high, oxidative muscles. In man, muscles may fall between these extremes (Mellander, 1981).

c. Arachidonic acid metabolites: Prostaglandins are released during exercise (Young and Sparks, 1979), but it is not known whether they originate from skeletal muscle cells or from vessel walls. Moreover, there is little evidence as to which of the prostaglandins are involved in the response. Exogenous thromboxane A_2 and $PGF_{2\alpha}$ cause skeletal muscle vasoconstriction, whereas most of the others produce vasodilatation. Prostaglandins contribute to the vasodilatation which follows ischemic exercise but not to that which occurs during either free flow or ischemic exercise (Sparks, 1980a).

d. Myogenic response: Resistance vessels dilate following the compression that occurs during skeletal muscle contraction. Such myogenic responses are noted after brief periods of intense tetanic contraction (Mohrman and Sparks, 1974b) and during twitch exercise (Bacchus *et al.*, 1981).

e. Potassium ion: Potassium ion is released with each action potential of skeletal muscle cells. At first the rate of release exceeds active transport back into muscle and hence K^+ concentration in interstitial fluid rises. Then, as intracellular Na^+ concentration increases and active transport is stimulated, extracellular K^+ returns to the control level (Hazeyama and Sparks, 1979). There are two characteristics of potassium-induced vasodilatation that suggest that the potential role of such a response in causing exercise vasodilatation is limited. First, K^+ does not cause maximum vasodilatation in skeletal muscle, and second, it produces transient but not steady state vasodilatation (Kjellmei, 1965). Increased interstitial K^+ concentration probably contributes to the augmented flow following brief tetanic contractions (Mohrman and Sparks, 1974a) and during the first few minutes of twitch exercise (Stowe *et al.*, 1975).

f. Adenosine triphosphate: There is some evidence for the belief that exercise is associated with an increase in release of adenosine triphosphate (ATP) into venous plasma draining active human skeletal muscles (Forrester, 1981). It is difficult to know whether the substance is released from skeletal muscle cells, endothelium, or formed elements in blood. Because of such uncertainty, it is difficult to evaluate the role of ATP in exercise vasodilatation.

g. Local neurogenic vasodilatation: Honig (1979) presented pharmacologic evidence that the vascular resistance changes associated with short periods of exercise can be reduced by anesthetizing neural elements within the arterial walls. However, the study should be repeated elsewhere using other preparations and different experimental conditions, in order to determine the generality of the results. Honig suggested that initiation of exercise vasodilatation is the result of a local reflex and that a sustained response results from the formation of vasodilator metabolites.

h. Acetate: It has been found that plasma acetate concentration, as well as skeletal muscle acetate content, increases during free flow exercise. Such a finding may have significance, since it has been reported that acetate infusion causes resistance vessel dilatation (Steffen *et al.*, 1981). More studies of this potentially important metabolite are in order.

i. Adenosine: This substance, a vasodilator of skeletal muscle resistance vessels, is released from skeletal muscle during free flow and ischemic exercise. Whether it is present in a sufficiently high concentration in the vicinity of vascular smooth muscle to play an important role has been studied by measuring tissue adenosine content and adenosine release into venous effluent. However, both of these approaches have shortcomings. For example, determination of tissue adenosine would be an accurate reflection of interstitial adenosine concentration if all measured were in the interstitial space. However, recent evidence suggests that in some tissues, a large proportion, perhaps as much as 90%, is not free in the interstitial space (Sparks and Fuchs, 1983). Also, in the case of adenosine release into venous effluent, the quantitative relationship between interstitial fluid and venous adenosine concentration is unknown. Currently, available evidence suggests that adenosine is at least partly responsible for ischemic exercise vasodilatation (Sparks, 1980b).

3. NERVOUS CONTROL

a. Sympathetic adrenergic nerves: Activation of sympathetic adrenergic nerves by direct stimulation reduces blood flow and blood volume of skeletal muscle, a response that is blocked by α-receptor antagonists such as phenoxybenzamine or phentolamine. Norepinephrine (NE) release from nerve terminals is influenced by nerve firing rate and also by a number of local factors, including metabolites, prostaglandins, acetylcholine, and NE itself (Vanhoutte *et al.*, 1981). The constrictor response to NE likewise is modulated by local factors (Sparks, 1980b). Although the constrictor effects mediated by α-receptor activation are most evident, vasodilatation mediated by β-receptors also may have a physiologic role.

Lundvall and Järhult (1976) hypothesized that both circulating and neurally released catecholamines cause dilatation of precapillary vessels. They found that sympathetic nerve stimulation produces decreased flow but increased capillary filtration coefficient (K_f), the latter suggesting a greater capillary surface area. It was of interest that the increase in K_f was blocked by β-receptor antagonists. Lundvall and Järhult (1974) proposed that increased capillary surface area is a physiologic compensation for the reduced flow associated with sympathetic vasoconstriction. β-Adrenergic receptor activation, to some extent, may compete with α-adrenergic actions on resistance vessels. The dilator effects of β-adrenergic receptor activation are evident when circulating catecholamines are elevated (Brick *et al.*, 1967), a response often observed during exercise in man (Häggendal *et al.*, 1970).

α-Receptor vasoconstriction must compete with vasodilatation initiated by metabolites released from exercising skeletal muscle (Blair *et al.*, 1961; Strandell and Shepherd, 1967; Bevegård and Shepherd, 1966). When sympathetic nerves are stimulated during exercise, there is a large transient fall in flow, followed by a much smaller steady state decrease (Rowlands and Donald, 1968). Potassium may be involved in the metabolic antagonism of sympathetic constriction. Burcher and Garlick (1975) demonstrated that sympathetic vasoconstriction was abolished with K^+ infusion, whereas the constrictor response to exogenous NE was not abolished. They concluded that the inhibiting effect of K^+ occurred at a prejunctional site. Beaty and Donald (1977) demonstrated that the attenuated constriction to sympathetic nerve stimulation was due to a high extracellular K^+/Ca^{2+} ratio. They proposed that increased K^+ levels during exercise reduced sympathetic vasoconstriction by interfering with Ca^{2+} movement across the smooth muscle membrane.

b. Sympathetic cholinergic nerves: In the presence of an adrenergic blocking agent, sympathetic nerve stimulation produces muscle vasodilatation that can be blocked by atropine. This response has been found in the cat, rabbit, and dog, but it appears to be absent in the rat and some primates, including man (Uvnäs, 1966). Cholinergic vasodilator nerves seem to be restricted to relatively large resistance vessels of skeletal muscles (Folkow and Neil, 1971).

Acetylcholine (Ach) causes contraction of isolated vascular strips devoid of endothelium, but when the latter is present, the opposite effect is produced (Furchgott *et al.*, 1981). Both types of responses are sensitive to atropine. Relaxation can be blocked by lipoxygenase inhibitors, phospholipase A_2 inhibitors, and anoxia (Furchgott *et al.*, 1981), and it has been proposed that Ach stimulates release of a vasodilator substance from endothelial cells that produces relaxation of vascular smooth muscle.

The cholinergic dilatation of skeletal muscle blood vessels observed in intact preparations could be the result of release of an endothelial dilator. An alternative explanation is that Ach inhibits NE release from adrenergic nerve terminals (Vanhoutte and Levy, 1980). The inhibitory action of Ach on NE release is mediated by activation of muscarinic receptors of adrenergic nerve endings (Van Hee and Vanhoutte, 1978). This explanation of cholinergic vasodilatation is weakened by the observation that vasodilatation produced neurogenically can exceed that observed when adrenergic nerves are cut. On the other hand, if the response is mediated by endothelium, the problem still exists of explaining how Ach released from nerve endings located at the adventitiomedial border can reach the endothelium.

The physiologic role of cholinergic vasodilatation is still unclear. It has been suggested that in skeletal muscle, the reaction is part of the hypothalamic alarm response (Abrahams *et al.*, 1964). Transient cholinergic vasodilatation may serve to raise muscle blood flow just before the increased metabolic activity of exercise takes over. In man, muscle dilatation also occurs in response to mental stress (Blair *et al.*, 1959; Barcroft *et al.*, 1960). However, this is more likely to be the result of increased circulating catecholamines than due to cholinergic mechanisms.

c. Histaminergic nerves: Baroreceptor stimulation results in hindlimb vasodilatation that exceeds what could be accounted for by inhibition of adrenergic tone (Brody, 1978). This response is not blocked by propranolol or atropine, is accompanied by an increase in histamine release (Tuttle, 1967), and is blocked by antihistamines (Heitz *et al.*, 1970). It appears that histamine release from cells associated with the vessel wall occurs when NE is withdrawn during inhibition of sympathetic tone (Brody, 1966).

D. PHARMACOLOGY OF BLOOD CIRCULATION

By H. V. Sparks, Jr., and L. P. Thompson

This section deals with the direct effects of a number of pharmacologic agents on skeletal muscle blood vessels (Table 19.I). With systemic administration, their direct actions may be masked by reflex compensations. However, such indirect effects will not be discussed.

1. Drugs Acting on Autonomic Nervous System

a. Cholinergic agents: Acetylcholine infused into cat skeletal muscle causes vasodilatation of resistance and capacitance vessels. In addition, capillary surface area is increased (Kjellmer and Odelram, 1965). Pre- to postcapillary resistance ratio is decreased, resulting in net filtration (Mellander, 1966). Acetylcholine has the same effects on the human forearm, except that capillary exchange is not increased (Dale, 1954). The longer-acting cogeners, e.g., methacholine, carbachol, and bethanechol, have similar actions, but they are less effective than acetylcholine (Taylor, 1980). Anticholinesterase agents cause vasodilatation, although the responses to them are mixed because they have ganglionic, as well as neuromuscular junctional, effects. Anticholinergics are not active except in the presence of a choline ester, in which case they block the vasodilatation.

b. Sympathetic agents: α_1-Adrenergic receptor agents, such as norepinephrine, phenylephrine, and methoxamine, cause constriction of resistance and capacitance vessels. Precapillary resistance rises more than

TABLE 19.I

PHARMACOLOGIC AGENTS AND THEIR VASCULAR EFFECTS[a]

Agent	Resistance	Capillary exchange	Capacitance
Cholinergic agents			
Acetylcholine	↓	↑	↑
Methacholine	↓		
Carbachol	↓		
Bethanechol	↓		
Sympathetic agents			
Norepinephrine	↑	↓	↓
Phenylephrine	↑		
Methoxamine	↑		
Clonidine	↓		
α-Methylnorepinephrine	↓		
Isoproterenol	↓	↑	↑
Amphetamine	↑		
Ephedrine	↑		
Tyramine	↑		
Dopamine	↑		
Isoxsuprine	↓		
Nylidrin hydrochloride	↓		
Autacoids			
Histamine	↓	↑	↑
Angiotensin II	↑	↓	○
Bradykinin	↓	↑	↑
Prostaglandins			
I_2	↓		
E	↓	↑	↑
A	↓	↑	↑
D	↓		
$F_{2\alpha}$	↑	↑,○,↓	↓
Cardiac drugs			
Cardiac glycosides	↑		
Lidocaine	↓		
Hydralazine	↓		
Nitroprusside	↓	○	↑
Methyldopa	↓		
Guanethidine	↓		
Reserpine	↓		
Thiazides	↓		
Saralasin	↓		
Captopril	↓		
Calcium entry blocker	↓		↑
Other drugs			
Methylxanthine	↓		
Morphine	↓		↑
α-Receptor antagonist prazosin	↓		↑

[a] ↑, Increased; ↓, decreased; ○, no effect. No indication means data are not available.

postcapillary resistance so that capillary reabsorption occurs (Mellander, 1966). Infusion of these agents causes a decrease in capillary exchange, but this is transient, probably because of countervailing metabolic effects. If the agonist has β-receptor activity, it may increase capillary exchange by this mechanism as well (Lundvall *et al.*, 1981).

α_2-Adrenergic receptor agents, such as clonidine and α-methylnorepinephrine, act primarily on prejunctional or presynaptic receptors limiting norepinephrine release. These agents cause vasodilatation, primarily because of their central action (Zelis and Flaim, 1981).

β_2-Adrenergic receptor agents, characterized by isoproterenol, cause vasodilatation of resistance vessels and increase capillary exchange. Capacitance vessels are not dilated as much. A decreased pre- to postcapillary resistance ratio produces filtration (Johansson and Öberg, 1968). Isoxsuprine (Vasodilan) and nylidrin hydrochloride (Arlidrin) cause vasodilatation of resistance vessels but are longer acting β_2-receptor stimulants than isoproterenol. Unlike the latter, however, the vasodilatation is resistant to blockade by propranolol.

Sympathomimetics with mixed action, such as amphetamine, ephedrine, and tyramine, act by a combination of release of norepinephrine and of their own inherent activity. They cause vasoconstriction but may be tachyphylactic due to exhaustion of neuronal stores of norepinephrine. Dopamine acts on α- and β-adrenergic receptors, as well as producing the release of norepinephrine via a prejunctional dopamine receptor. The net effect of these actions is mild limb vasoconstriction (Goldberg and Kohli, 1981).

α-Receptor antagonists cause skeletal muscle vasodilatation when adrenergic tone is present. They facilitate filtration of fluid because pre- to postcapillary resistance is lowered. Prazosin, a new postjunctional α_1-receptor blocker, produces a reduction in both vascular resistance and capacitance (Awan *et al.*, 1981).

2. AUTACOIDS

a. Histamine: This drug reduces resistance and increases capillary surface area. The combination of a high capillary hydrostatic pressure (the result of lowered pre- to postcapillary resistance ratio), increased capillary surface area, and raised microvascular permeability to macromolecules causes marked capillary filtration (Marciniak *et al.*, 1977). Catecholamines, acting via β-adrenergic receptors, antagonize the effect of histamine on increased permeability.

b. Polypeptides: Angiotensin II causes resistance vessel constriction but has little effect on capacitance vessels. This results in a fall in capillary hydrostatic pressure and in reabsorption of fluid. A portion of its vasoconstrictor effect is via the release of norepinephrine. Bradykinin produces relaxation of resistance and capacitance vessels; the rise in capillary hydrostatic pressure and surface area results in net filtration. Higher doses of the drug cause increased protein permeability, furthering net filtration. Its vasodilating action is partly the result of prostaglandin formation (Kline *et al.*, 1973).

c. Arachidonate metabolites: PGI_2, as well as the PGEs, PGAs, and PGDs, are vasodilators of both arteries and veins. $PGF_{2\alpha}$ has a weak vasoconstrictor effect (Dusting *et al.*, 1979). Prostaglandins have relatively little effect on capillary permeability.

3. Cardiac Drugs

a. Cardiac glycosides: These agents produce increased skeletal muscle resistance by inhibiting the electrogenic Na^+/K^+ active transport system of vascular smooth muscle.

b. Lidocaine: This agent, as well as other local anesthetics, causes dilation of resistance vessels.

c. Antihypertensives: Almost all drugs in this class produce decreased resistance by one of the following mechanisms: (1) a direct action on vascular smooth muscle (e.g., hydralazine or nitroprusside), (2) central effects on the sympathetic nervous system (e.g., methyldopa or clonidine), (3) peripheral effects on the sympathetic nervous system (e.g., guanethidine, reserpine, or prazosin), (4) increased salt and water excretion (e.g., thiazides), or (5) interference with the renin angiotensin system (e.g., saralasin or captopril) (Blaschke and Melmon, 1980).

d. Nitrates: These agents dilate resistance and capacitance vessels but do not increase capillary surface area or cause net filtration (Ablad and Mellander, 1963). They may act by interfering with catecholamine-induced Ca^{2+} entry into smooth muscle.

e. Calcium entry blockers: These drugs are vasodilators. They act by preventing the entry of Ca^{2+} into vascular smooth muscle via a voltage-sensitive membrane channel (Weiss, 1981).

4. Other Drugs

a. Methylxanthines: These agents cause resistance vessel vasodilatation by an unknown mechanism. Among their cellular effects are antagonism of adenosine receptors and inhibition of phosphodiesterase (at high concentrations) (see references in Fox and Kelly, 1978).

b. Morphine: This drug produces dilatation of resistance and capacitance vessels. The most likely mechanism is central reduction of sympathetic neural tone (Zelis *et al.*, 1977).

E. PATHOPHYSIOLOGY, PATHOGENESIS, AND PATHOLOGY OF BLOOD CIRCULATION

By D. I. ABRAMSON

1. PATHOPHYSIOLOGY

a. Intermittent claudication: The most common symptom originating in skeletal muscle in man is intermittent claudication, a pain experienced during periods of physical activity and relieved by rest. The complaint reflects a transient disparity between blood flow and nutrient supplied, relative to that required by the exercising muscle.

Substances such as lactic acid, potassium, serotonin, and stable chemical agents have all been implicated in the pathophysiology of intermittent claudication. In each case it has been proposed that the etiologic agent is formed locally and then accumulates in the interstitial tissues until a level sufficient to stimulate or irritate sensory nerve endings is reached.

Although the most common cause of intermittent claudication is stenosis or occlusion of main arteries supplying the intramuscular vessels, actually any condition that interferes with the normal transport of oxygen to the muscles may initiate symptoms. For example, it may also be experienced by patients suffering from severe anemia, congestive heart failure, chronic lung and congenital heart disorders associated with unsaturation of the blood, hyperthyroidism, and coarctation of the aorta.

b. Vascular changes in muscle atrophy: The reported alterations in the microcirculation produced by disuse atrophy do not consistently support any one view. There is some physiologic evidence to indicate an increase in local blood flow in dogs' limbs immobilized in casts (Imig *et al.*, 1953) and a similar type of response noted in the gastrocnemius of cats following tenotomy (Hudlická et al., 1964). However, microangiographic studies of rabbits' limbs have revealed no changes in the capillaries of skeletal muscle after 2 weeks of immobilization (Hulth and Olerud, 1961). It is possible that the reported vasodilator response subsequent to the development of atrophy of muscle is due to relaxation of both resistance and exchange vessels in the muscle mass (Hudlická and Renkin, 1968), unrelated to any anatomic change in microcirculation.

2. PATHOGENESIS AND PATHOLOGY

As one facet of their clinical pattern, a large number of unrelated systemic disorders manifest pathologic changes in the vascular tree of skeletal muscle.

a. Hypertension: In this disorder, peripheral vascular resistance is chronically elevated, the change occurring in all organs and tissues (Tobia et al., 1974a,b), with the major end organ being the microcirculation of the entire body (Bohlen, 1982). The mechanisms involved in producing such an alteration have not been identified. Among those suggested are vasoconstriction, lengthening of normal arterioles, a decrease in the number of vessels, hyperviscosity, or a combination of these factors. The possibility of vasoconstriction secondary to vessel wall hypertrophy, first proposed by Folkow et al. (1973), has received some recent physiologic experimental support.

In vitro histologic examination of skeletal muscle in the hypertensive rat and in the human with essential hypertension has revealed first arterial wall hypertrophy and then the appearance of similar changes in the microvasculature (Furuyama, 1962; Mulvany et al., 1978). Whether comparable changes occur in the blood vessels in other organs is still in doubt. The most promising data currently available to explain the mechanical origin of the increased vascular resistance in essential hypertension support the view that there is a decreased number of arterioles open to flow in the hypertensive animal, particularly in skeletal muscle (Bohlen, 1982). Those that are closed tend to open to blood flow upon denervation or disruption of vascular control (Bohlen, 1979). Another possibility is that in the hypertensive animal, there is a reduction in the total number of vessels. Bohlen (1979, 1982) has proposed that the hyper-responsiveness to norepinephrine observed in small arterioles in skeletal muscle of hypertensive animals is expressed *in vivo* as a strong tendency to constrict to closure rather than exhibiting a dose-dependent partial constriction, as occurs in normal arterioles.

b. Chronic occlusive arterial diseases: In the case of arteriosclerosis obliterans, a degenerative vascular disorder, the terminal arterial tree in the skeletal muscle of the lower limbs rarely demonstrates atherosclerotic changes, except possibly in the larger vessels. This is probably due to the fact that most of the intramuscular branches are small (less than 250 μm in diameter) and hence are not susceptible to plaque formation. However, there may be intimal and medial fibrosis, with thickening of the walls of smaller arteries and of arterioles. The basis for the consistent finding in arteriosclerosis obliterans of intermittent claudication (see Section E-1a, above), despite absence of stenotic or obstructive lesions in the terminal vessels, is the existence of such pathologic changes in the more proximal supplying channels: the main arteries in the abdomen and/or the large vessels of the lower limbs.

Pathologic changes have been found in the capillary bed of skeletal muscle in arteriosclerosis obliterans (Mäkitie, 1977), consisting of severe basement membrane thickening, degenerative changes, significantly decreased lumen size, and increased capillary density, possibly related to the chronic state of ischemia. The basement membrane also demonstrates replication to form many continuous and/or discontinuous layers around the capillary. Frequently observed are large vacuoles and protrusion of endothelial cells into the lumen of the vessel. The degree of pathologic changes appears to parallel the severity of the intermittent claudication.

Thromboangiitis obliterans, which falls into the category of vasculitides, affects the small- and medium-sized arteries in striated muscle of the lower limbs and, to a lesser degree, similar vessels of the upper limbs. Small- and medium-sized veins in striated muscle are not involved. The segmental occlusive process, characteristic of the disorder, also affects the large arteries of the limbs, generally at the level of the knee and distally and at the level of the elbow and below. Proximal vessels usually are not involved.

The histopathologic changes in the small- and medium-sized arteries in thromboangiitis obliterans consist initially of an acute arteritis, with edema and infiltration of all three coats of the vessel wall by polymorphonuclear leukocytes. Swelling and proliferation also are noted in the endothelium of the vasa vasorum. Invariably associated with such alterations is early thrombosis of the lumen of the vessels in the muscle, the clot being composed of inflammatory cells. In the subacute stage, mononuclear cells predominate, with multinucleated giant cells being found in varying numbers. In this phase, the thrombus shows signs of slowly being replaced by fibroblastic organization and collagen formation. The final, chronic stage is typified by the development of a fibrous and hyaline connective tissue mass occluding the lumen of the vessel, with prominent canalization.

In generalized arteriosclerosis associated with hypertension, the intramuscular arterioles commonly demonstrate thickening of their walls, due primarily to hypertrophy and hyperplasia of the media. To a lesser extent, the intima and adventitia are involved in the process.

c. Atherothrombotic or cholesterol emboli and miscellaneous agents: Minute debris, liberated from ulcerated atherosclerotic tissue in the infrarenal or terminal portion of the abdominal aorta and the common and external iliac arteries or from the interior of proximally located arterial aneurysms may be responsible for total obliteration of arterioles in skeletal muscle of the legs, as well as in the skin of the toes. The particulate material usually contains a variable combination of lipoid material and cholesterol crystals, together with adherent red blood cells and fibrin aggregates. In a gastrocnemius or quadriceps muscle biopsy, the condition may be identified by the presence of clefts of cholesterol crystals filling the lumen of involved vessels (Anderson and Richards, 1968). The resulting endarteritis is characterized by intimal fibrosis and foreign body reaction. At times areas of necrotizing angiitis may be present, resembling the pathologic changes of polyarteritis nodosa (see Section E-2e, below).

Among other agents that may obstruct the microcirculation in striated muscle are air emobli (caisson disease) and emboli composed of fat particles liberated following fractures of long bones or in the presence of a number of unrelated disease entities. Various infectious disorders, such as trichinosis, pneumonia, and meningococcic septicemia, also may be associated with thrombosis of small arteries and arterioles in skeletal muscle. The histopathologic manifestations consist of small vessel vasculitis, with infiltration by either polymorphonuclear or mononuclear cells, depending upon the type of underlying inflammatory agent.

d. Diabetes mellitus: In contrast to essential hypertension (see Section E-2a, above), in which the vascular pathology appears to be a hyperactive process not harming the majority of host tissues, in diabetes, it is of a degenerative nature, limiting or impairing the function of the host tissues (Bohlen, 1982). The process involves thickening of the capillary basement membrane (see below) and generalized atrophy of the arteriolar wall structure and loss of capillaries and arterioles. At the same time, the integrity of the remaining capillaries as a semipermeable barrier is impaired (Papachristodoulou and Health, 1977). Many possible causes have been offered for the vascular changes observed in diabetes mellitus, but none is supported by definitive proof.

Microangiopathy, a typical pathologic alteration in diabetes (Siperstein, 1972), is consistently noted in the capillaries of skeletal muscle. In fact, there is an apparent direct relationship between the prevalence of such a change and the clinical finding of retinopathy (Yodaiken et al., 1975), although recently some doubt has been cast on the validity of this view (Peterson et al., 1980). The abnormality in skeletal muscle is extensive, having been noted in the muscles of the abdominal and chest walls, the hand, midthigh, midleg, and foot (Vracko, 1970a). However, it is not always present, nor is it consistently found in every site (Zacks et al., 1962; Fuchs, 1964; Bencosme et al., 1966; Vracko, 1970a).

The typical histologic change in diabetes mellitus consists of early thickening of the basement membrane of capillaries, associated with concentric lamination (Zacks *et al.*, 1962). The degree of hypertrophy of the lamina appears to have no relationship to the age or weight of the patient. However, there is some suggestive evidence to indicate that when carbohydrate control is combined with an exercise program, a decrease in basement membrane thickening of muscle capillaries occurs (Peterson *et al.*, 1980). The average basement width in diabetes is approximately three times that found in normal subjects (2900 Å, as compared with 700 Å) (Siperstein *et al.*, 1968). An average basement membrane thickness of 3000 Å or more in the capillaries of skeletal muscle probably indicates the presence of widespread capillaropathology (Yodaiken *et al.*, 1975). The controversy regarding normal average widths of basement membrane is compounded by the fact that microvascular lesions can be found in aged nondiabetic patients that are indistinguishable from those in diabetics, except that the changes are rarely as severe (Kilo *et al.*, 1972) (see Section B-3a, above).

The cause of the basement membrane thickening in the capillaries of diabetics is unknown. One view proposes that the cyclic degeneration and regeneration of endothelium and pericytes are initiated by focal and intermittent ischemia. Another considers the possibility of an accelerated turnover rate of pericytes caused by genetic or metabolic defects. The mechanism responsible for the pathologic change may be repeated depositions and build-up of layers of successive generations of endothelial cells and pericytes, this type of response causing lamellation, redundancy, and eventually thickening of the basement membrane. It has been suggested, but not proved, that hyperglycemia produces the pathologic change by the increased incorporation of carbohydrate residues into the glycoprotein fibrils of the basement membrane.

Similar lesions to those noted in capillaries are also found in the arterioles, arteriovenous shunts, and venules of skeletal muscle in diabetes. There is cellular hyperplasia of the walls of the vessels, with PAS-positive staining of the basement membrane (Goldenberg *et al.*, 1959). The result is narrowing of the lumen and atrophy of the medial layer, at least, in the arterioles. The change is patchy, with many normal vessels being found in the vicinity of obviously thickened ones. Hyalinization and sclerosis of arterioles may be present, at times to the point of cessation of blood flow and actual necrosis of tissues supplied by the involved vessels. (For a review of the pathogenesis of atherosclerosis in diabetes mellitus, see Colwell *et al.*, 1981.)

e. Connective tissue disorders: A number of diseases falling into this category demonstrate pathologic changes in the vessels of skeletal muscle. Of these, polyarteritis nodosa often is associated with intense necrotizing vasculitis (Diaz-Perez and Winkelman, 1974), the changes occurring in approximately 40% of autopsied cases (Cupps and Fauci, 1981). Characteristically, the histopathologic alteration (pannecrosis) is found in the small- to intermediate-sized muscular arteries, generally at bifurcations. The lesions consist of marked distortions of the wall and segmental obliteration of vessel lumen. Small hemorrhages may be present in the muscle, but foci of infarct necrosis are rare.

Although hypersensitivity vasculitis is associated primarily with pathologic changes in the small vessels of the skin, biopsies of symptomatic skeletal muscle groups, particularly of the lower limbs, not infrequently also reveal vascular abnormalities. The most common of these is a polymorphonuclear leukocytic infiltrate of postcapillary venules, with leukocytoclasis (presence of nuclear debris), fibrinoid necrosis, endothelial swelling, and extravasation of erythrocytes into the perivascular areas (Cupps and Fauci, 1981). The damaged vessels are usually all at the same stage of involvement, suggesting a "burst" of immune complex deposition (Cupps and Fauci, 1981), in contrast to polyarteritis nodosa, in which lesions in all stages of development are found in a single section. Two cellular patterns have been described in hypersensitivity vasculitis. In one the change consists predominantly of a neutrophilic infiltration associated with hypocomplementemia, whereas the other is characterized by lymphocytic infiltration with normal complement. The possibility has also been advanced that the two patterns actually represent a continuous evolution from acute to chronic vasculitis and are not indicative of different mechanisms (Phanuphak et al., 1978; Andrews et al., 1979). Microinfarcts are typical of hypersensitivity vasculitides.

In patients with rheumatoid arthritis, examination of striated muscle obtained some distance from inflamed joints has revealed the presence of nonspecific arteritis (Sokoloff et al., 1951). The pathologic changes are found in very small arteries, and they resemble the alterations seen in association with rheumatic fever. There is also some similarity to the vascular abnormalities reported in noninfectious arteritis of other types. On occasion, the inflammatory changes affect all three coats of the involved vessel wall and hence the lesion can be termed a panarteritis; at other times, the adventitia is the only coat appreciably involved. Cellular exudate is varied in composition, with polymorphonuclear leukocytes predominating in some instances and mononuclear forms in others (presumably in older lesions). Accompanying the cellular infiltrate is proliferation of endothelial cells and fibroblasts. Although not possessing distinctive histologic characteristics, the vascular lesion in rheumatoid arthritis is believed to be a specific manifestation of the disease (Sokoloff et al., 1951).

A vasculitis or a perivasculitis may occasionally be seen in striated muscle in systemic lupus erythematosus. The pathologic change consists of a loose collection of mononuclear cells, resembling lymphocytes, which surrounds the wall of small intramuscular arteries, arterioles, venules, and veins. The endothelial cells are swollen and contain autophagic vacuoles and multivesicular bodies lying within the endoplasmic reticulum (Norton, 1970). The changes are considered to be nonspecific reactions, often noted, but not exclusively, in connective tissue disorders.

In cases of polymyositis and adult dermatomyositis, thickening and re-duplication of the basement membrane and swelling of the endothelial cells have been reported in the small intramuscular blood vessels. The endothelial cells may contain autophagic vacuoles and multivesicular bodies, as well as tubular inclusions which are apparently endoplasmic reticular formations (Norton, 1970). However, such changes are not specific for polymyositis or dermatomyositis, the basement membrane thickening being found in diabetes mellitus (see Section E-2d, above) and in other disorders and the endothelial changes being noted in systemic lupus erythematosus (see above).

In childhood dermatomyositis, the early vascular abnormalities in skeletal muscle consist of perivascular collections of inflammatory cells, arteritis, and phlebitis (Banker and Victor, 1966). Later in the disease, intimal hyperplasia of the arteries and veins may occur, followed by occlusion of many of these vessels by fibrin thrombi, with subsequent infarction of tissues.

f. Venous disorders: The most common involvement of the venous tree in striated muscles is acute occlusion of the venous plexuses in the calf by a bland thrombus. However, several hours after its formation, the clot causes a phlebitic reaction that lightly attaches the thrombotic process to the intima of the involved vessel. The venous plexuses most often affected are those located in the soleus muscle in the leg and those found in the plantar portion of the foot.

g. Congenital malformation: Congenital arteriovenous fistula is one of the relatively common vascular anomalies found in striated muscle. In this condition the voluntary muscle of a limb, as well as neighboring structures, is diffusely infiltrated by numerous abnormal vascular plexuses and communications between small arteries and small veins, with no intervening capillary bed. There are at least three types of fistulous tracts. In a relatively rare arrangement, a direct transverse axis shunt exists, similar to a patent ductus arteriosus. In the most frequent form, there are multiple communications in the muscle, the lesions resembling a hemangioma in vascular pattern. In the third group, the main artery directly joins the main vein through dilated channels, with a capillary bed being present. This type also is infrequently noted in striated muscle.

h. Miscellaneous disorders or states: Changes in the vascular tree in striated muscle have been reported in a relatively large number of other unrelated disorders or states. For example, intramuscular microangiopathy has been found in uremia in the form of prominent thickening of the basement membrane of capillaries and heavily vacuolated endothelial cells. The vacuoles often contain myelin-like figures or gray electron-dense material (Ahonen *et al.*, 1981). Capillary luminal size is reduced by numerous endothelial cell protrusions. An increase in pinocytotic vesicles in capillary endothelium is also present, as well as marked degeneration, with only remnants of basement membranes and endothelial cells remaining. The electron microscopic changes closely resemble those observed in diabetes mellitus (see Section E-2d, above).

In azotemic hyperparathyroidism, the pathologic changes in the vascular tree of striated muscle are a direct consequence of local ischemia resulting from intimal proliferation in small arteries. Associated findings are medial calcification and fibrous intimal thickening of intramuscular arteries, causing fulminant muscle necrosis and myoglobinuria (Richardson *et al.*, 1969).

Patients receiving steroid medication for protracted periods have been reported to develop marked thickening of the basement membrane of capillaries for reasons not clear at present (Afifi *et al.*, 1968).

F. LYMPHATIC SYSTEM

By D. R. Bell

1. Anatomy of Lymphatic System

Skeletal muscle has a lymphatic system that drains the interstitial space. Water, solutes, and blood cells may cross the blood capillary, distribute in portions of the interstitial space, and enter the lymphatic capillary, to be returned to the venous system through the thoracic duct. In 1913, Aagaard (cited by Drinker and Field, 1933) described a rich supply of lymphatics in a variety of striated muscles. The lymphatic capillaries were located in the fascial planes around muscle bundles, with none observed in the endomysium between muscle fibers. After injection of India ink into rat thigh muscle, Florey (1927) found that the particulate matter would appear in the femoral lymphatics only when the muscles were moved either passively or actively and not when they were quiescent. He proposed that muscle movement provides the hydrostatic forces necessary for transport of water and solutes surrounding the muscle fibers and capillaries into the lymphatics located in the perimysium.

a. Lymphatic capillaries: The endothelial cells lining lymphatic capillaries in the diaphragm resemble those found in blood capillaries (see Section A, Chapter 3). The basement membrane of terminal lymphatics appears incomplete, with large gaps between the collagen bundles (Casley-Smith and Florey, 1961). Large spaces have also been shown between the endothelial cells, with the width of intercellular junctions varying from 0.05 to 5 μm. Fluid and solutes may pass from the interstitium into the lymphatics through these intercellular junctions. Connective tissue filaments, found attached to the endothelial cells of the terminal lymphatics (Leak and Burke, 1968), may open the intercellular junctions during muscle movement or edema, resulting in increased transport from the interstitium into the lymphatics (Casley-Smith, 1964a, 1980). In the diaphragm, there appear to be more open intercellular junctions than in skin (Casley-Smith, 1964b), a finding that may be due to a greater tissue movement of this structure prior to the fixation process. It is possible that muscle movement promotes fluid and solute transport into the terminal lymphatics by increasing interstitial hydrostatic pressure, with the lymphatic filaments holding open the intercellular junctions during the period of tissue compression.

   b. Lymphatic trunks: The anatomy of the large lymphatics draining skeletal muscle has been studied in the lower limb of man (Pflug and Calnan, 1971) and in the hindlimb of the dog (Pflug and Calnan, 1969) and rabbit (Bach and Lewis, 1973). In all three species, there are two main systems of lymphatics draining the extremity: the superficial and the deep. The superficial lymphatics, located in loose subcutaneous tissue or close to superficial blood vessels, drain the skin of the limb and enter either the popliteal or the inguinal lymph nodes. The deep system of lymphatics is located along the deep blood vessels associated with muscles of the lower limb. It originates from the skeletal muscles and connects directly with the femoral lymphatics. In dogs, Pflug and Calnan (1969) found no evidence of Patent Blue dye being transported from one lymphatic system to another. After intramuscular injections of Evans blue into the hindlimbs of rabbits, Bach and Lewis (1973) noted that a large lymphatic arising from the body of the gastrocnemius muscle joined the efferent lymphatic from the popliteal node and then became the femoral lymphatic above the knee. It would appear that such an anatomic arrangement provides a means for collection of muscle lymph separately from that originating from skin.

2. PHYSIOLOGY OF LYMPHATIC SYSTEM

   At present there are few studies on skeletal muscle lymph flow and composition. In an investigation on cats and dogs, Jacobsson and Kjellmer (1964) cannulated femoral lymphatics and tied ligatures around the ankle joint to prevent lymph flow from the entire paw. With passive movement of the hindlimbs, ligation of the paw resulted in a decrease in lymph flow from 25 to 5 μl/min per limb. There was no significant change in the lymph total protein concentration during the 1 hr of lymph collection following paw ligation. The 5-fold reduction in lymph flow resulting from paw ligation indicates that under normal conditions, femoral lymph predominantly originates from skin. The lack of a change in lymph total protein is surprising, since most of the studies using micropipettes or wicks have shown that muscle has a higher interstitial fluid-to-plasma concentration ratio for total protein than skin (Renkin, 1979). The findings of Jacobsson and Kjellmer (1964) may have been influenced by their short lymph collection times or possibly by inclusion of skin lymph from between the ankle (where the ligature was applied) and the knee.

Bach and Lewis (1973) collected muscle lymph from the hindlimb of rabbits by cannulating the femoral lymphatics distal to the inguinal node and ligating the efferent lymphatic from the popliteal node. Under such circumstances, it was assumed that the lymph sample was derived from the deep muscle lymphatics, with little contamination from the superficial lymphatics. After ligation of the popliteal efferent, lymph flow in the cannulated femoral lymphatic vessel decreased by 5-fold, a result similar to the observation of Jacobsson and Kjellmer (1964) following ligation of the paw. In addition, Bach and Lewis collected lymph from a prepopliteal superficial lymphatic that drains hind paw skin. Lymph flow from the muscles of the lower limb was found to average 4.2 ± 0.4 μl/min, whereas that from paw skin averaged 3.5 ± 0.5 μl/min. The total protein concentration in muscle lymph was usually greater than that from skin. From localized dye injections and the anatomy of the lymphatics, Bach and Lewis estimated that the muscle mass from which they collected lymph was at least 10 times greater than that for skin. Thus, based upon tissue mass, they concluded that lymph flow from skin was much higher than that from muscle.

In a study also on rabbits, Bell and Mullins (1982) found that ligation of the popliteal efferent lymphatic resulted in a decrease in flow from 15 to 0.8 μl/min over a 2-hr period; it remained steady at the lower rate for 2–3 additional hours. The lymph : plasma concentration ratio for total protein increased from 0.44 to 0.56 over the initial 2-hr postligation period and remained constant thereafter. When subcutaneous injections of Evans Blue into the paw were used to check for skin lymph contamination, it was found that the initial lymph collections were blue, but they became clear 1 to 2 hr after ligation of the popliteal node. The time required to obtain clear lymph or a constant lymph protein concentration after ligation of the node was considered to be the time necessary to wash out the residual lymph in the large lymphatics between the muscles and the lymphatic cannula. This fluid probably represented skin lymph in great part. On the other hand, the lymph subsequently collected in the preparation was believed to originate mostly from the gastrocnemius and soleus muscle (Bach and Lewis, 1973; Bell and Mullins, 1982).

During normal, steady-state conditions, muscle lymph has a total protein concentration twice as high as skin lymph (Bell and Mullins, 1982). This could be explained by a difference in the capillary wall permeability to plasma proteins in the two tissues. However, the sieving curves for albumin, fibrinogen, and different-sized polyvinylpyrrolidone (PVP) molecules are similar for both tissues, indicating similar capillary permeabilities (Firrell *et al.*, 1982). An alternative explanation is that the net fluid filtration across muscle capillaries is less than that across skin capillaries, as suggested by the difference in lymph flows per unit tissue mass. This would result in a higher lymph protein concentration in muscle than in skin (Taylor *et al.*, 1981).

Alterations in lymph flow and protein concentration with changes in the microcirculation appear to be different in skeletal muscle from those in skin. For example, injection of dimethyl sulfoxide (DMSO) results in increased lymph flow with a decreased lymph protein concentration in muscle, but not in skin (Bach and Lewis, 1974). In addition, 4 hr of ischemia produces a 3-fold increase in lymph flow from muscle, with no change in flow from skin (Bach and Lewis, 1974). On the other hand, saline expansion, equivalent to 10% body weight, causes a greater increase in lymph flow from skin than from muscle (Mullins and Bell, 1982). The reasons for the differences in behavior between the two tissues are not clear and require further investigation.

REFERENCES*

Ablad, B., and Mellander, S. (1963). Comparative effects of hydrazaline, sodium nitrate and acetylcholine on resistance and capacitance blood vessels and capillary filtration in skeletal muscle in the cat. *Acta Physiol. Scand.* **58**, 319–329. (D)

Abrahams, V. C., Hilton, S. M., and Zbrozyma, A. W. (1964). The role of active muscle vasodilation in the alerting stage of the defense reaction. *J. Physiol. (London)* **171**, 189–202. (C)

Adolfsson, J., Ljungquist, A., Tornling, G., and Unge, G. (1981). Capillary increase in the skeletal muscle of trained young and adult rats. *J. Physiol. (London)* **310**, 529–535. (B)

Afifi, A. K., Bergman, R. A., and Harvey, J. C. (1968). Steroid myopathy: Clinical, histologic and cytologic observations. *Johns Hopkins Med. J.* **123**, 158–174. (E)

Ahonen, R. E., Mäkitie, J., and Kock, B. (1981). Striated muscle capillaries in uremic patients and in renal transplant recipients. *Arch. Intern. Med.* **141**, 867–869. (E)

Andersen, P., and Henriksson, J. (1977). Capillary supply of the quadriceps femoris muscle of man: Adaptive response to exercise. *J. Physiol. (London)* **270**, 677–690. (B)

Anderson, W. R., and Richards, A. M. (1968). Evaluation of lower extremity muscle biopsies in the diagnosis of atheroembolism. *Arch. Pathol.* **86**, 535–541. (E)

Andrews, B. S., Cains, G., McIntosh, J., Petts, V., and Penny, R. (1979). Circulating and tissue-immune complexes in cutaneous vasculitis. *J. Clin. Lab. Immunol.* **1**, 311–320. (E)

Awan, N., Everson, M., Needham, K., and Mason, D. (1981). Arteriolar and venous vasodilatation with prazosin in management of refractory congestive heart failure. *In* "Vasodilatation" (P. M. Vanhoutte and J. Leusen, eds.), pp. 511–523. Raven Press, New York. (D)

Bacchus, H., Gamble, G., Anderson, D., and Scott, J. (1981). Role of the myogenic response in exercise hyperemia. *Microvasc. Res.* **21**, 92–102. (C)

Bach, C. S., and Lewis, G. P. (1973). Lymph flow and lymph protein concentration in skin and muscle of the rabbit hind limb. *J. Physiol. (London)* **235**, 477–492. (F)

Bach, C. S., and Lewis, G. P. (1974). Flow and composition of skin and muscle lymph of the hind limb of the rabbit after injury. *Br. J. Pharmacol.* **52**, 359–365. (F)

*In the reference list, the capital letter in parentheses at the end of each reference indicates the section in which it is cited.

Baez, S. (1968). Vascular smooth muscle quantitation of cell thickness in the wall of arterioles in the living animal in situ. *Science* **159**, 536–538. (B)

Baez, S. (1973). An open cremaster muscle preparation for the study of blood vessels by *in vivo* microscopy. *Microvasc. Res.* **5**, 384–394. (B)

Baez, S. (1977). Skeletal muscle and gastrointestinal microvascular morphology. In "Microcirculation" (G. Kaley and B. M. Altura, eds.), Vol. 1, pp. 69–94. University Park Press, Baltimore, Maryland. (B)

Banker, B. O., and Victor, M. (1966). Dermatomyositis (systemic angiopathy) of childhood. *Medicine (Baltimore)* **45**, 261–289. (E)

Barcroft, H., Brod, J., Hejl, Z., Hirajarv, E. A., and Kitchin, A. H. (1960). The mechanism of the vasodilatation in the forearm muscle during stress (mental arithmetic). *Clin. Sci.* **19**, 1577–1586. (C)

Beaty, O., and Donald, D. E. (1977). Role of potassium in the transient reduction in vasoconstrictive responses of muscle resistance vessels during rhythmic exercise in dogs. *Circ. Res.* **41**, 452–460. (C)

Bell, D. R., and Mullins, R. J. (1982). Effects of increased venous pressure on albumin- and IgG-excluded volumes in muscle. *Am. J. Physiol. (Heart Circ. Physiol. 11)* **242**, H1044–H1049. (F)

Bencosme, S. A., West, R. O., Kerr, J. W., and Wilson, D. L. (1966). Diabetic capillary angiopathy in human skeletal muscles. *Am. J. Med.* **40**, 67–77. (B,E)

Bevegård, B. S., and Shepherd, J. T. (1966). Reaction in man of resistance and capacity vessels in forearm and hand to leg exercise. *J. Appl. Physiol.* **21**(1), 123–132. (C)

Blair, D. A., Glover, W. E., Greenfield, A. D. M., and Roddie, I. C. (1959). Excitation of cholinergic vasodilator nerves to human skeletal muscles during emotional stress. *J. Physiol. (London)* **148**, 633–647. (C)

Blair, D. A., Glover, W. E., and Roddie, I. C. (1961). Vasomotor responses in the human arm during leg exercise. *Circ. Res.* **9**, 264–274. (C)

Blaschke, T. F., and Melmon, K. L. (1980). Antihypertensive agents and the drug therapy of hypertension. In "The Pharmacological Basis of Therapeutics" (A. G. Gilman, L. S. Goodman, and A. Gilman, eds.), 6th ed., pp. 793–818. Macmillan, New York. (D)

Blomfield, L. B. (1945). Intramuscular vascular patterns in man. *Proc. R. Soc. Med.* **38**, 617–618. (B)

Bohlen, H. G. (1979). Arteriolar closure mediated by hyperresponsiveness to norepinephrine in hypertensive rats. *Am. J. Physiol.* **5**, H157–H164. (E)

Bohlen, H. G. (1982). Pathological expression in the microcirculation: Hypertension and diabetes. *Physiologist* **25**, 391–395. (E)

Brånemark, P. I., and Eriksson, E. (1971). Method for studying qualitative and quantitative changes of blood flow in skeletal muscle. *Acta Physiol. Scand.* **84**, 284–288. (B)

Brick, I., Hutchinson, K. J., and Roddie, I. C. (1967). The vasodilator properties of noradrenaline in the human forearm. *Br. J. Pharmacol. Chemother.* **30**, 561–567. (C)

Brody, M. J. (1966). Neurohumoral mediation of active reflex vasodilation. *Fed. Proc., Fed. Am. Soc. Exp. Biol.* **25**, 1583–1592. (C)

Brody, M. J. (1978). Histaminergic and cholinergic vasodilator systems. In "Mechanisms of Vasodilatation" (P. M. Vanhoutte and J. Leusen, eds.), pp. 266–277. Karger, Basel. (C)

Brown, M. D., Cotter, M. A., Hudlická, O., and Vrbová, G. (1976). The effects of different patterns of muscle activity on capillary density, mechanical properties and structure of slow and fast muscles. *Pfluegers Arch.* **361**, 241–250. (B)

Burcher, E., and Garlick, P. (1975). Effects of exercise metabolites on adrenergic vasoconstriction in the gracilis muscle of the dog. *J. Pharmacol. Exp. Ther.* **192**, 149–156. (C)

Campbell, J., and Pennefather, C. M. (1919). An investigation into the blood-supply of muscles, with special reference to war surgery. *Lancet* **1**, 294–296. (B)

Casley-Smith, J. R. (1964a). Endothelial permeability—The passage of particles into and out of diaphragmatic lymphatics. *Q. J. Exp. Physiol. Cogn. Med. Sci.* **49**, 365–383. (F)

Casley-Smith, J. R. (1964b). An electron microscopic study of injured and abnormally permeable lymphatics. *Ann. N.Y. Acad. Sci.* **116,** 803–830. (F)

Casley-Smith, J. R. (1980). Are the initial lymphatics normally pulled open by the anchoring filaments? *Lymphology* **13,** 120–129. (F)

Casley-Smith, J. R., and Florey, H. W. (1961). The structure of normal small lymphatics *Q. J. Exp. Physiol. Cogn. Med. Sci.* **46,** 101–106. (F)

Cassin, S. R., Gilbert, D., Bunnell, C. F., and Johnson, E. M. (1971). Capillary development during exposure to chronic hypoxia. *Am. J. Physiol.* **220,** 448–451. (B)

Coburn, R. F. (1977). Oxygen tension sensors in vascular smooth muscle. *In* "Tissue Hypoxia and Ischemia" (M. Reivich, P. F. Coburn, S. Lehiri, and B. Chance, eds.), pp. 101–115. Plenum, New York. (C)

Colwell, J. A., Lopes-Virella, M., and Halushka, P. V. (1981). Pathogenesis of atherosclerosis in diabetes mellitus. *Diabetes Care* **4,** 121–133. (E)

Cupps, T. R., and Fauci, A. S. (1981). "The Vasculitides" (L. H. Smith, Jr., ed.). Saunders, Philadelphia, Pennsylvania. (E)

Curtis, D. H., and Zalin, R. J. (1981). Regulation of muscle differentiation: Stimulation of myoblast fusion *in vitro* by catecholamines. *Science* **214,** 1355–1357. (A)

Dale, H. (1954). The action of certain esters of choline and their relation to muscarine. *J. Pharmacol. Exp. Ther.* **6,** 147–190. (D)

Diaz-Perez, J. L., and Winkelmann, R. K. (1974). Cutaneous periarteritis nodosa. *Arch. Dermatol.* **110,** 407–414. (E)

Drinker, C. K., and Field, M. E. (1933). "Lymphatics, Lymph and Tissue Fluid." Williams & Wilkins, Baltimore, Maryland. (F)

Duling, B. R., and Pittman, R. N. (1975). Oxygen tension: Dependent or independent variable in local control of blood flow? *Fed. Proc., Fed. Am. Soc. Exp. Biol.* **34,** 2020–2024. (C)

Dusting, G. J., Moncada, S., and Vane, J. R. (1979). Prostaglandins, their intermediates and precursors: Cardiovascular actions and regulatory roles in normal and abnormal circulatory systems. *Prog. Cardiovasc. Dis.* **21**(6), 405–430. (D)

Ericson, L. E., Eriksson, E., and Johanson, B. (1973). Morphological aspects of the microvessels in cat skeletal muscle. *In* "Advances in Microcirculation" (H. Harders, ed.), pp. 62–79. Karger, Basel. (B)

Eriksson, E., and Lisander, B. (1972). Changes in precapillary resistance in skeletal muscle vessels studied by intravital microscopy. *Acta Physiol. Scand.* **84,** 295–305. (B)

Eriksson, E., and Myrhage, R. (1972). Microvascular dimensions and blood flow in skeletal muscle. *Acta Physiol. Scand.* **86,** 211–222. (B)

Firrell, J. C., Lewis, G. P., and Youlten, L. J. F. (1982). Vascular permeability to macromolecules in rabbit paw and skeletal muscle: A lymphatic study with a mathematical interpretation of transport processes. *Microvasc. Res.* **23,** 294–310. (F)

Florey, H. (1927). Reactions of, and absorption by, lymphatics, with special reference to those of the diaphragm. *Br. J. Exp. Pathol.* **8,** 479–489. (F)

Folkow, B., and Neil, E. (1971). The principles of vascular control. *In* "Circulation," pp. 294–301. Oxford Univ. Press, London and New York. (C)

Folkow, B., Hallback, M., Lundgren, Y., Sivertsson, R., and Weiss, L. (1973). Importance of adaptive changes in vascular design for establishment of primary hypertension, studied in man and in spontaneously hypertensive rats. *Circ. Res.* **32,** Suppl. 1, 12–113. (E)

Forrester, T. (1981). Adenosine or adenosine triphosphate? *In* "Vasodilatation" (P. M. Vanhoutte and J. Leusen, eds.), pp. 205–229. Raven Press, New York. (C)

Fox, I. H., and Kelly, W. N. (1978). The role of adenosine and 2'-deoxyadenosine in mammalian cells. *Annu. Rev. Biochem.* **47,** 655–686. (D)

Fuchs, U. (1964). Elektronenmikroscopische untersuchungen menschlicher Muskelcapillaren bei Diabetes mellitus. *Frankf. Z. Pathol.* **73,** 318–327. (B,E)

Furchgott, R. F., Zawadzki, J. V., and Cherry, P. D. (1981). Role of endothelium in vasodilator response to acetylcholine. In "Vasodilatation" (P. M. Vanhoutte and J. Leusen, eds.), pp. 49–66. Raven Press, New York. (C)

Furuyama, M. (1962). Histometrical investigations of arteries in reference to arterial hypertension. Tohoku J. Exp. Med. 76, 388–414. (E)

Gaskell, W. H. (1877). On the changes of the bloodstream in muscles through stimulation of their nerves. J. Anat. 11, 360–402. (C)

Gellai, M., and Detar, R. (1974). Evidence in support of hypoxia but against high potassium and hyperosmolarity as possible mediators of sustained vasodilation in rabbit cardiac and skeletal muscle. Circ. Res. 35, 681–691. (C)

Goldberg, L. I., and Kohli, J. D. (1981). Specific dopamine receptors in vascular smooth muscle. In "Vasodilatation" (P. M. Vanhoutte and J. Leusen, eds.), pp. 131–140. Raven Press, New York. (D)

Goldenberg, S., Alex, M., and Joshe, R. A. (1959). Nonatheromatous peripheral vascular disease of the lower extremity in diabetes mellitus. Diabetes 8, 261–273. (E)

Gorzynski, R. J., and Duling, B. R. (1978). Role of oxygen in arteriolar functional vasodilation in hamster striated muscle. Am. J. Physiol. 235, H505–H515. (C)

Häggendal, J., Hartley, L. G., and Saltin, B. (1970). Arterial noradrenaline concentration during exercise in relation to the relative work levels. Scand. J. Clin. Lab. Invest. 26, 337–342. (C)

Hazeyama, Y., and Sparks, H. V. (1979). A model of potassium ion efflux during exercise of skeletal muscle. Am. J. Physiol. 236, R83–R90. (C)

Heitz, D. C., Schaffer, R. A., and Brody, M. J. (1970). Active vasodilatation evoked by stimulation of sinus nerve in the conscious dog. Am. J. Physiol. 218, 1296–1300. (C)

Henneman, E., and Olson, C. B. (1965). Relations between structures and functions in the design of skeletal muscle. J. Neurophysiol. 28, 581–598. (B)

Hermansen, L., and Wachtlová, M. (1971). Capillary density of skeletal muscle in well-trained and untrained men. J. Appl. Physiol. 30, 860–863. (B)

Heroux, O., and St. Pierre, J. (1957). Effect of cold acclimation on vascularization of ears, heart, liver and muscles of white rats. Am. J. Physiol. 188, 163–168. (B)

Holm, J., Björntorp, P., and Schérsten, T. (1972). Metabolic activity of human skeletal muscle. Effect of peripheral arterial insufficiency. Eur. J. Clin. Invest. 2, 321–325. (B)

Honig, C. R. (1979). Contributions of nerves and metabolites to exercise vasodilation: A unifying hypothesis. Am. J. Physiol. 236, H705–H719. (C)

Hudlická, O., and Renkin, E. M. (1968). Blood flow and blood-tissue diffusion of ^{86}Rb in denervated and tenotomized muscles undergoing atrophy. Microvasc. Res. 1, 147–151. (E)

Hudlická, O., Hník, P., and Stulcová, B. (1964). Changes in blood circulation in skeletal muscle undergoing atrophy. Physiologist 7, 163 (abstr.). (E)

Hulth, A., and Olerud, S. (1961). Disuse of extremities. II. A microangiographic study in the rabbit. Acta Chir. Scand. 120, 388–394. (E)

Imig, C. J., Randall, B. F., and Hines, H. M. (1953). Effect of immobilization on muscular atrophy and blood flow. Arch. Phys. Med. Rehabil. 34, 296–299. (E)

Ingjer, F., and Brodal, P. (1978). Capillary supply of skeletal muscle fibers in untrained and endurance trained women. Eur. J. Appl. Physiol. 38, 291–299. (B)

Jacobsson, S., and Kjellmer, I. (1964). Flow and protein content of lymph in resting and exercising skeletal muscle. Acta Physiol. Scand. 60, 278–285. (F)

Johansson, G., and Öberg, B. (1968). Comparative effects of isoprenaline and nitroglycerin on consecutive vascular sections in the skeletal muscle of the cat. Angiologica 5, 161–171. (D)

Kalsner, S. (1976). Intrinsic prostaglandin release: A mediator of anoxia-induced relaxation in an isolated coronary artery preparation. Blood Vessels 13, 155–166. (C)

Kilo, C., Volger, N., and Williamson, J. R. (1972). Muscle capillary basement membrane changes related to aging and to diabetes mellitus. Diabetes 21, 881–905. (B,E)

Kjellmer, I. (1965). The potassium ion as a vasodilator during muscular exercise. *Acta Physiol. Scand.* **63**, 460–468. (C)

Kjellmer, I., and Odelram, H. (1965). The effect of some physiological vasodilators on the vascular bed of skeletal muscle. *Acta Physiol. Scand.* **63**, 94–102. (D)

Kline, R. L., Scott, J. B., Haddy, F. J., and Grega, G. J. (1973). Mechanism of edema formation in canine forelimbs by locally administered bradykinin. *Am. J. Physiol.* **225**, 1051–1056. (D)

Krogh, A. (1929). "The Anatomy and Physiology of Capillaries." Yale Univ. Press, New Haven, Connecticut. (B)

Leak, L. V., and Burke, J. F. (1968). Ultrastructural studies on the lymphatic anchoring filaments. *J. Cell Biol.* **36**, 129–149. (F)

Lee, J. C.-Y. (1958). Vascular patterns in the red and white muscles of the rabbit. *Anat. Rec.* **132**, 597–611. (B)

Lundvall, J., and Järhult, J. (1974). β-Adrenergic microvascular dilatation evoked by sympathetic stimulation. *Acta Physiol. Scand.* **92**, 572–574. (C)

Lundvall, J., and Järhult, J. (1976). β-Adrenergic dilation component of the sympathetic vascular response in skeletal muscle. *Acta Physiol. Scand.* **96**, 180–192. (C)

Lundvall, J., Hillman, J., and Gustafsson, D. (1981). β-Adrenergic regulation of the capillary exchange and resistance functions. *In* "Vasodilatation" (P. M. Vanhoutte and J. Leusen, eds.), pp. 107–116. Raven Press, New York. (D)

Mai, J. V., Edgerton, V. R., and Barnard, R. J. (1970). Capillarity of red, white and intermediate muscle fibers in trained and untrained guinea pigs. *Experientia* **26**, 1222–1223. (B)

Mäkitie, J. (1977). Skeletal muscle capillaries in intermittent claudication. *Arch. Pathol. Lab. Med.* **101**, 500–503. (E)

Marciniak, D. L., Dobbins, D. E., Maciejko, J. J., Scott, J. B., Haddy, F. J., and Grega, G. J. (1977). Effects of systemically infused histamine on transvascular fluid and protein transfer. *Am. J. Physiol.* **233**, H148–H153. (D)

Mathes, S. J., and Nahai, F. (1981). Classification of the vascular anatomy of muscles: Experimental and clinical correlation. *Plast. Reconstr. Surg.* **67**, 177–187. (B)

Mellander, S. (1966). Comparative effects of acetylcholine, butyl-nor-synephrine (Vasculat), nor-adrenaline, and ethyl-adrianol (Effontil) on resistance, capacitance, and precapillary sphincter vessels and capillary filtration in cat skeletal muscle. *Angiologica* **3**, 77–99. (D)

Mellander, S. (1981). Differentiation of fiber composition, circulation and metabolism in limb muscles of dog, cat and man. *In* "Vasodilatation" (P. M. Vanhoutte and J. Leusen, eds.), pp. 243–254. Raven Press, New York. (C)

Minguetti, G., and Mair, W. G. P. (1979). Ultrastructure of human intramuscular blood vessels in development. *Arq. Neuro-Psiquiatr.* **37**, 127–137. (A)

Mohrman, D. E., and Sparks, H. V. (1974a). Role of potassium ions in the vascular response to a brief tetanus. *Circ. Res.* **35**, 384–390. (C)

Mohrman, D. E., and Sparks, H. V. (1974b). Myogenic hyperemia following brief tetanus of canine skeletal muscle. *Am. J. Physiol.* **227**, 531–535. (C)

Mohrman, D. E., Cant, J. R., and Sparks, H. V. (1973). Time course of vascular resistance and venous oxygen changes following brief tetanus of dog skeletal muscle. *Circ. Res.* **33**, 323–336. (C)

Müller, W. (1976). Subsarcolemmal mitochondria and capillarization of soleus muscle fibers in young rats subjected to an endurance training. *Cell Tissue Res.* **174**, 367–389. (B)

Mullins, R. J., and Bell, D. R. (1982). Permeability of rabbit skin and muscle microvasculature after saline infusion. *J. Surg. Res.* **32**, 390–400. (F)

Mulvany, M. J., Hansen, P. K., and Aalkjaer, C. (1978). Direct evidence that greater contractility of resistance vessels in spontaneously hypertensive rats is associated with a narrowed lumen, a thickened media, and an increased number of smooth muscle cell layers. *Circ. Res.* **43**, 854–864. (E)

Myrhage, R. (1977). Microvascular supply of skeletal muscle fibres. Thesis. *Acta Orthop. Scand. Suppl.* **168**, 1–46. (B)

Norton, W. L. (1970). Comparison of the microangiopathy of systemic lupus erythematosus, dermatomyositis, scleroderma, and diabetes mellitus. *Lab. Invest.* **22**, 301–308. (E)

Papachristodoulou, D., and Heath, H. (1977). Ultrastructural alterations during the development of retinopathy in sucrose-fed and streptozotocin-diabetic rats. *Exp. Eye Res.* **25**, 371–384. (E)

Parízková, J., Wachtlová, M., and Soukupová, M. (1972). The impact of different motor activity on body composition, density of capillaries and fibers in the heart and soleus muscles, and cell's migration in male rats. *Int. Z. Angew. Physiol. Einschl. Arbeitsphysiol.* **30**, 207–216. (B)

Peterson, C. M., Jones, R. L., Esterly, J. A., Wantz, G. E., and Jackson, R. L. (1980). Changes in basement membrane thickening and pulse volume concomitant with improved glucose control and exercise in patients with insulin-dependent diabetes mellitus. *Diabetes Care* **3**, 586–589. (E)

Petrén, T., Sjöstrand, T., and Sylvén, B. (1937). Der Einfluss des Trainings auf die Häufigkeit der capillaren in Herz-und Skeletmuskulatur. *Arbeitsphysiologie* **9**, 376–386. (B)

Pette, D., Smith, M. E., Staudte, H. W., and Vrbová, G. (1973). Effects of long-term electrical stimulation on some contractile and metabolic characteristics of fast muscles. *Pfluegers Arch.* **338**, 257–272. (B)

Pflug, J. J., and Calnan, J. S. (1969). Lymphatics: Normal anatomy in the dog hind leg. *J. Anat.* **105**, 457–465. (F)

Pflug, J. J., and Calnan, J. S. (1971). The normal anatomy of the lymphatic system in the human leg. *Br. J. Surg.* **58**, 925–930. (F)

Phanuphak, P., Kohler, P. F., Stanford, R. E., Thorne, E. G., and Claman, H. N. (1978). Value of skin biopsy in vasculitis. *Clin. Res.* **26**, 123A (abstr.). (E)

Renkin, E. M. (1979). Lymph as a measure of the composition of interstitial fluid. *In* "Pulmonary Edema" (A. P. Fishman and E. M. Renkin, eds.), pp. 145–159. *Am. Physiol. Soc.*, Bethesda, Maryland. (F)

Richardson, J. A., Herron, G., Reitz, R., and Layzer, R. (1969). Ischemic ulceration of skin and necrosis of muscle in azotemic hyperparathyroidism. *Ann. Intern. Med.* **71**, 129–138. (E)

Romanul, F. C. A. (1965). Capillary supply and metabolism of muscle fibres. *Arch. Neurol. (Chicago)* **12**, 497–509. (B)

Rowlands, D. J., and Donald, D. E. (1968). Sympathetic vasoconstrictive responses during exercise of drug-induced vasodilation: A time-dependent response. *Circ. Res.* **23**, 45–60. (C)

Senior, H. D. (1919). The development of the arteries of the human lower extremity. *Am. J. Anat.* **25**, 55–95. (A)

Senior, H. D. (1924). The description of the larger direct or indirect muscular branches of the human femoral artery: A morphogenic study. *Am. J. Anat.* **33**, 243–265. (A)

Sillau, A. H., and Banchero, N. (1977). Effects of hypoxia on capillary density and fiber composition in rat skeletal muscle. *Pfluegers Arch.* **370**, 227–232. (B)

Sillau, A. H., and Banchero, N. (1979). Effect of hypoxia on the capillarity of guinea pig skeletal muscle. *Proc. Soc. Exp. Biol. Med.* **160**, 368–373. (B)

Sillau, A. H., Aquin, L., Bui, M. V., and Banchero, N. (1980a). Chronic hypoxia does not affect guinea pig skeletal muscle capillarity. *Pfluegers Arch.* **386**, 39–45. (B)

Sillau, A. H., Aquin, L., Lechner, A. J., Bui, M. V., and Banchero, N. (1980b). Increased capillary supply in skeletal muscle of guinea pigs acclimated to cold. *Respir. Physiol.* **42**, 233–245. (B)

Siperstein, M. D. (1972). Capillary basement membranes and diabetic microangiopathy. *Adv. Intern. Med.* **18**, 325–344. (E)

Siperstein, M. D., Unger, R. H., and Madison, L. L. (1968). Studies of muscle capillary basement membranes in normal subjects, diabetic, and prediabetic patients. *J. Clin. Invest.* **47**, 1973–1999. (E)

Smaji, L., Zweifach, B. W., and Intaglietta, M. (1970). Micropressures and capillary filtration coefficients in single vessels of the cremaster muscle in the rat. *Microvasc. Res.* **2**, 96–110. (B)

Sokoloff, L., Wilens, S. L., and Bunim, J. J. (1951). Arteritis of striated muscle in rheumatoid arthritis. *Am. J. Pathol.* **27**, 157–173. (E)

Sparks, H. V. (1980a). Effect of local metabolic factors on vascular smooth muscle: vessel pO_2, K^+ and osmolarity. *In* "Handbook of Physiology" (D. F. Bohr, A. P. Somlyo, and H. V. Sparks, eds.), 2nd ed., Section 2, Vol. 2, pp. 475–513. Am. Physiol. Soc., Washington, D.C. (C)

Sparks, H. V. (1980b). Mechanism of vasodilation during and after ischemic exercise. *Fed. Proc., Fed. Am. Soc. Exp. Biol.* **39**, 1487–1490. (C)

Sparks, H. V., and Fuchs, B. D. (1983). Adenosine as a mediator of sustained exercise hyperemia. *In* "Frontiers of Exercise Biology" (K. Borer, D. W. Edington, and T. P. White, eds.), pp. 119–127. Human Kinetics Publ., Champaign, Illinois. (C)

Steffen, R. P., McKenzie, J. E., Yachnis, A. T., and Haddy, F. J. (1981). Correlation between skeletal muscle acetate content and vascular resistance. *Physiologist* **24**, 76 (abstr.). (C)

Stowe, D. F., Owen, T. L., Anderson, D. K., Haddy, F. J., and Scott, J. B. (1975). Interaction of O_2 and CO_2 in sustained exercise hyperemia of canine skeletal muscle. *Am. J. Physiol.* **229**, 28–33. (C)

Strandell, T., and Shepherd, J. T. (1967). The effect in humans of increased sympathetic activity on the blood flow to active muscles. *Acta Med. Scand., Suppl.* **472**, 146–167. (C)

Taylor, A. E., Parker, J. C., Granger, D. N., Mortillaro, N. A., and Rutilli, G. (1981). Assessment of capillary permeability using lymphatic protein flux: Estimation of the osmotic reflection coefficient. *In* "Microcirculation" (R. M. Effros, H. Schmid-Schonbein, and J. Ditzel, eds.), pp. 19–32. Academic Press, New York. (F)

Taylor, P. (1980). Cholinergic agonists. *In* "The Pharmacological Basis of Therapeutics" (A. G. Gilman, L. S. Goodman, and A. Gilman, eds.), 6th ed., pp. 91–99. Macmillan, New York. (D)

Tobia, A. J., Lee, J. Y., and Walsh, G. M. (1974a). Regional blood flow and vascular resistance in the spontaneously hypertensive rat. *Circ. Res.* **8**, 758–762. (E)

Tobia, A. J., Walsh, G. M., Tadepalli, A. S., and Lee, J. Y. (1974b). Unaltered distribution of cardiac output in the conscious young spontaneously hypertensive rat: Evidence for uniform elevation of regional vascular resistances. *Blood Vessels* **11**, 287–294. (E)

Tuttle, R. S. (1967). Physiological release of histamine ^{14}C in the pyramidal cat. *Am. J. Physiol.* **213**, 620–626. (C)

Uvnäs, B. (1966). Cholinergic vasodilator nerves. *Fed. Proc., Fed. Am. Soc. Exp. Biol.* **25**, 1618–1622. (C)

Valdivia, E. (1958). Total capillary bed in striated muscle of guinea pigs native to the Peruvian mountains. *Am. J. Physiol.* **194**, 585–589. (B)

Van Harn, G. L., Rubio, R., and Berne, R. M. (1977). Formation of adenosine nucleotide derivatives in isolated hog carotid artery strips. *Am. J. Physiol.* **233**, H299–H304. (C)

Van Hee, R. H., and Vanhoutte, P. M. (1978). Cholinergic inhibition of adrenergic neurotransmission in the canine gastric artery. *Gastroenterology* **74**, 1266–1270. (C)

Vanhoutte, P. M., and Levy, M. N. (1980). Prejunctional cholinergic modulation of adrenergic neurotransmission in the cardiovascular system. *Am. J. Physiol.* **7**, H275–H281. (C)

Vanhoutte, P. M., Verbreuren, T. J., and Webb, C. R. (1981). Local modulation of adrenergic neuroeffector interaction in the blood vessel wall. *Physiol. Rev.* **61**, 151–247. (C)

Vracko, R. (1970a). Skeletal muscle capillaries in diabetics: A quantitative analysis. *Circulation* **41**, 271–283. (E)

Vracko, R. (1970b). Skeletal muscle capillaries in nondiabetics: A quantitative analysis. *Circulation* **41**, 285–297. (B)

Warwick, R., and Williams, P. L., eds. (1973). "Gray's Anatomy," 35th Br. Ed. Saunders, Philadelphia, Pennsylvania. (A)

Weiss, G. B. (1981). Multiple Ca^{2+} sites and channels provide a basis for alterations in Ca^{2+} mobilization and vascular contractility. *In* "Vasodilatation" (P. M. Vanhoutte and J. Leusen, eds.), pp. 307–310. Raven Press, New York. (D)

Yodaiken, R. E., Menefee, M., Seftel, H. C., Kew, M. C., and McClaren, M. J. (1975). Capillaries of South African diabetics. IV. Relation to retinopathy. *Diabetes* **24**, 286–290. (E)

Young, E. W., and Sparks, H. V. (1979). Prostaglandins and exercise hyperemia of dog skeletal muscle. *Am. J. Physiol.* **238**, H190–H195. (C)

Zacks, S. I., Pegues, J. J., and Elliot, F. A. (1962). Interstitial muscle capillaries in patients with diabetes mellitus: A light and electron microscopic study. *Metab., Clin. Exp.* **11**, 381–393. (B,E)

Zelis, R., and Flaim, S. (1981). Hemodynamic effects of vasodilator drugs. *In* "Vasodilatation" (P. M. Vanhoutte and J. Leusen, eds.), pp. 441–449. Raven Press, New York. (D)

Zelis, R., Flaim, S. F., and Eisele, J. H. (1977). Effects of morphine on reflex arteriolar constriction induced in man by hypercaporia. *Clin. Pharmacol. Ther.* **22**, 172–178. (D)

Zweifach, B. W., and Metz, D. B. (1955). Selective distribution of blood through the terminal vascular bed of mesenteric structures and skeletal muscle. *Angiology* **6**, 282–290. (B)

Chapter 20
Locomotor System: Bones

A. ANATOMY OF BLOOD CIRCULATION
By L. A. WHITESIDE

1. VASCULAR ANATOMY OF THE DIAPHYSIS

a. Periosteum: The periosteum has a complex vascular network that interconnects with the overlying vessels of the skeletal muscle and the underlying vessels of the bone (Morgan, 1959; Whiteside *et al.*, in press). The superficial vascular plexus directly overlies the fibrous layer of the periosteum (Fig. 20.1). It is anastomosed at the gross and microscopic level with the vessels of the

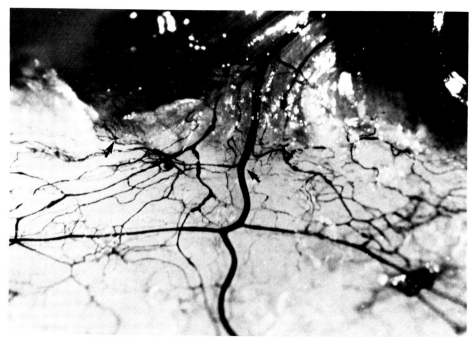

FIG. 20.1. Ink injection study of muscle attachment to the periosteum in an adult dog. Large and small vascular channels (arrows) spread out over the surface to form the superficial periosteal vascular plexus. ×15.

Blood Vessels and Lymphatics in Organ Systems
Copyright © 1984 by Academic Press, Inc.

FIG. 20.2. Ink injection study of the two layers of periosteum and underlying cortical bone in an immature dog. Both layers have been stripped to expose the bone in the foreground (B). The superficial layer (S) has been separated from the cambium layer (C). Vessels from the superficial layer anastomose with the cambium vessels, which, in turn, anastomose with the vascular plexus of the bone (arrows) ×26.

skeletal muscle and with the vascular plexus of the cambium layer of the periosteum. The finer vascular plexus of the cambium layer, in turn, blends with the anatomically similar plexus of the underlying cortical bone (Fig. 20.2). At skeletal maturity, the cambium layer disappears, and the superficial layer becomes thin and almost acellular, but its vascular plexus remains intact and connected with the vessels of the diaphyseal bone (Fig. 20.3). Although the density of periosteal vessels diminishes and the periosteal contribution to the circulation of bone decreases as bone ages, the periosteal vascular plexus remains an important collateral pathway for diaphyseal cortical bone in adulthood (Whiteside *et al.*, 1978; Trias and Ferg, 1979).

The studies of Zucman (1960-1961) demonstrate the importance of the periosteal vascular plexus as the collateral circulatory pathway for skeletal muscle. The geniculate arterial system is a specialization of the periosteal vascular plexus, connecting the periosteal circulation of the femur to that of the tibia. In man, these connections provide collateral circulation in case of occlusion of the major conduit arteries.

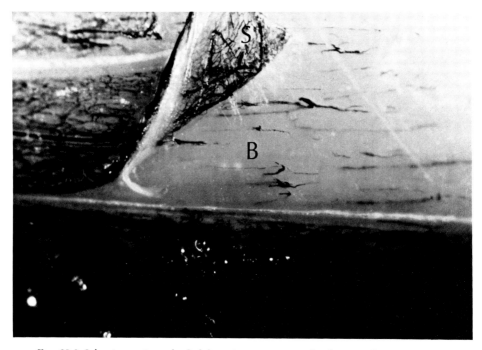

Fig. 20.3. Ink microangiograph of adult canine periosteum. The thin superficial layer (S), with its vascular plexus, has been stripped back from the diaphyseal bone (B). Anastomotic vessels penetrating into bone to combine with the vascular plexus of the diaphyseal cortical bone.

Since the periosteum and its vascular plexus form the extraosseous callus in fracture healing (Kolodney, 1923; Cavadias and Trueta, 1965), it is apparent that the source of this intrinsic periosteal blood supply is of major importance. Ink injection studies have shown that the blood supply to both layers of the periosteum is dependent upon vascular connections with adjacent skeletal muscles (Whiteside *et al.*, in press). The periosteum survives and even forms bone after subperiosteal elevation if the intrinsic vessels and attachments to muscle are preserved (King, 1976). The available evidence points to the extraosseous route as the single most important source of blood supply to the periosteum. It seems reasonable to speculate that the decreased periosteal bone formation noted after extraperiosteal exposure of the bone is caused by the ischemic effect on the periosteum.

b. Diaphyseal cortical bone: This structure is considerably less vascular than the metaphysis and epiphysis, but still it has a rich, extensively anastomosed circulation. The vascular anatomy is complex and difficult to study, so it is not surprising that this is a controversial area. Classically, the blood supply to the inner two-thirds of the cortex is thought to be supplied by medullary vessels and that to the outer one-third, by periosteal vessels (Johnson, 1972; Trueta and Cavadias, 1964). Opposed to such a concept was the observation by Stein *et al.* (1957) that relatively high pressures are present in the medullary canal—high enough to prevent cortical venous blood from returning to the medullary veins. Such findings led to the hypothesis that blood flows one way (from inside to outside), with the periosteal vessels serving to drain the diaphysis and metaphysis (Brookes, 1961; Rhinelander, 1968; Lopez-Cuerto *et al.*, 1980). However, Wilkes and Visscher (1975) improved the techniques for measuring intramedullary pressure and showed marrow interstitial pressures to be no higher than extraosseous venous pressure. Furthermore, the vital microscopy techniques utilized by Brånemark (1961) clearly demonstrated that arterial vessels enter the endosteal surface, and venous channels from the diaphyseal cortical bone empty into the medullary canal. Evidence from most of the microangiographic anatomic studies suggests that arterial and venous channels parallel one another and that venous vessels return to the medullary canal from the bone (Nelson and Kelly, 1960; Trias and Ferg, 1979). Trias and Ferg demonstrated two separate vascular systems of the diaphyseal cortical bone, with parallel arterial supply and venous drainage. A radial system originates in the medullary canal and branches toward the surface (Fig. 20.4), and venous drainage returns to the medullary canal and its nutrient vein. This radial system is characteristic of woven bone and predominates in immature animals. As the diaphysis matures and develops haversian systems, a longitudinal system of vessels forms and increases in importance. After skeletal maturity, both radial and longitudinal systems are found in the diaphysis. The anatomic details described by Trias are consistent with other reports of direct anastomoses between the arterial circulation of the periosteum and of the medullary canal (Danckwardt-Lilliestrom, 1969; Göthman, 1960). This concept of dual arterial and venous blood supply of the diaphyseal bone agrees closely with a study of the effects of surgical procedures on bone circulation (Whiteside and Lesker, 1978a,b; Whiteside *et al.*, 1978). Stripping the diaphysis of all periosteum in the canine and rabbit tibia did not impair venous or arterial blood flow (as measured by the hydrogen wash-out technique) in the middle layers of the diaphyseal cortical bone, regardless of whether the animal was mature or immature, a result possible only if arterial, as well as venous, channels crossed the endosteal surface of the diaphysis. Similarly, blood flow in the middle layers of the diaphyseal cortical bone was not significantly slowed by reaming the medullary canal, even in adult rabbits and dogs. Such findings suggest that the diaphysis can receive a major portion of its blood supply from either the periosteum or the medullary circulation and also that either the medullary canal or the periosteal circulation can provide channels for venous drainage.

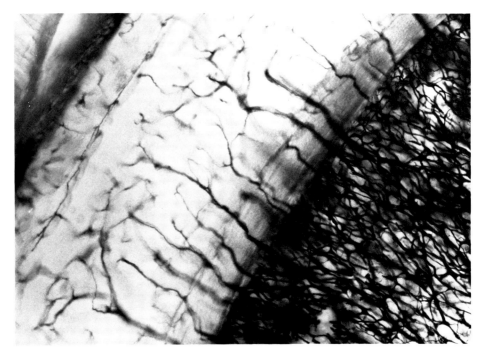

Fɪɢ. 20.4. A transverse section from the femur of the adult mongrel dog in which vessels and marrow are filled with India ink; those of the periosteum are only partially filled. The transverse vascular system radiates from the marrow to branch in the middle third of the cortex. ×19. (From Trias and Ferg, 1979; reproduced with permission from *Journal of Bone and Joint Surgery.*)

2. Bone Ends

a. Epiphysis: Before ossification, the vascular channels penetrate through the cartilaginous epiphysis, supplying nutrients to the entire structure. As ossification occurs, these cartilage channels persist as osseous vessels. Earlier studies suggested that there are free anastomoses between the metaphysis and epiphysis in infancy through cartilage channels across the growth cartilage (Trueta, 1957). However, the detailed study of circulation of the capital femoral epiphysis by Chung (1976) suggests that some of the concepts concerning epiphyseal circulation in infancy must be revised. Early in infancy, prior to ossification, the epiphyseal cartilage channels enter the epiphysis near the articular margin. Penetration of vessels across the growth cartilage between the epiphysis and metaphysis is an uncommon or rare occurrence. As growth progresses, one group of epiphyseal vessels (the lateral epiphyseal vessels of the femoral head) becomes predominant. The epiphysis and metaphysis fuse at maturity, allowing vascular channels to develop across the epiphyseal scar; however, the functional significance of these vessels is not clear. Trueta (1957)

FIG. 20.5. Ink injection study of an adult canine femoral head and neck, sagittal section. Anastomoses are numerous across the fused epiphysis. ×7. Rectangular area enlarged in Fig. 20.6.

reported a large number of interconnections between the old epiphyseal area and the metaphyseal area, but studies in human cadaver material demonstrate that extensive stripping of the capsule and retinaculum routinely devascularizes the human femoral head (Judet *et al.*, 1955; Sevitt and Thompson, 1965). In adult dogs, epiphyseal blood supply remains separate from the metaphyseal circulation, even in advanced age (Whiteside *et al.*, 1983). Although anastomoses exist at the capillary level, these small vessels do not appear to be adequate to reconstitute the epiphyseal circulation in the event of loss of the major epiphyseal vessels (Figs. 20.5 and 20.6). Brookes (1971) reported that the epiphyseal and metaphyseal marrow spaces retain separate blood supplies even in adulthood. Despite abundant anastomoses within the cancellous bone, in most cases, the subchondral epiphyseal arteries appear to be end-vessels.

b. Articular cartilage: Nutrition of the articular cartilage is generally considered to be dependent upon diffusion from the synovial fluid to the deep cell layers of the cartilage, but recent studies demonstrate that the nutrient pathways of subchondral bone and articular cartilage are more complex than simple diffusion. In infancy, epiphyseal vessels lie just under the articular cartilage. Radioactive tracer studies have shown that these vessels deliver nutrients to all

FIG. 20.6. Magnification of isolated portion of Fig. 20.5. Most of the anastomoses are capillaries.

layers of the cartilage (Ogata and Whiteside, 1979). After the bone has matured, subchondral bone and calcified cartilage act as a barrier to diffusion. In adult small animals, the calcified cartilage layer isolates the articular cartilage from the epiphyseal circulation. Thus, in the adult rabbit, the articular cartilage relies upon synovial fluid for its nutrition. However, in human weight-bearing joints, such as the hip, knee, and first metatarsal head, subchondral vessels are present even late in adulthood (Greenwald and Haynes, 1969). Haynes and Woods (1975) found that fluorescent dye injected into the femoral artery quickly penetrates through the substance of the articular cartilage from the subchondral vessels, whereas almost no dye penetrates from the articular surface. There is agreement among investigators that the subchondral vessels have an important function in articular cartilage nutrition in large weight-bearing joints. The relationship between these vessels and the pumping action of the weight-bearing joints in articular cartilage nutrition is not yet understood. This subject and its role in the pathophysiology of osteoarthritis should be an important area for future investigation.

  c. *Growth cartilage:* The blood supply of the growth cartilage comes from the epiphyseal arteries. Small vessels branch from the epiphyseal circulation

FIG. 20.7. Photomicrograph of thick unstained section of the upper end of tibia, visualized by ink injection study. Section shows terminal branches of the epiphyseal arteries supplying the epiphyseal plate. (From Morgan, 1959; reproduced with permission from *Journal of Bone and Joint Surgery, Oxford.*)

and arborize over the epiphysis, providing the sole source of nutrients to the proliferating cartilage cells (Fig. 20.7). Loss of the epiphyseal circulation results in death or deformity of the articular cartilage, thus resulting in growth disturbance. Metaphyseal vessels nourish the newly forming bone adjacent to the growth cartilage and participate in the calcification process. Loss of the metaphyseal vessels does not result in growth disturbance of the bone, but, instead, there is failure of ossification, thus leading to thickening of the epiphyseal plate (Trueta and Amato, 1960).

3. CONCLUSIONS

The diaphysis and metaphysis have dual circulations with abundant anastomoses and are not likely to be seriously injured by loss of one of their two major vascular pathways. The epiphysis, however, even in adulthood, is dependent upon the epiphyseal arteries entering at the articular margins. Loss of these vessels is likely to lead to significant avascular necrosis.

B. PHYSIOLOGY OF BLOOD CIRCULATION

By D. W. Lennox and D. S. Hungerford

Besides biomechanically providing a rigid framework for protecting vital organs and for the attachment of muscle and ligament, bone also possesses a remarkable capacity to modify its structure and function in response to injury, growth, aging, neoplasm, metabolic derangement, infection, surgery, and neurohumoral factors. Of utmost importance in maintaining the varied responsiveness of bone is the existence of a highly responsive system of bone blood flow to deliver necessary nutrients, cells, and possibly chemical mediators.

Although at the present time there is no clinically useful technique for the direct measurement of bone blood flow in human subjects, a number of procedures are available for this purpose in experimental animals. Several of these are considered in the following discussion.

1. METHODS OF STUDYING BONE BLOOD FLOW

a. Direct method: Direct measurement of total bone blood flow is nearly impossible due to the multiple afferent and efferent channels involved. Cumming (1960) used a direct technique to study flow to the rabbit femur, by cannulating the superficial femoral vein (having previously ligated the deep femoral and circumflex veins) and measuring flow from the cannulated vessel, considering this to represent femoral flow. A similar methodology was reported earlier by Drinker and Drinker (1916). Such an approach has more utility in investigating gross changes in circulation in response to any of a number of interventions than in determining precise values for blood flow.

b. Indirect methods: A number of indirect procedures exist for the study of blood flow in bone, one of which is red blood cell (RBC) labeling. White *et al.* (1964) utilized ^{51}Cr-tagged red cells for this purpose. The technique consists of injecting tagged red cells intravenously, and, after an appropriate time to permit mixing, blood is sampled to establish a reference radioactivity level per volume collected. The animal is then sacrificed and the bones under investigation are fixed, weighed, sectioned, and measured for radioactivity. Circulating red cell volume is then calculated and expressed as volume per net weight of specimen. This method assumes that the hematocrit in bone blood is the same as that in the blood sample.

682

A second approach, the radioisotope clearance method, consists of the use of both non-bone-seeking and bone-seeking isotopes to assess bone blood flow. With the non-bone-seeking isotopic method, a given amount of radioisotope is injected into the bone to be studied and the radioactivity level is followed over time. Rapid clearance is evidenced by a rapid diminution in radioactivity, which is considered to be proportional to blood flow. Isotopes such as $Na^{131}I$ have been employed for this purpose. Since the procedure is an invasive one, it is subject to artefacts produced by the injecting trocar. With the bone-seeking isotopic method, either ^{86}Rb (Kane, 1968), ^{45}Ca, or ^{85}Sr is employed to estimate lower limb blood flow.

With the use of the substances mentioned above, flow estimates are based upon the Fick principle. For example, in the case of ^{85}Sr the amount of the isotope fixed in bone in a given time is divided by the arteriovenous concentration difference in order to calculate flow. The flow value generated by such an approach is less than the true value since the extraction ratio for ^{85}Sr is less than unity (Shim et al., 1971).

The use of tracer microspheres is a relatively new method for measuring bone blood flow in experimental animals, the value of which has not been fully established. It employs 15 μm microspheres labeled with any of a number of different isotopes, including ^{125}I, ^{85}Sr, ^{46}Sc, ^{141}Ce, and ^{95}Nb. The availability of several different isotopes and their separability by spectroscopy create the possibility for repeated measurements in one animal. Arterial reference samples are obtained just prior to and for several minutes following the intracardiac microsphere injection. At the conclusion of the experiment, the animal is sacrificed and the bone cleaned and counted in a gamma counter. Microspheres are trapped in the bone microcirculation in numbers proportional to flow. The technique allows for analysis of flow to various anatomic regions of bone, as well as for overall bone blood flow. Although the method appears to be applicable to many studies with experimental animals, its validity in states of fracture healing, growth, or necrosis has not been established (Gross et al., 1981). Tothill and MacPherson (1980) have emphasized that due to removal of some microspheres in preosseous capillaries, the microsphere technique has limited applicability as a standard in determining extraction ratios. (For details of the technique, its assumptions, and discussion of the validity and applicability of the method, see Gross et al., 1979, 1981.)

2. BLOOD FLOW TO BONE

Recent measurements made with the microsphere technique indicate a heterogeneous pattern of flow to bone and bone marrow. In the anesthetized dog, low flow rates were reported for compact bone (2 ml/min/100 gm) in humeral and femoral diaphysis) and much higher values for hematopoietic cancellous bone and hematopoietic marrow (18–30 ml/min/100 gm) (Gross *et al.*, 1981). Morris and Kelly (1980) found similar values for compact and cancellous blood flow in the dog but recommended that, for optimal microsphere technique, the animal should be conscious with the reference catheter in the aorta. Whiteside *et al.* (1977a), utilizing the hydrogen washout technique, measured epiphyseal cancellous bone blood flow in the rabbit and found that for cancellous bone, the flow was 0.129 ± 0.015 ml/min/ml, with the corresponding value for cortical bone being 0.069 ± 0.002 ml/min/ml.

3. Skeletal Blood Flow

Estimates of total skeletal blood flow vary widely. Figures as high as 27.5% of cardiac output have been proposed (Brookes, 1967). Wootton *et al.* (1976b) found bone blood flow to be 4.1 ml/min/100 gm in eight normal male volunteers using an ^{18}F isotope technique. Shim *et al.* (1971) reported a reading of 2.4 ml/min/100 gm for man with an [85]Sr clearance method, a value similar to that previously found by Van Dyke *et al.* (1965). With the microsphere technique, Gross *et al.* (1981) estimated total skeletal blood flow in the dog to represent 11% of cardiac output. An earlier report by Shim *et al.* (1967) noted a rate of 7.3 ± 3.0% of cardiac output. Morris and Kelly (1980) estimated the percentage of cardiac output to bone tissue in the conscious dog to be 9.6% in the mature animal and 10.3% in the immature one. Of interest was their finding that in the mature animal, approximately 2% was flow to cortical bone and 8% to cancellous bone, whereas, in the immature dog, 7% was flow to cortical bone and 4% to cancellous bone. Such results were attributed to the fact that although cortical bone represents 80% of the skeleton by weight, its surface area is roughly equivalent to less voluminous cancellous bone; moreover, in the immature animal, appositional growth is greater, with possible shunting to cortical bone.

4. Altered Physiologic Parameters and Bone Blood Flow

Although it seems reasonable to assume that bone blood flow, constituting a significant portion of overall cardiac output, should respond and be sensitive to altered physiologic parameters, it is only recently that this expectation has gained experimental confirmation.

a. Response to hemorrhage: Syftestad and Boelkins (1980) subjected conscious rabbits to nonfatal reversible hemorrhage and analyzed the effects on marrow, bone, and a number of other tissues, utilizing the radioactive microsphere technique. There was no evidence for immediate shunting of blood from bone to marrow, but an increase in marrow blood flow did occur 16 hr after hemorrhage. This response was interpreted as a possible preparatory mechanism for increased erythropoietic activity.

In a related experiment, Gross *et al.* (1979) induced hypotension in the dog by arterial hemorrhage and followed changes in bone blood flow. The response to this state was a marked increase in vascular resistance and a decreased bone and marrow blood flow.

b. Response to exercise: In the dog, exercise induced by treadmill markedly increased blood flow to exercising skeletal muscles (Gross *et al.*, 1979), and, at the same time, vascular resistance in bone rose significantly and flow, both to bone and marrow, diminished. During exercise, nonexercising muscle (temporalis) exhibited increased vascular resistance.

c. Response to hypoxia: In the dog, systemic arterial hypoxia was found to reduce blood flow to bone and marrow and raise vascular resistance in skeletal muscle (Gross *et al.*, 1979). However, Adachi *et al.* (1976) had previously found no change under similar conditions. The differences in results may reflect variation in anesthesia technique or level of hypoxia.

d. Response to aging: Macpherson and Tothill (1978) presented evidence for decreased bone blood flow to the tibia, fibula, femur, pelvis, humerus, radius, ulna, and scapula with increase in age and weight of rats.

e. Response to growth: McInnis *et al.* (1977) reported a positive correlation between bone blood flow and percent of new bone formation in a standardized tibial defect in the dog. The finding that cortical bone blood flow is higher in the immature than in the mature animal (Pasternak *et al.*, 1966; Morris and Kelly, 1980) may be related to greater appositional growth and more extensive bone remodeling (Lee, 1964; Vanderhoeft *et al.*, 1962). Whiteside *et al.* (1977b) noted a positive correlation between bone blood flow and osteoblastic activity in the rabbit tibia.

f. Summary: Blood flow to bone and marrow is responsive to altered physiology. Vessels to bone and marrow appear to be involved in the overall circulatory adjustment to hypotension, hypoxia, and exercise. Neurohumoral factors also produce alterations in blood flow to bone. (For a discussion of neurohumoral factors, see Section C, below.)

C. PHARMACOLOGY OF BLOOD CIRCULATION

By D. W. Lennox and D. S. Hungerford

It has been known for many years that neurohumoral and metabolic factors regulate blood flow to bone (Drinker and Drinker, 1916), but only with the advent of sophisticated blood flow measurement techniques have the specific effects upon bone blood flow of a number of hormones and other neuropharmacologic agents been investigated.

1. Humoral Factors

Gross *et al.* (1979, 1981) demonstrated that, in the dog, norepinephrine infusion increases vascular resistance to bone and marrow by two times the baseline value, whereas the potent vasodilator, adenosine, on infusion, diminishes vascular resistance by one-third. Similarly, Driessens and Vanhoutte (1977, 1979) noted in dog tibias increased perfusion pressures in response to norepinephrine, a response that was blocked by phentolamine and inhibited by acetylcholine. In addition, the acetylcholine effect was abolished by atropine. Intraarterial acetylcholine was found to induce bone blood vessel dilatation by Michelsen (1968). Calcitonin produced dose-dependent elevations in perfusion pressure in dog tibias (Driessens and Vanhoutte, 1981) and a decrease in skeletal blood flow in patients with Paget's disease (Wootton *et al.*, 1976a, 1978). Bone blood vessel dilatation in response to parathyroid hormone was reported by Boelkins *et al.* (1976).

The effect of hydrocortisone on bone blood vessels was studied by Driessens and Vanhoutte (1981). They found that at low concentrations, the drug augmented the vasoconstrictor response to norepinephrine, whereas, at high concentrations, it had the opposite effect.

Dusting *et al.* (1978) investigated the action of arachidonic acid and several related metabolites on the femoral vasculature in the dog. They noted vasodilatation with injections of PGI_2, PGE_2, sodium arachidonate, and the endoperoxide PGH_2. Utilizing the femoral vascular bed of the dog as a model, Laubie *et al.* (1977) studied the effect of apomorphine and piribedil (ET 495) as dopamine agonists and, on the basis of their results, proposed the existence of dopamine receptors involved in mediating vasodilator and sympathoinhibitory effects.

Blood Vessels and Lymphatics in Organ Systems
Copyright © 1984 by Academic Press, Inc.

2. NEURAL EFFECTS

Tonic sympathetic vasoconstriction of blood vessels of bone in anesthetized dogs was demonstrated by Gross *et al.* (1979). On stimulation of carotid baroreceptors, a one-third reduction in vascular resistance in bone was noted, as compared with a corresponding 80% decrease in skeletal muscle vascular resistance under similar conditions. In cats, deafferentation of baroreceptors to activate sympathetic nerve discharge produced heightened vascular resistance. Increased perfusion pressure in the nutrient artery of the dog tibia following periarterial electrical stimulation was noted by Driessens and Vanhoutte (1979). This response was blocked by phentolamine and hence was presumed to be due to sympathetic activation. An augmentation in tibial blood flow following lumbar sympathectomy was reported by Trotman and Kelly (1963). In accord with such a finding is the earlier observation of vasoconstriction in bone blood vessels after sympathetic stimulation (Drinker and Drinker, 1916; Weiss and Root, 1959).

3. METABOLIC EFFECTS

Evidence exists for the control of bone blood flow by a number of metabolic factors, including acidosis and hypercapnea. Cumming (1962), studying the effect of rebreathing expired air, or breathing a low oxygen–high carbon dioxide gas mixture on the femoral nutrient vein outflow in the rabbit, reported a 20% increase in outflow under such conditions. This finding was confirmed by Shim and Patterson (1967), who noted increased blood flow in rabbit bone

TABLE 20.I

VARIATIONS IN BONE BLOOD FLOW IN RESPONSE TO A
NUMBER OF FACTORS

Agent/parameter	Flow
Norepinephrine	↓
Adenosine	↑
Acetylcholine	↑
Calcitonin	↓
PTH	↑
Sympathetic stimulation	↓
Sympathectomy	↑
Hypercapnea/acidosis	↑
Hemorrhage	↓
Exercise	↓
Hypoxia	↓

following rebreathing of expired air. Gross *et al.* (1979) studied the changes in bone and marrow blood flow and resistance in the baboon in response to arterial acidosis and hypercapnea induced by adding 10% CO_2 to inspired air while maintaining constant and normal arterial oxygen concentration. At arterial $pCO_2 = 65 \pm 1$ mm Hg and pH 7.14 ± 0.02, they noted a statistically significant increase in bone blood flow to sternum, rib, and femoral marrow and a marked decrease in calculated vascular resistance. Variations in bone blood flow in response to a number of factors are presented in Table 20.I.

4. SUMMARY AND AVENUES FOR FUTURE INVESTIGATION

Regulatory mechanisms for the control of blood flow to bone and marrow thus appear to be highly integrated and responsive, involving nervous, humoral, and metabolic factors. The development of new techniques of blood flow assessment has resulted in the discovery of additional pharmacologic data on bone blood flow.

Future advances will doubtless include study of the complex interplay among humoral, neural, and metabolic factors in regulating bone blood flow. As new pharmacologically active agents are discovered or previously known ones reassessed, pharmacologic intervention in altered bone blood flow states may become possible, as for example, in the case of bone neoplasms, osteoporosis, fractures, growth, and aging.

D. PATHOPHYSIOLOGY, PATHOGENESIS, AND PATHOLOGY OF BLOOD CIRCULATION

By D. W. LENNOX AND D. S. HUNGERFORD

Circulatory abnormalities of bone blood vessels underlie a number of clinical entities. Several of the better known of these are discussed below.

1. EPIPHYSITIDIES (APOPHYSITIDIES)

A number of eponymic disorders, including Legg–Calve–Perthes' disease, Osgood–Schlatter's disease, Sindig–Larrsen–Johannson disease, Kohler's dis-

ease, Freiberg's disease, Schuermann's disease, Panner's disease, and Thie-
mann's disease, were believed to originate from a similar vascular abnormality
of bone blood vessels. It now appears that those conditions about the patella or
patellar tendon represent avulsion fractures.

   a. Legg–Calve–Perthes' disease: One theory regarding the development of
this entity is that circulatory embarrassment from retinacular vessel disruption
in the region of the physis plays a very important role (Ponseti, 1956). In
comparing the effect of venous tamponade of the hip capsule on femoral head
blood flow in mature and immature dogs, Launder *et al.* (1981) reported a
significant rise in femoral head pressure in puppies following capsule inflation
but noted no significant change in either pressure or flow in adults. These
results were attributed to the absence of an intact intramedullary venous
drainage of the femoral head in the immature animal. The validity with which
these data can be applied to humans, as in the case of Legg–Calve–Perthes'
disease, is unclear. Heikkinen *et al.* (1980) reported delayed venous drainage
from the femoral neck to be indicative of a poor prognosis in this disorder.
Heikkinen *et al.* (1976) found that after treatment by intertrochanteric osteotomy, venous outflow patterns became normal at 4–15 months following surgery. Suramo *et al.* (1964) have demonstrated that venous drainage of the
femoral neck occurs via capsular veins in the normal child's hip, but via the
intramedullary route in children with Perthes' disease and with an intact
growth plate. In a study of intracapsular tamponade in puppies, Tachdjian and
Grana (1968) reported partial vascular obstruction at 80–100 mm Hg pressures
and avascular necrosis with pressures of 200 mm Hg maintained for 10 hr.
Utilizing a similar approach but measuring flow with the hydrogen washout
technique, Borgsmiller *et al.* (1980) eliminated epiphyseal flow at 150 mm
Hg pressure and on the basis of their findings concluded that Legg–Calve–
Perthes' disease resulted from arterial rather than venous tamponade. Woodhouse
(1962, 1964) found that in adult dogs, avascular necrosis developed after exposure to hip intracapsular pressures of 50 mm Hg for 12 hr following femoral
neck osteotomy. In puppies, a similar tamponade, but without osteotomy, also
produced femoral head necrosis at 12 hr. This worker concluded that avascular
necrosis resulted from venous tamponade, capillary engorgement, fluid extravasation, microcapillary sludging, and irreversible intravascular thrombosis.

2. FRACTURES AND DISLOCATIONS

   Trauma sufficient to produce a fracture or dislocation can result in damage
to the blood supply to an entire bone, e.g., the talus in subtalar dislocation, or a
portion of a bone, e.g., the femoral neck in femoral neck fracture. With severe
circulatory compromise, avascular (ischemic) necrosis may result. Particularly

vulnerable to the development of ischemia are intracapsular fractures, as occur in the hip and shoulder. In these locations, blood supply is marginal and damage to surrounding soft tissue may be sufficient to induce irreversible ischemia. The duration of the ischemia appears to be a critical factor since better results are obtained in cases of hip dislocation reduced within 12 hr than in those treated after that time period. In fractures of the femoral neck, bone scans have been recommended as diagnostic tools to determine the viability of the femoral head.

Avascular necrosis of the femoral head in children is a complication following reduction and cast immobilization, as in the treatment of congenitally dislocated hips. Schoenecker *et al.* (1978) demonstrated in dogs that a position of forced frog-leg abduction and internal rotation obliterated or drastically reduced circulation to the femoral head.

Duncan and Shim (1977) reported that in rabbits with trauma-induced dislocation of the hip, circulatory compromise was present in both adult and immature animals, but it was worse in immature rabbits, reaching a maximum effect after 24 hr of continued dislocation. Aseptic necrosis was evident in the specimens, particularly in the case of immature animals. Duncan and Shim found that in the adult rabbit, the anastomotic connection between epiphyseal and metaphyseal vessels afforded some protection to the femoral head from the insult to the extraosseous nutrient system, whereas in the immature animal, without such a vascular arrangement, damage to the system of supply and of drainage could result in necrosis.

3. Bone Infarcts

Bone infarcts, which are believed to arise from arterial obstruction, in many instances are asymptomatic, being noted incidentally in roentgenograms or bone scans performed for other reasons. Long bones are involved almost exclusively. Most commonly, lesions are a few millimeters in size but may vary to involve a large portion of the shaft; the ends of long bones are more often involved than other areas. Roentgenograms demonstrate areas of mottled sclerosis. In contrast, areas of infarction and necrosis involving a periarticular region are associated with a considerably different clinical pattern, roentgenographic picture, pathologic changes, and prognosis (see Section D-5, below).

4. Surgery

Orthopedic and/or vascular surgery may greatly alter circulation in bone. For example, internal fixation devices in the treatment of fractures can affect bone blood supply. In this regard, Rand *et al.* (1981) compared the vascular effects of open intramedullary nailing after reaming with those of compression plate fixation of a tibial fracture in the dog. Bone blood flow remained elevated for longer periods and reached a higher level overall in the intramedullary rod group than in that with compression plate fixation. However, the latter approach provided mechanical strength at the fracture site sooner than did rod fixation. Similar effects on bone healing were noted with both extraperiosteal and subperiosteal plate placement, with fracture healing occurring by different mechanisms in the two groups. Whereas with the intramedullary rod group, fracture healing was predominantly by periosteal callus, in the compression plate group, endosteal callus formation occurred. Both open intramedullary nailing after reaming and intramedullary rod insertion damage the medullary vasculature and produce avascularity of a portion of the cortical diaphyseal region (Trueta and Cavadias, 1955; Göthman, 1961; Rhinelander, 1974). Compression plate fixation damages cortical efferent blood flow beneath the plate (Olerud and Danckwardt-Lilliestrom, 1968).

Whiteside et al. (1978) studied the effects of periosteal stripping and medullary reaming on regional blood flow in the tibias of mature and immature rabbits. By stripping tibial epiphyseal periosteum, the epiphyseal circulation was eliminated, as measured by the hydrogen washout technique. In mature rabbits, the epiphyseal flow was markedly reduced by periosteal stripping, but no change in blood flow was noted following wide reaming of the epiphyseal center in both mature and immature animals. Diaphyseal and metaphyseal flow was unchanged following separate medullary reaming or periosteal dissection in both groups. When intramedullary reaming and periosteal stripping were performed, cortical diaphyseal flow ceased, but metaphyseal flow persisted. Whiteside et al. (1978) concluded that both venous drainage and arterial supply systems traverse endosteal and periosteal systems and that either can sustain adequate circulation.

Because a number of orthopedic implants, including total hip replacement and total knee replacement prostheses, utilize the polymer polymethylmethacrylate (PMMA) to secure components to bone, the vascular response elicited by this substance was studied by Brookes and Gallanaugh (1975). They implanted a plug of methylmethacrylate into the rat tibia and calculated blood flow using ^{51}Cr- and ^{59}Fe-tagged resin particles. At both 14 and 112 days postoperatively, both blood volume and blood flow were significantly depressed in the tibias in which the acrylic cement had been implanted. The extent to which this factor may be operative in the case of human joint replacement is uncertain. Theoretically, however, devascularized bone appears prone to infection, and this may be a contributing factor in the major problem of loosening in joint replacements.

5. ISCHEMIC NECROSIS OF THE FEMORAL HEAD

a. Theories of pathogenesis: A number of hypotheses have been proposed with regard to the pathogenesis of ischemic necrosis of the femoral head. One of these, the infarction theory, was espoused by Chandler (1948), who attributed the condition to compromise of the lateral retinacular vessel, with subsequent infarction of the anterolateral segment of the femoral head. According to him, with revascularization of the infarcted segment, the femoral head then becomes softened and finally fails mechanically. However, McFarland and Frost (1961) suggested that it is the accumulation of microfractures without repair that eventuates in macrofractures, causing the femoral head to collapse.

Another view, the embolization theory, is based on the assumption that fat emboli produce infarction and ischemic necrosis. Jones and Sakovich (1966) demonstrated that rabbits given intraarterial injections of Lipiodol were found to have fat droplets in a subchondral location in the femoral head. Of interest in this regard is the finding that hyperlipidemia is a common abnormality in ischemic necrosis patient populations (Cruess *et al.*, 1975; Fisher, 1978; Jones, 1971). Although fat emboli have been reported in ischemic necrosis of the femoral head, it is still not possible to state unequivocally that they are responsible for the disease and, if so, whether the basis is one of arterial or arteriolar infarction.

Finally, the progressive ischemia theory has been proposed to explain the mechanisms responsible for the development of ischemic necrosis of the femoral head. In this regard, Michelsen (1967) and Wilkes and Visscher (1975) have offered the analogy between circulation in bone and the functioning of a Starling resistor, with the rigid canister of the resistor corresponding to the rigidity of cortical bone. Thin-walled tissues, the vessels, traverse the canister but do not open into it. Under such conditions, structures outside the vascular space but within the container of bone (the elements of marrow) can modulate blood flow by changes in tissue pressure. Thus, an increase in pressure in the compartment (elevated bone marrow pressure) could collapse such thin-walled vessels as sinusoids and veins, increase peripheral resistance, and diminish flow.

The concept of bone and its circulation as functioning in a manner similar to a Starling resistor is supported by the common finding of elevated intramedullary pressure in a number of different pathologic circumstances, including ischemic necrosis of the femoral head (Arlet and Ficat, 1964; Ficat and Arlet, 1968; Hungerford, 1979; Hungerford and Zizic, 1978).

In the case of Gaucher's disease, proliferating reticuloendothelial cells could increase bone marrow pressure, with resultant diminished flow. With regard to caisson disease, the nitrogen bubbles generated by decompression could expand extravascularly, increase bone marrow pressure, and compromise flow. Essentially any circulatory injury that produces ischemia could result in fluid extravasation into the extravascular (bone marrow) space, with resultant increased bone marrow pressure and circulatory embarrassment.

Jacqueline and Rabinowitz (1973) reported no evidence in support of sudden infarction in an examination of 82 femoral heads studied at various times following femoral neck fracture. Multiple areas of ischemia were distributed throughout the femoral head after fracture. With time and the added stress of weight bearing, the ischemic areas were localized to that portion of the femoral head that assumed the weight-bearing load. Hungerford (1983) presented several examples of ischemic necrosis of the femoral head that were localized not to the anterolateral fracture but rather at a site that, although previously a non-weight-bearing region, was rotated into a weight-bearing position by the fracture. This argues for the role of biomechanical factors in determining the region of morphologic change in ischemic necrosis.

b. Diagnosis: Bone marrow pressure can be measured clinically and is elevated in all phases of ischemic necrosis, even at the preclinical stages in some patients (Hungerford, 1979). In this disorder, venograms, performed by injecting the radiopaque material intraosseously, demonstrate poor filling of the metaphyseal veins, stasis, and diaphyseal reflux on films taken 5 min following administration.

Arlet and Ficat (1964) and Ficat and Arlet (1968) studied necrosis of the femoral head utilizing bone marrow pressure measurements and venograms, together with a bone biopsy obtained from the femoral neck and head. They found that core decompression (removal of a core or plug of bone) resulted in pain reduction, and in some cases this procedure retarded radiologic progression of the condition in the early stages.

REFERENCES*

Adachi, H., Strauss, H. W., Ochi, H., and Wagner, H. N. (1976). The effect of hypoxia on the regional distribution of cardiac output in the dog. *Circ. Res.* **39**, 314–319. (B)

Arlet, J., and Ficat, P. (1964). Forage-biopsie de la tête femorale dans l'ostéonecrose primitive, observations histopathologiques portant sur huit cas. *Rev. Rhum. Mal. Osteo-Articulaires* **31**, 256–264. (D)

Boelkins, J. N., Mazarkiewicz, M., Mazur, P. E., and Mueller, W. J. (1976). Changes in blood flow to bone during the hypocalcemic and hypercalcemic phases of the response to parathyroid hormone. *Endocrinology* **98**, 403–412. (C)

Borgsmiller, W. K., Whiteside, L. A., Goldsand, E. M., and Lange, D. R. (1980). Effect of hydrostatic pressure in the hip joint on proximal femoral epiphyseal and metaphyseal blood flow. *Proc. Orthoped. Res. Soc.* **5**, 23. (D)

Brånemark, P. I. (1961). Experimental investigation of microcirculation in bone marrow. *Angiology* **12**, 239–253. (A)

Brookes, M. (1961). A new concept of capillary circulation in bone cortex. Same clinical applications. *Lancet* **1**, 1078–1081. (A)

Brookes, M. (1967). Blood flow rates in compact and cancellous bone, and bone marrow. *J. Anat.* **101**, 533–541. (B)

Brookes, M. (1971). "The Blood Supply of Bone," Chapter 7, p. 117. Butterworth, London. (A)

*In the reference list, the capital letter in parentheses at the end of each reference indicates the section in which it is cited.

Brookes, M., and Gallanaugh, S. C. (1975). Circulatory depression in bone after acrylic implantation. *Clin. Orthop. Relat. Res.* **107**, 274–276. (D)

Cavadias, A. X., and Trueta, J. (1965). An experimental study of the vascular contribution to the callus of fracture. *Surg., Gynecol. Obstet.* **119**, 731–748. (A)

Chandler, F. A. (1948). Coronary disease of the hip. *J. Int. Coll. Surg.* **11**, 34–36. (D)

Chung, S. M. K. (1976). The arterial supply of the developing proximal end of the human femur. *J. Bone Jt. Surg., Am. Vol.* **58-A**, 961–970. (A)

Cruess, R. L., Ross, D., and Crawshaw, E. (1975). The etiology of steroid induced avascular necrosis of bone. *Clin. Orthop. Relat. Res.* **113**, 178–183. (D)

Cumming, J. D. (1960). A method for studying the rate of blood flow through the bone marrow of a rabbit's femur. *J. Physiol. (London)* **152**, 39–40. (B)

Cumming, J. D. (1962). A study of blood flow through bone marrow by a method of venous effluent collection. *J. Physiol. (London)* **162**, 13–20. (C)

Danckwardt-Lilliestrom, G. (1969). Reaming of the medullary cavity and its effect on diaphyseal bone. *Acta Orthop. Scand., Suppl.* **128**, 134–144. (A)

Driessens, M., and Vanhoutte, P. M. (1977). Vascular reactivity in an isolated bone preparation. *Arch. Int. Pharmacodyn. Ther.* **230**, 330. (C)

Driessens, M., and Vanhoutte, P. M. (1979). Vascular reactivity of the isolated tibia of the dog. *Am. J. Physiol.* **236**, 904–908. (C)

Driessens, M., and Vanhoutte, P. M. (1981). Effect of calcitonin, hydrocortisone, and parathyroid hormone on canine bone blood vessels. *Am. J. Physiol.* **241**, H91–H94. (C)

Drinker, C. K., and Drinker, K. R. (1916). A method for maintaining an artificial circulation through the tibia of the dog, with a demonstration of the vasomotor control of the marrow vessels. *Am. J. Physiol.* **40**, 511–521. (B, C)

Duncan, C. P., and Shim, S. S. (1977). Blood supply to the head of the femur in traumatic hip dislocation. *Surg., Gynecol. Obstet.* **144**, 185–191. (D)

Dusting, G. J., Moncada, S., and Vane, J. R. (1978). Vascular actions of arachidonic acid and its metabolites in perfused mesenteric and femoral beds of the dog. *Eur. J. Pharmacol.* **49**, 65–72. (C)

Ficat, P., and Arlet, J. (1968). Diagnostic de l'ostéonecrose femoro-capitales primitive au stade I (stade pre-radiologique). *Rev. Chir. Orthoped.* **54**, 637–648. (D)

Fisher, D. E. (1978). The role of fat embolism in the etiology of corticosteroid-induced avascular necrosis. Clinical and experimental results. *Clin. Orthop. Relat. Res.* **130**, 68–80. (D)

Göthman, L. (1960). The normal arterial pattern of the rabbit's tibia. A microangiographic study. *Acta Chir. Scand.* **120**, 201–230. (A)

Göthman, L. (1961). Vascular reactions in experimental fractures. Microangiographic and radio-isotope studies. *Acta Chir. Scand., Suppl.* **284**, 1–34. (D)

Greenwald, A. S., and Haynes, D. W. (1969). A pathway for nutrients from the medullary cavity to the articular cartilage of the human femoral head. *J. Bone Jt. Surg., Br. Vol.* **51B**, 747–753. (A)

Gross, P. M., Heistad, D. D., and Marcus, M. L. (1979). Neurohumoral regulation of blood flow to bone and marrow. *Am. J. Physiol.* **237**, H440–H448. (B, C)

Gross, P. M., Marcus, M. L., and Heistad, D. D. (1981). Measurement of blood flow to bone and marrow in experimental animals by means of the microsphere technique. *J. Bone Jt. Surg., Am. Vol.* **63-A**, 1028–1031. (B, C)

Haynes, D. W., and Woods, C. G. (1975). Nutritional pathways for adult human articular cartilage. *Orthopaedics (Oxford)* **8**, 1–8. (A)

Heikkinen, E., Puranen, J., and Suramo, I. (1976). The effect of intertrochanteric osteotomy on the venous drainage of the femoral neck in Perthes' disease. *Acta Orthop. Scand.* **47**, 89–95. (D)

Heikkinen, E., Lanning, P., Suramo, I., and Puranen, J. (1980). The venous drainage of the femoral neck as a prognostic sign in Perthes' disease. *Acta Orthoped. Scand.* **51**, 501–503. (D)

Hungerford, D. S. (1979). Bone marrow pressure, venography, and core decompression in ischemic necrosis of the femoral head. *Proc. Am. Hip Soc. Meet. 7th*, St. Louis, pp. 218–237. (D)

Hungerford, D. S. (1983). Treatment of ischemic necrosis of the femoral head. *In* "Surgery of the Musculoskeletal System" (C. McCollister Evarts, ed.). Vol. 3, pp. 6:5–6:29. Churchill-Livingstone, Edinburgh and London. (D)

Hungerford, D. S., and Zizic, T. M. (1978). Alcohol associated ischemic necrosis of the femoral head. *Clin. Orthop. Relat. Res.* **130**, 144–153. (D)

Jacqueline, F., and Rabinowitz, T. M. (1973). Lesions de la hanche secondaires a la fracture du col du femur. *Proc. Int. Symp. Circ. Bone, 1st, 1973* p. 283. (D)

Johnson, R. (1927). A physiological study of the blood supply of the diaphysis. *J. Bone Jt. Surg.* **9**, 153–184. (A)

Jones, J. P., Jr. (1971). Alcoholism, hypercortisonism, fat embolism and osseous avascular necrosis. *In* "Idiopathic Ischemic Necrosis of the Femoral Head in Adults" (W. M. Zinn, ed.), pp. 112–132. Thieme, Stuttgart. (D)

Jones, J. P., Jr., and Sakovich, L. (1966). Fat embolism of bone: A roentgenographic and histological investigation with use of intraarterial lipiodol in rabbits. *J. Bone. Jt. Surg., Am. Vol.* **48-A**, 149–164. (D)

Judet, J., Judet, R., Legrange, G., and Dunoyer, G. (1955). A study of the arterial vascularization of the femoral neck in the adult. *J. Bone Jt. Surg., Am. Vol.* **37-A**, 663–680. (A)

Kane, W. J. (1968). The determination of blood flow to tissues of the human leg. *J. Bone Jt. Surg., Am. Vol.* **50-A**, 1070 (abstr.). (B)

King, K. F. (1976). Periosteal pedicle grafting in dogs. *J. Bone Jt. Surg., Br. Vol.* **58B**, 117–121. (A)

Kolodney, A. (1923). The periosteal blood supply and healing of fractures. Experimental study. *J. Bone Jt. Surg.* **21**, 698–711. (A)

Laubie, M., Schmitt, H., and Falg, E. (1977). Dopamine receptors in the femoral vascular bed of the dog as mediators of a vasodilator and sympathoinhibitory effect. *Eur. J. Pharmacol.* **42**, 307–310. (C)

Launder, W. J. Hungerford, D. S., and Jones, L. H. (1981). Hemodynamics of the femoral head. *J. Bone Jt. Surg., Am. Vol.* **63-A**, 442–447. (D)

Lee, W. R. (1964). Appositional bone formation in canine bone: A quantitative microscopic study using tetracycline markers. *J. Anat.* **98**, 665–677. (B)

Lopez-Cuerto, J. A., Bassingthwaite, J. B., and Kelly, P. J. (1980). Anatomy of the microvasculature of the tibial diaphysis of the adult dog. *J. Bone Jt. Surg., Am. Vol.* **62-A**, 1362–1369. (A)

McFarland, P. H., and Frost, H. M. (1961). A possible new cause for aseptic necrosis of the femoral head. *Henry Ford Hosp. Med. Bull.* **9**, 115–122. (D)

McInnis, J. C., Robb, R. A., and Kelly, P. J. (1977). Relationship of bone blood flow, mineral deposition, and endosteal new bone formation in healing canine tibial defects. *Orthoped. Trans.* **2**, 162 (abstr.). (B)

Macpherson, J. N., and Tothill, P. (1978). Bone blood flow and age in the rat. *Clin. Sci. Mol. Med.* **54**, 111–113. (B)

Michelsen, K. (1967). Pressure relationships in the bone marrow vascular bed. *Acta Physiol. Scand.* **71**, 16–29. (D)

Michelsen, K. (1968). Hemodynamics of the bone marrow circulation. *Acta Physiol. Scand.* **73**, 264–280. (C)

Morgan, F. D. (1959). Blood supply of growing rabbit's tibia. *J. Bone Jt. Surg., Br. Vol.* **41B**, 185–203. (A)

Morris, M., and Kelly, P. J. (1980). Use of tracer microspheres to measure bone blood flow in conscious dogs. *Calcif. Tissue Int.* **32**, 69–76. (B)

Nelson, G. E., and Kelly, P. J. (1960). Blood supply of the human tibia. *J. Bone Jt. Surg., Br. Vol.* **42B**, 625–636. (A)

Ogata, K., and Whiteside, L. A. (1979). Barrier to material transfer at the bone-cartilage interface. *Clin. Orthop. Relat. Res.* **145**, 273–276. (A)

Olerud, S., and Danckwardt-Lilliestrom, G. (1968). Fracture healing in compression osteosynthesis in the dog. *J. Bone Jt. Surg., Br. Vol.* **50B**, 844–851. (D)

Pasternak, N. S., Kelly, P. J., and Owen, C. A., Jr. (1966). Estimation of oxygen consumption, and carbon dioxide production and blood flow of bone in growing and mature dogs. *Mayo Clin. Proc.* **41**, 831–835. (B)

Ponseti, I. V. (1956). Legg-Perthes' disease: Observations on pathological changes in two cases. *J. Bone Jt. Surg., Am. Vol.* **38-A**, 739–750. (D)

Rand, J. A., An, Kai Nan, Chao, E. Y., and Kelly, P. J. (1981). A comparison of the effect of open intramedullary nailing and compression-plate fixation on fracture site blood flow and fracture union. *J. Bone Jt. Surg., Am. Vol.* **63-A**, 427–441. (D)

Rhinelander, F. W. (1968). The normal microinculation of diaphyseal cortex and its response to fractures. *J. Bone Jt. Surg., Am. Vol.* **50-A**, 784–800. (A)

Rhinelander, F. W. (1974). Tibial blood supply in relation to fracture healing. *Clin. Orthop. Relat. Res.* **105**, 34–81. (D)

Schoenecker, P. D., Bitz, D. M., and Whiteside, L. A. (1978). The acute effect of position of immobilization on capital femoral epiphyseal blood flow. *J. Bone Jt. Surg., Am. Vol.* **60-A**, 899–904. (D)

Sevitt, S., and Thompson, R. G. (1965). The distribution and anastomoses of arteries supplying the head and neck of the femur. *J. Bone Jt. Surg., Br. Vol.* **47B**, 560–573. (A)

Shim, S. S., and Patterson, E. P. (1967). A direct method of qualitative study of bone blood circulation. *Surg., Gynecol. Obstet.* **125**, 261–268. (C)

Shim, S. S., Copp, D. H., and Patterson, F. P. (1967). An indirect method of bone–blood flow measurement based on the bone clearance of a circulating bone-seeking radioisotope. *J. Bone Jt. Surg., Am. Vol.* **49-A**, 693–702. (B)

Shim, S. S., Mokkhavesa, S., McPherson, G. D., and Schweigel, J. F. (1971). Bone and skeletal blood flow in man measured by a radioisotopic method. *Can. J. Surg.* **14**, 38–41. (B)

Stein, A. H., Morgan, H., and Reynolds, F. C. (1957). Variations in normal bone marrow pressures. *J. Bone Jt. Surg., Am. Vol.* **39-A**, 1129–1140. (A)

Suramo, I., Puranen, J., Heikkinen, E., and Vuorinen, P. (1964). Disturbed patterns of venous drainage of the femoral neck in Perthes' disease. *J. Bone Jt. Surg., Br. Vol.* **56B**, 448–453. (D)

Syftestad, G. T., and Boelkins, J. N. (1980). Effect of hemorrhage on blood flow to marrow and osseous tissue in conscious rabbits. *Am. J. Physiol.* **238**, H360–H364. (B)

Tachdjian, M. O., and Grana, L. (1968). Response of the hip joint to increased intra-articular hydrostatic pressure. *Clin. Orthop. Relat. Res.* **6**, 199–212. (D)

Tothill, P., and MacPherson, J. M. (1980). Limitations of radioactive microspheres as tracers for bone blood flow and extraction ratio studies. *Calcif. Tissue Int.* **31**, 261–265. (B)

Trias, A., and Ferg, A. (1979). Cortical circulation of long bones. *J. Bone Jt. Surg., Am. Vol.* **61-A**, 1052–1059. (A)

Trotman, N. M., and Kelly, W. D. (1963). The effect of sympathectomy on blood flow to bone. *JAMA, J. Am. Med. Assoc.* **183**, 123–124. (C)

Trueta, J. (1957). The normal vascular anatomy of the human femoral head during growth. *J. Bone Jt. Surg., Br. Vol.* **39B**, 358–394. (A)

Trueta, J., and Amato, V. P. (1960). The vascular contribution to osteogenesis. III. Changes in the growth cartilage caused by experimentally induced ischemia. *J. Bone Jt. Surg., Br. Vol.* **42B**, 571–590. (A)

Trueta, J., and Cavadias, A. X. (1955). Vascular changes caused by the Kuntscher type of nailing. An experimental study in the rabbit. *J. Bone Jt. Surg., Br. Vol.* **37B**, 492–505. (D)

Trueta, J., and Cavadias, A. X. (1964). A study of the blood supply of long bones. *Surg., Gynecol. Obstet.* **118**, 485–499. (A)

Vanderhoeft, P. J., Kelly, P. J., and Peterson, L. F. A. (1962). Determination of growth rates in canine bone by means of tetracycline-labeled patterns. *Lab. Invest.* **11**, 714–726. (B)

Van Dyke, D., Anger, H. O., Yano, Y., and Bozzini, C. (1965). Bone blood flow shown with ^{18}F and the position camera. *Am. J. Physiol.* **209**, 65–70. (B)

Weiss, R. A., and Root, W. S. (1959). Innervation of the vessels of the marrow cavity of certain bones. *Am. J. Physiol.* **197**, 1255–1257. (C)

White, N. B., Ter-Pogossian, M. M., and Stein, A. H. (1964). A method to determine the rate of blood flow in long bone and selected soft tissues. *Surg., Gynecol. Obstet.* **119**, 535–540. (B)

Whiteside, L. A., and Lesker, P. A. (1978a). The effects of extraperiosteal and subperiosteal dissection. I. On blood flow in muscle. *J. Bone Jt. Surg., Am. Vol.* **60-A**, 23–26. (A)

Whiteside, L. A., and Lesker, P. A. (1978b). The effects of extraperiosteal and subperiosteal dissection. II. On fracture healing. *J. Bone Jt. Surg., Am. Vol.* **60-A**, 26–30. (A)

Whiteside, L. A., Lesker, P. A., and Simmons, D. J. (1977a). Measurement of regional bone and bone marrow blood flow in the rabbit using the hydrogen washout technique. *Clin. Orthop. Relat. Res.* **122**, 340–346. (B)

Whiteside, L. A., Simmons, D. J., and Lesker, P. A. (1977b). Comparison of regional blood flow in areas with differing osteoblastic activity in the rabbit tibia. *Clin. Orthop. Relat. Res.* **124**, 267–269. (B)

Whiteside, L. A., Ogata, K., Lesker, P., and Reynolds, F. C. (1978). The acute effect of periosteal stripping and medullary reaming on regional bone blood flow. *Clin. Orthop. Relat. Res.* **131**, 266–272. (A, D)

Whiteside, L. A., Lange, D. R., and Fraser, B. (1983). The effects of surgical procedures on blood supply to the femoral head. *J. Bone Jt. Surg. Am. Vol.* **65-A**, 1127–1133. (A)

Whiteside, L. A., Lange, D. R., and Fraser, B. (in press). The sources of periosteal blood supply. A microanatomical study. *J. Bone Jt. Surg.*

Wilkes, C. H., and Visscher, M. B. (1975). Some physiological aspects of bone marrow pressure. *J. Bone Jt. Surg., Am. Vol.* **57-A**, 49–57. (A, D)

Woodhouse, C. F. (1962). Anoxia of the femoral head. *Surgery* **52**, 55–62. (D)

Woodhouse, C. F. (1964). Dynamic influences of vascular occlusion affecting the development of avascular necrosis of the femoral head. *Clin. Orthop. Relat. Res.* **32**, 119–129. (D)

Wootton, R., Reeve, J., and Veall, N. (1976a). Measurement of skeletal blood flow in normal man and in patients with Paget's disease of bone. *Calcif. Tissue Res.* **21**, Suppl., 380–385. (C)

Wootton, R., Reeve, J., and Veall, N. (1976b). The clinical measurement of skeletal blood flow. *Clin. Sci. Mol. Med.* **50**, 261–268. (B)

Wootton, R., Reeve, J., Spellacy, E., and Tellez-Yudilevich, M. (1978). Skeletal blood flow in Paget's disease of bone and its response to calcitonin therapy. *Clin. Sci. Mol. Med.* **54**, 69–74. (C)

Zucman, J. (1960–1961). Studies on the vascular connections between periosteum, bone and muscle. *Br. J. Surg.* **40**, 324–328. (A)

Chapter 21
Hemopoietic System: Spleen and Bone Marrow

A. BLOOD VESSELS AND LYMPHATICS OF THE SPLEEN

By R. S. McCuskey and F. D. Reilly

The mammalian spleen is a lymphoreticular organ that serves as a selective filter interposed in the blood vascular system. The structure of its unique microvasculature functions to filter the blood, thereby removing abnormal or senescent blood cells, particulates, and antigenic materials from the circulation. The spleen also sequesters, stores, and releases blood cells and platelets; its lymphoid tissue participates in immune responses; and it can serve as a hemopoietic organ (Crosby, 1959; Moore *et al.*, 1964; Weiss and Tavassoli, 1970; Weiss, 1977a).

1. ANATOMY OF BLOOD CIRCULATION

Considerable controversy exists in the literature as to whether blood flow through the red pulp of the spleen is contained within channels lined by endothelium ("closed" circulation) or whether it perculates through the reticular meshwork of the red pulp cords ("open" circulation) (Tischendorf, 1969; Weiss, 1977a). In recent years, however, the question about the "intermediate" circulation of the spleen appears to have been resolved through scanning and transmission electron microscopy (Weiss and Tavassoli, 1970; Irino *et al.*, 1977; Weiss, 1973, 1977a; Blue and Weiss, 1981a–d), as well as through high resolution *in vivo* microscopic methods (McCuskey and McCuskey, 1977; McCuskey and Meineke, 1977). As a result, although considerable differences in splenic morphology exist among various species (Snook, 1950; Tischendorf, 1969), a basic pattern of vascular organization is recognizable (Fig. 21.1).

a. Intrasplenic vasculature: Blood enters the spleen at its hilus through branches of the splenic artery, which then course in the connective tissue extensions of the splenic capsule as trabecular arteries. Twigs of the trabecular arteries penetrate the connective tissue to enter the pulp of the spleen as central arteries. Each of the latter is surrounded by a periarterial lymphatic

698

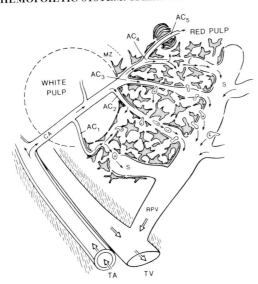

FIG. 21.1. Diagrammatic illustration of the splenic microvasculature. TA, trabecular artery; TV, trabecular vein; CA, central artery; AC_1 and AC_2, "arterial" capillaries terminating in the marginal zone (MZ); AC_3 and AC_4, "arterial" capillaries terminating in the red pulp; AC_5, sheathed "arterial" capillary terminating in the red pulp; S, venous sinus; RPV, red pulp venule. Arrows indicate direction of blood flow. Solid black lines indicate the endothelial lining of vessels.

sheath (PALS), composed principally of T lymphocytes but also containing lymphatic nodules (principally B lymphocytes). Collectively, this accumulation of lymphatic tissue is known as the white pulp. As the central artery courses through the white pulp, it branches and gives rise to arterioles and "arterial" capillaries that course radially through the lymphatic tissue to terminate in the surrounding marginal zone. Some of these "arterial" vessels penetrate the marginal zone and terminate in the red pulp, as do the terminal branches of the main stem of the central artery. The terminal branches, known as penicillar arterioles, may lose their smooth muscle investment and become "arterial" capillaries before terminating in the red pulp. The "arterial" capillaries may bear a sheath (ellipsoid) of reticular cells and macrophages. The extent to which penicillar arterioles and sheathed "arterial" capillaries are developed varies considerably in different species (Snook, 1950; Tischendorf, 1969). In most species, however, the endothelium of these arterial vessels protrudes into the lumen and contains numerous microfilaments suggestive of contractile elements. *In vivo* microscopic studies have demonstrated that the endothelium is very responsive to vasoactive substances and nerve stimulation (McCuskey and Meineke, 1977; Reilly and McCuskey, 1977a,b). Hence, it is possible that the "arterial" capillaries play a key role in regulating blood flow through the spleen.

b. Vasculature of marginal zone and red pulp: The marginal zone borders the PALS and consists of a meshwork of cell processes of reticular cells and associated reticular fibers that form channels through which the blood flows and in which macrophages, lymphocytes, platelets, and some erythrocytes are entrapped. The reticular meshwork blends into the cords of the surrounding red pulp, which differs only in having somewhat larger channels and a higher concentration of entrapped erythrocytes. Where "arterial" capillaries empty into the marginal zone and red pulp, the processes of the reticular cells are in close apposition to the endothelium of these vessels, thereby forming an attenuated adventitial coat. Near the termination of the capillaries, discontinuities develop in their walls that permit blood cells and plasma to exit laterally into the surrounding reticular meshwork. Finally, the endothelium of these vessels is lost, and the blood flows into channels within the marginal zone or red pulp formed solely by the cytoplasmic processes of reticular cells. As a result, blood entering two adjacent channels frequently is separated by only a single cytoplasmic process of a reticular cell that is dimensionally similar to endothelium. When viewed by light microscopy, this relationship often creates the appearance of a vascular channel completely lined by endothelium, especially when the "arterial" capillary terminates in close proximity to a venous sinus (see below) and the continuing channel formed by reticular cell processes provides a direct pathway for blood flow into the sinus. Such a channel appears to act as an arteriovenous (AV) shunt and is functionally significant during conditions when the red pulp is congested with blood cells, e.g., in polycythemia and during active hemopoiesis (McCuskey and Meineke, 1977).

c. Apertures in endothelium of venous sinuses: Blood leaves the reticular meshwork of the red pulp cords by passing through apertures in the endothelial lining of venous sinuses. Such openings are formed as gaps between contiguous endothelial cells at sites where the latter are not effaced on their abluminal surface by a basal lamina or adventitial processes of reticular cells. Most of the apertures are of such size that blood cells are deformed during their passage from the cords into the sinus. As a result, intracellular inclusions in erythrocytes are pinched off during passage and "pitted" from the cells (Weiss and Tavossoli, 1970; Klausner *et al.*, 1975). Rigid erythrocytes, such as spherocytes and sickle cells, frequently are trapped in such sites and are removed from circulation. The transit time through the apertures for leukocytes is slower than for pliable erythrocytes, these cells appearing to traverse the apertures by diapedesis (McCuskey and McCuskey, 1977). A number of the openings, however, are large enough to allow passage of blood cells without being deformed in the process; these usually are associated with the functional AV shunts, described in the previous paragraph. The apertures appear to be dynamic structures, the diameter of which can be modified by vasoactive substances (McCuskey and Meineke, 1977). The structural mechanism for effecting this may reside in the sinus endothelial cells that contain microfilaments suggestive of a contractile mechanism (Chen and Weiss, 1973; DeBruyn and Cho, 1974). Thus, the apertures contained in the sinus walls provide an additional site for regulating intrasplenic blood flow. While some species contain poorly developed sinuses (mouse, cat) in comparison with others (rat, human, rabbit), the basic patterns of circulation are similar. Nevertheless, the functional significance of the wide variation of intrasplenic structure should not be dismissed—especially when interpreting physiologic and pharmacologic data obtained from different species. While a few "arterial" capillaries may be continuous with the venous sinuses or red pulp venules, such an arrangement appears to be rare, and most of the blood flows through the red pulp, as described above.

The venous sinuses drain into venules within the red pulp which, in turn, enter trabecular veins. The latter return the blood to the hilus where it leaves the spleen via the splenic vein.

2. NERVOUS INNERVATION OF BLOOD VESSELS

Innervation of the spleen appears to be relatively sparse. Whereas adrenergic (sympathetic) nerves have been demonstrated in the spleens of many mammals, there is limited evidence for cholinergic (parasympathetic) or sensory innervation (Utterback, 1944; Dahlström and Zetterström, 1965; Tranzer and Thoenen, 1967; Fillenz, 1970; Reilly et al., 1976, 1979; Heusermann and Stutte, 1977; Kudoh et al., 1979; Blue and Weiss, 1981b,d).

There is considerable species variation in the distribution of adrenergic nerves. They are found to innervate smooth muscle of arteries and arterioles of the human and murine spleen, and in the rat, cat, and dog, they are also distributed to the smooth muscle of the capsule and trabeculae. In addition, there is evidence of adrenergic axons innervating reticular cells of the white pulp of the mouse (Reilly et al., 1979) and red pulp of the canine spleen (Blue and Weiss, 1981d). Moreover, some of these axons have been reported to be contiguous with blood cells in the white pulp of the mouse (Reilly et al., 1979) and the red pulp of the dog (Zetterström et al., 1973). However, a functional relationship between such nerves and the reticulum or blood elements has yet to be established. Since it has been suggested that the reticular cells of the spleen are contractile (Saito, 1977; Blue and Weiss, 1981d), it is possible that they participate, along with the smooth muscle in the capsule and trabeculae, in producing the large decrease in splenic volume observed following adrenergic stimulation (Davies and Withrington, 1973).

3. PHYSIOLOGY OF BLOOD CIRCULATION

a. Regulation of blood flow: Splenic blood flow is principally controlled by splenic arterial resistance. In some species, but not man, smooth muscle in the capsule and trabeculae contributes to the regulation of flow. Dynamic, steady-state adjustments in muscular tone affect the rate(s) of filtration, synthesis, storage, and release of blood elements in the red and white pulps of the spleen. Vasoconstrictor mechanisms, primarily of an α-adrenergic origin, predominate and exert relatively strong influences on governing blood flow to and from, as well as within and between, the red and white pulps. In contrast, vasodilator mechanisms in the spleen are poorly developed in most mammalian species.

   b. Variability of splenic blood flow: In vivo microscopic studies of the spleen reveal that the patterns of blood flow in the red pulp are complex and variable from moment to moment. For example, blood cells may flow rapidly through the channels in the red pulp with minimal or no distortion. A few seconds later, the flow may be reduced to only a few highly-distorted erythrocytes slowly pursuing a tortuous pathway through the red pulp, which may have many of its channels congested with blood cells. During such periods of congestion, the main flow usually continues in the direct pathways (AV shunts), with most of the blood bypassing the more tortuous channels in the red pulp. Under all conditions, there are a number of formed elements of the blood trapped in the reticular meshwork, adherent to macrophages and very resistant to dislodgment by blood flowing past them. Such observations support the existence of a three compartment system predicted by the erythrocyte washout studies of Groom (1980) and Cilento *et al.* (1980).

4. Pharmacology of Blood Circulation

   Epinephrine, norepinephrine, neurostimulation, isoproterenol, vasopressin, angiotensin II, and bradykinin all produce vasoconstriction and contraction of the splenic capsule in dogs, cats, mice, guinea pigs, rabbits, monkeys, and man (Ayers *et al.*, 1972; Davies and Withrington, 1973; Reilly and McCuskey, 1977a,b). These responses generally are dosage- or frequency-dependent, and in the larger mammals, they have been shown to decrease splenic arterial inflow and volume; concomitantly, splenic arterial and venous pressures increase, as do venous outflow and hematocrit (Cilento *et al.*, 1980; Groom, 1980). In man, capsular contractions are weak or nonexistent (Ayers *et al.*, 1972). The degree of contraction is more on the order of that seen following pharmacologic or neural stimulation of the spleen in mice and rats. These species differences are attributed to the sparse distribution of smooth muscle and innervation in the trabeculae and capsule (Reilly *et al.*, 1976; Heusermann and Stutte, 1977; Kudoh *et al.*, 1979).

   In contrast to the findings in rats, guinea pigs, rabbits, monkeys, and man, low concentrations of epinephrine or norepinephrine in the dog and cat or of isoproterenol in the dog, cat, and mouse produce vasodilatation and an increase in splenic volume, venous pressure, and blood flow by activating β-adrenoceptors. At these concentrations, the capsule contracts, this structure being more sensitive than the vasculature to such chemicals (Davies and Withrington, 1973; Reilly and McCuskey, 1977a). Bradykinin or acetylcholine in the dog (Davies and Withrington, 1973) and histamine or adenosine in the mouse (Reilly and McCuskey, 1977a) also provoke vasodilatation and increase splenic blood flow. However, these responses are not modified by β-adrenoceptor blockers. The magnitude of the responses elicited by such chemicals appears to be directly proportional to vascular and capsular tone, which, in turn, is dependent on the degree of basal sympathetic innervation.

An interaction of acetylcholine, histamine, serotonin, bradykinin, lactic acid, or prostaglandins with adrenergic mechanisms has been reported in several species (Davies and Withrington, 1973; Reilly and McCuskey, 1977a,b; Malik, 1978; Einzia et al., 1980; Kondo et al., 1980; Peskar et al., 1980). Acetylcholine, histamine, serotonin, lactic acid, and bradykinin excite, or release neurotransmitter from, sympathetic nerve endings. These effects on stored catecholamine result in α-mediated vasoconstriction and capsular contraction. Prostaglandins A_1, D_2; high doses of E_1, E_2, I_2, and $F_{2\alpha}$; or thromboxane B_2 amplifies α-mediated contractions of capsular and/or vascular smooth muscle to nerve stimulation or injections of epinephrine or norepinephrine. However, prostaglandins A_2, $F_{1\alpha}$, G_2 and lower doses of E_1, E_2, $F_{2\alpha}$, or I_2 have the opposite effect. Based on these findings, it has been hypothesized that prostaglandins modulate adrenergic vasoconstrictor responses by interacting directly with splenic nerves and/or smooth muscle.

In vivo microscopic studies of transilluminated spleens by MacKenzie et al. (1941), Fleming and Parpart (1958), and Reilly and McCuskey (1977a,b) indicate that arteries, arterioles, and "arterial" capillaries of the splenic red and white pulps of mice, rats, guinea pigs, and rabbits are the vascular segments responsive to pharmacologic or neural stimulation. Although the venules and veins do not contract or dilate following the use of such agents, some vasoactive substances appear to influence the permeability of the venous sinuses to blood cells (McCuskey and Meineke, 1977).

5. Lymphatic System of the Spleen

The white pulp of the spleen contains lymphatic vessels (Janout and Weiss, 1972). However, there is considerable species variability in their extent; for example, they are more prevalent in man and hedgehogs than in rats and mice. Lymphatic capillaries originate in the lymphatic tissue associated with the PALS and drain into larger lymphatics that course in the trabeculae and exit the spleen at the hilus. While lymphatics also are found in the capsule, none has been demonstrated in the red pulp.

6. FUTURE AVENUES FOR INVESTIGATION

The principal limitation to the various vascular studies of the spleen has been the technical difficulties encountered and the few studies performed in this field. As a result, caution must be applied to the interpretation of pharmacodynamic and neurophysiologic data on blood flow and other parameters. Different experimental protocols impose limitations on the nature and validity of the information derived from them. For example, binding of drugs in the bloodstream or to vascular receptors peripheral to the spleen must be considered as sources of error when interpreting experimentally derived data, since the results may not reflect the potential activity of substances when they are released *in situ*. Conflicting results also may be related to differences in the experimental techniques used or the different concentrations of drugs or vasoactive agents at the effector site(s) within the spleen. Investigations are required to ascertain whether compounds are released in sufficient quantities to cause changes in blood flow during conditions of optimal circulation or of low-flow states provoked by hemorrhage, arterial occlusion, shock, or other disease processes. Up to now only circulating levels of catecholamines, vasopressin, and angiotensin II in dogs and cats have been demonstrated to accumulate in quantities sufficient to modify splenic blood flow during pathophysiologic conditions, such as hemorrhagic, cardiogenic, or traumatic shock (Zetterström *et al.*, 1964; Greenway and Stark, 1970; Stark *et al.*, 1971; Beckman *et al.*, 1981). It also must be remembered that discrepancies in responses to pharmacologic and neural stimulation may reflect real species variations in structure and function.

Even more ambiguous are the factors controlling the regulation of blood flow through the splenic microvasculature. Most information on microvascular regulation has been extrapolated from investigations of humoral effects on innervated vessels larger than 300 μm or from measurements of flow, resistance, and clearance following arterial or venous injections of vasoactive substances or tracers. These studies fail to elucidate directly the sensitivity of the microvascular system to such substances *in situ*; the discrete site(s) of chemical interaction within the "resistance" vessels of the spleen; or the role of hormones, the autonomic nervous system, or other stimuli in primary (direct) and secondary (indirect) microvascular responses. Indirect effects may be caused by such chemicals modifying the responses of vascular and capsular smooth muscle to vasoactive agents or producing the release of stores of neurotransmitter(s) from postganglionic nerves. Since microvascular innervation appears to be sparse, a strong theoretical case can be made for the role of humoral agents other than norepinephrine in the local regulatory mechanisms that maintain homeokinesis. Yet to be elucidated is the role of resident cells in the spleen in secreting vasoactive substances that influence intrasplenic blood flow. Macrophages, for example, produce prostaglandins; and hemopoietic elements also may release vasoactive substances (McCuskey and Meineke, 1977).

B. BLOOD CIRCULATION OF BONE MARROW

By S. S. ADLER

Galen believed that bone marrow is the nutritive tissue of bones. By 1868, however, its hemopoietic function had been realized. Serious study of structure of bone marrow dates back to the last quarter of the nineteenth century, at which time only occasional investigators viewed the marrow vascular system as a closed one; now, this concept is universally accepted. During the late 1950's and early 1960's, the circulatory pathway of blood in bone marrow was defined, and, since then, studies have concentrated on the ultrastructural, pathophysiologic, and functional aspects of the marrow vasculature. Most of them have been performed on specimens taken from rodents or lagomorphs because of the practical problems encountered in using bones from larger mammals.

1. GENERAL DESCRIPTION OF MARROW VASCULATURE

a. Pattern of blood flow: This subject has been investigated primarily on the marrow of long bones. As depicted in Fig. 21.2, the main afferent vessel, the nutrient artery, enters about midshaft through the nutrient foramen and divides into ascending and descending limbs. Each of the latter gives off radial branches that extend toward the periphery as arterioles and capillaries, penetrating the endosteum and anastomosing with capillaries in the cortical canalicular system. The cortical capillary system also is comprised of vessels derived from interfascicular vessels from muscles and of ramifications of periosteal twigs. Efferent vessels from the cortical bone capillary system feed into the complex of marrow sinuses (Figs. 21.3 and 21.4), the latter draining into a large longitudinal central sinus (Figs. 21.2–21.4). A small amount of blood may enter marrow sinuses directly from the radial arterioles without first passing through cortical capillaries (DeBruyn *et al.*, 1970). The central venous sinus joins the emissary or comitant vein, which exits through the nutrient foramen and merges with the systemic venous circulation.

It has been estimated that the velocities of blood flow in marrow arterioles, capillaries, sinuses, and venules are about 1–1.5, 0.5, 0.2, and 0.1–0.3 mm/sec, respectively (Brånemark, 1961).

b. General anatomy of marrow vessels: Although the structure of medullary arteries, arterioles, capillaries, and the emissary vein is similar to that of such vessels in other parts of the body, the marrow also contains specialized

Blood Vessels and Lymphatics in Organ Systems
Copyright © 1984 by Academic Press, Inc.

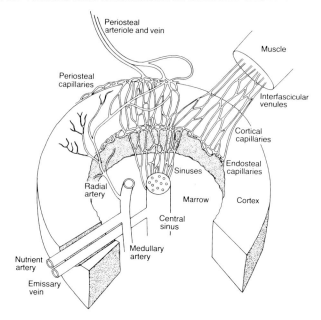

Fig. 21.2. Schematic drawing of circulation of marrow in long bones. The medullary arterial system joins the cortical bone vascular system which then merges with the marrow sinusoidal system. (From Lichtman, 1981; reproduced with permission from Plenum Publishing Co.)

thin-walled radial artery twigs (Tavassoli, 1974) that branch at right angles, rather than at acute ones, and are more numerous at the epiphyseal ends than elsewhere. Whether these vessels serve to regulate intramedullary pressure or whether they play a role in regulation of blood cell egress from the marrow, as suggested by some investigators, is purely a matter of speculation.

The central venous vessel of the marrow is a sinus consisting of endothelial cells with scattered flocculent ground substance on its abluminal surface. The highly branched network of endothelial-lined channels of the marrow sinus system drains into this vessel (Figs. 21.3 and 21.4). Parasinal fibroblastic cells may have their cell bodies or cytoplasmic extensions closely applied to the abluminal surface of the sinus endothelial cells (Figs. 21.4 and 21.6). Hemopoietic tissue is located extravascularly between the sinuses (Figs. 21.4 and 21.5). The marrow, like other organs with extensive sinus systems, has no lymphatics.

   c. *Function of marrow vasculature—an overview:* Because of the interconnections between the marrow radial arteries and cortical bone capillaries, it has been postulated that, aside from serving as a simple conduit for systemic blood, the marrow vasculature provides for one or more of the following: (1) transpor-

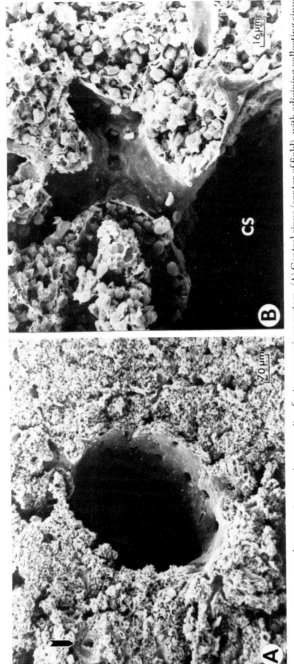

FIG. 21.3. Scanning electron microscopic composite of rat marrow sinus system. (A) Central sinus (center of field), with adjoining collecting sinus (at 12:30 o'clock) and many collecting sinus orifices. Note medullary artery at arrow. (B) Collecting sinus (CS) shown in A, at higher magnification. Note extrasinal hemopoietic elements. (From Lichtman et al., 1978; reproduced with permission from Plenum Publishing Co.)

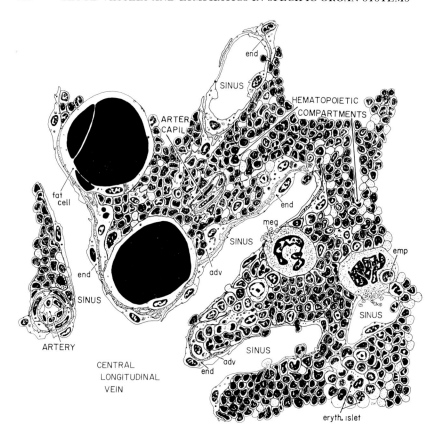

FIG. 21.4. Schematic drawing of marrow structure in cross section near central sinus [central longitudinal vein (CLV), term used by some investigators]. Shown are sinusoids entering the central sinus and blood cells emigrating from the intersinal marrow space into sinus lumens. Sinus walls are composed of endothelial cells (end) and some have parasinal, adventitial cell bodies (adv) or processes applied to their abluminal surfaces. Megakaryocytes (meg) shed platelets from processes which they extend into sinusoidal lumens. Cells wandering through megakaryocytes are depicted (emperiopolesis; emp). (From Weiss, 1977b; reproduced with permission from McGraw-Hill Book Co.)

tation of bone or endosteal-related nutrients and hemopoietic factors, (2) adjustment of pH and/or P_{O_2} for optimal hemopoiesis, and (3) support and anchorage of intramedullary contents. Since these possibilities have not been critically evaluated, they do not warrant further discussion. The marrow vasculature plays a definite role in the regulation of cell egress and some of the cells which comprise the marrow vascular system may also be involved in hemopoiesis; these functions will be considered. However, a full discussion of hemopoiesis is beyond the purview of this section; thus, hemopoietic functions

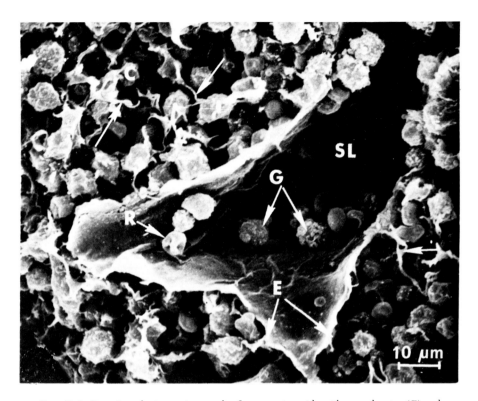

FIG. 21.5. Scanning electron micrograph of mouse sinusoid, with granulocytes (G) and re-ticulocytes (R) entering the lumen (SL) through endothelial cell (E) wall. Note fine processes of parasinal cells in hemopoietic cords (arrows). The large number of cells seen in transit is the result of erythropoietin stimulation. (From Chamberlain *et al.*, 1975; reproduced with permission from Springer-Verlag, New York.)

will only be mentioned if the cells which are involved in this process are constituents of the marrow vasculature.

The cells lining the medullary sinus are, at once, a barrier to the movement of immature blood cells into the circulation and a passageway for mature cells. By means of their pinocytotic apparatus, they may function as part of the body's "reticuloendothelial system," removing substances from circulating blood (see Section B-3c, below).

2. EMBRYOLOGIC AND OTHER DEVELOPMENTAL CONSIDERATIONS

Ontologic, embryologic, and marrow regeneration studies have docu-mented that a close relationship exists between the formation of marrow vasculature and the development of hemopoietic function, a critical amount of vasculature being a prerequisite for the development of hemopoietic marrow.

In this regard, three species of amphibians were studied by Tanaka (1976). In one, a primitive urodele, neither a marrow sinus system nor marrow hemopoiesis was found. In another, a primitive anuran, a subendosteal venous vascular system was detected, but no central vein was seen; in this species, some marrow hemopoiesis was observed. In the third type, an advanced anuran, (*Rana catesbiana*), both the marrow vascular system and the hemopoietic compartment were found to be well-developed.

In the developing human fetus, periosteal-derived vascular buds grow in toward the future marrow cavities of long bones. Together with actively growing capillaries in the vascular bud, osteoclasts and chondroclasts are present, and these break the periosteal bone down, permitting further invasion of the vascular bud.

In developing vertebral cartilage, blood vessels penetrate from dorsal and ventral sides, with the result that the future bone marrow space becomes occupied by blood vessels situated among scattered mesenchymal and mononuclear cells. A large central cavity, formed by the action of chondroclasts, is occupied by a dilated sinus. As the marrow cavity expands, the initially dilated round central sinus develops an irregular contour and receives tributaries from the periphery of the marrow (Chen and Weiss, 1975).

In both long bone and vertebral marrow, hemopoiesis is not found until a fairly well-developed vascular system is in place. Similarly, hemopoietic marrow regeneration, such as is noted in bone marrow after high-dose irradiation (Patt and Maloney, 1970) or after subcutaneous implantation of femurs in mice (Tavassoli and Weiss, 1971), does not occur until there is at least partial restoration of marrow sinusoidal structures.

3. MARROW SINUS SYSTEM

a. Structure: The medullary sinusoidal system is a highly branched interconnected network of endothelial cell-lined polygonal channels that traverses the hemopoietic compartment and connects with the central venous sinus (Figs. 21.3 and 21.4). The endothelial cells of this structure are broad, flat, and pavement-type, forming an uninterrupted layer (Figs. 21.4 and 21.6). Their cytoplasm may just touch, overlap, or interdigitate. Most investigators believe that junctional structures, e.g., zonulae adherens, may exist between endothelial cells (Weiss, 1977b; DeBruyn, 1981; Lichtman, 1981); however, recently, based on studies using freeze-fracture techniques and colloidal lanthanum impregnation, the existence of such structures has been questioned (Tavassoli and Shaklai, 1979). The marrow sinuses vary in size with the species: in mice, they may be 5–30 μm (Lichtman, 1981), but in humans, they are probably somewhat larger.

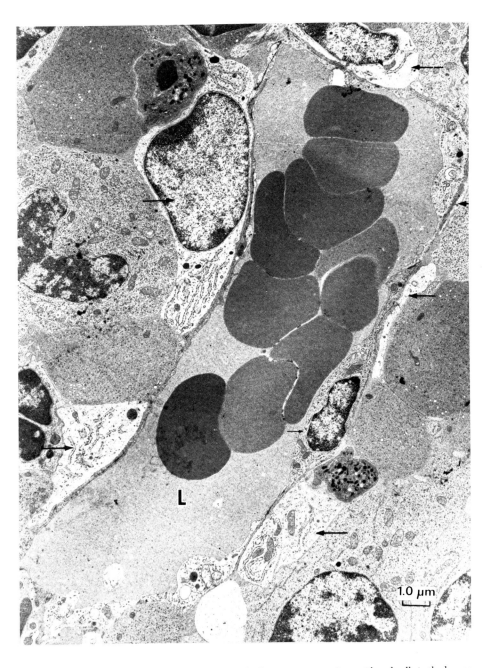

FIG. 21.6. Transmission electron micrograph of mouse marrow sinus with red cells in the lumen (L). Parasinal cell bodies and processes (both depicted by large arrows) are applied to portions of the sinus wall. Note endothelial cell body (small arrow). (From Lichtman, 1981; reproduced with permission from *Experimental Hematology*.)

The endothelial cell lining the sinus contains a nucleus, mitochondria, endoplasmic reticulum, ribosomes, centrioles, microtubules, microfilaments, a Golgi apparatus, lysosomes, and bristle-coated pits and vesicles (DeBruyn, 1981). The nuclei usually bulge into the sinus lumens, but they may also protrude into the extravascular space. Both sialic acid residues and a neuraminidase-resistant, anionic substance, with a pK_a higher than that of sialic acid (perhaps carboxyl-containing moieties), are detectable on the membranes of the endothelial cells. However, sialic acid is not found at sites of specialized membrane structures (e.g., bristle-coated pits and diaphragmed fenestrae) (DeBruyn, 1981). Portions of the cell lining the sinus may be markedly attenuated, forming structures known as fenestrated diaphragms that appear as small beads of cytoplasm separated from each other by plasma membrane. Marrow sinus endothelial cells have no true basal lamina except where they directly abut against arterioles; ground substance and the cell bodies and processes of parasinal cells cover portions of the abluminal side of the sinus (Figs. 21.4 and 21.6).

b. Cell egress through blood marrow-blood barrier (Figs. 21.4–21.7): Mature blood cells enter the sinusoidal lumen via transient migration pores (DeBruyn, 1981; Lichtman, 1981), which usually form close to, but not at, cell junctions (Lichtman, 1981). Cells do not emigrate from the marrow through cell junctions. Pore formation occurs by fusion of the luminal and abluminal portions of endothelial cell membranes; frequently, early during migration, diaphragmed fenestrae are found close to the sites of migration pores (DeBruyn, 1981). Whether hemopoietic cell–endothelial cell interaction creates the diaphragmed fenestrae or whether the latter are formed first and hemopoietic cells are then attracted to those sites (perhaps by chemoattractants entering through the fenestrae from the circulation) is not known. Increased numbers of diaphragmed fenestrae are not regularly seen at parajunctional sites when blood cells are not in the process of emigration. Cells fit extremely tightly in migration pores, apparently not even permitting leakage of plasma (DeBruyn, 1981). Circumferential and radial filaments, found by some investigators near migration pores, have raised the question as to whether these structures play some role in the formation or closure of the pores (Lichtman, 1981). In advance of the migrating cell is a dense accumulation of an undefined anionic substance, which is rapidly dissipated after the major portion of the cells is delivered through the migration pore (DeBruyn, 1981). The role this substance plays in cell migration is unknown.

Although white and red blood cells enter the sinus lumen, megakaryocytes, as a rule, do not. The latter abut against and extend tentacle-like proplatelet processes through pores; teardrop-shaped platelets then bud off from the processes (Figs. 21.4 and 21.7) (DeBruyn, 1981).

Unanswered is the question of what causes selective release of mature blood

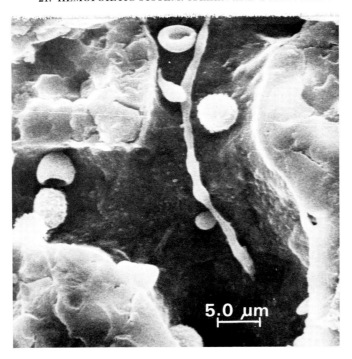

Fig. 21.7. Scanning electron micrograph of proplatelet process in sinus lumen. Teardrop-shaped platelet buds are shown. (From De Bruyn, 1981; reproduced with permission from Grune & Stratton.)

elements. Cell deformability and other mechanical explanations cannot alone account for such selectivity. There is, as yet, no evidence to suggest that cell passage is receptor mediated.

c. *Transport system of sinus endothelial cells:* Transport of plasma-contained substances through marrow endothelial cells, from luminal to abluminal surfaces, or vice versa, may occur via diaphragmed fenestrae. Although electron micrographs fail to reveal any pores in these fenestrae, it is clear that even carbon particles can migrate through them (DeBruyn, 1981). How this takes place remains a mystery.

Bristle-coated pits are found on the surface of sinus epithelial cells, and ferritin, horseradish peroxidase, and carbon and latex particles have been found to be incorporated into bristle-coated endocytotic vesicles forming from these pits. The bristles are made of clathrin (a protein with a molecular weight of about 180,000), found on endothelial cells of diverse origins. In other systems, this protein has been associated with receptor-mediated pinocytosis, recycling of cell membranes, and stabilization of vesicles. Both large (mean diameter,

1400 Å; range, 1000–2100 Å) and small (mean diameter, 400 Å; range, 300–560 Å) bristle-coated vesicles are found in sinus endothelial cells (DeBruyn, 1981).

The large bristle-coated vesicles clearly form from surface pits after pinocytoses. They may remain intact or fuse with other similar vesicles, with loss of the clathrin coat and the formation of a phagosome still containing the unaltered ingested matter. Either of these fates is possible for vesicles containing a particulate such as carbon. On the other hand, large bristle-coated vesicles may fuse with another type of cell organelle, the transfer tubule. The latter is a smooth-walled organelle (240–500 Å) that may contain acid phosphatase; it is similar to the expanded lateral portion of the Golgi apparatus from which it may be derived (DeBruyn, 1981). After transfer tubules fuse with bristle-coated vesicles, they may go on to fuse with dense bodies, resulting in formation of secondary lysosomes. This scenario is probably most common for vesicles containing proteinaceous materials. There is no evidence to suggest that transcellular transport is mediated by large bristle-coated vesicles; rather it may be that the latter subserve the "reticuloendothelial" function of scavenging substances from the circulation.

The small bristle-coated vesicles do not seem to originate as bristle-coated pits on the cell surface (DeBruyn, 1981); instead they may be part of the Golgi–lysosomal system, rather than part of a scavenging system. One could speculate that these vesicles transport such substances as complex polysaccharides from the Golgi system to other areas of the cell. Unlike capillary endothelial cells, sinus endothelial cells do not contain smooth micropinocytotic vesicles.

4. MARROW PARASINAL (ADVENTITIAL OR RETICULUM) CELL

a. Morphology: The parasinal cell is a fibroblast or fibroblastoid cell with long cytoplasmic processes protruding into the hemopoietic tissue for which it provides structural support (Figs. 21.3 and 21.5). Parasinal cells contain the regular complement of cellular organelles, and at least a portion of them seem to possess alkaline phosphatase activity (Westen and Bainton, 1979). In the unperturbed state, about 60% of the abluminal surface of the sinus lining is covered by cell bodies or processes of parasinal cells; apparently two-thirds of this cover may be withdrawn in response to endotoxin, erythropoietin, hemorrhage, and possibly other stimuli (Chamberlain *et al.*, 1975).

b. Function: Contrary to statements appearing in the older literature, the marrow parasinal or reticulum cell is not and cannot become a hemopoietic stem cell. There is convincing cytogenetic evidence that indicates that marrow

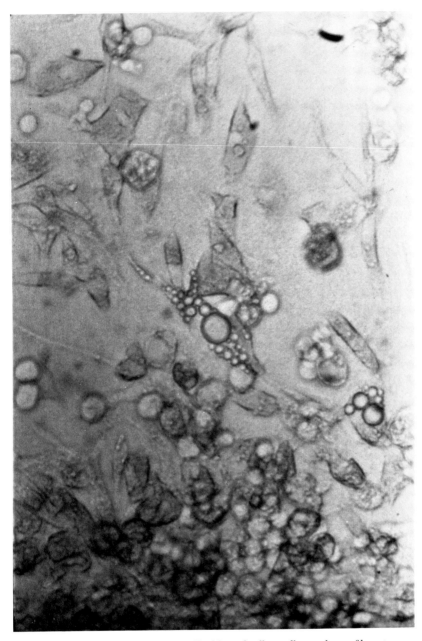

FIG. 21.8. Photomicrograph of marrow fibroblastoid cells in adherent layer of long-term mouse marrow culture containing hydrocortisone. Fat globules are present in one of the fibroblast-like cells. Marrow fat-containing cells develop fat globules under the influence of hydrocortisone, unlike adipose cells from other organs. [Original magnification × 10; final magnification approximately 214 (length) × 150 (width).]

fibroblasts and hemopoietic elements are derived from different precursor cells (Golde *et al.*, 1980; Friedenstein *et al.*, 1978). Parasinal cells provide support for developing hemopoietic cells and, perhaps, also for the sinus endothelial cells. They may help quantitatively to regulate hemopoietic cell egress into the sinusoidal lumen, for, although hemopoietic cells migrate through sinus endothelial cells, as a rule, they do not migrate through parasinal cells. Therefore, areas under parasinal cell cover are not conducive to cell migration.

The parasinal cell probably is the primary fibroblast of the marrow. Tissue culture studies have suggested that at least some marrow fibroblasts are capable of taking up fat; hence, these cells may be the adipose cells of the marrow as well (Fig. 21.8).

Marrow fibroblasts may be capable of inducing formation of hemopoietic support tissue even in extramedullary sites (Friedenstein *et al.*, 1978) and probably are of primary importance during regeneration of marrow after severe injury. Recently it has been suggested that at least some marrow fibroblasts (perhaps the parasinal cells) are involved in support or regulation of granulopoiesis (Westen and Bainton, 1979), much as macrophages are in erythropoiesis.

Although parasinal cells may not be part of the marrow vascular system *sensu strictu*, most investigators do consider them to be so, for the following reasons: they are closely associated with the endothelial cells; endothelial cell function as a gateway for hemopoietic elements is shared with them; and they seem to buttress the meager lining of the vascular sinus.

5. Marrow Vasculature in Pathologic States

Although marrow tissue is not usually used to diagnose systemic vascular diseases, marrow vessels may be involved in those that affect blood vessels, such as thrombotic thrombocytopenic purpura. Rarely, findings diagnostic of giant cell arteritis have been found in marrow biopsy specimens (Enos *et al.*, 1981). In agnogenic myeloid metaplasia (myelofibrosis with myeloid metaplasia), there is an increase in the number of intramedullary capillaries (up to 4 times normal) and of sinusoids (1.5–1.7 times normal); this may be related to activity of new bone formation (Lennert *et al.*, 1975). In a small study on aplastic anemia, no ultrastructural abnormality of marrow sinusoidal structure was detected (Samsom *et al.*, 1972). Marrow necrosis is a well-documented entity and may occur in patients with sickle cell anemia, leukemia, and other myeloproliferative disorders, infections, and, occasionally, in other disorders (Conrad and Carpenter, 1979). Experimentally, marrow necrosis can be induced by occlusion of vessels feeding the marrow or by injection of endotoxin; the pathophysiology of marrow necrosis in the above disease states is not clear but may be related to occlusion of or damage to small blood vessels (Conrad and Carpenter, 1979).

REFERENCES*

Ayers, A. B., Davies, B. N., and Withrington, P. G. (1972). Responses of the isolated, perfused human spleen to sympathetic nerve stimulation, catecholamines and polypeptides. *Br. J. Pharmacol.* **44**, 17–30. (A)

Beckman, C. B., Ziazi, Z., Dietzman, R. H., and Lillehei, R. C. (1981). Protective effects of epinephrine tolerance in experimental cardiogenic shock. *Circ. Shock* **8**, 137–149. (A)

Blue, J., and Weiss, L. (1981a). Periarterial macrophage sheaths (ellipsoids) in cat spleen. An electron microscopic study. *Am. J. Anat.* **161**, 115–134. (A)

Blue, J., and Weiss, L. (1981b). Vascular pathways in nonsinusal red pulp. An electron microscopic study of the cat spleen. *Am. J. Anat.* **161**, 135–168. (A)

Blue, J., and Weiss, L. (1981c). Species variation in the structure and function of the marginal zone. An electron microscopic study of the cat spleen. *Am. J. Anat.* **161**, 169–187. (A)

Blue, J., and Weiss, L. (1981d). Electron microscopy of the red pulp of the dog spleen including vascular arrangements, periarterial macrophage sheaths (ellipsoids) and the contractile, innervated reticular meshwork. *Am. J. Anat.* **161**, 189–218. (A)

Brånemark, P.-I. (1961). Experimental investigation of microcirculation in bone marrow. *Angiology* **12**, 293. (B)

Chamberlain, J. K., Leblond, P. F., and Weed, R. I. (1975). Reduction of adventitial cell cover: An early direct effect of erythropoietin of bone marrow ultrastructure. *Blood Cells* **1**, 655–674. (B)

Chen, L.-T., and Weiss, L. (1973). The role of the sinus wall in the passage of erythrocytes through the spleen. *Blood* **41**, 529–537. (A)

Chen, L.-T., and Weiss, L. (1975). The development of vertebral bone marrow of human fetuses. *Blood* **46**, 389–407. (B)

Cilento, E. V., McCuskey, R. S., Reilly, F. D., and Meineke, H. A. (1980). Compartmental analysis of circulation of erythrocytes through the rat spleen. *Am. J. Physiol.* **239**, H272–H277. (A)

Conrad, M. E., and Carpenter, J. T. (1979). Bone marrow necrosis. *Am. J. Hematol.* **7**, 181–189. (B)

Crosby, W. H. (1959). Normal functions of the spleen relative to red blood cells: A review. *Blood* **14**, 399–408. (A)

Dahlström, A. B., and Zetterström, B. E. M. (1965). Noradrenaline stores in nerve terminals of the spleen: Changes during hemorrhagic shock. *Science* **147**, 1583–1585. (A)

Davies, B. N., and Withrington, P. G. (1973). The actions of drugs on the smooth muscle of the capsule and blood vessels of the spleen. *Pharmacol. Rev.* **25**, 373–412. (A)

De Bruyn, P. P. H. (1981). Structural substrates of bone marrow function. *Semin. Hematol.* **18**, 179–193. (B)

De Bruyn, P. P. H., and Cho, Y. (1974). Contractile structures in endothelial cells of splenic sinusoids. *J. Ultrastruct. Res.* **49**, 24–33. (A)

De Bruyn, P. P. H., Breen, P. C., and Thomas, T. B. (1970). The microcirculation of the bone marrow. *Anat. Rec.* **168**, 55–68. (B)

Einzia, S., Rao, G. H., and White, J. G. (1980). Differential sensitivity of regional vascular beds in the dog to low-dose prostacyclin infusion. *Can. J. Physiol. Pharmacol.* **58**, 940–946. (A)

Enos, W. F., Pierre, R. V., and Rosenblatt, J. E. (1981). Giant cell arteritis detected by bone marrow biopsy. *Mayo Clin. Proc.* **56**, 381–383. (B)

---

*In the reference list, the capital letter in parentheses at the end of each reference indicates the section in which it is cited.

Fillenz, M. (1970). The innervation of the cat spleen. *Proc. R. Soc. London, Ser. B.* **174**, 459–468. (A)

Fleming, W. W., and Parpart, A. K. (1958). Effects of topically applied epinephrine, norepinephrine, acetylcholine, and histamine on the intermediate circulation of the mouse spleen. *Angiology* **9**, 294–302. (A)

Friedenstein, A. J., Ivanov-Smolenski, A. A., Chajlakjan, R. K., Gorskaya, U. F., Kuralesova, A. I., Latzinik, N. W., and Gerasimow, U. W. (1978). Origin of bone marrow stromal mechanocytes in radio-chimeras and heterotopic transplants. *Exp. Hematol.* **6**, 440–444. (B)

Golde, D. W., Hocking, W. G., Quan, S. G., Sparkes, R. S., and Gale, R. P. (1980). Origin of human bone marrow fibroblasts. *Br. J. Haematol.* **44**, 183–187. (B)

Greenway, C. V., and Stark, R. D. (1970). The vascular response of the spleen to intravenous infusions of catecholamines, angiotensin and vasopressin in the anesthetized cat. *Br. J. Pharmacol.* **38**, 583–592. (A)

Groom, A. C. (1980). Microvascular transit of normal, immature, and altered red blood cells in spleen versus skeletal muscle. In "Erythrocyte Mechanics and Blood Flow" (G. R. Cokelet, H. J. Meiselman, and D. E. Brooks, eds.), pp. 229–259. Alan R. Liss, Inc., New York. (A)

Heusermann, U., and Stutte, H. J. (1977). Electron microscopic studies of the innervation of the human spleen. *Cell Tissue Res.* **184**, 225–236. (A)

Irino, S., Murakami, T., and Fujita, T. (1977). Open circulation in the human spleen. Dissection scanning electron microscopy of conductive-stained tissue and observation of resin vascular casts. *Arch. Histol. Jpn.* **40**, 297–304. (A)

Janout, V., and Weiss, L. (1972). Deep splenic lymphatics in the marmot: An electron microscopic study. *Anat. Rec.* **172**, 197–209. (A)

Klausner, M. A., Hirsch, L. H., Leblond, P. F., Chamberlain, J. K., Klemperer, M. R., and Segal, G. B. (1975). Contrasting splenic mechanisms in the blood clearance of red blood cells and colloidal particles. *Blood* **46**, 965–976. (A)

Kondo, K., Okuno, T., Suzuki, H., and Saruta, T. (1980). The effects of prostaglandins E_2 and I_2, and arachidonic acid on vascular reactivity to norepinephrine in isolated rat mesenteric artery, hindlimb, and splenic artery. *Prostaglandins Med.* **4**, 21–30. (A)

Kudoh, G., Hoshi, K., and Murakami, T. (1979). Fluorescence microscopic and enzyme histochemical studies of the innervation of the human spleen. *Arch. Histol. Jpn.* **42**, 169–180. (A)

Lennert, K., Nagai, K., and Schwarze, E. W. (1975). Patho-anatomical features of the bone marrow. *Clin. Haematol.* **4**, 331–351. (B)

Lichtman, M. A. (1981). The ultrastructure of the hemopoietic environment of the marrow: A review. *Exp. Hematol.* **9**, 391–410. (B)

Lichtman, M. A., Chamberlain, J. K., and Santillo, P. A. (1978). Factors thought to contribute to the regulation of egress of cells from marrow. In "Year of Hematology" (R. Silber, J. Le Bue, and A. S. Gordon, eds.), pp. 243–279. Plenum, New York. (B)

McCuskey, R. S., and McCuskey, P. A. (1977). *In vivo* microscopy of the spleen. *Bibl. Anat.* **16**, 121–125. (A)

McCuskey, R. S., and Meineke, H. A. (1977). Erythropoietin and the hemopoietic microenvironment. In "Kidney Hormones" (J. W. Fisher, ed.), Vol. 2, pp. 311–327. Academic Press, New York. (A)

MacKenzie, D. W., Whipple, A. O., and Wintersteinner, M. P. (1941). Studies on the microscopic anatomy and physiology of living transilluminated mammalian spleens. *Am. J. Anat.* **68**, 397–456. (A)

Malik, K. U. (1978). Prostaglandin-mediated inhibition of the vasoconstrictor responses of the isolated perfused rat splenic vasculature to adrenergic stimuli. *Circ. Res.* **43**, 225–233. (A)

Moore, R. D., Mumaw, V. R., and Schoenberg, M. D. (1964). The structure of the spleen and its functional implications. *Exp. Mol. Pathol.* **3**, 31–50. (A)

Patt, H. M., and Maloney, M. A. (1970). Reconstitution of bone marrow in a depleted medullary cavity. In "Hemopoietic Cellular Proliferation" (F. Stohlman, Jr., ed.), pp. 56–66. Grune & Stratton, New York. (B)

Peskar, B. A., Förstermann, U., and Simmet, T. (1980). Effect of prostaglandins and thromboxane A₂ on the contractility of rabbit splenic capsular smooth muscle. *Artery* **8**, 1–6. (A)

Reilly, F. D., and McCuskey, R. S. (1977a). Studies of the hematopoietic microenvironment. VI. Regulatory mechanisms in the splenic microvascular system of mice. *Microvasc. Res.* **13**, 79–90. (A)

Reilly, F. D., and McCuskey, R. S. (1977b). Studies of the hematopoietic microenvironment. VII. Neural mechanisms in splenic microvascular regulation in mice. *Microvasc. Res.* **14**, 293–302. (A)

Reilly, F. D., McCuskey, R. S., and Meineke, H. A. (1976). Studies of the hemopoietic microenvironment. VIII. Adrenergic and cholinergic innervation of the murine spleen. *Anat. Rec.* **185**, 109–117. (A)

Reilly, F. D., McCuskey, P. A., Miller, M. L., and McCuskey, R. S. (1979). Innervation of the periarteriolar lymphatic sheath of the spleen. *Tissue Cell* **11**, 121–126. (A)

Saito, H. (1977). Fine structure of the reticular cells in the rat spleen, with special reference to their fibro-muscular features. *Arch. Histol. Jpn.* **40**, 333–345. (A)

Samsom, J. P., Hulstaert, C. E., Molenaar, I., and Nieweg, H. O. (1972). Fine structure of the bone marrow sinusoidal wall in idiopathic and drug-induced panmyelopathy. *Acta Hematol.* **48**, 218–226. (B)

Snook, T. (1950). A comparative study of the vascular arrangements in mammalian spleens. *Am. J. Anat.* **87**, 31–65. (A)

Stark, R. D., McNeill, J. R., and Greenway, C. V. (1971). Sympathetic and hypophyseal roles in the splenic response to haemorrhage. *Am. J. Physiol.* **220**, 837–840. (A)

Tanaka, Y. (1976). Architecture of the marrow vasculature in three amphibian species and its significance in hematopoietic development. *Am. J. Anat.* **145**, 485–498. (B)

Tavassoli, M. (1974). Arterial structure of bone marrow in the rabbit with special reference to the thin-walled arteries. *Acta Anat.* **90**, 608–616. (B)

Tavassoli, M., and Shaklai, M. (1979). Absence of tight junctions in endothelium of marrow sinuses: Possible significance for marrow cell egress. *Br. J. Haematol.* **41**, 303–307. (B)

Tavassoli, M., and Weiss, L. (1971). The structure of developing bone marrow sinuses in extramedullary autotransplant of the marrow in rats. *Anat. Rec.* **171**, 477–494. (B)

Tischendorf, F. (1969). Die Milz. In "Möllendorffs Handbuch der mikroskopischen Anatomie der Menschen" (W. Bargmann, ed.), Vol. VI, Part 6, pp. 498–820. Springer-Verlag, Berlin and New York. (A)

Tranzer, J. P., and Thoenen, H. (1967). Elektronenmikroskoposche Untersuchungen am peripheren sympathischen nervensystem der Katze; physiologische und pharmakologische Aspekte. *Naunyn-Schmiedebergs Arch. Pharmakol. Exp. Pathol.* **257**, 73–75. (A)

Utterback, R. A. (1944). The innervation of the spleen. *J. Comp. Neurol.* **81**, 55–66. (A)

Weiss, L. (1973). A scanning electron microscopic study of the spleen. *Blood* **43**, 665–691. (A)

Weiss, L. (1977a). The spleen. In "Histology" (L. Weiss and R. Greep, eds.), 4th ed., pp. 445–476. McGraw-Hill, New York. (A)

Weiss, L. (1977b). "The Blood Cells and Hematopoietic Tissues." McGraw-Hill, New York. (B)

Weiss, L., and Tavassoli, M. (1970). Anatomical hazards to the passage of erythrocytes through the spleen. *Semin. Hematol.* **7**, 372–380. (A)

Westen, H., and Bainton, D. F. (1979). Association of alkaline-phosphatase-positive reticulum cells in bone marrow with granulocytic precursors. *J. Exp. Med.* **150**, 919–937. (B)

Zetterström, B. E. M., Palmerio, C., and Fine, J. (1964). Protection of function and vascular integrity of the spleen in traumatic shock by denervation. *Proc. Soc. Exp. Biol. Med.* **117**, 373–376. (A)

Zetterström, B. E. M., Hökfelt, T., Norberg, K.-A., and Olsson, P. (1973). Possibilities of a direct adrenergic influence on blood elements in the dog spleen. *Acta Chir. Scand.* **139**, 117–122. (A)

Index

A

Abdominal aorta
  aneurysms, 49, 65–66
  testicular artery from, 544
Abdominal pressure, Valsalva maneuver, 88
Abductor hallucis, 640
ABO antigens, endothelial cells, 27
Acceleration, convective and local, 74–75
Accessory hepatic artery, 464
Accessory internal pudendal artery, 555
Accessory renal arteries, 508
Acetaldehyde, and skin flushing, 612
Acetate, and muscle blood flow, 649
Acetylcholine
  and bladder blood flow, 565
  and bone blood flow, 686, 687
  and cerebral blood flow, 195
  and cutaneous blood flow, 605
    in rosacea, 611
    sweating, 613
  and eye blood flow, 252
  and gastric blood flow, 419–420
  and hepatic blood flow, 486
  and inner ear circulation, 240
  and intestinal blood flow, 448
  and microcirculation, 116, 117
  and muscle blood flow, 650–653
  nasal mucosa, 245, 246
  and ocular vessels, 251
  and pancreatic blood flow, 299, 301
  and pituitary blood flow, 273
  and pulmonary blood flow, 386–387, 390
  and renal blood flow, 527
  and splenic blood flow, 703
  and vasodilatation, endothelium-dependent,
    37
Achalasia, 423
Acidosis, and bone blood flow, 687–688
Acinar microvascular units
  agglomerate, 469, 475
  complex, 469–474
  simple, 469, 474–475
Acinar plexus, pancreas, 297
Acinulus, 472
Acral areas, glomus bodies in, 599–600
Acrocyanosis, 615–616
ACTH
  and adrenal blood flow, 283

and adrenal medullary function, 287
and epinephrine biosynthesis, 286
Actin, 185
  erythrocyte, 81
  lymphatic endothelial cells, 145
  smooth muscle, 17–20, 22
Actomyosin, *see also* Myosin
  ATPase, 23–24
  cross bridges, 66
Adductor digiti minimi, 640
Adamkiewicz, artery of, 213
Adenine nucleotides and derivatives
  and adipose tissue blood flow, 626
  and bladder blood flow, 566
  and bone blood flow, 687
  and cerebral blood flow, 189, 195
  and coronary thrombosis, 346
  and cutaneous blood flow, axon reflex dil-
    atation, 609, 610
  and gastric blood flow, 422
  and intestinal blood flow, 448
  and microcirculation, 116, 117
  and muscle blood flow, 648, 650
  and splenic blood flow, 703
Adenohypophysis, *see also* Pituitary gland
  brain and, 260–261
  microcirculation, 269
Adipose tissue, blood circulation in, 622–632
  anatomy, 622–624
  arteries, 7
  physiology, 624–631
    and catecholamines, 629–630
    central nervous system regulation,
      626–628
    hormonal regulation, 629–631
    local regulation, 625–626
    peripheral sympathetic regulation,
      628–629
    vascular dimensions, 624–625
  pathology, 631
Adrenal arteries, 280, 561
Adrenal gland, blood circulation in, 280–287
  anatomy, 280–283
  physiology, 283–287
Adrenal gland, lymphatics, 287
  medulla–cortex relationship, 286–287
Adrenal medulla, and cutaneous blood flow,
  604

in arteriosclerosis obliterans, 658
    steroids and, 663
    uremia and, 662–663
pituitary vessels, 269
Basilar artery, 182, 233, 234
Basilar membrane, 239
Basilovertebral system, embryonic development, 175–179
Batson's plexus, 219, 550
Benzopyrones, and pancreatic lymph flow, 304
Bergmeister's papilla of the eye, 248
Bernoulli equation, 75
Berry aneurysms, 202, 225
Beta-blockers, *see* Adrenergic regulation; specific substances
Bethanechol
    and bladder blood flow, 566
    and inner ear blood flow, 240
    and intestinal lymph flow, 454
    and muscle blood flow, 652, 653
    and pancreatic blood flow, 300
Bicarbonate, and cerebral blood flow, 188–189
Biceps brachii, 642
Biceps femoris, 640
Bilateral longitudinal neural artery, 176
Bile
    and lymph, 500–501
    transsinusoidal pressure and, 481
Bile canaliculi, 476, 477
Bile ducts, 462, 470, 476
Bile salt, and gastric mucosal blood flow, 417
Binding sites, endothelial cells, vasoactive agents, 116
Biochemistry, smooth muscle, *see* Smooth muscle, mechanochemistry
Biomechanics, *see* Blood flow; Hemodynamics; Mechanics
Bladder, *see* Urinary bladder
Blood-vessel wall interactions, 76–77
Blood-air barrier, 365, 367
Blood-aqueous barrier, 251
Blood-brain barrier, 155, 187, 192
    carbon dioxide permeability, 187–188
    pineal gland, 309–310
Blood cells, *see also* Erythrocytes; Leukocytes
    in marrow sinusoids, 712–713
    skin capillaries, 597
Blood flow, *see also* specific glands and organ systems

in adipose tissue, 624–630
in adrenal, 16, 283–287
and adventitia, vein, 15
coronary, stenosis and, 341
and critical closure, 66
and endothelium, 3–5
and hemodynamics, 70–77
in microcirculation
    exchange processes, 107–110
    patterns of, 105–106
    regulation of, 110–113
in nose veins, 243
and pancreas, 298
and pineal gland, 310
and pituitary, 270–271
and placenta, 574–577
    and exchange, 591–592
    and fetal, 591
    and maternal, 588–591
pulmonary, 376–379, *see also* Pulmonary system
renal, *see* Kidney, blood circulation in
rheology, 77–85
spinal cord, 219–220
thyroid gland, 275, 277
Blood gases, *see also* Oxygen tension
    and bone blood flow, 687–688
    and cerebral blood flow, 188–189
Blood pressure, 14, *see also* Hypertension
    analgesics and, 115
    and autoregulation, spinal cord, 220
    drug withdrawal and, 115
    and exchange processes, 109
    and intestinal blood flow, 445
    and blood vessel mechanics, 64–70
    regulation of
        cutaneous circulation and, 608
        skin and, 599
    and spinal cord blood flow, 221
Blood reservoir
    cutaneous, 608
    hepatic, 479, 481, 487
Blood-retinal barrier, 251, 252
    drugs and, 253
    fluorescein angiography, 248
Blood volume
    and capacitance, 87
    and cardiac output, 86
    hepatic, 479–481, 487–488
    and intestinal blood flow, 445, 446
    spleen, 703

Splanchnic nerves
  and adrenal medullary function, 284, 287
  and hepatic blood flow, 479
Spleen, blood circulation in, 698–703, *see also*
    Hemopoietic system
  anatomy, 698–701
  and hepatic portal flow, 479
  innervation, 701–702
  pharmacology, 702–703
Spleen, lymphatics, 703
Splenic artery, 410, 411, 438
  aneurysms, 50
  hepatic artery collaterals, 464
Splenic lymph nodes, pancreatic drainage,
    301
Splenic veins, and portal vein, 466
Splenomegaly, portal hypertension and, 490
Spongiosal artery, 553, 555–556
SRS-A, and pulmonary circulation, 388
Stapedial artery, 175
Starling resistor model
  bone circulation, 692
  pulmonary vascular system, 377, 378
  renal circulation, 512
Steal phenomenon, 221–222, 228
Stellate ganglion, and blood flow, 240, 328
Stenosis, vascular
  and aneurysms, 50
  coronary artery, 340–342
  flow through, 75–76
  intestinal, 423
  renal artery, 528
Steroids
  and cutaneous blood flow, 613
  and liver, 496
  and muscle vessels, 664
  and pulmonary blood flow, 386
  and renal blood flow, 522
  and spinal cord blood flow, 223
Stiffness, arterial, 67–68
Stomach, *see* Esophagus and stomach, blood
    circulation in
Storage diseases, and liver, 496
Straight sinus, cerebral circulation, 181
Streptokinase, in coronary thrombosis, 346
Stress
  and pituitary circulation, 272
  stress ulcers, 417, 425
Stresses, vascular, *see also* Mechanical
  compressive, 70
  shear, 3–4, 6

tensile, 8, 64–69
and thrombus formation, 46
traction, 69
Stressed volume, 87
Striated muscle, vs. vascular smooth muscle,
  17–20, *see also* Skeletal muscle, blood
  circulation in
Stria vascularis, 234–237, 239–240
Stroke, sympathetic nerves and, 187
Subarachnoid hemorrhage, spinal cord, 227
Subarachnoid space, 127
Subcapsular plexuses
  adrenal, 280
  thyroid, 279
Subcardinal vein, 507
Subclavian artery
  cervicothoracic spinal cord supply, 213
  thyroid supply, 273
Subclavian lymphatic trunk, 132–133
Subendosteal venous vascular system, 710
Submucosal glands, intestinal, 440
Submucosal plexus, gastric, 412
Subserous plexus, gastric, 411–412
Substance P
  and cutaneous blood flow, axon reflex vaso-
    dilatation, 610
  in vitro studies, 117
  and microcirculation, 116, 117
  nasal mucosa, 245–246
  and pulmonary circulation, localization of,
    393
Suction theory, lymphatic, 163
Sulfonurea, and pancreatic blood flow, 301
Sulpiride, and renal response to dopamine,
  525
Superficial petrosal nerve, nasal blood vessels,
  246
Superficial plexus, pulmonary lymphatics,
  397–399
Superfusion, 193
Superior adrenal artery, 280
Superior cervical ganglia
  and blood flow, 240
  and pineal blood flow, 309, 310
  and pituitary blood flow, 271
Superior mesenteric artery, 411, 438
  embryology, 437
  hepatic artery collaterals, 464
  right hepatic artery from, 463, 464
Superior mesenteric vein
  large intestine drainage by, 439